T0320961

Basic Algebraic Topology

Anant R. Shastri

Indian Institute of Technology Bombay Mumbai,
Maharastra, India

CRC Press
Taylor & Francis Group
Boca Raton   London   New York

CRC Press is an imprint of the
Taylor & Francis Group, an **informa** business

A CHAPMAN & HALL BOOK

CRC Press
Taylor & Francis Group
6000 Broken Sound Parkway NW, Suite 300
Boca Raton, FL 33487-2742

© 2014 by Taylor & Francis Group, LLC
CRC Press is an imprint of Taylor & Francis Group, an Informa business

No claim to original U.S. Government works

Printed on acid-free paper
Version Date: 20130812

International Standard Book Number-13: 978-1-4665-6243-1 (Hardback)

This book contains information obtained from authentic and highly regarded sources. Reasonable efforts have been made to publish reliable data and information, but the author and publisher cannot assume responsibility for the validity of all materials or the consequences of their use. The authors and publishers have attempted to trace the copyright holders of all material reproduced in this publication and apologize to copyright holders if permission to publish in this form has not been obtained. If any copyright material has not been acknowledged please write and let us know so we may rectify in any future reprint.

Except as permitted under U.S. Copyright Law, no part of this book may be reprinted, reproduced, transmitted, or utilized in any form by any electronic, mechanical, or other means, now known or hereafter invented, including photocopying, microfilming, and recording, or in any information storage or retrieval system, without written permission from the publishers.

For permission to photocopy or use material electronically from this work, please access www.copyright.com (http://www.copyright.com/) or contact the Copyright Clearance Center, Inc. (CCC), 222 Rosewood Drive, Danvers, MA 01923, 978-750-8400. CCC is a not-for-profit organization that provides licenses and registration for a variety of users. For organizations that have been granted a photocopy license by the CCC, a separate system of payment has been arranged.

Trademark Notice: Product or corporate names may be trademarks or registered trademarks, and are used only for identification and explanation without intent to infringe.

Library of Congress Cataloging-in-Publication Data

Shastri, Anant R.
   Basic algebraic topology / Anant R. Shastri.
      pages cm
   Includes bibliographical references and index.
   ISBN 978-1-4665-6243-1 (hardback)
   1. Algebraic topology--Textbooks. I. Title.

QA612.S535 2013
514'.2--dc23
                                                   2013025264

Visit the Taylor & Francis Web site at
http://www.taylorandfrancis.com

and the CRC Press Web site at
http://www.crcpress.com

Contents

Foreword

While the subject of algebraic topology began long before H. Poincaré's *Analysis Situs*, the discipline started to take shape only in the 1930s during which the foundation of modern algebraic topology was laid. Fundamental concepts such as manifolds, fiber spaces, higher homotopy groups, and various homology and cohomology theories were firmly established. Meanwhile, obstruction theory, cohomology operations, and spectral sequences were among some of the powerful tools developed as the subject rapidly grew. By the 1960s (see [Dieudonné, 1989]), algebraic topology was already a well-established discipline and together with differential topology dominated much of mathematics at the time.

Applications to analysis and other fields were some of the motivating factors in the early development of algebraic topology. For instance, the Lusternik–Schnirelmann (LS-) category cat(X) of a topological space X was first introduced in the early 1930s as a means to obtain information about the critical points of a functional. This subject was later taken up and advanced by R. Palais and S. Smale (1963–64). The homotopy approach to the LS-category by T. Ganea (1971) revived the subject. The so-called Ganea conjecture claiming that $\operatorname{cat}(X \times S^n) = \operatorname{cat}(X) + 1$ for any sphere S^n with $n > 0$ attracted much attention until a counter-example was given by N. Iwase in 1998. The study of the classical LS-category and its many variants and their applications to analysis continues to be an active area of current research. The classical Borsuk-Ulam theorem (first conjectured by S. Ulam and later proved by K. Borsuk in 1933) is another example which has generated many new and interesting problems with diverse applications in other fields such as combinatorics and economics (see [Matousek, 2003]), among many others.

One of the deepest and most important theorems in homotopy theory is J. F. Adams' work ([Adams, 1960]) on the Hopf invariant one problem, which asserts that the sphere S^n is an H-space exactly when n is 0, 1, 3, or 7. In 1966, an alternate proof by Adams and M. Atiyah was given using Adams operations and topological K-theory. The study of stable homotopy theory (see [Adams, 1974]) recently saw a major breakthrough when M. Hopkins, M. Hill, and D. Ravenel resolved the so-called Kervaire-invariant one problem except in dimension 126. The existence of smooth framed manifolds of Kervaire-invariant one has been a long-standing problem in differential and algebraic topology. Through the work of W. Browder (1969), the original problem is equivalent to a problem in stable homotopy groups of spheres and it is known that such framed manifolds can only exist in dimension $n = 2^{j+1} - 2$. The recent achievement of Hopkins et al. states that such manifolds exist only in dimensions 2, 6, 14, 30, 62, and possibly 126.

Nowadays, every student in his or her first year of a Ph.D. program in mathematics must take basic graduate courses in algebra, analysis, and geometry/topology. Algebraic topology constitutes a significant portion of such basic knowledge a practicing mathematician should know in geometry/topology. As suggested by the title, Professor Shastri's book covers the most basic and essential elements in algebraic topology. Similar to his other well-written textbook [Shastri, 2011] on differential topology, Professor Shastri's book gives a detailed introduction to the vast subject of algebraic topology together with an abundance of carefully chosen exercises at the end of each chapter. The content of Professor Shastri's book furnishes the necessary background to access many major achievements such as the results cited above, to explore current research work as well as the possible applications to other branches of mathematics of modern algebraic topology.

Peter Wong
Lewiston, Maine

Preface

This book is intended for a two-semester first course in algebraic topology, though I would recommend not to try to cover the whole thing in two semesters. A glance through the contents page will tell the reader that the selection of topics is quite standard whereas the sequencing of them may not be so. The material in the first five chapters is very basic and quite enough for a semester course. A teacher can afford to be a little choosy in selecting exactly which sections she may want to teach. There is more freedom in the choice of material to be taught from latter chapters. It goes without saying that the material in later chapters demands much higher mathematical maturity than the first five chapters. Also, this is where some knowledge of differential manifolds helps to understand the material better.

The book can be adopted as a text for M.Sc./B.Tech./M.Tech./Ph.D. students. We assume that the readers of this book have gone through a semester course each in real analysis and point-set-topology and some basic algebra. It is desirable that they also have had a course in differential topology or are concurrently studying such a course, but that is necessary only for a few sections. There are exercises at the end of many sections or within a section, which involve a single theme of that particular section. There are Miscellaneous Exercises at the end of most of the chapters, which may normally involve themes studied thus far. Most of these exercises are part of the main material and working through them is an essential part of learning. However, it is not necessary that a student get the right answers before proceeding further. Also, it is not a good idea to get stuck with a problem for too long—keep going further and come back to it later. There is a hint/solution manual at the end of the book for some selected exercises, especially for those which are being used in a later section, so as to make this book self-readable by any interested student. However, peeping into the solutions at the beginning is like reading the last section of a thriller first. You will notice that the number of exercises goes down as the chapters proceed for the simple reason that more and more 'routine verifications' of claims in the main text are left to the reader as exercises.

In the first chapter, we begin with a general discussion of what algebraic topology is and what to expect from this book, and then go on to introduce one of the very basic algebraic topological invariants, viz., the fundamental group. We then give a quick introduction to some set topological results such as function spaces and quotient spaces, which are crucial to understanding homotopy theory. The concept of relative homotopy, basics of cofibrations and fibrations, and an introduction to the language of category theory, etc., make up the rest of this chapter. No doubt the material here is used throughout the rest of the book and a teacher/reader may choose only a part of it and go ahead with other chapters preferring to come back later, to whatever is wanted.

In Chapter 2, we begin with an introduction to basics of convex polytopes laying down a foundation for the study of simplicial complexes/polyhedral topology. We take the view that simplicial complexes are a very special type of CW-complexes, with additional combinatorial structure but with the same point-set-topological and homotopy theoretic behaviour. If nothing else, this point of view saves us some time. Simplicial approximation theorem is one of the milestone results here. We give a number of applications of this. A simple proof of Brouwer's invariance of domain via Sperner's lemma is one such.

Chapter 3 deals with the notion of covering spaces, along with the study of discontinuous group actions and the relationship with fundamental group. We then give yet another powerful tool of computation of the fundamental groups, viz., Seifert–Van Kampen theorems. Grothendieck's idea of G-coverings is introduced especially for this purpose.

In Chapter 4, we start the study of homology theory. With singular homology taking centre stage, we also introduce CW-homology, simplicial homology, etc. Standard applications to results such as Brouwer's and Lefschetz's fixed point theorems, hairy ball theorem, Jordan-Brouwer separation theorem, Brouwer's invariance of domain, etc., are included. We also give the result which relates fundamental group with the first homology group, paving the way for a more general result known as Hurewicz's isomorphism theorem to be discussed in Chapter 10. The emphasis here is to get familiar with the tools so as to start using them rather than the theory and the proofs. So, most of the long and pedagogically less important proofs have been clubbed together in one section.

In Chapter 5, we introduce topological manifolds, the central objects of study in topology. This chapter also contains a topological classification of compact surfaces by first showing that they are all triangulable. We also include some preparatory materials on vector bundles and fibrations.

Chapter 6 contains more algebraic tools which help us to develop homology with coefficients and study homology of product spaces, etc. (Method of acyclic models should not be postponed any more.) Künneth formula is an important result here.

In Chapter 7, we develop cohomology algebra, carry out some computations and applications, and discuss cohomology operations. Steenrod squares are constructed and their fundamental properties are verified but the proof of the uniqueness is omitted. Similarly, though we discuss Adem's relations to some extent and verify them on finite product of infinite real projective spaces, further discussion is postponed to Chapter 10.

In Chapter 8, we return to the study of manifolds. Poincaré duality theorem is the central result here. We include a number of variants of it such as Alexander duality and Lefschetz's duality. Bootstrap lemma which plays the central role in the proof here is taken from [Bredon, 1977]. Various applications of duality are included. The notion of degree and the index of a $4n$-dimensional smooth manifold, etc., are discussed. This chapter ends with another important result, viz., de Rham's Theorem which relates the singular cohomology with that of cohomology of differential forms on a smooth manifold. The proof here does not use sheaf cohomology.

Chapter 9 contains more topics on cohomology. We introduce the important concept of sheaves and basics of sheaf cohomology, and Čech cohomology of sheaves. As an application we present the standard proof of de Rham's theorem.

Chapter 10 is the heart of the book. With a somewhat digressive note on H-spaces and co H-spaces in Section 10.1, we quickly reintroduce higher homotopy groups (which have been introduced in the Miscellaneous Exercises to Chapter 1) and verify their basic properties, in Section 10.2. In Section 10.3, we thoroughly discuss the effect of change of base points on homotopy groups. In Section 10.4, we present Hurewicz's isomorphism theorem, Whitehead's theorem, etc. In Section 10.5, we are able to address one of the central problems that we had posed in Section 1.1, through obstruction theory. In Section 10.6, we give a number of applications to extension and classification problems such as Eilenberg classification and Hopf-Whitney theorem. As a natural fall-out, the homotopy theoretic building blocks, viz., the Eilenberg–Mac Lane spaces are introduced in Chapter 10.7. As an application, we continue our discussion on Steenrod squares and show how to prove Adem's relations modulo a technical result of Serre on the structure of cohomology algebra of $K(\mathbb{Z}_m; \mathbb{Z}_2)$-spaces. In Section 10.8, we present a method of breaking up spaces into these building blocks, viz., Moore-Postnikov decomposition. We then carry out some elementary computations with the homotopy groups of classical groups in Section 10.9. The chapter ends with the section on homology with local coefficients.

In Chapter 11, we return again to the study of homology. Here the theme is to relate the homology of the total space of a fibration with that of the base and the fibre under special conditions. We first consider the case when the fibre is a sphere. After establishing the celebrated Thom isomorphism theorem, and as a consequence the Gysin exact homology sequences, we present a generalization of this, viz., Leray Hirsch theorem. We then consider fibrations in which the base is a sphere. Since the technique involved uses only the fact that spheres are suspensions, we treat the broader class of fibrations over suspensions. Wang homology exact sequence and Freudenthal's homotopy suspension theorem are two important results here. We give an application to computation of the integral homology of the Eilenberg-Mac Lane space of type $(\mathbb{Z}, 3)$. We then compute the cohomology algebra of some of the classical groups. As a necessity, we include Borel's structure theorem for Hopf algebras.

Chapter 12 is a quick introduction to characteristic classes of vector bundles. In Section 12.1, we discuss orientation and Euler class. The relation between Euler class and the Euler characteristic is the main result here. In section 12.2, we give constructions of Steifel–Whitney classes and Chern classes, treating both of them simultaneously. Section 12.3 contains discussion of standard properties of these characteristic classes and applications to non-existence of division algebras and un-oriented cobordism theory. Section 12.4 contains the splitting principle and the proof of uniqueness of characteristic classes. In section 12.5, we study complex vector bundles and Pontrjagin classes and give some applications to oriented cobordism theory. All in all, our treatment of this subject here is merely a glimpse of the theory of characteristic classes and is far from being complete.

The last chapter introduces spectral sequences. After some brief discussion of generalities, we concentrate on one particular spectral sequence, viz., the Leray–Serre spectral sequence of a fibration. We first give the construction of homology spectral sequence, give some immediate applications. For instance, we show how to derive both Gysin sequence and Wang sequence from spectral sequence. We then discuss transgression in homology. In Section 13.7, we discuss cohomology spectral sequences with product structure without proof. (Theorem 13.7.4 is one of the few results in the book that we have used without proving it.) This is immediately applied in obtaining the structure of cohomology algebra of the loop space ΩX under two different types of assumptions on the cohomology algebra of X. In Section 13.8, we introduce "Serre class of abelian groups" and generalize several homotopy theoretic results of Chapter 10. For example, an immediate consequence here is that all homotopy groups of all the spheres are finitely generated. The book concludes with a presentation of Serre's celebrated results on higher homotopy groups.

According to Ahlfors, no teacher should follow any single book in toto. There is a certain amount of comprehensiveness in the early chapters, which is time consuming but deliberate. For instance, a lot of material in the later chapters can be understood without the knowledge of Van Kampen theorem. So, I have included a 'section-wise dependence tree' which may help a teacher to make his/her pick-and-choose course plan and then tell the students to read the book for the rest of the stuff.

Acknowledgments

I have benefited mainly from [Spanier, 1966] and [Whitehead, 1978]. In addition, the books [Bredon, 1977], [Fulton, 1995], [Hatcher, 2002], [Husemoller, 1994], [McCleary, 2001], [Milnor–Stasheff, 1974], [Mosher–Tangora, 1968], [Ramanan, 2004], [Seifert–Threlfall, 1990], etc., were also used whenever I needed extra help or have found an irresistibly beautiful presentation. The bibliography contains the list of all of these from which I may have borrowed something or the other.

This book grew out of regular courses that I have taught to M.Sc. and Ph.D. students since 1989 at the Indian Institute of Technology, Bombay. Initially, I was mostly following the classic book [Spanier, 1966] from which I have myself learned algebraic topology. Invariably, most of the students were finding the course difficult and so I started writing my own notes in Chi-writer. When Allen Hatcher's book on the subject arrived, it was a big relief and writing my own notes came to an end. With some younger faculty at the department willing to teach algebraic topology, I was not teaching the course so regularly any more.

The interest in writing the notes was revived when we started the Advanced Training in Mathematics (ATM) schools under the aegis of the National Board for Higher Mathematics, DAE, Govt. of India. However, the old Chi-writer notes were lost since the old floppies which had those files had become unreadable. So, the present version has grown out of these notes for the Annual Foundation Schools and Advanced Instructional Schools of ATM schools. The revision efforts were supported *twice* by the Curriculum Development Programme of Indian Institute of Technology, Bombay.

In the first year of my graduation at Tata Institute of Fundamental Research, I received a lot of help and encouragement from Anand Doraswami with whom I was sharing my office, while working through exercises in [Spanier, 1966]. M. S. Raghunathan, R. R. Simha and Gopal Prasad have educated me through 'coffee table discussions.' Interaction with several students in the department as well as at ATM schools have helped me in understanding and presenting the material in a better way. Many friends such as Parameswaran Sankaran, Basudev Datta, Goutam Mukherjee, Mahuya Datta, Keerti Vardhan, and students B. Subhash and K. Ramesh have gone through various parts of these notes and pointed out errors, and have suggested improvements in presentation. Discussions with colleague Gopal Srinivasan were always informative. My heartfelt thanks to all these people. The errors which still persist are all due to my own limitations. Readers are welcome to report them to me so that I can keep updating the corrections on my website

http://www.math.iitb.ac.in/~ars/

Thanks to Prof. Peter Wong for providing a friendly foreword. Finally, my thanks to CRC Press for publishing these notes and for doing an excellent job of converting it into a book.

Anant R. Shastri
Department of Mathematics
Indian Institute of Technology, Bombay
Powai, Mumbai

List of Symbols and Abbreviations

$\mathbb{N}$	set of natural numbers	$\mathbb{Z}$	ring of integers
$\mathbb{Q}$	field of rational numbers	$\mathbb{R}$	field of real numbers
$\mathbb{C}$	field of complex numbers	$\mathbb{H}$	skew field of quaternions
$\mathbb{Z}_m$	ring of integers modulo m	$\mathbb{I}$	unit interval $[0,1] \subset \mathbb{R}$
$\mathbb{D}^n$	unit disc in $\mathbb{R}^n$	$\mathbb{S}^{n-1}$	unit sphere in $\mathbb{R}^n$
$\mathbb{P}^n$	real projective space of dim. n	$\mathbb{CP}^n$	complex projective space of dim. n
$\mathbb{R}^\infty$	countable infinite sum of $\mathbb{R}$	$\mathbb{S}^\infty$	unit sphere in $\mathbb{R}^\infty$
$\mathbb{P}^\infty$	infinite real projective space	$\mathbb{CP}^\infty$	infinite complex projective space
CX	cone over X	SX	suspension of X
C_f	mapping cone of f	M_f	mapping cylinder of f
LHS	left hand side	RHS	right hand side
WLOG	without loss of generality	NDR	neighbourhood deformation retract
DR	deformation retract	SDR	strong deformation retract
HED	homotopy extension data	HEP	homotopy extension property
HLD	homotopy lifting data	HLP	homotopy lifting property
UCT	universal coefficient theorem	UPL	unique path lifting property
PID	principal ideal domain	♠	end of the proof

Sectionwise Dependence Tree

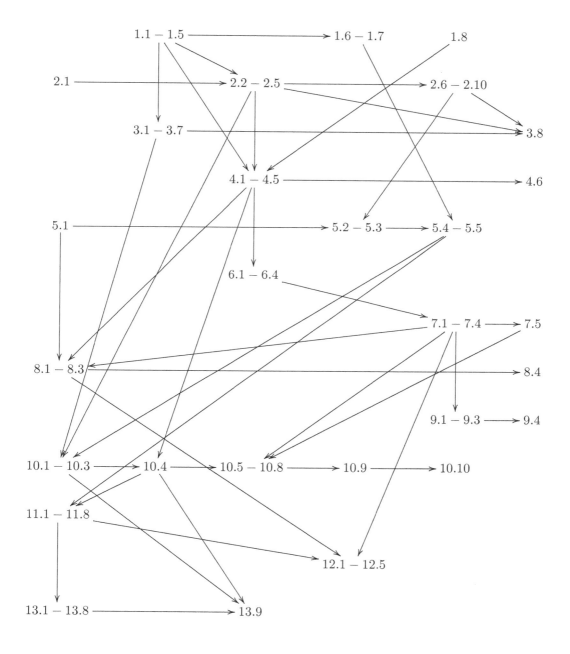

Chapter 1

Introduction

We shall assume that the readers of this book have had a course in general point-set topology and are familiar with some basic notions such as connectedness, path connectedness, local path-connectedness, compactness, Hausdorffness, etc. Some of the slightly more advanced topics such as function spaces, quotient spaces, etc., will be recalled as a ready reference.

Throughout this exposition, we shall use the word 'space' to mean a topological space. Similarly, we shall use the word 'map' to mean a continuous function between topological spaces. This however, does not forbid us from using terminologies such as 'linear map' or 'simplicial map', etc., wherein we may not really be bothered about the function being continuous, the emphasis being on something else.

In Section 1.1, we begin with an attempt to describe what algebraic topology is and what to expect from this book, and discuss an experiment with the Möbius band.

In Section 1.2, as a typical motivating example of tools of algebraic topology, we introduce the concept of fundamental group of a topological space, establish some basic properties and compute it in the case of the circle. Applications to (2-dimensional) Brouwer's fixed point theorem, Borsuk-Ulam theorem, etc., are included, which illustrate the power of categorical constructions in general, and the fundamental group in particular.

In Section 1.3, we shall quickly introduce the compact open topology and quotient spaces. These are fundamental point-set-topological background needed to understand homotopies and constructions in algebraic topology. Section 1.4 will plunge the reader into technicalities of relative homotopy. In Section 1.5, we give certain basic constructions which keep cropping up repeatedly in algebraic topology. In Sections 6 and 7, we introduce the reader to cofibrations and fibrations, respectively. In Section 1.7, the language of category theory is introduced.

This chapter, like many others, will end with a large number of doable and challenging exercises. It is not necessary that the reader solve all of them before proceeding with the book but she is expected to give a good try. The joy that one gets after cracking a problem on one's own is perhaps the best motivation for many of us for doing mathematics.

1.1 The Basic Problem

A central problem in topology is to determine whether two given topological spaces are homeomorphic or not. For instance, we all know that any two open intervals in $\mathbb{R}$ are homeomorphic to each other, since we can actually write down a homeomorphism in each case. On the other hand, we also know that a closed interval and an open interval are not homeomorphic to each other because the former is compact whereas the latter is not.

In general, displaying such homeomorphisms between topological spaces becomes very difficult. On the other hand, it is fruitful and easier to find out that there is no homeomorphism between two given specific spaces X and Y. The standard method is to look for a suitable 'topological invariant' such as compactness, connectedness, etc., which is present in one of the two spaces and absent in the other.

Let us consider an example. Let us show that $\mathbb{R}$ and $\mathbb{R}^2$ are not homeomorphic. If

$f : \mathbb{R} \to \mathbb{R}^2$ were a homeomorphism then the restriction map $f : \mathbb{R} \setminus \{0\} \to \mathbb{R}^2 \setminus \{f(0)\}$ is also a homeomorphism. Now the domain of f is not connected whereas the range is. Since this is absurd, we conclude that $\mathbb{R}$ and $\mathbb{R}^2$ are not homeomorphic. However, note that the connectivity could not be directly applied to the map $f : \mathbb{R} \to \mathbb{R}^2$ to arrive at this conclusion. Again, this method may not work very far. For example, it is not effective if the problem is to prove that $\mathbb{R}^n$ and $\mathbb{R}^m$ are not homeomorphic to each other for $m > n > 1$. Of course, there are purely point-set-topological proofs of this result as well but they are not so easy. So, one looks for other topological invariants, which are perhaps not so demanding.

Taking up a different thread, let us take an example from complex analysis of 1-variable. Look at Figure 1.1 in which oriented closed smooth curves C_1, C_1', C_2, C_2' are drawn around the origin (with varying thicknesses),

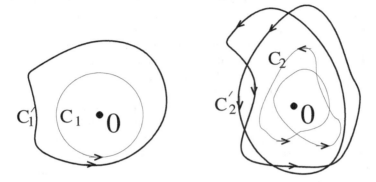

FIGURE 1.1. Why do the integrals take the same value on C_j and C_j'?

The reader may recall that by Cauchy's integral formula,

$$\int_{C_j} \frac{dz}{z} = \int_{C_j'} \frac{dz}{z} = (2\pi\imath)j, \quad j = 1, 2.$$

Certainly, we can observe that the curves C_1, C_1' are homeomorphic (indeed diffeomorphic) to each other, though this really does not help us here. Even this information is not available for C_2 and C_2', as the two curves are not homeomorphic as subsets of $\mathbb{C} \setminus \{0\}$. What makes the integrals have the same value? It is the property that one curve can be 'deformed' into the other while remaining all the time inside $\mathbb{C} \setminus \{0\}$. The other technical terms that are used in this contexts in complex analysis are 'homologous' and 'homotopic'. Also, the reader may recall that several variants of the notion of simple connectedness were used in complex analysis. One may call this the starting point of algebraic topology, wherein 'quantitative' invariants such as the integral, which is a number, were introduced. Complex analysis of 1-variable may be considered as the birthplace of algebraic topology (and also of many other branches of modern mathematics).

When we start the study of a discipline, it is good to have at least some idea of what the fundamental problems in that discipline are. In most of the cases, these problems remain unsolved. In some cases, some day one may find that the fundamental problem cannot be solved. However, that does not mean that we have come to the end of the road—there will always be some related problems or modified problems demanding our attention.

And this is the case with topology in general and algebraic topology in particular. The central objects of study in topology are 'manifolds'. Roughly speaking, an n-dimensional manifold is a topological space in which each point has a neighbourhood system consisting of open sets which are homeomorphic to open sets in a n-dimensional Euclidean space.

(For a formal definition of a manifold, see 5.1.1.) The central problem then is to determine whether any two given manifolds are homeomorphic or not. It is known that this problem is not solvable. Even this negative result is quite valuable and let us take a few seconds to see how this result was established.

To each space X one 'associates' a group called the fundamental group $\pi_1(X)$. This association has the property that for any map $f : X \to Y$, there is the homomorphism of groups $f_\# : \pi_1(X) \to \pi_1(Y)$. The association is 'natural' in the sense that if $g : Y \to Z$ is another map, then $(g \circ f)_\# = g_\# \circ f_\#$ and for the identity map $Id : X \to X$, we have $Id_\# : \pi_1(X) \to \pi_1(X)$ is the identity homomorphism. (See Section 1.2 for more details.) In particular, it now follows that if $f : X \to Y$ is a homeomorphism then $f_\#$ is an isomorphism. Thus, in order that two given spaces X and Y are homeomorphic, first of all, their fundamental groups must be isomorphic. The next step is to construct a manifold such that $\pi_1(M)$ is isomorphic to a given group G. Indeed, given a group G with finitely many generators and relations, (i.e., a *finitely presented group*) it can be shown that there is a compact 4-dimensional manifold M with $\pi_1(M) = G$. (See Exercise 5.5.11.) The net result is that now the homeomorphism problem for compact 4-dimensional manifolds implies the isomorphism problem for finitely presented groups. This latter problem goes under the name 'word problem' and nowadays is a very specialized branch of group theory and mathematical logic. The non solubility of the word problem was established in 1955 by P.S. Novikov [Novikov, 1955].

This, however does not close the subject altogether—topology is still a very lively subject extending a helping hand in solving problems from several areas of mathematics. The process of associating the fundamental group to a space X as considered above is called constructing a functor (see Section 1.8 for more). Algebraic topology may be described at the outset as the study of such functors.

Coming back to the fundamental problem, there are many interesting related problems. For instance before finding a homeomorphism $f : X \to Y$, we may want to find some map which may be defined on a part of X or may be defined all over X but does not have all the properties that we demand. This, in turn, raises many other questions. Instead of listing all these questions, we shall begin with two of the important ones:

(i) the lifting problem and
(ii) the extension problem.

The underlying themes in these two problems occur repeatedly throughout this book. So, it may be worthwhile to get some familiarity with these concepts.

Consider a triangle of maps as represented in Figure 1.2.

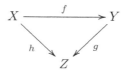

FIGURE 1.2. Triangle of maps

Such a diagram is called a commutative diagram, if $h = g \circ f$. Given any two of the three maps, we can consider the problem of finding one or all maps which fit the third arrow in the diagram. Naturally, this problem can be broken up into three cases, out of which one case is too easy, viz., if f, g are given then h can be taken to be $g \circ f$ and nothing else. So, we consider the other two cases which we reformulate as follows:

Q. I Given maps, $p : E \to B$ and $f : X \to B$, does there exist $g : X \to E$ such that, $p \circ g = f$? The map g is called a *lift* of f through p and this problem goes under the name

$$(I) \qquad\qquad\qquad\qquad (II)$$

FIGURE 1.3. Two fundamental problems

lifting problem.
Q. II Given maps $\eta : A \to X$ and $f : A \to Y$ does there exist $\hat{f} : X \to Y$ such that $\hat{f} \circ \eta = f$?

Question II has two important special cases:
(a) η is a quotient map; this goes under the name 'factorization problem'.
(b) η is an inclusion map; this goes under the name 'extension problem'.
It turns out that (a) has an easy answer (see Section 1.3) and so we need to bother about (b) only. Clearly, these are purely point-set-topological problems which may or may not have satisfactory solutions. In analysis, often we need to approximate a given function by nicer functions. On the other hand, in physics for example, one is interested in properties which are 'stable', i.e., if the initial position is disturbed a little bit, the system returns to the original position on its own. A primitive mathematical notion which helps to describe these ideas is the notion of homotopy. The modern definition of homotopy is due to Brouwer.

Definition 1.1.1 Let X, Y be topological spaces and $\mathbb{I}$ denote the closed interval $[0, 1]$. Then any continuous function $H : X \times \mathbb{I} \to Y$ is called a *homotopy*. We set up the notation $h_t(x) = H(x, t)$. Often the family of maps, $H = \{h_t : X \to Y\}_{t \in \mathbb{I}}$ is called a homotopy from h_0 to h_1. Given two maps f and g, we say f *is homotopic to* g if there exists a homotopy $\{h_t\}$ such that $h_0 = f$ and $h_1 = g$. Symbolically, we express this as:

$$f \simeq g. \tag{1.1}$$

Given a homotopy H from f to g, consider the map G defined by

$$G(x, t) = H(x, 1 - t).$$

Then G is a homotopy from g to f. Therefore, $\simeq$ is a symmetric relation. On the other hand, if H' is a homotopy from g to h, then one can define a homotopy F obtained by juxtaposing H and G :

$$F(x, t) = \begin{cases} H(x, 2t), & \text{if} \quad 0 \le t \le 1/2, \\ H'(x, 2t - 1), & \text{if} \quad 1/2 \le t \le 1. \end{cases}$$

(Why is F continuous?)
Clearly, F is a homotopy from f to h. This shows that $\simeq$ is transitive. Of course, the relation is easily seen to be reflexive. Thus, the relation $\simeq$ is an equivalence relation on the set of all maps from X to Y. The set of homotopy equivalence classes of maps from X to Y will be denoted by $[[X, Y]]$. These are the basic objects of study in algebraic topology. An element of $[[X, Y]]$ represented by a map $f : X \to Y$ is denoted usually by $[[f]]$.

The following lemma says that homotopy is well behaved under composition of maps.

Lemma 1.1.2 Let f and g be homotopic maps from X to Y, $p : W \to X$ and $q : Y \to Z$ be any maps. Then $f \circ p$ is homotopic to $g \circ p$ and $q \circ f$ is homotopic to $q \circ g$.

Proof: If H is the homotopy between f and g, then $H \circ (p \times Id_{\mathbb{I}})$ and $q \circ H$ are the required homotopies. ♠

Remark 1.1.3

1. The assignment $(X, Y) \rightsquigarrow [[X, Y]]$ has the following three properties:

   (a) Given a map $\alpha : Y \to Z$, there is a function $\alpha_\# : [[X, Y]] \to [[X, Z]]$ defined by post-composing with α, viz., $f \mapsto \alpha \circ f$. If $\beta : Z \to W$ is another map, then we have $(\beta \circ \alpha)_\# = \beta_\# \circ \alpha_\#$.

   (b) Given a map $\gamma : W \to X$, there is a function $\gamma^\# : [[X, Y]] \to [[W, Y]]$, defined by pre-composing with γ. If $\mu : V \to W$ is a map, then $(\gamma \circ \mu)^\# = \mu^\# \circ \gamma^\#$.

   (c) Observe that $Id_\# = Id; Id^\# = Id$.

   The 'workspace' for algebraic topology consists of
   (i) the collection of 'all' topological spaces and
   (ii) for each pair (X, Y) of topological spaces, the set $[[X, Y]]$ of homotopy classes $[[f]]$ of maps $f : X \to Y$.

   Modern mathematics is so full of such collections and assignments which satisfy the above three properties that they deserve a name and a full discussion. This will be done in Section 1.8, *Categories and Functors*.

2. If we take $X = \{\star\}$, a singleton space, then what is the set [[X,Y]]? It is nothing but the set of all path components of the space Y. This is a topological invariant, i.e., if Y_1, Y_2 are two topological spaces homeomorphic to each other then they have the same 'number' of path components. This is different from other topological invariants which we have come across, such as compactness, Hausdorffness, metrizability, etc., in the sense that it is quantitative rather than qualitative. In algebraic topology, we produce and study various such invariants of topological spaces which are not only sets but with more algebraic structures on them such as groups, rings, modules and so on. This basic idea of assigning algebraic invariants such as groups and rings instead of numbers goes back to Emmy Noether. We shall see a large number of such examples in this book.

We shall now reformulate the above two fundamental questions, one by one, using the notion of homotopy, which makes these questions more accessible.

Q. Ia: Homotopy Lifting Property (HLP)

Given $f : X \to B$, does there exist $g : X \to E$ such that, $p \circ g \simeq f$?
Here, we are asking: *can f be lifted through p up to homotopy?* Obviously, if **Q. I** has an affirmative answer then **Q. Ia** also has an affirmative answer. This simple observation is often used in the negative sense, viz., if we know that the answer to Q. Ia is in the negative then so is it for Q. I. Now, often it turns out that the converse does not hold, viz., though we can lift a certain map up to homotopy, it may not be possible to actually lift the same map. We do not like to get stuck up in such a situation. So, we make an axiom, the so called *homotopy lifting property*.

Definition 1.1.4 Consider the following situation for a map $p : E \longrightarrow B$. We are given a homotopy $F : X \times \mathbb{I} \to B$, and a map $g : X \to E$ such that, $p \circ g(x) = F(x, 0)$, $\forall\, x \in X$.

This is called *homotopy lifting data* for p. We say p satisfies the *homotopy lifting property* if for each homotopy lifting data as above, there exists a homotopy $G : X \times \mathbb{I} \to E$ with the property $p \circ G = F$ and $G(x, 0) = g(x)$, $\forall\, x \in X$. Schematically, this is represented in the following commutative diagram, (see Figure 1.4) where, the space X is identified with the subspace $X \times 0$ of $X \times \mathbb{I}$, the dotted arrow indicating the existence of the corresponding map.

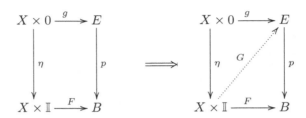

FIGURE 1.4. The homotopy lifting property

Remark 1.1.5 A special class of maps called 'covering projections' which we shall study soon, possess HLP. The notion of HLP is important enough to prompt the following definition.

Definition 1.1.6 If the map p satisfies the HLP for all spaces X, then it is called a *Hurewicz fibration* or simply a *fibration*.

Remark 1.1.7 Hurewicz was the one who first recognized the importance of HLP and studied it extensively. The reader will have to wait for a while before we can tackle this problem head-on. On the face of it, it looks like we are merely avoiding the 'difficult' situation by rephrasing it as an axiom or a definition. The point is that Hurewicz fibrations occur in abundance, and indeed, up to homotopy, every map can be replaced by a Hurewicz fibration. (See Theorems 1.7.12, 1.7.13 and 1.7.8.) We shall now take the homotopy version of Q. II, which is in some sense 'dual' to the concept of HLP.

Q. IIa: Homotopy Extension Property (HEP)

Definition 1.1.8 Given a map $\eta : A \longrightarrow X$, consider the following data: A homotopy $F : A \times \mathbb{I} \longrightarrow Y$ and a map $g : X \longrightarrow Y$ such that $g(\eta(a)) = F(a, 0)$ for all $a \in A$. Such a data is called a *homotopy extension data* for η. We say η has *homotopy extension property* with respect to Y if for each such data, there exists a homotopy $H : X \times \mathbb{I} \longrightarrow Y$ such that $H \circ (\eta \times Id) = F$, and $H(x, 0) = g(x)$, $\forall\, x \in X$. If η has HEP with respect to all spaces Y then it is called a *cofibration*. Often the situation is such that A is a subspace of X and η is the inclusion map, in which case we say that the pair (X, A) has HEP with respect to Y. (See the Figure 1.5.)

We shall keep visiting these two concepts time and again. We need to start developing the notion of homotopy. Since maps which are homotopy equivalent to each other are going to be treated on par, it is natural to have the next definition.

Definition 1.1.9 Let X and Y be any two topological spaces. A map $f : X \to Y$ is called a *homotopy equivalence,* if there exists a map $g : Y \to X$ such that, $g \circ f \simeq Id_X$ and $f \circ g \simeq Id_Y$. In this case, f and g are said to be *homotopy inverses* of each other. If there exists a homotopy equivalence $f : X \longrightarrow Y$, we say X is homotopy equivalent to Y or X and Y have the same homotopy type.

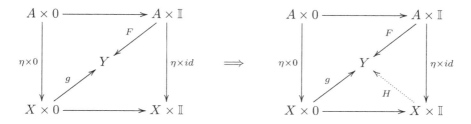

FIGURE 1.5. Homotopy extension property

Remark 1.1.10 Of course, using Lemma 1.1.2, it is fairly easy to verify that, 'homotopy type' defines an equivalence relation on the collection of all topological spaces. Using Remark 1.1.3, verify that, a homotopy equivalence $f : X \to Y$ induces bijection of sets

$$[[Y, A]] \longleftrightarrow [[X, A]], \quad [[B, X]] \longleftrightarrow [[B, Y]].$$

Since point-spaces have hardly any non trivial properties, it is appropriate to make the following definitions.

Definition 1.1.11 A topological space X which is homotopy equivalent to a singleton space is called a *contractible space*. A map $f : X \longrightarrow Y$ which is homotopic to a constant map will be called *null homotopic*.

Remark 1.1.12 If Y is path connected any two constant maps $X \longrightarrow Y$ will be homotopic to each other. Hence, the terminology 'null homotopic' is unambiguous.

The following lemma gives some of the other easy ways to know whether a space is contractible or not.

Lemma 1.1.13 The following conditions on a space X are equivalent:
(1) X is homotopy equivalent to a singleton space, i.e., X is contractible.
(2) For every space Y, every map $h : Y \to X$ is null homotopic.
(3) The identity map of X is null homotopic.
(4) For every space Z, every map $h : X \to Z$ is null homotopic.

Proof: $(1) \Longrightarrow (2)$ Let $f : X \to \{*\}$ and $g : \{*\} \to X$ be the homotopy inverses of each other, where $\{*\}$ is any singleton space. Then we have, $g \circ f \simeq Id_X$ and hence, for any map $h : Y \to X$, $g \circ f \circ h \simeq Id_X \circ h = h$. But, since $g \circ f$ is a constant map so is $g \circ f \circ h$.
$(2) \Longrightarrow (3)$ Take $X = Y$ and $h = Id_X$.
$(3) \Longrightarrow (1)$ Given that Id_X is homotopic to a constant map $c : X \to X$, let the image of c be denoted by $\{*\}$. Then we can view c as a map from X to $\{*\}$. Let $g : \{*\} \to X$ be the inclusion map. Clearly, $c \circ g = Id_{\{*\}}$. Moreover, we are given $g \circ c \simeq Id_X$. Thus c is a homotopy equivalence from X to $\{*\}$ which means X is contractible.
$(3) \Longleftrightarrow (4)$ Easy. ♠

Example 1.1.14 The Euclidean space $\mathbb{R}^n$ is contractible. Any convex subspace of $\mathbb{R}^n$ is also contractible. More generally, call a subset S of $\mathbb{R}^n$ *star-shaped* if there is a point $s_0 \in S$ such that, for every $s \in S$, the line segment $[s_0, s]$ is contained in S. In this case, s_0 is called an *apex* of S. Then define $H : S \times \mathbb{I} \to S$, by $H(s, t) = ts_0 + (1-t)s$ to obtain a homotopy of Id_S with a constant map. This shows that, every star-shaped subset of $\mathbb{R}^n$ is contractible. (See Figure 1.6.)

FIGURE 1.6. A star-shaped subset of $\mathbb{R}^2$

Remark 1.1.15 Clearly, a homeomorphism is a homotopy equivalence and hence spaces which are homeomorphic to each other have the same homotopy type. Thus, if we can somehow show that two given spaces X and Y are not of the same homotopy type, then we can conclude that X and Y are not homeomorphic to each other. This is one of the most effective and typical ways in which tools of algebraic topology will be employed in general. Now, what are the means to see that X and Y are not of the same homotopy type? Algebraic topology addresses this problem in a variety of ways, by investigating properties which are preserved under homotopy. These properties are called *homotopy invariants*. Of course, every homotopy invariant is a topological invariant, i.e., preserved under a homeomorphism. However, there are plenty of topological invariants which are not homotopy invariants. One would expect that if two spaces which are of the same homotopy type and share 'all' known topological invariants, then they are homeomorphic to each other. As an example, we have the celebrated, century old, 3-dimensional Poincaré conjecture[1], which states that a 3-dimensional topological manifold which has the same homotopy type of the 3-dimensional sphere $\mathbb{S}^3$ is homeomorphic to $\mathbb{S}^3$. The same question can be asked in any dimension, viz., given an n-dimensional topological manifold M which has the same homotopy type as $\mathbb{S}^n$, is M homeomorphic to $\mathbb{S}^n$. For $n = 1, 2$ it is a classical fact not difficult to prove. (See Chapter 5.) However, for $n \geq 3$ this becomes a very difficult problem. Somewhat surprisingly, the problem becomes a little more tractable for $n \geq 5$ and a positive answer was provided by Smale in the differential manifold case, by Stallings in the piecewise linear case and by Zeeman in the topological case. For, $n = 4$, the problem is even harder. Only the topological version is known to be true due to very deep work of Freedman who was awarded Field's medal for it. We will not be able to discuss the Poincaré conjecture in this book.

The purpose of this book is to lead the reader to the doorsteps of such great results in topology. Algebraic tools have been invented and sharpened by masters while attempting to solve topological problems. This requires the reader to master a formidable amount of technical tools even before understanding what the master is trying to do. We have tried to minimise this with shortcuts without missing out on important points which have certain permanent value. For instance, consider the following classical result due to Brouwer.

Theorem 1.1.16 ((Brouwer's Invariance of domain)) *Let U, V be some subspaces of*

---

[1]In 2002, G. Perelman proved this conjecture using as well as developing deep results in differential geometry. He was awarded Field's medal in 2006 and the Millennium prize in 2010 both of which he has declined.

the Euclidean space $\mathbb{R}^n$. Suppose U is homeomorphic to V. Then U is open in $\mathbb{R}^n$ iff V is open in $\mathbb{R}^n$.

An easy consequence of this is:

Corollary 1.1.17 For $n \neq m$, $\mathbb{R}^n$ is not homeomorphic to $\mathbb{R}^m$.

The standard method of proof of Theorem 1.1.16 is to obtain it as a 'not-too-difficult' consequence of the singular homology theory. We shall present this in Chapter 4. On the other hand, we shall obtain a proof of Theorem 1.1.16 as a consequence of simplicial approximation and Sperner lemma (see Chapter 2), without using any homotopy invariants such as fundamental group or homology groups. There are purely point-set-topological proofs of the invariance of domain which are much too long and difficult. Notice that mere homotopy equivalence is not able to detect the fact that $\mathbb{R}^n$ and $\mathbb{R}^m$ are not homeomorphic for $n \neq m$, since both are contractible. It should be noted that any known proof of the purely set-topological invariance of domain is not too easy (see [Engleking, 1968] or [Hurewicz–Wallman, 1948] for a proof).

In what follows, we shall keep sharpening our tools so as to solve problems mentioned in Q. I and Q. II above and many other related ones.

Example 1.1.18 An Experiment with Möbius Band

Before winding up this section, let us carry out an experiment. We need to equip ourselves with several, long, rectangular strips of waste paper and a glue-stick. First identify the pair of (shorter) opposite sides of a strip of paper without introducing any twists in the strip. (In practice, this is done by 'gluing' the two sides with a small overlap.) The resulting object is easily seen to be a homeomorphic copy of the cylinder $C_0 := \mathbb{S}^1 \times \mathbb{I}$. Now let us take another strip and do the same thing except that before gluing, we give the paper a half-twist, (i.e., through an angle π), to the paper. Call this object C_1. Schematically, we have represented these two operations Figure 1.7 with arrows indicating what 'identification' we have to carry out.

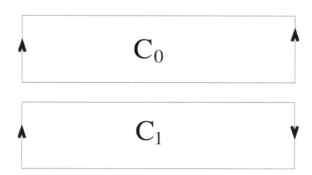

FIGURE 1.7. Paper schemes for the cylinder and the Möbius band

We may carry on and make some more bands, each time giving different numbers of half-twists to the strip and then identifying the edges and call the resulting surfaces $C_2, C_3, \cdots$, etc.

Observe each piece carefully and note as many similarities and differences between them,

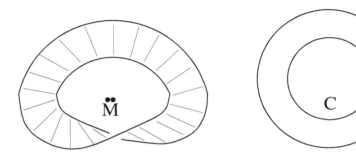

FIGURE 1.8. The Möbius band and the cylinder

such as number of boundary components, number of sides, etc. Now try to answer the following questions:

Q.1 Is each C_i homeomorphic to C_j? or are they all homeomorphically different?

Q.2 If you cut C_i along the central circle, how many pieces would you get. How does each piece look? How is each piece related to the other piece?

Do not actually perform the cut, until you have thought enough on these questions. After giving enough thought, you may take a pen-knife and cut the bands as instructed and verify the results.

The object C_1 is called a Möbius band. We shall now give a formal definition of this. Consider the space $\mathbb{I} \times \mathbb{I}$. Introduce the relation $(0, y) \sim (1, 1 - y)$ for all $y \in \mathbb{I}$. The Möbius band is the quotient of $\mathbb{I} \times \mathbb{I}$ by this relation. We shall denote it by $\ddot{\mathrm{M}}$. It is the simplest example of a surface that is not 'orientable'. Locally you can see that the surface has two sides. However, if you try to colour one side of it, say, green and the other red, you will not succeed.

There are only two homeomorphism classes amongst all the surfaces C_i that we have constructed. All C_{2n} are homeomorphic to C_0 which is easily seen to be a cylinder $\mathbb{S}^1 \times \mathbb{I}$. All C_{2n+1} are homeomorphic to $C_1 = \ddot{\mathrm{M}}$. The difference is that each C_i is 'embedded' in a different way in the Euclidean 3-space.

Finally, we shall give a hypothetical (not a practical) method of constructing a Möbius band, obviously different from the method described above. Consider the space $X = \mathbb{S}^1 \times \mathbb{I}$. Identify $(x, 0)$ with $(-x, 0)$ for each $x \in \mathbb{S}^1$. Denote the quotient space by Y. We claim Y is homeomorphic to a Möbius band. In practice, it is not easy to carry out this antipodal identification on the bottom circle of the cylinder.

So, what we suggest is the following: Cut the cylinder back into a strip. We will have to put some indicator arrows along the resulting edges so as not to loose track of where we have done the cuttings, since ultimately we need to carry out identification along these cuts. Indeed in the first place, we have obtained the cylinder from a rectangular strip by identifying a pair of opposite sides. So, take the strip and just do not perform this operation of identification yet. Now check whether we can carry out the antipodal identifications $(x, 0)$ with $(-x, 0)$? (See Figure 1.9.)

Even this step is difficult since all the identification is taking place within the bottom side. So, we perform one more cut, viz., along the vertical middle line to obtain two rectangular pieces. Now it is possible to carry out the 'antipodal identifications', since it has got converted into identifying one side of a strip with a side of another strip. After this identification is carried out, we obtain one rectangular strip with indicators on its boundary

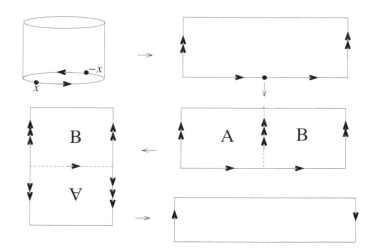

FIGURE 1.9. Another scheme for the Möbius band

for further identifications for the two cuts that we have performed. It is easy to see that these two indicators can be combined to one single indicator and what we have is nothing but a scheme for the Möbius band.

This entire process above has been perfected into a method called the 'cut and paste' technique in the study of low-dimensional topology. We shall use this technique extensively in the proof of the classification of surfaces in Chapter 5.

We conclude this introductory section with the following words:
> ''... algebraic topology does not consist solely of the juggling of categories and functors and the like, but it has some genuine geometric content.''
> —G.W. Whitehead

Exercise 1.1.19

(i) Show that a contractible space is path connected.

(ii) List half a dozen topological properties not preserved under homotopy. Justify your list.

(iii) Show that composite of two homotopy equivalences is a homotopy equivalence.

(iv) Show that homotopy equivalence amongst spaces is an equivalence relation.

(v) Show that a map which is homotopic to a homotopy equivalence is a homotopy equivalence.

(vi) Let $f : X \longrightarrow Y$, $g : Y \longrightarrow Z$ be such that f and $g \circ f$ are homotopy equivalences. Show that g is a homotopy equivalence.

(vii) Let $f : X \longrightarrow Y$, $g : Y \longrightarrow X$ be maps such that $f \circ g$ and $g \circ f$ are homotopy equivalences. Then show that f and g are homotopy equivalences. (Observe that this does not mean that g is the homotopy inverse of f.)

1.2   Fundamental Group

This section contains the definition of the fundamental group and its functorial properties. We shall also introduce two best known 'methods' of computing the fundamental group and use them to compute the fundamental group of the spheres $\mathbb{S}^n, n \geq 1$. Extensive study of these methods will then be taken up in Chapter 3.

Recall that a path in a space can be thought of as the track of a moving point. The fact that we may 'move' from one point to another point in a continuous way within a space is described by saying that the space is path connected. We know that the set of path components of a space is an important topological invariant. This can be viewed as $[[*, X]]$, the set of homotopy classes of maps from a point space $*$ to X.

Given a path connected space X, we are now interested in looking at various 'different ways' in which two given points may be joined in X. For example, suppose X is the disc $\mathbb{D}^2$. Then given any two points in X, the natural way to join them is to take the line segment between them. If we are not so economical, there will be a lot of nearby paths but they would all be the 'same' in the sense that they are all homotopic. On the other hand, suppose the space is $\mathbb{S}^1$. Then given any two points in $\mathbb{S}^1$, the natural way is to trace the circular arc from one point to the other. Obviously there are two choices here. We can say that the shorter one is a better choice. But then if the two points are antipodal there will be two distinct choices to make and common sense tells us that mathematically also we should distinguish them. Since a path in X is described by a continuous function $\mathbb{I} \to X$, and since $\mathbb{I}$ is contractible, it follows that any two paths are homotopic. So, homotopy as considered in the previous section is not exactly the tool that is going to help us here.

So, we fix two points $x_0, x_1 \in X$ and look at the space $\Omega(X, x_0, x_1)$ of all possible paths in X from x_0 to x_1, with the compact-open-topology. We may then look at the path components of this space. These turn out to be nothing but the classes of paths which are homotopy equivalent to each other by a homotopy which *keeps the end-points fixed*. This is the modification in the concept of homotopy that is going to play the crucial role.

Following the simple common sense rule of tracing one curve until its end-point and then tracing another curve which begins at the end-point of the first curve, we get a binary operation on the set of all loops at a given point in a space. The constant loop is expected to play the role of a two-sided identity and tracing a given loop in the opposite direction should play the role of taking the inverse. A moment's reflection tells us that this is not exactly the case. However, our expectations are met when we pass onto the homotopy classes of loops—we obtain a very powerful notion, viz., the fundamental group of a space, which is going to play a very important role in the study of topological behaviour of a space. Let us lay down a sound foundation for this important notion.

Definition 1.2.1 By a *path* in X, we mean a continuous function $\mathbb{I} \longrightarrow X$. If $\omega : \mathbb{I} \longrightarrow X$ is a path, then $\omega(0)$ is called the *initial point* of ω and $\omega(1)$ is called the *terminal point* of ω. These two points are also called *end-points* of ω. When they coincide, the path ω is called a *loop* in X based at $\omega(0) = \omega(1)$.

Definition 1.2.2 Let $\omega, \tau : \mathbb{I} \to X$ be any two paths with the same end-points, i.e., $\omega(0) = \tau(0) = x_0, \omega(1) = \tau(1) = x_1$. By a *path-homotopy* from ω to τ in X, we mean a map $H : \mathbb{I} \times \mathbb{I} \longrightarrow X$ such that

$$H(0, s) = x_0, \; H(1, s) = x_1; \; \& \; \omega(t) := H(t, 0), \quad \tau(t) := H(t, 1), 0 \leq t \leq 1.$$

If there exists such a path homotopy, we say that the two paths ω, τ are *path-homotopic in* X and write this

$$\omega \simeq \tau. \tag{1.2}$$

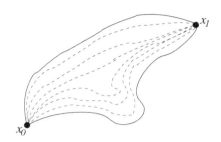

FIGURE 1.10. A path homotopy

Now fixing two points $x_0, x_1 \in X$, on the set of all paths in X with initial point x_0 and end-point x_1, it is easily seen that path homotopy is an equivalence relation. We shall denote the equivalence class represented by a path ω by $[\omega]$. Notice that path homotopy is more restrictive than the homotopy of maps which we have defined in the previous section.

Example 1.2.3 A typical example of homotopic paths is obtained by taking any path and re-parameterizing it: Given a path $\omega : \mathbb{I} \longrightarrow X$, by a re-parameterisation of ω we mean a path $\omega \circ \alpha$ where $\alpha : \mathbb{I} \longrightarrow \mathbb{I}$ is any map such that $\alpha(0) = 0$ and $\alpha(1) = 1$. Observe that $A(t, s) = (1 - s)\alpha(t) + st$ gives a homotopy of α with the identity map, relative to $\{0, 1\}$. Therefore, it follows that all re-parameterisations of a given path are path homotopic to each other. (*It may be noted here that in differential geometry, wherein we are concerned about the concept of length, etc., a parameterisation α is required to be smooth and satisfy $\alpha'(t) > 0$ for all $t \in \mathbb{I}$.*)

Definition 1.2.4 Let X be a topological space. Let $\omega, \tau : \mathbb{I} \longrightarrow X$ be two paths such that the terminal point $\omega(1)$ of ω is the same as the initial point $\tau(0)$ of τ. We then define $\omega * \tau : \mathbb{I} \longrightarrow X$ by the formula

$$\omega * \tau(t) = \begin{cases} \omega(2t), & 0 \leq t \leq 1/2, \\ \tau(2t - 1), & 1/2 \leq t \leq 1. \end{cases} \tag{1.3}$$

The usefulness of this operation is essentially due to its path-homotopy invariance:

Lemma 1.2.5 If $\omega_1 \simeq \omega_2$ and $\tau_1 \simeq \tau_2$ and $\omega_1 * \tau_1$ is defined then $\omega_2 * \tau_2$ is defined and we have, $\omega_1 * \tau_1 \simeq \omega_2 * \tau_2$.

Proof: Given a path-homotopy $H : \mathbb{I} \times \mathbb{I} \longrightarrow X$ between ω_1 and ω_2 and a path-homotopy $G : \mathbb{I} \times \mathbb{I} \longrightarrow X$ between τ_1 and τ_2, first observe that the initial point of τ_2 is the same as the initial point of τ_1 which is the same as the terminal point of ω_1 which is equal to the terminal point of ω_2. Therefore $\omega_2 * \tau_2$ is defined. Now consider

$$F(t, s) = \begin{cases} H(2t, s), & 0 \leq t \leq 1/2, \\ G(2t - 1, s), & 1/2 \leq t \leq 1, \end{cases} \tag{1.4}$$

and check that this gives a homotopy of the composites as required. ♠

The next lemma tells us about some basic algebraic properties of the path-composition.

Lemma 1.2.6
(i) **Associativity:** If $\omega * (\tau * \lambda)$ is defined then so is $(\omega * \tau) * \lambda$ and the two are path-homotopic.

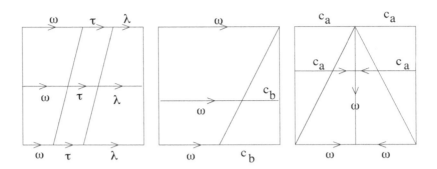

FIGURE 1.11. Group laws for the fundamental group

(ii) **Identity:** Let c_x denote the constant path at $x \in X$. Then for any path ω, we have $\omega * c_b \simeq \omega \simeq c_a * \omega$ where $a = \omega(0), b = \omega(1)$.

(iii) **Inverse:** For any path ω, let $\underline{\omega}$ be the path given by $\underline{\omega}(t) = \omega(1 - t)$. Then $\omega * \underline{\omega} \simeq c_a$ and $\underline{\omega} * \omega \simeq c_b$ where a, b are initial and terminal points of ω, respectively.

Proof: The first two statements follow from Example 1.2.3, once we observe that the path on the right side is a re-parameterisation of the path on the left.

(i) Observe that in defining the LHS, we first divide the interval $\mathbb{I}$ into two parts and the first half is divided into two parts to be shared by ω and τ. On the other hand, for the RHS we first divide $\mathbb{I}$ into two parts and then divide the second half into two parts to be shared by τ and λ. So, the re-parameterisation map α has to be taken such that $0 \mapsto 0, 1/4 \mapsto 1/2, 1/2 \mapsto 3/4$ and $1 \mapsto 1$. We extend this linearly in each subinterval. If γ_1 and γ_2 denote the LHS and RHS, respectively, then check that $\gamma_1 = \gamma_2 \circ \alpha$.

(ii) Here consider the maps α_1, α_2 defined by

$$\alpha_1(t) = 2t, 0 \leq t \leq 1/2; \; \alpha_1(t) = 1, \; 1/2 \leq t \leq 1.$$

$$\alpha_2(t) = 0, 0 \leq t \leq 1/2; \; \alpha_2(t) = 2t - 1, \; 1/2 \leq t \leq 1.$$

Note that $\omega * c_b = \omega \circ \alpha_1$ and $c_a * \omega = \omega \circ \alpha_2$. The first diagram in Figure 1.11 indicates the required homotopy. (iii) In this case, we write down the homotopy. Define

$$H(t, s) = \begin{cases} \omega(0), & 0 \leq t \leq \frac{s}{2}, \\ \omega(2t - s), & \frac{s}{2} \leq t \leq \frac{1}{2}, \\ \omega(2 - 2t - s), & \frac{1}{2} \leq t \leq \frac{2-s}{2}, \\ \omega(1), & \frac{2-s}{2} \leq t \leq 1. \end{cases}$$

Then H is the required homotopy from $\omega \circ \underline{\omega}$ to c_a. Schematically, these homotopies are represented in Figure 1.11.

For instance consider case (iii). At time s, the curve $H(-, s)$ remains idle at the point $\omega(0)$ until $t = s/2$ and then starts tracing the curve ω until it reaches $\omega(1 - s)$ and then retraces it back to $\omega(0)$ and then takes a rest for the remaining time. By symmetry, the homotopy between $\underline{\omega} * \omega$ with c_b also follows. ♠

Remark 1.2.7

(a) It is interesting to note that during these homotopies the entire action is taking place in the domain itself and so the proofs that the composition satisfies associativity, etc., do not depend upon the actual paths.

(b) Because of property (iii) in the above lemma, many authors use the notation $\underline{\omega}^{-1}$ for $\underline{\omega}$. We shall also make use of this notation. However, do not confuse this for the inverse of ω as a function.

(c) Intuitively, any continuous map $\omega : [a, b] \to X$ should be called a path in X, where $[a, b]$ is any closed interval. Our definition of a path as a map from the closed interval $\mathbb{I} = [0, 1]$ causes a minor irritation: if you restrict a path $\omega : [0, 1] \to X$ to a closed subinterval, it is no longer a path which is contrary to our intuition. Notice that the composition law had to be defined after re-parameterisation only, even if we adopt the more general definition. Strict associative law fails in either case. Similarly, the constant path is not a strict unit. Indeed, there are a few different ways to avoid some of these difficulties but they will acquire other difficulties. See Exercise 1.9.40 for one such.

The definition we have adopted is not at all restrictive once we pass on to path homotopy classes. The crux of the matter is that we can use the standard parameterisation of any closed interval $[a, b]$ by the interval $[0, 1]$ and think of a map $\gamma : [a, b] \to X$ as a path in X, viz.,

$$\hat{\gamma}(t) := \gamma((b - a)t + a).$$

The following elementary result answers all the above objections satisfactorily and therefore we shall stick to our definition of a path.

Theorem 1.2.8 (Invariance under subdivision) *Let* $\gamma : [0, 1] \to X$ *be a path, and* $0 < t_1 < \cdots < t_n < 1$. *Then* γ *is path homotopic to* $\widehat{\gamma|_{[0,t_1]}} * \cdots * \widehat{\gamma|_{[t_n,1]}}$.

Proof: It is enough to prove this for $n = 1$ with $t_1 = a$. By repeated application of this we get the general case. So, for $0 < a < 1$, consider the following parameterisation of the unit interval: $\alpha : [0, 1] \to [0, 1]$ given by

$$\alpha(t) = \begin{cases} 2at, & 0 \le t \le 1/2, \\ 2t - 1 + (2 - 2t)a, & 1/2 \le t \le 1. \end{cases}$$

Check that $\widehat{\gamma|_{[0,a]}} * \widehat{\gamma|_{[a,1]}}) = \gamma \circ \alpha$. Therefore from Example 1.2.3, it follows that it is path homotopic to γ. ♠

Definition 1.2.9 The Fundamental Group Let X be any topological space and $x_0 \in X$. By the above two lemmas it follows that the set $\pi_1(X, x_0)$ consisting of all path-homotopy classes of loops in X based at x_0 forms a group under the path composition. We call this group *the fundamental group* of X at x_0.

Note that from property (iii) in the above lemma, it follows that for a loop ω, the inverse of $[\omega]$ in the group $\pi_1(X, x_0)$ is $[\underline{\omega}]$ or in the new notation $[\omega^{-1}]$, i.e., $[\omega]^{-1} = [\omega^{-1}]$.

Remark 1.2.10 Since a loop and its homotopies will always be contained in the same path component of a space X, it follows that $\pi_1(X, x_0)$ is equal to $\pi_1(C, x_0)$ where C is the path component of X containing the point x_0. Because of this reason, while discussing π_1 (and also most other algebraic topology questions) we may as well assume that the space is path connected by restricting our attention to the path component that is involved. The following theorem describes what happens when we change base points within a path component.

Theorem 1.2.11 *Let* τ *be a path in a space* X *with* a, b *as its initial and terminal points, respectively. Then the assignment* $[\omega] \mapsto [\tau^{-1} * \omega * \tau]$ *defines an isomorphism* $h_{[\tau]} : \pi_1(X, a) \longrightarrow \pi_1(X, b)$.

Proof: Verify directly, that $h_{[\tau]}$ is a homomorphism. Use Lemma 1.2.6 (iii) to verify that the inverse of $h_{[\tau]}$ is nothing but the homomorphism $h_{[\tau^{-1}]}$. ♠

Remark 1.2.12 Thus while dealing with a path connected space, we often need not mention the base point at which the fundamental group is being taken, since any such choice leads to the same group up to isomorphism. It should be noted that the isomorphism depends upon the path τ in general and hence it is dangerous to identify $\pi_1(X, a)$ with $\pi_1(X, b)$ blindly. Even though there are many situations wherein it is enough to know that the two groups are isomorphic, there are situations wherein the isomorphism chosen may play a crucial role. Exercise 1.2.33(v) tells you about a situation in which we need not worry about this.

Definition 1.2.13 Let X be a path connected space. We say X is simply connected if $\pi_1(X, x_0) = (1)$ for some point $x_0 \in X$ (and hence for every point $x_0 \in X$).

Remark 1.2.14 Recall that the map $\theta \mapsto e^{2\pi i \theta}$ may be used to identify the quotient space $\mathbb{I}/\{0, 1\}$ with the unit circle $\mathbb{S}^1$. It follows from the properties of quotient spaces that this identification, in turn, induces an identification of $\pi_1(X, a)$ with the set $[(\mathbb{S}^1, 1); (X, a)]$ of relative homotopy classes of base point preserving maps from $\mathbb{S}^1$ to X. This description of π_1 comes often very handy. As an illustration we have:

Corollary 1.2.15 If X is a contractible space then X is simply connected.

Proof: We have already seen that every contractible space is path connected. Given a loop $\omega : (\mathbb{S}^1, 1) \to (X, a)$ since X is contractible, we know that ω is null homotopic (see Theorem 1.1.13). Therefore, from Theorem 1.5.5 it follows that ω is null homotopic relative to 1. ♠

Remark 1.2.16 While studying complex functions of one variable, the reader may have come across the definition of 'simple connectivity' which is quite different from the one that we have given here. Consider, for instance, a domain Ω in $\mathbb{C}$ which satisfies the property that its complement in the extended complex plane $\widehat{\mathbb{C}}$ is connected. Complex function theory offers a proof that this property is equivalent to the simple connectedness of the domain. However the proof is not easy and one has to go through a deep theorem, viz., the Riemann mapping theorem that says every simply connected domain in $\mathbb{C}$ is bi-holomorphic to $\mathbb{C}$ itself or to the open unit disc. In particular, it follows that every simply connected domain is contractible. Certainly, the reader will soon see that such is not the case, even for domains in $\mathbb{R}^n, n \geq 3$, let alone for arbitrary topological spaces. We know that open subsets of the real line do not display any too complicated topological properties. The case in dimension 2 is a little more complicated. So, the reader may expect that the complications increase as the dimension increases.

Remark 1.2.17 The importance of the assignment $(X, a) \rightsquigarrow \pi_1(X, a)$ is enhanced by the fact that it is functorial. Thus, given a map $f : (X, a) \longrightarrow (Y, b)$, we have a homomorphism $f_\# : \pi_1(X, a) \longrightarrow \pi_1(Y, b)$ defined by

$$f_\#[\omega] = [f \circ \omega]. \tag{1.5}$$

This is well defined for if $\omega \simeq \tau$ then $f \circ \omega \simeq f \circ \tau$. Since $f \circ (\omega * \tau) = (f \circ \omega) * (f \circ \tau)$, it follows that $f_\#$ is a homomorphism. Moreover, we have the following two properties which are verified directly.
(i) For any space X, if $Id : X \longrightarrow X$ denotes the identity map then the induced homomorphism on the fundamental groups is the identity homomorphism, i.e., $Id_\# = Id$.
(ii) Given maps $f : (X, a) \longrightarrow (Y, b)$ and $g : (Y, b) \longrightarrow (Z, c)$ we have $(g \circ f)_\# = g_\# \circ f_\#$.

This property is going to play a very crucial role throughout the study of fundamental group. For instance this immediately implies that if two spaces are homeomorphic then they have isomorphic fundamental groups.

Further, observe that if f and g are homotopic maps relative to the base point, then they induce the same homomorphism on the fundamental groups.

Remark 1.2.18 As yet, we do not know any example of a space for which π_1 is non trivial. We shall now provide such an example, viz., $\mathbb{S}^1$. In some sense, this is the simplest example and the actual computation is an illustration of a powerful notion called *covering spaces* that we are going to study later.

We shall exploit the map $\exp : \mathbb{R} \longrightarrow \mathbb{S}^1$ defined by $\theta \mapsto e^{2\pi i\theta}$, and the fact that $\pi_1(\mathbb{R}, r) = (1)$ to compute $\pi_1(\mathbb{S}^1, 1)$. Recall that $\exp$ is a surjective map and $\exp(t_1) = \exp(t_2)$ iff $t_1 - t_2$ is an integer. (See Figure 1.12.) In particular, we have

Lemma 1.2.19 For every point $z \in \mathbb{S}^1$, the open set $\exp^{-1}(\mathbb{S}^1 \setminus \{z\})$ is a disjoint union of intervals and $\exp$ restricted to each one of these intervals is a homeomorphism onto $\mathbb{S}^1 \setminus \{z\}$.

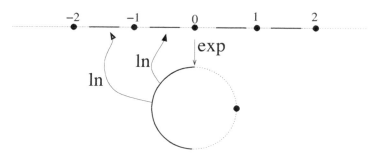

FIGURE 1.12. The exponential function and its local inverses

Remark 1.2.20 An inverse of $\exp$ defined on any sub-arc of $\mathbb{S}^1$ is called a branch of the logarithm. Maximal sub-arcs on which a branch of logarithm may be defined are $\mathbb{S}^1 \setminus \{z\}$ for some z. In what follows, we will use branches of logarithm defined on open arcs $\mathbb{S}^1 \setminus \{\pm 1\}$.

In the following lemma, we begin to relate maps into $\mathbb{R}$ with those into $\mathbb{S}^1$ via $\exp$.

Lemma 1.2.21 Let X be any connected space and $f_1, f_2 : X \longrightarrow \mathbb{R}$ be any two continuous functions such that $\exp \circ f_1 = \exp \circ f_2$. Then there exists an integer n such that $f_1(x) - f_2(x) = n$ for all $x \in X$.

Proof: The map $g := f_1 - f_2 : X \longrightarrow \mathbb{R}$ has the property that $\exp(g(x)) = 1$ for all x. Therefore, $g : X \longrightarrow \mathbb{R}$ is a map which takes only integral values. Since X is connected, this must be a constant function $g(x) = n$ for some n and for all x. ♠

Proposition 1.2.22 Let $f : \mathbb{I} \longrightarrow \mathbb{S}^1$ be any map such that $f(0) = 1$. Then there exists a unique map $g : \mathbb{I} \longrightarrow \mathbb{R}$ such that $g(0) = 0$ and $\exp \circ g = f$.

Proof: From the previous lemma, the uniqueness follows. So, we have to show only the existence of g. Let $Z = \{t \in \mathbb{I} : g$ is defined in $[0, t]\}$. Observe that by the very definition, Z is a subinterval of $\mathbb{I}$ and contains 0. Let t_0 be the least upper bound of Z. It is enough to show that $t_0 \in Z$ and $t_0 = 1$. Consider the open set $V = \mathbb{S}^1 \setminus \{-f(t_0)\}$. For $0 < \epsilon < 1$ put $I_\epsilon = [t_0 - \epsilon, t_0 + \epsilon] \cap \mathbb{I}$. Then by continuity, there exists $\epsilon > 0$ such that $f(I_\epsilon) \subset V$. Now use Lemma 1.2.19. Let $\ln : V \longrightarrow U$ be the inverse of $\exp$ where U is the interval containing $g(t_0 - \epsilon)$ and contained in $\exp^{-1}(V)$. Take $h = \ln \circ f$ on I_ϵ. Then, we have $g(t_0 - \epsilon/2) = h(t_0 - \epsilon/2)$ and $\exp \circ g = \exp \circ h$ on the interval $[t_0 - \epsilon, t_0)$ Hence, by the uniqueness again, we have $g = h$ on this interval. Therefore, we can extend g continuously on $Z \cup I_\epsilon$. This first of all implies that $t_0 \in Z$. Secondly, if $t_0 < 1$, then this interval will contain numbers larger than t_0, which will be absurd. Therefore $t_0 = 1$. ♠

Definition 1.2.23 Given a map $f : \mathbb{I} \longrightarrow \mathbb{S}^1$ such that $f(0) = 1$, we take the unique map $g : \mathbb{I} \longrightarrow \mathbb{R}$ as in the above proposition. Further, if f is a loop, i.e., $f(1) = 1$, then it follows that $g(1)$ is an integer. We call this integer the *degree* of f and write deg f for it. In particular, given any map $f : \mathbb{S}^1 \to \mathbb{S}^1$, we can view it as a loop via the parameter $t \mapsto e^{2\pi i t}$, i.e., $t \mapsto f(e^{2\pi i t})$, take the corresponding map $g : \mathbb{I} \to \mathbb{R}$ and call $g(1)$ the degree of f.

Remark 1.2.24 The justification for this terminology is in the following example. Take $f(z) = z^n$. Then the map g is nothing but $g(t) = nt$ and hence $g(1) = n$ which coincides with the degree of f. The important thing about the degree is that it is a homotopy invariant and respects the group laws. In Chapter 4, we shall generalize this concept to all spheres and then in Chapter 8, to all manifolds.

Proposition 1.2.25 If $f_1 \simeq f_2$ then deg $f_1 =$ deg f_2.

Proof: Let $H : \mathbb{I} \times \mathbb{I} \longrightarrow \mathbb{S}^1$ be a homotopy of f_1 to f_2 relative to $\{0, 1\}$. Put $U_\pm = \mathbb{S}^1 \setminus \{\pm 1\}$. Then $\{U_+, U_-\}$ forms an open cover for $\mathbb{S}^1$. Therefore, by the compactness of $\mathbb{I} \times \mathbb{I}$ there exists $\delta > 0$ such that if S is any sub-square of $\mathbb{I} \times \mathbb{I}$ of side less than δ then $H(S)$ is contained in U_+ or U_-. We can now subdivide $\mathbb{I}$ into intervals $0 < \frac{1}{n} < \cdots < \frac{n-1}{n} < 1$ such that the corresponding sub-squares of $\mathbb{I} \times \mathbb{I}$ are all mapped into U_+ or U_- by H.

Let $G : \mathbb{I} \times \mathbb{I} \longrightarrow \mathbb{R}$ be a function such that
(i) $\exp \circ G = H$ and
(ii) for all $t \in \mathbb{I}$ the function $s \mapsto G(t, s)$ is continuous and $G(t, 0) = 0$.
Such a function exists by Proposition 1.2.22. We claim that G is actually a continuous function on the whole of $\mathbb{I} \times \mathbb{I}$. For this it is enough to prove that it is so restricted to each sub-square $S_{k,l} := [\frac{k}{n}, \frac{k+1}{n}] \times [\frac{l}{n}, \frac{l+1}{n}]$. This we do by induction on l. For $l = 0$, consider $S_{k,0}$. Clearly $H(S_{k,0}) \subset \mathbb{S}^1 \setminus \{-1\}$. Therefore $G(S_{k,0})$ is contained in the disjoint union

$$\coprod_n \left(n - \frac{1}{2}, n + \frac{1}{2} \right)$$

Since $G(t, 0) = 0$ for all t, it follows that $G(t \times [0, \frac{1}{n}]) \subset (-\frac{1}{2}, \frac{1}{2})$ by continuity of $G|_{t \times \mathbb{I}}$. Thus $G(S_{k,0}) \subset (-\frac{1}{2}, \frac{1}{2})$ on which $\exp$ is a homeomorphism. Hence $G = \ln \circ H$ on $S_{n,0}$ where $\ln$ is the inverse map of $\exp$ on $\mathbb{S}^1 \setminus \{-1\}$. In particular, G is continuous on $S_{k,0}$.

Inductively, assume that we have proved the continuity of G on $S_{k,l}$. In particular, this implies that $G([\frac{k}{n}, \frac{k+1}{n}] \times \{\frac{l+1}{n}\})$ is contained in an interval of the type $(n - \frac{1}{2}, n + \frac{1}{2})$ or of the type $(n, n+1)$ for some integer n. As in the case $l = 0$ above, this then implies that $G(t \times [\frac{l+1}{n}, \frac{l+2}{n}])$ are contained in the same interval for all $t \in [\frac{k}{n}, \frac{k+1}{n}]$. That is $G(S_{k,l+1})$ is contained in an interval on which $\exp$ is a homeomorphism. Therefore $G = \ln \circ H$ for a suitably chosen branch of the logarithm. This proves the continuity of G. In particular, $G(-, 1) : \mathbb{I} \longrightarrow \mathbb{R}$ is continuous. But this is an integer valued function. Hence it is a constant. In particular deg $f_1 = G(0, 1) = G(1, 1) =$ deg f_2. ♠

Remark 1.2.26 Thus we have a well-defined function

$$\deg : \ \pi_1(\mathbb{S}^1, 1) \longrightarrow \mathbb{Z}, \quad [f] \mapsto \deg \ f.$$

In what follows we shall see that this is indeed an isomorphism.

Theorem 1.2.27 *The function* deg $: \pi_1(\mathbb{S}^1) \longrightarrow \mathbb{Z}$ *is an isomorphism.*

Proof: To prove that deg is a homomorphism, for $i = 1, 2$, given two loops $f_i : \mathbb{I} \longrightarrow \mathbb{S}^1$, bases at 1, let g_i be such that $\exp \circ g_i = f_i$ and $g_i(0) = 0$. Put $h(t) = g_2(t) + $ deg $f_1 =$

$g_2(t) + g_1(1)$. Then observe first that $\exp \circ h = f_2$ and then that $\exp \circ (g_1 * h) = f_1 * f_2$. Moreover, $g_1 * h(0) = g_1(0) = 0$. Therefore $\deg (f_1 * f_2) = (g_1 * h)(1) = h(1) = g_2(1) + g_1(1) = \deg f_2 + \deg f_1$. This proves that deg is a homomorphism.

It is easily checked that $\deg Id = 1$. Therefore deg is surjective. Finally suppose $\deg f = 0$. This means that we have a map $g : \mathbb{I} \longrightarrow \mathbb{R}$ such that $\exp \circ g = f$ and $g(0) = g(1) = 0$. Consider the homotopy $G(t, s) = sg(t)$. Clearly, it is a homotopy of the 0 map with g, relative to the end-points. Then $\exp \circ G$ defines a homotopy of the constant function 1 with f relative to $\{0, 1\}$. This proves the injectivity of deg. ♠

One can give a number of applications. Here are just two samples. Some others are included in the exercises.

Corollary 1.2.28 The boundary of the disc $\mathbb{D}^2$ is not a retract of $\mathbb{D}^2$.

Proof: If $r : \mathbb{D}^2 \to \mathbb{S}^1$ is a retraction, and $\iota : \mathbb{S}^1 \to \mathbb{D}^2$ is the inclusion map, then we have $r \circ \iota = Id_{\mathbb{S}^1}$. Therefore, on the fundamental groups, we have

$$Id_{\mathbb{Z}} = r_\# \circ \iota_\# : \pi_1(\mathbb{S}^1) \to \pi_1(\mathbb{D}^2) \to \pi_1(\mathbb{S}^1)$$

which is absurd, since $\pi_1(\mathbb{D}^2) = (1)$. ♠

Corollary 1.2.29 (Brouwer's fixed point theorem for $\mathbb{D}^2$) Every continuous function $f : \mathbb{D}^2 \to \mathbb{D}^2$ has a fixed point.

Proof: Suppose there is a map $f : \mathbb{D}^2 \to \mathbb{D}^2$ such that $f(x) \neq x$ for any x. Extend the unique line segment $[f(x), x]$ in the direction from $f(x)$ to x so as to meet the circle in a unique point $g(x)$.

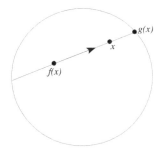

FIGURE 1.13. The sphere is not a retract of the disc

Indeed, the line is parameterised by

$$t \mapsto (1 - t)x + tf(x), \quad t \in \mathbb{R},$$

and $g(x) = (1 - t_0)x + t_0 f(x)$ where t_0 is the root of the quadratic equation in t

$$t^2 \|v\|^2 + 2tv \cdot x + \|x\|^2 - 1 = 0$$

such that $t_0 \leq 0$. Here $v = f(x) - x$. Since $\|x\|^2 - 1 \leq 0$, it follows that the discriminant of this quadratic is non negative and identically zero iff $\|x\|^2 = 1$ and $v \cdot x = 0$. But then $\|f(x)\|^2 = \|x\|^2 + \|v\|^2 > 1$ which is absurd. Therefore, the discriminant is strictly positive and hence the two roots are continuous. In particular, t_0 is a continuous function of the variable x.

Hence g is a continuous function which coincides with x if $\|x\| = 1$. Thus $g : \mathbb{D}^2 \to \mathbb{S}^1$ is a retraction, contradicting the above corollary. ♠

Remark 1.2.30 The method adopted above in the computation of the fundamental group of $\mathbb{S}^1$ leads to the notion of covering spaces and its relation with the fundamental group. We shall end this section by initiating another powerful method of computing the fundamental group which goes under the name Seifert–van Kampen Theorems. Both these methods will be taken up again in later chapters for further investigation.

Theorem 1.2.31 (Seifert–van Kampen theorem version-1) *Let* $X = U \cup V$ *where* U *and* V *are open subsets of* X *and* $U \cap V$ *is path connected. Suppose further that for some* $x_0 \in U \cap V$, *the inclusion maps* $i : U \to X, j : V \to X$ *induce homomorphisms* $i_\# : \pi_1(U, x_0) \to \pi_1(X, x_0)$ *and* $j_\# : \pi_1(V, x_0) \to \pi_1(X, x_0)$ *which are trivial. Then* $\pi_1(X, x_0) = (1)$.

Proof: Let $\gamma : \mathbb{I} \to X$ be a loop at x_0. It follows that $\{\gamma^{-1}(U), \gamma^{-1}(V)\}$ is an open covering for the interval $[0, 1]$. Let $\delta > 0$ be the Lebesgue number of this cover. Choose a partition $0 < t_1 < \cdots < t_n < 1$ such that $|t_{i+1} - t_i| < \delta/2$ so that $[t_i, t_{i+1}]$ is contained in one of the two open sets. Without loss of generality, we may assume $[0, t_1] \subset \gamma^{-1}(U)$. By dropping some of the points t_i, we can further assume that two consecutive intervals are not contained in the same open set. Thus it follows that $\gamma[t_i, t_{i+1}] \subset U$ for i even and is contained in V for i odd.

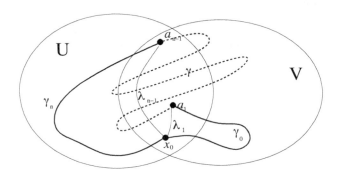

FIGURE 1.14. A simple version of Seifert–van Kampen Theorem

Our aim is to prove that γ is path homotopic to the constant loop c at x_0 in X. We induct on the number n of divisions required to express γ in the above form. If $n = 1$, this already implies that γ itself is contained in U and by the hypothesis that $i_\#$ is the trivial homomorphism, the conclusion follows.

Now assume that $n \geq 2$ and that the claim holds whenever we can express a path γ in the above form with fewer than n divisions. Put $\gamma_i = \gamma|_{[t_i, t_{i+1}]}$ so that we have

$$\gamma \simeq \gamma_0 * \gamma_1 * \cdots * \gamma_n$$

and γ_i are alternatively inside U and V. (See Figure 1.14.)

Note that $a_i := \gamma(t_i), i = 1, \ldots, n-1$ are all in $U \cap V$. Since $U \cap V$ is path connected, we can choose paths λ_i, in $U \cap V$ joining x_0 to a_i. Then the loop $\lambda_{n-1} * \gamma_n$ based at x_0 is completely contained in U, or V. By the hypothesis that $i_\#, j_\#$ are the trivial homomorphisms, it follows that this loop is homotopic to the constant loop c at x_0 in X. Therefore,

$$\begin{aligned}
\gamma & \simeq & \gamma_0 * \gamma_1 * \cdots * \gamma_n \\
& \simeq & \gamma_0 * \gamma_1 * \cdots * \gamma_{n-1} * \lambda_{n-1}^{-1} * (\lambda_{n-1} * \gamma_n) \\
& \simeq & \gamma_0 * \gamma_1 * \cdots * \gamma_{n-1} * \lambda_{n-1}^{-1} * c \\
& \simeq & \gamma_0 * \gamma_1 * \cdots * \gamma_{n-1}'
\end{aligned}$$

where we have put $\gamma'_{n-1} = \gamma_{n-1} * \lambda_{n-1}^{-1} * c$ which is a path completely contained in U or V. By induction hypothesis, it follows that γ is path homotopic the constant loop. ♠

Corollary 1.2.32 $\pi_1(\mathbb{S}^n) = (1), n \geq 2$.

Proof: Write $\mathbb{S}^n = U \cup V$, where $U = \mathbb{S}^n \setminus \{(0, \ldots, 0, 1)\}$, $V = \mathbb{S}^n \setminus \{(0, \ldots, 0, -1\}$, Then by stereographic projection we know that both U and V are homeomorphic to $\mathbb{R}^n$ and hence contractible. Also, it is clear that U, V are both open and $U \cap V$ is connected. (This is where you need the hypothesis that $n \geq 2$. It follows that all the hypotheses in the above theorem are satisfied and hence $\pi_1(\mathbb{S}^n) = (1)$. ♠

Exercise 1.2.33

(i) Let ω and τ be two paths with same end-points, in a space X. Show that $\omega \simeq \tau$ iff $\omega * \tau^{-1}$ is null homotopic.

(ii) Show that any homomorphism $\alpha : \mathbb{Z} \longrightarrow \mathbb{Z}$ can be thought of as induced by a map $f : \mathbb{S}^1 \longrightarrow \mathbb{S}^1$ on the fundamental group. Show that such a map is unique up to homotopy. [Remark: Such a result is not true for arbitrary spaces. However, analogous results hold for all spheres (see Remark 10.2.25).]

(iii) Let $A \subset X$ and $a \in A$. Show that the inclusion induced homomorphism $\pi_1(A, a) \longrightarrow \pi_1(X, a)$ is surjective iff any loop in X based at a can be homotoped to a loop in A. Further, if A is path connected, show that this is equivalent to saying that every path in X with its end-points in A can be homotoped to a path in A.

(iv) Show that $\pi_1(X \times Y, (x_0, y_0)) \approx \pi_1(X, x_0) \times \pi_1(Y, y_0)$. In particular, compute the group $\pi_1(\mathbb{S}^1 \times \mathbb{S}^1)$.

(v) Consider the map $f : \mathbb{S}^1 \times \mathbb{S}^1 \to \mathbb{S}^1 \times \mathbb{S}^1$ given by $f(z_1, z_2) = (z_1 z_2, z_2)$. Compute the induced homomorphism $f_\#$ on the fundamental group.

(vi) Recall that the Möbius band is defined as the quotient space of $\mathbb{I} \times \mathbb{I}$ by the relation $(0, y) \sim (1, 1 - y), y \in \mathbb{I}$. Let C be the central circle in $\ddot{\mathrm{M}}$ which is the image of $\mathbb{I} \times \{1/2\}$. Show that C is SDR of $\ddot{\mathrm{M}}$. Deduce that $\pi_1(\ddot{\mathrm{M}}) \approx \mathbb{Z}$.

(vii) Let $\ddot{\mathrm{M}}$ be the Möbius band and B its boundary circle. Compute the inclusion induced homomorphism $\iota_\# : \pi_1(B) \longrightarrow \pi_1(\ddot{\mathrm{M}})$. Deduce that B is not a retract of $\ddot{\mathrm{M}}$.

(viii) In later chapters, we shall see that there are many spaces with their fundamental group non abelian. Assuming this, show that the fundamental group of the bouquet of two circles (the figure-8) is non abelian.

(ix) Show that any map $f : X \longrightarrow \mathbb{S}^n$ which is not surjective is null homotopic.

(x) Let $f, g : X \to \mathbb{S}^n$ be any two maps such that $f(x) \neq -g(x)$ from any x. Show that f is homotopic to g.

(xi) Let $n \geq 2$. Given a map $f : (\mathbb{S}^1, 1) \longrightarrow (\mathbb{S}^n, p)$ and a point $q \neq p$ in $\mathbb{S}^n$, show that f can be homotoped to a map g relative to 1 such that q is not in the image of g. Deduce that $\pi_1(\mathbb{S}^n, p)$ is trivial. (This gives an alternative proof of Corollary 1.2.32.)

(xii) Show that $\mathbb{S}^1$ is not homotopy type of $\mathbb{S}^n$ for any $n \geq 2$.

(xiii) Show that $\mathbb{R}^2$ is not homeomorphic to $\mathbb{R}^n$ for any $n \geq 2$.

(xiv) Suppose $f, g : X \to Y$ are homotopic maps. Given $x_0 \in X$, show that there is an isomorphism $\phi : \pi_1(Y, f(x_0)) \to \pi_1(Y, g(x_0))$ such that $\phi \circ f_\# = g_\#$ on $\pi_1(X, x_0)$.

(xv) Show that homotopy equivalent spaces have isomorphic fundamental groups.

1.3   Function Spaces and Quotient Spaces

We have seen that thinking of a homotopy $H : X \times \mathbb{I} \to Y$ as a path in the space of all maps from $X \to Y$ has some advantages. Therefore it is advantageous to get some familiarity with the topology of function spaces. We have also seen that it is advantageous to think of a cylinder or a Möbius band as obtained by identifying certain points in the unit square, which demands certain familiarity with quotient space construction. Therefore, before going further with the algebraic topology, we shall take a break here and pay some attention to these basic concepts, which, in the long run, saves us a lot of time.

In this section, we shall quickly present some basic facts about function spaces and quotient topology, which are going to be useful for us throughout the study of algebraic topology. The reader who is eager to go ahead with algebraic topology and finds this section a bit of a drag may skip it for the time being and return to it if and when the need arises.

Given two topological spaces X, Y we denote by Y^X the set of all continuous functions (maps) from X to Y. This set is topologised by the so-called **compact-open-topology,** viz., the topology with the subbase as the collection of all sets of the form

$$\langle K, U \rangle = \{f \in Y^X \ : \ f(K) \subset U\}$$

where $K \subset X$ is compact and $U \subset Y$ is open.

It is necessary to get familiar with this topology in order to verify various claims of continuity one may make at various places. The central results that we shall need about compact open topology are the following:

Theorem 1.3.1 (Exponential correspondence) *Let X be a locally compact Hausdorff space and Y, Z be any two topological spaces.*
(a) The evaluation map $E : Y^X \times X \to Y$ given by $E(f, x) = f(x)$ is continuous.
(b) A function $g : Z \to Y^X$ is continuous iff the composite $E \circ (g \times Id_X) : Z \times X \to Y$ is continuous.
(c) If Z is also locally compact and Hausdorff, then the function

$$\psi : (Y^X)^Z \to Y^{Z \times X}$$

defined by

$$\psi(g) = E \circ (g \times Id_X)$$

is a homeomorphism.

Proof: (a) Given an open set $U \subset Y$ we have to show that $E^{-1}(U)$ is open in $Y^X \times X$. Let $(f_0, x_0) \in E^{-1}(U)$. This implies that $f_0 : X \to Y$ is a continuous function and $f_0(x_0) \in U$. By local compactness of X there exists a compact neighbourhood K of x_0 such that $f_0(K) \subset U$. This means that $f_0 \in \langle K, U \rangle$. Since $\langle K, U \rangle$ is an open subset of Y^X, we get a neighbourhood $\langle K, U \rangle \times K$ of the point (f_0, x_0) and clearly $E(\langle K, U \rangle \times K) \subset U$.
(b) From (a) we need to prove only one part here, viz., that if $E \circ (g \times Id_X)$ is continuous then so is g. For this it is enough to show that $g^{-1}\langle K, U \rangle$ is open whenever $K \subset X$ is compact and $U \subset Y$ is open. Let $z_0 \in Z$ be such that $g(z_0) \in \langle K, U \rangle$. Then for every point $k \in K$ we have $E \circ (g \times Id_X)(z_0, k) = g(z_0)(k) \in U$. So, there exist open sets V_k, W_k of Z and X, respectively, such that $z_0 \in V_k, k \in W_k$ and $E \circ (g \times Id_X)(V_k \times W_k) \subset U$. Since K is compact, we can pass on to a finite cover $K \subset \cup_{i=1}^n W_{k_i} =: W$ and put $V = \cap_i V_{k_i}$. Then $E \circ (g \times Id_X)(V \times W) \subset U$ which implies that $g(V) \subset \langle K, U \rangle$. Since V is a neighbourhood of z_0, we are through.
(c) Note that from (b) it follows that ψ is well defined and is a bijection. Applying (b) with

$(Y^X)^Z$ in place of Z and $Z \times X$ is place of X, continuity of ψ is the same as the continuity of $E \circ (\psi \times Id_{Z \times X}) : (Y^X)^Z \times (Z \times X) \to Y$ given by

$$(\lambda, z, x) \mapsto \lambda(z)(x).$$

This latter map is the composite of the two maps

$$((Y^X)^Z \times Z) \times X \to Y^X \times X \to Y$$

given by

$$((\lambda, z), x) \mapsto (\lambda(z), x) \mapsto \lambda(z)(x).$$

The continuity of these two maps follows from (a).
The continuity of ψ^{-1} is checked in a similar fashion. ♠

Remark 1.3.2 Note that in (b) we do not need the local compactness of X to prove the continuity of $g : Z \to Y^X$ from the continuity of the corresponding map $Z \times X \to Y$. The local compactness of X is needed in the other implication only, because it is needed in (a).

Next, we take up quotient spaces. Recall the following purely set theoretic fact: Let X be a set. A partition of X is a collection of non empty subsets $\{A_y : y \in Y\}$ of X which are pair-wise disjoint such that $X = \cup_{y \in Y} A_y$. We can then consider the surjective function $f : X \to Y$ such that $f(x) \in A_{f(x)}$. Conversely, given any surjective function $g : X \to Y$ we note that the fibres of g, viz., $\{g^{-1}(y) : y \in Y\}$ forms a partition of X. These two correspondences are inverses of each other. Next given a partition on X as above, we can define an equivalence relation R on X by the rule $x \sim_R x'$ iff x, x' are in the same subset A_y. Conversely, given any equivalence relation on X, the equivalence classes define a partition on X. Once again these two correspondences are inverses of each other. Thus, a surjective function $f : X \to Y$ will be called a quotient function, Y can be thought of as the indexing set of a partition on X and fibres of f can be thought of as equivalence classes of an equivalence relation. While talking about quotient topology, we shall implicitly use all these three different avatars of the same set theoretic concept.

Let $q : (X, \tau) \to (Y, \tau')$ be a surjective map (i.e., continuous function) of topological spaces.

Lemma 1.3.3 The following statements are equivalent.
(i) $U \in \tau'$ iff $q^{-1}(U) \in \tau$.
(ii) A function $g : (Y, \tau') \to (Z, \tau'')$ is continuous iff $g \circ q$ is continuous.
(iii) For a fixed τ, τ' is the maximal topology on Y such that q is continuous.

Definition 1.3.4 *Under the above conditions, we say (Y, τ') is a quotient space of (X, τ) and the map q is called a quotient map.*

From now on, we shall revert back to the practice of using simplified notation X, Y, etc., for topological spaces, instead of $(X, \tau), (Y, \tau')$, etc.

Remark 1.3.5
(a) Quotient maps occur aplenty in mathematics. For instance, suppose $f : X \to Y$ is a surjective open map (or a closed map). Then f is a quotient map and the topology on Y is a quotient topology of the topology on X. The projection map $\pi : X \times Y \to X$ is a surjective open mapping. Therefore it is a quotient map. On the other hand, often it is not a closed map. For example, the map $(x, y) \mapsto x$ from $\mathbb{R}^2 \to \mathbb{R}$ is not a closed map. In general, a quotient map need not be an open map nor a closed map, as illustrated by the following example.

(b) Consider the quotient space of $\mathbb{R}$ in which we collapse the open interval $(0, 1)$ to a single point and also the closed interval $[2, 3]$ to a single point. If $q : \mathbb{R} \to Y$ is the resulting quotient map then q is neither an open map (the image of the open interval $(2, 3)$ is not open) nor a closed map (the image of the closed set $\{1/2\}$ is not closed). The quotient space is not even a T_1 space since $\{q(0, 1)\}$ is not closed. Luckily, we hardly need to deal with such pathological situations.

(c) Naturally, one would like to know whether a given topological property of X holds for the quotient space Y. For example, if X is compact (or connected) then so is Y. *However, all separation properties such as Hausdorffness, regularity, etc., are not passed onto the quotient space in general.*

(d) It is easy to see that Y is a T_1 space iff all the fibres of q are closed subsets of X, irrespective of whether X is T_1 or not.

(e) In general, there are no easy criteria to determine when a quotient space is Hausdorff. Indeed even if X is Hausdorff, it is unlikely that Y is also Hausdorff. Therefore, especially in the study of manifolds, where we make the blanket assumption that all manifolds are Hausdorff, we need to pay attention whenever quotient topology comes into the picture.

(f) Recall that a space X is Hausdorff iff the diagonal $\Delta_X \subset X \times X$ is a closed subset of the product space. Now suppose $X \to Y$ is a quotient map, given by an equivalence relation $R \subset X \times X$. Then $(q \times q)^{-1}(\Delta_Y) = R$. If R is a closed subset of $X \times X$ and if q is an open mapping, then it follows that $q \times q$ is an open mapping and hence Δ_Y is a closed subset of $Y \times Y$. Thus for a surjective open mapping $q : X \to Y$ given by a closed relation, Y is a Hausdorff space, irrespective of whether X is Hausdorff or not. However, we need to deal with quotient maps which are not open and then the above result is useless.

(g) If a group G acts on a Hausdorff topological space X, the orbits of the action define a partition on X. The set of orbits X/G can then be given the quotient topology. If the action is properly discontinuous and if G is finite, then the quotient topology is Hausdorff. Finiteness condition cannot be ignored as illustrated by the action of $\mathbb{Z}$ on $\mathbb{R}^2 \setminus \{(0, 0)\}$ via

$$(n, (x, y)) \mapsto (2^n x, 2^{-n} y).$$

(h) However, all standard constructions in homotopy theory such as mapping cones, mapping cylinders, suspensions, reduced suspensions, joins, etc., do not destroy Hausdorffness. This may be attributed to a single fact that in all these cases, equivalence classes are separated by open sets which are themselves a union of equivalence classes. Most often, we will be dealing with maps which are called 'cofibrations' which ensure such a situation (see Theorem 1.6.9 and Exercises 1.6.15.(vii) and (viii)).

(i) Consideration of a subspace $A \subset X$ such that $q_A : A \to Y$ is surjective often helps to understand the topology of Y better. For example consider the quotient map defining the real projective space: $p : \mathbb{R}^{n+1} \setminus \{0\} \to \mathbb{P}^n$ given by

$$(x_0, \ldots, x_n) \sim (\lambda x_0, \ldots, \lambda x_n,), \quad \lambda \in \mathbb{R}^*.$$

From this, it is not at all clear why $\mathbb{P}^n$ is compact. However, we can restrict p to the unit sphere $\mathbb{S}^n$ to see that p is surjective and then it immediately follows that $\mathbb{P}^n$ is compact. Once again, restricting p further to just the closed upper-hemisphere, it follows that $\mathbb{P}^n \setminus \mathbb{P}^{n-1}$ is an open n-cell. Also, when $n = 1$, the closed upper-hemisphere is homeomorphic to a closed interval and the identification now reduces to identifying the end-points. From this picture, you may immediately deduce that $\mathbb{P}^1$ is homeomorphic to $\mathbb{S}^1$.

These examples naturally lead us to the following question: *Given a quotient map $q : X \to Y$ a subspace $A \subset X$, is the restriction map $q_A : A \to q(A)$ a quotient map?* In general, the answer is of course in the negative. (Look out for a counter-example from (b) above.) However, here is a satisfactory affirmative answer.

Theorem 1.3.6 *Let $q : X \to Y$ be a quotient map and $A \subset X$. Then the restriction map $q_A : A \to q(A) =: B$ is a quotient map if the following conditions are satisfied:*
(a) B is the intersection of an open set and a closed set in Y.
(b) For $C \subset B$, $q_A^{-1}(C)$ is closed in A (respectively open in A) iff $q^{-1}(C)$ is closed (respectively open) in $q^{-1}(B)$.

We shall leave the proof of this theorem as an exercise to the reader.

Example 1.3.7 Let $n \geq 2$ be an integer. Consider the map $\eta_n : \mathbb{S}^1 \to \mathbb{S}^1$ given by $\eta_n(z) = z^n$. By fundamental theorem of algebra, it follows that η_n is surjective. This map is indeed a group homomorphism with $\operatorname{Ker} \eta_n$ being the subgroup consisting of all n^{th}-roots of unity. It follows that the fibres of this map are nothing but $\{\zeta^k z \ : \ 1 \leq k \leq n\}$ where ζ is a primitive n^{th}-root of unity. (In case $n = 2$ this just means that the fibres are $\{z, -z\}$.) Therefore, if we take the quotient space Y as the space of equivalence classes, then η_n induces a bijective continuous map $\bar{\eta}_n : Y \to \mathbb{S}^1$.

Being the continuous image of $\mathbb{S}^1$, Y is compact. Since $\mathbb{S}^1$ is also Hausdorff, it follows that $\bar{\eta}_n$ is a homeomorphism. In other words, one could say that η_n itself is a quotient map. Thus a given space can be its own quotient in so many ways.

A natural question that arises with respect to quotient topology is the following: If $q_i : X_i \to Z_i$ are quotient maps, $i = 1, 2$, is the product map $q_1 \times q_2 : X_1 \times X_2 \to Z_1 \times Z_2$ a quotient map? Since the composite of two quotient maps is a quotient map, and since

$$q_1 \times q_2 = (q_1 \times Id) \circ (Id \times q_2)$$

the problem reduces to the case $q_2 = Id$. A satisfactory answer then is:

Theorem 1.3.8 *If Y is a locally compact Hausdorff space, then for any quotient map $q : X \to Z$, the product $q \times Id_Y$ is a quotient map.*

Proof: Clearly $q \times Id_Y$ is surjective and continuous. What we need to prove is the following: For any space W and any function $g : Z \times Y \to W$, if $f = g \circ (q \times Id_Y) : X \times Y \to W$ is continuous, then g is continuous. By Theorem 1.3.1 (b), the map $\hat{f} : X \to W^Y$ given by $\hat{f}(x)(y) = f(x, y)$ is continuous. This factors down through q to give a continuous function $\hat{g} : Z \to W^Y$ such that $\hat{f} = \hat{g} \circ q$. But then $g = E \circ (\hat{g} \times Id_Y)$ and hence is continuous. ♠

Corollary 1.3.9 *If $q_i : X_i \to Z_i, i = 1, 2$ are quotient maps such that X_1, Z_2 or Z_1, X_2 are locally compact Hausdorff spaces, then $q_1 \times q_2$ is a quotient map.*

Example 1.3.10 (Munkres) Here is an example when the conclusion of Theorem 1.3.8 fails without the locally compactness condition on Y. We take $X = \mathbb{R}$ and Z to be the quotient of $\mathbb{R}$, where all positive integers are identified to a single point $*$. Let $q : X \to Z$ be the quotient map. Let $\mathbb{Q}$ denote the set of rational numbers with the subspace topology. We know $\mathbb{Q}$ is not locally compact and take $Y = \mathbb{Q}$. We claim that $q \times Id : X \times Y \to Z \times Y$ is not a quotient map. Let $c_n = \sqrt{2}/n$ and

$$U_n = \{(x, y) \in X \times Y \ : \ x^2 - y^2 < n^2 - c_n^2 \ \& \ n - 1/4 < x < n + 1/4\}$$

Then U_n is open and contains $\{n\} \times Y$ since c_n is not rational. Put $U = \cup_{n \geq 1} U_n$. Since

$\mathbb{Z}^+ \times \mathbb{Q} \subset U$, it follows that $U = (q \times Id)^{-1}(q \times Id)(U)$. Our claim is completed if we show that $(q \times Id)(U)$ is not open in the product topology of $Z \times Y$. Suppose on the contrary that this set is open. Since $(*, 0) \in (q \times Id)(U)$, there exists an open set V in Z and $\delta > 0$ such that $V \times I_\delta \subset (q \times Id)(U)$, where $I_\delta = (\delta, \delta) \cap \mathbb{Q}$. This is the same as saying $q^{-1}(W) \times I_\delta \subset U$. Choose $n >> 0$ so that $c_n < \delta$. Then for some $0 < \epsilon < 1/4$ it follows that

$$(n - \epsilon, n + \epsilon) \times I_\delta \subset W \times I_\delta \subset U.$$

On the other hand, one can verify that $(n + \epsilon/2, 1/n) \in W \times I_\delta$ but not in U, which is absurd.

Exercise 1.3.11

(i) Given an example of a function $H : \mathbb{I} \times \mathbb{I} \to \mathbb{I}$ such that $H(-, s)$ and $H(t, -)$ are all continuous and H is not continuous.

(ii) Consider the right-half disc

$$G := \{(x_1, x_2) \in \mathbb{D}^2 \; : \; x_1 \geq 0\}.$$

Let Y be the quotient space of G by the identification $(0, x_2) \sim (0, -x_2)$. Show that Y is homeomorphic to $\mathbb{D}^2$.

(iii) Let X denote the right hemisphere in $\mathbb{S}^2$:

$$X = \{(x_1, x_2, x_3) \in \mathbb{S}^2 \; : \; x_1 \geq 0\}.$$

(a) Let Y be the quotient space of X obtained by the identification

$$(0, x_2, x_3) \sim (0, x_2, -x_3)$$

Show that Y is homeomorphic to $\mathbb{S}^2$.
(b) Let Z be the quotient space of X by the identification

$$(0, x_2, x_3) \sim (0, -x_2, -x_3).$$

Show that Z is homeomorphic to $\mathbb{P}^2$.

(iv) Consider the orbit space of the $\mathbb{Z}_2$ action on $\mathbb{S}^1 \times \mathbb{S}^1$ given by $(x, y) \mapsto (x^{-1}, y^{-1})$. Show that it is homeomorphic to $\mathbb{S}^2$.

(v) Give an example to show that the evaluation map $E : Y^X \times X \to Y$ need not be continuous if we do not assume X is locally compact as in Theorem 1.3.1.(a). [Hint: Use Munkres' Example 1.3.10.]

(vi) Prove Theorem 1.3.6.

1.4 Relative Homotopy

We now return to the homotopy theory. In Section 1.2 we have seen that a meaningful study of the paths can be carried out with a modified notion of homotopy, which is called path-homotopy. This was nothing but a homotopy with an additional property that the two end-points were fixed. We would like to generalize this notion now.

In order to study homotopy properties of spaces, we need to work out from smaller pieces of the space to larger chunks. This demands that the information that we have on smaller spaces is not lost when we move on to larger spaces and so, the notion of homotopy needs to be strengthened by allowing us to exercise control over some smaller part of a given space. This is formalized in the notion of relative homotopy. This section will only make a small beginning with a few basic definitions and observations.

Definition 1.4.1 Let $A \subset X$, and $f, g : X \longrightarrow Y$ be any two maps such that $f(a) = g(a)$, $\forall\, a \in A$. We say f is homotopic to g *relative to* A if there is a homotopy H from f to g such that $H(a, t) = f(a)$, $\forall\, a \in A$. We write this

$$f \simeq g, \quad \text{rel } A.$$

Of course if $A = \emptyset$, then this notion coincides with the usual homotopy. All the earlier properties that we have discussed for homotopy hold good for relative homotopy as well.

Definition 1.4.2 Two topological pairs $(X, A), (Y, A)$ are said to have the same homotopy type if there exist maps $f : (X, A) \to (Y, A), g : (Y, A) \to (X, A)$ such that $f_A = Id_A = g_A$ and

$$g \circ f \simeq Id_X \text{ rel } A; \quad f \circ g \simeq Id_Y \text{ rel } A.$$

It is easily checked that the above relation is an equivalence relation among all topological pairs (X, A), where A is fixed.

Definition 1.4.3 By a retraction $r : X \longrightarrow A$ of a space X to a subspace $A \subset X$, we mean a map r such that $r(a) = a$, $\forall\, a \in A$. If such a retraction exists, then we call A a *retract* of X. A map $f : X \longrightarrow X$ which is homotopic to Id_X is called a *deformation* of X. If r is a retraction and is homotopic to Id_X, then it is called a *deformation retraction*. Again, if such a map $r : X \longrightarrow A$ exists, then we call A a *deformation retract* (DR) of X. Further, if the homotopy from Id_X to r is relative to A, then r is called a *strong deformation retraction* and A is called a *strong deformation retract* (SDR) of X. Finally, if the inclusion map $\iota : A \hookrightarrow X$ is a homotopy equivalence then we say A is a *weak deformation retract* (WDR) of X.

Note that SDR implies DR and DR implies WDR.

Example 1.4.4

1. Let X be any convex subspace of $\mathbb{R}^n$. Then every point $x \in X$ is a SDR of X. This follows from the observation that parameterisation of a line segment is continuous in terms of its end-points, viz., we simply write down, $(x, t) \mapsto (1 - t)x + tx_0$ which shows that Id_X is homotopic to c_{x_0} relative to $\{x_0\}$. (See Figure 1.15.)

2. For any space X, the subspace $X \times 0$ is a SDR of $X \times \mathbb{I}$. More generally, if Y is a contractible space, then $X \times \{y\}$ is a deformation retract of $X \times Y$ for any point $y \in Y$. Generalize this statement further.

3. The unit circle $\mathbb{S}^1$ in $\mathbb{C}$ is a SDR of $\mathbb{C} \setminus \{0\}$. Consider $H : (\mathbb{C} \setminus \{0\}) \times \mathbb{I} \longrightarrow \mathbb{C} \setminus \{0\}$ defined by

$$H(x, t) = (1 - t)x + \frac{tx}{\|x\|}.$$

In fact, this map can be used to show that $\mathbb{S}^1$ is SDR of $X \setminus \{0\}$, where X is any subspace of $\mathbb{C}$ containing $\mathbb{S}^1$ and is 'annulus-shaped' around $\mathbb{S}^1$, i.e., for each $x \in X$, the line segment $\left[x, \dfrac{x}{\|x\|}\right] \subset X$. Suitable modification of the above map may be used to prove similar statements about circles of different radii and different centres.

FIGURE 1.15. Every point of a convex set is a SDR

4. Less obvious is the fact that this is true for $S = \mathbb{D}^1 \times \mathbb{D}^1 \setminus \{(0,0)\}$ since we are not going to get a single formula for a retraction of S onto the boundary $\partial(\mathbb{D}^1 \times \mathbb{D}^1)$. There are many ways to do this. Here is one concrete way: Cut S into four triangles as in Figure 1.16 and then write down the formulae for the radial projection on each triangles:

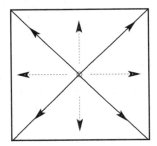

FIGURE 1.16. Square minus the centre retracts onto the boundary

The actual 'formula' for r is as follows:

$$r(x,y) = \begin{cases} (1, y/x), & |y| \le x; \\ (-1, -y/x), & |y| \le -x; \\ (x/y, 1), & |x| \le y; \\ (-x/y, -1), & |x| \le -y. \end{cases}$$

We leave it you to see that r is a continuous function and defines the required retraction. The homotopy of r with the identity map is easy to obtain, viz.,

$$(z, t) \mapsto (1-t)z + tr(z).$$

Remark 1.4.5

1. Every singleton subspace of a space is a retract of the space.

2. Let X be a Hausdorff space. If a subspace A of X is a retract, then check that A is a closed subspace of X. Thus, in a Hausdorff space subsets which are not closed will not be retracts.

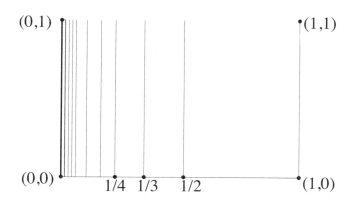

FIGURE 1.17. The combspace

3. Note that, if A is a deformation retract of X, then the inclusion map $\iota : A \to X$ is a homotopy equivalence; the retraction r as above is its homotopy inverse. In this terminology, to say that X is contractible is equivalent to saying that, every singleton subspace of X is a deformation retract of X. The following example shows that an arbitrary singleton subset need not be a strong deformation retract of a contractible Hausdorff space.

Example 1.4.6 Combspace
Consider the following subspace E of the Euclidean space $\mathbb{R}^2$ given by (see Figure 1.17):

$$E = \{(x,y) \in \mathbb{I} \times \mathbb{I} \ : \ y = 0 \text{ or } x = 0 \text{ or } x = 1/n, \ n \in \mathbb{N}\}.$$

E is called the *combspace*. This space and certain clever modifications of it serve as counter examples to many pathological questions in elementary algebraic topology. Its role may be compared to the topologist's sine-curve which you may have come across in point set topology (see Exercise 1.9.19). Show that each singleton $\{(x,0)\}$ is a strong deformation retract of E for $0 \le x \le 1$. Also show that the point $\{(0,1)\}$ is not a strong deformation retract of E. (Hint: Solve Exercise 15 from the Miscellaneous Exercises at the end of the chapter first and use the fact that E is not locally path connected. Of course, there are other ways to solve this exercise.) What about other singleton subsets?

Do not worry if you are not getting the answers immediately. We have relegated this exercise to Miscellaneous section, so as to give you enough time to solve it. On the other hand, obtaining solutions to these exercises is not at all necessary to go further.

1.5 Some Typical Constructions

We shall now study a few important constructions in algebraic topology. If the reader faces some difficulty with the notion of quotient spaces, then she should go back to Section 1.3 and get familiar with the quotient spaces and then return here.

How does one construct homotopies? There seems to be just one single source for this viz., 'convexity'. In the simple form, it is just that we are dealing with a subset A of a

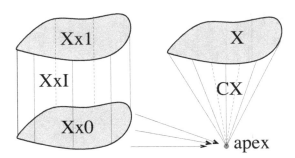

FIGURE 1.18. The cone

vector space V over $\mathbb{R}$ with the property that if $u, v \in A$ then the line segment joining u, v, is contained in A. Convexity can occur in some slightly disguised form in other situations, such as geodesic convexity in geometry, in simplexes and more generally in cells (which are nothing but homeomorphic copies of discs). Convexity presents subtly in some other situations such as cones and suspensions and path spaces. We shall discuss some of them here through examples.

We begin with an example which works like a mother of all constructions of deformations. Using the mapping cylinder, we show how to replace, up to homotopy, any arbitrary map by a cofibration. We also consider the result dual to this, viz., how to replace an arbitrary map by a fibration by using the 'universal' fibration, viz., the mapping-path-fibration. See Exercises (ix), 3 of 1.2.33 for examples where geodesic convexity is used.

The Cone

Definition 1.5.1 By a cone CX over X, we mean the quotient space of $X \times I$ obtained by identifying the subspace $X \times 0$ to a single point. (See Figure 1.18.) This point is called the *vertex* or *apex* of the cone and is denoted simply by $\star$. The space X itself can then be identified with the image of the subspace $X \times 1$ in the quotient space, and this is then referred to as the *base* of the cone.

Remark 1.5.2
(a) An arbitrary point of CX can be represented in the form $[x, t]$ for some $x \in X$ and $t \in \mathbb{I}$, the representation being unique, if $t \neq 0$. Also, $[x, 0]$ represents the vertex $\star$, $\forall x \in X$. Though CX may not have any linear structure, it makes sense to talk about *'line segments'* through any point $[x, t]$ of CX and $\star$, viz., set of all points $[x, st], 0 \leq s \leq 1$. Thus the entire cone CX is *'star-shaped'* in this sense, with $\star$ as its vertex. Observe that the map $([x, t], s) \mapsto [x, st]$ defines a homotopy of the constant map $\star$ with the identity map of CX. Therefore by Lemma 1.1.13, CX is contractible.
(b) Given any map $f : X \to Y$ there is a map $C(f) : CX \to CY$ defined by $C(f)[x, t] = [f(x), t]$. It has the property that $C(f)|_X = f$ where we have identified X with the subspace $X \times 1$ of CX and Y with the subspace $Y \times 1$ of CY. Note that $C(f)$ is a homeomorphism iff f is a homeomorphism (exercise).

Theorem 1.5.3 *Any topological space X is contractible iff it is a retract of the cone CX.*

Proof: Let $q : X \times \mathbb{I} \to CX$ denote the quotient map. Given a homotopy $H : X \times \mathbb{I} \to X$ such that $H(x, 0) = x_0$ and $H(x, 1) = x$, there is a continuous map $r : CX \to X$ defined by $r([x, t]) = H(x, t)$, i.e., $r \circ q(x, t) = H(x, t)$. Clearly $r(x, 1) = x$ and since we have identified X with $X \times 1$ this means $X \times 1 \subset CX$ is a retract of CX. Conversely, given any retraction $r : CX \to X$ the formula $H(x, t) = r([x, t])$ gives a homotopy of the constant map with the identity map of X and hence X is contractible. ♠

Example 1.5.4 The cone over $\mathbb{S}^{n-1}$ is homeomorphic to $\mathbb{D}^n$. For $n = 2$ this follows from polar coordinates. Indeed for $n > 2$, it is the generalization of polar coordinates. Consider the map $P : \mathbb{S}^{n-1} \times \mathbb{I} \to \mathbb{D}^n$ given by $(x, t) \mapsto tx$. The map is clearly a continuous bijection and one-to-one except that all points of $\mathbb{S}^{n-1} \times 0$ are mapped to a single point $0 \in \mathbb{D}^n$. Therefore it induces a continuous surjection $\bar{P} : C\mathbb{S}^{n-1} \to \mathbb{D}^n$. Since the domain is compact and the range is Hausdorff, $\bar{P}$ is a homeomorphism.

We shall now put the observations made in the above examples to good use.

Theorem 1.5.5 *Let X be any topological space and $f : \mathbb{S}^{n-1} \to X$ be any map. The following statements are equivalent:*
(i) $f : \mathbb{S}^{n-1} \longrightarrow X$ is null homotopic;
(ii) f can be extended to a map $\bar{f} : \mathbb{D}^n \to X$;
(iii) f is null homotopic relative to any given point $p \in \mathbb{S}^{n-1}$.

Proof: (i) $\Longrightarrow$ (ii) Let $H : \mathbb{S}^{n-1} \times \mathbb{I} \to X$ be a homotopy of f to a constant map. Then H defines a map $\bar{H} : C\mathbb{S}^{n-1} \to X$ such that $\bar{H} \circ q = H$, where $q : \mathbb{S}^{n-1} \times \mathbb{I} \to C\mathbb{S}^{n-1}$ is the quotient map. Now consider the homeomorphism $\bar{P}^{-1} : \mathbb{D}^n \to C\mathbb{S}^{n-1}$, as in the above example. The composite $\bar{f} = \bar{H} \circ \bar{P}^{-1}$ will do the job.
(ii) $\Longrightarrow$ (iii) Let $F : \mathbb{D}^n \times \mathbb{I} \to \mathbb{D}^n$ be the relative homotopy of the identity map with the constant map p viz., $(x, t) \mapsto (1 - t)x + tp$. Take the composite $\bar{f} \circ F : \mathbb{S}^{n-1} \times \mathbb{I} \to X$.
(iii) $\Longrightarrow$ (i) Obvious. ♠

Remark 1.5.6 There is nothing sacrosanct about identifying $X \times \{0\}$ to a single point in the definition of the cove over X. We could do as well by identifying $X \times \{1\}$ to a single point. The map $(x, t) \mapsto (x, 1 - t)$ induces a homeomorphism of the resulting quotient space with CX. Indeed, we could have started the product of X with any closed interval $[a, b], a < b \in \mathbb{R}$ and identified $X \times \{a\}$ to a single point to obtain a 'copy' of CX.

The Adjunction Space

Definition 1.5.7 Let X be a closed subspace of a space Z and $f : X \to Y$ be a continuous map. The *adjunction space* $Z \cup_f Y =: A_f$ *of f* is defined to be the quotient of the disjoint union $Y \sqcup Z$ by the identification: $x \sim f(x) \ \forall \ x \in X$. This is also called the *space obtained by attaching Z to Y via the map f*. Note that, Y can be identified with a closed subspace of A_f via $j : y \mapsto [y]$. If the image of f is closed in Y, then the image of Z is closed in A_f. Also, if f is injective then Z can be identified with its image in A_f.

(The reader is requested to verify the validity of each and every 'claim' made in the above paragraph. For instance, why is j an embedding, why is $j(Y)$ a closed subset of A_f, etc. This is the kind of point-set of topological background that we assume on the part of the reader.)

Adjunction space construction is a very wide notion which encompasses several interesting special cases. We now consider some of them here, one by one.

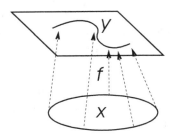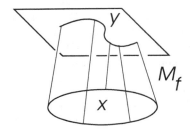

FIGURE 1.19. The mapping cylinder

Definition 1.5.8 The Mapping Cylinder Given a map $f : X \longrightarrow Y$ consider the quotient space of $Y \sqcup X \times \mathbb{I}$ by the identification $(x, 1) \sim f(x)$, $\forall\, x \in X$. This is called the mapping cylinder of f and is denoted by M_f. (See Figure 1.19.) The reader may think of this as a special case of adjunction space construction (by taking $Z = X \times \mathbb{I}$ and identifying X with the subspace $X \times 1$). There are maps

$$j : Y \longrightarrow M_f; \quad i : X \longrightarrow M_f; \text{ and } \hat{f} : M_f \longrightarrow Y$$

defined by

$$j(y) = [y]; \quad i(x) = [x, 0]; \quad \hat{f}[x, t] = f(x), \hat{f}[y] = y. \tag{1.6}$$

Clearly i and j are embeddings and we use them to identify X, Y as subspaces of M_f. Also, $\hat{f}$ defines a continuous function on M_f such that $\hat{f} \circ i(x) = \hat{f}(x, 0) = f(x)$, $\forall\, x \in X$. Also, check that $\hat{f} \circ j = Id_Y$. We shall soon see that in a very strong sense, the mapping cylinder can be used as a device to replace the continuous function $f : X \longrightarrow Y$ by the inclusion map $i : X \longrightarrow M_f$.

Definition 1.5.9 The mapping cone In the construction of adjunction space above, putting $Z = CX$, the cone over X where $f : X \longrightarrow Y$ is a given map, we get the definition of the mapping cone of f. Let us denote this by C_f. Observe that X is identified with a subspace of Z, viz., $X \times 1$ and is called the base of the cone CX. Clearly Y is a subspace of C_f. The image of $X \times 1$ in C_f itself may not be homeomorphic to X but there are other copies of X, viz., $X \times \{t\}, t \neq 0, 1$ in C_f.

Remark 1.5.10 Suppose that $\phi : Z_1 \longrightarrow Z_2, \psi : Y_1 \longrightarrow Y_2$ are homeomorphisms such that $\phi(X_1) = X_2$, where $X_i \subset Z_i$ are closed subsets. Let $f_1 : X_1 \to Y_1$ be any map. Put $f_2 = \psi \circ f_1 \circ \phi^{-1}$. Then the homeomorphism $\phi \amalg \psi : Z_1 \amalg Y_1 \longrightarrow Z_2 \amalg Y_2$ defines a homeomorphism of the adjunction spaces $A_{f_1} \longrightarrow A_{f_2}$. In particular, we may say that the homeomorphism type of the adjunction space depends on the 'ambient homeomorphism' class of the adjoining map $f : X \longrightarrow Y$. A similar statement holds for mapping cylinders and mapping cones also.

Example 1.5.11 Construction of a typical SDR We shall now illustrate a typical construction which helps to produce deformation retractions in many contexts. We first carry this out for $\mathbb{I} \times \mathbb{I}$. Consider the subspace $A = \mathbb{I} \times \{0\} \cup \{1\} \times \mathbb{I}$. Then A is a strong deformation retract of $\mathbb{I} \times \mathbb{I}$. There are several ways of writing down such a strong deformation. We find the one given below and illustrated in Figure 1.20 to be quite useful and geometric.

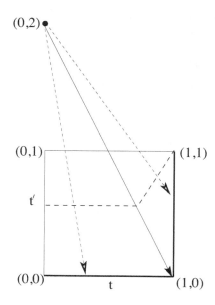

FIGURE 1.20. A typical strong deformation retract

We fix some point P in the plane lying somewhere above the line $y = 1$ and on the left side of the line $x = 1$. For definiteness let $P = (0, 2)$. We observe that for each point $z \in \mathbb{I} \times \mathbb{I}$, the line through P and z meets A in a unique point $r(z)$. We can then simply take the map which 'pushes' the point z to $r(z)$ in a unit time as our homotopy. The idea is over. Often many expositions in algebraic and differential topology stop at this stage. A student of topology is expected to appreciate this as the final proof and others are expected to swallow it in good faith. Let us prepare ourselves by working out a few such ideas into proofs so that we can appreciate such 'ideas' as proofs later in many other situations.

Of course, the very first thing we must verify is that the map $z \mapsto r(z)$ is continuous. To see this, we divide the square $\mathbb{I} \times \mathbb{I}$ into two parts by the line joining $(1, 0)$ with $P = (0, 2)$. If $z = (t, t')$ is in the lower part, viz., $t' \leq 2(1 - t)$, then $r(z) = \left(\dfrac{2t}{2 - t'}, 0 \right)$. On the other hand, if $t' \geq 2(1 - t)$ then $r(z) = \left(1, \dfrac{t' - 2(1 - t)}{t} \right)$. Therefore r is continuous. The homotopy S between Id and r is then given simply by

$$(z, t'') \mapsto (1 - t'')z + t'' r(z).$$

We can rewrite $S : \mathbb{I} \times \mathbb{I} \times \mathbb{I} \longrightarrow \mathbb{I} \times \mathbb{I}$ as follows:

$$S(t, t', t'') = \begin{cases} \left((1 - t'')t + \dfrac{2tt''}{2 - t'}, (1 - t'')t' \right), & t' \leq 2(1 - t) \\[3mm] \left((1 - t'')t + t'', (1 - t'')t' + \dfrac{t''(t' - 2(1 - t))}{t} \right), & t' \geq 2(1 - t) \end{cases} \tag{1.7}$$

Observe that the map S_t 'pushes' the square along the lines passing through the point

$(0,2)$. The union of the two line segments shown by the thick dotted broken lines is the image of the segment $t' = 1$ at the time $t'' = 1 - s$.

Now, think of $\mathbb{I}$ as a cone over the point $\{1\}$. Then the construction above can be generalized to the case when the point space $\{1\}$ is replaced by any topological space X using the polar coordinates for CX. Thus consider $H : X \times \mathbb{I} \times \mathbb{I} \times \mathbb{I} \longrightarrow X \times \mathbb{I} \times \mathbb{I}$ defined by,

$$H(x, t, t', t'') = (x, S(t, t', t'')).$$

Let us use the notation $[x, t], [x, a, b]$ to denote the image of $(x, t) \in X \times \mathbb{I}, (x, a, b) \in X \times \mathbb{I} \times \mathbb{I}$, etc., under q or $q \times Id$, etc., where $q : X \times \mathbb{I} \to CX$ is the quotient map. The function H respects the equivalence relation defining the cone because, the t coordinate of $S(0, t', t'')$ is always zero. Hence there is a well-defined function $\bar{H} : CX \times \mathbb{I} \times \mathbb{I} \longrightarrow CX \times \mathbb{I}$, viz.,

$$\bar{H}([x, t], t', t'') = [x, S(t, t', t'')]. \tag{1.8}$$

It can be directly verified that this map is continuous. The continuity at points of the form $([x, t], t', t'')$ for $t \neq 0$ is trivial. At all points of the form, $([x, 0], t', t'')$, it follows from the fact that $\bar{H}$ is uniformly continuous in x. Alternatively, we can use Theorem 1.3.8 to see that that the topology on $CX \times \mathbb{I}$ is the quotient topology from $X \times \mathbb{I} \times \mathbb{I}$ and hence continuity of H automatically implies the continuity of $\bar{H}$.

Thus we have proved:

Theorem 1.5.12 *For any topological space X, $CX \times 0 \cup X \times \mathbb{I}$ is a SDR of $CX \times \mathbb{I}$ where we identify X with the image in CX of the space $X \times 1$.*

Combining with Proposition 1.6.1, we obtain:

Corollary 1.5.13 *For any topological space, the inclusion map $x \mapsto [x, 1]$ of X into CX is a cofibration.*

Some immediate applications of this are the following.

Theorem 1.5.14 *For any $t \in \mathbb{I}$ the inclusion map $t \hookrightarrow \mathbb{I}$ is a cofibration.*

Proof: The method above can be used to construct a retraction $\mathbb{I} \times \mathbb{I} \to \mathbb{I} \times 0 \cup \{t\} \times \mathbb{I}$. ♠

Corollary 1.5.15 *For any space Z and any point $t \in \mathbb{I}$ the inclusion map $Z \times t \hookrightarrow Z \times \mathbb{I}$ is a cofibration.*

Proof: Use Exercise 1.6.15.(iii).

Remark 1.5.16 The essence of the types of constructions in the above example is that the space parameter X does not play any role. Figure 1.21 depicts the cases for $X = \mathbb{S}^0$ and $\mathbb{S}^1$, respectively, in Theorem 1.5.12. The second case may be named as 'bath-tub construction'.

We shall now give an application of this in obtaining a powerful tool in homotopy theory which 'replaces' any map by an inclusion map which is a cofibration.

Theorem 1.5.17 *Given any map $f : X \longrightarrow Y$, let $i, j, \hat{f}$, etc., be as in (1.6). Consider the following homotopy commutative diagram:*

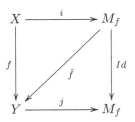

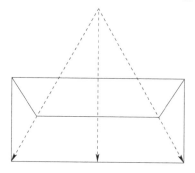
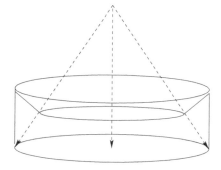

FIGURE 1.21. The bath-tub construction

We have,

(a) $\hat{f} \circ i = f$.

(b) $j \circ \hat{f}$ *homotopic to* Id_{M_f} *relative to* Y, *i.e.,* $\hat{f}$ *is a strong deformation retraction of* M_f *to* Y. *In particular,* Y *is a SDR of* M_f.

(c) $M_f \cup X \times \mathbb{I}$ *is a strong deformation retract of* $M_f \times \mathbb{I}$.

(d) i *is a cofibration.*

(e) $j \circ f$ *is homotopic to* i.

Proof: We have already seen (a). Consider the maps $G_1 : X \times \mathbb{I} \times \mathbb{I} \longrightarrow M_f$ and $G_2 : Y \times \mathbb{I} \longrightarrow M_f$ defined by

$$G_1(x, t, s) = [x, (1-s)t + s]; \quad \text{and} \quad G_2(y, s) = [y].$$

These maps fit together and induce a homotopy $G : M_f \times \mathbb{I} \longrightarrow M_f$ from Id_{M_f} to $j \circ r$ relative to the subspace Y. This proves (b).

For the proof of (c), the construction (1.8) comes to help. Define $H_1 : X \times \mathbb{I} \times \mathbb{I} \times \mathbb{I} \longrightarrow M_f \times \mathbb{I}$ and $H_2 : Y \times \mathbb{I} \times \mathbb{I} \longrightarrow M_f \times \mathbb{I}$ by

$$H_1(x, t, t', t'') = [x, S(t, t', t'')]; \quad H_2(y, t', t'') = [y, 1 - t'],$$

with similar notation as in (1.8). Verify that they fit together to define a map $H : M_f \times \mathbb{I} \times \mathbb{I} \longrightarrow M_f \times \mathbb{I}$ which is a strong deformation retraction of $M_f \times \mathbb{I}$ into $M_f \cup X \times \mathbb{I}$.

In view of Proposition 1.6.1, the conclusion (d) is obvious using (c). Finally, (e) follows from (a) and (b). ♠

Remark 1.5.18 Thus, the mapping cylinder is a device that enables us to replace an arbitrary map $f : X \longrightarrow Y$ by an inclusion map which is a cofibration up to homotopy. Observe that M_f contains 'lots' of copies of X and a copy of Y. Moreover, (b) of the above theorem tells us that Y is SDR of M_f. Also the mapping cone C_f is called the *cofibre* of the cofibration $X \hookrightarrow M_f$. In the next section we shall give a number of applications of mapping cylinder.

Exercise 1.5.19

(i) Given any two points p, q in the interior of the disc $\mathbb{D}^n$ show that there is a homeomorphism $f : \mathbb{D}^n \to \mathbb{D}^n$ such that $f(p) = q$ and $f|_{\mathbb{S}^{n-1}} = Id$.

(ii) Let X be a convex polygon in $\mathbb{R}^2$ with n vertices $n \geq 3$.
   (a) Show that X is homeomorphic to the cone over ∂X, the boundary of X.

(b) Choose any distinct n points $a_1, \ldots, a_n$ on the circle $\mathbb{S}^1$. Construct a homeomorphism $f : \partial X \to \mathbb{S}^1$ so that the vertices of X are mapped onto $a_1, \ldots, a_n$.
(c) Construct a homeomorphism $g : X \to \mathbb{D}^2$ which extends f.
(d) Do the same thing as in (b) and (c) with the right-half disc G with three of the points on the boundary being $a_1 = (0, 1), a_2 = (0, 0), a_3 = (0, -1)$.
(e) Assume that $n \geq 4$ and let A_1, A_2, A_3 be any three consecutive vertices of X. Let Y be the quotient space of X obtained by identifying the points on the edge $A_1 A_2$ with the points in $A_3 A_2$ by the rule $tA_2 + (1 - t)A_1 \sim tA_2 + (1 - t)A_3, 0 \leq t \leq 1$. Show that Y is homeomorphic to $\mathbb{D}^2$.

(iii) Show that there is a homeomorphism $\Psi : \mathbb{I} \times \mathbb{I} \to \mathbb{I} \times \mathbb{I}$ which takes $0 \times \mathbb{I} \cup \mathbb{I} \times \dot{\mathbb{I}}$ onto $\mathbb{I} \times 0$.

(iv) Let $f : X \to Y$ be any map.
   (a) If $g : W \longrightarrow C_f$ is any map such that the image of g does not intersect Y then g is null homotopic.
   (b) Show that if X and Y are locally contractible then so is C_f.
   (c) Show that a map $f : X \to Y$ is a homeomorphism iff $c(f) : C(X) \to C(Y)$ is a homeomorphism.
   (d) Show that a map $f : X \to Y$ is null homotopic iff it extends to a map $\hat{f} : C(X) \to Y$.

(v) Let $H : \mathbb{I} \times \mathbb{I} \to X$ be a map. Put

$$\omega_j(t) = H(t, j), \lambda(j, s) = H(j, s), j = 0, 1, t, s \in \mathbb{I}.$$

Show that $\omega_0 * \lambda_1 * \omega_1^{-1} * \lambda_0^{-1}$ is a null-homotopic loop at $x_0 = H(0, 0)$.

(vi) Establish a bijection between $\pi_1(X, a)$ and $[(\mathbb{S}^1, 1); (X, a)]$, the set of all relative homotopy classes of maps of pointed spaces.

(vii) Show that any map $\mathbb{S}^1 \longrightarrow X$ is null homotopic iff $\pi_1(X, x_0)$ is trivial for each point $x_0 \in X$.

(viii) Let X be a path connected space. Show that the set of free homotopy classes $[[\mathbb{S}^1, X]]$ is equal to the set of conjugacy classes of elements of $\pi_1(X, x_0)$. Deduce the previous exercise from this.

(ix) Suppose X is path connected. Prove that $\pi_1(X, a)$ is abelian for some $a \in X$ iff for each $b \in X$ and for all paths τ from a to b in X the isomorphisms $h_{[\tau]} : \pi_1(X, a) \longrightarrow \pi_1(X, b)$ are the same.

1.6   Cofibrations

We begin this section with an easy reformulation of HEP. After giving some immediate consequences, we then go on to give yet another reformulation of HEP in terms of the so-called 'neighbourhood deformation retracts' thereby laying down a firm foundation for the study of cofibrations. We give just one application here to homotopy invariance of adjunction space. The result is quoted several times throughout this book.

Proposition 1.6.1 Let A be a closed subspace of X. Then (X, A) has HEP with respect to every space, i.e., $A \hookrightarrow X$ is a cofibration iff the subspace $Z = A \times \mathbb{I} \cup X \times 0$ is a retract of $X \times \mathbb{I}$. (See Figure 1.22.)

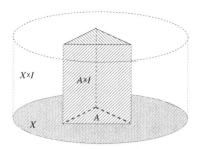

FIGURE 1.22. A retract and homotopy extension property

Proof: Assume that there is a map $r : X \times \mathbb{I} \to Z$ such that, $r|_Z = id_Z$. Then given a homotopy extension data for (X, A) (see Definition 1.1.8), we have a map $\theta : Z \to Y$ such that, $\theta|_{A \times \mathbb{I}} = F$, and $\theta|_{X \times 0} = g$. Consider $G = \theta \circ r$. Then, G will be as required by the HEP.

Conversely, suppose (X, A) has HEP with respect to any space. Then, take $Y = Z$, F and g as the corresponding inclusion maps. Then the HEP gives a map $G : X \times \mathbb{I} \to Z$ which is obviously a retraction. ♠

Remark 1.6.2 The above proposition comes extremely handy in determining whether an inclusion map $A \hookrightarrow X$ is a cofibration or not. It reduces the practically impossible task of verifying the HEP with respect to every space Y to just one task of checking whether the subspace $Z = X \times 0 \cup A \times \mathbb{I} \hookrightarrow X \times \mathbb{I}$ is a retract or not. Here is an immediate application of this.

Corollary 1.6.3 If $A \hookrightarrow X$ is a cofibration then $Z \times A \hookrightarrow Z \times X$ is also a cofibration.

Proof: If $r : X \times \mathbb{I} \to X \times 0 \cup A \times \mathbb{I}$ is a retraction, take $R(z, x, t) = (z, r(x, t))$ to see that $Z \times X \times 0 \cup Z \times A \times \mathbb{I}$ is a retract of $Z \times X \times \mathbb{I}$. ♠

The following useful results are some immediate consequences of the homotopy extension property.

Theorem 1.6.4 *Let $A \hookrightarrow X$ be a cofibration.*

(a) *A is a WDR of X iff it is a DR of X.*

(b) *If A is contractible, then the quotient map $q : X \longrightarrow X/A$ is a homotopy equivalence.*

(c) *If $\{a_0\} \hookrightarrow A$ is a SDR, then the inclusion $(X, a_0) \to (X, A)$ is a homotopy equivalence.*

Proof: (a) Given a homotopy inverse $r_0 : X \to A$ of the inclusion $\eta : A \to X$, and a homotopy $F : A \times \mathbb{I} \to A$ from $r_0 \circ \eta$ to Id_A, put $Y = A$ and $g = r_0$ in Figure 1.5 to obtain a homotopy $H : r_0 \sim r_1$ such that $r_1 \circ \eta = Id_A$. This also implies $\eta \circ r_1 \sim \eta \circ r_0 \sim Id_X$.

(b) Let $F : A \times \mathbb{I} \to A$ be a homotopy of the identity map with the constant map at a_0. Put $Y = X, g = Id_X$ in Figure 1.5 to obtain a homotopy $H : X \times \mathbb{I} \to X$ such that $H(a, t) = F(a, t), a \in A$ and $H(x, 0) = x, x, \in X$. Since the map $h_1(x) = H(x, 1)$ has the property $h_1(a) = a_0$ for all $a \in A$, it factors down through $q : X \to X/A$ and defines a map $g_1 : X/A \to X$, i.e., $h_1 = g_1 \circ q$. Also, the homotopy $q \circ H : X \times \mathbb{I} \to X/A$ factors down to define a homotopy $G : (X/A) \times \mathbb{I} \to X/A$, (i.e., $G \circ (q \times Id) = q \circ H$) from $Id_{(X/A)}$ to $q \circ g_1$.

(c) In the above discussion, if we start with a homotopy $F : A \times \mathbb{I} \to A$ which is relative to $a_0 \in A$, it follows that $h_1 : (X, A) \to (X, a_0)$ is the relative homotopy inverse of the inclusion map $(X, a) \hookrightarrow (X, A)$ as needed. ♠

Corollary 1.6.5 A map $f : X \longrightarrow Y$ is a homotopy equivalence iff X is a deformation retract of M_f.

Proof: If X is a deformation retract of M_f then clearly the inclusion map $i : X \to M_f$ is a homotopy equivalence. From (a) of Theorem 1.5.17, it follows that f is a homotopy equivalence. Conversely, if f is a homotopy equivalence, then again from (a) of Theorem 1.5.17, $i : X \to M_f$ is a homotopy equivalence. Now we combine (d) of Theorem 1.5.17 with Theorem 1.6.4 to conclude that X is a deformation retract of M_f.   ♠

Corollary 1.6.6 Two spaces X, Y are homotopy equivalent iff there is another space W which contains both of them as deformation retracts.

Proof: If $f : X \to Y$ is a homotopy equivalence, we take $W = M_f$. The converse is obvious.
    We need to strengthen the cofibration property a little bit further by introducing another equivalent concept.

Definition 1.6.7 A topological pair (X, A) is called a *neighbourhood deformation retract (NDR) pair* iff there exist maps $u : X \to [0, 1]$; $h : X \times \mathbb{I} \to X$ such that
(i) $A = u^{-1}(0)$;
(ii) $h(x, 0) = x$, $x \in X$;
(iii) $h(x, t) = x$, $(x, t) \in A \times \mathbb{I}$;
(iv) $h(x, 1) \in A$, wherever $u(x) < 1$.
We sometimes also say that A is a neighbourhood retract in X.

Remark 1.6.8 Condition (i) implies that A is a closed G_δ-set. If $U = u^{-1}([0, 1))$, then h defines a relative homotopy of a retraction of U onto A inside X, i.e., A is a DR of one of its neighbourhoods. This justifies the name NDR for this definition. The point of introducing this definition is in the following theorem.

Theorem 1.6.9 *Let A be a closed subset of X. Then the following conditions are equivalent.*
(a) $A \hookrightarrow X$ *is a cofibration.*
(b) $X \times 0 \cup A \times \mathbb{I}$ *is a retract of $X \times \mathbb{I}$.*
(c) $X \times 0 \cup A \times \mathbb{I}$ *is a deformation retract of $X \times \mathbb{I}$.*
(d) (X, A) *is an NDR pair.*

Proof: We have already seen the equivalence of (a) and (b) and the implication (c) $\Longrightarrow$ (b) is obvious. Putting
$$H(x, t, s) = (h(x, t), t(1 - su(x))),$$
we get (d) $\Longrightarrow$ (c). It takes some effort to prove (b) implies (d).
    Let then $r : X \times \mathbb{I} \to X \times 0 \cup A \times \mathbb{I}$ be a retraction. Let $p_1 : X \times \mathbb{I} \to X$ be the projection. Put $h = p_1 \circ r$. Then h satisfies (ii) and (iii) of the above definition. So, it remains to construct $u : X \to \mathbb{I}$ satisfying (i) and (iv). Let $p_2 : X \times \mathbb{I} \to \mathbb{I}$ be the projection to the second factor. For $n \geq 0$, define

$$\phi_n(x) = \text{Min} \left\{ \frac{1}{2^n}, p_2 \circ r \left(x, \frac{1}{2^n} \right) \right\}.$$

Then each $\phi_n : X \to [0, \frac{1}{2^n}]$ is continuous and the series $\sum_n \phi_n(x)$ is uniformly convergent on the whole of X. Now put

$$u(x) = 1 - \sum_{n=1}^{\infty} \phi_0(x)\phi_n(x) = \sum_{n=1}^{\infty} \left(\frac{1}{2^n} - \phi_0(x)\phi_n(x) \right).$$

If follows that $u : X \to \mathbb{I}$ is continuous. If $x \in A$, then $p_2 r(x, 1/2^n) = p_2(x, 1/2^n) = 1/2^n$ for each n and therefore $u(x) = 0$. If $x \in X \setminus A$, by continuity of r, we can choose $\epsilon > 0$ and a neighbourhood V of x in X such that $r(V \times [0, \epsilon)) \subset (X \setminus A) \times 0$. But then for some $n >> 0$, $(x, 1/2^n) \in V \times [0, \epsilon)$ which implies by the second formula for u above, that $u(x) > 0$. This proves (i).

Now let $u(x) < 1$. This implies $\phi_0(x) \neq 0$ which in turn gives $p_2 r(x, 1) > 0$. Therefore $r(x, 1) \in A \times \mathbb{I}$. This then implies $h(x, 1) = p_1 \circ r(x, 1) \in A$. This proves the implication (b) $\implies$ (d). ♠

Theorem 1.6.10 *Suppose $X \hookrightarrow Z$ is a cofibration, where $X \subset Z$ is a closed subspace. If $f, g : X \longrightarrow Y$ are homotopic, then the adjunction space pairs (A_f, Y) and (A_g, Y) are homotopy equivalent.*

Proof: Let $H : X \times \mathbb{I} \longrightarrow Y$ be a homotopy from f to g. By (c) of the previous theorem, there is $r_0 : Z \times \mathbb{I} \to Z \times 0 \cup X \times \mathbb{I}$, a deformation retraction. This induces a deformation retraction $\bar{r}_0 : A_H \to Y \cup_H (Z \times 0 \cup X \times \mathbb{I})$. The latter space is a quotient of $Y \sqcup Z \times 0 \cup X \times \mathbb{I}$ and the quotient map restricted to $Y \sqcup Z \times 0$ is surjective. Therefore, from Theorem 1.3.6, we get,

$$Y \cup_H (Z \times 0 \cup X \times \mathbb{I}) = Y \cup_f (Z \times 0) = A_f.$$

Changing t to $(1 - t)$, there is a deformation retraction $r_1 : Z \times \mathbb{I} \to Z \times 1 \cup X \times \mathbb{I}$ also, which in turn yields a deformation retraction $\bar{r}_1 : A_H \to A_g$. From Corollary 1.6.6, it follows that (A_f, Y) and (A_g, Y) are homotopy equivalent. ♠

Remark 1.6.11 Authors such as Steenrod (see [Steenrod, 1967]) have proposed that the category of NDR pairs (X, A) where X is a topological space which is Hausdorff and compactly generated, is the most suitable for doing algebraic topology. The above result is central to this theme. We prefer to keep the treatment somewhat less sophisticated than this and stick to taking closed subspaces A such that $A \hookrightarrow X$ is a cofibration.

One of the main point-set-topological difficulties is that quotients of Hausdorff spaces are not necessarily Hausdorff. The above result comes handy in almost all quotient space constructions which we come across in algebraic topology. (See Exercises 1.6.15 (vii),(viii) for some immediate usefulness of the above theorem.)

Example 1.6.12 Here are some examples to illustrate how all this theory helps us to see whether two spaces are homotopic to each other without actually constructing a homotopy, under certain situations.

It is easy to see that any point in $\mathbb{R}^n$ or $\mathbb{S}^n$ is an NDR. So is any proper arc of a great circle in $\mathbb{S}^n$. (Indeed, the tubular neighbourhood theorem of differential topology tells us that (M, N), where N is any smooth submanifold of M is a NDR pair. We shall see many more examples of NDR pairs in the next chapter.) As a consequence, it follows that if $A \subset \mathbb{S}^n$ is an arc, then collapsing A to a point we get a space $\mathbb{S}^n/A$ which is homotopy equivalent to $\mathbb{S}^n$.

Next, let us consider the space X which is the union of $\mathbb{S}^2$ and one of its diameters. By collapsing one of the great arcs joining the end-points of the diameter we immediately see that X has the same homotopy type of $Y = \mathbb{S}^2 \vee \mathbb{S}^1$ the one-point union. (See Figure 1.23.)

A topologist may describe this result by saying that 'we can move one of the end-points of the diameter slowly to coincide with its other end-point along one of the arcs of a great circle'. To a beginner in topology or an outsider, such statements do look too heuristic.

However, this is precisely the homotopy invariance of the adjunction space: X can be thought of as the adjunction space of $\mathbb{S}^2$ and $[-1, 1]$. The attaching map $\mathbb{S}^0 \to \mathbb{S}^2$ of the adjunction space is homotopic to a constant map. This implies X has the same homotopy

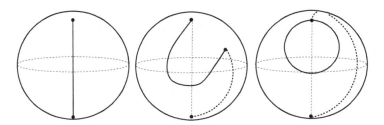

FIGURE 1.23. Same homotopy type

type as the adjunction space obtained by attaching $[-1, 1]$ to $\mathbb{S}^2$, where the attaching map is a constant map; and this latter space is nothing but Y.

Remark 1.6.13 The above results about cofibration may encourage one to ask a bold question such as the one below:

Let (X, A) be a pair such that $A \hookrightarrow X$ is a cofibration, and let two maps $f, g : X \to X$ be such that $f_A = g_A$. Suppose f is homotopic to g. Is f homotopic to g relative to A?

[Of course, as seen in Example 1.4.6, the answer to this question is in the negative, without the hypothesis $A \hookrightarrow X$ is a cofibration.]

Here is a simple counter example. We take $X = \mathbb{S}^1 \times \mathbb{I}$ and $A = \mathbb{S}^1 \times \{0, 1\}$. Clearly $A \hookrightarrow X$ is a cofibration (see Exercise 1.6.15.(iv)). Take $f(z, t) = (e^{2\pi i t} z, t)$ and $g = Id$. $H(z, t, s) = (e^{2\pi i t s} z, t)$ gives a homotopy between the two maps. To see that f and g are not homotopic relative to A takes a little more effort. Note that if $q : X \to \mathbb{S}^1 \times \mathbb{S}^1$ is the quotient map which identifies $(z, 0)$ with $(z, 1)$ for each $z \in \mathbb{S}^1$, then f and g induce maps $F, G : \mathbb{S}^1 \times \mathbb{S}^1 \to \mathbb{S}^1 \times \mathbb{S}^1$ where G is the identity map. A homotopy from f to g relative to A would induce a homotopy from F to G relative to $\mathbb{S}^1 \times \{1\}$. In particular, they induce the same homomorphism on the fundamental group $\pi_1(\mathbb{S}^1 \times \mathbb{S}^1) \approx \mathbb{Z} \oplus \mathbb{Z}$. Now using Exercises (iv) and (v) of 1.2.33, you will see that this is absurd.

Therefore it is quite remarkable that the following theorem is true. Since we have no use of it, we shall merely state it here and refer the reader to [Hatcher, 2002] (Proposition 0.19) for a tricky proof of it.

Theorem 1.6.14 *Suppose $(X, A), (Y, A)$ satisfy homotopy extension property and $f : X \to Y$ is a homotopy equivalence, then f is a homotopy equivalence relative to A.*

Exercise 1.6.15

(i) Show that any non empty open interval is not a retract of any larger interval.

(ii) Let $X \subset \mathbb{R}^n$ be homeomorphic to $\mathbb{I}^k$. Show that X is a retract of $\mathbb{R}^n$. (Hint: Tietze extension theorem.)

(iii) Let $A \hookrightarrow X$ be a cofibration. Show that for any space Z, $Z \times A \hookrightarrow Z \times X$ is a cofibration.

(iv) Let A be any finite subset of $\mathbb{I}$. Show that $A \hookrightarrow \mathbb{I}$ is a cofibration. Deduce that $\mathbb{S}^1 \times \{0, 1\} \hookrightarrow \mathbb{S}^1 \times \mathbb{I}$ is a cofibration.

(v) Let C be a space such that $C = A \cup B$, and $A \cap B \hookrightarrow A$ be a cofibration. Show that $B \hookrightarrow C$ is a cofibration.

(vi) Given $A \subset X$ note that CA is a subspace of CX in a natural way. Consider the subspace $X \times 0 \cup CA \subset CX$. If $A \subset X$ is a cofibration, show that $CA \subset X \times 0 \cup CA$ is a cofibration.

(vii) Given any map $f : X \to Y$, show that the mapping cone C_f (respectively, mapping cylinder M_f) is Hausdorff iff X, Y are Hausdorff.

(viii) More generally, suppose $X \hookrightarrow Z$ is a cofibration and $f : X \to Y$ is a map. Then the adjunction space A_f is Hausdorff iff both Z and Y are Hausdorff.

(ix) Obtain the following version of Theorem 1.5.12, viz., for any two topological spaces X, Y, $CX \times CY \times 0 \cup X \times CY \times \mathbb{I} \cup CX \times Y \times \mathbb{I}$ is a SRD of $CX \times CY \times \mathbb{I}$. [Hint: Make a 3-dimensional version of the picture in Example 1.5.11.]

1.7 Fibrations

We now consider a construction dual to mapping cylinder which tells us how to replace an arbitrary map by a fibration.

Definition 1.7.1 Given $p : E \to B$ and $f : B' \to B$, the *pullback* of p via f is defined to be the map $p' : E' \to B'$ given as follows:

$$E' = \{(b', e) \in B' \times E : p(e) = f(b')\}; \quad p'(b', e) = b'.$$

The map p' is also called the *fibred product* of p and f. (See Figure 1.24.) Let us denote the second projection restricted to E' by f'. Then we have $p \circ f' = f \circ p'$. We also use the notation $f^*(p)$ for the map p'.

The following 'universal property' gives a characterization of the pullback. (see Example 1.8.18 for more.)

Lemma 1.7.2 Given maps $\alpha : X \to B', \beta : X \to E$ such that $f \circ \alpha = p \circ \beta$, there is a unique map $\gamma : X \to E'$ such that $p' \circ \gamma = \alpha$ and $f' \circ \gamma = \beta$.

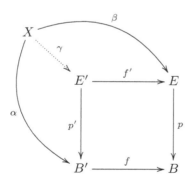

FIGURE 1.24. The fibred product

The pullbacks retain a lot of topological properties of p in general. The point of our interest is the following result, which is immediate from the above lemma.

Lemma 1.7.3 If $p : E \to B$ is a fibration so is its pullback under any map $f : B' \to B$.

The reader is advised to make an appropriate diagram to see the proof.

We shall now consider a very important and natural fibration over a space, called the path fibration. For any topological space X we consider the path space $X^{\mathbb{I}}$ of all paths in X with the compact open topology. There is an obvious map $p : X^{\mathbb{I}} \to X$ given by $p(\omega) = \omega(0)$, which is of prime importance (see Exercise 1.9.43).

Lemma 1.7.4 The map $p : X^{\mathbb{I}} \to X$ is a fibration.

Proof: Given a homotopy $H : Y \times \mathbb{I} \to X$ and a map $G : Y \to X^{\mathbb{I}}$ such that $G(y)(0) = H(y,0)$, we have to find $\hat{H} : Y \times \mathbb{I} \to X^{\mathbb{I}}$ such that $\hat{H}(y,t)(0) = H(y,t)$ and $\hat{H}(y,0) = G(y)$. In other words, for each $(y,t) \in Y \times \mathbb{I}$, we must find a path which starts at $H(y,t)$ and such that for $t = 0$ this path is the given path $G(y)$. So, consider

$$\hat{H}(y,t)(s) = \begin{cases} H(y, t - 2s), & 0 \le s \le t/2, \\ G(y)\left(\frac{2s-t}{2-t}\right), & t/2 \le s \le 1. \end{cases}$$

Verify the details. (What we have done is the following: for each fixed $(y,t) \in Y \times \mathbb{I}$, we have traced the path $H(y,-)$ from $H(y,t)$ to $H(y,0)$ and then traced the path $G(y)$ to obtain $\hat{H}(y,t)$. (See Figure 1.25.) ♠

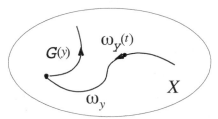

FIGURE 1.25. The path fibration

Remark 1.7.5
(a) There is nothing very special about taking the end-point map $X^{\mathbb{I}} \to X$. For any $t \in \mathbb{I}$, if $p_t(\omega) = \omega(t)$ then $p_t : X^{\mathbb{I}} \to X$ is a fibration. We only need to make a small modification in the proof of the above lemma.
(b) In particular, it is easily seen that the map $\omega \mapsto \omega(1)$ is also a fibration.
(c) Consider the subspace $P(X, x_1) \subset X^{\mathbb{I}}$ of all paths which end at a given point x_1. Then the map p restricts to a fibration $P(X, x_1) \to X$. The proof of the above lemma works verbatim. By interchanging the two end-points as in Remark (b), we obtain the following base-point version of the above lemma.

Proposition 1.7.6 Given a pointed space (X, x_0), let PX denote the space of all paths in X which start at x_0. Then the map $p : \omega \mapsto \omega(1)$ is a fibration.

Definition 1.7.7 The above fibration is also called the *path fibration* for the pointed space (X, x_0). For any map $f : (Y, y_0) \to (X, x_0)$ the pullback fibration $p_f : P_f \to Y$ from the path fibration $PX \to X$ is called the *principal fibration* induced by f.

This will re-enter our study a little later. Right now we are interested in something else.

We shall now prove a statement somewhat 'dual' to Theorem 1.5.17 which replaces any given map by a fibration up to homotopy. Given any map $f : X \to Y$ consider the pullback fibration $p' : M^f \to X$ from the fibration $p = p_0 : Y^{\mathbb{I}} \to Y$ where $p_0(\omega) = \omega(0)$, viz.,

$$M^f := \{(x, \omega) \in X \times Y^{\mathbb{I}} \;:\; f(x) = \omega(0)\}, \quad p'(x, \omega) := x.$$

Theorem 1.7.8 *For any map $f : X \to Y$, the map $p^f : M^f \to Y$, given by $p^f(x, \omega) = \omega(1)$, is a fibration. Moreover, there is a section $s : X \to M^f$ of the fibration p' such that*
(i) $p^f \circ s = f$;   (ii) $s \circ p' \sim Id_{M^f}$;   (iii) $f \circ p' \sim p^f$.

Proof: Let $s : X \to M^f$ be defined by $s(x) = (x, c_{f(x)})$ where c_y is the constant path at $y \in Y$. Then $p' \circ s = Id_X$ and s satisfies (i). Let us prove the two homotopies: Consider $\alpha(x, \omega, t') = (x, \omega_{t'})$, where $\omega_{t'}(t) = \omega(tt')$. Then $\alpha : s \circ p' \sim Id_{M^f}$. Likewise consider $\beta(x, \omega, t') = \omega(t')$. Then $\beta : f \circ p' \sim p^f$.

To prove that p^f has HLP, let $H : Z \times \mathbb{I} \to Y$ and $g : Z \to M^f$ be maps such that $p^f \circ g(z) = H(z, 0)$. Writing $g(z) = (g_1(z), g_2(z))$, it follows that $f g_1(z) = g_2(z)(1) = p^f \circ g(z) = H(z, 0)$. We should get maps $G_1 : Z \times \mathbb{I} \to X, G_2 : Z \times \mathbb{I} \to Y^{\mathbb{I}}$ such that $p_0 \circ G_2(z, t') = G_2(z, t')(0) = f \circ G_1(z, t')$, and such that $p_1 \circ G_2(z, t') = G_2(z, t')(1) = H(z, t')$ so that the map $G = (G_1, G_2) : Z \times \mathbb{I} \to M^f$ is such that $p^f \circ G = H$ and $G(z, 0) = g(z)$. This is obtained by merely reversing the arrows in Figure 1.25. So, we take $G_1(z, t') = g_1(z)$, for all $(z, t') \in Z \times \mathbb{I}$. We define

$$G_2(z, t')(t) = \begin{cases} g_2(z)(2t/2 - t'), & 0 \le 2t \le 2 - t' \le 2, z \in Z; \\ H(z, 2t + t' - 2), & 1 \le 2 - t' \le 2t \le 2, z \in Z. \end{cases}$$

The required property of G is verified easily. Therefore p^f is a fibration. ♠

Definition 1.7.9 The fibration $p^f : M^f \to Y$ is called the *mapping path fibration* of f.

Remark 1.7.10

1. Notice that the homotopy in (ii) above does not move the section s. In other words, s defines X as a subspace of M^f which is SDR of M^f.

2. In particular, if f is the inclusion $X \hookrightarrow Y$, then M^f can be identified with the subspace of $Y^{\mathbb{I}}$,
$$P(X, Y) := \{\omega \in Y^{\mathbb{I}} : \omega(0) \in X\}.$$

3. Thus, we have shown that up to homotopy every map is a fibration. Nevertheless, one would like to have maps which are actually fibrations and not up to homotopy. Practically there is just one good source of this. We shall merely state this result. Readers familiar with a bit of differential topology will be able to appreciate this immediately.

Definition 1.7.11 A map $p : E \to B$ is called locally trivial, if there is a space F and an open cover $\{U_\alpha\}$ of B and homeomorphisms $\phi_\alpha : U_\alpha \times F \to p^{-1}(U_\alpha)$ such that $p \circ \phi_\alpha = p_1$ the projection to the first factor.

Theorem 1.7.12 *If B is paracompact then any locally trivial map $p : E \to B$ is a fibration.*

Proof: See [Spanier, 1966].

Theorem 1.7.13 *Let $p : M \to N$ be a smooth surjective submersion of manifolds with each fibre $p^{-1}(x)$ being compact, $x \in N$. Then p is locally trivial and hence is a fibration.*

Proof: See [Shastri, 2011].

Exercise 1.7.14

(i) Show that a map $f : W \to X$ is null homotopic iff it can be lifted to the path fibration $p : P(X, x_0) \to X$ of Proposition 1.7.6.

(ii) Let $p : E \to B$ be a fibration with E path connected and B simply connected. Show that for any $b \in B$ the fibre $p^{-1}(b)$ is path connected.

(iii) Show that for any space X, the space PX of all paths in X starting at a given point x_0 is contractible.

1.8 Categories and Functors

This section is a quick introduction to the language of *Categories and Functors*. No need to master this section in one go. You may keep coming back to it now and then as and when the need arises. However, by the time you start with Chapter 6 or so, we hope you have mastered this section.

All sciences are essentially the study of patterns. This is more so in the case of mathematics. The fundamental concepts such as number and space combine to produce enumerable patterns that we come across in day-to-day life. All mathematical concepts involve a certain family of objects which fit into a pattern and the study involves relations among these objects.

Let us consider some examples. In topology we study the so-called topological properties of spaces. The objects that we consider here are topological spaces. The relation between two topological spaces is studied by considering continuous functions from one to the other. A fundamental property of continuous functions is that the composite of two continuous functions if defined, is again continuous. Also, for every space X, there is at least one continuous function, viz., the identity map $Id_X : X \longrightarrow X$. Similar statements are also true for groups and homomorphisms, viz., the objects that we study here are groups and the relations are group homomorphisms; composite of two homomorphisms, if defined, is a homomorphism and there is always the identity homomorphism $Id_G : G \longrightarrow G$ for any group G. We can go on listing similar properties in other contexts such as vector spaces and linear maps between them, modules over a ring and linear maps between them and so on.

Category theory is the language that sums up this aspect of mathematics in a technically precise way. Not so surprisingly, it can prove theorems too.

Definition 1.8.1 A category $\mathcal{C}$ consists of a family $Obj(\mathcal{C})$ of *objects* with the following properties:
(i) To each ordered pair (A, B) of objects with $A, B \in Obj(\mathcal{C})$, there is a set denoted by $M(A, B)$; $M(A, B), M(A', B')$ are disjoint unless $A = A'$ and $B = B'$ in $\mathcal{C}$. The elements of $M(A, B)$ are called *morphisms* with *domain* A and *codomain* B.
(ii) For each triple (A, B, C) of objects, there is a binary operation

$$\circ : M(A, B) \times M(B, C) \longrightarrow M(A, C) \qquad (f, g) \mapsto g \circ f;$$

these are, collectively associative in the following sense: for $f \in M(A, B), g \in M(B, C), h \in M(C, D)$, we have

$$h \circ (g \circ f) = (h \circ g) \circ f.$$

(iii) For each object A in $\mathcal{C}$, there exists a (unique) $Id_A \in M(A, A)$ such that for any $f \in M(A, B)$ and $g \in M(C, A)$ we have,

$$f \circ Id_A = f; \quad Id_A \circ g = g.$$

A morphism $f : A \longrightarrow B$ in $\mathcal{C}$ is called an *equivalence* or *invertible* if there exists a morphism $g : B \longrightarrow A$ such that $f \circ g = Id_B$ and $g \circ f = Id_A$. Then g is said to be an inverse to f. If there exists an equivalence $f : A \longrightarrow B$ then we say A and B are equivalent objects in $\mathcal{C}$.

Remark 1.8.2
(i) Often when more than one category is involved in the discussion, we write $M_{\mathcal{C}}(X, Y), \operatorname{Mor}(X, Y)$ or $\operatorname{Hom}_{\mathcal{C}}(X, Y)$ to denote the set of all morphisms from X to Y, where X and Y are considered as objects in a particular category $\mathcal{C}$.
(ii) It follows easily that the identity morphism Id_A is unique in $M(A, A)$ for all A. Note that $M(A, B)$ is a set, may or may not be empty, whereas objects of $\mathcal{C}$ need not be sets.

Neither is the collection of objects a set, in general. You need not bother about these purely 'logical' problems at this stage.

(iii) It also follows that if f is invertible, then its inverse is unique and hence we can write it as f^{-1}.

(iv) The equivalence of any two objects defines an equivalence relation on the family of all objects of a category. The central theme in any categorical study is to determine this set.

(v) The condition of disjointness of $M(A, B)$ and $M(A', B')$ unless $A = A', B = B'$ is important especially in the categorical language. We are aware of the common practice of denoting a restriction of a function $f : X \to Y$ to a subset $A \subset X$ by the same notation, and often we too will follow this 'convenient' practice. For example the functions $\sin : \mathbb{R} \to [-1, 1], \sin : \mathbb{R} \to \mathbb{R}, \sin : \mathbb{C} \to \mathbb{C}$ are all denoted by $\sin$, (which is a sin according to the modern practice). The justification is that we are cutting down on clumsy notations which may obscure the mathematical ideas that we want to present. On the other hand, one of the basic objects of the language of category theory is to provide us rigor without being too verbose.

Example 1.8.3

1. **The category of all sets and functions, Ens** There is the category **Ens** whose objects are all sets and whose morphisms are all functions, with the usual law of composition. This is the 'mother' (see Example 1.8.6) of many categories that we deal with usually. That is one reason for expressing $f \in M(X, Y)$ by $f : X \longrightarrow Y$. The short form **Ens** is taken from the French word *ensembles*.

2. **The singleton category P** Consider a category with a single object denoted by $\star$ and with the morphism set $M(\star, \star)$ consisting of the single element, viz., the identity morphism. In some sense this category is the smallest of all categories which are 'non empty'.

3. **The opposite category** Given any category $\mathcal{C}$, there is another category called the *opposite* of $\mathcal{C}$ and denoted by $\mathcal{C}^{\mathrm{op}}$. Its objects are the same as that of $\mathcal{C}$. The morphisms are defined by $M_{\mathcal{C}^{\mathrm{op}}}(A, B) := M_{\mathcal{C}}(B, A)$ and the composition law is also the same. Note that $(\mathcal{C}^{\mathrm{op}})^{\mathrm{op}} = \mathcal{C}$.

4. **The category of topological spaces, Top** As discussed in Remark 1.1.3 and in the beginning of this section, there is a category whose objects are topological spaces, for each pair X, Y of spaces, $M(X, Y)$ is the set of all continuous functions from X to Y and the composition is the usual composition of functions as the binary operations. This is called the category of topological spaces and is usually denoted by **Top**. An equivalence in this category is called a homeomorphism. The equivalence class of objects in this category are called homeomorphism types.

5. **The homotopy category of topological spaces, T** Consider the category in which objects are all topological spaces and $M(X, Y)$ is the set of all homotopy class of maps from X to Y. This category is called the *homotopy category* which we shall denote by $\mathcal{T}$. (Notice that the objects in this category are the same as objects in **Top**. In some literature, this category is denoted by **Top** instead of the one we have introduced in Example 1 above.) More generally, given any topological category, we may retain the objects as they are but change the morphisms to homotopy class of maps to obtain the associated homotopy category.

6. **The category of groups, Gr** Likewise, we have the category of all groups and group homomorphisms which is denoted by **Gr**. Here the composition of two homomorphisms

is defined in the usual way. An equivalence in this category is an isomorphism. The equivalence classes are isomorphism classes of groups.

7. **The category of abelian groups, Ab** Consider the category **Ab** whose objects are abelian groups and $M(G, H)$ is taken to be all homomorphisms from G to H. Of course we compose two such homomorphisms in the usual way. In some sense this category is a subcategory of **Gr** which we shall define soon.

8. **The category of open sets on** X, $\mathcal{U}_X$ Given a topological space X consider the category whose objects are open sets $U \subset X$. For open sets U, V, if $U \subset V$ then take $M(U, V) = \{\iota\}$, the singleton set consisting of the inclusion map $U \hookrightarrow V$; otherwise take $M(U, V) = \emptyset$. In particular, note that $M(U, U) = \{Id_U\}$. We shall denote this category by $\mathcal{U}_X$.

9. **The fundamental groupoid** Let X be a fixed topological space. Let us try to define a category $\mathcal{P}_X$ whose objects are points of X and for any two points $x, y \in X$, let $M(x, y)$ be the set of all paths in X from x to y. We can concatenate a path from x to y with a path from y to z. However, this binary operation fails to satisfy the associativity condition. (Also the identity condition.) So, we rectify this by taking $M(x, y)$ to be the set of all path homotopy classes of paths from x to y. Then part (i) and (ii) of Lemma 1.2.6 essentially states that P_X becomes a category. So, we shall not prove this here. For each $x \in X$ the class of the constant path at x forms the identity element. Observe that $M(x, y)$ is non empty iff x and y are in the same path component of X. Every morphism in this category is an equivalence. (See Part (iii) of Lemma 1.2.6 again.) More generally, any category in which every morphism is an equivalence is called a *groupoid*. The set $M(x, x)$ forms a group which is nothing but the fundamental group $\pi_1(X, x)$ of X at x. (See Definition 1.2.9.)

10. **The semi-group** By a semigroup M we mean a set M with an associative binary operation and a two-sided identity. In any category $\mathcal{C}$, for any object A in $\mathcal{C}$, $M(A, A)$ is a semigroup in two different ways: either by taking the operation $f \star g$ as $f \circ g$ or $g \circ f$ (the two semigroups being the opposite of one another). Conversely, given a semigroup M, there are essentially two different ways of thinking M as $M_\mathcal{C}(A, A)$: Define the category $\mathcal{L} = \mathcal{L}_M$ with a single object $A = M$ and $M_\mathcal{L}(A, A) = \{L_x \ : \ x \in M\}$ where $L_x : M \to M$ denotes the left-multiplication by x. Likewise we can define $\mathcal{R}_M$ using right-multiplication in M. We can almost recover the semigroup M from $\mathcal{L}_M$ or $\mathcal{R}_M$ except for the fact that we cannot tell whether it is the original semigroup or the opposite one. The reason is $\mathcal{L}_M$ is the opposite of $\mathcal{R}_M$.

A semigroup is called a *monoid* if it is cancellative, i.e., $xy = xz \implies y = z$. This does not necessarily mean that the morphisms in $\mathcal{L}_M$ are invertible. In any case, notice that each morphism in $M(A, A)$ is invertible iff the semigroup is actually a group.

11. **The poset as a category** Let $(X, \leq)$ be a partially ordered set. Let us define a category associated to this. Take elements of X as objects of this category. For any two $x, y \in X$ take $M(x, y)$ to be a singleton set if $x \leq y$, and $= \emptyset$ otherwise. The binary operations are defined in an obvious way, due to the transitivity condition. For each x, the unique element in $M(x, x)$ plays the role of two-sided identity. These steps can be reversed. Starting with a category whose family of objects is a set and for each pair of objects (A, B), the set of morphisms $M(A, B)$ is either empty or a singleton and if $M(A, B) \neq \emptyset, A \neq B$ then $M(B, A) = \emptyset$, we obtain a partial order on the set of objects.

The above three examples illustrate the fact that there are many interesting categories in which morphisms are not necessarily functions. We shall see a little later an explicit example in which objects are not sets.

12. **The smooth category Diff** Let the objects be smooth manifolds and morphisms be all smooth functions. This category is called the smooth (differential) category denoted by **Diff**.

13. **The simplicial and CW categories** Let the objects be all simplicial complexes (CW-complexes) and morphisms be the set of all simplicial (cellular) maps. This category is called the simplicial (cellular) category. These categories will be denoted by **Simp, CW** respectively (see Chapter 2 for details).

14. **The category of vector spaces** Let **k** be a field. Consider the family of all (finite dimensional) vector spaces over **k** with $M(V, U)$ being the set of all linear maps from V to U. This forms a category called the category of all (finite dimensional) vector spaces over **k**. We shall denote it by **Vect$_k$** (**FVect$_k$**, respectively).

15. **The category of modules** More generally, if R is a commutative ring, we can take the objects to be all modules over R with $M(A, B) = \mathrm{Hom}_R(A, B)$ to be the set of all R-linear homomorphisms from A to B, we get a category called the category of R-modules denoted by **R-mod**.

Definition 1.8.4 Let $\mathcal{C}$ be a category. By a *subcategory* $\mathcal{D}$ of $\mathcal{C}$, we mean a category $\mathcal{D}$ each of whose objects are objects of $\mathcal{C}$ and for each pair of objects $A, B \in \mathcal{D}$ we have the set of all morphisms from A to B is a subset of the set of all $\mathcal{C}$-morphisms from A to B, i.e.,

$$M_{\mathcal{D}}(A, B) \subset M_{\mathcal{C}}(A, B) \tag{1.9}$$

and the binary operations in $\mathcal{D}$ are just the restrictions of the same in $\mathcal{C}$. If equality holds in (1.9), for all possible pairs of objects in $\mathcal{D}$, then we say $\mathcal{D}$ is a *full subcategory* of $\mathcal{C}$.

Example 1.8.5 We have already indicated one such example of a subcategory of the category of all groups, viz., the category of all abelian groups. Indeed, **Ab** is a full subcategory of **Gr**. On the other hand, **Diff** is a subcategory of **Top**, which is not a full subcategory.

Example 1.8.6 List all categories occurring in the above examples which are subcategories of **Ens**. You will find that most of the categories that we come across are subcategories of **Ens**. This explains why we called **Ens** as the 'mother' in Example 1.8.3.1.

We shall now consider the concept of relations between two categories, which is more general than the concept of subcategory.

Definition 1.8.7 Let $\mathcal{C}, \mathcal{D}$ be any two categories. By a covariant (respectively, contravariant) functor $\mathcal{F} : \mathcal{C} \rightsquigarrow \mathcal{D}$ we mean
(i) as association denoted by $\mathcal{F}$ itself

$$\mathcal{F} : Obj\,\mathcal{C} \longrightarrow Obj\,\mathcal{D}; \quad A \mapsto \mathcal{F}(A), \text{ and}$$

(ii) to each pair of objects A, B in $\mathcal{C}$ a function, again denoted by

$$\mathcal{F} : M_{\mathcal{C}}(A, B) \longrightarrow M_{\mathcal{D}}(\mathcal{F}(A), \mathcal{F}(B)); f \mapsto \mathcal{F}(f)$$

$$(\text{respectively } M_{\mathcal{C}}(A, B) \longrightarrow M_{\mathcal{D}}(\mathcal{F}(B), \mathcal{F}(A)); \quad f \mapsto \mathcal{F}(f))$$

satisfying the following properties:
(a) $\mathcal{F}(f \circ g) = \mathcal{F}(f) \circ \mathcal{F}(g);$ (respectively $\mathcal{F}(f \circ g) = \mathcal{F}(g) \circ \mathcal{F}(f)$);
(b) $\mathcal{F}(Id_A) = Id_{\mathcal{F}(A)};$
for all objects A, B, C of $\mathcal{C}$ and all morphisms $g \in M_{\mathcal{C}}(A, B); f \in M_{\mathcal{C}}(B, C)$.

Remark 1.8.8

1. The difference between covariance and contravariance is simply in the fact that covariance preserves the direction of the arrow whereas contravariance reverses it. However, in practice, it turns out that contravariance has more mathematical structure in it whereas covariance is more geometrical and easy to understand.

2. If $\mathcal{F}_1 : \mathcal{C}_1 \longrightarrow \mathcal{C}_2$ and $\mathcal{F}_2 : \mathcal{C}_2 \longrightarrow \mathcal{C}_3$ are covariant functors then there is an obvious way to define the composite functor $\mathcal{F}_2 \circ \mathcal{F}_1$ which is again a covariant functor from $\mathcal{C}_1$ to $\mathcal{C}_3$. Similarly for contravariant functors. Of course, a composite of two contravariant functors will be covariant.

3. Suppose $\mathcal{F}$ is a functor from $\mathcal{C}$ to $\mathcal{D}$. If two objects A, B are equivalent in $\mathcal{C}$ then $\mathcal{F}(A)$ and $\mathcal{F}(B)$ are equivalent in $\mathcal{D}$. (Verify this.) This is one of the most effective ways functors are exploited, viz., if we know that $\mathcal{F}(A)$ and $\mathcal{F}(B)$ are inequivalent for some functor then we know that A and B are inequivalent. We will have several illustrations of this. For a quick one see the second example below.

Example 1.8.9

1. One of the most important class of functors are the so called *forgetful* functors from a category whose objects may also be considered as objects of a larger category. For instance, from **Ab** $\rightsquigarrow$ **Ens** we have the forgetful functor which associates to each abelian group its underlying set and to each homomorphism $f : G \to H$ of abelian groups, the corresponding function. The study of inter-relations between a category and its subcategory is a common feature in all mathematics and that is how the forgetful functors are important. The most widely studied forgetful functor in topology is the one from **Diff** to **Top** with other categories interspersed between them.

2. Consider the category **Top**. The set of connected components $\operatorname{conn}(X)$ of a topological space defines a covariant functor from **Top** to **Ens**. Let us illustrate the importance of such functors in this simple example. Suppose now that we are given two topological spaces X and Y with the underlying sets having different cardinality, i.e., $\#(c(X)) \neq \#(C(Y))$. Then from Remark 3 above, it follows that X and Y cannot be equivalent in **Top**.

3. Let $\mathcal{C}$ be any category and $\hat{\mathcal{C}}$ denote the category whose single object is $\mathcal{C}$ itself and the set of morphisms $M_{\hat{\mathcal{C}}}(\mathcal{C}, \mathcal{C}) = \{Id\}$, the singleton consisting of the identity functor. This is a simple example of a category whose objects may not be sets.

4. Let $\mathcal{C}$ be any category and $A \in Obj\,\mathcal{C}$. Then there is a covariant and a contravariant functor, $\operatorname{Hom}(A, -), \operatorname{Hom}(-, A) : \mathcal{C} \rightsquigarrow$ **Ens** defined as follows:

$$\operatorname{Hom}(A, -)(B) = M(A, B); \quad \operatorname{Hom}(-, A)(B) = M(B, A)$$

and for any $f \in M(B, C)$, $f_* := \operatorname{Hom}(A, f) : M(A, B) \to M(A, C)$ and $f^* := \operatorname{Hom}(f, A) : M(C, A) \to M(B, A)$ are given as follows: For any $g \in M(A, B), h \in M(C, A)$ take
$$f_*(g) = f \circ g; \qquad f^*(h) = h \circ f.$$

These are called representable functors. They are very important, since understanding these functors is equivalent to understanding the category $\mathcal{C}$ itself.

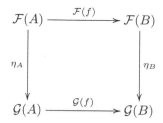

FIGURE 1.26. Naturality

5. Given a surjective group homomorphism $f : G \to H$, the first isomorphism theorem says that the quotient group $G/\operatorname{Ker} f$ is canonically isomorphic to H. Likewise given a finite dimensional vector space V, you are told that its double-dual V^{**} is canonically isomorphic to V. Perhaps, you were not told the true meaning of the adverb 'canonically' in these situations. We shall explain this using category theory.

Definition 1.8.10 Let $\mathcal{F}, \mathcal{G} : \mathcal{C} \rightsquigarrow \mathcal{D}$ be any two covariant functors. By a *natural transformation of functors* $\eta : \mathcal{F} \rightsquigarrow \mathcal{G}$, we mean the following: For each $A \in Obj\mathcal{C}$, there is a morphism $\eta_A \in M_{\mathcal{D}}(\mathcal{F}(A), \mathcal{G}(A))$ such that for each $f \in M_{\mathcal{C}}(A, B)$, the following diagram is commutative.

In addition, if each η_A is an equivalence, then η is called an *equivalence of functors*. When we have such an equivalence, we say $\mathcal{F}$, $\mathcal{G}$ are naturally equivalent or naturally isomorphic functors.

The naturality is of utmost importance and cannot be brushed away as merely a rhetoric of pure mathematicians. Let us consider one more important notion which generalizes the notion of equivalence.

Definition 1.8.11 Let $\mathcal{F} : \mathcal{C} \to \mathcal{D}$ and $\mathcal{G} : \mathcal{D} \to \mathcal{C}$ be functors. We say $\mathcal{F}$ is a *left-adjoint of* $\mathcal{G}$ (and equivalently, $\mathcal{G}$ is a right-adjoint of $\mathcal{F}$) if there is a natural isomorphism

$$\eta : \operatorname{Hom}_{\mathcal{C}}(-, \mathcal{G}(-)) \approx \operatorname{Hom}_{\mathcal{D}}(\mathcal{F}(-), -).$$

This just means that for every pair (C, D) with C in $\mathcal{C}$ and D in $\mathcal{D}$, there is an isomorphism

$$\eta_{C,D} : \operatorname{Hom}_{\mathcal{C}}(C, \mathcal{G}(D)) \to \operatorname{Hom}_{\mathcal{D}}(\mathcal{F}(C), D)$$

such that for every pair of morphisms $\phi : C \to C'$ and $\psi : D \to D'$ in the respective categories, the diagram

$$
\begin{array}{ccc}
\operatorname{Hom}_{\mathcal{C}}(C', \mathcal{G}(D)) & \xrightarrow{\eta_{C',D}} & \operatorname{Hom}_{\mathcal{D}}(\mathcal{F}(C'), B) \\
{\scriptstyle \operatorname{Hom}_{\mathcal{C}}(\phi, \mathcal{G}(\psi))} \downarrow & & \downarrow {\scriptstyle \operatorname{Hom}_{\mathcal{C}}(\mathcal{F}(\phi), \psi)} \\
\operatorname{Hom}_{\mathcal{C}}(C, \mathcal{G}(D')) & \xrightarrow{\eta_{C,D'}} & \operatorname{Hom}_{\mathcal{D}}(\mathcal{F}(C), D')
\end{array}
$$

commutes.

Remark 1.8.12 We shall leave it to the reader to verify that any two left adjoints of a functor $\mathcal{G}$ are naturally equivalent. Similarly, any two right adjoints of a functor $\mathcal{F}$ are naturally equivalent.

Example 1.8.13

1. Consider the category $\mathbf{Vect_k}$. Let V^* denote the 'dual' space of all linear maps from $V \to \mathbf{k}$. Then the assignment

$$V \rightsquigarrow V^*$$

defines a contravariant functor $\mathbf{Vect_k} \to \mathbf{Vect_k}$ where for a linear map $f : V \to W$ we take $f^* : W^* \to V^*$ defined by the formula

$$f^*(\phi) = \phi \circ f.$$

Take $\mathcal{C} = \mathbf{Vect_k} = \mathcal{D}$ and $\mathcal{F}$ to be the identity functor and $\mathcal{G}$ to be the double dual: $T : V \rightsquigarrow V^{**}$. Define a natural transformation $\eta : \mathcal{F} \rightsquigarrow \mathcal{G}$ as follows: given a vector space V over $\mathbf{k}$, we have to first define a linear map $\eta_V : V \longrightarrow V^{**}$. This is done by taking

$$\eta_V(v)(\phi) = \phi(v), \quad v \in V; \phi \in V^*.$$

It follows that $f^{**} : V^{**} \to W^{**}$ has the property

$$f^{**}(\psi) = \psi \circ f^*, \quad \psi \in V^{**}.$$

It is straightforward to verify that the following diagram is commutative.

$$
\begin{array}{ccc}
V & \xrightarrow{\ f\ } & W \\
{\scriptstyle \eta_V}\downarrow & & \downarrow{\scriptstyle \eta_W} \\
V^{**} & \xrightarrow{\ f^{**}\ } & W^{**}
\end{array}
$$

Thus η is a natural transformation of functors.

Observe that there is a subcategory $\mathbf{FVect_k}$ of $\mathbf{Vect_k}$ whose objects are finite dimensional vector spaces over k. If V is finite dimensional then so is V^{**}. Thus, η restricts to a natural transformation of functors Id and T on $\mathbf{FVect_k}$. Further, we know that η_V is an isomorphism whenever V is finite dimensional. Therefore η is an equivalence of the two functors on $\mathbf{FVect_k}$. This is the explanation that we promised of the word 'canonical' in the Example 1.8.9.5.

In the next example, we explain the word 'canonical' occurring in the first isomorphism theorem in group theory.

2. Consider the category $\mathcal{C}$ whose objects are surjective homomorphisms $f : G \to G'$ of groups and whose morphisms are pairs (α, β) of homomorphisms which make up a commutative diagram:

$$
\begin{array}{ccc}
G_1 & \xrightarrow{\ f_1\ } & G_1' \\
{\scriptstyle \alpha}\downarrow & & \downarrow{\scriptstyle \beta} \\
G_2 & \xrightarrow{\ f_2\ } & G_2'
\end{array}
$$

Note that because of the commutativity of the diagram, $\alpha(\mathrm{Ker}\, f_1) \subset \mathrm{Ker}\, f_2$. Therefore α induces a homomorphism $\bar{\alpha} : G_1/\mathrm{Ker}\, f_1 \to G_2/\mathrm{Ker}\, f_2$.

On the category $\mathcal{C}$, we consider two functors: $\mathcal{Q}$ and $\mathcal{I}$ defined as follows:

$$\mathcal{Q}(f) = G/\mathrm{Ker}\, f, \quad \mathcal{Q}(\alpha, \beta) = \bar{\alpha}$$

$$\mathcal{I}(f) = \text{Image of } f; \quad \mathcal{I}(\alpha, \beta) = \beta.$$

The first isomorphism theorem yields an isomorphism: $\phi_f : G/\mathrm{Ker}\, f \to G'$ such that the following diagrams are commutative:

$$
\begin{array}{ccc}
G_1/\mathrm{Ker}\, f_1 & \xrightarrow{\ \phi_{f_1}\ } & G_1' \\
{\scriptstyle \bar{\alpha}}\downarrow & & \downarrow{\scriptstyle \beta} \\
G_2/\mathrm{Ker}\, f_2 & \xrightarrow{\ \phi_{f_2}\ } & G_2'
\end{array}
$$

Thus ϕ defines a natural equivalence of the two functors. This is the meaning of the word 'canonical' occurring in the statement of the first isomorphism theorem.

3. The definitions and various properties needed to understand these examples will be given in a later chapter. Here we merely record them. The category $\mathcal{T}_0$ of all pointed topological spaces whose objects consist of all pairs (X, x_0) where X is any topological space and $x_0 \in X$ is the chosen base point; morphisms in this category, are continuous functions which take base points to base points. On this category, we have the functor π_1 which associates the fundamental group $\pi_1(X, x_0)$. We also have the first integral homology group $H_1(X, \mathbb{Z})$ of X defining another functor. To each (X, x_0) the Hurewicz map $\phi : \pi_1(X, x_0) \longrightarrow H_1(X)$ which is nothing but the abelianisation of $\pi_1(X, x_0)$ now defines a natural transformation of functors $\varphi : \pi_1 \rightsquigarrow H_1$.

Example 1.8.14 Here are a few examples of adjoint functors. We shall meet more of them as we proceed deeper into algebraic topology.

1. Let $\mathcal{F} : \mathbf{Ens} \to \mathbf{Ab}$ be the functor that assigns to each set S the free abelian group over it: $\mathcal{F}(S) = F(S)$. A right adjoint $G : \mathbf{Ab} \to \mathbf{Ens}$ is the forgetful functor which assigns the underlying set to a given abelian group. Verify this and imitate this to obtain many other examples.

2. Let B be an A-algebra, where A is a commutative ring. Let $S : \mathbf{B}\text{-mod} \to \mathbf{A}\text{-mod}$ be the forgetful functor that associates to each B-module, the underlying A-module. Then the functor $M \rightsquigarrow M \otimes_A B$ is a left-adjoint to S and the functor $N \rightsquigarrow \mathrm{Hom}_A(B, N)$ is a right adjoint to S. Verify these statements.

We shall end this section with one more useful concept.

Definition 1.8.15 Let A be an object in a category $\mathcal{C}$. We say A is an *initial object* in $\mathcal{C}$ if for every object $B \in \mathcal{C}$, $M(A, B)$ consists of a single element. Likewise A is called a *terminal object* in $\mathcal{C}$ if for every $B \in \mathcal{C}$, $M(B, A)$ is a singleton. The object A is called a *zero object* if it is both an initial object and a terminal object.

Example 1.8.16 In the category $\mathbf{Ens}$, the emptyset $\emptyset$ is an initial object but not a terminal object; also any singleton set $\{p\}$ is a terminal object but not an initial object. Categories such as $\mathbf{Gr}, \mathbf{Ab}, \mathbf{Vect_k}, \mathbf{R}\text{-mod}$, etc., all have zero objects.

Remark 1.8.17 In any category $\mathcal{C}$, it is easy to see that any two initial (terminal) objects in a category (if they exist) are equivalent. Likewise, it is not difficult to see that any two zero objects in a category $\mathcal{C}$ are equivalent (isomorphic). Moreover, if there are zero objects in $\mathcal{C}$ then for every $A, B \in \mathcal{C}$, there is a special morphism $0_{A,B}$ which is the composite of the unique morphisms:

$$A \to 0 \to B.$$

These morphisms have the additional property that for any morphisms $f : D \to A, g : B \to E$ in $\mathcal{C}$ we have

$$0_{A,B} \circ f = 0_{D,B}; \quad g \circ 0_{A,B} = 0_{A,E}.$$

Example 1.8.18 Recall the definition of pull-back from 1.7.1. Here we shall describe the same thing in any category $\mathcal{C}$. Let $f \in M(X, B), p \in (E, B)$ be any two morphisms. There is a category $\mathcal{C}(f, p)$ whose objects are commutative diagrams

$$
\begin{array}{ccc}
Z & \xrightarrow{\ \alpha\ } & E \\
{\scriptstyle\beta}\downarrow & & \downarrow{\scriptstyle p} \\
X & \xrightarrow{\ f\ } & B
\end{array}
$$

We shall leave it to the reader to figure out what are morphisms in this category. A terminal object in this category is called the pullback of p under f and is denoted by $f^*p : f^*E \to X$. The uniqueness of the pullback according to this definition is obvious. Lemma 1.7.2 relates this definition with the constructive definition 1.7.1. This construction is not available in a general category and even if available may not serve to exhibit the pullback.

Example 1.8.19 Direct limits and inverse limits In example 1.8.3.11 we have seen how to view a poset as a small category. A *directed set* $(J, \leq)$ is a poset such that for any two elements $i, j \in J$, there is $k \in J$ such that $i \leq k$ and $j \leq k$. It is not difficult to reformulate this condition in terms of category theory: given $i, j \in J$, there is $k \in J$ such that $M(i, k)$ and $M(j, k)$ are non empty.

We define a *directed system* in a category $\mathcal{C}$ to be a covariant functor $\mathcal{F} : (J, \leq) \to \mathcal{C}$ where $(J, \leq)$ is a directed set viewed as a category. We shall denote the objects $\mathcal{F}(j) := F_j, j \in J$ and $\mathcal{F}(M(i, j)) = \{f_{ij}\}$ whenever $i < j$. Associated to a directed system $\mathcal{F}$ one defines another category $\bar{\mathcal{F}}$ as follows: the objects are pairs $(A, \{\alpha_j\}_{j \in J})$, where A is an object in $\mathcal{C}$ and $\alpha_j : F_j \to A$ are morphisms in $\mathcal{C}$ such that $\alpha_i = \alpha_j \circ f_{ij}$ whenever f_{ij} exists. A morphism $\tau : (A, \{\alpha_j\}_{j \in J}) \to (B, \{\beta_j\}_{j \in J})$ in $\bar{\mathcal{F}}$ is a morphism $\tau : A \to B$ in $\mathcal{C}$, such that $\tau \circ \alpha_j = \beta_j, j \in J$.

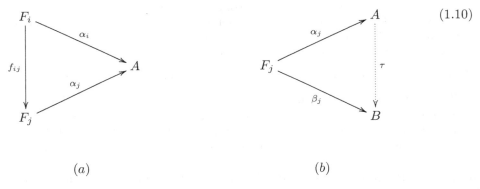

$$(1.10)$$

$$(a) \qquad\qquad\qquad\qquad (b)$$

By the *direct limit* of the directed system $\mathcal{F}$ we mean an initial object in the category $\bar{\mathcal{F}}$. Note that if $\mathcal{C}$ has initial objects then the category $\bar{\mathcal{F}}$ is non empty; however, there is no guarantee that it will have initial objects. Of course, if an initial object exists, we have seen that it is unique up to equivalence. Thus the same holds for the direct limit also. We denote this object by $\mathrm{dlim}_{\rightarrow} F_j$.

Thus the direct limit of a directed system $\{F_i\}$ in $\mathcal{C}$ is an object A in $\mathcal{C}$ together with a collection of morphisms $\alpha_j : F_j \to A$ satisfying the compatibility condition (1.10(a)) such that for every other compatible family of morphisms $\beta_j : F_j \to B$ there is a *unique* morphism $\tau : A \to B$ making up a commutative diagram as in (1.10(b)).

The simplest thing is to see that in the category **Ens** every directed system has a direct limit: Take B to be the quotient of the disjoint union $\coprod_{j \in J} F_j$ by the relation $x \sim f_{ij}(x)$

for all $x \in F_i$ whenever f_{ij} makes sense and $\beta_i : F_i \to B$ as the composite of the inclusion followed by the quotient map. Similar construction works in many other algebraic categories such as **Ab**, **Vect$_k$**, **Gr**, **R-mod**, etc. In **Top** also, this works; we have to take the obvious disjoint union topology on the disjoint union and the quotient topology on the quotient. A typical special case of this is when F_i are subsets of a given set F and f_{ij} are inclusion maps whenever $F_i \subset F_j$. The direct limit of this system is nothing but $F' := \cup_i F_i$ with the coinduced topology: $A \subset F'$ is open iff $A \cap F_i$ is open for every i.

An important observation which often helps in computations is the following. Let $(J, \leq)$ be a sub-poset of a directed set $(I, \leq)$. We say J is 'final' in I if to each $i \in I$ there exists some $j \in J$ such that $i \leq j$. In this situation, given a directed system $\{F_i\}_{i \in I}$, we may consider the subsystem $\{F_j\}_{j \in J}$. If one of the two direct limits exists then the other exists and the two are equal:

$$\underset{\to \ i \in I}{\mathrm{dlim}} \ F_i = \underset{\to \ j \in J}{\mathrm{dlim}} \ F_j.$$

Verification of this statement is straightforward and left to the reader as an exercise.

If we replace 'covariant' by 'contravariant' and 'initial' by 'terminal' in the above passages, we obtain the notion of inverse systems and inverse limits and corresponding conclusions. However, about the existence of inverse limits, one needs to be careful. In the category of **Ens**, one may take the subset of $\prod_i F_i$ consisting of those elements (x_i) such that $f_{ij}(x_j) = x_i$ whenever $i \leq j$. Once again, the same construction goes through in many other familiar categories as well. For more details, the reader may refer to [Hilton–Stommbach, 1970].

Exercise 1.8.20

(i) List all possible category-pairs where one is a 'sub' of the other, occurring in the above set of Examples 1.8.3.

(ii) List all the pairs of categories where one is a full subcategory of the other, occurring in the above set of Examples 1.8.3.

(iii) In any category $\mathcal{C}$ and for any object A in $\mathcal{C}$, let Aut A denote the set of all invertible morphisms in $M_{\mathcal{C}}(A, A)$. Verify that this forms a group under the composition law coming from $\mathcal{C}$. Now, let $\alpha : A \to B$ be an invertible morphism in $\mathcal{C}$. Define $\hat{\alpha} :$ Aut $A \to$ Aut B by $\hat{\alpha}(f) = \alpha \circ f \circ \alpha^{-1}$. Show that $\hat{\alpha}$ is an isomorphism of groups.

(iv) Let $\mathcal{C}$ be any category, $f \in M_{\mathcal{C}}(A, X), g \in M_{\mathcal{C}}(A, Y)$ be any two morphisms. There is a category $(f, g)\mathcal{C}$ whose objects are commutative diagrams of morphisms in $\mathcal{C}$:

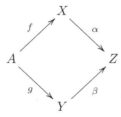

An initial object in this category is called the *pushout* of (g, f) and is denoted by $X_f \cup_g Y$. Show that pushouts exist in the categories **Ens**, **Ab**. It exists in the category **Gr**. But it may be better to read Section 3.6 before trying to prove this.

(v) Show that the direct limit of the following system

$$\mathbb{Z} \xrightarrow{.2} \mathbb{Z} \xrightarrow{.2} \mathbb{Z} \xrightarrow{.2} \cdots$$

is isomorphic to $\mathbb{Z}[1/2]$, the group of rational numbers of the form $\frac{m}{2^n}, m, n \in \mathbb{Z}$.

1.9   Miscellaneous Exercises to Chapter 1

1. Construct an explicit strong deformation retraction in each of the following cases:

   (a) $\mathbb{S}^{n-1} \subset \mathbb{R}^n \setminus \{0\}$.

   (b) $\partial \mathbb{I}^n \subset \mathbb{I}^n \setminus \{p\}$ where $p \in \mathbb{I}^n$ is any interior point.

   (c) $\mathbb{S}^1 \times 1 \cup 1 \times \mathbb{S}^1 \subset \mathbb{S}^1 \times \mathbb{S}^1 \setminus \{(-1, -1)\}$.

   (d) $\mathbb{S}^1 \times \{0, 1\} \cup 1 \times \mathbb{I} \subset \mathbb{S}^1 \times \mathbb{I} \setminus \{(-1, 1/2)\}$.

   (e) $A \cup \partial \ddot{\mathrm{M}} \subset \ddot{\mathrm{M}} \setminus \{[1/2, 1/2]\}$, where $\ddot{\mathrm{M}}$ denotes the Möbius band as a quotient of $\mathbb{I} \times \mathbb{I}$ by the relation:
   $$\ddot{\mathrm{M}} := \frac{\mathbb{I} \times \mathbb{I}}{(t, 0) \sim (1 - t, 1)},$$
   A is the image of $\mathbb{I} \times 0$ and $\partial \ddot{\mathrm{M}}$ is the boundary $\ddot{\mathrm{M}}$.

2. Let $\emptyset \neq K \subset \operatorname{int} \mathbb{D}^n$ be a convex subset. Show that $\partial \mathbb{D}^n$ is a SDR of $\mathbb{D}^n \setminus K$.

3. Let $X = \mathbb{S}^n \times \mathbb{S}^n \setminus \{(x, -x) \ : \ x \in \mathbb{S}^n\}$. Show that the subspace $\{(x, x) \ : \ x \in \mathbb{S}^n\} \subset X$ is a deformation retract.

4. Let X and Y be closed subspaces of a topological space Z. Suppose that $X \cap Y$ is a strong deformation retract of Y. Then X is a strong deformation retract of $X \cup Y$.

5. Show that the group $O(n)$ of all orthogonal $n \times n$ real matrices is a strong deformation retract of the group $GL(n; \mathbb{R})$ of all invertible $n \times n$ real matrices. (Hint: Use polar decomposition of invertible matrices OR Gram-Schmidt.)

6. Show that $GL(n, \mathbb{R})$ has precisely two path connected components. Similarly, show that the real orthogonal group $O(n)$ of all $n \times n$ matrices A such that $AA^t = Id$ has precisely two path connected components.

7. Consider the mappings $(x_1, \ldots, x_n) \mapsto (\pm x_1, \pm x_2, \ldots, \pm x_n)$ from $\mathbb{S}^{n-1} \to \mathbb{S}^{n-1}$ and put them into homotopy equivalent classes. How many classes are there?

8. Suppose $f : \mathbb{S}^n \to \mathbb{S}^n$ is a mapping such that $f(x) \neq x$ for any $x \in \mathbb{S}^n$. Then show that f is homotopic to the antipodal map. Similarly, if $f(x) \neq -x$ for any x then show that f is homotopic to the identity map.

9. Let $A \subset X$, and let $f : A \longrightarrow Y$ be a continuous function.
   (a) Suppose $A \subset B \subset X$, are closed subsets and B is a SDR of X. Then show that the adjunction space $Y \cup_f B$ is a SDR of the adjunction space $Y \cup_f X$.
   (b) Show that $f : A \longrightarrow Y$ can be extended to $\hat{f} : X \longrightarrow Y$ continuously iff Y is a retract of the adjunction space $Y \cup_f X$.
   (c) Show that (X, A) has homotopy extension property with respect to a space W iff $(Y \cup_f X, Y)$ has homotopy extension property with respect to W.

10. Let $A \subset B \subset X$ be subspaces of a space X and let $f : A \to Y$ be a map. Suppose B is a retract of X. Then show that $B \cup_f Y$ is a retract of $X \cup_f Y$.

11. For any map $f : X \to Y$, let M_f be the mapping cylinder as in Example 1.5.8. Show that the subspace $M_f \times \{0\} \cup X \times \{0\} \times \mathbb{I}$ is a retract of $M_f \times \mathbb{I}$. Deduce that the inclusion map $X \times \{0\} \hookrightarrow M_f$ is a cofibration.

12. Let X be a locally (path) connected space and $A \subset X$ be a retract of X. Show that A is locally (path) connected.

13. If X is a Hausdorff space and the inclusion map $A \longrightarrow X$ is a cofibration then show that A is a closed subspace of X.

14. Show that the combspace is not a retract of $\mathbb{I} \times \mathbb{I}$. Deduce that $A \subset \mathbb{I}$ is not a cofibration where $A = \{0, \frac{1}{2}, \frac{1}{3}, \frac{1}{4}, \cdots\}$.

15. Suppose $p \in X$ is a SDR of X. Show that for each open neighbourhood U of p in X, there exists an open neighbourhood V of p such that the inclusion map $V \hookrightarrow U$ is null homotopic.

16. **The zig-zag zebra** Consider the following closed subspace of $\mathbb{R}^2$ as depicted in Figure 1.27 Let for any integer n,

$$A = \left\{ (t+s, 2s) \ : \ 0 \leq t \leq 1, \ t \text{ rational and } 0 \leq s \leq \frac{1-t}{2} \right\}$$

and

$$B = \left\{ (t+s, 1-2s) \ : \ 0 \leq t \leq 1, \ t \text{ rational and } 0 \leq s \leq \frac{1-t}{2} \right\}.$$

Let

$$X = \cup_{n \in \mathbb{Z}} (A + (n, 0)) \cup (B + (n + \frac{1}{2}, 0)).$$

Let Y be the union of all the line segments

$$\left\{ (s, 2s) \ : \ 0 \leq s \leq \frac{1}{2} \right\} + (n, 0)$$

and

$$\left\{ (s, 1-2s) \ : \ 0 \leq s \leq \frac{1}{2} \right\} + (n + \frac{1}{2}, 0).$$

The subspace Y is shown by thicker line segments in Figure 1.27.

FIGURE 1.27. Every point of the zig-zag is degenerate

Show that
(a) For each $x \in \mathbb{R}$, there is a unique $\alpha(x) = (x+1, q(x)) \in Y$ and the function q is continuous.
(b) Define $p : X \to Y$ by $p(x, y) = \alpha(x)$. Construct a homotopy $H : X \times \mathbb{I} \to X$ such that $H(z, 0) = z$ and $H(z, 1) = p(z)$. Hence conclude that p is a homotopy equivalence.
(c) Deduce that X is contractible.
(d) Show that X fails to be locally connected at each and every point.
(e) Deduce that no point of X is a SDR of X.

17. **The double combspace** Consider the following double combspace (see Figure 1.28): For $v \in \mathbb{R}^2$ let $T_v : \mathbb{R}^2 \to \mathbb{R}^2$ be the translation: $T_v(u) = u + v$. Take $E_1 = T_{(0,-1)}(E)$

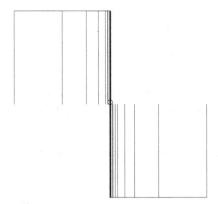

FIGURE 1.28. The double combspace

where E is the combspace defined in Example 1.4.6. Let $E_2 = -E_1$ be the reflection of E_1 in the point $(0,0) = p$. Take $D = E_1 \cup E_2$.

Show that D is not contractible. This gives an example where the bouquet of two contractible spaces fails to be contractible. (See Exercise 32 below. Also see Exercise 5 in 4.7 and Exercise (ii) in 10.4.17.)

18. If we stretch the bad point $(0,0)$ in the above example then the space behaves slightly better but still is not very good in the sense it still has certain pathological properties: Consider $E_1' = T_{(0,-2)}(E)$, $E_2' = -E_1'$. Let J denote the line segment between $(0,1)$ and $(0,-1)$. Put $D' = E_1' \cup E_2' \cup J$.
    (a) Show that D' is contractible.
    (b) Show that there is a surjective map $p : D' \to D$ such that fibres of p are all contractible. (Yet p is not a homotopy equivalence.)

19. **Topologist's sine loop** Consider the following subspaces of $\mathbb{R}^2$:

$$A_1 = \left\{ \left(x, \sin \frac{\pi}{x} \right) \ : \ 0 < x \leq 1 \right\}; A_2 = \{(0,y) \ : \ -1 \leq y \leq 1\};$$

$$A_3 = \{(x, 1 + \sqrt{x - x^2}) \ : \ 0 \leq x \leq 1\}; A_4 = \{(1,y) \ : \ 0 \leq y \leq 1\};$$

$$S = A_1 \cup A_2; \quad C = A_1 \cup A_2 \cup A_3 \cup A_4.$$

S is called the **topologist's sine curve** whereas C is called the **topologist's sine loop** or the **Warsaw circle**. (See Figure 1.29.)

(a) Show that S is connected but not path connected. Also show that S is not locally connected.
(b) Show that $\pi_1(C) = (1)$.

20. Show that the inclusion map $\mathbb{S}^1 \longrightarrow \mathbb{C}^\star$ is a homotopy equivalence.

21. Let $q_1(z) = z^n$ and $q_2(z) = \bar{z}$. Compute the degrees of $q_1, q_2 : \mathbb{S}^1 \longrightarrow \mathbb{S}^1$.

22. Let $p(z)$ be a monic polynomial of degree n in z with complex coefficients and let $q(z) = z^n$. For $r > 0$, let S_r denote the circle of radius r and centre 0. Show that for large enough r the maps $p|_{S_r} : S_r \longrightarrow \mathbb{C}^\star$ and $q|_{S^r} : S^r \longrightarrow \mathbb{C}^\star$ are homotopic to each other.

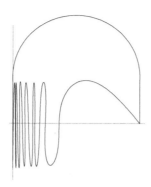

FIGURE 1.29. The Warsaw circle

23. Prove the fundamental theorem of algebra using some of the exercises above. [Hint: If p has no zeros, then $p|_{S_r}$ extends to the disc $\{z : |z| \leq r\}$.]

24. (a) Let $f : \mathbb{S}^1 \longrightarrow \mathbb{S}^1$ be a map with the property that $f(-x) = -f(x)$ for all x. Show that deg f is odd.
(b) Deduce that there is no continuous function $f : \mathbb{S}^2 \longrightarrow \mathbb{S}^1$ with the property $f(-x) = -f(x)$ for all x.
(c) **Borsuk–Ulam theorem** Show that given any map $g : \mathbb{S}^2 \longrightarrow \mathbb{R}^2$ there exists $x \in \mathbb{S}^2$ such that $g(-x) = g(x)$. [Hint: consider $g(x) - g(-x)$ and use (b).]

25. **Lusternik–Schnirelmann** Let $\mathbb{S}^2 = F_1 \cup F_2 \cup F_3$ where each F_i is a closed subset of $\mathbb{S}^2$. Show that one of the F_i contains a pair of antipodal points. [Hint: First consider two of the functions $d(x, F_i)$ and apply the previous result.]

26. **Ham–sandwich problem** A ham-sandwich consists of two pieces of bread and a piece of ham placed in between. Show that we can cut all of them into two equal halves by a single stroke of the knife.

27. **The topological join $X * Y$** : Let X and Y be any two nonempty topological spaces. We define the *topological join* $X * Y$ of X and Y to be the quotient space of $X \times \mathbb{I} \times Y$ by the relations:$(x_1, 1, y) \sim (x_2, 1, y)$, $\forall x_1, x_2 \in X, y \in Y$ and $(x, 0, y_1) \sim (x, 0, y_2)$, $\forall x \in X$ and $y_1, y_2 \in Y$. Intuitively, the topological join is the space formed by the union of all line segments $\{[x, t, y], 0 \leq t \leq 1\}$ with the two end-points $x \in X$ and $y \in Y$. If Y is empty, we define $X * Y$ to be X itself. Prove:

(a) $X * Y$ contains both X and Y as subspaces. (In fact, $x \mapsto [x, 0, y]$ for any y defines a canonical inclusion of X into $X * Y$. Similarly, Y can be identified with the subspace consisting of $[x, 1, y]$ for any fixed x and all $y \in Y$.)

(b) The topological join is commutative and associative, i.e., for any topological spaces X, Y, Z, there are homeomorphisms

$$X * Y \simeq Y * X; \quad X * (Y * Z) \simeq (X * Y) * Z.$$

(c) If X is a singleton space, then $X * Y$ is the same as the cone CY.

(d) The assignment $X \rightsquigarrow CX$ is functorial.

(e) If X and Y are homeomorphic, then so are CX and CY.

28. **The Suspension** SX We define the suspension of a space X to be $SX := \mathbb{S}^0 * X$ (see Figure 1.30).

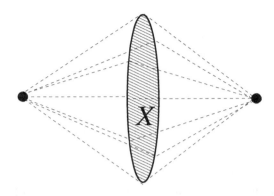

FIGURE 1.30. The unreduced suspension

Given a map $f : X \longrightarrow Y$ there is an obvious way to define a map $S(f) : SX \longrightarrow SY$. Prove the following statements:

(a) SX is the quotient of $\mathbb{I} \times X$ by the identifications given by the two sets of relations $(x, 0) \sim (x', 0), x, x' \in X$ and $(x, 1) \sim (x', 1), x, x' \in X$.

(b) $X \rightsquigarrow SX$ is a functor.

(c) If $f : X \longrightarrow Y$ is a homeomorphism then so is $Sf : SX \longrightarrow SY$.

(d) $S(\mathbb{S}^n) = \mathbb{S}^{n+1}$.

(e) SX can be obtained by gluing two copies $C_{\pm}X$ of CX along X and the homeomorphism type is independent of the gluing homeomorphism $f : X \longrightarrow X$.

(f) There is a relative homeomorphism $(SX, C_-X) \to (C_+X, X)$ and a homeomorphism $SX/C_-X \to C_+X/X$.

(g) If there is a point $x_0 \in X$ such that $\{x_0\} \hookrightarrow X$ is a cofibration, then the quotient map $(SX, C_-X) \to (SX/C_-X, \{C_-X\})$ is a homotopy equivalence.

(h) If we glue two copies of n-dimensional discs $\mathbb{D}^n$ along their boundary via a homeomorphism the resulting space is homeomorphic to the n-dimensional sphere $\mathbb{S}^n$.

29. Show that $\mathbb{S}^m * \mathbb{S}^n$ is homeomorphic to $\mathbb{S}^{m+n+1}$.

Pointed spaces and base points

30. A special case of relative homotopy occurs when the subspace that we take is a single-ton. More precisely, by a *pointed space* we mean a pair (X, a) where X is a topological space and $a \in X$. The point a is called the *base point* of the pointed space. A map $f : (X, a) \longrightarrow (Y, b)$ is a continuous function $f : X \longrightarrow Y$ such that $b = f(a)$. The product of two pairs $(X, a) \times (Y, b)$ is defined to be the pair $(X \times Y, (a, b))$. Let (X, A) be a topological pair. Given a point $a \in A$ we say a is *non degenerate base point* for the pair (X, A) iff the inclusion map $(a, a) \hookrightarrow (X, A)$ is a cofibration. Similarly, in the absolute case, $a \in X$ is a non degenerate base point if $\{a\} \hookrightarrow X$ is a cofibration. A point is called *degenerate* if it is **not** non degenerate. There is a category of pointed topological pairs, with non degenerate base points and base point preserving maps of pairs. The condition of non degeneracy of the base point is a technical necessity when we do homotopy theory on this category.
    (a) Show that if X is contractible and $p \in X$ is non degenerate, then p is a SDR of X.
    (b) Show that the point $(0, 1)$ is a degenerate point of the combspace.
    (c) Locate all examples of degenerate points that you have come across (but were not aware of).

31. **Reduced cones and cylinders** The notion of mapping cylinder mapping cone, sus-pension, etc., all have to be correspondingly modified in this context.
    (a) Given a pointed space (X, a) the *reduced cone* $C(X, a)$ is defined to be the pointed space which is the quotient of the ordinary cone CX by the relation $(a, t) \sim (x, 0)$ for all $0 \le t \le 1$ and $x \in X$ and where the class $[a, t]$ is taken as the base point.
     (b) Similarly the reduced suspension $S(X, a)$ is defined to be the quotient of the ordi-nary suspension in which all points on the two lines $[-1, t, a]$ and $[1, t, a]$ are identified to a single point denoted by $[a]$.
    (c) Given $f : (X, a) \longrightarrow (Y, b)$ the *reduced mapping cylinder* $\hat{M}(f)$ of f is defined to be the pointed space which is the quotient space of the ordinary mapping cylinder by the relation $(a, t) \sim b \ \forall \ 0 \le t \le 1$, and in which the class $[a, t] = [b]$ taken as the base point. Show that
    (i) For any pointed space, the reduced cone is contractible.
    (ii) $C(\mathbb{S}^{n-1}, p)$ is homeomorphic to $(\mathbb{D}^n, p)$ for any $p \in \mathbb{S}^{n-1}$ and $S(\mathbb{S}^{n-1}, p)$ is home-omorphic to $(\mathbb{S}^n, q)$ for any $p \in \mathbb{S}^{n-1}, q \in \mathbb{S}^n$.
    (iii) $f : (X, a) \longrightarrow (Y, b)$ is homotopic to the constant map relative to a iff f can be extended continuously to $C(X, a) \longrightarrow (Y, b)$.

32. Given a family (X_α, a_α) of pointed spaces, we define the *bouquet* or the *one-point-union* $\vee_\alpha X_\alpha$ to be the quotient space of the disjoint union $\coprod_\alpha X_\alpha$ by the identification $a_\alpha \sim a_\beta$ for any two indices α, β. The base point for this space is taken to be the class $[a_\alpha]$. Show that
    (i) Each (X_α, a_α) is a retract of $(\vee_\alpha X_\alpha, [a_\alpha])$.
    (ii) Given a family of maps $f_\alpha : (X_\alpha, a_\alpha) \longrightarrow (Y, b)$, there is a unique map $\vee_\alpha f_\alpha : (\vee_\alpha X_\alpha, [a_\alpha]) \longrightarrow (Y, b)$ which extends each f_α.
    (iii) Given a family of maps $f_\alpha : (X_\alpha, a_\alpha) \longrightarrow (Y_\alpha, b_\alpha)$, there is a unique map $\prod_\alpha f_\alpha : (\vee_\alpha X_\alpha, [a_\alpha]) \longrightarrow (\vee_\alpha Y_\alpha, b_\alpha)$, which extends each f_α. (This is nothing but the restriction of the product map.)
    (iv) The bouquet of the family of reduced cones (suspension) of pointed spaces is homeomorphic to the reduced cone (suspension) over the bouquet of the family of pointed spaces.
    (v) The bouquet of a family of pointed spaces can be identified with a subspace of the product space of the family in a natural way.

33. Let (X, a) be any pointed space and $\tilde{S}X = S(X, a)$ denote its reduced suspension. (See Figure 1.31.) Show that $\tilde{S}X$ can be thought of as the quotient space of $X \times \mathbb{I}$ in which all points in $X \times 0 \cup X \times 1 \cup a \times \mathbb{I}$ are identified to a single point.

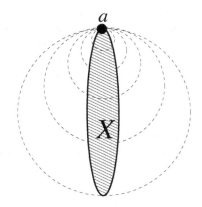

FIGURE 1.31. The reduced suspension

This point is then the base point of $\tilde{S}X$ and is again denoted by a. For $(x, t) \in X \times \mathbb{I}$, let us denote by $[x, t]$ the corresponding element in $\tilde{S}X$ under the quotient map.

Define the *co-multiplication* $\mu : \tilde{S}X \longrightarrow \tilde{S}X \vee \tilde{S}X$ by the formula:

$$\mu[x, t] = \begin{cases} ([x, 2t], a), & 0 \leq t \leq \frac{1}{2} \\ (a, [x, 2t - 1]), & \frac{1}{2} \leq t \leq 1. \end{cases} \tag{1.11}$$

Let Id denote the identity map of $\tilde{S}X$ and let c_a, the constant map $z \mapsto a$. base point. Show that the following homotopies hold relative to the base point.
(i) (**Identity**) $(Id \vee c_a) \circ \mu \sim Id \sim (c_a \vee Id) \circ \mu$.
(ii) (**Associativity**) $(Id \vee \mu) \circ \mu \sim (\mu \vee Id) \circ \mu$.
(iii) (**Inverse**) $(Id \vee \eta) \circ \mu \sim c_a$. [Hint: see the proof of Lemma 1.2.6]

34. Use Exercise 30 (ii) to define $\mu_n : \mathbb{S}^n \longrightarrow \mathbb{S}^n \vee \mathbb{S}^n$ for $n \geq 1$, as the iterated suspension of $\mu : \tilde{S}X \longrightarrow \tilde{S}X \vee \tilde{S}X$ where $X = \mathbb{S}^0$. Verify that homotopies similar to those in the previous exercise hold for μ_n also.

35. For $n = 2$, points of $S^2 X$ can be parameterised by $[[x, t], s]$ as a quotient of $(X \times \mathbb{I}) \times \mathbb{I}$. We then have two different co-multiplications on $S^2 X$ using the two different time-slots t, s as in 1.11:

$$\mu([[x, t], s]) = \begin{cases} ([[x, 2t], s], a), & 0 \leq t \leq \frac{1}{2}; \\ (a, [[x, 2t - 1], s]), & \frac{1}{2} \leq t \leq 1. \end{cases} \tag{1.12}$$

$$\nu([x, t], s]) = \begin{cases} ([[x, t], 2s], a), & 0 \leq s \leq \frac{1}{2}; \\ (a, [[x, t], 2s - 1]), & \frac{1}{2} \leq s \leq 1. \end{cases} \tag{1.13}$$

(i) Show that μ is the suspension of the co-multiplication on $\tilde{S}X$.
(ii) Show that μ and ν are mutually distributive: i.e., $(\mu, \mu) \circ \nu = (\nu, \nu) \circ \mu$.
(iii) Deduce that μ and ν are homotopic to each other.
(iv) Show that μ is homotopy commutative, i.e., if $T : \tilde{S}X \vee \tilde{S}X \longrightarrow \tilde{S}X \vee \tilde{S}X$ is the

map which interchanges the factors then $\mu \sim T \circ \mu$ relative to the base point. (This is rather tricky if you try to prove it directly. However, read further.)

36. Let $*$ and $\circ$ be any two binary composition laws, on a set S. Suppose there exists an element $e \in S$ which serves as two-sided identity for both these compositions, i.e., for all $s \in S$ we have $e * s = s * e = e \circ s = s \circ e = s$ and that the two compositions are mutually distributive, i.e, for $p, q, r, s \in S$, we have,

$$(p * q) \circ (r * s) = (p \circ r) * (q \circ s).$$

Then show that the two compositions are the same and that this common composition is associative and commutative. (Hint: Put $q = e = r$, to obtain $* = \circ$. Next put $p = e = s$ to get commutativity. Prove the associativity by yourself.)

37. Given any pointed space (Y, b), consider $[(\tilde{S}X, a); (Y, b)]$ the set of all base point preserving homotopy classes of base point preserving maps. Define a binary operation on this set by

$$[f] \circ [g] = [(f, g) \circ \mu]$$

where μ is the co-multiplication as in (1.11).
(i) Show that $[(\tilde{S}X, a); (Y, b)]$ becomes a group under this operation.
(ii) If (X, a) itself is a suspension of some space (Z, a) then show that the above group is abelian.
(iii) For any pointed space (Y, b) and $n \geq 1$, we define $\pi_n(Y, b)$ as the group on the set $[(\mathbb{S}^n, p); (Y, b)]$ as above. Deduce that for $n \geq 2$, $\pi_n(Y, b)$ is abelian.

38. Show that for any two path connected based spaces $(X, x_0), (Y, y_0)$ and for all integers $n \geq 1$, $\pi_n(X \times Y, (x_0, y_0)) \approx \pi_n(X, x_0) \times \pi_n(Y, y_0)$.

39. Prove that $\pi_1(G, e)$ is commutative for a connected topological group. (Hint: The group law in G can be used to obtain another composition '$\circ$' of homotopy classes of loops at e. Compare this with the standard composition in π_1.)

40. For any topological space with a base point x_0, we define the space of *measured paths* as a subspace:

$$P^*(X) = \{(r, \omega) \in [0, \infty) \times X^{[0,\infty)} \ : \ \omega(0) = x_0, \omega(t) = \omega(r), t \geq r\}.$$

(a) Show that $F(X)$ is contractible.
(b) Let $p^*(r, \omega) = \omega(r)$. Show that p is a fibration.
(c) Let $\Omega^*(X) = \{(r, \omega) \in P^*(X)$ such that $\omega(r) = x_0\}$. Given any path $\omega : \mathbb{I} \to X$ such that $\omega(0) = x_0$ there is an obvious extension $\hat{\omega} : [0, \infty) \to X$ given by $\omega(t) = \omega(1)$ for all $r > 1$. Show that $\omega \mapsto (1, \hat{\omega})$ defines an embedding $h : P(X) \to F(X)$.
(d) h is a fibre homotopy equivalence from the fibration $p : P(X) \to X$ to the fibration $p^* : F(X) \to X$.
(e) h defines a homotopy equivalence of $\Omega(X)$ with the space $\Omega^*(X)$ of measured loops.
(f) $\Omega^*(X)$ has a composition which is associative and the constant loop is a two sided identity. Thus $\Omega^*(X)$ is a *strictly associative* H-space.
(g) The association $X \rightsquigarrow \Omega^*(X)$ is a functor.

41. Let $\{X_i\}_{i \in I}$ be a directed system of path connected spaces. Show that the direct limit space is path connected.

42. **Solenoids** Fix any infinite sequence of primes $P = (p_1, \ldots, p_n, \ldots)$. Define the directed system of topological spaces $\{(G_n, f_n)\}$ where $G_n = \mathbb{S}^1$ for all n and $f_n : G_n \to G_{n+1}$ is given by $f_n(z) = z^{p_n}$. Other maps $G_n \to G_{n+k}$ are taken to be composites of these homomorphisms. Denote the direct limit by $\mathbb{S}_P^1$.

   (a) For the constant sequence $P = (2, 2, 2, \ldots)$, we denote the solenoid by S_2 and call it dyadic solenoid. Compute the fundamental group of S_2. (Hint: Compare Exercise 1.8.20.(vi)).

   (b) What are the sequences P which will give $\pi_1(\mathbb{S}_P^1) \approx \mathbb{Q}$?

43. Consider the fibration $P = P_X : X^{\mathbb{I}} \to X$ given by $P_X(\omega) = \omega(0)$. (See Lemma 1.7.4.) Show that the assignment $X \rightsquigarrow (P_X : X^{\mathbb{I}} \to X)$ defines covariant functor on **Top** to a suitable category (which you have to describe).

44. Verify that there is a covariant functor from the category of pointed spaces and base-point-preserving maps to the category of fibrations, which assign to a map $f : X \to Y$, the principal fibration $p_f : P_f \to X$ induced by f, as in Definition 1.7.7

45. **The category of compactly generated spaces** We shall work inside the category of Hausdorff topological spaces. Given any space X consider the weak topology on X generated by the family of compact subsets of X. We shall denote X with this topology by $k(X)$. Prove the following statements.
   (a) $Id : k(X) \to X$ is continuous, i.e., the topology of $k(X)$ finer than the topology of X.
   (b) If X is locally compact then $Id : k(X) \to X$ is a homeomorphism.
   (c) If $f : X \to Y$ is continuous, then so is $f : k(X) \to k(Y)$.
   (d) The assignment $X \rightsquigarrow k(X)$ is a covariant functor.
   (e) The family of compact subsets of X and that for $k(X)$ coincide.
   (f) If $q : X \to Y$ is a quotient map and X is compactly generated then so is Y. We shall denote $k(X \times Y)$ by the symbol $X \times_w Y$. We shall also denote $k(Y^X)$ by the symbol $\mathrm{Map}(X, Y)$. Now prove the following version of Theorem 1.3.1

   (g) Show that the evaluation map $E : \mathrm{Map}(X, Y) \times_w X \to Y$ is continuous. (Compare with Theorem 1.3.1 and observe that locally compactness hypothesis on X is removed.)
   (h) Show that a function $g : Z \to M(X, Y)$ is continuous iff the composite $E \circ (g \times Id) : Z \times_w X \to Y$ is continuous.
   (i) Show that $\psi : \mathrm{Map}(Z, \mathrm{Map}(X, Y)) \to \mathrm{Map}(Z \times X, Y)$ given by $\psi(g) = E \circ (g \times Id)$ is a homeomorphism.

Chapter 2

Cell Complexes and Simplicial Complexes

In this chapter we shall introduce two important classes of topological spaces. To begin with, by an open (closed) n-cell we mean a topological space which is homeomorphic to the open (closed) unit disc in $\mathbb{R}^n$. Being contractible, these are among the simplest objects from the point of view of algebraic topology. On the other hand, from the point of view of differential topology, they are among the richest objects. The interior of these objects, viz., the open cells are the building blocks for manifolds and manifolds are the most suitable objects on which we can do calculus. The closed cells are going to be the building blocks for a large class of topological spaces called *cell complexes,* though the process of 'building-up' is quite different here from the one that is employed in defining manifolds. Originally named 'CW-complexes', introduced and studied extensively by J. H. C. Whitehead [Whitehead, 1939], cell-complexes are best suited for the study of algebraic topology.

In the first section, we quickly recall some basics of convex polytopes partly to illustrate the nature of 'geometry' that is behind simplicial complexes and cell complexes and partly to give a small step toward PL-topology which is getting more attention in recent years. After establishing the two basic descriptions of a convex polytope, we introduce the classical concept of Euler characteristic. We give a completely elementary proof of the Euler formula convex polytopes. This is then used in the proof of the classification of regular polytopes. Our treatment of this topic just stops when it starts becoming more and more combinatorial in nature. Interested readers may look into books such as [Coxeter, 1973], [Grunbaum, 1967], [Brondsted, 1982].

In Sections 2.2 and 2.3, we shall begin a study of cell complexes. This automatically takes care of all the point-set topological and homotopical aspects of simplicial complexes that we are going to discuss in Sections $2.4 - 2.8$. The most important result here seems to be the simplicial approximation theorem, from which we will be able to deduce Brouwer fixed point theorem and the mild version of Brouwer invariance of domain, viz., $\mathbb{R}^n$ and $\mathbb{R}^m$ are not homeomorphic for $n \neq m$. Note that we are getting these results without computing any kind of algebraic invariants such as fundamental group or homology groups.

2.1 Basics of Convex Polytopes

The simplest maps $\mathbb{R}^n \to \mathbb{R}^m$ to deal with are linear maps. Calculus helps us to study differentiable maps via their linear approximation, viz., the derivative. A polygonal path approximating a smooth curve gives a lot of information on the curve and at the same time makes it much easier to handle. In this section, we shall briefly introduce the basics of the theory of convex polytopes laying down motivation for the study of simplicial complexes and polyhedral topology on the one hand and a foundation to the combinatorial study of convex polytopes on the other. The contents of this section are not quite necessary to understand the rest of the book with a few exceptions and so, if you prefer you may skip it or merely browse through it and return to it only when necessary.

Notation Let us denote by $\mathbb{R}^d$, the d-dimensional vector space over the reals with the stan-

dard addition and scalar multiplication and with the standard basis. We identify $\mathbb{R}^n$ with
the subspace of $\mathbb{R}^{n+1}$ consisting of vectors with their last coordinate zero. Let $\mathbb{R}^\infty$ denote
the vector space of sequences $(x_0, x_1, \ldots, x_n, \ldots)$ of real numbers, which vanish eventually.
Let $\mathbf{e}_i = (0, \ldots, 1, 0, \ldots)$ be the sequence with all entries 0 except at the ith instance at which
the entry is 1. Then the set $\{\mathbf{e}_i \; : \; i \geq 0\}$ forms a basis for the vector space $\mathbb{R}^\infty$. We shall
denote by

$$|\Delta_n| = \{(x_0, x_1, \ldots, x_n) \in \mathbb{R}^{n+1} \; : \; 0 \leq x_0 \leq 1, \sum_0^n x_i = 1\}.$$

This object is going to be the central object of study in this chapter and is called the
standard n-simplex.

We assume that the reader is familiar with a fair amount of linear algebra but perhaps
not with the notion of 'affine' geometry. The so-called *affine structure* on $\mathbb{R}^d$ can be thought
of as the vector space *without any specified origin*.

Definition 2.1.1 Let $x_1, x_2, \ldots, x_n$ be points in $\mathbb{R}^d$. By an *affine combination* of these
points we mean a linear combination $\sum \lambda_i x_i$ where $\sum \lambda_i = 1$. Observe that $n \geq 1$. By an
affine subspace A of $\mathbb{R}^d$ we mean a subset A with the property that every affine combination
of points in A is again in A. (This allows the empty set as an affine subspace.) Given a subset
M of $\mathbb{R}^d$, by the *affine hull* of M is meant the collection of all affine combinations of points
in M and this is denoted by aff M. We say that a subset S of $\mathbb{R}^d$ is *affinely independent* if

$$\sum_i \lambda_i x_i = 0, \quad x_i \in S \text{ and } \sum_i \lambda_i = 0 \implies \lambda_i = 0 \; \forall \; i = 1, 2, \ldots, n.$$

A function $f : A \to B$ where A, B are any two affine subspaces of Euclidean spaces, is said
to be an *affine transformation* if $f(tx + (1-t)y) = tf(x) + (1-t)f(y)$ for all $x, y \in A$ and
for all $t \in \mathbb{R}$.

Observe that any singleton set and any 2-set are affinely independent. There is a close
relation between affine independence and linear independence but they should not be confused with each other. The key result is the following lemma, the proof of which is very
easy.

Lemma 2.1.2 A function $f : \mathbb{R}^r \to \mathbb{R}^s$ is an affine transformation iff the function $F(x) :=
f(x) - f(0)$ is a linear transformation.

Exercise 2.1.3

(i) Show that an affine combination of points of aff M is again a point of aff M.

(ii) Show that an affine subspace A of $\mathbb{R}^d$ is a vector subspace iff $0 \in A$.

(iii) Show that an affine subspace A of $\mathbb{R}^d$ is nothing but a translate of a linear subspace
of $\mathbb{R}^d$, i.e., there is a vector subspace V and a vector $x \in \mathbb{R}^d$ such that $A = V + x$.

  For any affine subspace A of $\mathbb{R}^d$ we define dim A to be the dimension of the linear
  subspace $A - x$, where $x \in A$ is a point.

(iv) Show that $\{x_1, x_2, \ldots, x_n\} \subset \mathbb{R}^d$ is affinely independent iff $\{x_1 - x_n, x_2 - x_n, \ldots, x_{n-1} -
x_n\}$ is linearly independent. In particular, this implies that $n \leq d + 1$.

(v) Let X denote the $n \times (d+1)$ matrix with its ith row equal to $(1, x_{i1}, \ldots, x_{id})$. Using
the standard coordinates in $\mathbb{R}^d$, let us write $x_i = (x_{i1}, x_{i2}, \ldots, x_{id})$. Show that the
subset $\{x_1, x_2, \ldots, x_n\} \subset \mathbb{R}^d$ is affinely independent iff the matrix X has rank n.

(vi) For any point x in $\mathbb{R}^d$, let us consider the point $(1, x)$ in the space $\mathbb{R} \times \mathbb{R}^d = \mathbb{R}^{d+1}$. Show that $\{x_1, x_2, \ldots, x_n\}$ is affinely independent in $\mathbb{R}^d$ iff the set

$$\{(1, x_1), (1, x_2), \ldots, (1, x_n)\}$$

is linearly independent in $\mathbb{R}^{d+1}$.

(vii) **General position theorem:** Let A be any subset of $\mathbb{R}^d$. We say that A is in *general position* if every k-subset of A is affinely independent for all $k \leq d+1$. (Thus for example, a subset $A \subset \mathbb{R}^3$ is in general position if no three distinct points of A are collinear and no four distinct points of A are coplanar. Notice that the definition is stronger than saying that every $(d+1)$-subset of A is affine independent. These two conditions are equivalent only if A has at least $d+1$ elements.) Let $A = \{v_1, \ldots, v_n\}$ be a subset of $\mathbb{R}^d$. Prove the following statements:

(a) Given any $\epsilon > 0$, there exist $w_1, \ldots, w_n \in \mathbb{R}^d$ such that $\|v_i - w_i\| < \epsilon \; \forall \; i$ and the set $\{w_1, \ldots, w_n\}$ is in general position in $\mathbb{R}^d$.

(b) If A is in general position, then there exists $\epsilon > 0$ such that any set $\{w_1, \ldots, w_n\}$ such that $\|v_i - w_i\| < \epsilon, \; \forall \; i$, is in general position.

[*Hint*: In the affine space of dimension $(d+1)n$ of $(d+1) \times n$ matrices over $\mathbb{R}$, the set of those matrices with rank less than $d+1$ is contained in the union of finitely many hyperplanes.]

Statements (a) and (b) together constitute what is known as the *General Position Theorem*.

Definition 2.1.4 Let $x_1, x_2, \ldots, x_n$ be points in $\mathbb{R}^d$. By a *convex combination* of these points we mean a finite linear combination $\sum \lambda_i x_i$ where $\sum \lambda_i = 1,$ and each $\lambda_i \geq 0$. The *convex hull* of a set M is the set of all convex combinations of points in M and is denoted by conv M. Observe that conv $M \subset$ aff M. A subset A of $\mathbb{R}^d$ is said to be *convex* if conv $A = A$. The *relative interior* of a convex set A, denoted by ri A, is the interior of A as a subset of aff A. Similarly, the *relative boundary* of A is the complement of ri A in the (topological) closure of A,

$$\text{rb } A := \text{cl } A \setminus \text{ri } A.$$

The following classical theorem is of fundamental importance in the entire theory of convex sets.

Theorem 2.1.5 (Carathéodory) *For any $M \subset \mathbb{R}^d$, the convex hull* conv M *has the property that every element of* conv M *can be expressed as a convex combination of at most $d+1$ points of M.*

Proof: Let $x = \sum_{i=1}^{k} \lambda_i x_i$ be a convex combination of points in M. Assuming that $k \geq d+2$ we shall express x as a convex combination of $k-1$ points and that is enough.

Since $k \geq d+2$ (see 2.1.3.((iv))), it follows that the set $\{x_1, x_2, \ldots, x_k\}$ is not affinely independent. Hence, there exists a non trivial relation $\sum_{i=1}^{k} \mu_i x_i = 0$ with $\sum_i \mu_i = 0$. Of course, by changing the sign, if necessary, we may assume that $\mu_k > 0$ and also by rearranging, if necessary, that $\lambda_i / \mu_i \geq \lambda_k / \mu_k$ whenever, $\mu_i > 0$. Put $\gamma_i = \lambda_i - \lambda_k \mu_i / \mu_k$ and verify that $x = \sum_{i=1}^{k-1} \gamma_i x_i$ is a convex combination as required. ♠

Corollary 2.1.6 If M is compact then conv M is compact.

Proof: Let $\phi : M^{d+1} \times |\Delta_d| \to \mathbb{R}^d$ be defined by $((x_0, \ldots, x_d), (\alpha_0, \ldots, \alpha_d)) \longmapsto \sum \alpha_i x_i$. Here $|\Delta_d| = \{(\alpha_0, \ldots, \alpha_d) \in \mathbb{R}^{d+1} : 0 \leq \alpha_i \leq 1, \sum \alpha_i = 1\}$ and so is a compact subset of $\mathbb{R}^{d+1}$. Since M is compact it follows that $M^{d+1} \times |\Delta_d|$ is compact. Observe that ϕ is continuous and Im $\phi = $ conv M and hence, conv M is compact. ♠

Definition 2.1.7 By a geometric n-simplex we mean the convex hull of any $n+1$ affinely independent points $\{v_1, \ldots, v_{n+1}\}$ in $\mathbb{R}^d$. The elements v_i are called the vertices of the simplex.

Thus a 0-simplex is nothing but a point and 1-simplex is a line segment, a 2-simplex is a triangle, a 3-simplex is a tetrahedron. The important consequence of Carathéodory's theorem about a geometric simplex A is that every element of the conv A is a unique convex combination $\sum_i t_i v_i$ of its vertices. Often the functions $t_1, \ldots, t_{n+1}$ are called *barycentric coordinates* of the points of the simplex. The point $\frac{\sum_i v_i}{n+1}$ is called the *barycentre* of A.

Theorem 2.1.8 *Let A, B be any two geometric simplices. There exists an affine isomorphism $f : \operatorname{conv} A \to \operatorname{conv} B$ such that $f(A) = B$ iff $\dim A = \dim B$.*

Proof: Choose some labeling $A = \{a_1, \ldots a_k\}, B = \{b_1, \ldots, b_k\}$ where $k = \dim A + 1$. Take $f(a_i) = b_i$ and extend linearly. ♠

Definition 2.1.9 By a half-space in $\mathbb{R}^d$ we mean a subset of the form

$$H^+ = \{x \in \mathbb{R}^d : L(x) \geq 0\}$$

where L is a non trivial affine linear map, say,

$$L(x) = \sum_{i=1}^d l_i x_i + l_0.$$

H^+ is called a *supporting half-space* for a convex set A if $A \subseteq H^+$.

A *supporting hyperplane* for a convex set A in $\mathbb{R}^d$ is a hyperplane

$$H = \{x \in \mathbb{R}^d : L(x) = 0\}$$

such that $H \cap A \neq \phi$ and $A \subseteq H^+$. Further, if $A \not\subseteq H$, then we call H a *proper supporting hyperplane*. Denoting the vector, $(l_1, \ldots, l_d) =: y$ and (say) $l_0 = \alpha$ we could use the following more descriptive notation for H, H^+ and H^- :

$$H(y, \alpha) := \{x \; : \; \langle y, x \rangle + \alpha = 0\}$$
$$H^+(y, \alpha) := \{x \; : \; \langle y, x \rangle + \alpha \geq 0\}$$
$$H^-(y, \alpha) := \{x \; : \; \langle y, x \rangle + \alpha \leq 0\}.$$

Here $\langle, \rangle$ denotes the standard inner product on $\mathbb{R}^d$.

Exercise 2.1.10

(i) Let A be a closed convex subset of $\mathbb{R}^d$. Show that $H(y, \alpha)$ is a supporting hyperplane for A iff

$$\alpha = \max_{x \in A}\{\langle x, y \rangle\} \quad \text{or} \quad \alpha = \min_{x \in A}\{\langle x, y \rangle\}.$$

Moreover, $H(y, \alpha)$ is a proper one iff $H \cap \operatorname{ri} A = \emptyset$.

(ii) Let A be a non empty convex set. Show that $\operatorname{ri} A \neq \emptyset$. If $x_0 \in \operatorname{ri} A$ and $x_1 \in \operatorname{cl} A$ then show that the line segment $[x_0, x_1) \subseteq \operatorname{ri} A$.

(iii) For a convex subset $A \subset \mathbb{R}^d$, show that $\operatorname{ri} A = \operatorname{int} A$ iff $\dim A = d$.

(iv) For a convex set A, show that its topological closure $\operatorname{cl} A$ is also convex.

Theorem 2.1.11 *Let A be a non empty closed convex set in $\mathbb{R}^d$. Then we have:*
(a) For every point $x \notin A$, there exists a hyperplane H which separates A and x.
(b) Every point on the relative boundary of A is contained in a supporting hyperplane for A.
(c) A is the intersection of its supporting half-spaces.

Proof: (a) Since A is a closed set there exists a point $y \in A$ which is nearest to x. In fact such a point y is unique also (use convexity of A). Now take the mid point z of the line segment $[x, y]$ and take H to be the hyperplane orthogonal to this line segment and passing through z.

(b) Without loss of generality we may assume that $\dim A = d$. If $d = 0, 1$, then there is nothing to prove. Consider the case $d = 2$. Consider a circle C with centre at x. (See Figure 2.1 (a).) Let C' be the set of points $w \in C$ such that the line segment $(x, w]$ intersects A. Observe that

(i) C' is non empty.
(ii) if for some w, the line segment $(x, w]$ intersects ri A then its antipodal point w' on C (with respect to x) cannot be in C'. In particular, $C' \neq C$.
(iii) C' is connected since A is convex.
Thus C' is an arc. If w_1 is one of the end-points of C' then let H be the line through x and w_1. Verify that this will do.

(Observe that it may happen that the other end-point of C' also lies on this line.)

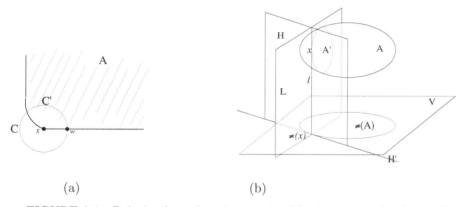

(a)	(b)

FIGURE 2.1. Relative boundary is contained in the supporting hyperplane

We shall now induct on the dimension d of A. Assume the validity of the statement for lower values of d and let $d > 2$. (See Figure 2.1(b).) Let L be any 2-dimensional affine subspace passing through x and some point of ri A. Then $A' = L \cap A$ is a 2-dimensional convex set with $x \in \mathrm{bd}\, A$. Therefore, we get a line l through x that supports A'. Let V be the hyperplane orthogonal to this line and let $\pi : \mathbb{R}^d \longrightarrow V$ be the orthogonal projection. Then $\pi(x)$ is in the boundary of the convex set $\pi(A)$ and $\dim \pi(A) = d - 1$. Therefore by induction hypothesis, there is a supporting hyperplane H' for $\pi(A)$ in V and passing through $\pi(x)$. Take $H = \pi^{-1}(H')$. Verify that this is the required hyperplane.
(c) This is immediate from (a). ♠

For the simplicity of the exposition we shall, from now on, implicitly assume that for the given convex set A, $\dim A = d$, the dimension of the ambient affine space, unless specifically mentioned otherwise. Of course this can always be arranged by merely taking aff A as the ambient affine space.

Definition 2.1.12 Let A be a convex set. $F \subseteq A$ is called a *face* of A if
(i) F is convex, and
(ii) for any two points $y \neq z \in A$, if $(y, z) \cap F \neq \emptyset$ then $[y, z] \subseteq F$.

We call a point $x \in A$ an *extreme point* of A if $\{x\}$ is a face. (In other words, for every pair of points $y, z \in A$, $x \in [y, z]$ implies $x = y = z$. This is also equivalent to saying that $A \setminus \{x\}$ is convex.) The set of extreme points in A is denoted by $\text{ext}\, A$.

Remark 2.1.13

1. The empty set $\emptyset$ and the entire set A are also considered as faces of A. These are called *improper faces*. Other faces are called *proper faces*.

2. 0-dimensional faces are extreme points of A.

3. By a *facet* of A we mean a $(d-1)$-dimensional face of A. More generally a *k-face* is a face of dimension k.

4. Every proper face of A is contained in the boundary of A.

5. The set of all faces of a convex set A forms a partially ordered set (poset) under the obvious inclusion relation. This poset is called the *face poset* of A. We shall denote it by $\mathcal{F}(A)$. One may say that the combinatorial information about A is coded in this poset.

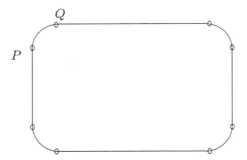

FIGURE 2.2. Extreme points in an oval

Example 2.1.14 All points of the boundary of a closed disc are extreme points. On the other hand, consider the set A of points bounded by the oval shown in Figure 2.2. This forms a compact convex subset of $\mathbb{R}^2$. It is easily verified that any point in the interior of any of the two horizontal and vertical segments of the boundary is not an extreme point. Of course, the points in the interior of the curved portions of the boundary are extreme points. The boundary points of the curved portions such as P and Q are also extreme points.

Theorem 2.1.15 *Let A be a convex set. Then the intersection of any arbitrary family of faces of A is a face of A. In particular, the set $\mathcal{F}(A)$ of all faces of A forms a complete lattice, under the operations:*

$$\inf \mathcal{G} \quad := \quad \cap \{F \ : \ F \in \mathcal{G}\}$$
$$\sup \mathcal{G} \quad := \quad \cap \{G \ : \ G \in \mathcal{F}(A) \ \textit{such that for all } F \in \mathcal{G}, \ F \subseteq G\}$$

Thus the poset $\mathcal{F}(A)$ is indeed a lattice. So it is also called the *face lattice* of A. The proof of the above theorem is straightforward. (Recall that a partially ordered set is called a *lattice* if inf and sup exist for every finite subset. If this happens for all subsets then it is called a *complete lattice*.)

Remark 2.1.16 (1) Every face F of a (closed convex set) A is closed and convex.
(2) Let $F \subset G \subset H$ be convex subsets and G be a face of H. Then F is a face of G iff F is face of H. Thus, a face of a face is a face.
(3) If $F \neq A$ is a face of A then $F \subset \operatorname{bd} A$. As a consequence, $\dim F < \dim A$.
(4) Suppose $x \in F$ and F is a face of A. Then F is the smallest face containing x iff $x \in \operatorname{ri} F$.

Exercise 2.1.17 Determine the face lattice of A, where A is the closed convex set given in Example 2.1.14.

Theorem 2.1.18 (Minkowski) *Let A be a compact convex set in $\mathbb{R}^d$. Let M be any subset of A. Then the following conditions are equivalent.*
(a) $A = \operatorname{conv} M$.
(b) $\operatorname{ext} A \subset M$.
In particular, $A = \operatorname{conv} \operatorname{ext} A$.

Proof: We first observe that the equivalence of (a) and (b) implies that $A = \operatorname{conv} \operatorname{ext} A$, by taking $\operatorname{ext} A = M$. This we shall use in the inductive step involved in the proof of the theorem.
(a) $\Rightarrow$ (b). Suppose $x \in \operatorname{ext} A \setminus M$. Then $M \subset A \setminus \{x\}$. $A \setminus \{x\}$ is convex and so $\operatorname{conv} M \subset A \setminus \{x\}$, a contradiction.
(b) $\Rightarrow$ (a). Enough to show $A \subset \operatorname{conv} \operatorname{ext} A$.

We prove this by induction on d, the dimension of A. For $\dim C = -1, 0$, there is nothing to prove. Even for $\dim A = 1$, this is obvious. Suppose that the statement is true for all smaller values of d and now $d > 1$. Let $x \in A$. If x is not an extreme point already, then there exist segments in A containing x in their interior and amongst these we choose one which is of the largest size, say, $[y_1, y_2]$. Clearly, $y_1, y_2 \in \operatorname{rb} A$. Let F_1 and F_2 be the smallest faces containing y_1 and y_2, respectively. Then both F_1 and F_2 are proper faces of A and so we can apply the induction hypothesis. Hence, $y_i \in \operatorname{conv} \operatorname{ext} F_i$, for $i = 1, 2$. Therefore, $x \in (y_1, y_2) \subset \operatorname{conv}(\operatorname{ext} F_1 \cup \operatorname{ext} F_2)$. Since extreme points are 0-dimensional faces, and a face of a face is a face, $\operatorname{ext} F_1 \cup \operatorname{ext} F_2 \subseteq \operatorname{ext} A$. Therefore, $x \in \operatorname{conv} \operatorname{ext} A$. ♠

Definition 2.1.19 By a *convex polytope* in $\mathbb{R}^d$ we mean the convex hull of a finite subset of $\mathbb{R}^d$.

Let K be a convex polytope. By Minkowski's theorem, it follows that $\operatorname{ext} K$ is a finite set. We also call extreme points of K *vertices* of K. It follows that each face F of K is again a convex polytope of appropriate dimension, and $\operatorname{ext} F = (\operatorname{ext} K) \cap F$. Observe that each face is determined by its extreme set which is necessarily a subset of $\operatorname{ext} K$. Hence it follows that $\mathcal{F}(K)$ is finite.

Theorem 2.1.20 *Let K be a convex polytope, F be a face of K. Then $F = K \cap H$ where H is a supporting hyperplane for K.*

Proof: By induction on $d = \dim K$. If $d = -1, 0$ or 1, the statement is obvious. We prove the statement for $d = 2$ separately (since the inductive hypothesis is not going to be of any use at this stage). Let F be a face of K and $\dim K = 2$. If $\dim F = 1$, we can take $H = \operatorname{aff} F$, and then clearly, $F = H \cap K$. Let $\dim F = 0$, i.e., F is a singleton set, $F = \{x\}$. Proceed exactly as in the proof of Theorem 2.1.11 (b). We have seen that C' is a proper, connected subset of C. The important difference is that now we can show that C' is closed also. Indeed, each of the two end-points of C' must lie on one of the segments $[x, y]$ where y runs over the finite set ext $K \setminus \{x\}$. Let us say, $[x, y_i] \cap C$, $i = 1, 2$, are the end-points of C'. Moreover, it is also clear that the two line segments do not lie on the same affine line, for otherwise, it follows that $x \in (y_1, y_2)$ and hence cannot be an extreme point of K. Now, take $y = x + (y_1 - y_2)/2$ and $H = \text{aff } \{x, y\}$. Verify that $H \cap K = \{x\}$. (See Figure 2.3.)

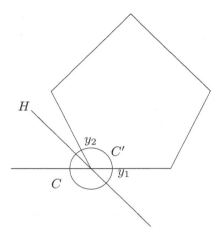

FIGURE 2.3. Every face is the intersection with a supporting hyperplane

Inductively, suppose that $\dim K \geq 3$. Let F be a proper face of K, and let $x \in \text{ri } F$. Let H be a supporting hyperplane of K through x (see the proof of theorem 2.1.11 again). Then it follows that $F \subseteq H \cap K$. If $F = H \cap K$ then we are through. Suppose that F is a proper face of $H \cap K$. Note that $H \cap K$ is itself a polytope of dimension $< d$. Indeed, $H \cap K = conv(\text{ext } K \cap H)$. Hence, by induction, there is a hyperplane H' in H such that $H' \cap K = F$. It remains to extend this hyperplane to a hyperplane H_1 of $\mathbb{R}^d$ such that $H_1 \cap K = F$ and this is where we are going to use the result for $d = 2$.

Let B be any (2-dimensional) plane in $\mathbb{R}^d$ orthogonal to H' and $\pi : \mathbb{R}^d \to B$ be the projection map. Then clearly, $\pi(K)$ is a convex polygon in B. Since $\pi(H') = \{y\}$ is a singleton set so is $\pi(F) = \{y\}$. Observe that y must be a vertex of $\pi(K)$. [If this is not the case, $\exists y_1, y_2 \in K$, such that $\pi(y_1) \neq \pi(y_2)$, and $y = \lambda \pi(y_1) + (1 - \lambda)\pi(y_2), \lambda \in (0, 1)$. But then $u = \lambda y_1 + (1 - \lambda)y_2 \in K$ and $\pi(u) = y$. Therefore $u \in K \cap H' = F$. Hence $y_1, y_2 \in F \Rightarrow \pi(y_1) = \pi(y_2)$, a contradiction.] So let L be a line in B such that $L \cap \pi(K) = \{y\}$. Take $H_1 = \pi^{-1}(L)$. Then it follows that $H_1 \cap K = F$. ♠

Theorem 2.1.21 *Each d-polytope K is the intersection of a finite number of closed half-spaces defined by the hyperplanes which are the affine hulls of facets of K. Conversely, if K is a bounded subset of $\mathbb{R}^d$ and is the intersection of finitely many closed half-spaces, then K is a convex polytope.*

Proof: The first part is immediate from Theorem 2.1.20. To prove the converse, we observe that K is a compact convex set. So it is enough to prove that ext K is finite. This we do by induction on the dimension d of K. For $d \leq 1$, there is nothing to prove. Let $\{H_i\}$ be the set of finitely many supporting hyperplanes and let $K_i = K \cap H_i$. Then observe that each K_i is a bounded intersection of finitely many closed half-spaces in H_i. Also, since K_i is a proper face of K, $\dim K_i < d$. Hence, by induction, ext K_i is finite. Since, ext $K = \cup \text{ ext } K_i$, we are done. ♠

Remark 2.1.22 Below we sum up a few important consequences of what we have done so far. The details are left to the reader as exercises.

(1) Each face of a convex polytope is a convex polytope.
(2) If $f_k(K)$ denotes the number of k-faces of K, then clearly for all k,

$$f_k(K) \leq \binom{f_0(K)}{k+1}.$$

(3) Intersection of finitely many convex polytopes is a convex polytope. Intersection of a convex polytope with an affine subspace or with a polyhedral set is a convex polytope.
(4) If K is a d-polytope, each $(d-2)$-face F of K is contained in precisely two facets F_1 and F_2 of K, and $F = F_1 \cap F_2$. [The proof of this statement for $d = 2$ is almost contained in the proof of Theorem 2.1.20.]
(5) Let h, k be integers such that $-1 \leq h < k \leq d-1$. Let K be a d-polytope. Then, each h-face of K is the intersection of the family of k-faces of K containing it.
(6) Let K be a d-polytope and F be a k-face of K. Then there exists a $(d-k-1)$-face $\tilde{F}$ of K such that $\dim \operatorname{conv}(F \cup \tilde{F}) = d$ and $F \cap \tilde{F} = \emptyset$.
(7) Let K be a convex polytope in $\mathbb{R}^d$ and $f : \mathbb{R}^d \longrightarrow \mathbb{R}^s$ be an affine linear map. Then $f(K)$ is a convex polytope.
(8) If $K_i \subset \mathbb{R}^{d_i}$ $i = 1, 2$ are convex polytopes, then so is $K_1 \times K_2$ in $\mathbb{R}^{d_1} \times \mathbb{R}^{d_2}$. [Hint: If $\{u_1, u_2, \ldots, u_n\}$ spans K_1 and $\{v_1, v_2, \ldots, v_m\}$ spans K_2 then we have,

$$\left(\sum_i t_i u_i, \sum_j s_j v_j \right) = \sum_{i,j} t_i s_j (u_i, v_j).]$$

Definition 2.1.23 Given a polytope P, let $\mathcal{F}(P)$ denote the poset of all (proper as well as improper) faces of P. Two polytopes P and Q are said to be *'equivalent'* or *'of the same combinatorial type'* if there is an inclusion preserving bijection $\psi : \mathcal{F}(P) \longrightarrow \mathcal{F}(Q)$. We write this in the form $P \approx Q$.

The following observations are immediate.
(1) $P \approx Q \implies \dim P = \dim Q$. Also if F is a face of P and $\psi : \mathcal{F}(P) \longrightarrow \mathcal{F}(Q)$ is an equivalence as above, then ψ restricts to an equivalence $\mathcal{F}(F) \longrightarrow \mathcal{F}(\psi(F))$.
(2) If $T : \mathbb{R}^d \longrightarrow \mathbb{R}^d$ is a non singular affine transformation, then $P \approx T(P)$. Also if T is a non singular *projective transformation* permissible for P, then $P \approx TP$.
(3) Notice that either of the conditions in (2) is too strong and hence we do not have the converse. However, it is true that if $P \approx Q$ then there are simplicial subdivisions P' and Q' of P and Q and a homeomorphism $f : |P| \to |Q|$ which restricts to linear isomorphism on each simplex of P'. Often this condition is taken as the definition of combinatorial equivalence and is also generalized to larger classes of polyhedrons.

The following result is what makes the study of convex polytopes so fundamental from a topological point of view.

Theorem 2.1.24 *Every convex polytope of dimension d is homeomorphic to the unit disc $\mathbb{D}^d$ in $\mathbb{R}^d$. Its boundary is homeomorphic to the sphere $\mathbb{S}^{d-1}$.*

Proof: Take any point z_0 in the interior of a convex polytope K. Let $B = B_\delta(z_0)$ be a ball of radius $\delta > 0$ around z_0 such that $B \subset \operatorname{int} K$. The crucial point here is that given a point y on any hyperplane the expression for the point of intersection of the line segment $[z_0, y]$ with the sphere ∂B is a root of a quadratic equation in the coordinates of the point y. (Do you remember the so-called *stereographic projection?*) Thus the assignment $y \mapsto \partial B \cap [z_0, y]$ defines a continuous map $h : \partial K \longrightarrow \partial B$. It is easily seen that this is a bijection. Now use the fact that every point of K is a unique convex combination of z_0 and a point on ∂K to write down a homeomorphism

$$(1 - t)z_0 + ty \mapsto (1 - t)z_0 + th(y)$$

of K onto B. Since, B is homeomorphic to $\mathbb{D}^d$, we are done.   ♠

One is led to consider the convex polytopes as the basic units or *building blocks* for more general types of spaces. The study of such spaces is called *polyhedral topology*. Below, we consider one single little step in this direction.

Definition 2.1.25 A finite family $\mathcal{C}$ of polytopes in $\mathbb{R}^d$ (not necessarily, all of them d-polytopes) will be called a *polyhedral complex* or polyhedral presentation if
(i) every face of a member of $\mathcal{C}$ is itself a member of $\mathcal{C}$;
(ii) the intersection of any two members of $\mathcal{C}$ is a face of each of them.

Members of $\mathcal{C}$ are called *faces* of $\mathcal{C}$, or the *cells* in $\mathcal{C}$. Note that each member of $\mathcal{C}$ is a genuine convex polytope in $\mathbb{R}^d$. The geometric carrier $|\mathcal{C}|$, associated to $\mathcal{C}$ is the underlying topological subspace of $\mathbb{R}^d$ which is the union of all the members of $\mathcal{C}$.

Example 2.1.26
(i) The collection of all faces of dimension $\leq k$ of a given convex polytope P forms a polyhedral complex. It is called the *k-skeleton* of P. If $k = \dim P = d$ then this coincides with $\mathcal{F}(P)$. For $k = d - 1$ this gives the *boundary* complex $\mathcal{B}(P)$ of P.
(ii) In the following picture (Figure 2.4) the rectangle is represented in three different ways, in (a) as a single convex polytope (and its faces) and in (b) and (c) as the union of two convex polytopes in two different ways.

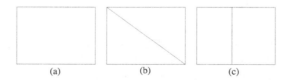

$$(a) \qquad\qquad (b) \qquad\qquad (c)$$

FIGURE 2.4. Three different combinatorial presentations of a rectangle

Definition 2.1.27 Given a complex $\mathcal{C}$ and a face $F \in \mathcal{C}$, the *Star* of F, *Anti-star* of F and the *Link* of F are defined and denoted respectively as follows:

$$
\begin{aligned}
St\ F &:= \{G \in \mathcal{C} : \exists\, H \in \mathcal{C}, F \subseteq H,\ G \subseteq H\}, \\
Anst\ F &:= \{G \in \mathcal{C} : F \cap G = \emptyset\}, \\
Lk\ F &:= Anst\ F \cap St\ F.
\end{aligned}
$$

Theorem 2.1.28 *Let P be a d-polytope and v be a vertex of P. Then the topological space covered by $St_{\mathcal{B}(P)}(v)$ and $Anst_{\mathcal{B}(P)}(v)$ are both homeomorphic to the $(d - 1)$-dimensional Euclidean ball and the one covered by $Lk_{\mathcal{B}(P)}(v)$ is homeomorphic to the sphere $\mathbb{S}^{d-2}$.*

In the pretext of presenting a proof of this theorem, we will study another important geometric notion with convex polytopes.

Definition 2.1.29 Let K be a convex polytope, F be a facet of K, defined by a supporting hyperplane H and let x be any point not on H. We say that x is *beneath* (respectively *beyond*) F with respect to K if x and K lie on the same side of H (respectively, on the opposite sides of H). Further if $x \notin K$, then we say that F is *invisible* (respectively, *visible*) from x, if x is beneath (respectively, beyond) F with respect to K, accordingly.

It is best to imagine that K is a solid made up of some opaque matter, in order to understand the above definition. Observe that the notion of visibility can be extended, by common sense, to a point x lying on H itself. In this case we take F to be not visible from x by convention. Similarly, for any point x which lies in K, we take that every facet of K is invisible from x.

Lemma 2.1.30 Let F be a facet of a convex polytope K, and let x be a point in $\mathbb{R}^d$. Then F is visible from x iff for every $y \in F$ the open line segment (y, x) does not meet K.

The proof of this lemma is quite easy and is left to the reader as an exercise. Note that this also gives a criterion for a face to be invisible. We introduce the notation $\mathcal{V}(K, x)$, (and $\mathcal{I}(K, x)$) to denote the polyhedral complex generated by the facets of K that are visible (respectively, invisible) from x. Often we will denote the underlying topological space also by the same notation. Observe that both these complexes are subcomplexes of the boundary complex $\mathcal{B}(K)$ of the polytope K. Moreover, since every facet is either visible or invisible from a given point we see that $\mathcal{V}(K, x) \cup \mathcal{I}(K, x) = \mathcal{B}(K)$.

Definition 2.1.31 We say that a point $x \in \mathbb{R}^d$ is an *admissible point* for the convex polytope K if $x \notin K$ and x does not lie on any aff F for any facet F of K.

Since there are only finitely many such hyperplanes, it follows that the set of all admissible points is open and dense in $\mathbb{R}^d \setminus K$.

Definition 2.1.32 Let K be a polytope, and v be a vertex of it. Let H be a hyperplane that separates v from the rest of the vertices of K. Consider the convex polytope $P = K \cap H$. The combinatorial type of P is easily seen to be independent of the choice of H. We call P a vertex diagram of K at the vertex v.

Lemma 2.1.33 Suppose that x is an admissible point for the d-dimensional convex polytope K. Consider the convex polytope $K_1 = \text{conv}(K \cup \{x\})$. Let X be the underlying topological space of a vertex diagram of K_1 at v. Then X, $\mathcal{V}(K, x)$, $\mathcal{I}(K, x)$ are all homeomorphic to the closed unit disc $\mathbb{D}^{d-1}$.

Proof: We shall show that X and $Y := \mathcal{V}(K, x)$ are homeomorphic. The same argument shows that X and $Z = \mathcal{I}(K, x)$ are homeomorphic. Since, X is the underlying space of a $(d - 1)$-polytope, we will be through.

Let H be the hyperplane in which lies the vertex diagram X at x for K_1. We claim that given any point a in X the line aff $\{x, a\}$ meets Y in a unique point $\phi(a)$. (Similarly, this line meets $\mathcal{I}(K, x)$ also in a unique point and so on.) By the definition, every point of X looks like $a = (1 - t)x + tw$ for some $w \in K$. It follows that aff $\{x, a\}$ meets K. Hence, it meets Y also. If it met Y in two different points then, it follows that x is a point of aff F for some F in Y. That contradicts the admissibility of x. Now it is elementary to see that ϕ is a continuous map. Since it is a bijection of compact metric spaces, it is a homeomorphism. This completes the proof of the lemma. ♠

Remark 2.1.34 Observe that the intersection of Y and Z is mapped onto the boundary of X under ϕ.

Proof of Theorem 2.1.28: Let $\epsilon > 0$ be such that the ball $B_\epsilon(v)$ does not meet any of the supporting hyperplanes of K which do not pass through v. Then for any facet F of K not incident at v, every point of $B_\epsilon(v)$ is beneath F. Let $y \in B_\epsilon(v) \cap int\, K$, and let x be *antipodal* to y with respect to v, i.e., take $x = 2v - y$. Then x is an admissible point for K. Also observe that the set of facets that are visible from x is precisely those in $St_{\mathcal{B}(K)}(v)$. Hence, the geometric complex generated by them is precisely equal to $St_{\mathcal{B}(K)}(v)$. Thus the underlying topological space is homeomorphic to the disc. Similarly the set of facets that are invisible generate $Anst_{\mathcal{B}(K)}(v)$. The underlying topological space being Z, is also homeomorphic to a disc. Finally since $Lk_{\mathcal{B}(K)}(v) = St_{\mathcal{B}(K)}(v) \cap Anst_{\mathcal{B}(K)}(v)$, it follows that the topological space underlying the link is homeomorphic to $\mathbb{S}^{d-2}$. This completes the proof of the theorem.   ♠

Remark 2.1.35 The techniques that we have developed can be used in proving one of the most interesting results, viz., *that the boundary complex of any convex polytope is shellable.*

Exercise 2.1.36 We say a convex polytope is *simplicial* if each of its facets is a simplex. Let K be a simplicial convex d-polytope and F be a k-face in K. Let $\pi : \mathbb{R}^d \longrightarrow \mathbb{R}^{d-k}$ be the affine linear projection with aff $F = \pi^{-1}(v)$. Let $K' = \pi(K)$, $v = \pi(F)$. Then prove that $Lk_{\mathcal{B}(K)}(F)$ is isomorphic to $Lk_{\mathcal{B}(K')}(v)$.

Exercise 2.1.37 Prove the following *Radon's Theorem*: If M is a n-subset of $\mathbb{R}^d$ for $n \geq d + 2$, then there exist two disjoint subsets M_1, M_2 of M such that $M_1 \cup M_2 = M$ and conv $M_1 \cap$ conv $M_2 \neq \emptyset$.

One of the earliest topological notions is the so-called Euler characteristic

$$\chi(K) = \sum_{0}^{d} (-1)^i f_i(K) \tag{2.1}$$

where $f_i(K)$ denotes the number of faces of dimension i of a convex polytope K. It has far-reaching generalizations and has remained central in the topological studies. We shall make a beginning of understanding this concept with a geometric proof of Euler's formula

$$\chi(K) = 1. \tag{2.2}$$

This proof can be found in [Grunbaum, 1967]. I do not know any simpler proof in the general case. For dimension ≤ 3 of course, one can give simpler proofs. The reader familiar with some Morse theory may notice the striking similarity of the idea involved here with that in Morse theory. Indeed, this may be taken as the starting point of what is known as PL-Morse theory. Unfortunately, we will not be able to touch upon this aspect here.

    The proof is by induction on the dimension d. For $d = 1$ it is obvious. (Even for $d = 2$, it is obvious but we do not need this.) Now begin with a convex polytope K of dimension d in $\mathbb{R}^d$. Tilt it so that no two vertices are in the same horizontal level. (Strictly speaking, what we are doing is to choose a direction such that any hyperplane perpendicular to this direction contains at most one vertex of K and treat this direction as the last coordinate axis.) The idea is to break K into 'simpler' polytopes by using horizontal planes, prove the formula for each of the pieces and then put the results together.

Lemma 2.1.38 Let H be a hyperplane cutting K into two polytopes K_1 and K_2 and not passing through any vertex of K. Let $H \cap K = K_0$. Then we have,

$$\chi(K) = \chi(K_1) + \chi(K_2) - \chi(K_0).$$

Proof: (See Figure 2.5(a).) Observe that

$$f_0(K) = f_0(K_1) + f_0(K_2) - 2f_0(K_0).$$

Note that $f_d(K) = f_d(K_1) = f_d(K_2) = 1 = f_{d-1}(K_0)$. Further for each $0 \leq i \leq d-1$, there is a bijection between the i-faces of K_0 and the 'vertical' $(i+1)$-faces of K, K_1 and K_2.

Therefore, we have

$$f_{i+1}(K) = f_{i+1}(K_1) + f_{i+1}(K_2) - 2f_{i+1}(K_0) - f_i(K_0), \ 0 \leq i \leq d-1.$$

Adding all these with appropriate signs gives the result. ♠

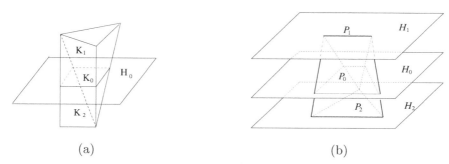

(a)   (b)

FIGURE 2.5. Euler's formula being verified

Lemma 2.1.39 Let P_i, be convex polytopes lying in two distinct horizontal hyperplanes $H_i, i = 1, 2$, respectively. Let K be the convex hull of $P_1 \cup P_2$. Let H_0 be another horizontal hyperplane between H_1 and H_2 and let $P_0 = H_0 \cap K$. Then

$$\chi(K) = \chi(P_1) + \chi(P_2) - \chi(P_0).$$

Proof: (See Figure 2.5(b).) Observe that P_i are faces of K and contain all the vertices of K. Therefore, $f_0(K) = f_0(P_1) + f_0(P_2)$. Moreover, there is a bijective correspondence between the k-faces of P_0 and $(k+1)$-faces of K that are not contained in either P_1 or P_2. Hence, for $k \geq 1$, we have, $f_k(K) = f_k(P_1) + f_k(P_2) + f_{k-1}(P_0)$. Now take the alternating sum of these identities to complete the proof. ♠

Lemma 2.1.40 Let P_1, P_2, etc., be as in the above lemma except that K is now the convex hull of P_1, P_2 and a point v lying between H_1 and H_2 so that the vertex set of K precisely consists of v and those in P_i. Let H_0 be the plane parallel to H_i and passing through v and let $P_0 = H_0 \cap K$ and Q_i be the portion of K lying between H_0 and H_i. Then

$$\chi(K) = \chi(Q_1) + \chi(Q_2) - \chi(P_0).$$

Proof: See Figure 2.6. Observe that

$$f_0(K) = f_0(Q_1) + f_0(Q_2) - 2f_0(P_0) + 1.$$

Next, observe that each k-face of P_0 other than v corresponds to a unique $(k+1)$-face that is vertical in K (and Q_1 and Q_2.) Hence we have,

$$f_1(K) = f_1(Q_1) + f_1(Q_2) - 2f_1(P_0) - f_0(P_0) + 1$$

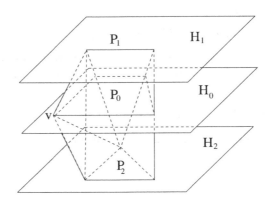

FIGURE 2.6. Proof of Euler formula

and

$$f_k(K) = f_k(Q_1) + f_k(Q_2) - 2f_k(P_0) - f_{k-1}(P_0), \ 2 \leq k \leq d.$$

Taking the alternating sum yields the result.   ♠

We can now state and complete the proof of this classical result:

Theorem 2.1.41 *Let K be a convex polytope. Then $\chi(K) = 1$.*

Proof: We induct on the dimension d of K. For $d = 0, 1$, this is obvious. Assume this to be true for all convex polytopes of dimension less than d. Then, it follows that for all convex polytopes K of dimension d and satisfying the hypothesis of Lemma 2.1.39 $\chi(K) = 1$. From this it follows that $\chi(K) = 1$ for all convex polytopes satisfying the hypothesis of Lemma 2.1.40 also.

Finally, given any convex polytope K of dimension d, as indicated above, choose a direction such that any plane perpendicular to this direction contains at most one vertex of K. Cut K into a finite number of polytopes K_i, by these parallel planes not passing through any vertex of K and such that each K_i contains exactly one vertex of K. It follows that each K_i satisfies the hypothesis of Lemma 2.1.40. Hence $\chi(K_i) = 1$, for each i. Now use Lemma 2.1.38, iteratively to conclude that $\chi(K) = 1$.   ♠

We shall end this section with a proof of the classification of regular 3-polytopes. For full details, [Coxeter, 1973] is strongly recommended.

Classification of Regular 3-Polytopes

Definition 2.1.42 Let p, q be any two positive integers. By a regular 3-polytope of type (p, q) we mean a 3-dimensional convex polytope P with the property that
(i) every 2-face of P has precisely p vertices and
(ii) the number of 2-faces incident at any given vertex is q.

Familiar examples of regular 3-polytopes are the tetrahedron and the cube. Not so familiar ones are the so-called octahedron, the icosahedron and the dodecahedron. These five polytopes are called *Platonic solids*.

There are slightly different definitions of a regular polytope. The more familiar one is the most geometric which demands that each facet should be a regular polygon of the same size as well. The above definition, allows objects such as any parallelepiped, etc., as regular polytope, which are combinatorially equivalent to a cube. However, the classification result is the same. The key is Euler's formula, which we have proved in Theorem 2.1.41.

Theorem 2.1.43 *There are only five combinatorial types of regular 3-polytopes and these are represented by the five Platonic solids.*

Proof: The proof is presented in three steps.

Step I List of all possible patterns: We begin with the following simple observations which will be used repeatedly in the proof:

(i) Every convex polygon has at least 3-vertices, i.e., $p \geq 3$.

(ii) For any 3-polytope P, the link in ∂P of any vertex is homeomorphic to a circle, (see Theorem 2.1.28) which will have at least three edges. Therefore every vertex is incident at least in three faces. Therefore $q \geq 3$.

(iii) Likewise, every edge is present in precisely two faces.

(iv) Because of regularity, each 2-face being a convex polygon with p vertices, has precisely p edges.

If v, e, f denote, respectively, the number of vertices, edges and the faces of a 3-polytope P, we have $1 = \chi(P) = v - e + f - 1$ and therefore,

$$v - e + f = 2. \tag{2.3}$$

Since each edge is incident at exactly two vertices and exactly at two faces, by regularity, it follows that

$$2e = pf = qv. \tag{2.4}$$

Substituting for v, f from this in (2.3), we get

$$e\left(\frac{2}{q} - 1 + \frac{2}{p}\right) = 2. \tag{2.5}$$

Equivalently,

$$\frac{e}{2} = \frac{pq}{2p + 2q - pq}. \tag{2.6}$$

This implies

$$0 < 2p + q(2 - p) \leq 2p + 3(2 - p) \leq 6 - p.$$

Therefore $3 \leq p < 6$. Also, (2.6) can be rewritten as

$$\frac{2}{q} = \frac{2}{e} + \frac{p-2}{p}$$

and hence $q < \frac{2p}{p-2}$. Substituting $p = 3, 4, 5$, respectively, in this relation, it follows that the only possibilities for the pair (p, q) are

$$(3,3), (3,4), (3,5), (4,3), (5,3).$$

In each case, the number of edges is determined by (2.6) and then the number of vertices and faces are determined by (2.4). Thus the five possible cases give the number of faces $4, 8, 20, 6$ and 12, respectively.

Step-II Realisation Each of these cases is represented by the five Platonic solids. The names are according to the number of faces in them.

Type	Name	Vertices	Edges	Faces
(3,3)	Tetrahedron	4	6	4
(4,3)	Cube	8	12	6
(3,4)	Octahedron	6	12	8
(5,3)	Dodecahedron	20	30	12
(3,5)	Icosahedron	12	30	20

Clearly the **tetrahedron** is represented as a convex hull of the points

$$(1, 1, 1), (1, -1, -1), (-1, 1, -1), (-1, -1, 1)$$

and the cube is represented by the convex hull of

$$\{(\pm 1, \pm 1, \pm 1)\}.$$

By taking the barycentres of the faces we get the vertices of the dual polytope octahedron:

$$(\pm 1, 0, 0), (0, \pm 1, 0), (0, 0, \pm 1).$$

By certain elementary geometric considerations, it is possible to show that the twelve vertices of an icosahedron can be obtained by dividing the twelve edges of an octahedron in the golden ratio. Instead, we shall merely display the vertex set of an icosahedron

$$(0, \pm \tau, , \pm 1), (\pm 1, 0, \pm \tau), (\pm \tau, \pm 1, 0)$$

where τ is a root of $x^2 - x - 1 = 0$ and leave it to the reader to verify that the convex hull of the above set is indeed a regular polytope. For instance two of its triangular faces are the convex hulls of

$$\{(0, \tau, 1), (1, 0, \tau), (-1, 0, \tau)\}; \{0, \tau, 1), (1, 0, \tau), (\tau, 1, 0)\}.$$

We shall give a little more description of the **dodecahedron.** By taking the barycentres of the faces of the twenty faces of the icosahedron, we obtain the vertices of the dodecahedron. Using the relation $x^2 = x + 1$ satisfied by τ and by scaling we get a simplified form for these vertices:

$$(0, \pm \tau^{-1}, \pm \tau), (\pm \tau, 0, \pm \tau^{-1}), (\pm \tau^{-1}, \pm \tau, 0), (\pm 1, \pm 1, \pm 1).$$

Note that from the vertex $(1, 1, 1)$, there are three edges, respectively, to the vertices $(0, \tau^{-1}, \tau)$, $(\tau, 0, \tau^{-1})$, and $(\tau^{-1}, \tau, 0)$. Similarly, we can locate edges at other vertices. The length of each edge is $= \sqrt{6 - 2\sqrt{5}} = \sqrt{5} - 1$. Of course, all these vertices lie on the sphere $\{x^2 + y^2 + z^2 = 3\}$. We can get the model of the dodecahedron inscribed in the unit sphere by scaling by a factor of $1/\sqrt{3}$. It is a standard result that the symmetry group of the dodecahedron is generated by a permutation (123) (of order 3) of the coordinate axes and the rotation through $180°$ in the plane perpendicular to a vector which bisects one of the edges, say, the vector $(1, 1 + \tau^{-1}, 1 + \tau)$. (See more of it in Section 3.8.) It is an interesting exercise to write down the 3×3 matrix that represents this rotation using quaternion algebra. Think of $\mathbb{S}^2$ as the purely imaginary part $x\mathbf{i} + y\mathbf{j} + z\mathbf{k}$ with $x^2 + y^2 + z^2 = 1$. Then the unit vector in the above direction is given by $\mathbf{v} = \frac{1}{2}(\tau^{-1}\mathbf{i} + \mathbf{j} + \tau\mathbf{k})$ and the rotation through an angle π in the plane perpendicular to this vector is given by conjugation by the same element in the quaternion algebra. Since $\mathbf{v}^{-1} = -\mathbf{v}$ we have only to compute the numbers $-\mathbf{viv}, -\mathbf{vjv}, \mathbf{vkv}$ which comes out to be

$$T = \frac{1}{2} \begin{pmatrix} -\tau & \tau^{-1} & 1 \\ \tau^{-1} & -1 & \tau \\ 1 & \tau & \tau^{-1} \end{pmatrix}.$$

Check that this is an element of order 2. Write S for the permutation matrix (123), viz.,

$$S = \begin{pmatrix} 0 & 0 & 1 \\ 1 & 0 & 0 \\ 0 & 1 & 0 \end{pmatrix}$$

which is clearly of order 3. Check that TS is of order 5 and corresponds to a rotation through an angle $2\pi/5$ about the line which joins the centres of two of the opposite faces of the dodecahedron. It can be shown that any group generated by two elements T, S and satisfying the relations

$$T^2 = S^3 = (TS)^5 = 1$$

is isomorphic to the group of symmetries of the dodecahedron. By mapping $T \to (14)(25)$ and $S \to (123)$ one can check that this group is isomorphic to the alternating group A_5 on 5 letters and hence is of order 60.

Step III Uniqueness The distinct number of 2-faces in each case clearly implies that these five Platonic solids are of distinct combinatorial type. It remains to see that each case represents a single combinatorial type. (Since many popular expositions leave out the proof of this, we shall include it here.)

Assuming P, P' are two regular convex polytopes of dimension 3 of the same type (p, q), we shall write down a vertex map $V(P) \to V(P')$ which will induce a lattice isomorphism $\mathcal{F}(P) \to \mathcal{F}(P')$.

In what follows we use the standard convention of cyclic labeling—for instance, in the cyclic labeling $x_1, x_2, \ldots, x_5$ of vertices of a pentagon, by x_0 and x_6 we mean, respectively, x_5 and x_1.

Case $(p, q) = (3, 3)$. This is the simplest case where the number of vertices $v = 4$ and they are therefore affinely independent. Therefore any bijection of vertices can be extended to an affine isomorphism of the polytopes $P \to P'$.

Case $(p, q) = (3, 4)$. Here $v = 6$. We select any one vertex a in P. Now a is incident exactly at four edges and hence is joined to precisely four other vertices. Label these vertices x_1, x_2, x_3, x_4 in any cyclic order, i.e., so that $\{a, x_i, x_{i+1}\}, i = 1, 2, 3, 4$ are the four triangles incident at a. Now there is just one more vertex left; label it b. Label the vertices of P' as $a', x_0', \ldots, x_4', b'$ in the same manner. The bijection

$$a \mapsto a'; \quad b \mapsto b'; \quad x_i \mapsto x_i'; \quad i = 1, \ldots, 4$$

is easily seen to induce a lattice isomorphism $\mathcal{F}(P) \to \mathcal{F}(P')$.

Case $(p, q) = (4, 3)$. This case is dual to the above case. We begin with any 2-face F_0 of P and label its vertices $x_1, \ldots, x_4$ in a cyclic order. Each vertex x_i is joined to precisely one more vertex other than any of the x_j. Let us call it y_i. There is one face F_i incident at the edge $x_i x_{i+1}$ which is not equal to F_0. Clearly each of these F_i have to be distinct. It also follows that the vertices of F_i are $x_i x_{i+1} y_{i+1} y_i$ in the cyclic order. In particular, this implies $y_i \neq y_{i+1}$, for all i. We claim that $y_i \neq y_{i+2}$. For, otherwise, the link at y_{i+1} (as well as y_{i-1}) fails to be homeomorphic to a circle no matter how we fit the third 2-face at y_{i+1}. Therefore y_1, y_2, y_3, y_4 are distinct. That takes care of all eight vertices and five faces. It turns out that the sixth face has to be $y_1 y_2 y_3 y_4$. Label the vertices of P' exactly in the same manner as x_i', y_i' and take the bijection

$$x_i \mapsto x_i'; \quad y_i \mapsto y_i'; \quad i = 1, 2, 3, 4.$$

Case $(p, q) = (3, 5)$. Here the number of vertices is 12. As usual, start with any vertex x of P. There are precisely 5 edges incident at this vertex which give us vertices a_1, a_2, a_3, a_4, a_5 so that $\{x, a_i, a_{i+1}\}$ are the five triangles incident at x. Now at each of the edges $a_i a_{i+1}, i = 1, \ldots, 5$ we have precisely one new triangle; label the new vertex of this triangle b_i. Clearly b_i is distinct from x, a_i and a_{i+1}. Each b_i is distinct from b_{i+1} for otherwise we cannot have five triangles incident at a_{i+1} (respectively, at a_{i-1}). Next each b_i is also distinct from a_{i-1} and a_{i+2} since we cannot have two edges between two vertices. We need to know why b_i is distinct from a_{i-2} (which is anyway equal to a_{i+3}). If, say, $b_1 = a_4$ then at this vertex there are already six distinct edges (viz., to all vertices $a_j, j \neq i+3$ and b_{i+1}, x_0) which is absurd. Therefore b_i are all distinct from $a'_j s$. This implies that at each of the vertices a_i we already have five distinct edges. Next, we claim that b_i are themselves distinct. If not suppose some $b_i = b_{i+2}$. Then the link at this vertex includes the four vertices $a_j, j \neq i-1$ whereas together they contribute only two triangles incident at $b_i = b_{i+2}$. We can have only one more vertex joined to b_i which can create at most two more triangles at b_i. This means that there can be only four triangles incident at b_i which is a contradiction. Therefore $b_i \neq b_{i+2}$. The proof that $b_i \neq b_{i-2}$ is similar. This completes the proof that x, a_i, and b_i are all distinct.

So far, we have accounted for four triangles at each of the vertices a_i. It follows that the missing triangle at each of a_i is nothing but $a_i b_i b_{i-1}$. This takes care of 11 vertices and 15 triangles as depicted in the Figure 2.7 (a).

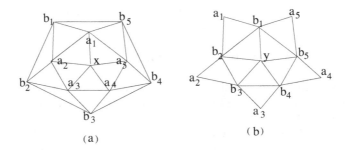

(a)   (b)

FIGURE 2.7. The icosahedron

There is just one more vertex. Call it y. The five more triangles will all be incident at this vertex y. See Figure 2.7 (b).

Label the vertices of P' as x', a'_i, b'_i, y', etc., exactly in the same manner and take the bijection

$$x \mapsto x'; \quad y \mapsto y'; \quad a_i \mapsto a'_i; \quad b_i \mapsto b'_i; \quad i = 1, \ldots, 5.$$

Case $(p,q) = (5,3)$. Here the number of vertices is 20 and the number of faces is 12. (This case is of course dual to the above case. Nevertheless, we shall write down a detailed proof.)

We begin with any one face F_0 of P and label its vertices $a_1 \ldots, a_5$ in a cyclic order. Pick up the unique vertex not accounted so far and joined to a_i and label it b_i. Let F_i be the unique face incident at $a_i a_{i+1}$ and not equal to F_0. It follows that four of the five vertices of F_i are $b_i, a_i, a_{i+1}, b_{i+1}$. Call the fifth vertex c_i. Consequently, it follows that none of the b_i are equal to a_i because any two F_i, F_j can have at most two vertices in common. For the same reason, each c_i is distinct from any a_j and $c_i \neq c_{i+1}$. Clearly $b_i \neq b_{i+1}$ for any i. If $b_i = b_{i+2}$ then at this vertex all the four F_j's $(j \neq i-2)$ will be incident, which is absurd. Therefore $b_i \neq b_{i+2}$. This proves that all the b_i' s are distinct.

Let us show that $c'_i s$ are different from $b'_j s$. Clearly each c_i is different from b_i and b_{i+1}. Suppose some c_i is equal to b_{i+2}. For the sake of definiteness let us say, $c_1 = b_3$. This will imply that at this common vertex the three faces that are incident are F_1, F_2 and F_3. This will imply that $b_1 = c_2$ and $b_2 = c_3$. For the same reason now, the three faces which are incident at $c_2 = b_1$ are F_1, F_3 and F_5. Therefore, $b_3 = c_5$ which in turn implies $c_1 = b_5$. But then $b_1 = b_5$, which is absurd.

Therefore, $c_i \neq b_{i+2}$. The same argument gives $c_i \neq b_{i-1}$ also. Finally suppose $c_i = b_{i+3}$. This implies that the three faces that are incident at this vertex are F_i, F_{i+2} and F_{i+3}. This will force b_i to be equal to c_{i+3} or c_{i+2}. The former is ruled out because we have already seen that c_{i+3} is different from $b_{i+3+2} = b_i$. In the latter case, viz., if $b_i = c_{i+2} = b_{i+2+3}$ which will in turn imply $b_{i+2} = c_{i+4} = b_{i+4+3}$, which in turn implies that $b_{i+4} = c_{i+4+2} = c_{i+1}$ and so on, each c_j is equal to b_{j+3}.

Thus, we obtain a genuine surface of type $(5,3)$ with precisely 10 vertices, 15 edges and 6 faces as depicted in Figure 2.8. With the remaining faces we can get a carbon copy of the same space.

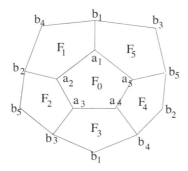

FIGURE 2.8. A PL representation of the projective space $\mathbb{P}^2$

However, this is not allowed for the simple reason that the resulting surface is not connected. [The Euler characteristic of each of these components is equal to 1 and therefore neither of them is the boundary of a 3-polytope. Indeed, what we have obtained is a presentation of the projective space as a regular surface of type $(5,3)$. (Notice that one could have discarded this case in other ways as well. For instance, the pair of identifications $c_i = b_{i+3}, b_i = c_{i+2}$ makes the surface non orientable. But we may not want to use an unfamiliar concept which is not properly established before hand.] Thus we have established that c_i are distinct from $a_i, b_i, i = 1, \ldots, 5$.

We need to see why c_i are themselves distinct. We have already seen $c_i \neq c_{i+1}$. Suppose for some i, $c_i = c_{i+2}$. At this common vertex we will have two pentagons, F_i, F_{i+2} which will intersect only in this vertex. That means there would be at least two more pentagons incident at this vertex which contradicts the fact $q = 3$. Therefore $c_i \neq c_{i+2}$. In the cyclic notation this also proves $c_i \neq c_{i+3}$. Therefore, all the c_i are distinct.

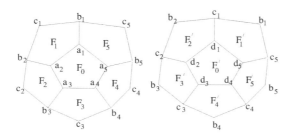

FIGURE 2.9. The dodecahedron

So far, we have taken care of 6 faces and 15 vertices. At each of the vertices c_i, we have only two edges so far and so each c_i is joined precisely to one more vertex which we label d_i. At each of the vertex b_i there are two faces so far, viz., F_i and F_{i-1}. So, there is yet another face incident at this vertex and we label it F'_i. Three of its vertices are c_{i-1}, b_i, c_i. Therefore, it cannot be any of the faces F_j. Nor any $F'_i = F'_j, i \neq j$. We name the vertices of F'_i as $c_{i-1}b_ic_id_id_{i-1}$. Since all the vertices a_i, b_i, c_i already have three edges incident at them, each d_j is distinct from any a_i, b_i, c_i. Clearly $d_i \neq d_{i+1}$. If, on the other hand, $d_i = d_{i+2}$, for some i, accounting for the number of edges at this vertex will force us to conclude that $d_{i+1} = d_{i+3}$ and so on, it follows that all the d_i are equal to a single vertex, which is absurd. Therefore, $d_i \neq d_{i+2}$. The same argument gives $d_i \neq d_{i-2}$. Thus the vertices $d'_i s$ are also distinct. (See Figure 2.9.)

This accounts for 11 faces and 30 edges and all 20 vertices. The missing face must be $d_1 d_2 d_3 d_4 d_5$.

Label the vertices of P' also in the same manner as a'_i, b'_i, c'_i, d'_i and take the bijection

$$a_i \mapsto a'_i; \quad b_i \mapsto b'_i; \quad c_i \mapsto c'_i; \quad d_i \mapsto d'_i, \quad i = 1, \ldots, 5.$$

The procedure we have used in this labeling ensures that this bijection defines an isomorphism of the lattices.   ♠

Exercise 2.1.44 Write down expressions for the vertices of a regular n-simplex inscribed inside $\mathbb{S}^{n-1}$.

2.2   Cell Complexes

In this section, we shall first introduce the most important class of topological spaces, viz., the CW-complexes. We shall study some fundamental point-set topological properties of these spaces.

Definition 2.2.1 Let $k \geq 1$ be an integer and for each index $\alpha \in \Lambda$, let D^k_α denote a copy of the closed unit ball $\mathbb{D}^k$ in $\mathbb{R}^k$. Given two spaces X and Y, we say X is obtained by attaching k-cells to Y if there exists a family of maps $f_\alpha : \mathbb{S}^{k-1} \longrightarrow Y$, $\alpha \in \Lambda$, such that X is the quotient space of the disjoint union

$$Y \sqcup_{\alpha \in \Lambda} D^k_\alpha \tag{2.7}$$

by the relation $x \sim f_\alpha(x)$ for each $x \in \partial D^k_\alpha$, and for each α.

The maps $\{f_\alpha\}$ are called *attaching maps for the cells*. Let us denote the quotient map restricted to a cell D^k_α by ϕ_α. Observe that $\phi_\alpha | \partial D^k_\alpha = f_\alpha$ and ϕ_α is injective in the interior of D^k_α. Thus ϕ_α defines a homeomorphism of the interior of D^k_α onto the image. The image $\phi_\alpha(\text{int}\,(D^k_\alpha))$ is called an *open cell* in X. The maps ϕ_α are called *characteristic maps of the cells*. Observe that the image of each D^k_α is a compact subspace of X. We call them the *closed k-cells* in (X, Y) and denote them by e^k_α.

The following statement is immediate from the definition:

Lemma 2.2.2 Let X, Y, etc., be as in the definition above. Then
(a) a subset A of X is closed in X iff $A \cap Y$ is closed in Y and $A \cap e^k_\alpha$ is closed in e^k_α for each $\alpha \in \Lambda$;
(b) Y is a closed subset of X.

Proof: This is a direct consequence of the definition of quotient topology: Given a subset $A \subset X$, observe that
$$q^{-1}(A) = A \cap Y \sqcup_\alpha \phi_\alpha^{-1}(A)$$

is closed iff each of these disjoint sets is closed in the corresponding decomposition (2.7). Since each characteristic map $\phi_\alpha : \mathbb{D}^k \to e^k_\alpha$ is itself a quotient map, $A \cap e^k_\alpha$ is closed iff $\phi_\alpha^{-1}(A)$ is closed. This proves statement (a). Now we use (a) to prove (b): $Y \cap e^k_\alpha$ is closed in e^k_α because, $\phi_\alpha^{-1}(Y) = f_\alpha^{-1}(Y) = \partial D^k_\alpha$ is closed in D^k_α.   ♠

Remark 2.2.3

(a) Observe that in the definition above, the family $\{f_\alpha\}$ may be empty also. Of course this is the most uninteresting case. However, we should include this case for technical reasons.

(b) If the family has just one member f, it may be noted that X can be identified with the mapping cone of f. In this case, it is customary to denote the resulting space by

$$Y \cup_f e^k. \tag{2.8}$$

(c) In general, e_α^k need not be a closed subset of X. (Exercise. Illustrate this by an example). However, if Y is a Hausdorff space then being a compact subset, $f_\alpha(\mathbb{S}^{k-1})$ is a closed subset of Y and hence it follows that e_α^k is also a closed subset of X.

(d) Also, in general, e_α^k is not homeomorphic to $\mathbb{D}^k$. However, its interior is homeomorphic to int $(\mathbb{D}^k)$.

(e) In Figure 2.10 below, $k = 1$ and $\Lambda = \{1, 2, 3\}$. Observe that f_1 and f_2 are injective and f_3 is not injective.

FIGURE 2.10. Attaching maps need not be injective

The quotient topology retains a large number of topological properties and hence we can expect a certain Euclidean behavior in the spaces obtained by attaching cells. In order to put this to some good use, we need to prepare ourselves a bit. The proof of following lemma is obvious (see Figure 2.11).

Lemma 2.2.4 Consider a subset $A \subset \mathbb{S}^{n-1}$ and let $0 < \epsilon < 1$. Let us put

$$N_\epsilon(A) = \{x \in \mathbb{D}^n \ : \ \|x\| > \epsilon \ \& \ \frac{x}{\|x\|} \in A\}.$$

Then

(i) $N_\epsilon(A) \cap \mathbb{S}^{n-1} = A$;

(ii) $N_\epsilon(A)$ is an open subset of $\mathbb{D}^n$ iff A is an open subset of $\mathbb{S}^{n-1}$.

(iii) $(x, t) \mapsto (1 - t)x + \dfrac{tx}{\|x\|}$ defines a SDR of $N_\epsilon(A)$ onto A.

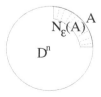

FIGURE 2.11. Extending neighbourhoods

Proposition 2.2.5 Let X be a space obtained by attaching a family of k-cells $\{e_\alpha^k : \alpha \in \Lambda\}$ to Y with characteristic maps $\{\phi_\alpha : \alpha \in \Lambda\}$ and let $q : Y \sqcup_{\alpha \in \Lambda} D_\alpha^k \longrightarrow X$ be the quotient map. Let $\epsilon : \Lambda \longrightarrow (0,1)$ be any set function. For any subset A of Y define

$$N_\epsilon(A) := q(A \cup \{N_{\epsilon(\alpha)}(\phi_\alpha^{-1}(A)) : \alpha \in \Lambda\}).$$

Then
(i) $N_\epsilon(A) \cap Y = A$;
(ii) $N_\epsilon(A)$ is open in X iff A is open in Y;
(iii) A is a SDR of $N_\epsilon(A)$.

Proof: We simply appeal to Lemma 2.2.4: Parts (i) and (ii) follow directly from the corresponding parts of this lemma. To obtain (iii) we have to put together all the deformations given by (iii) of 2.2.4, for each $\alpha \in \Lambda$. ♠

Corollary 2.2.6 If Y is T_1, T_2, or normal, then so is X.

Proof: Exercise.

Definition 2.2.7 A relative CW-complex (or a relative cell complex) (X, A) consists of a topological space X and a closed subspace A together with a sequence of closed subspaces $\{X^{(n)}\}, n \geq 0$ such that
(i) $A \subset X^{(0)}$ and $X^{(0)} \setminus A$ is a discrete space;
(ii) $X^{(k)}$ is obtained by attaching k-cells to $X^{(k-1)}$ for all $k \geq 1$;
(iii) $X = \cup_k X^{(k)}$;
(iv) The topology on X is the weak topology with respect to the family $\{X^{(k)}\}$, i.e., a subset $F \subset X$ is closed in X iff $F \cap X^{(k)}$ is closed in $X^{(k)}$ for every $k \geq 0$.

Remark 2.2.8
(a) An interesting case is when $A = \emptyset$. Then we say X is a *CW-complex*. It follows that $X^{(0)}$ is a discrete space (possibly empty). Thus, in building up a CW-complex, we start off with a discrete topological space, then attach 1-cells to this space, then attach 2-cells to the space obtained, and so on, to obtain the CW-complex X. We call elements of $X^{(0)} \setminus A$ the 0-cells of (X, A). Note that a 0-cell is both an open 0-cell as well as a closed 0-cell.
(b) It may happen that $X = X^{(k)}$ for some k. In that case, condition (iv) is automatically satisfied. (Keep using (a) of Lemma 2.2.2.) Further if $X^{(k-1)} \neq X$, then we say (X, A) is k-dimensional. We also write $\dim(X, A) = k$.
(c) The terminology that we introduced in Definition 2.2.1, such as *open cells, characteristic maps*, etc., holds good here also. Caution: open cells in a CW-complex need not be open subsets!
(d) Note that we allow the case when $X^{(0)} = \emptyset$. Of course, it then follows that $X^{(k)} = \emptyset$ for all k and $X = \emptyset = A$.

Definition 2.2.9 By a subcomplex (Y, B) of a relative CW-complex (X, A), we mean a relative CW-complex where, $Y \subset X, B \subset A$ and each cell in Y is also a cell in X with precisely the same attaching map.

Remark 2.2.10
(a) If (Y, B) is a sub-pair of a relative CW-complex (X, A) and Y is the union of B and some closed cells in X then (Y, B) is a subcomplex of (X, A).
(b) If Y is a subcomplex of X then (X, Y) is a relative CW-complex. More generally, if (Y, B) is a subcomplex of (X, A) then $(X, Y \cup A)$ is a relative CW-complex.
(c) For all k, $X^{(k)}$ is a subcomplex of any CW-complex X. It is called the kth-skeleton of X. For a relative CW-complex (X, A), observe that $(X^{(k)}, A)$ is a subcomplex.

Example 2.2.11

(i) Any discrete space can be thought of as a CW-complex with only 0-cells.

(ii) The n-dimensional unit sphere $\mathbb{S}^n$ is a CW-complex with a single 0-cell and a single n-cell. The attaching map of the n-cell is the constant map. This follows from the fact that the quotient space $\mathbb{D}^n/\mathbb{S}^{n-1}$ is homeomorphic to $\mathbb{S}^n$. Observe that with these CW-structures, even though $\mathbb{S}^{n-1}$ is a subspace of $\mathbb{S}^n$ via the equatorial inclusion, it is not a subcomplex. (See Figure 2.12(a).)

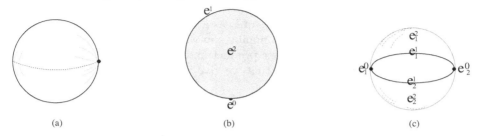

FIGURE 2.12. Standard examples of CW-complexes

(iii) Fix some CW-structure on $\mathbb{S}^n$, for instance given as in Figure 2.12(a). Then $\mathbb{D}^{n+1}$ itself can be viewed as a CW-complex obtained by attaching one $(n+1)$-cell to $\mathbb{S}^n$ via the identity map $\mathbb{S}^n \longrightarrow \mathbb{S}^n$. It follows that $\mathbb{S}^n$ is then a subcomplex of $\mathbb{D}^{n+1}$. The case $n = 1$ is illustrated in Figure 2.12(b).

(iv) We can have different CW-structures on the same topological space. For instance, take $\mathbb{S}^n$, $n \geq 1$ and consider the usual equatorial inclusions

$$\mathbb{S}^0 \subset \mathbb{S}^1 \subset \cdots \subset \mathbb{S}^{n-1} \subset \mathbb{S}^n.$$

Then each $\mathbb{S}^k, k \geq 1$ can be obtained from $\mathbb{S}^{k-1}$ by attaching two k-cells, viz., the upper and lower hemispheres. (For instance, $\mathbb{S}^1$ has two 0-cells and two 1-cells. See Figure 2.12(c).) This cell decomposition of $\mathbb{S}^n$ is more useful than the earlier simpler one given in (2). It immediately gives us a cell structure for the real projective space, which we shall discuss now.

(v) Recall that the real projective space $\mathbb{P}^n$ is the quotient space of $\mathbb{S}^n$ under the antipodal action $x \sim -x$. By the definition of quotient topology, if $q : \mathbb{S}^n \longrightarrow \mathbb{P}^n$ denotes the quotient map then a subset U of $\mathbb{P}^n$ is open iff its inverse image $q^{-1}(U)$ is open in $\mathbb{S}^n$. We first observe that the quotient map q is both an open mapping as well as a closed mapping. This follows easily from the fact that for any subset $F \subset \mathbb{S}^n$, $F \cup (-F)$ is open (closed) if F is open (closed, respectively). From this, many of the topological properties of $\mathbb{S}^n$ pass onto the quotient space $\mathbb{P}^n$. For instance, using the openness of q we can easily conclude that $\mathbb{P}^n$ is II-countable. Indeed given any base $\mathcal{B}$ for the topology of $\mathbb{S}^n$, it follows that $\{q(U) \ : \ U \in \mathcal{B}\}$ is a base for $\mathbb{P}^n$. Since $\mathbb{S}^n$ is compact, it follows that $\mathbb{P}^n$ is also compact.

The CW-structure of $\mathbb{S}^n$ as described in (iv) is compatible with this action in the sense that the action preserves each skeleton and merely permutes the various cells. In such a situation, the quotient space acquires a natural CW-structure: We begin with $X^{(0)}$ as a single 0-cell for $\mathbb{P}^n$, which is the image of $\mathbb{S}^0$ under the quotient map. Inductively, having defined $X^{(k-1)}$ whose underlying space happens to be $\mathbb{P}^{k-1}$, we attach a single k-cell to $X^{(k-1)}$ which could be either the upper or the lower hemisphere of $\mathbb{S}^{k-1}$

to get $X^{(k)}$. For definiteness, let us choose the upper-hemisphere of $\mathbb{S}^k$ and take the attaching map to be q restricted to $\mathbb{S}^{k-1}$. The space so obtained is indeed equal to $\mathbb{P}^k$. Thus, $\mathbb{P}^n$ has a CW-structure with one cell for each dimension $0 \leq k \leq n$. Indeed, this is really the first non trivial example of a CW-complex.

(vi) The infinite sphere $\mathbb{S}^\infty$ which is the union of all spheres

$$\mathbb{S}^0 \subset \mathbb{S}^1 \subset \cdots \subset \mathbb{S}^n \subset \cdots$$

is a CW-complex with two cells in each dimension. It is infinite dimensional. Indeed, let $\mathbb{R}^\infty$ denote the infinite sum of copies of $\mathbb{R}$ as a vector space. Elements of $\mathbb{R}^\infty$ can be written as infinite sequences of real numbers in which all but finitely many entries are zero. The standard Euclidean inner product extends to $\mathbb{R}^\infty$ and makes it into a Hilbert space. However, the topology that we give it is the *weak topology* coinduced from the family

$$\mathbb{R} \subset \mathbb{R}^2 \subset \mathbb{R}^3 \subset \cdots,$$

viz., a subset $F \subset \mathbb{R}^\infty$ is closed iff $F \cap \mathbb{R}^n$ is closed in $\mathbb{R}^n$ for every n. Note that the weak topology is finer than the metric topology and so all vector space operations are continuous. The unit sphere $\mathbb{S}^\infty$ inherits this weak topology.

The cell structure on $\mathbb{S}^\infty$ is compatible with the antipodal action and hence we get an induced cell structure on the infinite real projective space $\mathbb{P}^\infty$, with exactly one cell in each dimension $0 \leq k < \infty$.

(vii) The case of complex projective spaces is not much different. Recall that $\mathbb{CP}^n$ is defined as the quotient space of $\mathbb{C}^{n+1} \setminus \{0\}$ by the relation

$$(z_0, \ldots, z_n) \sim \lambda(z_0, \ldots, z_n), \quad \lambda \in \mathbb{C} \setminus \{0\}.$$

Under the standard inclusions

$$\mathbb{C} \subset \mathbb{C}^2 \subset \cdots \mathbb{C}^n \subset \cdots$$

the above relations are compatible and hence we can use the same notation q for all the quotient maps $q : \mathbb{C}^{n+1} \setminus \{0\} \to \mathbb{CP}^n$.

As before, restricted to the unit sphere $\mathbb{S}^{2n+1}$, q is surjective and hence we can view $\mathbb{CP}^n$ as the quotient space of $\mathbb{S}^{2n+1}$ modulo the same relation as above wherein λ is now restricted to being of unit length. Clearly, $\mathbb{CP}^0$ is a singleton. Next, the quotient map $q : \mathbb{S}^3 \to \mathbb{CP}^1$ sends the circle $z_1 = 0$ to a single point which is our $\mathbb{CP}^0$. The subspace

$$e^2 = \{(z_0, r) \ : \ |z_0|^2 + r^2 = 1, r \geq 0\}$$

is clearly homeomorphic to $\mathbb{D}^2$ with its boundary being the circle $z_1 = 0$. Note that $q(e_2) = \mathbb{CP}^1$ and q restricted to the interior of the 2-cell e^2 is injective. Hence $\mathbb{CP}^1$ is nothing but a 2-disc with its boundary collapsed to a single point and hence is homeomorphic to $\mathbb{S}^2$. Indeed the map $q : \mathbb{S}^3 \to \mathbb{S}^2$ is the familiar Hopf fibration. Inductively, having established that $\mathbb{CP}^k$ is obtained by attaching a $2k$-cell to $\mathbb{CP}^{2k-1}$ via the quotient map $\mathbb{S}^{2k-1} \to \mathbb{CP}^{k-1}$, we see that the subspace $e_{2k+2} \subset \mathbb{S}^{2k+3}$ given by

$$e^{2k+2} = \{(z_0, \ldots, z_{k+1}, r) \ : \ \sum_{i=0}^{k+1} |z_i|^2 + r^2 = 1, r \geq 0\}$$

is homeomorphic to $\mathbb{D}^{2k+2}$ with its boundary being the sphere given by $z_{k+2} = 0$. One merely checks that $q(e^{2k+2}) = \mathbb{CP}^{k+1}$ and restricted to the interior of the cell e^{2k+2} is injective. Therefore, $\mathbb{CP}^{2k+2}$ is obtained by attaching the $(2k+2)$-cell e^{2k+2} to $\mathbb{CP}^k$ via the map $q : \mathbb{S}^{2k+1} \to \mathbb{CP}^k$.

Thus the infinite complex projective space $\mathbb{CP}^\infty$ has a CW-structure one cell for each even dimension.

We shall now study some of the topological properties of a CW-complex. Proofs of corresponding properties for relative CW-complexes, with appropriate modifications, will be left to the reader as exercises.

Proposition 2.2.12 Let X be a CW-complex.
(a) Each skeleton $X^{(k)}$ is a closed subset of X.
(b) Each closed cell is also a closed subset of X.
(c) A subset S of X is closed iff S intersects each closed cell e^k of X in a closed subset of e^k.
(d) The topology on X is compactly generated.

Proof: (a) By Lemma 2.2.2 (b), each $X^{(k)}$ is closed in $X^{(k+1)}$ and hence each $X^{(k)}$ is closed in $X^{(r)}$ for $r > k$. This means $X^{(k)} \cap X^{(r)}$ is closed in $X^{(r)}$ for every r. Hence by part (iv) of Definition 2.2.7, $X^{(k)}$ is closed in X.
(b) By Corollary 2.2.6, applied inductively, it follows that each skeleton $X^{(k)}$ is a Hausdorff space. Now if e^k is a closed cell, then it is a compact subset of $X^{(k)}$ and hence is a closed subset of $X^{(k)}$. Since each $X^{(k)}$ is closed in X, we are done.
(c) By (b), if S is closed in X then $S \cap e$ is closed in X for each closed cell e. Conversely, suppose $S \cap e$ is closed in X for every closed cell e. Inductively, we shall prove that $S \cap X^{(k)}$ is closed for every k. Again for $k = 0$, there is nothing to prove since $X^{(0)}$ is discrete. Having proved this for some k, we appeal to Lemma 2.2.2 to conclude that $S \cap X^{(k+1)}$ is closed.
(d) Recall that a topology is *compactly generated if it is coherent with the family of its compact subsets, i.e., a subset S of X is closed iff $S \cap K$ is closed for every compact subset K of X.* In this situation, since each closed cell is compact, $S \cap e$ is closed for each cell e and hence we are through by (c). ♠

Just as in Corollary 2.2.6, it is not difficult to prove the following:

Theorem 2.2.13 *Let (X, A) be a relative CW-complex. If A is T_1, T_2 or normal so is X.*

Lemma 2.2.14 *Every CW-complex is the disjoint union of its open cells.*

Proof: By definition, we know that

$$X = \coprod_{k \geq 0} (X^{(k)} \setminus X^{(k-1)}).$$

Here $X^{(-1)} = \emptyset$ and $X^{(0)}$ is the union of 0-cells which are, by convention open cells. And for $k \geq 1$ each $X^{(k)} \setminus X^{(k-1)}$ is the disjoint union of open k-cells. ♠

Lemma 2.2.15 *Let S be a subset of X such that S contains at most one point from each open cell of X. Then S is a closed discrete subset of X.*

Proof: To show the discreteness of S, let $x \in S$ be a point such that it belongs to an open k-cell. In this cell, we can take a small neighbourhood U_x of x such that $\bar{U}_x$ does not intersect the boundary of the cell. Then clearly, the image of U_x is open in $X^{(k)}$ which we shall denote by V_k. Also observe that $V_k \cap S = \{x\}$. Suppose we have constructed an open set $V_m \subset X^{(m)}$ such that $V_m \cap S = \{x\}$ where $m \geq k$. Put W_{m+1} to be the union of all open

$(m+1)-$ cells except for the points in S. Take $V_{m+1} = V_m \cup W_{m+1}$. Clearly $V_{m+1} \subset X^{(m+1)}$ is open and $V_{m+1} \cap S = V_m \cap S = \{x\}$. Then it follows that $V = \cup_{m \geq k} V_m$ is open in X and $V \cap S = \{x\}$.

To show that S is a closed set, clearly $S \cap X^{(0)}$ is closed in $X^{(0)}$. Assume that $S \cap X^{(k)}$ is closed in $X^{(k)}$. Then for any closed $(k+1)-$cell e, $e \cap S$ is either equal to $e \cap S \cap X^{(k)}$ or has one extra point. And in either case this is a closed subset of $X^{(k+1)}$. Therefore, $S \cap X^{(k+1)}$ is closed in $X^{(k+1)}$. Again by condition (iv) of Definition 2.2.7, we are through.   ♠

Theorem 2.2.16 *Every compact subset K of a CW-complex X is contained in the union of finitely many open cells.*

Proof: If not, K will intersect infinitely many open cells which, we know are disjoint. Selecting one point in each such intersection, we will get a subset S of K which is, by Lemma 2.2.15, a discrete subset of the compact set K. But S is infinite?!   ♠

Corollary 2.2.17 The closure of every cell in X meets only finitely many open cells.

Remark 2.2.18 This property of a CW-complex is known as 'closure finiteness' and it explains the presence of the letter 'C' in this mysterious nomenclature, 'CW-complex'. Historically, this property was an additional part of the definition itself. We have just proved that we don't need to put this extra condition in the definition itself. By the way, the other letter 'W' in the name corresponds to the 'weak topology' on X. So, it might have been appropriate to call these spaces W-complexes. However, we shall stick to the present-day fashion of calling them *cell complexes* or the old name CW-complexes.

One of the salient features of the coherent topology that we take on a CW-complex is that it provides a practical method to construct continuous functions as well as a method to verify whether a given function is continuous or not. We shall have many instances of this. Let us begin with an illustration by showing the existence of partition of unity on CW-complexes. We assume that the reader is familiar with the concept of partition of unity and related results. However, let us recall the definition.

Definition 2.2.19 Let X be a topological space. By a partition of unity on X, we mean a family $\{\theta_\alpha : \alpha \in \Lambda\}$ of continuous function $\theta_\alpha : X \to \mathbb{I}$ such that
(i) each θ_α has compact support, i.e., the closure of $\{x \in X : \theta_\alpha(x) \neq 0\}$ is compact.
(ii) for each $x \in X$ there exists a neighborhood N_x of x in X, such that only finitely many of θ_α are non zero on N_x; (this property is called 'locally finiteness');
(iii) $\sum_\alpha \theta_\alpha(x) = 1$, for all $x \in X$.

Given an open covering $\mathcal{U} = \{U_\beta\}$ of X, the family $\{\theta_\alpha\}$ is said to be subordinate to $\mathcal{U}$ if for each α there exists $\beta = \beta(\alpha)$ such that support of θ_α is contained in U_β.

It is a standard result that over any subspace of a Euclidean space, there is always a partition of unity subordinate to a given open cover. Similar to Proposition 2.2.5, the key result that we need is the following:

Lemma 2.2.20 Let $\mathcal{U}$ be an open covering of $\mathbb{D}^n$, $\{\eta_\alpha\}$ be a partition of unity on $\mathbb{S}^{n-1}$ subordinate to the restricted covering $\mathcal{U}|_{\mathbb{S}^{n-1}} = \{U \cap \mathbb{S}^{n-1} : U \in \mathcal{U}\}$. Then there is an extension of $\{\eta_\alpha\}$ to a partition of unity on $\mathbb{D}^n$ subordinate to $\mathcal{U}$.

Proof: Since $\mathbb{S}^{n-1}$ is compact and since the partition of unity is locally finite, it follows that only finitely many η_α are non zero. For each such α let the support of η_α be contained in $U_\alpha \cap \mathbb{S}^{n-1}$, say. By Tietze's extension theorem there exist extensions $\eta'_\alpha : \mathbb{D}^n \to \mathbb{I}$ of η_α with support contained in U_α. It follows that there exists an open set V in $\mathbb{D}^n$ such that $\mathbb{S}^{n-1} \subset V$ and $\sum_\alpha \eta'_\alpha > 0$ on V. Choose any finite subcover $\{U_1, \ldots, U_k\} \subset \mathcal{U}$ of $\mathbb{D}^n$ and let

$\{\varphi_1, \ldots, \varphi_k\}$ be a partition of unity on $\mathbb{D}^n$ subordinate to this cover. Let $\theta : \mathbb{D}^n \to \mathbb{I}$ be a map such that $\theta \equiv 0$ on $\mathbb{S}^{n-1}$ and $\equiv 1$ on $\mathbb{D}^n \setminus V$. Put

$$\Phi(x) = \left(\sum_\alpha \eta'_\alpha \right) + \theta \sum_i \varphi_i.$$

Put $\hat{\eta}_\alpha = \eta'_\alpha / \Phi$, and $\lambda_i = \theta \varphi_i / \Phi$. It follows that $\{\hat{\eta}_\alpha\} \cup \{\lambda_1, \ldots, \lambda_k\}$ is a partition of unity subordinate to $\mathcal{U}$ and which is an extension of $\{\eta_\alpha\}$. ♠

Theorem 2.2.21 *Let X be a CW-complex and $\mathcal{U}$ be an open covering. Then there is partition of unity subordinate to $\mathcal{U}$.*

Proof: Define $p_{\alpha,0} : X^{(0)} \to \mathbb{I}$ to be 1 if $x \in U_\alpha$ and otherwise, to be 0. Since $X^{(0)}$ is discrete, this is clearly a partition of unity on $X^{(0)}$. Inductively, having defined a partition of unity $\{p_{\alpha,n-1}\}$ on $X^{(n-1)}$, we can extend each $p_{\alpha,n-1}$ restricted to ∂e^n_j inside the n-cell e^n_j for each n-cell as in the above lemma. Then we can put them all together to get a partition $\{p_{\alpha,n}\}$ on $X^{(n)}$ which extends $\{p_{\alpha,n-1}\}$. It is easily verified that $\{p_{\alpha,n}\}$ is subordinate to $\mathcal{U}|_{X^{(n)}}$.

Finally we define p_α on X by taking $p_\alpha|_{X^{(n)}} = p_{\alpha,n}$. ♠

Corollary 2.2.22 Every CW-complex is paracompact.

Proof: This follows from the fact that for a Hausdorff space, paracompactness is equivalent to the property that to every open cover $\mathcal{U}$ of X, there is a partition of unity subordinate to $\mathcal{U}$. ♠

Exercise 2.2.23

   (i) Show that the weak topology on $\mathbb{R}^\infty$ is finer than the metric topology, by giving an example of a subset $F \subset \mathbb{R}^\infty$ which is closed in the weak topology but not closed in the metric topology.

   (ii) Let X be a paracompact space and $\mathcal{U}$ be an open cover for X. Show that there exists a family of continuous functions $\theta_j : X \to \mathbb{I}$ subordinate to $\mathcal{U}$ and such that for every $x \in X$, $\max_j \theta_j(x) = 1$.

2.3 Product of Cell Complexes

We shall now study finite Cartesian product of CW-complexes. Experience shows that Cartesian product neither misbehaves totally nor behaves all too well with respect to several topological/geometric structures. CW-structure is one such instance of this. In this section, we shall study how to give CW-structure to product spaces and obtain conditions under which this construction yields the standard product topology.

Example 2.3.1 Let us begin with an example. Let us fix a CW-structure on $\mathbb{S}^1$ with one 0-cell and one 1-cell as in the previous example (see Figure 2.12(a)). Let us denote them by σ_0, σ_1. Similarly, let us fix a CW-structure on the unit interval $\mathbb{I}$ with two 0-cells and one 1-cell. We would like to obtain a CW-structure on the cylinder $\mathbb{S}^1 \times \mathbb{I}$. For the 0-skeleton we take the obvious candidate, viz., $\{(\sigma_0, 0), (\sigma_0, 1)\}$, being the product space of the 0-skeleton of the two spaces. What should be the 1-skeleton. We observe that the product of a 1-cell and a 0-cell is again a 1-cell and there are three different possibilities of obtaining a 1-cell from this process, viz., $\sigma_1 \times \{0\}, \sigma_1 \times \{1\}, \sigma_0 \times \mathbb{I}$. Finally the product of a 1-cell and a 1-cell yields a 2-cell, viz., $\sigma_1 \times \mathbb{I}$. (See Figure 2.13.) This process can actually be generalized completely, at least when both the CW-complexes are finite (hence compact).

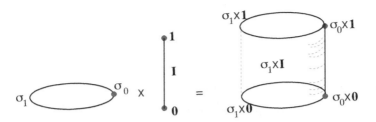

FIGURE 2.13. Product of CW-complexes

Definition 2.3.2 We fix once for all a set of homeomorphisms $h_{m,n} : \mathbb{D}^{n+m} \longrightarrow \mathbb{D}^n \times \mathbb{D}^m$. Given two CW-complexes X, Y, we define the product CW-complex $X \times_w Y$ as follows: The underlying set is the product set $X \times Y$. For each cell σ in X and a cell τ in Y with characteristic maps ϕ and ψ, respectively, we take a product cell $\sigma \times \tau$ in $X \times Y$ with its characteristic map $(\phi \times \psi) \circ h_{m,n} : \mathbb{D}^{m+n} \longrightarrow X \times Y$. In particular, observe that the 0-skeleton $(X \times_w Y)^{(0)} = X^{(0)} \times Y^{(0)}$. Inductively define the k-skeleton $(X \times_w Y)^{(k)}$ of $(X \times_w Y)$ to be the space obtained from the $(k-1)$-skeleton by attaching all possible product cells $\sigma \times \tau$ where σ, τ range over cells of X, Y, respectively, with the condition $\dim \sigma + \dim \tau = k$. Finally give the weak topology on $X \times_w Y = \cup_{k \geq 0} (X \times_w Y)^{(k)}$, viz., a subset S is closed iff $S \cap (X \times_w Y)^{(k)}$ is closed in $(X \times_w Y)^{(k)}$.

Example 2.3.3 Let us consider the cell structure for $\mathbb{S}^1 \times \mathbb{S}^1$. To begin with we must fix some cell structure on $\mathbb{S}^1$ explicitly. We consider $\mathbb{S}^1$ as the standard subspace of $\mathbb{C}$. Take the 0-cell to be the point 1 and the characteristic map of the 1-cell to be $\phi : \mathbb{I} \to \mathbb{C}$ where $\phi(t) = e^{2\pi i t}$.

Now we look at the product CW-structure on the subspace $\mathbb{S}^1 \times \mathbb{S}^1 \subset \mathbb{C}^2$. There is just one 0-cell, viz., the point $(1, 1)$, whereas there are two 1-cells given by $\phi_1(t) = (\phi(t), 1)$ and $\phi_2(t) = (1, \phi(t))$. Finally there is just one 2-cell given by $\phi \times \phi : \mathbb{I} \times \mathbb{I} \to \mathbb{C} \times \mathbb{C}$ which fills up the subspace $\mathbb{S}^1 \times \mathbb{S}^1$. It is interesting to note that the attaching map of this 2-cell traces the first circle $\mathbb{S}^1 \times 1$ then the second circle $1 \times \mathbb{S}^1$ then again the first circle but in the opposite direction and finally the second circle, again in the opposite direction. Therefore in the fundamental group this loop represents the element $xyx^{-1}y^{-1} \in \pi_1(\mathbb{S}^1 \vee \mathbb{S}^1)$. (See Corollary 3.8.12 for more details.)

Remark 2.3.4 Returning to the study of $X \times Y$ in general, it is not difficult to see that the images of the product cells cover $X \times Y$ as σ ranges over all cells of X and τ ranges over all cells of Y and hence the underlying set of $X \times_w Y$ is $X \times Y$. Our major concern now is to compare the weak topology with the product topology. Note that for each 0-cell x_0 of X, the $x_0 \times Y$ is a subcomplex of $X \times_w Y$. Similarly, for each 0 cell y_0 of Y, $X \times y_0$ is a subcomplex $X \times_w Y$. A product cell can be thought of as obtained by attaching $\mathbb{D}^m \times \mathbb{D}^n$ with the characteristic map as the product of the two characteristic maps, instead of going to the round disc $\mathbb{D}^{m+n}$. (This is what we have done in the above example.) In particular, all the closed cells of $X \times_w Y$ are closed and compact subsets of the product space $X \times Y$. As it turns out, there is a much closer relation between the product topology on $X \times Y$ and the weak topology on $X \times_w Y$ though, in general, they may not be the same. To understand this, we begin with recalling one of the fundamental properties of co-induced topology. Not being familiar with it is going to cause problems for us at some stage or other. A reader who does not want to be bothered about this problem may skip this, take Theorem 2.3.10 for granted and go ahead.

Definition 2.3.5 Let $\mathcal{F} = \{(X_\alpha, \tau_\alpha) : \alpha \in \Lambda\}$ be a family of topological spaces, $X = \cup_\alpha X_\alpha$, so that for every pair (α, β) of indices, the two induced topologies on $X_\alpha \cap X_\beta \subset X_\alpha$ and $X_\alpha \cap X_\beta \subset X_\beta$ are the same. Then there is a topology τ on X such that each (X_α, τ_α) is a subspace and τ is the smallest topology on X with this property. It is called the *co-induced topology on X*. Let us denote it temporarily by $X_\mathcal{F}$.

Lemma 2.3.6
(a) The coinduced topology $X_\mathcal{F}$ is characterised by the property that a subset U of X is open in X iff $U \cap X_\alpha$ is open in X_α.
(b) If X is a topological space of which each X_α is a subspace, then $X_\mathcal{F}$ is finer than the topology on X.
(c) If X is a topological space such that $\mathcal{F}$ is an open covering of X or a closed covering which is locally finite, then $X_\mathcal{F}$ coincides with the topology on X.
(d) Given subsets $A \subset V \subset X$, V is a neighbourhood of A in X iff $V \cap X_\alpha$ is a neighbourhood of $A \cap X_\alpha$ in X_α.

Proof: The proofs of (a), (b) (c) are all routine. We shall only prove (d) here.

The 'only if' part of (d) is obvious. Therefore, we need to prove the 'if' part only. We first prove this for the case when $\Lambda = \{1, 2\}$. Then by a simple application of Zorn's lemma the general result follows.

So, consider the case $\Lambda = \{1, 2\}$. Let U_1 be open in X_1 such that $A \cap X_1 \subset U_1 \subset V \cap X_1$. This implies $U_1 \cap (X_1 \cap X_2) = U_1 \cap X_2$ is open in $X_1 \cap X_2$ and therefore there is an open set W_2 in X_2 such that $U_1 \cap X_2 = W_2 \cap X_1$. Put $W_2' = W_2 \cap \text{int}_{X_2}(V \cap X_2)$. Then $U_1 \cap X_2 \subset \text{int}_{X_1 \cap X_2}(V \cap X_1 \cap X_2)$, being open in $X_1 \cap X_2$ and contained in $V \cap X_1 \cap X_2$, and hence $U_1 \cap X_2 \subset \text{int}_{X_2}(V \cap X_2)$. Therefore

$$W_2' \cap X_1 = W_2 \cap X_1 \cap \text{int}_{X_2}(V \cap X_2) = U_1 \cap X_2.$$

Put $T = X_2 \setminus W_2'$. Then $V \cap T$ is a neighbourhood of $A \cap T$ in T and hence there exists an open set W_2'' in T such that

$$A \cap T \subset W_2'' \subset V \cap T \subset V \cap X_2.$$

Put $U_2 = W_2' \cup W_2''$. Then U_2 is open in X_2 and

$$A \cap X_2 = (A \cap W_2') \sqcup (A \cap T) \subset W_2' \sqcup W_2'' = U_2 \subset V \cap X_2.$$

Put $U = U_1 \cup U_2$. Verify that $U \cap X_i = U_i, i = 1, 2$. In particular, this proves U is open in X. Verify also that $A \subset U \subset V$.

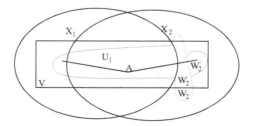

FIGURE 2.14. Neighbourhoods in coherent topology

Now for the general case, consider the family $\mathcal{F}$ of pairs (Y, U) where Y is a union of members of $\{X_\alpha\}$ (with the co-induced topology) and U is an open subset of Y such that

$A \cap Y \subset U \subset V \cap Y$. By the hypothesis, for each α, there exists U_α such that $(X_\alpha, U_\alpha) \in \mathcal{F}$ and hence this family is non empty. We partially order this family by $(Y_1, U_1) < (Y_2, U_2)$ iff $Y_1 \subset Y_2$ and $U_1 = Y_1 \cap U_2$. It is easily checked that every chain in this partially ordered family has an upper bound. By Zorn's lemma there exists a maximal element (Y^*, U^*). Clearly, it is enough to show that $Y^* = X$. If this is not true, then there exists α such that $X_\alpha \not\subset Y^*$. Put $X_1 = Y^*, U_1 = U^*, X_2 = X_\alpha$ and apply the above case to obtain an element $(X_1 \cup X_2, U_1 \cup U_2) \in \mathcal{F}$. Indeed, from the proof of the above case, it follows that $(U_1 \cup U_2) \cap X_1 = U_1$ and hence $(Y^*, U^*) < (X_1 \cup X_2, U_1 \cup U_2)$ which is a contradiction to the maximality of (Y^*, U^*). ♠

Lemma 2.3.7 Given any topological space X, let X_w denote X with the topology induced from the collection of compact subsets of X, i.e., a subset S of X is closed in X_w iff $S \cap K$ is closed in K for each compact subset K of X. Then
(a) The identity map $Id : X_w \longrightarrow X$ is continuous.
(b) Id is a homeomorphism iff X is compactly generated.
(c) Id defines a bijection of compact sets in the two topologies.

Proof: (a), (b) are straightforward, being just a rewording of the definition. In (c), again if K is compact in X_w, by continuity of Id, K is compact in X. To see the converse, let $\mathcal{F}$ be a family of subsets of K closed in X_w with finite intersection property. Then it is a family of closed sets in X as well. Since K is compact in X, it follows that $\mathcal{F}$ has non empty intersection. This proves (c). ♠

Lemma 2.3.8 The identity map $Id : (X \times Y)_w \longrightarrow X \times Y$ is a homeomorphism if one of the following holds:
(a) X and Y are compact.
(b) X is compactly generated and Y is compact.
(c) X is compactly generated and Y is locally compact.

Proof: (a) Since X and Y are compact, so is $X \times Y$. For any compact space, the compactly generated topology coincides with the given topology.
(b) Anyway this follows easily from (c) and so we shall directly prove (c).
(c) Let U be an open subset of $(X \times Y)_w$ and $(x_0, y_0) \in U$. Let K be a compact neighbourhood of y_0 in Y. Then $U \cap \{x_0\} \times K$ is open in $\{x_0\} \times K$ and hence there exists a compact neighbourhood K' of y_0 such that $\{x_0\} \times K' \subset U$. Let $V = \{x \in X : \{x\} \times K' \subset U\}$.

Let L be a compact subset of X, such that $x_0 \in L$. Then $U \cap L \times K'$ is a neighbourhood of $x_0 \times K'$ in $L \times K'$. Hence by Wallman's theorem, there is an open subset W of L such that $x_0 \in L$ and $W \times K' \subset U$. This means $W \subset V$. Therefore, $V \cap L$ is a neighbourhood of x_0 in L. By Lemma 2.3.6 (d), it follows that V is a neighbourhood of x_0 in X. Since $V \times K' \subset U$ we have proved that U is a neighbourhood of (x_0, y_0) in $X \times Y$. Therefore, U is open in $X \times Y$. ♠

Returning to the case when X and Y are CW-complexes, we now have:

Theorem 2.3.9 *Let X, Y be any two CW-complexes. We have:*
(a) *$(X \times_w Y) = (X \times Y)_w$.*
(b) *The identity map $Id : X \times_w Y \longrightarrow X \times Y$ is continuous i.e., the CW-topology is finer than the product topology.*
(c) *Id is a homeomorphism iff the product topology is compactly generated.*
(d) *The identity map $X \times_w Y \to X \times Y$ defines a bijection of compact subsets in the two topologies.*

Proof: (a) The family $\mathcal{F}$ of product cells $\{\sigma \times \tau\}$ as σ, τ runs over all cells of X and Y, respectively, is a cover of $X \times Y$ consisting of compact subsets. Therefore the coinduced topology of $X \times_w Y$ from this family is finer than the weak topology on $(X \times Y)_w$. The equality follows from the observation that every compact subset of $X \times Y$ is covered by finitely many members of $\mathcal{F}$. Details are left to the reader.

Other conclusions now follow from Lemma 2.3.7. ♠

We can now give a number of instances when the product topology coincides with the weak topology for the product complex $X \times_w Y$.

Theorem 2.3.10 *Let X, Y be any two CW-complexes. The product topology coincides with the CW-topology on $X \times Y$ in the following instances.*
(1) *X, Y are finite.*
(2) *If either X or Y is finite.*
(3) *If X or Y is locally compact (or equivalently, locally finite).*
(4) *If both X and Y have countably many cells.*
(5) *If X and Y are both locally countable.*

Proof: (1), (2), (3) directly follow from (a), (b), (c) of Lemma 2.3.8. To prove (4), let W be an open subset of $X \times_w Y$ and $(x_0, y_0) \in W$ be an arbitrary point. It is enough to produce neighbourhoods U of x_0 in X and V of y_0 in Y such that $U \times V \subset W$. Write $X = \cup_k K_k, Y = \cup_k L_k$ as the increasing union of finite subcomplexes such that $x_0 \in K_0, y_0 \in L_0$. Then $W \cap (K_k \times L_k)$ is a neighbourhood of (x_0, y_0) in $K_k \times L_k$. So, we can choose neighbourhoods U_k, V_k of x_0, y_0 in K_k, L_k, respectively, such that $U_k \cap K_{k-1} = U_{k-1}, V_k \cap L_{k-1} = V_{k-1}$, and $U_k \times V_k \subset W \cap (K_k \times L_k)$. Put $U = \cup_k U_k, V = \cup_k V_k$. Then it follows from Lemma 2.3.6(d) that U and V are neighbourhoods of x_0 and y_0 in K, L, respectively. Clearly, $U \times V \subset W$ and we are through.

Finally to prove (5), recall that a CW-complex is said to be locally countable, if every open cell meets only countably many closed cells. This means that every point of the complex is contained in a countable subcomplex which is a closed neighbourhood. (If $x \in \text{int } e^n$ then there are countably many closed cells which meet $\text{int } e^n$ and each of these closed cell is contained in a finite subcomplex. Therefore, the union of all these subcomplexes will be a countable subcomplex as required.) Hence (5) immediately reduces to the situation of (4). ♠

Example 2.3.11 (Dowker [Dowker, 1951].) This example tells us that we may not be able to generalize the results in the above theorem any further. Let $\mathbb{N}$ denote the set of natural numbers and $\mathcal{L}$ denote the set of all functions $\phi : \mathbb{N} \to \mathbb{N}$. Consider the real vector spaces $\mathbb{R}^{\mathbb{N}}$ and $\mathbb{R}^{\mathcal{L}}$ (direct sums) with the topology co-induced by finite dimensional vector subspaces. Let $\mathbf{v}_n, \mathbf{u}_\phi$ denote the standard basis elements in $\mathbb{R}^{\mathbb{N}}$ and $\mathbb{R}^{\mathcal{L}}$, respectively. Define

$$X = \{r\mathbf{v}_n \ : \ n \in \mathbb{N}, 0 \leq r \leq 1\} \subset \mathbb{R}^{\mathbb{N}}; \quad Y = \{r\mathbf{v}_\phi \ : \ \phi \in \mathcal{L}, 0 \leq r \leq 1\} \subset \mathbb{R}^{\mathcal{L}}.$$

Then both X and Y are 1-dimensional CW-complexes which are nothing but the 1-point-union of edges indexed by $\mathbb{N}$ and $\mathcal{L}$, respectively. We claim that the product CW-topology on $X \times Y$ is strictly finer than the product topology, i.e., considered as a subspace of $\mathbb{R}^n \times \mathbb{R}^{\mathcal{L}}$. Define

$$P = \left\{ \left(\frac{\mathbf{v}_n}{\phi(n)}, \frac{\mathbf{u}_\phi}{\phi(n)} \right) \ : \ n \in \mathbb{N}, \phi \in \mathcal{L} \right\}.$$

Observe that P consists of precisely one element from the interior of each 2-cell in $X \times Y$ and hence is a discrete closed subset of $(X \times Y)_w$. (See Lemma 2.2.15.) However, we shall see that the origin 0 of $\mathbb{R}^{\mathbb{N}} \times \mathbb{R}^{\mathcal{L}}$ is in the closure of P in the product topology. So, let U, V be neighbourhoods of the origin in $\mathbb{R}^{\mathbb{N}}$ and $\mathbb{R}^{\mathcal{L}}$, respectively. Then for each n and each ϕ there exist $r_n, s_\phi \in (0, 1]$ such that

$$\{\lambda\mathbf{v}_n \ : \ 0 \leq \lambda \leq r_n\} \subset U; \quad \{\lambda\mathbf{u}_\phi \ : \ 0 \leq \lambda \leq s_\phi\} \subset V.$$

Consider $\psi : \mathbb{N} \to \mathbb{N}$ given by

$$\psi(n) = \max\left\{n, \left\lfloor \frac{1}{r_n} \right\rfloor\right\} + 1.$$

Clearly $\psi(n) \to \infty$ as $n \to \infty$ and hence we can choose m such that $\psi(m) > \frac{1}{s_\psi}$. It follows that $(\frac{1}{\psi(m)}\mathbf{v}_m, \frac{1}{\psi(m)}\mathbf{u}_\psi) \in P \cap (U \times V)$. This proves that P is not closed in the product topology.

2.4 Homotopical Aspects

We shall now study the homotopical aspects of cell complexes, which is central to our study. This section will contain only the basics of this aspect on which the entire edifice of algebraic topology will be built.

Theorem 2.4.1 *Let X be a CW-complex. A function $f : X \times \mathbb{I} \longrightarrow Y$ is continuous iff the restricted functions $f : X^{(k)} \times \mathbb{I} \longrightarrow Y$ are all continuous for $k \geq 0$.*

Proof: Since $\mathbb{I}$ is compact, $Id : (X \times \mathbb{I})_w \to X \times \mathbb{I}$ is a homeomorphism. Therefore f is continuous iff $f|_{(X \times \mathbb{I})^{(k)}}$ is continuous for each k. Observe that for each k, $(X \times_w \mathbb{I})^{(k)} \subset X^{(k)} \times \mathbb{I} \subset (X \times_w \mathbb{I})^{(k+1)}$ are closed subsets. The claim follows. ♠

The following lemma is central to the homotopical aspects of CW-complexes.

Lemma 2.4.2 *Let U be an open (or a closed) subset of X where (X, A) is a relative CW-complex. Put $U_{-1} = U \cap A$; and $U_n = U \cap (X, A)^{(n)}, n \geq 0$. Suppose U_{n-1} is a SDR of U_n for each n. Then U_{-1} is SDR of U.*

Proof: Let $F_n : U_n \times \mathbb{I} \longrightarrow U_n$ be a homotopy such that $F_n(x, t) = x$ for all $x \in U_{n-1}, t \in \mathbb{I}$ and $F_n(x, 0) = x$ for all $x \in U_n$. Put $f_n(x) = F_n(x, 1)$. Then clearly, $f_n : U_n \longrightarrow U_{n-1}$ is a SDR for each n. Therefore taking the composites, viz., $g_n = f_0 \circ f_1 \circ \cdots \circ f_n$, we get a SDR $g_n : U_n \longrightarrow U_{-1}$. Observe that $g_{n+1}|_{U_n} = g_n$ for all n. Take $g(x) = g_n(x)$ whenever $x \in U_n$. Then $g : U \longrightarrow U_{-1}$ is well-defined, continuous and a retraction. However, to show that it is a SDR needs a little more effort.

Let us define $G_n : U_n \times \mathbb{I} \longrightarrow U_n$ inductively as follows:

$$G_0(x, t) = \begin{cases} x, & 0 \leq t \leq \dfrac{1}{2}; \\ F_0(x, 2t - 1), & \dfrac{1}{2} \leq t \leq 1. \end{cases}$$

For $n \geq 1$, having defined G_{n-1}, now define

$$G_n(x, t) = \begin{cases} x, & 0 \leq t \leq \dfrac{1}{n+2}; \\ F_n(x, (n+1)[(n+2)t - 1]), & \dfrac{1}{n+2} \leq t \leq \dfrac{1}{n+1}; \\ G_{n-1}(f_n(x), t), & \dfrac{1}{n+1} \leq t \leq 1. \end{cases}$$

Inductively, we now verify that each G_n is a SDR of U_n into U_{-1}. Moreover, $G_n|_{U_{n-1}} \times \mathbb{I} = G_{n-1}$. Therefore, there is a well-defined map $G : U \times \mathbb{I} \longrightarrow U$ given by $G(x, t) = G_n(x, t)$ whenever $x \in U_n$. If V is an open subset of U then $G^{-1}(V) \cap (X, A)^{(n)} = G^{-1}(V) \cap U_n = G_n^{-1}(V \cap (X, A)^{(n)})$ and hence is open in U_n for each n. This means that $G^{-1}(V)$ is open in U. Therefore, G is continuous. It is easily verified that G is a SDR of U into U_{-1}. ♠

Theorem 2.4.3 *Every CW-complex is locally contractible.*

Proof: Given $x \in V \subset X$, where V is an open set, we shall construct a neighbourhood $U \subset V$ of x which is contractible. Using the previous lemma, this construction is done inductively. To begin with, there is a unique open cell e^k in X to which x belongs. First choose a contractible neighbourhood U_k of x in e^k so that the closure $\bar{U}_k$ is contained in $V \cap e^k$. For $n > k$ having constructed a neighbourhood U_n so that the closure $\bar{U}_n \subset X^{(n)} \cap V$ is compact and such that U_{n-1} is a SDR of U_n, the inductive step is carried out as follows. Let Λ be the indexing set of all $(n+1)$-cells. For each $\alpha \in \Lambda$, it follows that $\phi_\alpha^{-1}(\bar{U}_n)$ is a compact subset of $\mathbb{S}^n$ contained in the open set $\phi_\alpha^{-1}(V)$. Therefore (by Wallman's theorem) there exists $0 < \epsilon(\alpha) < 1$ such that

$$\overline{N_{\epsilon(\alpha)}(\phi_\alpha^{-1}(\bar{U}_n))} \subset \phi_\alpha^{-1}(V).$$

(See, Remark 2.2.4.) With this choice of $\epsilon : \Lambda \longrightarrow (0,1)$, let $U_{n+1} = N_\epsilon(U_n)$ as defined in Proposition 2.2.5. Now we take $U = \cup_n U_n$. From Lemma 2.4.2, it follows that U is a contractible neighbourhood of x in V. ♠

Lemma 2.4.4 Let X be obtained from Y by attaching k-cells, then $X \times 0 \cup Y \times \mathbb{I}$ is a strong deformation retract of $X \times \mathbb{I}$.

Proof: From Example 1.5.11, we know that $\mathbb{D}^n \times 0 \cup \mathbb{S}^{n-1} \times \mathbb{I}$ is a SDR of $\mathbb{D}^n \times \mathbb{I}$. Therefore, a similar statement holds for the disjoint union of a family of copies of them. Now $X \times \mathbb{I}$ can be thought of as the adjunction space of the map $f : \sqcup_\alpha D^n_\alpha \times \mathbb{I} \longrightarrow Y \times \mathbb{I}$ where $f(x,t) = (f_\alpha(x),t)$ for $x \in D^n_\alpha$. Now use Exercise 1.9.9, to see that $X \times 0 \cup Y \times \mathbb{I}$ is a SDR of $X \times \mathbb{I}$. ♠

Theorem 2.4.5 *If (X,A) is a relative CW-complex, then $A \hookrightarrow X$ is a cofibration.*

Proof: By Proposition 1.6.1, it is enough to see that $X \times 0 \cup A \times \mathbb{I}$ is a retract of $X \times \mathbb{I}$. Applying the above lemma successively to various skeletons of X, we get retractions $r_k : X^{(k)} \times \mathbb{I} \longrightarrow X^{(k)} \times 0 \cup X^{(k-1)} \times \mathbb{I}$. Now define $r : X \times \mathbb{I} \longrightarrow X \times 0 \cup A \times \mathbb{I}$ by the formula $r|_{X^{(k)} \times \mathbb{I}} = r_0 \circ r_1 \circ \cdots \circ r_k$. Verify that r is well defined continuous and is identity on $X \times 0 \cup A \times \mathbb{I}$. ♠

Example 2.4.6 Let $A = \Omega_0$ be the first uncountable ordinal (See [Joshi, 1983] for details) with the order topology. For each $\alpha \in A$, define $\phi_\alpha : \mathbb{S}^0 \to A$ by the rule $\phi_\alpha(-1) = \alpha, \phi_\alpha(1) = \alpha + 1$ where $\alpha + 1$ denotes the immediate successor to α in the well order on A. Let (X,A) be the 1-dimensional relative CW-complex obtained by attaching 1-cells $\{e^1_\alpha\}$ with $\{\phi_\alpha\}$ as attaching maps. It is easily seen that as a set, X can be identified with the set $\Omega_0 \times [0,1)$. With the usual order on the open interval $[0,1)$ coming from real numbers, we can then give $X = \Omega_0 \times [0,1)$, the lexicographic ordering and then take the order topology $\mathcal{O}$ on it. This gives the so-called *long-line* (see [Shastri, 2011] Example 5.1.1) which is a connected '1-dimensional Hausdorff manifold' except that it fails to satisfy the II-countability condition. Check that the adjunction space topology on X (which is the same as CW-topology) is finer than $\mathcal{O}$ (look at the neighbourhoods of $\alpha \in \Omega_0$ where α is an infinite ordinal). Indeed, in the adjunction space topology, X is connected but not locally connected. Nor it is path connected.

Exercise 2.4.7

(i) Show that a CW-complex X is locally compact iff each closed cell meets finitely many other closed cells. CW-complexes satisfying this condition are called locally finite. (Compare corollary 2.2.17.)

(ii) If X and Y are CW-complexes then show that

$$\pi_n(X \times_w Y) \cong \pi_n(X \times Y) \cong \pi_n(X) \times \pi_n(Y), \text{ for all } n \geq 1.$$

2.5 Cellular Maps

In mathematical studies, the study of objects always goes hand in hand with the study of appropriate maps between them. In this section, we introduce the cellular maps from one CW-complex to another, thereby completing the introduction to the category of CW-complexes. The most important property of cellular maps stems from the fact that every continuous map can be approximated by a cellular map. This result goes under the name 'cellular approximation theorem' and is a part of the tool-kit in the study of algebraic topology. We shall indicate only a few applications here.

Definition 2.5.1 Let $f : (X, A) \to (Y, B)$ be a continuous function of CW-pairs. We say f is cellular, if $f((X, A)^{(q)}) \subset (Y, B)^{(q)}$ for all $q \geq 0$.

Remark 2.5.2 Thus a cellular map takes each q-skeleton of the domain inside the q-skeleton of the codomain. It is easy to check that the composite of cellular maps is again cellular and hence CW-complexes and cellular maps form a category as mentioned in Section 1.8.

Theorem 2.5.3 (Cellular approximation theorem) *Let $f : (X, A) \to (Y, B)$ be any continuous map of CW-pairs. Suppose (X', A) is a subcomplex on which f is cellular. Then there exists a cellular map $g : (X, A) \to (Y, B)$ such that $g = f$ on (X', A) and $g \simeq f$ (rel X').*

We shall give a proof of this based on smooth approximations and Sard's theorem which you may have learnt in your calculus course (see [Shastri, 2011] Theorem 1.7.2 and 2.2.1). In Section 2.5, we shall indicate how to modify this proof so that we can use the simplicial approximation theorem in place of calculus.[1] The basic idea is in the following lemma:

Lemma 2.5.4 Let $\alpha : (\mathbb{D}^n, \mathbb{S}^{n-1}) \to (Y, B)$ be a continuous map, where Y is got by attaching a single cell e^m to B. If $m > n$, then there is a homotopy $H : \mathbb{D}^n \times \mathbb{I} \to Y$ such that $H(x, 0) = \alpha(x), x \in \mathbb{D}^n, H(x, t) = \alpha(x), x \in \mathbb{S}^{n-1}, 0 \leq t \leq 1$, and $H(x, 1) \in B$, for all $x \in \mathbb{D}^n$.

Proof: Suppose the image of α misses a point $z \in \text{int } e^m$. If $r : Y \setminus \{z\} \to B$ is a strong deformation retraction (which exists), then α is homotopic to $r \circ \alpha$ and H can be taken to be such a homotopy. Therefore, it is enough to prove that there is a homotopy H of α to a map α_1 relative to $\mathbb{S}^{n-1}$ such that image of α_1 misses a point in the interior of e^m. Let $\phi : \mathbb{D}^m \to Y$ be the characteristic map of e^m. Choose $0 < r < 1/5$ and put

$$B_j = jr\mathbb{D}^m, \quad C_j = \phi(B_j), \quad K_j = \alpha^{-1}(C_j), \quad j = 1, 2, 3, 4.$$

We then have compact subsets $C_1 \subset \cdots \subset C_4 \subset \text{int } e^m$ and $K_1 \subset \cdots \subset K_4 \subset \text{int } \mathbb{D}^n$. Let $\beta = \phi^{-1} \circ \alpha : K_4 \to \mathbb{D}^m$. Let $\eta : \mathbb{D}^n \to \mathbb{I}$ be a smooth map such that $\eta \equiv 1$ on K_2 and $\equiv 0$ outside $\text{int } K_3$. Let $\gamma : \text{int } K_4 \to \mathbb{D}^m$ be a smooth r-approximation to β. Consider the homotopy $t \mapsto (1 - t\eta)\beta + t\eta\gamma$ from β to $\beta_1 = (1 - \eta)\beta + \eta\gamma$. On K_2, β_1 is equal to γ and outside K_3, it is equal to β. Hence β_1 can be extended over all of K_4 by β. Indeed, the same

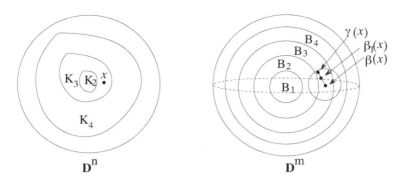

FIGURE 2.15. Homotoping away from a point of B_1

holds for the entire homotopy as well and hence we get a homotopy of β with the map β_1 on K_4. (See Figure 2.15.)

We claim that $\beta_1(K_4)$ does not contain some points in $r\mathbb{D}^m$. First of all, any point x outside K_2 is mapped by β outside of the ball $2r\mathbb{D}^m$. Since γ is a r-approximation to β, $\gamma(x)$ is in a ball of radius r with centre $\beta(x)$ which misses the ball $r\mathbb{D}^m$. But then $\beta_1(x)$ belongs to the line segment $[\beta(x), \gamma(x)]$ and hence cannot be in $r\mathbb{D}^m$. Therefore, $\beta_1(K_4) \cap r\mathbb{D}^m = \beta_1(K_2)$. Now β_1 is a smooth map on K_2 and since $n < m$, by Sard's theorem $\beta_1(K_2)$ is of measure zero. Therefore, $r\mathbb{D}^m \setminus \beta_1(\mathbb{D}^n) = r\mathbb{D}^m \setminus \beta_1(K_2) \neq \emptyset$.

Composing back with ϕ we obtain a homotopy $G : K_4 \times \mathbb{I} \to Y$ of α with a map $\alpha' = \phi \circ \beta_1$ which is constant outside K_3 and hence can be extended by α outside K_4. This however does not affect its image inside $\phi(r\mathbb{D}^m)$, i.e., $\alpha'(\mathbb{D}^n)$ misses a point of int e^m. ♠

Proof of Theorem 2.5.3: For simplicity put $X^{(k)} = (X, X')^{(k)}$; $Y^{(k)} = (Y, B)^{(k)}$. We shall construct inductively a family $H_n : X \times \mathbb{I} \to (Y, B)$ such that
(i) $H_0(x, 0) = f(x), x \in X$;
(ii) $H_n(x, 0) = H_{n-1}(x, 1), x \in X, n \geq 1$;
(iii) $H_n(x, t) = H_{n-1}(x, 1)$ for all $x \in X^{(n-1)}, 0 \leq t \leq 1$;
(iv) $H_n(X^{(n)} \times 1) \subset Y^{(n)}$.

Then as in the proof of Lemma 2.4.2, we can piece them together to get a homotopy H as required, viz.,

$$H(x, t) = \begin{cases} H_{n-1}(x, (n+1)(nt-n-1)), & \frac{n-1}{n} \leq t \leq \frac{n}{n+1}; \\ H_n(x, 1), & x \in X^{(n)}. \end{cases}$$

Note that $X^{(-1)} = (X', A)$ and we take H_0 to be identically f on it. If $x \in X^{(0)} \setminus X^{(-1)}$, choose a path ω_x from $f(x)$ to some point in $Y^{(0)}$ and define $H_0(x, t) = \omega_x(t)$. This gives us a map $X \times 0 \cup X^{(-1)} \times \mathbb{I} \to Y$. Since $X^{(-1)} \subset X$ is a cofibration, this map extends to $H_0 : X \times \mathbb{I} \to Y$ as required. Inductively suppose we have defined H_{n-1} with properties as specified. For each n-cell e^n in $X^{(n)}$, we need to define homotopies $h_n : e^n \times \mathbb{I} \to Y$ such that $h_n|_{\partial e^n \times \mathbb{I}} = H_{n-1}|_{\partial e^n \times \mathbb{I}}$, $h_n(x, 0) = H_{n-1}(x, 0)$ and $h_n(e_n \times 1) \subset Y^{(n)}$. For once we have done this for each n-cell in $X^{(n)}$, we can put all h_n together, use the cofibration property of $X^{(n)} \subset X$ and get the map H_n as desired.

---

[1]A teacher/student who does not like to use smooth approximation theorem may post-pone the proof of this theorem till Section 2.9.

Now since $H_{n-1}(e^n \times 1)$ is a compact set it meets the interior of only finitely many cells in Y. If all these cells are of dimension $\leq n + 1$, there is nothing to prove. So, suppose e^m is a cell, $m > n$ and $H_{n-1}(e^n \times 1)$ meets the interior of e^m.

Composing with the characteristic map of the n-cell e^n in X and taking $Z = Y \setminus \text{int } e^m$ makes this problem precisely the same as in Lemma 2.5.4. Therefore we can find a homotopy of $H_{n-1}(-, 1)$ so as to avoid the interior of e^m. The required homotopy h_n is then the composite of all these finitely many homotopies.

Putting all these h_n together and using the cofibration property as mentioned, this completes the definition of H_n and thereby the proof of Theorem 2.5.3. ♠

Remark 2.5.5 Since composite of cellular maps is cellular, there is a category **CW** of CW-complexes and cellular maps. This category is of prime importance from the point of view of algebraic topology. We shall meet CW-complexes all over this book.

Following the important Lemma 2.5.4 we now make a definition.

Definition 2.5.6 Let $n \geq 0$ be an integer. A topological pair (X, A) is said to be n-connected, if for each $0 \leq k \leq n$, every map $f : (\mathbb{D}^k, \mathbb{S}^{k-1}) \to (X, A)$ is homotopic relative to $\mathbb{S}^{k-1}$ to a map $g : \mathbb{D}^k \to A$. (Here by convention, for $n = 0$, $\mathbb{D}^0$ is a singleton space and $\mathbb{S}^{-1} = \emptyset$.) In particular, if $A = \{x_0\}$, where x_0 is the base point then we say X is n-connected.

Remark 2.5.7 Note that a based space is n-connected iff it is path connected and every map $\alpha : \mathbb{S}^k \to X$ is null-homotopic for every $1 \leq k \leq n$. (See Theorem 1.5.5.)

Theorem 2.5.8 *The pair* $(\mathbb{D}^{n+1}, \mathbb{S}^n)$ *and the space* $\mathbb{S}^{n+1}$ *are* n-*connected, for all* $n \geq 0$.

Proof: Each $\mathbb{S}^{k-1}$ is a CW-complex with a single 0-cell and a single $(k-1)$-cell. Each $\mathbb{D}^k$ is a CW-complex of dimension k obtained by attaching a single n-cell to the CW-complex $\mathbb{S}^{k-1}$. The theorem follows immediately from the Lemma 2.5.4. ♠

The CW-approximation theorem has many applications. Here is just a sample.

Theorem 2.5.9 *Let X be a connected CW-complex. Then the inclusion induced homomorphism $\iota_\# : \pi_1(X^{(2)}) \to \pi_1(X)$ is an isomorphism.*

Proof: We choose one of the vertices $\star$ as the base point for X. Given a map $\alpha : (\mathbb{S}^1, 1) \to (X, \star)$, by the cellular approximation theorem, there is a homotopy $(\mathbb{S}^1, 1) \times \mathbb{I} \to (X, \star)$ relative to 1 of α to a map $\beta : (\mathbb{S}^1, 1) \to (X^{(1)}, \star)$. This clearly implies that $\iota_\#([\beta]) = [\alpha]$ and hence the surjectivity of $\iota_\#$. The injectivity is proved similarly, by starting with a map $f : (\mathbb{D}^2, \mathbb{S}^1, 1) \to (X, X^{(1)}, \star)$. ♠

Exercise 2.5.10

(i) Show that a CW-complex X is connected iff its 1-skeleton $X^{(1)}$ is connected.

(ii) Let X be a locally finite CW-complex which is connected. Show that X has countably many cells.

2.6 Abstract Simplicial Complexes

We shall now take up the study of spaces which are built-up from convex polyhedrons. We are all familiar with the concept of a polygonal curve in a Euclidean space. The spaces that we are going to consider now are unions of convex polyhedrons, two of them intersecting along some lower dimensional common face which may be empty as well. Such a space is called a *polyhedron* and the study of such spaces is called *polyhedral topology*. The important point here is that we need not necessarily begin with a total space which is a subspace of some Euclidean space and would like to work in a more abstract set-up. On the other hand, to keep the discussion simple, we shall use, as building blocks, only the simplest of the convex polyhedrons called *simplices,* viz., those which have their vertex-set affinely independent. The result of such a restriction is that, to a large extent, the topological study becomes elementary combinatorics. This is reflected right in the abstract definition with which we begin. In this section, we shall introduce the concept of abstract simplicial complex. The reader should wait till the next section for the underlying geometric motivations.

Definition 2.6.1 By a *simplicial complex* K we mean a pair $(V, \mathcal{S})$, where V is a set and $\mathcal{S}$ is a collection of finite subsets of V such that
(i) $\forall\ v \in V,\ \{v\} \in \mathcal{S}$;
(ii) $F \in \mathcal{S}$, and $F' \subset F \Longrightarrow F' \in \mathcal{S}$.

Elements of V are called *vertices* of K, and those of $\mathcal{S}$ are called *simplices* of K. If $F \in \mathcal{S}$ and $\#(F) = q + 1$, then F is called a *q-simplex* (or a simplex of *dimension* q). Thus the vertices of K are also the 0-simplices of K. If $F' \subset F$ for $F \in \mathcal{S}$, then F' is called a *face* of F. Often a simplex is displayed by enumerating its 0-faces. Also, by abuse of notation, we may simply say $F \in K$ when we actually mean that F is a face of K, i.e., $F \in \mathcal{S}$. Observe that we have allowed the empty subset also as a simplex in every simplicial complex. This is rather unusual in topology, but quite a convenient convention in a combinatorial set-up. We define the dimension of the empty face to be -1. If V is a finite set, then K is called a finite simplicial complex.

The *dimension,* $\dim K$, of a simplicial complex is defined to be equal to the supremum of n such that K has a n-simplex. Thus, if K is finite then $\dim K < \infty$. A simplex in K is said to be *maximal* if it is not contained in another larger simplex. If the dimension of K is $n < \infty$ then there are simplices in K of dimension n and all simplices of dimension n are maximal. However, not all maximal simplices need be of dimension n. If this happens such a simplicial K is called a *pure* simplicial complex. This concept is very useful in combinatorial algebra.

By a *simplicial map* $\varphi : K_1 \to K_2$ from one simplicial complex to another, we mean a set theoretic function on the vertex sets, $\varphi : V_1 \to V_2$, such that for each simplex F in K_1, $\varphi(F)$ is a simplex in K_2. Composite of simplicial maps is defined in an obvious way. There is a category of simplicial complexes and simplicial maps.

By a *simplicial isomorphism* we mean a simplicial map with a simplicial inverse. Note that a simplicial isomorphism is a bijection on the vertex set. However, the converse is not true, viz., any simplicial map which is a bijection on the vertex set need not be a simplicial isomorphism (exercise).

By a *subcomplex* $K' = (V', \mathcal{S}')$ of a simplicial complex K, denoted by $K' \subset K$, we mean a simplicial complex K', such that $V' \subset V$ and $\mathcal{S}' \subset \mathcal{S}$). Note that in this situation, the inclusion map $K' \subset K$ is a simplicial map.

Example 2.6.2

(i) Given a set V, let S be the set of all finite subsets of V. Then (V, S) is a simplicial complex. This simplicial complex contains as subcomplexes, all simplicial complexes whose vertex set is a subset of V. The isomorphism type of this simplicial complex depends only on the cardinality of V.

(ii) Consider the special case where V itself is finite with $n + 1$ elements. We get an important simplicial complex whose simplexes are all subsets of V. More specifically, if this set V consists of the $n+1$ basic unit vectors $\{\mathbf{e}_1, \ldots, \mathbf{e}_{n+1}\}$ in $\mathbb{R}^{n+1} \subset \mathbb{R}^\infty$, we then denote the corresponding simplicial complex by the symbol Δ_n. Recall that we had already used this symbol to denote the convex hull of these $n+1$ points. This over-use of the same symbol is deliberate. These simplicial complexes are the building blocks of all other simplicial complexes. This point will ne explained time and again.

(iii) Any simplicial complex of finite dimension can be described by declaring all the maximal simplices in it, viz., take the collection of all subsets of all of these maximal simplices. This is especially effective while describing a finite simplicial complex—we need to merely list all its maximal simplices.

(iv) If F is a simplex of a simplicial complex, then the set of all faces of F forms a simplicial subcomplex denoted also by F. The set of all proper faces of F also forms a subcomplex denoted by $\mathcal{B}(F)$; clearly $\mathcal{B}(F) \subset F$. If the dimension of F is n then F is a carbon copy of Δ_n. This should already justify to some extent, the claim that the simplicial complexes Δ_n's are the building blocks of other simplicial complexes.

(v) If K is a simplicial complex, define $K^{(q)}$, its *q-dimensional skeleton,* to be the simplicial complex consisting of all p-simplices of K for all $p \le q$. If F is a q-dimensional face of K then $\mathcal{B}(F)$ is the $(q-1)$-skeleton of the subcomplex F.

(vi) Let $K_i = (V_i, \mathcal{S}_i)$, $i = 1, 2$ be any two simplicial complexes. Then the *join* $K_1 * K_2 = (V_1 \amalg V_2, \mathcal{S})$ is defined by taking

$$\mathcal{S} = \{F_1 \cup F_2 \ : \ F_i \in \mathcal{S}_i, \ i = 1, 2\}.$$

In particular, if K_2 is a singleton set $\{v\}$, then the join $K_1 * \{v\} =: K_1 * v$ is called a *cone* over K_1. Note that $\dim(K_1 * K_2) = \dim K_1 + \dim K_2 + 1$.

(vii) **Nerve of a covering:** Let $\mathcal{U}$ be a collection of non empty subsets of a non empty set X. We get a simplicial complex $K(\mathcal{U})$ by taking $\mathcal{U}$ as the set of vertices and finite subsets $\{U_0, \ldots, U_k\} \ : \ U_i \in \mathcal{U}$ with the property $\cap_{i=1}^k U_i \ne \emptyset$ as k-simplices. This simplicial complex is called the *nerve* of $\mathcal{U}$. When $\mathcal{U}$ happens to be an open cover for a topological space X, its nerve $K(\mathcal{U})$ plays a central role in topological dimension theory, which we shall not discuss. (See [Hurewicz–Wallman, 1948] or exercises in Chapter 3 of [Spanier, 1966].) For us, it is important because we are going to use them in the study of Čech cohomology.

(viii) Many interesting examples of simplicial complexes arise while studying various mathematical problems. For instance, somewhat dual to the above example, let V be the set of k-subsets of a $(k+n)$-set and

$$\mathcal{S} = \{\{v_0, ..., v_r\} \ : \ v_i \in V \text{ and } v_i \cap v_j = \emptyset \text{ for each } i \ne j\}.$$

Then $K = (V, S)$ is a simplicial complex which arises in the study of Kneser's conjecture (see [Lovasz, 1978]).

2.7 Geometric Realization of Simplicial Complexes

We are familiar with the notion of a graph and drawing diagrams of them by placing some dots and joining these dots by some lines. The dots represent the vertices. And two vertices v_1, v_2 are joined by a line if $\{v_1, v_2\}$ is an edge in the graph. Now consider a 1-dimensional simplicial complex K. To keep the discussion simple, assume that the number of vertices is finite. We can then select as many distinct points in some Euclidean space $\mathbb{R}^N$ to represent distinct vertices of K. We now join those pairs of vertices v_i, v_j for which the set $\{v_i, v_j\}$ is a simplex of K by the line segment between v_i and v_j. The only snag in this is that two such line segments may intersect each other at points other than the vertices. For the moment, we shall assume that we can avoid this. So far, the difference between the representation of a graph and the representation of K is only in the fact that the edges have to be straight line segments in the latter case. (This seems to be rather a peripheral issue and let us come back to it later.) In any case, given K we have been able to assign a collection of line segments in a Euclidean space, so that the line segments meet in a specific manner only at their end-points.

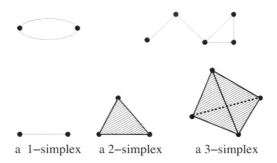

a 1–simplex   a 2–simplex   a 3–simplex

FIGURE 2.16. Graphs and simplices

Now suppose K has 2-dimensional simplices also. Say $\{v_1, v_2, v_3\}$ is such a 2-simplex. We then take care that the three vertices are not chosen on a straight line. This then enables us to fill up the triangle formed by the three vertices. We do this to all the 2-simplices in K. We have to ensure that two distinct triangles do not overlap. This gives us a subspace of the Euclidean space which is the union of a number of triangles and line segments. The Figure 2.16 depicts a graph which is not a simplicial complex, a graph, and also a 1-simplex, a 2-simplex and a 3-simplex.

This idea can be easily generalized to any finite simplicial complex. The problem is that there are too many choices involved: the choice of the Euclidean space to begin with and then the choice of vertices. As such we are not even sure whether we can always do this successfully. Let us see how we can avoid some of these ambiguities.

Observe that in any Euclidean space, the moment two points p, q are given, the line segment $[p, q]$, viz., the set of points $tp + (1 - t)q, 0 \leq t \leq 1$, is well defined. If three non collinear points p, q, r are given then they define a triangle, viz., it is the set of points

$$\{\alpha p + \beta q + \gamma r \; : \; 0 \leq \alpha, \beta, \gamma \leq 1, \alpha + \beta + \gamma = 1.\}$$

Observe that there is a one-one correspondence between this set and the set

$$\{(t_1, t_2, t_3) \in \mathbb{I}^3 \; : \; \sum_i t_i = 1\}.$$

Thus a point in the triangle can be thought of as a function

$$t : \{1, 2, 3\} \longrightarrow \mathbb{I}$$

such that $t(1) + t(2) + t(3) = 1$. These are some of the ideas that go into making up the following 'abstract' definition of the geometric realization of a simplicial complex, which, indeed, removes all ambiguities involved and brings functoriality.

Definition 2.7.1 Let $K = (V, \mathcal{S})$ be a simplicial complex. Let $|K|$ denote the set of all functions $\alpha : V \to \mathbb{I}$ such that
(i) *supp* $\alpha := \{v \in V : \alpha(v) \neq 0\}$ is a simplex in K.
(ii) $\sum_{v \in V} \alpha(v) = 1$.

Note that by (i), the summation in (ii) involves only finitely many non zero terms and hence makes sense. The set of all functions $\alpha : V \longrightarrow \mathbb{I}$ can be identified with the Cartesian product $\mathbb{I}^V$, and given the product topology where each copy of $\mathbb{I}$ is given the usual topology. Since $|K| \subset \mathbb{I}^V$, this gives us ideas for topologising $|K|$.

At this stage note that we have two definitions of $|\Delta_n|$: the first one being a mere notation for a topological subspace of $\mathbb{R}^{n+1}$ as given in Section 2.1 and the second one given by the above definition, where we treat Δ_n as a simplicial complex. Check that the two definitions coincide.

The special case when K is finite Suppose K is a finite simplicial complex with $\#(V) = N$. We then simply take the subspace topology on $|K|$ induced from $\mathbb{I}^V = \mathbb{I}^N$ and call it the *geometric realization* of K. If $V = \emptyset$, then by convention, $|K| = \emptyset$. Verify that $|K|$ is a closed subset of $\mathbb{I}^N$ (exercise) and so is compact.

For each $F \in \mathcal{S}$, the *closed simplex* $[F]$ is defined to be the subspace

$$[F] := \{\alpha \in |K| : \alpha(v) = 0, \ \forall \ v \in K \setminus F\} = \{\alpha \in |K| : \operatorname{supp} \alpha \subset F\}.$$

Clearly $[F]$ is a closed subset of $|K|$. Note that $[F]$ has a natural convex structure coming from that of $\mathbb{I}^V$: if $\alpha, \beta \in [F]$ and $0 \leq t \leq 1$, then $t\alpha + (1-t)\beta$ is again an element of $[F]$. We may identify $[F]$ with the convex subset of $\mathbb{I}^F$ defined by condition (ii). The relative interior of $[F]$ is given by

$$\operatorname{int}([F]) = \{\alpha \in |K| : \alpha(v) \neq 0 \text{ iff } v \in F\}.$$

The boundary of $[F]$ is then given by $[F] \setminus \operatorname{int}[F]$ and is the union of all $[G]$ such that G is a proper face of F.

Moreover, $[F]$ is homeomorphic to the geometric realization $|F|$, of the simplicial complex F. Each map $\alpha : F \to \mathbb{I}$ can be identified with a map $\alpha : V \to \mathbb{I}$ which is zero outside of F, yielding the embedding $[F] \to |K|$ onto the subspace $|F|$.

Using this property, we shall get rid of the notation $[F]$ and write $|F|$ instead to mean the closed simplex underlying F as well.

Further if $F_1, F_2 \in K$ then $|F_1| \cap |F_2| = |F_1 \cap F_2|$ and the two subspace topologies on $|F_1 \cap F_2|$ from $|F_1|$ and $|F_2|$ coincide. Also, the family of closed sets

$$\{|F| : F \text{ is a face of } K\}$$

is a finite covering for $|K|$. Clearly, $G \subset |K|$ is closed iff $G \cap |F|$ is closed in $|F|$ for every $F \in K$. This is the same as saying that the topology on $|K|$ is the weak topology with respect to the above covering. This property gives us an idea how to get a good topology on $|K|$ when K is infinite.

The case when K is not finite: We then take the topology on $|K|$ to be the weak topology with respect to the closed simplices $\{|F|, F \in K\}$: a set $A \subset |K|$ is closed if and only if $A \cap |F|$ is closed in $|F|$, $\forall F \in K$.

Remark 2.7.2 Note that even if K is finite, the above definition holds good, because the collection $\{|F|, \; : \; F \in \mathcal{S}\}$ forms a finite closed covering. Why then we do not simply take the subspace topology on $|K|$ from the product topology of $\mathbb{R}^{\#(V)}$ even when V is infinite? One good reason is that constructing continuous functions on $|K|$ becomes simpler and is coherent with our theme that the simplices are the building blocks of simplicial complexes. Indeed we have:

Definition 2.7.3 By a *triangulation* of a topological space X we mean a pair (K, f) where K is a simplicial complex and $f : |K| \to X$ is a homeomorphism. If X has a triangulation then it is called a *triangulable space*. If we select a specific triangulation on it then we call it a *simplicial polyhedron* (or a *polyhedron*).

Definition 2.7.4 A CW-complex is called regular, if the attaching map of each cell in it is an embedding (equivalently, the characteristic map of each cell is an embedding).

Theorem 2.7.5 *Any triangulation (K, f) of a topological space X defines a regular CW-structure on X such that the nth-skeleton is given by $X^{(n)} = f(|K^{(n)}|), n \geq 0$.*

Proof: Identify X with $|K|$ and $X^{(k)}$ with $|K^{(k)}|$ for each k, via f. It is enough to show that $|K| = \cup_n |K^{(n)}|$ defines a CW-structure on $|K|$.

Fix a total order on the vertex set V of K and fix homeomorphisms $h_n : \mathbb{D}^n \to |\Delta_n|$, for each $n \geq 1$. For each $F \in S$, let $\psi_F : |\Delta_n| \to |F|$ be the simplicial isomorphism given by the vertex map which is order preserving. Put $\phi_F = \iota \circ \psi_F \circ h_n : \mathbb{D}^n \to |K^{(n)}|$. It is not difficult to see that $|K^{(n)}|$ is obtained by attaching the cells $\{|F| \; : \; F \in S, \; \#F = n+1\}$ to $|K^{(n-1)}|$ with $\phi_F|_{\partial \mathbb{D}^n}$ as attaching maps and ϕ_F as characteristic maps. The fact that the topology on $|K|$ is the CW-topology follows from Proposition 2.2.12 and the definition of the topology on $|K|$.

Finally, since each $|F| \subset |K|$ is homeomorphic to $\mathbb{D}^n$, it follows that $|K|$ is a regular CW-complex. ♠

Remark 2.7.6

(1) Thus, all topological properties of CW-complexes are shared by polyhedrons. Note that in the proof the CW-structure depends on the choice of a total order on the vertex set. However, any two of these structures are related by a cellular homeomorphism. In particular, since a polyhedron, being a CW-complex, is compactly generated, locally contractible, Hausdorff, normal, paracompact, etc., we see that a topological space has to satisfy a lot of restrictions in order to be triangulable, which is the same as being a polyhedron. However, the class of triangulable spaces is quite large and encompasses many spaces that we come across in mathematics. For instance, all regular CW-complexes are triangulable. Regularity is not quite necessary for triangulability (see [Lundell–Weingram, 1969]). All algebraic varieties, analytic sets etc. are triangulable (see [Hardt, 1977]).

(2) The reader and the teacher who have chosen to skip Sections 2.2 and 2.3 are now advised to stick to finite simplicial complexes K so that the topology on $|K|$, being a closed subspace of a Euclidean space is familiar to them. They can then go through the rest of this chapter without any hurdles. We shall record one of the most important properties below.

Theorem 2.7.7 *Let X be a topological space. A function $f : |K| \to X$ is continuous if and only if the restriction map $f|_{|F|} : |F| \to X$ is continuous $\forall F \in K$. Likewise a function $H : |K| \times \mathbb{I} \to X$ is continuous iff $H|_{|F| \times \mathbb{I}}$ is continuous for every face $F \in K$.*

Definition 2.7.8 Let $K_i, i = 1, 2$ be simplicial complexes with V_i as their vertex sets. Given a simplicial map $\varphi : K_1 \to K_2$, we define $|\varphi| : |K_1| \to |K_2|$ as follows :

$$|\varphi|(\alpha)(v_2) = \sum_{\varphi(v_1) = v_2} \alpha(v_1).$$

Note that, given $\alpha \in |K_1|$, though $\{v_1 \; : \; \varphi(v_1) = v_2\}$ may be infinite, $\alpha(v_1) \neq 0$ for only finitely many $v_1 \in V_1$. Therefore the summation on the right above is finite.

Also
$$supp \, |\varphi|(\alpha) = \varphi(supp \, \alpha)$$

is a simplex of K_2. Hence $|\varphi|$ is well defined.

Remark 2.7.9
(1) Restricted to any closed simplex $|F|$, it is easily seen that $|\varphi|$ is *an affine linear map,* i.e.,
$$|\varphi|(t\alpha + (1 - t)\beta) = t|\varphi|(\alpha) + (1 - t)|\varphi|(\beta), \; (\alpha, \beta \in |F|, \; 0 \leq t \leq 1).$$

In particular, $|\varphi||_F$ is continuous. Hence $|\varphi| : |K_1| \to |K_2|$ is continuous.
(2) One easily checks that if $\phi : K_1 \longrightarrow K_2, \psi : K_2 \longrightarrow K_3$ are simplicial maps, then

$$|\psi \circ \phi| = |\psi| \circ |\phi|; \quad |Id| = Id.$$

Thus the geometric realization defines a covariant functor $K \rightsquigarrow |K|$ from the category of simplicial complexes and simplicial maps to the category of topological spaces and continuous maps. Also note that if $L \subset K$, then $|L| \subset |K|$. In fact, this functor takes values inside the subcategory of CW-complexes and cellular maps.

Example 2.7.10

(i) The convex hull of the points $\{\mathbf{e}_0, \ldots, \mathbf{e}_n\} \subset \mathbb{R}^\infty$ is easily seen to be the geometric realization of the standard simplex Δ_n introduced earlier. We shall from now on use the symbol $|\Delta_n|$ to denote this subspace of $\mathbb{R}^\infty$ also.

(ii) All spheres and discs are triangulable. For $n = 0$ clearly both $\mathbb{D}^0$ and $\mathbb{S}^0$ being a singleton and a double-ton (respectively) are triangulable. The simplest polyhedron is $|F|$ for any q-simplex F. We shall now identify it topologically. By choosing an order on the vertex set $F = \{v_0, \ldots, v_q\}$, elements $\alpha \in |F|$ can be written as $(\alpha_0, \ldots, \alpha_q)$, where $\alpha_i = \alpha(v_i)$. Thus $|F|$ is a subspace of $\mathbb{I}^{q+1}$ consisting of elements $(\alpha_0, \ldots, \alpha_q)$ with $\sum \alpha_i = 1$. It is not difficult to see that this is homeomorphic to the unit disc $\mathbb{D}^q$ in $\mathbb{R}^q$. (For instance, first shift the origin to centroid $(1/q + 1, \ldots, 1/q + 1)$. Now rotate through an angle of $\pi/4$ so that the complex is now mapped inside $\mathbb{R}^q \times 0$. Finally use the map $\mathbf{x} \mapsto \mathbf{x}/\|\mathbf{x}\|$ to get a homeomorphism of $|\mathcal{B}(F)|$ with $\mathbb{S}^{q-1}$ and extend it over the interior of the cells via cone construction.) Furthermore the subspace $|\mathcal{B}(F)|$ of $|F|$ goes homeomorphically onto $\mathbb{S}^{q-1} = \partial \mathbb{D}^q$ under this homeomorphism.

Thus we have indeed shown that the pair $(\mathbb{D}^q, \mathbb{S}^{q-1})$ is homeomorphic to $(|F|, |\mathcal{B}(F)|)$, i.e., a polyhedral pair.

(iii) Here is another interesting way to triangulate the spheres. We have just seen that the topological cone over $\mathbb{S}^{n-1}$ is homeomorphic to $\mathbb{D}^n$. It follows that if (K, f) triangulates X then this triangulation extends to a triangulation $(K * \{\star\}, F)$ of the cone CX over X. Since the suspension can be thought of as a double cone, it follows that we get a triangulation (SK, Sf) of SX.

Now recall that $\mathbb{S}^n$ can be thought of as a n-fold suspension of $\mathbb{S}^0$. In particular, beginning with the obvious triangulation of $\mathbb{S}^0$, by taking successive suspension, we obtain a triangulation of $\mathbb{S}^n$. We shall refer to this triangulation of $\mathbb{S}^n$ by *S-triangulation.* It is worthwhile to note that the antipodal map $\alpha : \mathbb{S}^n \longrightarrow \mathbb{S}^n$ (i.e., given by $\alpha(x) = -x$) is a simplicial isomorphism with respect to the S-triangulation.

(iv) Consider the following 1-dimensional simplicial complex K whose vertex set is the set of integers $\mathbb{Z}$ and whose edges are $\{n, n+1\}, n \in \mathbb{Z}$. The map which sends the vertex $n \in K$ to the integer n extends linearly to define a continuous map $f : |K| \to \mathbb{R}$. Check that f is a homeomorphism. This way we obtain a triangulation of $\mathbb{R}$. (Why can we not choose K to be a finite simplicial complex here?)

(v) **Join of two complexes** Recall how we have defined the topological join $X * Y$ of two spaces X, Y from Exercise 1.9.27. If K_1, K_2 are any two simplicial complexes, then $|K_1 * K_2|$ is homeomorphic to $|K_1| * |K_2|$. Let us just construct one homeomorphism: Let
$$\phi : |K_1| \times \mathbb{I} \times |K_2| \to |K_1 * K_2|$$
be defined by $(\alpha, t, \beta) \mapsto (1-t)\alpha + t\beta$. Clearly, $\phi(\alpha, 0, \beta) = \alpha$ and $\phi(\alpha, 1, \beta) = \beta$ for all α, β. Therefore, ϕ factors through the quotient and defines a map $\hat{\phi} : |K_1| * |K_2| \to |K_1 * K_2|$. Verify that this is a continuous bijection. If $F_j \subset K_j$ are some faces, then it is easily verified that the restriction map $\phi : |F_1| * |F_2| \to |F_1 * F_2|$ is a homeomorphism. From this, it follows that ϕ is a homeomorphism.

In particular, the geometric realization $|K * \{v\}|$ of a cone over a simplicial complex K is in fact the topological cone over $|K|$. It follows that $|K * \{v\}|$ is contractible. In particular, if $|K_j| = \mathbb{S}^{p_j}, j = 1, 2$, then it follows that $|K_1 * K_2| = \mathbb{S}^{p_1} * \mathbb{S}^{p_2} = \mathbb{S}^{p_1+p_2+1}$. (See Exercise 1.9.29.)

Example 2.7.11 A CW-complex which is not triangulable: This example is perhaps, one of the simplest of its kind. We are going to construc a CW-complex X of dimension 3. Take $\mathbb{D}^2$ with a 0-cell, a 1-cell and a 2-cell. This is the 2-skeleton of X. Consider the loop $\gamma : (-1, 1) \longrightarrow \mathbb{D}^2$ given by $\gamma(t) = (-t, 0), \ -1 \leq t \leq 0$; and $g(t) = (t, t\sin(\pi/t)), 0 \leq t \leq 1$. Let C denote the image of this loop. Attach a 3-cell to $\mathbb{D}^2$ now via the map $\eta : \mathbb{S}^2 \longrightarrow \mathbb{D}^2$ where $\eta(x, y, z) = \gamma(x)$ to obtain the 3-dimensional CW-complex X. (See Figure 2.17.)

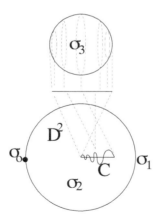

FIGURE 2.17. A non-triangulable CW-complex

Suppose X is triangulable. Since it is compact, the simplicial complex K which triangulates it will be finite. Let A be the union of all 3-simplices in K. Then $\bar{A}$ the closure of A should be contained in the closure of the 3-cell in X and will be a subcomplex of K. (This, rather intuitively clear fact needs to be proved. It is an easy consequence of the celebrated result of Brouwer's *invariance of domain* 2.9.16 that we shall prove later in this chapter and right now we shall assume this.) For similar reasons, $\mathbb{D}^2$ is also a subcomplex of K

and hence $\mathbb{D}^2 \cap \bar{A}$ is a subcomplex. Since the attaching map is onto C it follows easily that $\mathbb{D}^2 \cap \bar{A} = C$. But $\bar{A}$ is a finite CW-complex whereas C has infinitely many loops in it?!

Example 2.7.12 The prism construction: The simplicial structures are not 'well behaved' under Cartesian products. For instance, what should be the definition of $\Delta_1 \times \Delta_1$? Whatever it is, the first requirement is that its geometric realization should be homeomorphic to $|\Delta_1| \times |\Delta_1| \cong \mathbb{I} \times \mathbb{I}$. We are tempted to take the set of vertices to be the four element set $\{0,1\} \times \{0,1\}$ and the set of simplices to be $F \times G$ where F and G range over simplices of Δ_1. It turns out that we were lucky in the choice of vertices but completely off the track once we come to higher dimensional simplices—already in dimension 1, there are too many 1-simplices and that we cannot accommodate them all. So, we are forced to choose only some of them and Figure 2.18 shows two distinct choices, both of them equally good.

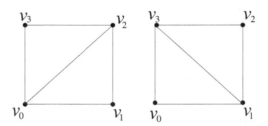

FIGURE 2.18. Two choices for triangulating a square

There are quite a few different ways to resolve this problem, each one being suitable for some particular situation. (See Exercises at the end of this section and 2.11.(1).) Our aim is to find a 'canonical' way to triangulate $|K| \times \mathbb{I}$ rather than do it economically. Figure 2.19 depicts this for $\sigma \times \mathbb{I}$ where σ is a simplex of dimension $0, 1$, and 2, respectively.

Inductively, assume that for each $r < n, |K^{(r)}| \times \mathbb{I}$ is the polyhedron of a simplicial complex $K^{(r)} \times \mathbb{I}$ such that:

(a) the inclusion maps η_0 and η_1 from $|K^{(r)}|$ to $|K^{(r)}| \times \mathbb{I}$ are simplicial maps, where, $\eta_t(\alpha) = (\alpha, t),\ t = 0, 1$;

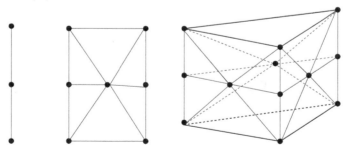

FIGURE 2.19. The prism constructions

(b) if L is a subcomplex of $K^{(r)}$, there exists a subcomplex $L \times \mathbb{I}$ of $K^{(r)} \times \mathbb{I}$ such that

$$|L \times \mathbb{I}| = |L| \times \mathbb{I}.$$

(Note that for $n = 0$, the hypothesis is vacuous). Now if F is a n-simplex of K, the boundary of $|F| \times \mathbb{I}$, viz.,

$$|F| \times 0 \cup |\mathcal{B}(F)| \times \mathbb{I} \cup |F| \times 1$$

is the polyhedron of $F \times 0 \cup \mathcal{B}(F) \times \mathbb{I} \cup F \times 1$. We shall denote it by $\partial(F \times \mathbb{I})$ temporarily. Let $F = \{v_0, ..., v_q\}$. The *barycentre* of F is defined to be the element $\widetilde{F} \in |K|$ such that

$$\widetilde{F}(v) = \begin{cases} 1/(q+1), & \text{if } v = v_i, \ i = 0, ..., q; \\ 0, & \text{otherwise.} \end{cases}$$

Clearly $\widetilde{F} \in |F|$. If $F = \{v_0\}$, then $\widetilde{F} = \{v_0\} = F$.

We now define

$$K^{(n)} \times \mathbb{I} = K^{(n-1)} \times \mathbb{I} \cup \{\partial(F \times \mathbb{I}) * \{\widehat{F}\} \ : \ F \in K, \ \& \ \dim F = n\}.$$

where $\widehat{F} = (\widetilde{F}, 1/2)$. Then the inductive hypothesis is checked easily. If the dimension of K is m then take $K \times \mathbb{I} = K^{(m)} \times \mathbb{I}$. Finally in the general case, observe that the product topology on $|K| \times \mathbb{I}$ coincides with the weak topology from the family $\{|K^{(m)} \times \mathbb{I}|\}$ and hence the inductive process defines a triangulation on $|K| \times \mathbb{I}$. We shall call this triangulation of $|K| \times \mathbb{I}$ the *prism construction* and write $K \times \mathbb{I}$ for it. Thus $|K \times \mathbb{I}| = |K| \times \mathbb{I}$. You will meet other variants of this construction later.

Example 2.7.13 Triangulation of a torus Recall that we can obtain the torus $\mathbb{S}^1 \times \mathbb{S}^1$ as the quotient of $\mathbb{I} \times \mathbb{I}$ wherein the opposite sides are identified by the rule:

$$(t, 0) \sim (t, 1); \quad (0, s) \sim (1, s), \quad t, s \in \mathbb{I}.$$

Figure 2.20 depicts a triangulation of the torus with 9 vertices. It is of course not a very economical way of triangulating a torus–there is one with just 7 vertices! Can you find it?

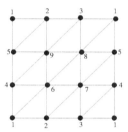

FIGURE 2.20. Triangulation of a torus

Exercise 2.7.14

(i) In Definition 2.7.1, we have claimed that $|K| \subset \mathbb{I}^V$ is a closed set when V is finite. Verify this.

(ii) For any simplicial complex K, triangulate $K \times \mathbb{I}$ with exactly twice as many vertices as in K and such that $K \times \{0\}$ and $K \times \{1\}$ are subcomplexes. [Hint: Choose an order on the vertices of K.]

(iii) Give a triangulation of $\mathbb{R}^n$.

(iv) Give a triangulation of $\mathbb{P}^n$. [Hint: Use S-triangulation of the sphere.]

(v) Let K be a simplicial complex. Show that the following conditions are equivalent:
 (a) Each vertex of K belongs to a finite number of edges of K.
 (b) Each vertex of K belongs to a finite number of simplices of K.
 (c) $|K|$ is locally compact.
 We say K is *locally finite* if the above conditions are satisfied.

(vi) Suppose $f : |K| \to \mathbb{R}^n$ is a topological embedding. Show that
 (a) K is locally finite.
 (b) K is countable, i.e., the vertex set of K is countable.
 (c) $\dim K \leq n$.

(vii) A map $f : |K| \to \mathbb{R}^n$ is said to be *linear* if restricted to each simplex $F \in K$, $f : |F| \to \mathbb{R}^n$ is affine linear.
 Suppose f is such a linear map.
 (a) Show that $f|_{|F|}$ is injective iff $f(F)$ is an affinely independent set in $\mathbb{R}^n$.
 (b) Assume f is injective on $K^{(0)}$. Suppose further that f is injective on $|F|$ and $|G|$ where $F, G \in K$ and the set $f(F) \cup f(G)$ is affinely independent. Then show that $f(|F \cap G|) = f(|F|) \cap f(|G|)$.
 (c) Let $f : |K| \to \mathbb{R}^n$ be an injective linear mapping. Show that f is an embedding iff f is proper, i.e., inverse image of a compact set in $\mathbb{R}^n$ is compact in $|K|$.
 (d) Recall that a subset P of $\mathbb{R}^n$ is said to be in general position, if every $(n+1)$-subset of V is affinely independent. Show that there exists a countably infinite closed discrete subset P of $\mathbb{R}^n$ which is in general position.
 (e) Given a locally finite, countable simplicial complex of dimension at most $\lfloor \frac{n-1}{2} \rfloor$ show that there is a linear embedding $f : |K| \to \mathbb{R}^n$.

2.8   Barycentric Subdivision

Often in real analysis, we come across the situation when an interval has to be subdivided into a finitely many subintervals suitably. This happens in the study of polyhedrons as well. The mid-point of an interval divides the given interval symmetrically into two parts. We can keep repeating this process for each of the new intervals until we have reached a stage when each interval is of sufficiently small length for the problem at hand to have some nice solution. Of course, there are other not so systematic ways of subdividing an interval. We find that the obvious generalization of the process of taking mid-points repeatedly is quite a powerful way to handle many of the problems. The process goes under the name *iterated barycentric subdivisions*.

Definition 2.8.1 Let K be a simplicial complex and F be a non empty face of it. Recall the definition of the barycentre $\widetilde{F}$ of F from Example 2.7.12. We now define a new simplicial complex $\operatorname{sd} K$, called the *barycentric subdivision of K* as follows : the vertex set of $\operatorname{sd} K$ is given by

$$V(\operatorname{sd} K) = \{\widetilde{F} \ : \ F \in K\},$$

the set of all barycentres $\widetilde{F}$ for all simplices $F \in K$; the simplices of $\operatorname{sd} K$ are sets of the form

$$\{\widetilde{F}_0, \widetilde{F}_1, ..., \widetilde{F}_q\}$$

where $F_0 \subset F_1 \subset ... \subset F_q$ is a chain of simplices in K. The reader is urged to verify that this actually defines a simplicial complex. In Figure 2.21, (b) represents the barycentric subdivision of (a).

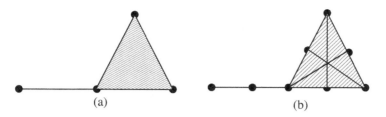

(a) (b)

FIGURE 2.21. (b) is the barycentric subdivision of (a)

Remark 2.8.2

(1) If we are given a simplicial map $f : K \longrightarrow L$ then we have a simplicial map sd $f :$ sd $K \longrightarrow$ sd L defined as follows:

$$(\text{sd } f)(\widetilde{F}) = \widetilde{f(F)}. \tag{2.9}$$

It is not difficult to verify that this is actually a simplicial map. In this way sd can be viewed as a functor on the category of simplicial complexes to itself. In particular, if $K \subset L$ then check that sd $K \subset$ sd L.

(2) Observe that the inclusion map sd $\mathcal{B}(F) \to$ sd F extends to a simplicial isomorphism of $(\text{sd } \mathcal{B}(F)) * \{\widetilde{F}\} \xrightarrow{\approx}$ sd F.

(3) The next theorem assures that topologically, the barycentric subdivision does not affect any change.

Theorem 2.8.3 *The inclusion map of the vertex set of* sd K *into* $|K|$ *extended linearly on each closed simplex of* sd K *defines a homeomorphism* $h : |\text{sd } K| \to |K|$. *This homeomorphism is canonical in the following weak sense: Given* $K \hookrightarrow L$ *the following diagram is commutative:*

$$\begin{array}{ccc} |\text{sd } K| & \hookrightarrow & |\text{sd } L| \\ \downarrow h & & \downarrow h \\ |K| & \hookrightarrow & |L| \end{array} \tag{2.10}$$

Proof: We use (2.9). Observe that for any simplex $F \in K$ we have the canonical inclusion sd $F \subset$ sd K which is simplicial and the map defined above satisfies the canonical property for any face $G \subset F$. Therefore the canonical property in the general case also follows.

Consider the following four statements:

(A_n) For all k-simplices such that $k \leq n$, $h : |\text{sd } F| \longrightarrow |F|$ is a homeomorphism.

(A) For all k-simplices F, $h : |\text{sd } F| \longrightarrow |F|$ is a homeomorphism.

(B_n) For all simplicial complexes K of dimension $\leq n$, $h : |\text{sd } K| \longrightarrow |K|$ is a homeomorphism.

(B) For all simplicial complexes K, $h : |\text{sd } K| \longrightarrow |K|$ is a homeomorphism.

We shall first show that $(A_n) \Longrightarrow (B_n)$. So, let dim $K \leq n$. Given any $\beta \in |K|$, clearly there exists a simplex $F \in K$ such that $\beta \in |F|$. Therefore, it follows from (A_n) that h is surjective. Given $\alpha \in |\text{sd } K|$ check first that $h(\alpha) \in |F|$ for some $F \in K$ iff $h(supp\,\alpha) \subset |F|$. Therefore, if α, $\alpha' \in |\text{sd } K|$ are such that $h(\alpha) = h(\alpha') = \beta$, then there exists $F \in K$ such that both $h(supp\,\alpha), h(supp\,\alpha')$ are contained in $|F|$. Again from (A_n), it follows that $\alpha = \alpha'$. Hence h is injective.

Clearly h is continuous also. To see that it is a closed map, let $G \subset |\operatorname{sd} K|$ be a closed set. We must verify that $h(G) \cap |F|$ is closed in $|F|$ for each $F \in K$. But $h(G) \cap |F| = h(G \cap |\operatorname{sd} F|)$. Since G is closed, $G \cap |\operatorname{sd} F|$ is closed in $|\operatorname{sd} F|$ from $((A_n)$ we conclude that $h(G) \cap |F|$ is closed in $|F|$.

To prove $(A) \Longrightarrow (B)$, we merely replace (A_n) everywhere by (A) in the above argument. Thus, in order to complete the proof of the theorem, we have only to prove (A_n) for all n, which we shall do by induction.

If dimension of F is 0, then both F and $\operatorname{sd} F$ coincide and there is nothing to prove. Suppose we have proved (A_n) for all simplices of dimension $\leq n$, for some $n \geq 0$. The theorem is then true for all simplicial complexes of dimension $\leq n$. Let now $\dim F = n + 1$.

$$
\begin{array}{ccc}
|\operatorname{sd} \mathcal{B}(F)| & \xrightarrow{\ \ h\ \ } & |\mathcal{B}(F)| \\
\downarrow & & \downarrow \\
|(\operatorname{sd} \mathcal{B}(F) * \{\widetilde{F}\}| & \xrightarrow{\ c|h|\ } & c(|\mathcal{B}(F)|) \\
{\scriptstyle |\lambda|}\downarrow & & \downarrow{\scriptstyle \mu} \\
|\operatorname{sd} F| & \xrightarrow{\ \ h\ \ } & |F|
\end{array}
$$

In the above diagram, the top horizontal map $h : |\operatorname{sd} \mathcal{B}(F)| \longrightarrow |\mathcal{B}(F)|$ is a homeomorphism by induction. Upon taking cones, it follows that the middle horizontal map is a homeomorphism. The vertical arrow $|\lambda|$ represents the homeomorphism induced by the canonical simplicial isomorphism as seen in Remark 4.5(2). The vertical map μ on the right is the homeomorphism obtained by mapping the apex of the cone to the barycentre $\widetilde{F}$ and extending linearly. This proves (A_{n+1}).   ♠

Example 2.8.4 Here is an example to show that the above homeomorphism is not canonical in the usual sense, viz., we cannot replace the inclusion map in (2.10) by an arbitrary simplicial map. Take $K = \Delta_2$, $L = \Delta_1$ and let $\varphi : K \to L$ be defined by $\varphi(\mathbf{e}_1) = \varphi(\mathbf{e}_2) = \mathbf{e}_1, \varphi(\mathbf{e}_3) = \mathbf{e}_2$. Then

$$|\varphi|(\mathbf{e}_1 + \mathbf{e}_2 + \mathbf{e}_3/3) = (2\mathbf{e}_1 + \mathbf{e}_2)/3 \neq (\mathbf{e}_1 + \mathbf{e}_2)/2 = \operatorname{sd}\varphi((\mathbf{e}_1 + \mathbf{e}_2 + \mathbf{e}_3)/3).$$

Remark 2.8.5 More generally, by a *subdivision* of a simplicial complex K we mean a simplicial complex K' such that the vertices of K' are elements of $|K|$ satisfying the following condition:
(i) for every simplex F' of K' there is a (unique) simplex F of K such that $F' \subseteq |F|$,
(ii) the inclusion map of the vertices of K' into $|K|$ extended linearly on each $|F'|$ defines a homeomorphism of $|K'|$ onto $|K|$.

We shall not have much to do with this general concept and so will not study it elaborately. A slight generalization of barycentric subdivision which is quite useful in combinatorial set-up will be discussed in exercises at the end of the chapter.

Definition 2.8.6 Let K be any simplicial complex and v be any vertex in K. We define

$$\operatorname{st} v = \{\alpha \in |K| \ : \ \alpha(v) \neq 0\}.$$

By the very definition, it follows that $\{\operatorname{st} v\}_{v \in V}$ is an open cover of $|K|$. A simplicial complex K is *finer* than an open covering $\mathcal{U}$ of $|K|$ if the covering $\{\operatorname{st} v\}_{v \in V}$ is finer than $\mathcal{U}$. Recall that this means that for each $v \in V$, there exists $U_v \in \mathcal{U}$ such that $st\, v \subset U_v$.

Now let K be finite. Using the definition of $|K|$ as a subspace of $\mathbb{I}^{\#(V)}$, we can take the restriction of the Euclidean metric of $\mathbb{I}^{\#(V)}$ on $|K|$, viz.,

$$d(x, y) = ||x - y|| = \sqrt{\sum_i (x_i - y_i)^2}.$$

The special property of this metric that is easily verified and crucial to us is:

Definition 2.8.7 A metric on $|K|$ is called a *linear metric* if restricted to each $|F|$; it is given by a norm function.

Lemma 2.8.8 Let K be a finite simplicial complex and d be a linear metric on $|K|$. Then d is a linear metric on $|\operatorname{sd} K| = |K|$ and for any $F' \in \operatorname{sd} K$ such that $F' \subset |F|$, where $F \in K$ is a q-simplex, we have the inequality of the diameters:

$$\operatorname{diam}|F'| \leq \frac{q}{q+1} \operatorname{diam}|F|. \tag{2.11}$$

Proof: Let $F = \{v_0, ..., v_q\}$. Any point $\alpha \in |F|$ can be regarded as a convex combination $\alpha = \sum t_i v_i$, $\sum t_i = 1$, $0 \leq t_i \leq 1$. Then for any $\beta \in |F|$, we have

$$
\begin{aligned}
d(\alpha, \beta) &= d\left(\sum_i t_i v_i, \beta\right) & &= d\left(\sum_i t_i v_i, \left(\sum_i t_i\right)\beta\right) \\
&= \left\|\sum_i t_i v_i - \left(\sum_i t_i\right)\beta\right\| & &= \left\|\sum_i (t_i(v_i - \beta))\right\| \\
&\leq \sum_i t_i \|v_i - \beta\| & &\leq \sum_i t_i d(v_i, \beta)
\end{aligned}
$$

Hence

$$d(\alpha, \beta) \leq \sup_i \{d(v_i, \beta)\}. \tag{2.12}$$

Since this is true for all $\alpha, \beta \in |F|$, we can put $\beta = v_j$ and take $\alpha = \beta$ in (2.12) to get

$$d(\beta, v_j) \leq \sup_i \{d(v_i, v_j)\} \tag{2.13}$$

Thus

$$\operatorname{diam}|F| \leq \sup_{i, j} \{d(v_i, v_j)\}. \tag{2.14}$$

Now, let u_1, u_2 be any two vertices of a simplex F' in $\operatorname{sd} F$. Without loss of generality we may write,

$$u_1 = \frac{v_0 + \cdots + v_k}{k+1}, \quad u_2 = \frac{v_0 + \cdots + v_l}{l+1}$$

with $k < l \leq q$. Then $d(u_1, u_2) \leq \sup_{0 \leq s \leq k} d(v_s, u_2)$ and

$$
\begin{aligned}
d(v_s, u_2) &= \left\|v_s - \frac{\sum_{r=0}^{l} v_r}{l+1}\right\| & &= \frac{1}{1+l}\left\|(1+l)v_s - \sum_{r=0}^{l} v_r\right\| \\
&\leq \frac{1}{l+1}\sum_{r=0}^{l} \|v_s - v_r\| & &\leq \frac{l}{l+1} \sup_{0 \leq r \leq l} \{d(v_s, v_r)\}.
\end{aligned}
$$

Hence

$$d(u_1, u_2) \leq \frac{l}{l+1} \operatorname{diam} |F| \leq \frac{q}{q+1} \operatorname{diam} |F|. \tag{2.15}$$

From this (2.11) follows easily.   ♠

Theorem 2.8.9 *If K is a finite simplicial complex and $\mathcal{U}$ is an open covering of $|K|$ then there exists N such that for all $n \geq N$, $\operatorname{sd}^n K$ is finer than $\mathcal{U}$.*

(Here $\operatorname{sd}^n K = \operatorname{sd}(\operatorname{sd}^{n-1} K)$ is the n-iterated barycentric subdivision).

Proof: Let c be the Lebesgue number of the covering. Choose N, such that for all $F' \in \operatorname{sd}^N K$, we have $\operatorname{diam} |F'| < c/2$. This is possible, by Lemma 2.8.8, since $(q/q+1)^n \to 0$ as $n \to \infty$.   ♠

Remark 2.8.10 Clearly, subdivisions give the same topological information on the original triangulation of a space. In some sense, we may say that they give the same combinatorial information as well. Based on the fact that two partitions of an interval have a common refinement, we can ask the question: Given any two triangulations K_1, K_2 are there subdivisions K_i' of K_i such that K_1' is isomorphic to K_2'? It is convenient to make a definition here: Two given triangulations K_1, K_2 of a space X are said to be combinatorially equivalent if there exists a common subdivision to both of them. The above question can be reformulated as follows: On a triangulable space are there more than one combinatorially inequivalent triangulations? This apparently simple question remained unsolved for a long time until Milnor gave an example of a 6-dimensional space with two combinatorially inequivalent triangulations. For this he had to invent a new combinatorial invariant which is obviously not a homeomorphic invariant [Milnor, 1961]. The study of these questions on manifolds goes by the name 'hauptvermutung'. For more, the reader may see Section 5.2.

Exercise 2.8.11

(i) Obtain a homeomorphism of $|\partial \Delta_n|$ with $\mathbb{S}^{n-1}$ so that for any face $F \in \partial \Delta_n$, the antipode of $\beta(F)$ goes to the antipode of $\beta(F')$ where F' is the complement of F in the vertex set of Δ_n. Use this to show that the barycentric subdivision of $\partial \Delta_n$ quotients down to define a triangulation of the real projective space $\mathbb{P}^{d-1}$.

(ii) Let K denote the boundary complex of Δ_n. Compute the nerve $K(\mathcal{U})$ of the open covering $\mathcal{U} = \{\operatorname{st}_K v \, : \, v \in K\}$.

(iii) Let X be a finite CW-complex. We write $f_k := f_k(X)$ for the number of k-cells of X. The alternate sum

$$\chi(X) := \sum_{i \geq 0} (-1)^i f_i(X)$$

is called the *Euler characteristic* of X. This is a very interesting number in topology and geometry. However, we cannot do anything with it right now. So, we take the special case when $X = K$ is a finite simplicial complex and make a beginning in the study of $\chi(K)$. [2]

(a) Let K_1, K_2 be two subcomplexes of K. Show that

$$\chi(K_1 \cup K_2) = \chi(K_1) + \chi(K_2) - \chi(K_1 \cap K_2).$$

---

[2]Caution: In combinatorics, we also set $f_{-1} = 1$ always and take $\chi(K) = \sum_{i \geq -1} f_i$. Therefore, the combinatorial Euler characteristic is always one less than the topological Euler characteristic.

(b) Show that $\chi(K \star v) = 1$, i.e., the Euler characteristic of any cone is always equal to 1. In particular the Euler characteristic of any simplex is equal to 1.

(c) Show that for any face F of K, $\chi(St_k(F)) = 1$.

(d) Show that $\chi(\text{sd } K) = \chi(K)$ and hence $\chi(\text{sd}^n K) = \chi(K)$ for all n.

(e) More generally, for any subdivision K' of K, it is true that $\chi(K') = \chi(K)$. Can you prove this at this stage? [3]

2.9   Simplicial Approximation

Given two simplicial complexes K, L, how big is the space of all simplicial maps inside the space of all continuous maps? Approximating certain functions with better behaved ones is an age-old game which happens to be quite rewarding. Approximating continuous functions by 'piecewise linear' ones is one of the motives of introducing simplicial complexes. The difference here is that the formulation of the problem is not in terms of convergent sequences.

Definition 2.9.1 Let K_1 and K_2 be simplicial complexes and $f : |K_1| \to |K_2|$ be a continuous map. A simplicial map $\varphi : K_1 \to K_2$ is called a *simplicial approximation* if

$$f(\alpha) \in |F_2| \Rightarrow |\varphi|(\alpha) \in |F_2| \text{ for } \alpha \in |K_1| \text{ and } F_2 \in K_2.$$

Remark 2.9.2 If $L_1 \subset K_1$ is a subcomplex and f and φ are as above and if φ is a simplicial approximation to f then $\varphi|_{L_1}$ is a simplicial approximation to $f|_{|L_1|}$; also if f is already induced by a simplicial map $\psi : K_1 \to K_2$, then $\varphi = \psi$. This follows from the simple observation that if for $v \in V_1$, $f(v)$ is a vertex, then $\varphi(v) = f(v)$. The importance of simplicial approximation stems from the following lemma.

Lemma 2.9.3 Suppose $\varphi : K_1 \to K_2$ is a simplicial approximation to $f : |K_1| \to |K_2|$ such that $|\varphi|(a) = f(a)$, $a \in A \subset |K_1|$. Then $|\varphi|$ is homotopic to f relative to A.

Proof: Define $h(\alpha, t) = tf(\alpha) + (1 - t)|\varphi|(\alpha), \alpha \in |K_1|, \quad 0 \leq t \leq 1$. ♠

Remark 2.9.4 At this stage, we hope that the reader is able to figure out the kind of details required to complete the proof of the above lemma and also supply them on her own. Take this as an exercise (see Exercise 2.9.21.(i)) and then check the answer that is provided at the end of the book.

The lemma below plays a key role in constructing simplicial approximations.

Lemma 2.9.5 Let $\varphi : V(K_1) \to V(K_2)$ be a vertex function, where K_1 and K_2 are any two simplicial complexes. Let $f : |K_1| \to |K_2|$ be a continuous function. Then the following conditions are equivalent:
(a) for every $v \in K_1$, $f(\text{st } v) \subset \text{st } \varphi(v)$;
(b) for every $\alpha \in |K_1|$, $\phi(\text{supp } \alpha) \subset \text{supp } f(\alpha)$.
(c) ϕ is a simplicial approximation to f.

---

[3]Indeed, Euler characteristic is a topological invariant which makes it very useful. This will be proved in Chapter 4 through a homological definition. For compact smooth manifolds, there are many other avatars of the Euler characteristic (see [Shastri, 2011]).

Proof: (a) $\Longrightarrow$ (b): $v \in \operatorname{supp} \alpha \Longrightarrow \alpha(v) > 0 \Longrightarrow \alpha \in \operatorname{st} v \Longrightarrow f(\alpha) \in \operatorname{st} \phi(v)$ (by (a))
$\Longrightarrow f(\alpha)(\phi(v)) > 0 \Longrightarrow \phi(v) \in \operatorname{supp} f(\alpha)$.
(b) $\Longrightarrow$ (a): $\alpha \in \operatorname{st} v \Longrightarrow \alpha(v) > 0 \Longrightarrow v \in \operatorname{supp} \alpha \Longrightarrow \phi(v) \in \operatorname{supp} f(\alpha)$ (by (b))
$\Longrightarrow f(\alpha)(\phi(v)) > 0 \Longrightarrow f(\alpha) \in \operatorname{st} \phi(v)$.
(c) $\Longrightarrow$ (a): Fix $v \in K_1$ and $\alpha \in \operatorname{st} v$. We have to show that $f(\alpha)(\varphi(v)) \neq 0$. Let $F_2 = \operatorname{supp} f(\alpha)$. Clearly $f(\alpha) \in |F_2|$. By the hypothesis, this implies that $|\varphi|(\alpha) \in |F_2|$. By definition of $|\varphi|$, $|\varphi|(\alpha)(\varphi(v)) = \alpha(v) +$ some non negative terms $\geq \alpha(v) > 0$. Therefore $\varphi(v) \in F_2$. Since F_2 is the support of $f(\alpha)$ this implies $f(\alpha)(\varphi(v)) \neq 0$.
(a) $\Longrightarrow$ (c) First to show that φ is a simplicial map, let $\{v_0, ..., v_q\} = F_1$ be a simplex in K_1. Then clearly $\widetilde{F_1}$, the barycentre, belongs to $\operatorname{st} v_i$ for each i. Hence $\emptyset \neq f(\cap_i \operatorname{st} v_i) \subset \cap_i f(\operatorname{st} v_i) \subset \cap_i \operatorname{st} \varphi(v_i)$. Say, $\beta \in \cap_i \operatorname{st} \varphi(v_i)$. Then $\varphi(v_i) \in \operatorname{supp} \beta$ and hence $\varphi(F_1)$ is a simplex of K_2. To show that φ is a simplicial approximation to f, assume $\alpha \in |K_1|$ is such that $f(\alpha) \in |F_2|$. This means that $\operatorname{supp} f(\alpha) \subset F_2$. Since we have proved (a)$\Longrightarrow$(b), we can use condition (b) from which it follows that

$$\operatorname{supp} |\varphi|(\alpha) = \varphi(\operatorname{supp} \alpha) \subset \operatorname{supp} f(\alpha) \subset F_2.$$

This just means that $|\varphi|(\alpha) \in |F_2|$ as desired.   ♠

Lemma 2.9.6 A map $f : |K_1| \to |K_2|$ admits simplicial approximations iff K_1 is finer than $\mathcal{U} = \{f^{-1}(\operatorname{st} v)\}_{v \in V_2}$.

Proof: By the above lemma, it is necessary and sufficient to find for each vertex v of K_1 a vertex $\varphi(v)$ of K_2 such that $\operatorname{st} v \subset f^{-1}(\operatorname{st} \varphi(v))$. This is assured by the statement that K_1 is finer than $\mathcal{U}$. The converse is a direct consequence of the lemma above.   ♠

Theorem 2.9.7 *Let* $f : |K_1| \to |K_2|$ *be any continuous map and* K_1 *be finite. Then there exists an integer* N, *such that for all* $n \geq N$, *there are simplicial approximations* $\varphi : \operatorname{sd}^n K_1 \to K_2$ *to* f.

Proof: Combine Lemma 2.9.6 and Lemma 2.8.9.   ♠

Remark 2.9.8 Simplicial approximations have applications similar to smooth approximations, which we shall not take trouble to list. (For instance, we can prove Theorem 2.5.8 directly from the simplicial approximation theorem by using the standard triangulation of discs and spheres.) Obviously, they have wider applicability than smooth approximations but perhaps narrower than CW-approximations. To some extent, say, in a combinatorial sense, they are better behaved than CW-approximations. The main drawback is that we need to keep going to subdivisions, which is not the case with CW-approximation.

Below we shall indicate an application in proving the cellular approximation theorem itself. We shall then prove the Sperner lemma which in turn has many applications.

An alternative proof of cellular approximation theorem

The difference in the present proof and the proof of given in 2.5.3 is in the fact that, there we have used smooth approximation and Sard's theorem (Lemma 2.5.4), whereas, here we are going to use only simplicial approximation theorem. As in Lemma 2.5.4, let K_1 and K_4 be given by $K_1 = \alpha^{-1}(C_1)$ and $K_4 = \alpha^{-1}(C_4)$. By uniform continuity, there exists $\delta > 0$ such that whenever $\|x - y\| < \delta, x, y \in K_4$ we have $\|\beta(x) - \beta(y)\| < r$. Now we use what is known in analysis as Runge's trick: We cut $\mathbb{R}^n$ into a mesh of n-cubes by planes parallel to the axes and at a distance less than $\frac{\delta}{2\sqrt{n}}$ (i.e., so that the diameter of each cube is less than δ). Let K_2 be the union of all those cubes which intersect K_1 and K_3 be the union of all those cubes which intersect K_2. Both are a finite union of cubes and hence triangulable. Choose any triangulation of K_3 so that K_2 is a subcomplex and let γ be a simplicial approximation to β on K_3. Since $\dim K_2 = n < m$, it follows that image of K_2 under γ cannot be the whole of B_1. The rest of the proof of Lemma 2.5.4 can now be reproduced verbatim. ♠

As an immediate corollary, we have:

Corollary 2.9.9 Every map $f : \mathbb{S}^n \to \mathbb{S}^{n+k}$, $k \geq 1$ is null homotopic.

Proof: Think of f as a map from $|\Delta_n| \to |\Delta_{n+k}|$. Let $\phi : \Delta'_n \to \Delta_{n+k}$ be a simplicial approximation to f, where Δ'_n is a subdivision of Δ_n. We know that f is homotopic to $|\phi|$. On the other hand, since ϕ is simplicial, its image is contained in the nth-skeleton of Δ_{n+k}. In particular, $|\phi|$ is not surjective. We know that any map into a sphere which is not surjective is null homotopic. This means f is null homotopic. ♠

Remark 2.9.10 Simplicial approximation theorem is valid over arbitrary simplicial complexes also, though we have proved it only over finite simplicial complexes, since we hardly need the general result. However, in the general case we cannot always do it by an iterated barycentric subdivision. We have indicated a proof using the notion of 'star-subdivision' in the exercises below.

We shall now present a proof of Sperner lemma and some of its consequences which illustrates the power of combinatorial methods in topology.

Theorem 2.9.11 (Sperner lemma) *Let Δ'_n be a subdivision of the standard simplicial complex Δ_n and let $\phi : \Delta'_n \to \Delta_n$ be a simplicial map which, when restricted to the boundary subcomplex $\mathcal{B}(\Delta_n)'$, is a simplicial approximation to the identity map. Then the number of n-simplices of Δ'_n mapped onto Δ_n is odd.*

As a step toward the proof of this, we shall first prove another lemma which is a little more general and elaborate result.

Lemma 2.9.12 Let Δ'_n be a subdivision of the standard simplicial complex Δ_n and $\phi : \Delta'_n \to \Delta_n$ be a simplicial map. Let $L = \Delta_{n-1}$. For any n-simplex F of Δ'_n, let $\alpha(F)$ denote the number of $(n-1)$-faces of F which are mapped onto L. Put $s_1 = \sum_F \alpha(F)$. Let s_2 denote the number of n-simplices of Δ'_n which are mapped onto Δ_n by ϕ and and s_3 be the number of $(n-1)$-simplices of $\mathcal{B}(\Delta_n)'$ mapped onto L by ϕ. Then

$$s_1 \equiv s_2 \equiv s_3 \mod 2. \tag{2.16}$$

Proof: First note that $\alpha(F) = 0, 1$ or 2. Also $\alpha(F) = 1$ iff ϕ restricted to F is a bijection onto Δ_n. Thus the collection of all n-simplices F of Δ'_n is divided into three groups A_0, A_1, A_2 according to the value of $\alpha(F)$. It follows that working modulo 2, we have

$$s_1 = \sum_F \alpha(F) \equiv \sum_{F \in A_1} \alpha(F) = s_2.$$

Now note that s_1 also counts the number of $(n-1)$-faces G of Δ'_n which are mapped onto L except that G is counted twice iff it is an interior $(n-1)$-face. Cutting down all these entries leaves us with only those G which are in $\mathcal{B}(\Delta_n)'$ and hence with the sum s_3. This proves $s_1 \equiv s_3$. ♠

We can now prove Sperner lemma by induction: Let $C_n, n \geq 1$ denote the statement of Sperner lemma. Accordingly we shall temporarily denote the numbers $s_1, s_2,$ and s_3 by $s_1(n), s_2(n)$ and $s_3(n)$. We need to show that $s_2(n) \equiv 1 \mod 2$. The case $n = 1$ is easy, viz., $s_3(1) = 1$ since ϕ is identity map on $\partial \Delta'_1 = \partial \Delta_1$. Assume that $n \geq 2$, and we have proved C_{n-1}. Since ϕ is a simplicial approximation to the identity when restricted to the boundary, each $(n-1)$-face of G', where G is an $(n-1)$-face of of the boundary will be mapped inside G itself. In other words, only some of the $(n-1)$-faces of L' are mapped onto L. Therefore, it follows that $s_3(n) = s_2(n-1)$. Now by appealing to (2.16) and the induction hypothesis we obtain $s_2(n) \equiv 1 \mod 2$. ♠

Here are some important applications of Sperner lemma to two of the celebrated results of Brouwer—the fixed point theorem and the theorem of invariance of domain. The proof of the fixed point theorem is standard and very quick. Equally quick is the proof of a mild version of invariance of domain, viz., that $\mathbb{R}^n$ and $\mathbb{R}^m$ are not homeomorphic to each other for $n \neq m$. The proof of the main version takes only a little bit more time.

Theorem 2.9.13 *For any integer $n \geq 1$, the following three statements are equivalent and each of them is true:*
(a) **(Brouwer's fixed point theorem)** *Every continuous map $f : \mathbb{D}^n \to \mathbb{D}^n$ has a fixed point, i.e., there is $x \in \mathbb{D}^n$ such that $f(x) = x$.*
(b) $\mathbb{S}^{n-1}$ *is not a retract of $\mathbb{D}^n$.*
(c) $\mathbb{S}^{n-1}$ *is not contractible.*

Proof: (a) $\Longrightarrow$ (b): If $r : \mathbb{D}^n \to \mathbb{S}^{n-1}$ is a retraction consider the composite f of the three maps

$$\mathbb{D}^n \xrightarrow{\ r\ } \mathbb{S}^{n-1} \xrightarrow{\ \alpha\ } \mathbb{S}^{n-1} \xrightarrow{\ \eta\ } \mathbb{D}^n$$

where $\alpha(x) = -x$ and η is the inclusion map. Then f has no fixed point, contradicting (a).
(b) $\Longrightarrow$ (a) The proof here is exactly the same as that we wrote for the case $n = 2$ in the proof of Corollary 1.2.29.
(b) $\Longleftrightarrow$ (c): We have seen that a space X is contractible iff X is a retract of the cone CX. (See Theorem 1.5.3.) Since $C\mathbb{S}^{n-1}$ is homeomorphic to $\mathbb{D}^n$ we are done.

Finally, to prove that each of the above statements is true, we prove (b). Assuming on the contrary, using the fact that $|\Delta_n|$ is homeomorphic to $\mathbb{D}^n$, we obtain a map $f : |\Delta_n| \to |\mathcal{B}(\Delta_n)|$ which is a retraction. If $\phi : \Delta'_n \to \mathcal{B}(\Delta_n)$ is a simplicial approximation to f then restricted to the boundary, ϕ is a simplicial approximation to the identity map. Now we can treat ϕ as a simplicial map $\Delta'_n \to \Delta_n$ and use Sperner lemma to conclude that the number of n-simplices of Δ'_n mapped onto Δ_n is odd. But that is absurd since we know that this number is zero in this case. ♠

Theorem 2.9.14 *For $n \neq m$, $\mathbb{S}^n$ is not homotopy equivalent to $\mathbb{S}^m$; in particular, $\mathbb{S}^n, \mathbb{S}^m$ are not homeomorphic to each other.*

Proof: Suppose $f : \mathbb{S}^n \to \mathbb{S}^m$, $n < m$ is a homotopy equivalence. By Corollary 2.9.9, we know that f is null homotopic. By pre-composing with the homotopy inverse $g : \mathbb{S}^m \to \mathbb{S}^n$ of f, this implies that the identity map of $\mathbb{S}^m$ is null homotopic. This is the same as saying $\mathbb{S}^m$ is contractible and contradicts the above theorem. ♠

By taking one-point compactification we obtain

Corollary 2.9.15 *For $n \neq m$, $\mathbb{R}^n$ is not homeomorphic to $\mathbb{R}^m$.*

The above corollary may be called a weak version of Brouwer's invariance of domain. We shall now embark upon proving the main version of the same as stated below.

Theorem 2.9.16 (Brouwer's invariance of domain) *Let X, Y be any two subsets of $\mathbb{R}^n$ and $h : X \to Y$ be a homeomorphism. If X is open in $\mathbb{R}^n$ then so is Y.*

The key steps are the Lemma 2.9.17 below leading to a point-set-topological result, Theorem 2.9.20.

Lemma 2.9.17 Let K be a finite simplicial complex of dimension $< m$, and A be a closed subset of $|K|$. Then given any map $f : (|K|, A) \to (\mathbb{D}^m, \mathbb{S}^{m-1})$, there exists a homotopy $H : |K| \times \mathbb{I} \to \mathbb{D}^m$ such that

$$H(x,0) = f(x), x \in |K|; \quad H(a,t) = f(a), a \in A, t \in \mathbb{I}; \quad H(x,1) \in \mathbb{S}^{m-1}, x \in |K|.$$

Corollary 2.9.18 Let $A \subset \mathbb{S}^n$ be a closed subset and $m - 1 \geq n$. Then every map $\alpha : A \to \mathbb{S}^{m-1}$ can be extended to a map $g : \mathbb{S}^n \to \mathbb{S}^{m-1}$.

Proof: By Tietze's extension theorem, there is a map $f : \mathbb{S}^n \to \mathbb{D}^m$ such that $f(a) = \alpha(a), a \in A$. Take $K = \partial \Delta^{n+1}$ and identify $|K|$ with $\mathbb{S}^n$. Let H be the homotopy given by the above lemma and put $g(x) = H(x,1)$. ♠

Let us now recall a definition from point set topology.

Definition 2.9.19 Let $X \subset Z$ where Z is a topological space. A point $x \in Z$ is called a relative interior point of X in Z if there exists an open subset U of Z such that $x \in U \subset X$. A point $x \in Z$ is called a relative boundary point of X in Z if it is not a relative interior point of X nor a relative interior point of $Z \setminus X$.

Thus a relative boundary point of X has the property that every neighbourhood of this point will intersect both X and $Z \setminus X$. A subset X is open in Z iff all its points are relative interior points.

The following theorem which characterizes intrinsically, the relative boundary points of a subset $X \subset \mathbb{R}^n$ is perhaps the strongest form of Brouwer's invariance of domain.

Theorem 2.9.20 *Let $X \subset \mathbb{R}^n$ be compact. A point $x \in X$ is a relative boundary point of X iff there exist arbitrarily small neighbourhoods U of x in X such that every continuous function $f : X \setminus U \to \mathbb{S}^{n-1}$ has a continuous extension over X.*

Proof: $\implies$: Put $U = B_r(x) \cap X$ for arbitrary $r > 0$. It is enough to prove that every $f : X \setminus U \to \mathbb{S}^{n-1}$ can be extended over X. Consider the restriction of f to $\partial B_r(x) \cap X = A$. This gives a map $f' : A \to \mathbb{S}^{n-1}$ and A is closed in $\partial B_r(x)$. By Corollary 2.9.17 above, there is a map $g : \partial B_r(x) \to \mathbb{S}^{n-1}$ extending f'. Take a point $p \in B_r(x) \setminus X$. (Such a point exists, because x is a boundary point of X.) Let $\eta : X \to \partial B_r(x)$ be the radial projection from the point p. (See Figure 2.22.) Put

$$h(x) = \begin{cases} f(x), & x \in X \setminus U, \\ g(\eta(x)), & x \in \bar{U}. \end{cases}$$

Then h is the required extension of f.

$\impliedby$: Let x be a relative interior point X. Choose $r > 0$ so that $U := B_r(x) \subset X$. Let $f : X \setminus \{x\} \to \mathbb{S}^{n-1}$ be defined by $f(y) = \dfrac{y - x}{\|y - x\|}$. If $g : X \to \mathbb{S}^{n-1}$ is an extension of $f|_{X \setminus U}$, then we can take $\phi(v) = g(rv + x) : \mathbb{D}^n \to \mathbb{S}^{n-1}$ which is a retraction, contradicting Theorem 2.9.13. Now for any neighbourhood V of x such that $V \subset B_r(x)$, f is defined and continuous on $X \setminus V$ and cannot be extended over the whole of X. ♠

Proof of Theorem 2.9.16 from Theorem 2.9.20: Given a relative interior point x of X, it is enough to prove that $h(x)$ is a relative interior point of Y. If not, this means that it is a relative boundary point of Y and hence has arbitrarily small neighbourhoods V in Y such that every continuous map $g : V \to \mathbb{S}^{n-1}$ can be extended over Y. Composing with h, we get the same property for $x \in X$ which means x is a relative boundary point of X?! ♠

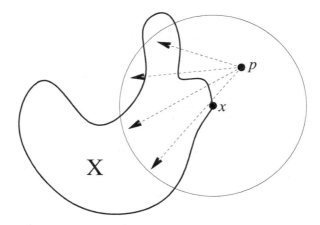

FIGURE 2.22. Characterization of interior points

Notice that if we tried to formulate the above proof in terms of relative interior points only, we will have to be extra careful— it is not enough to show that $h(x)$ has a neighbourhood such that continuous functions defined on the complement cannot be extended.

So, it remains to prove Lemma 2.9.17, which has the same flavour as that of Theorem 2.5.3.

First, we observe that it is enough to prove that f is homotopic to a map g relative to A such that g does contain an interior point q of $\mathbb{D}^m$. For then we can compose this with the standard deformation retraction $\mathbb{D}^m \setminus \{q\} \to \mathbb{S}^{m-1}$.

Next we also see that we can replace the pair $(\mathbb{D}^m, \mathbb{S}^{m-1})$ with the pair $(J^m, \partial J^m)$, where J is the closed interval $[-1, 1]$.

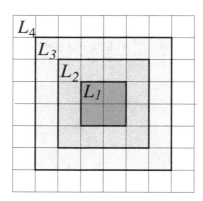

FIGURE 2.23. Homotoping away from an interior point

Cut the cube J^m by planes parallel to the coordinate planes at intervals of length $1/4$ and then take a triangulation L of J^m such that each of these little cubes is a subcomplex. Choose a subdivision K' of K so that there is a simplicial approximation $\alpha : K' \to L$ to f. Put $L_j = [-j/4, j/4]^m$; $K_j = f^{-1}(L_j)$ for $j = 1, 2, 3, 4$. Then check that each K_i is compact and

$$K_1 \subset \text{int } |K_2| \subset |K_2| \subset \cdots \subset K_4 = \mathbb{D}^n.$$

Observe that $A \cap K_2 = \emptyset$ and if $x \notin K_2$, then $f(x)$ and $|\alpha$ are contained in a simplex $|\sigma|$ of L which does not intersect L_1. (See Figure 2.23.)

Let $\eta : |K'| \to \mathbb{I}$ be a continuous map such that $\eta \equiv 0$ outside int $|K_3|$ and $\equiv 1$ on $|K_2|$.

Consider the homotopy $t \mapsto (1 - t\eta)f + t\eta|\alpha|$ from f to $g = (1 - \eta)f + \eta|\alpha|$. On K_2, $g = |\alpha|$ and outside K_3, it is equal to f and hence can be extended over all of $|K|$ by f. Indeed, the same holds for the entire homotopy as well and hence we get a homotopy of f with the map g on $|K|$. We claim that the image of g does not contain some points of L_1. First of all, since α is a simplicial map of a simplicial complex of dimension n, it follows that $|\alpha|(|K|)$ is contained in the nth-skeleton of L. In particular so is $|\alpha|(K_2)$. On the other hand, if $x \notin K_2$ then we know the line segment $[f(x), |\alpha|]$ does not intersect L_1 and since $g(x) \in [f(x), |\alpha|(x)]$, it follows that $g(x) \notin L_1$. Therefore, $g(|K|)$ is contained in $|L^{(n)}| \cup (|L| \setminus L_1)$ which does not cover L_1. This proves the claim and hence the lemma. ♠

This completes the proof of the Theorem 2.9.20 and hence that of Theorem 2.9.16.

Exercise 2.9.21

(i) Why does h, as given in the proof of Lemma 2.9.3, make sense? Why is it independent of the choice of F_2? Why is the map h so defined continuous on $|K| \times \mathbb{I}$?

(ii) Show that the composite of simplicial approximations is a simplicial approximation to the composite.

(iii) Consider the map $f(z) = z^2$. Show that for any simplicial complex K such that $|K| = \mathbb{S}^1$, there is no simplicial approximation $\phi : K \to K$ to f. (This simple example illustrates the need to subdivide only the domain of the function in order to get simplicial approximations.)

(iv) **Stellar-subdivision** There is a subdivision 'slower' than the barycentric subdivision which is quite useful in combinatorial aspects. Let F be a simplex of a simplicial complex. For each simplex G of K which contains F, let $B(G, F)$ denote the union of all the faces of G which do not contain F.
(a) Show that $B(G, F)$ is a subcomplex and $|B(G, F)|$ is homeomorphic to $\mathbb{D}^{n-1}$ or $\mathbb{S}^{n-1}$ where $n = \dim G$ depending upon F is a proper face of G or $F = G$.
(b) For each simplex G in K such that $F \subset G$, replace G by $cB(G, F)$ the cone over $B(G, F)$. (Keep simplices which do not contain F undisturbed.) Show that this gives a subdivision of K where the apex of the cones $cB(G, F)$ is mapped onto the barycentre $\beta(F)$ of F. This is called a *stellar subdivision of K obtained by adding just one extra vertex $\beta(F)$*. Figure 2.24 shows the stellar subdivisions of a tetrahedron into two or three tetrahedrons.

(c) Show that the barycentric subdivision of any finite simplicial complex can be obtained as an iterated stellar subdivision.

(v) **Star-subdivision** Here is another way to subdivide simplicial complexes. It is a slight generalization of barycentric subdivision in which we are allowed to omit some of the barycentres.

Let F be a simplex and L be a subdivision of ∂F, and v be a point in the interior of $|F|$. By *starring F at v* we mean the cone $L' = L \star \{v\}$ thought of as a subdivision of F, obtained by extending linearly, the identity map on K and the apex v being mapped to the point v itself. By a *star subdivision* of a simplex we mean a subdivision of F obtained by starring some faces of F finitely many times. By a star-subdivision of a simplicial complex K we mean a subdivision which is obtained by star-subdivision of some of its faces. In Figure 2.25, the first two pictures are star-subdivisions of a 2-simplex; the third one is not a star-subdivision and the fourth one is not a subdivision

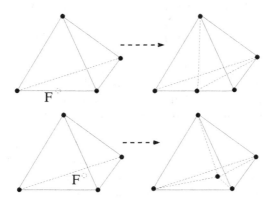

FIGURE 2.24. Two different stellar subdivisions of the tetrahedron

at all. (Note that starring some simplex of K does not always produce a subdivison of K. That is why we impose this condition in the definition itself.)

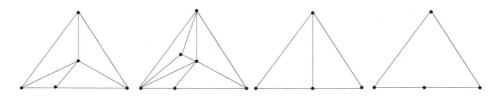

FIGURE 2.25. Which of them are star-subdivisions?

(a) If F is a maximal simplex of K, show that starring F defines a subdivision of K.

(b) Show that barycentric subdivision is a star-subdivision.

(c) Show that any subdivision of a 1-dimensional simplicial complex is a star-subdivision.

(d) Show that any convex polyhedron can be triangulated by starring its faces.

(vi) By a partial subdivision of a simplicial complex K, we mean a subdivision of only some of the simplices of K such that whenever $F_1 \subset F_2$ are two faces which have been subdivided, then the subdivision of F_1 induced from that of F_2 coincides with the given subdivision of F_1. Given a simplex F a subcomplex L of ∂F a subdivision L' of L and an open cover $\mathcal{U}$ of $|F|$ such that L' is finer than $\mathcal{U}$, show that there is a subdivision of F which is finer than the open covering and which restricts to L' on $|L|$. [Keep starring at the barycentres of all simplices not contained in L.]

(vii) Given any finite simplicial complex K, a subcomplex L and an open covering $\mathcal{U}$ of $|K|$ such that L is finer than $\mathcal{U}$, show that there is a subdivision of K which is finer than $\mathcal{U}$ and which restricts to L on $|L|$. (Hint: Induct on the skeletons $(K, L)^{(r)}$.)

(viii) Prove the following generalization of Theorem 2.9.7: *Let $f : |K_1| \to |K_2|$ be a continuous map, L_1 be a subcomplex of a finite simplicial complex K_1, and let L_1' be a subdivision of L_1 such that $f|_{|L_1|}$ is a simplicial map. Then there exists a subdivision K_1' of K_1 which extends the subdivision L_1' on L_1 and such that there are simplicial approximations $\phi : K_1' \to L$ such that $\phi|_{|L_1'|} = f$.*

2.10 Links and Stars

In this section, we shall introduce the 'combinatorial' study of simplicial complexes. We will have to wait until Chapter 5 to see the usefulness of the results here, when we take up the study of triangulated manifolds.

Definition 2.10.1 Let K be a (finite) simplicial complex and F be any simplex in K. Then the *open simplex* $<F>$ is the subset of $|K|$ consisting of all $\alpha \in |K|$ such that $\alpha(v) \neq 0$ iff $v \in F$. In general, an open simplex need not be an open subset of $|K|$. In fact, $<F>$ is open in $|K|$ iff F is a maximal simplex in K. Also note that $<F> = |F|$ iff $F = \{v\}$. An important thing to note is that the open simplices in a simplicial complex form a partition of $|K|$. Many topological properties of $|K|$ can be derived from this fact.

Definition 2.10.2 Let L be any subcomplex of K and let F be a simplex of L. Then the collection $\{H | H \cup F$ is a simplex of $L\}$ forms a subcomplex of L. (Verify.) This is called the *star* of F in L and is denoted by $St_L(F)$. When $L = K$, we use the simpler notation $St(F)$ for $St_L(F)$. Observe that $|St(F)|$ is naturally identified with the subspace of $|K|$ which is the union of all closed simplices $|G|$ in K that contain F. This subspace is referred to as the *closed star of F in K*. It is a closed subspace of $|K|$ containing $|F|$.

In contrast, consider now the subspace $st(F)$ defined as the union of all open simplices $<G>$ such that G is a simplex of K containing F. This is called the *open star* of F in K. Clearly it is an open subspace of $|K|$ containing $<F>$. Observe that $st(F)$ need not always contain $|F|$.

Definition 2.10.3 The *link*, $Lk_L(F)$ of F in L is defined to be the subcomplex of $St_L(F)$ consisting of simplices which are disjoint from F. Again when $L = K$ we denote it merely by $Lk(F)$. The space $|Lk(F)|$ is naturally identified with the subspace of $|K|$ which is the union of all closed simplices $|G|$ in $St(F)$ which are disjoint from $|F|$. Observe that

$$\dim(St(F)) = \dim(Lk(F)) + \dim(F) + 1.$$

Indeed, we have,

Lemma 2.10.4 In any simplicial complex and any face F we have, $St\,F = Lk(F) * F$. In particular, $|St(F)|$ is homeomorphic to the join $|Lk(F)| * |F|$.

Proof: The first assertion follows from the obvious fact that every simplex G in $St(F)$ can be written in a unique way as a disjoint union $G = F_1 \amalg H$, where $F_1 \subset F$ and $H \in Lk(F)$. Now for the second part. Given $\alpha \in |St(F)|$, $\alpha \in |G|$ for some $G \in St(F)$. Write $G = F_1 \amalg H$ with $H \subset Lk(F)$. We can write $\alpha = ta + (1-t)b$, for some $a \in |F|$, $b \in |Lk(F)|$ and $0 \leq t \leq 1$. Define $h : |St(F)| \to |Lk(F)| * |F|$ by $h(\alpha) = [a, 1 - t, b]$.

It is not hard to verify that this h is well defined and is indeed a homeomorphism as required. ♠

As a consequence, we have

Corollary 2.10.5 For any $x \in <F>$, we have,

$$|Lk(F)| * |\mathcal{B}(F)| \subset (|Lk(F)| * |F|) \setminus \{x\} = |St(F)| \setminus \{x\}$$

is a deformation retract of $|St(F)| \setminus \{x\}$.

Proof: Fix a homeomorphism $|\mathcal{B}(F)| * \{x\} \approx |F|$. This then yields

$$(|Lk(F)| * |\mathcal{B}(F)|) * \{x\} \approx |Lk(F)| * (|\mathcal{B}(F)| * \{x\}), \quad \text{(by associativity)}$$
$$\approx |Lk(F)| * |F| \approx |St(F)|.$$

On the other hand, we know that for any space Y the base $Y \times 0$ of the cone $Y * \{x\}$ is a deformation retract of $Y * \{x\} \setminus \{x\}$. Taking $Y = |Lk(F)| * |\mathcal{B}(F)|$, we get the required result. ♠

Theorem 2.10.6 *Let K be a simplicial complex, F, G be any two disjoint faces of a face $F \cup G$ in K. Then $Lk_{St(G)}(F) = St_{Lk(F)}(G)$.*

Proof: First observe that, $F \cap G = \emptyset$, and $F \cup G \in K$ implies that $G \in Lk(F)$. Thus under the given hypothesis we have,

$$
\begin{aligned}
L \in St_{Lk(F)}(G) &\Leftrightarrow L \in Lk(F), \& L \cup G \in Lk(F) \\
&\Leftrightarrow L \cup F \in K; \ L \cap F = \emptyset; \ L \cup G \cup F \in K \ \& \ (L \cup G) \cap F = \emptyset \\
&\Leftrightarrow L \cup G \in K; \ L \cap F = \emptyset \ \& \ L \cup F \cup G \in K \\
&\Leftrightarrow L \in St(G), \ L \cap F = \emptyset \ \& \ L \cup F \in St(G) \\
&\Leftrightarrow L \in Lk_{St(G)}(F).
\end{aligned}
$$

This completes the proof. ♠

Remark 2.10.7 In Chapter 5, (see Theorem 5.2.19) we shall use this theorem to sketch a proof of Poincaré duality, and a result due to Munkres on certain local-homological conditions on simplicial complexes (see Theorem 5.2.27). The latter result was a key step in the proof of Reisner's theorem, which itself was an important link in the proof of upper bound conjecture settled positively by Stanley (see [Stanely, 1975]). Just to understand that Theorem 2.10.6 is a non trivial result, try this:

Exercise 2.10.8 Given F, $G \in K$ such that $F \cup G \in K$, prove or disprove that

$$
St_{St(F)}(G) = St(F) \cap St(G).
$$

2.11   Miscellaneous Exercises to Chapter 2

1. (a) Triangulate a convex polytope without introducing extra vertices.
   (b) Triangulate $|\Delta_m| \times |\Delta_n|$ without introducing any extra vertices and such that for every pair of faces F, G of dimension k, l the restricted triangulation on $|F| \times |G| \subset |\Delta_m| \times |\Delta_n|$ coincides with that of $|\Delta_k| \times |\Delta_l|$.
   (c) Let K_1, K_2 be simplicial complexes with n_1, n_2 vertices. Triangulate $K_1 \times K_2$ with $n_1 n_2$ vertices.

2. Let K be a 1-dimensional simplicial complex. Show that K is path connected iff any two of its vertices can be joined by a sequence of edges. [Hint: Use simplicial approximation.]

3. Let K be any simplicial complex. Consider the inclusion maps $\iota_1 : |K^{(1)}| \longrightarrow |K|$ and $\iota_2 : |K^{(2)}| \longrightarrow |K|$. Show that they induce a bijection of path components. Show that the second one also induces an isomorphism of fundamental groups, when $|K|$ is connected. [Hint: Use simplicial approximation.] Deduce that
   (a) $|K|$ is path connected iff $|K^{(1)}|$ is.   (b)) $|K|$ is simply connected iff $|K^{(2)}|$ is.

4. Consider the quotient space D of the 2-simplex obtained by the identification of the three edges as shown in Figure 2.26. It is called the *dunce hat*.
   (a) Show that D has a CW-structure with a single cell in dimensions $0, 1$ and 2 each and a single 2-cell.
   (b) Show that D is triangulable.

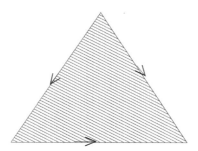

FIGURE 2.26. The dunce hat

(c) Show that D has the same homotopy type of the disc $\mathbb{D}^2$ and hence contractible.

(d) Try to write down an explicit contraction of D.

[Hint: For (c) use Theorem 1.6.10. For (d), go through the steps in the proof of this theorem.]

5. Given a simplicial map $\varphi : K \longrightarrow L$ show that it induces a simplicial map $\mathrm{sd}^n(\varphi) : \mathrm{sd}^n K \longrightarrow \mathrm{sd}^n L$.

6. Let $\varphi : K \longrightarrow L$ be a simplicial approximation to $f : |K| \longrightarrow |L|$. Is it necessary that $\mathrm{sd}\,\phi : \mathrm{sd}\,K \longrightarrow \mathrm{sd}\,L$ is a simplicial approximation to f?

7. If $f : \mathbb{D}^n \longrightarrow \mathbb{D}^n$ is a continuous map such that $\|f(x) - x\| < \epsilon$ for all $x \in \mathbb{S}^{n-1}$ where $0 < \epsilon < 1$ then the open ball $B_{1-\epsilon}(0)$ is contained in the image of the open ball $B_1(0)$ under f.

8. **Quotient complex** Let $K = (V, S)$ be a simplicial complex and R be an equivalence relation on the vertex set V. We define the quotient complex $K/R = (V', S')$ as follows: V' is the quotient set V/R and $F' \subset V'$ is in S' iff there is $F \in V$ such that $F/R = F'$. Clearly the quotient map $q : V \to V' = V/R$ becomes a simplicial map.

   (a) Use the above construction and the prism construction to define the mapping cylinder M_φ of a simplicial map $\varphi : K \to L$.

   (b) Show that the geometric realization $|M_\varphi|$ is homeomorphic to the mapping cylinder of $|\varphi| : |K| \to |L|$, thereby proving that the mapping cylinder of a simplicial map is triangulable.

9. Let $\lambda : \Delta_n \to \Delta_m$ be a surjective map. Show that the mapping cylinder $M_{|\lambda|}$ is homeomorphic to the convex hull of $\Delta_n \times \{0\} \cup \Delta_m \times \{1\}$ in $\mathbb{R}^{n+1} \times \mathbb{R}$. Use this to prove that the mapping cylinder of a simplicial map is triangulable.

10. Let $f : K \longrightarrow L$ be a simplicial map. Give a triangulation of the mapping cylinder $M_{|f|}$ of $|f| : |K| \longrightarrow |L|$ in such a way that $|K| \times 0 \subset M_{|f|}$ and $|L| \subset M_{|f|}$ are subcomplexes and the simplicial structure on them coincides with $\mathrm{sd}\,K$ and $\mathrm{sd}\,L$, respectively. [Hint: Follow the method somewhat similar to the prism construction.]

11. Show that every CW-complex is the homotopy type of a simplicial complex.

12. **Invariance of dimension** Show that if K is any triangulation whatsoever of $\mathbb{S}^n$, then $\dim K = n$. More generally, it is true that K_1 and K_2 are triangulations of the same topological space, then show that $\dim K_1 = \dim K_2$.

13. **Steifel manifolds and Grassmann manifolds:** For integers $1 \leq k \leq n$, a k-tuple $(\mathbf{v}_1, \ldots, \mathbf{v}_k)$ of vectors $\mathbf{v}_j \in \mathbb{R}^n$ is called an orthonormal k-frame if

$$\langle \mathbf{v}_i, \mathbf{v}_j \rangle = \delta_{ij}.$$

The subspace of $\mathbb{R}^{n \times k}$ consisting of all orthonormal k-frames in $\mathbb{R}^n$ is denoted by $V_{k,n}$ and is called the Steifel manifold of type (k, n). They are also called Steifel varieties, since they occur as the roots of polynomial equations and are studied extensively in algebraic geometry. Let $G_{k,n}$ denote the set of all k-dimensional linear subspaces of $\mathbb{R}^n$. They are called Grassmann manifolds, or Grassmann varieties. There is an obvious surjective map $\psi_k : V_{k,n} \to G_{k,n}$ sending $(\mathbf{v}_1, \ldots, \mathbf{v}_k)$ to the linear span $L(\mathbf{v}_1, \ldots, \mathbf{v}_k)$. We give the quotient topology to $G_{k,n}$ from $V_{k,n}$. The coordinate inclusion maps $\mathbb{R}^n \subset \mathbb{R}^{n+1}$ induce corresponding inclusions

$$\cdots V_{k,n} \subset V_{k,n+1} \subset \cdots$$

and we denote the infinite union by $V_k := V_{k,\infty}$ which is nothing but the space of orthonormal k-frames in $\mathbb{R}^\infty$.

Note that V_k can be thought of as a closed subspace of $\mathbb{S}^\infty \times \cdots \times \mathbb{S}^\infty$ (k factors).

(a) Write down an explicit homotopy $H : \mathbb{S}^\infty \times \mathbb{I} \to \mathbb{S}^\infty$ of $Id_{\mathbb{S}^\infty}$ with a constant map.

(b) Do the same thing for V_k.

(c) Show that the projection map $\tau_n : V_{k,n} \to \mathbb{S}^{n-1}$, viz., $\tau(\mathbf{v}_1, \ldots, \mathbf{v}_k) = \mathbf{v}_k$ is a (locally trivial) fibration with fibre $V_{k-1,n}$.

(d) Put the fibrations τ_n's together to obtain a fibration $V_k \to \mathbb{S}^\infty$ with fibre V_{k-1}.

(e) Show that the map $\psi_n : V_{k,n} \to G_{k,n}$ is a fibration with fibre $O(k)$, the group of orthogonal transformations of $\mathbb{R}^n$.

(f) Put these fibrations together to get a fibration

$$O(k) \subset V_k \xrightarrow{\psi} G_k,$$

where $G_k = G_{k,\infty}$ is the space of all k-dimensional linear subspaces of $\mathbb{R}^\infty$.

(g) Let $\rho : \mathbb{R}^\infty \to \mathbb{R}^\infty$ denote the right-shift operator given by $\rho(\mathbf{e}_i) = \mathbf{e}_{i+1}$ for every i. Let ρ^r denote $\rho \circ \rho \circ \cdots \circ \rho$ (r times). Show that each T^r induces an embedding of $V_k \to V_k$ which is isotopic to identity map $V_k \to V_k$. These in turn induce embeddings of $G_k \to G_k$ as well. Show that $T^r(X) = (\mathbf{e}_1, \ldots, \mathbf{e}_r, \rho(X))$ defines an embedding of $V_k \to V_{k+r}$ for each r and each k. In particular, they give a sequence of embeddings

$$V_1 \subset V_2 \subset V_3 \subset \cdots V_k \subset V_{k+1} \subset \cdots$$

which in turn induce embeddings

$$G_1 \subset G_2 \subset G_3 \subset \cdots G_k \subset G_{k+1} \subset \cdots$$

(h) Show that $V_{k,n}, G_{k,n}$ are smooth manifolds of dimension $nk - k(k+1)/2$ and $k(n-k)$, respectively.

14. **CW-structure on Grassmann manifolds** Solve the following sequence of exercises to obtain a CW-structure on each $G_k = G_k(\mathbb{R}^\infty)$ such that each $G_{k,n}$ is a subcomplex and such that each standard inclusion $G_k \hookrightarrow G_{k+1}$ is cellular.

(a) By a *Schubert symbol* σ we mean a finite sequence of integers, $\sigma := (\sigma_1, \sigma_2, \ldots, \sigma_k)$ satisfying

$$1 \leq \sigma_1 < \sigma_2 < \cdots < \sigma_k.$$

Show that there is a one-one correspondence s between the set $(\mathbb{Z}^+)^k$ of all sequences $\rho = (\rho_1, \ldots, \rho_k)$ of non negative integers and the set of all Schubert symbols of length k given by

$$s(\rho_1, \ldots, \rho_k) = (1 + \rho_1, 2 + \rho_1 + \rho_2, \ldots, k + \sum_1^k \rho_i)$$

(b) Given $L \in G_k$, show that there is a unique Schubert symbol $\sigma(L) = (\sigma_1, \ldots, \sigma_k)$ with the property:

$$\dim(L \cap \mathbb{R}^{\sigma_i}) = i, \ \& \ \dim(L \cap \mathbb{R}^{\sigma_i - 1}) = i - 1, \ \forall \ i = 1, 2, \ldots, k.$$

(c) Let $L \in G_k$. Show that $\sigma(L) = (\sigma_1, \ldots, \sigma_k)$ iff there exists a basis $\{v_1, \ldots, v_k\}$ of L, of column vectors such that $v_{\sigma_i, i} = 1$ and $v_{j,i} = 0$ for $j > \sigma_i$.

(d) Let $L \in G_k$. Show that L has a unique orthonormal basis $\{u_1, \ldots, u_k\}$ such that $u_i \in H^{\sigma_i}$, where H^n denotes the open upper-half subspace of $\mathbb{R}^n$ consisting of $(r_1, \ldots, r_n)$ with $r_n > 0$.

(e) Given two unit vectors $u \neq v \in \mathbb{R}^\infty$, let $T(u, v)$ denote the orthogonal transformation of $\mathbb{R}^\infty$ which fixes the subspace orthogonal to uv-plane and rotates u onto v. For any two elements $X, Y \in V_k$, let $T(X, Y)$ denote the composite

$$T(X, Y) = T(x_1, y_1) \circ T(x_2, y_2) \circ \cdots \circ T(x_k, y_k).$$

Given a Schubert symbol σ, let $u(\sigma) \in V_k$ be such that its ith vector has its σ_i^{th}-coordinate 1 (and all other coordinates zero). Given $n > \sigma_k$, let

$$D(\sigma, n) = \{u \in H^n \ : \ (u(\sigma), u) \in V_{k+1}\}.$$

Show that D is homeomorphic to a closed disc of dimension $n - k - 1$.

(f) Given a Schubert symbol σ, let

$$E'(\sigma) = V_k \cap (H^{\sigma_1} \times H^{\sigma_2} \times \cdots \times H^{\sigma_k})$$

and $\bar{E}'(\sigma)$ its closure in V_k. Define

$$f : \bar{E}'(\sigma) \times D(\sigma, n) \to \bar{E}'(\sigma_1, \ldots, \sigma_k, n)$$

by $f(X, u) = (X, T(u(\sigma), X)(u))$. Show that f is a homeomorphism.

(g) Show that $\bar{E}'(\sigma)$ is homeomorphic to $\mathbb{D}^{d(\sigma)}$, where $d(\sigma) = \sum_i \sigma_i - k(k+1)/2$ and $E'(\sigma)$ is its interior.

(h) Let $E(\sigma)$ be the subspace of G_k consisting of $X \in G_k$ such that $\sigma(X) = \sigma$. $E(\sigma)$ is called an *open Schubert cell*. Its closure G_k is called a *Schubert variety*. Show that the quotient map $\psi : V_k \to G_k$ maps $e(\sigma)$ onto $E(\sigma)$ homeomorphically.

(i) Show that the closures $\bar{E}(\sigma)$ as σ varies over all possible Schubert symbols of length k gives a CW-decomposition of G_k with $\psi : \bar{E}'(\sigma) \to G_k$ as characteristic maps.

(j) Show that the embedding $G_k \subset G_{k+1}$ induced by the shift operator ρ as in Exercise 13g above, is cellular and we have $G_{k+l}^{(k)} = G_k^{(k)}$, for all $l \geq 1$.

(k) By a partition of an integer $d \geq 0$, we mean a set $\{r_1, r_2, \ldots, r_s\}$ of positive integers such that $\sum_i r_i = d$. The number of partitions of d is denoted by $p(d)$. For example, $p(0) = 1$ (corresponding to the emptyset), $p(1) = 1, p(2) = 2, p(3) = 3, \ldots, p(10) = 42$, etc. Given a Schubert symbol σ, the sequence $(\sigma_1 - 1, \sigma_2 - 2, \sigma_k - k)$ gives a partition of $d(\sigma)$ after cancelling all possible occurrence of zeros in the beginning of the sequence. Show that the number of d-dimensional cells in $G_{k,n}$ is equal to the number of partitions of d into at most k integers in which each r_i is $\leq n - k$.

15. Let K be a simplicial complex. A n-simplex F in K is said to be free if there exists a $(n-1)$-face G of F which is not a face of any other simplex. In this situation, we remove both G and F from K to get a subcomplex K_1 of K and say that K_1 is obtained from K by an elementary collapsing. If there is a sequence of subcomplexes $K_1 \supset K_2 \supset \cdots \supset K_n = \{*\}$ a singleton, then we say K is collapsible. Show that if K is collapsible, then $|K|$ is contractible.

16. Show that the dunce hat is not collapsible.

17. **Duplex Igloo** Show that the two 'dimensional' subspace X of $\mathbb{R}^3$ shown in Figure 2.27 can be triangulated as a finite 2-dimensional simplicial complex. Also show that any such triangulation will not have any free simplex. Hence X is not collapsible. However, show that X is contractible. (See Exercise 4.)

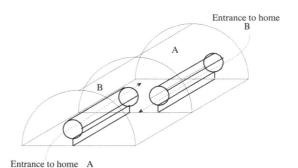

FIGURE 2.27. The duplex igloo

Chapter 3

Covering Spaces and Fundamental Group

We shall now study one of the most basic concepts in algebraic topology, viz., the *covering spaces*. They are closely related to the study of fundamental groups on the one hand and to the study of *the discontinuous groups* on the other. Having met the notion of fundamental groups, it is time to study the theory of covering spaces and their relation with fundamental groups. We shall also study a little bit about the 'discontinuous groups', vis-a-vis covering spaces and fundamental group. Classically however, these concepts occurred in the reverse order. The study of discontinuous groups goes back to the time of Gauss and occurred in the theory of elliptic functions and then in the theory of modular forms. During the time of Riemann, the notion of covering space started taking shape in what is today known as the theory of Riemann surfaces. The fundamental group appeared for the first time in the third installment of the celebrated papers ANALYSIS SITUS of Poincaré, around the turn of this century. Nowadays, these three notions have taken deep root in almost all branches of mathematics. They have been found useful and, in any case, make a very delightful subject of study.

3.1 Basic Definitions

In this section, we shall introduce the concept of covering projection, discuss some immediate properties and a few examples.

Definition 3.1.1 Let $p : \overline{X} \to X$ be a surjective map. We say an open subset V of X is *evenly covered* by p, if $p^{-1}(V)$ is a disjoint union of open subsets of $\overline{X}$:

$$p^{-1}(V) = \coprod_i U_i$$

where, each U_i is mapped homeomorphically onto V by p. In this case, we shall refer to each U_i as a *sheet for p over V*. If the space X can be covered by open subsets which are evenly covered by p, then we say that, p is a *covering projection*; the space $\overline{X}$ is called a *covering space* of X.

Remark 3.1.2
(a) Strictly speaking each time we mention the word 'covering space', we should not only mention the two spaces $\overline{X}$ and X but also the covering projection p therein. However, this will often be clear from the context and so, for simplicity of language, we will merely say, '$\overline{X}$ is a covering space of X'.[1] We also say, $\overline{X}$ is the *total space* and X is the *base space* of the covering projection p.
(b) Every covering projection is a local homeomorphism. (Recall that p is a local homeomorphism if $\forall \ \bar{x} \in \overline{X}$, there exists an open neighbourhood U of $\bar{x}$ such that $f|_U$ is a

---

[1]This is quite a common practice in mathematics and we refer to it as 'abuse of language'.

homeomorphism of U onto an open subset of X.) Indeed, $\overline{X}$ and X share all local topological properties of each other. For example, X is locally compact (respectively, locally connected, locally path connected, T_1, locally contractible, locally Euclidean, etc.) iff the same holds for $\overline{X}$. However, we will have to be careful with Hausdorffness, regularity, etc., which are, in any case, not local properties. (See Exercises (i), (ii) in 3.1.7.)

(c) Every local homeomorphism is an open map and so is every covering projection. In general, given a map $f : X \to Y$, and a point $y \in Y$, we call the set $f^{-1}(y)$, *fibre of f over y*. If f is a local homeomorphism then the fibres of f are discrete, i.e., the subspace topology on $f^{-1}(y)$ is discrete. In particular, the fibres of a covering projection are discrete. This fact is going to play a very important role in what follows.

(d) Clearly, given a covering projection, the cardinality of $p^{-1}(x)$ is a constant as x varies inside an evenly covered open set. Suppose X is connected, then since a locally constant function on X is a constant, it follows that the cardinality of $p^{-1}(x)$ is independent of $x \in X$. This common cardinality is referred to as the '*number of sheets*' of p. This terminology is borrowed from the theory of Riemann surfaces, where the notion of covering spaces has its roots. If this cardinality is finite, then we say that '*p is a finite covering*'. The map $z \mapsto z^n$ of $\mathbb{S}^1$ is a typical example of a finite covering, where the total space and base space are the same.

Example 3.1.3

(a) Any homeomorphism is a covering projection.

(b) A typical example of a covering projection is already familiar to us, viz., $\exp : \mathbb{R} \to \mathbb{S}^1$. For any fixed $\theta : 0 \leq \theta < 2\pi$, if we consider $U = \mathbb{S}^1 \setminus \{e^{i\theta}\}$, then $(\exp)^{-1}(U)$ is the union of disjoint intervals, $\theta + 2n\pi < t < \theta + (2n+2)\pi$. Restricted to any of these intervals, exp is a homeomorphism. (See Figure 1.12.)

In a similar way, it is not hard to see that the map $z \mapsto z^n$ defines a covering projection of $\mathbb{C}^\star$ onto itself. Here, n is any positive integer, and $\mathbb{C}^\star$ denotes the space of non zero complex numbers. This map restricted to the subspace $\mathbb{S}^1$ of complex numbers of modulus 1 defines a covering projection of $\mathbb{S}^1$ onto itself.

(c) If Y is any subspace of X, and if p is a covering projection onto X, then p restricts to a covering projection on $p^{-1}(Y)$ to Y. To see this, take the intersection of an evenly covered open subset V of X with Y to obtain an evenly covered open subset of Y. Such a result is not true for subspaces of the domain of a covering projection (see the theorem below).

(d) It is easy to construct examples of local homeomorphisms which are not covering projections. If we take the restriction of a covering projection : $\overline{X} \to X$ to an open set U of $\overline{X}$, it will continue to be a local homeomorphism. However, by choosing U appropriately we can arrange to destroy the covering space property. For instance, take $U = \overline{X} \setminus \bar{x}$ for any $\bar{x} \in \overline{X}$.

Remark 3.1.4 Path connectivity plays a very important role in algebraic topology. Most often, the discussion can be reduced to the case when the space is path connected. Since the nature of the covering projection is local–global, you can anticipate that even local path connectedness is important. This is illustrated in the following theorem.

Theorem 3.1.5 *Let $p : \overline{X} \longrightarrow X$ be a continuous map, where X is locally path connected.*
(1) *The map p is a covering projection iff for each component C of X, the restriction map $p : p^{-1}(C) \longrightarrow C$ is a covering projection.*
(2) *If p is a covering projection then for each component $\overline{C}$ of $\overline{X}$, the map $p : \overline{C} \longrightarrow p(\overline{C})$ is a covering projection and $p(\overline{C})$ is a component of X.*

Proof: (1) We have already seen that if p is a covering projection then for any subspace Y of X, the restriction map $p : p^{-1}(Y) \longrightarrow Y$ is a covering projection. Conversely, let

$p : p^{-1}(C) \longrightarrow C$ be a covering projection for each path component C of X. The crucial thing is that since X is locally path connected, each C is open. Therefore, for any point $x \in X$, consider the path component C which contains x and an open set $U \subset C$ containing x and evenly covered by $p : p^{-1}(C) \to C$. Then U is also an evenly covered neighbourhood of x in X for the map $p : \overline{X} \to X$.

(2) First of all $p(\overline{C}) =: C$ is open. Given $x \in C$, let V be a connected open neighbourhood of x which is evenly covered by p. Let $p^{-1}(V) = \coprod U_i$. Then each U_i is connected and hence, either $U_i \subset \overline{C}$ or $U_i \cap \overline{C} = \emptyset$. From this, it follows that V is evenly covered by $p|_{\overline{C}}$ also.

Finally to show that C is a component, let x be a point in its closure and V be an open neighbourhood of x as above. Then, one of the U_i has to intersect $\overline{C}$, which in turn means that, this U_i is contained in $\overline{C}$, and hence, $V \subset C$. This shows that C is open as well as closed. ♠

Remark 3.1.6

1. This theorem shows that, for locally path connected spaces, we can study covering spaces by restricting the given the covering projection to each of the path components of the total space, one at a time. So, from now onwards, in this section, **we shall assume that both the base and the total space of a covering projection are locally path connected and connected, unless specified otherwise or clear from the context.**

2. In algebraic geometry, the terminology 'covering projection' is used in a slightly different sense. Consider the following example. The map $\eta_n : \mathbb{C} \to \mathbb{C}$ given by $z \mapsto z^n$ is not a covering projection precisely at the point $z = 0$. Indeed, $\eta_n : \mathbb{C}^* \to \mathbb{C}^*$ is an n-sheeted covering, i.e., for every point $z \neq 0$, there are precisely n solutions of the equation $\eta_n(w) = z$. If we count these solutions with multiplicity, then this is true for $z = 0$ also. The point $z = 0$ is called a *branch point* or a *ramification point*. So, if we stick to the terminology of algebraic topology, a natural way would have been to call such maps covering projections with ramifications. In algebraic geometry, especially in the study of curves, we come across this type of map all the time and it would be too inconvenient to mention the word 'ramifications' all the time. So, the standard practice is to call this larger class of maps as covering projections and call the smaller class of better behaved ones *unramified covering projections*. So, do not apply results in algebraic topology about covering projections directly in algebraic geometry, but make this small modification beforehand.

Exercise 3.1.7

(i) Let $p : \overline{X} \to X$ be a covering projection. Show that if X is Hausdorff then $\overline{X}$ is Hausdorff. Further, if p is finite-to-one, then show that if $\overline{X}$ is Hausdorff then X is Hausdorff. (Caution: Hausdorffness is **not** a local property.)

(ii) Here is an example to show that we cannot relax the condition 'finite-to-one' on p in the previous exercise (see Remark 1.3.5 (g)). Consider the following equivalence relation on $\mathbb{R}^2 \setminus \{(0,0)\}$:

$$(x, y) \sim \left(\frac{1}{2^n} x, 2^n y \right), \text{ for all non negative integers } n.$$

Show that the quotient map $q : \mathbb{R}^2 \setminus \{(0,0)\} \to X$ is actually a covering projection. Show that the points $[(1,0)], [(0,1)] \in X$ cannot be separated by disjoint open sets and hence X is not Hausdorff.

(iii) Check that in the previous example the base space X is T_1. Hence it cannot be regular, nor normal. However the total space is actually Euclidean.

(iv) Let $p : \overline{X} \longrightarrow X$ be a continuous mapping. A continuous map $s : X \longrightarrow \overline{X}$ such that $p \circ s = Id_X$ is called a *section* of p. Suppose $\overline{X}$ is connected also. Suppose that
(a) p is a local homeomorphism and $\overline{X}$ is Hausdorff OR
(b) p is a covering projection.
Show that any section of p is a homeomorphism onto $\overline{X}$.
(Hint: Show that $s(X)$ is both open and closed in $\overline{X}$.) (Some books assume that X is locally path connected also in these exercises.)

3.2   Lifting Properties

We shall now embark upon showing that a covering projection has homotopy lifting property (HLP) (see Section 1.1). HLP asserts that a certain map 'exists'. As elsewhere in mathematics, there is a 'uniqueness' result which goes hand in hand with it. Indeed, often the uniqueness result can be put to use in proving the 'existence result', which is the case here. So we shall first have this uniqueness result. Recall that given $f : Y \to X$ a map $g : Y \to \overline{X}$ is called a lift (through $p : \overline{X} \to X$) if $p \circ g = f$.

Theorem 3.2.1 *Let $p : \overline{X} \to X$, be a covering projection, Y be any connected space and $f : Y \to X$ be any map. Let $g_1, g_2 : Y \to \overline{X}$ be any two lifts of f, such that, for some point $y \in Y$, $g_1(y) = g_2(y)$. Then $g_1 = g_2$.*

Proof: Let $Z = \{y \in Y : g_1(y) = g_2(y)\}$. It is given that, Z is non empty. Thus, if we show that, Z is open and closed then from the connectivity of Y, it follows that, $Z = Y$, i.e., $g_1 = g_2$.

Let $y \in Z$ and let V be an evenly covered open neighbourhood of $f(y)$ in X. Let U be an open subset of $\overline{X}$ mapped homeomorphically onto V by p and let $g_1(y) = g_2(y) \in U$. Choose W, an open neighbourhood of y in Y such that, $g_j(W) \subset U$ for $j = 1, 2$. Then, $p \circ g_1(z) = f(z) = p \circ g_2(z)$, $\forall z \in W$. Since $p|_U$ is injective, this implies that, $g_1(z) = g_2(z)$, $\forall z \in W$ and hence $W \subset Z$. Therefore Z is open.

[If $\overline{X}$ were Hausdorff, then the closedness of Z follows easily. We would like to prove it without the Hausdorffness assumption, just to emphasis the fact that the closedness of Z in this context has nothing to do with the Hausdorffness of $\overline{X}$.]

So, let z be a point in Y such that, $g_1(z) \neq g_2(z)$. Let V be an evenly covered open neighbourhood of $p \circ g_1(z) = p \circ g_2(z)$. For $i = 1, 2$, we can find an open neighbourhood U_i of $g_i(z)$ on which p is a homeomorphism and such that $U_1 \cap U_2 = \emptyset$. By continuity of g_1, g_2, there is an open neighbourhood W of z such that $g_i(W) \subseteq U_i$, $i = 1, 2$. It follows that W is an open neighbourhood of z not intersecting Z. Hence Z is closed as required. ♠

The next step is to prove the HLP of a covering projection for singleton spaces. This can also be termed 'path lifting property' (PLP).

Theorem 3.2.2 (Path lifting property) *Let $p : \overline{X} \to X$ be a covering projection. Then given a path $\omega : \mathbb{I} \to X$ and a point $\bar{x} \in \overline{X}$ such that, $p(\bar{x}) = \omega(0)$, there exists a path $\bar{\omega} : \mathbb{I} \to \overline{X}$ such that, $p \circ \bar{\omega} = \omega$ and $\bar{\omega}(0) = \bar{x}$.*

Proof: Let $Z = \{t \in \mathbb{I} : \bar{\omega} \text{ is defined in } [0, t]\}$. Observe that Z is a subinterval of $\mathbb{I}$ and contains 0. Let t_0 be the least upper bound of Z. It is enough to show that $t_0 \in Z$ and $t_0 = 1$.

Let V be an evenly covered open neighbourhood of $\omega(t_0)$. For $0 < \epsilon < 1$, put $I_\epsilon = [t_0 - \epsilon, t_0 + \epsilon] \cap \mathbb{I}$. Choose ϵ so that $\omega(I_\epsilon) \subset V$. Let U_i be the open neighbourhood of $\overline{\omega}(t_0 - \epsilon/2)$, that is mapped homeomorphically onto V. Then $\lambda = p^{-1} \circ \omega$ is a lift of ω on I_ϵ. Observe that $\lambda(t_0 - \epsilon/2) = \overline{\omega}(t_0 - \epsilon/2)$. Therefore, by the uniqueness theorem, $\lambda(t) = \overline{\omega}(t), \forall t \in [t_0 - \epsilon, t_0 - \epsilon/2]$. Therefore the two lifts can be patched up. That is, $\overline{\omega}$ can be extended to a lift of ω on the interval $[0, t_0 + \epsilon] \cap \mathbb{I}$. By the definition of t_0, we must then have $[0, t_0 + \epsilon] \cap \mathbb{I} = \mathbb{I}$ which means that $t_0 = 1$ and $t_0 \in Z$. ♠

Now we are ready to prove HLP of covering projections.

Theorem 3.2.3 *Every covering projection is a fibration.*

Proof: Let $p : \overline{X} \to X$ be a covering projection, Y be a topological space, $H : Y \times \mathbb{I} \to X$, $g : Y \to \overline{X}$ be the maps such that, $p \circ g(y) = H(y, 0), \forall y \in Y$. By the PLP of p, it follows that there is a unique function $G : Y \times \mathbb{I} \to \overline{X}$ such that $p \circ G = H$, $G|_{Y \times 0} = g$ and $G|_{y \times \mathbb{I}}$ is continuous for all $y \in Y$. It remains to prove that G is continuous as a function on $Y \times \mathbb{I}$.

Given any point $y \in Y$, we shall first construct an open neighbourhood W_y of y and a partition of $\mathbb{I} : 0 = t_0 < t_1 \cdots < t_n = 1$, with the property that each $H(W_y \times [t_i, t_{i+1}])$ is contained in an evenly covered open subset of X as follows: First choose an evenly covered open subset $V_{y,t}$ around $H(y, t), \forall t \in \mathbb{I}$ and use the compactness of $y \times \mathbb{I}$ to find finitely many of these $V_{y,t}$'s to cover $H(y \times \mathbb{I})$. Next choose a partition

$$0 = t_0 < t_1 < t_2 < \cdots < t_m = 1$$

of $\mathbb{I}$ such that $H(y \times [t_i, t_{i+1}])$ is contained in an evenly covered open set, say, $V_{y,i}, \forall i$. Again using compactness of $\mathbb{I}$ (Wallace theorem), find open neighbourhood $W_{y,i}$ of y such that $H(W_{y,i} \times [t_i, t_{i+1}])$ is contained in $V_{y,i}$. Now take $W_y = \cap_{i=1}^n W_{y,i}$. Check that the open neighbourhood W_y and the partition of $\mathbb{I}$ are as required.

[In Figure 3.1, we have deliberately taken Y as the closed interval $\mathbb{I}$. The proof of this theorem is very similar to that of Proposition 1.2.22. In the proof of Proposition 1.2.22, first we replace the exponential map by the given covering map and make corresponding changes in the proof. (This is what we have already done.) Next, we replace $\mathbb{I}$ by Y with very little modification to obtain the proof of the present theorem. Here are the details.]

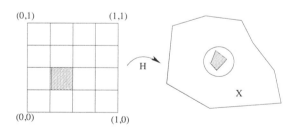

FIGURE 3.1. Subdivision finer than an even covering

Setting $W = W_y$, we shall prove that if $G|_{W \times \{t_i\}}$ is continuous then $G|_{W \times [t_i, t_{i+1}]}$ is continuous. Since $G|_{W \times 0} = g|_W$, successive application of this implication will produce that $G|_{W \times [t_i, t_{i+1}]}$ is continuous for all i. Since these subsets form a finite closed cover of $W \times \mathbb{I}$, it would follow that, $G|_{W \times \mathbb{I}}$ is continuous. Since $\{W_y \times \mathbb{I}\}$ form an open cover of $Y \times \mathbb{I}$, that will establish the continuity of G.

Let us consider an arbitrary point $z \in W$. By continuity of $G|_{W \times \{t_i\}}$, there exists an open neighbourhood A of z in W such that $G(A \times \{t_i\})$ is contained in U, say, which is mapped homeomorphically onto an evenly covered open set V. By the uniqueness of the lifts of paths, it follows that $G(A \times [t_i, t_{i+1}]) \subset U$. But then on $A \times [t_i, t_{i+1}]$, G, being equal to $p^{-1} \circ H$ is continuous. Since subsets of the form $A \times [t_i, t_{i+1}]$ cover $W \times [t_i, t_{i+1}]$, it follows that $G|_{W \times [t_i, t_{i+1}]}$ is continuous, as required. ♠

In subsequent sections, we shall apply this theorem to obtain several interesting properties of covering projections. Here is a warm-up exercise.

Exercise 3.2.4 Let $p : Y \to X \times \mathbb{I}$ be a covering projection where X is a locally path connected and path connected space. Show that $p : p^{-1}(X \times \{t\}) \to X \times \{t\}$, the restriction of p is a covering projection, for each $t \in I$. Moreover, for each $t, s \in \mathbb{I}$ show that there are homeomorphisms $\Theta(t, s)$ which make the following diagram commutative.

$$
\begin{array}{ccc}
p^{-1}(X \times t) & \xrightarrow{\Theta(t,s)} & p^{-1}(X \times s) \\
\downarrow{\scriptstyle p} & & \downarrow{\scriptstyle p} \\
X \times t & \xrightarrow{\quad Id \quad} & X \times s
\end{array}
$$

3.3   Relation with the Fundamental Group

Having established the HLP for covering projections, we continue our study of lifting maps with respect to a covering projection. Let us fix a base point $x_0 \in X$ and put $F := p^{-1}(x_0)$. To begin with, we know that every path can be lifted. What about loops? Given a loop $\omega : \mathbb{I} \longrightarrow X$ at x_0 say, we can lift it at $\bar{x} \in F$ as a path. By the ULP, if this path is not a loop, then we are helpless, in the sense that there is no loop at $\bar{x}$ which is a lift of ω. Maybe there is one at a different point $\bar{y} \in F$. We observe that if the lift of ω were a loop then it represents an element in $\pi_1(\overline{X}, \bar{x})$ which is mapped onto the element $[\omega]$ in $\pi_1(X, x_0)$. Thus we are led to study the inter-relationship between fundamental groups and covering projections. We begin with two lemmas which are immediate consequences of unique path lifting property.

Lemma 3.3.1 Let $p : \overline{X} \to X$ be a covering projection, $x_0 \in X, \bar{x}_0 \in \overline{X}$ and $p(\bar{x}_0) = x_0$. Let ω be a path in X with $\omega(0) = x_0$, and let $\bar{\omega}$ be a lift of ω at $\bar{x}_0$.
(i) The end-point of $\bar{\omega}$, viz., $\bar{\omega}(1)$ depends only on the path homotopy class of ω in X and not on actual representative path.
(ii) $\bar{\omega}$ is a loop in $\overline{X}$ if and only if ω is a loop in X such that $[\omega] \in p_{\#}(\pi_1(\overline{X}, \bar{x}))$. In this case, lift of any member of $[\omega]$ at $\bar{x}_0$ is a loop.

Proof: (i) Suppose ω' is a path in X path-homotopic to ω. (Note that this implies $\omega(0) = \omega'(0) = x_0$ and $\omega(1) = \omega'(1) = x_1$, say. Let H be a path homotopy from ω to ω' and $\overline{H}$ be a lift of H such that $H(0, 0) = \bar{x}_0$. By the uniqueness of the lifts, it follows that $\bar{H}(-, 0) = \bar{\omega}, \bar{H}(-, 1) = \bar{\omega}'$. It follows that $\overline{H}(0, s) \subset p^{-1}(x_0)$ and $\overline{H}(1, s) \subset p^{-1}(x_1)$. Hence, by the discreteness of the fibres and the connectedness of $\mathbb{I}$, $\bar{H}(0, s) = \bar{x}_0, \bar{H}(1, s) = \bar{x}_1, \forall s \in \mathbb{I}$ where $\bar{x}_1$ is a single element such that $p(\bar{x}_1) = x_1$. In particular, $\bar{\omega}(1) = \bar{H}(1, 0) = \bar{x}_1 = \bar{H}(1, 1) = \bar{\omega}'(1)$.
(ii) Easy. ♠

Lemma 3.3.2 Suppose ω_1, ω_2 are paths in X, with initial point x and end-point y. Suppose that, $\omega_1 * \omega_2^{-1}$ lifts to a loop at $\bar{x}$, where $p(\bar{x}) = x$. Let $\overline{\omega}_1, \overline{\omega}_2$ be the lifts of ω_1, ω_2, respectively, at $\bar{x}$. Then $\overline{\omega}_1(1) = \overline{\omega}_2(1)$.

Proof: Let γ be the loop at $\bar{x}$ such that, $p \circ \gamma = \omega_1 * \omega_2^{-1}$. By the uniqueness of the lift, it follows that, $\overline{\omega}_1(t) = \gamma(t/2) \; \forall \; t \in \mathbb{I}$. It also follows that, $\gamma(1 - t/2) = \overline{\omega}_2(t), \; \forall \; t \in \mathbb{I}$. In particular, $\overline{\omega}_2(1) = \gamma(1/2) = \overline{\omega}_1(1)$, as claimed. ♠

Combining the above two lemmas we have,

Theorem 3.3.3 *Let $p : \overline{X} \to X$ be a covering projection of path connected spaces, $x \in X, \bar{x} \in \overline{X}$ be such that $p(\bar{x}) = x$ and let $F = p^{-1}(x)$. A loop ω at x in X can be lifted to a loop at $\bar{x} \in F$ iff the element $[\omega]$ belongs to $p_\#(\pi_1(\overline{X}, \bar{x}))$. There is a loop in $\overline{X}$ which is a lift of ω iff some conjugate of $[\omega]$ belongs to the subgroup $p_\#(\pi_1(\overline{X}, \bar{x}))$.*

Proof: The first part is just Lemma 3.3.1. To see the second part, suppose that ω can be lifted to a loop $\bar{\omega}$ at $\bar{x}_1$. Choose a path λ from $\bar{x}$ to $\bar{x}_1$ in $\overline{X}$. Put $\tau = [p \circ \lambda] \in \pi_1(X, x)$. Check that $p_\#([\lambda \bar{\omega} \lambda^{-1}]) = \tau[\omega]\tau^{-1}$. Conversely, if there exists $\tau \in \pi_1(X, x)$ such that $\tau[\omega]\tau^{-1} \in p_\#(\pi_1(\overline{X}, \bar{x}))$, let θ be a loop in $\overline{X}$ at $\bar{x}$ such that $[p \circ \theta] = \tau[\omega]\tau^{-1}$. Let λ be the lift at $\bar{x}$ of a loop γ representing τ. It follows that $\theta = \lambda * \bar{\omega} * \lambda_1$, where $\bar{\omega}$ is the lift of ω at $\bar{x}_1 = \lambda(1)$ and λ_1 is a lift of γ^{-1} at $\bar{\omega}(1) = \bar{x}_2$. But now $\lambda_1 * \lambda$ is the lift of $\gamma^{-1} * \gamma$ which represents the trivial element. Therefore $\lambda_1 * \lambda$ is a loop, by Lemma 3.3.3. This means $\bar{x}_1 = \bar{x}_2$ which implies that $\bar{\omega}$ is a loop. ♠

Now the relation between fundamental group and covering space starts revealing itself. The subgroup $p_\#(\pi_1(\overline{X}, \bar{x}))$ of $\pi_1(X, x)$ has a special role to play in lifting property of the covering projection p. Obviously, we would then like to know how this subgroup is related to $\pi_1(\overline{X}, \bar{x})$ itself.

Theorem 3.3.4 *Let $p : \overline{X} \to X$ be a covering projection and $\bar{x} \in \overline{X}$ be such that, $p(\bar{x}) = x$. Then the induced homomorphism $p_\sharp : \pi_1(\overline{X}, \bar{x}) \to \pi_1(X, x)$ is injective. Moreover, there is a surjection $\Theta : \pi_1(X, x) \longrightarrow F = p^{-1}(x)$ which defines a bijection of right cosets of $K := p_\#(\pi_1(\overline{X}, \bar{x}))$ with F.*

Proof: Let $\bar{\omega}$ be a loop at $\bar{x}$ and H be a homotopy of $p \circ \bar{\omega}$ to the constant loop at x, relative to the end-points, i.e., $H : \mathbb{I} \times \mathbb{I} \to X$ be such that, $H(t, 0) = p \circ \bar{\omega}(t)$, $H(t, 1) = x$, $\forall \, t \in \mathbb{I}$ and $H(0, s) = H(1, s) = x$, $\forall \, s \in \mathbb{I}$. Let $\overline{H}$ be the lift of H such that, $\overline{H}(t, 0) = \bar{\omega}(t)$, $\forall \, t \in \mathbb{I}$. Then, $\overline{H}(0, s) \in p^{-1}(x)$. Hence, as seen above, by the discreteness of the fibre and the connectedness of $\overline{H}(0 \times \mathbb{I})$, it follows that, $\overline{H}(0, s) = \bar{x}$. For the same reason, it also follows that $\overline{H}(1, s) = \bar{x} = \overline{H}(t, 1)$, $\forall \, t, s \in \mathbb{I}$. Thus, $\overline{H}$ is a homotopy of $\bar{\omega}$ to the constant loop, relative to the end-points. This proves the first part.

Given a loop class $[\omega] \in \pi_1(X, x)$, lift ω to a path at $\bar{x}$ and let $\Theta([\omega]) = \omega(1)$. By lemma 3.3.1 (i), it follows that Θ is well defined. Given any $z \in F$ take a path τ from $\bar{x}$ to z in $\overline{X}$ and then see that $p \circ \tau$ is a loop at x and $\Theta([p \circ \tau]) = z$. Now $\Theta[\omega] = \Theta[\lambda]$ iff the lifts of ω and λ at $\bar{x}$ have the same end-points iff $\omega * \lambda^{-1}$ lifts to a loop at $\bar{x}$ iff $[\omega][\lambda]^{-1} \in K$ iff $K[\omega] = K[\lambda]$. ♠

Remark 3.3.5 We shall now investigate the effect of changing the base point in $\overline{X}$ without, of course, changing the base point in X, i.e., remaining within $p^{-1}(x_0) = F$. Remember that since $\overline{X}$ is path connected, the isomorphism class of $\pi_1(\overline{X})$ is not affected by change of base point. Can we then say that as a subgroup of $\pi_1(X)$ also, there is no change?

Proposition 3.3.6 For various points $\bar{x} \in p^{-1}(x_0)$, the subgroups $p_\#(\pi_1(\overline{X}, \bar{x}))$ of $\pi_1(X, x_0)$ are conjugate to each other.

Proof: Take a path ω in $\overline{X}$ from $\bar{x}_1$ to $\bar{x}_2$. We then know that

$$[\alpha] \mapsto [\omega^{-1} * \alpha * \omega]$$

defines an isomorphism of $\pi_1(\overline{X}, \bar{x}_1)$ onto $\pi_1(\overline{X}, \bar{x}_2)$. Observe that $p \circ \omega$ is a loop at x. Let $\tau = [p \circ \omega] \in \pi_1(X, x)$. When we pass onto the base space, the above isomorphism becomes the conjugation by the element τ^{-1}. ♠

The above result naturally leads us to investigate the case when K is a normal subgroup of $\pi_1(X, x)$.

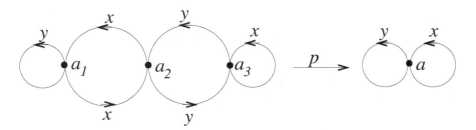

FIGURE 3.2. A covering which is not normal

Definition 3.3.7 The covering projection $p : \overline{X} \longrightarrow X$ is called a *normal covering* (or a *Galois covering* or a *regular covering*) if the subgroup $p_\#(\pi_1(\overline{X}, \bar{x}))$ is normal in $\pi_1(X, x)$.

Remark 3.3.8 It is immediate from this definition that if $\pi_1(X, x_0)$ is abelian, then every covering projection $p : \overline{X} \longrightarrow X$ is normal. However, we shall soon see that there are very many interesting spaces with $\pi_1(X, x)$ not abelian. So, having some topological criterion for a normal covering is quite desirable. Combining Proposition 3.3.6 with Theorem 3.3.3, we immediately obtain the following criterion for normal coverings.

Theorem 3.3.9 *The covering projection $p : \overline{X} \longrightarrow X$ is normal covering iff given a loop ω at x, all its lifts to $\overline{X}$ are either loops or none is.*

Example 3.3.10 We shall use the above criterion to show that the bouquet of two circles (see Exercise 1.9.32) has fundamental group non abelian. (Compare Exercise 1.2.33.(viii)) Figure 3.2 indicates a 3-fold covering projection map $p : \overline{X} \to X$ where X is the bouquet of two circles.

Consider the loop xyx^{-1} in X at the point a and three of its lifts at the points a_1, a_2, a_3, respectively. The second one is a loop, whereas the other two are not loops. Therefore p is not a normal covering and hence $p_\#(\pi_1(\overline{X}, a_i))$ are not normal subgroups of $\pi_1(X, a)$. (See Exercise 3.4.25.(vi) for another way to see that $\pi_1(\mathbb{S}^1 \vee \mathbb{S}^1)$ is non commutative.)

Remark 3.3.11 Having solved the lifting problem for loops satisfactorily, we now take up the problem of lifting maps defined on more general spaces. However, the nature of our investigation does not allow complete arbitrariness. We need to restrict ourselves to locally path connected spaces. Since the problem can always be studied component-wise, we can further assume that the spaces are connected. So, begin with a space Y which is path connected and a map $f : Y \longrightarrow X$. Let $y \in Y$ and $f(y) = x$ be the base point in X. Taking $\bar{x} \in F$ as the base point in $\overline{X}$, we ask: Is there a map $\tilde{f} : Y \longrightarrow \overline{X}$ such that $\tilde{f}(y) = \bar{x}$ and $p \circ \tilde{f} = f$? The following theorem, which gives a complete answer to this question, is an important milestone in our journey. It is also a typical example of how a topological problem is converted into an algebraic one.

Theorem 3.3.12 (Lifting criterion) *Let $p : \overline{X} \longrightarrow X$ be a covering projection of locally path connected and connected spaces. Let Y be a locally path connected and connected space and $f : Y \longrightarrow X$ be a map. Given $y \in Y, \bar{x} \in \overline{X}$, such that $p(\bar{x}) = f(y)$, there exists a map $\bar{f} : Y \longrightarrow \overline{X}$ such that $p \circ \bar{f} = f$ and $\bar{f}(y) = \bar{x}$ iff $f_\#(\pi_1(Y, y)) \subseteq p_\#(\pi_1(\overline{X}, \bar{x}))$.*

Proof: Given $\bar{f}$, as above, we have, $f_\sharp = (p \circ \bar{f})_\sharp = p_\sharp \circ \bar{f}_\sharp$, and hence, $f_\sharp(\pi_1(Y, y)) = p_\sharp(\bar{f}_\sharp(\pi_1(Y, y))) \subseteq p_\sharp(\pi_1(\overline{X}, \bar{x})) =: K$.

Conversely, suppose $f_\sharp(\pi_1(Y,y)) \subset K$. Given any point $z \in Y$, choose a path ω from y to z in Y. Let $\bar{\omega}$ be the lift of $f \circ \omega$ at $\bar{x}$ and let $\bar{f}(z) = \bar{\omega}(1)$.

We have to first show that, $\bar{f}$ is well-defined, i.e., it is independent of the choice of the path ω joining y to z. So, let γ be another such path, $\bar{\gamma}$ be the lift of $f \circ \gamma$ at $\bar{x}$. Then, $\omega * \gamma^{-1}$ is a loop at y. Since $f_\sharp[\omega * \gamma^{-1}]$ is an element of K, by Lemma 3.3.2, the lifts of $f \circ \omega$ and $f \circ \gamma$ should have the same end-point. Thus $\bar{f}$ is well-defined.

Clearly, $\bar{f}$ is a lift of f. It remains to show that $\bar{f}$ is continuous. So, let z be any point of Y, U be an open neighbourhood of $\bar{f}(z)$ mapped homeomorphically onto an evenly covered open neighbourhood V of $f(z)$. By the continuity of f, and the local path connectivity of Y, we can get a path connected open neighbourhood W of z in Y such that $f(W) \subset V$. Let ω be the path from y to z chosen to define $\bar{f}(z)$. For each point $a \in W$, we can choose a path γ_a inside W joining z to a, and then use the path $\omega * \gamma_a$ to define $\bar{f}(a)$. If $\bar{\gamma}_a$ is the lift of $f \circ \gamma_a$ at $\bar{f}(z)$, then, clearly, $\bar{\omega} * \bar{\gamma}_a$ is the lift of $f \circ (\omega * \gamma_a)$. Hence, $\bar{f}(a) = \bar{\gamma}_a(1)$. On the other hand, since V is evenly covered, and p maps U homeomorphically onto V, it follows that the entire path $\bar{\gamma}_a$ is contained in U. In particular, $\bar{\gamma}_a(1) = \bar{f}(a) \in U$. Thus, we have proved that $\bar{f}(W) \subseteq U$, thereby completing the proof of the continuity of $\bar{f}$. ♠

Apart from application within covering space theory itself, which we shall study in the next section, this result has many applications elsewhere also. Here is just a sample.

Corollary 3.3.13 Let Y be a locally path connected and simply connected space and $p : \overline{X} \to X$ be a covering projection. Then every map $f : Y \to X$ has a lift $\hat{f} : Y \to \overline{X}$. In particular, every map $f : Y \to \mathbb{S}^1$ is null-homotopic.

Proof: The first part is obvious. In order to prove the latter part, consider the covering projection $\exp : \mathbb{R} \to \mathbb{S}^1$. First get a lift $g : Y \to \mathbb{R}$ of $f : Y \to \mathbb{S}^1$. Now, use the fact that $\mathbb{R}$ is contractible to conclude that g is null-homotopic. A null-homotopy of g composed with $\exp$ would then yield the required null-homotopy of $f : Y \to \mathbb{S}^1$. ♠

Example 3.3.14 Let us construct some non trivial coverings, $p : \overline{X} \to X$. Starting with X, it is a consequence of the above results that we need to 'break' some loops in the base space X into non-loops. Obviously, these loops must be representing non trivial elements in the fundamental group of the base. For instance, consider the space X which is the union of $\mathbb{S}^2$ and the diameter $[-p, p]$ where $p = (1, 0, 0)$. Note that $\mathbb{S}^2 \subset X \setminus \{0\}$ is a deformation retract and hence $X \setminus \{0\}$ is simply connected. Because of this, as a consequence of lifting criterion, it follows that in any covering space of X we will have various copies of $X \setminus \{0\}$. To cover the missing point $\{0\}$, we take a neighbourhood of this point, say, the open interval $(-p/2, p/2)$. We now equip ourselves with several copies of $X \setminus \{0\}$ and the interval $(-p/2, p/2)$ and start gluing them systematically to construct various coverings of X. Figure 3.3 depicts three such examples: a 2-sheeted cover, a 3-sheeted cover and an infinite sheeted cover.

Exercise 3.3.15 Let $p : \overline{X} \to X$ be a simply connected covering of a path connected space X, and $A \subset X$ be a path connected subset.
(a) Show that the inclusion induced homomorphism $i_\# : \pi_1(A) \to \pi_1(X)$ is surjective, iff $p^{-1}(A)$ is path connected.
(b) Show that $i_\# : \pi_1(A) \to \pi_1(X)$ is injective iff each path component of $p^{-1}(A)$ is simply connected.

3.4 Classification of Covering Projections

We continue to assume that X is connected and locally path connected and all covering spaces over X are also connected. The relation between subgroups of $\pi_1(X)$ and covering projections over X will be investigated further.

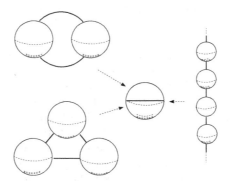

FIGURE 3.3. Three different coverings of a 2-sphere with a diameter attached

We shall now introduce the notion of equivalence of covering projections.

Definition 3.4.1 Two covering projections, $p_i : \overline{X}_i \to X$, $i = 1, 2$, are said to be *equivalent* if there is a homeomorphism $f : \overline{X}_1 \to \overline{X}_2$ such that $p_2 \circ f = p_1$.

Remark 3.4.2 Clearly, this defines an equivalence relation amongst the covering projections over X. The following is an immediate consequence of Theorem 3.3.12. In Exercise 3.2.4, we came across such a situation. Given a covering $p : Y \to X \times I$, we get various covering projections over X by taking restriction of p over $X \approx X \times \{t\} \subset X \times I$. The exercise precisely says that these are all equivalent to each other.

Proposition 3.4.3 Two covering projections of X are equivalent iff they define the same subgroup of $\pi_1(X)$ up to conjugation.

Proof: Given two covering projections $p_i : \overline{X}_i \longrightarrow X$, $i = 1, 2$, suppose $f : \overline{X}_1 \longrightarrow \overline{X}_2$ is an equivalence. If $\bar{x}_i$ are the base points such that $f(\bar{x}_1) = \bar{x}_2$, then, since f is a homeomorphism, we have $f_\#(\pi_1(\overline{X}_1, \bar{x}_1)) = \pi_1(\overline{X}_2, \bar{x}_2)$. Therefore, $p_{1\#}(\pi_1(\overline{X}_1, \bar{x}_1)) = p_{2\#}(\pi_1(\overline{X}_2, \bar{x}_2))$.

Conversely, suppose $p_{1\#}(\pi_1(\overline{X}_1, \bar{x}_1)) = \tau^{-1}(p_{2\#}(\pi_1(\overline{X}_2, \bar{x}_2)))\tau$, for some element $\tau \in \pi_1(X, x)$ where $x = p_1(\bar{x}_1) = p_2(\bar{x}_2)$. Let λ be a path in $\overline{X}_2$ such that $\lambda(0) = \bar{x}_2$, $[p_2 \circ \lambda] = \tau$ and $\hat{x}_2 = \lambda(1)$. Then we have $[\lambda]^{-1}\pi_1(\overline{X}_2, \bar{x}_2)[\lambda] = \pi_1(\overline{X}_2, \hat{x}_2)$. Therefore, it follows that

$$p_{2\#}(\pi_1(\overline{X}_2, \hat{x}_2)) = \tau^{-1}p_{2\#}(\pi_1(\overline{X}_2, \bar{x}_2))\tau = p_{1\#}(\pi_1(\overline{X}_1, \bar{x}_1)).$$

By applying the lifting criterion, either way, we get maps $f : \overline{X}_1 \longrightarrow \overline{X}_2$ and $g : \overline{X}_2 \longrightarrow \overline{X}_1$ such that $p_2 \circ f = p_1$ and $p_1 \circ g = p_2$ and $f(\bar{x}_1) = \hat{x}_2$, $g(\hat{x}_2) = \bar{x}_1$. Now $p_2 \circ f \circ g = p_2$ and $f \circ g(\hat{x}_2) = \hat{x}_2$. Therefore, by ULP, we have $f \circ g = Id_{\overline{X}_2}$. Likewise, we see $g \circ f = Id_{\overline{X}_1}$. Therefore f (and g) defines an equivalence of p_1 and p_2. ♠

Remark 3.4.4 Thus, we see that each equivalence class of a connected covering space corresponds to a unique conjugacy class of a subgroup of the fundamental group. To complete the picture, given any subgroup K of $\pi_1(X)$, we should also be able to tell whether or not there exists a covering projection $p : \overline{X} \to X$ such that $p_\#(\pi_1(\overline{X}, \bar{x}))$ is conjugate to K. In particular, taking $K = (e)$, we ask: does there exist a covering projection for which the total space is simply connected? Our next goal will be to answer these questions reasonably well.

Definition 3.4.5 Let $p : \overline{X} \to X$ be a covering projection. By a covering transformation of $p : \overline{X} \to X$, we mean a homeomorphism $f : \overline{X} \to \overline{X}$ such that, $p \circ f = p$. It is easily seen that the set $\mathbf{G}(p)$ of all covering transformations of p forms a subgroup of all homeomorphisms of $\bar{X}$ under the usual composition of maps. This group is called the *Deck transformation group* of p. It is also called the Galois group of p.

Lemma 3.4.6 There is an injective mapping Φ of $\mathbf{G}(p)$ to the set of right cosets of K in $\pi_1(X, x)$, where, $K = p_\#(\pi_1(\overline{X}, \bar{x}))$. Moreover, K is a normal subgroup iff this map Φ is an isomorphism of groups. In any case, the cardinality of $\mathbf{G}(p)$ is less than or equal to the number of sheets of p.

Proof: Given $f \in \mathbf{G} = \mathbf{G}(p)$, let γ be a path from $\bar{x}$ to $f(\bar{x})$. Let $\Phi(f) = K[p \circ \gamma]$. Check that, Φ is well defined. To show that Φ is injective, let $g \in \mathbf{G}$ be another element such that $\Phi(f) = \Phi(g)$. If τ is a path joining $\bar{x}$ to $g(\bar{x})$, we have, $K[p \circ \gamma] = K[p \circ \tau]$. This implies, from Lemma 3.3.2, that γ and τ have the same end-point, i.e., $f(\bar{x}) = g(\bar{x})$. By the uniqueness of the lift, it follows that $f = g$.

Suppose that Φ is surjective. Recall that from Theorem 3.3.4, there is a bijection between the right cosets of K and the fibre $p^{-1}(x)$. From this, it follows that to each $z \in p^{-1}(x)$ there exists $g \in \mathbf{G}(p)$ such that $g(\bar{x}) = z$. Now given any $[\omega] \in \pi_1(X, x)$ let $\bar{\omega}$ be a lift of ω at $\bar{x}$. Then the lift of ω at z is given by $g \circ \bar{\omega}$. Clearly $g \circ \bar{\omega}$ is a loop iff $\bar{\omega}$ is. This proves the normality of the cover p and hence that of K.

Conversely, suppose that the subgroup K is normal in $\pi_1(X, x)$. Then the right-cosets of K form the quotient group $K \backslash \pi_1(X, x)$. Given $K[\omega]$, lift the loop ω to a path at $\bar{x}$, and let z be the end-point. Then, $p_\#(\pi_1(\overline{X}, z))$ is conjugate to K and hence, by normality, is equal to K. Apply the lifting criterion to the map p itself, to get $f : \overline{X} \to \overline{X}$ such that, $f(\bar{x}) = z$ and $p \circ f = p$. Now use the lift of ω at $\bar{x}$, to join $\bar{x}$ and z and thereby to see that, $\Phi(f) = K[\omega]$. This shows that, Φ is onto. It remains to see that Φ is a homomorphism.

Let ω and τ be paths from $\bar{x}$ to $f(\bar{x})$ and $g(\bar{x})$, respectively. Then $g \circ \omega$ is a path joining $g(\bar{x})$ and $g \circ f(\bar{x})$. Hence, we have,

$$
\begin{aligned}
\Phi(g \circ f) &= K[p \circ (\tau * g \circ \omega)] \\
&= K[(p \circ \tau) * (p \circ \omega)] \\
&= K[p \circ \tau] K[p \circ \omega] = \Phi(g)\Phi(f).
\end{aligned}
$$

The last assertion in the lemma follows from Theorem 3.3.4. ♠

As an immediate consequence we have:

Theorem 3.4.7 *Let* $p : \overline{X} \to X$ *be a connected normal covering,* $\bar{x} \in \overline{X}$ *be such that* $p(\bar{x}) = x \in X$. *Then we have an exact sequence of groups and homomorphisms:*

$$
1 \longrightarrow \pi_1(\overline{X}, \bar{x}) \xrightarrow{p_\#} \pi_1(X, x) \xrightarrow{\Psi} G(p) \longrightarrow 1.
$$

Proof: Put $\Psi = \Phi^{-1} \circ q$ where $q : \pi_1(X, x) \to \pi_1(X, x)/\pi_1(\overline{X}, \bar{x})$ is the quotient map. ♠

Corollary 3.4.8 Let $p : \overline{X} \to X$ be a covering projection, where $\overline{X}$ is connected, and simply connected. Then $\Phi : \mathbf{G}(p) \to \pi_1(X, x_0)$ is an isomorphism of the group of covering transformations to the fundamental group of X. The covering projection p has the following universal property: given any connected covering projection $q : Z \to X$, there exists a map $f : \overline{X} \to Z$ such that, $p \circ f = q$.

Proof: The first statement is a direct consequence of the above lemma. The second one follows from a simple application of the lifting criterion.

Definition 3.4.9 Let $p : \overline{X} \to X$ be a covering projection of connected spaces. Fix $x \in X$ and $\bar{x} \in \overline{X}$ such that $p(\bar{x}) = x$. We say $p : \overline{X} \to X$ is *universal*, if for any given connected covering projection $q : Z \to X$, and an element $z \in Z$ such that $q(z) = x$, there is a unique map $f : \overline{X} \to Z$ such that, $p = q \circ f$ and $f(\bar{x}) = z$.

Lemma 3.4.10 Given two universal covering projections $p_i : \overline{X}_i \longrightarrow X, i = 1, 2$, there exists a homeomorphism $\psi : \overline{X}_1 \longrightarrow \overline{X}_2$ such that $p_2 \circ \psi = p_1$. A universal covering space, if it exists, is unique in this sense.

Proof: Apply the universal property of p_1 to obtain the map $\psi : \overline{X}_1 \longrightarrow \overline{X}_2$ and the universal property of p_2 to obtain a map $\xi : \overline{X}_2 \longrightarrow \overline{X}_1$ such that $p_2 \circ \psi = p_1$, $p_1 \circ \xi = p_2$ and $\psi(\bar{x}_1) = \bar{x}_2, \xi(\bar{x}_2) = \bar{x}_1$. It follows that $\psi \circ \xi$ and $\xi \circ \psi$ are covering homeomorphisms of $\overline{X}_2$ and $\overline{X}_1$, respectively, which fix the points $\bar{x}_2$ and $\bar{x}_1$, respectively. Therefore, they are identity transformations. Hence, ψ is also a homeomorphism.   ♠

Remark 3.4.11 Thus a connected universal covering projection over X, if it exists, is unique up to equivalence of covering projections. The above corollary says, in particular, that any simply connected covering projection is a universal covering projection. We can now complete the answer to the question of existence of covering projections corresponding to other subgroups, assuming that simply connected coverings exist.

Theorem 3.4.12 *Let X be connected, locally path connected space, admitting a simply connected covering space, $p : \overline{X} \to X$. Then for every subgroup K of $\pi_1(X)$ there corresponds a covering projection, $q : Z \to X$ such that, $q_\#(\pi_1(Z)) = K$.*

Proof: By Corollary 3.4.8, the function $\Phi : \mathbf{G}(p) \to \pi_1(X, x_0)$ as in the Lemma 3.4.6 is an isomorphism. Let $K' = \Phi^{-1}(K)$. We then take Z as the quotient space of $\overline{X}$ by the relations:

$$z_1 \sim z_2 \text{ iff } z_2 = \phi(z_1), \phi \in K'.$$

Clearly, there is a commutative diagram

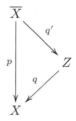

where q' is the quotient map. If V is a connected open subset evenly covered by p, we claim that, it is also evenly covered by q. This follows from the fact that, if $p^{-1}(V) = \sqcup U_i$, then any covering transformation of p maps each U_i homeomorphically to another U_j. (Incidentally, it turns out that q' is also a covering projection but we shall not need this here. See Section 3.5 for more.)

Given $[\omega] \in \pi_1(Z, z_0)$, we want to show that $q_\#([\omega]) \in K$. So, we take a lift $\bar{\omega}$ of $q \circ \omega$ in $\overline{X}$ at $\bar{x}_0$. Let $f \in \mathbf{G}(p)$ be the unique element such that $f(\bar{x}_0) = \bar{\omega}(1)$. Then by definition of Φ, we have, $\Phi(f) = q_\#([\omega])$. Since $q \circ q' \circ \bar{\omega} = p \circ \bar{\omega} = q \circ \omega$, and $\omega(0) = z_0 = q' \circ \bar{\omega}(0)$, by the uniqueness of the lifts in Z, we have $q' \circ \bar{\omega} = \omega$. In particular, $q' \circ \bar{\omega}(1) = \omega(1) = z_0 = q'(\bar{x}_0)$. This implies that $f \in K'$ and hence $\Phi(f) \in K$. This proves $q_\#\pi_1(Z, z_0) \subset K$. Proof of the other way inclusion is simialr and easier.   ♠

Remark 3.4.13 The existence of a simply connected covering projection over a given space X requires some more hypothesis than local path connectivity. Say, $p : \overline{X} \longrightarrow X$ is a covering projection and $\overline{X}$ is simply connected. Given a point $x \in X$, if U is an evenly covered open neighbourhood of x then there are copies of U in $\overline{X}$, i.e., we have open subsets V of $\overline{X}$ such that $p : V \longrightarrow U$ is a homeomorphism. Then we can write $i : U \hookrightarrow X$ as a composite of

$$U \xrightarrow{p^{-1}} V \hookrightarrow \overline{X} \xrightarrow{p} X$$

and therefore, it follows that $i_{\#} : \pi_1(U, x) \longrightarrow \pi_1(X, x)$ is the trivial homomorphism.

It turns out that this local condition on X is sufficient also for the existence of $\overline{X}$. So, we shall first make a few definitions.

Definition 3.4.14 Let X be a locally path connected and connected space. We say X is *semi-locally simply connected* if each point x of X has a path connected open neighbourhood U such that the inclusion induced homomorphism $\pi_1(U) \longrightarrow \pi_1(X)$ is trivial.

Definition 3.4.15 Let X be a topological space. We say X is *locally contractible* if for each point in X, there is a fundamental system of neighbourhoods consisting of contractible open subsets of X. Similarly, if for each point $x \in X$, we have a fundamental system of neighbourhoods which are all simply connected then we say that the space X is locally simply connected.

Remark 3.4.16 Any locally contractible space is semi-locally simply connected. Any locally simply connected space is semi-locally simply connected. In particular, all manifolds belong to this class. All CW-complexes and simplicial complexes are locally contractible. Hence the entire covering space theory is applicable to them. We shall now take up the task of proving the following theorem.

Theorem 3.4.17 *Over a connected, locally path connected and semi-locally simply connected space X, there exists a simply connected covering space.*

Remark 3.4.18 Understanding the proof of this theorem is not all too necessary to master the basics of covering space theory and may be skipped, especially by a reader who is not familiar with the function space topology. Also a reader who is quite conversant with the notion of compact-open-topology may even directly read the proof of the theorem below, skipping the two lemmas which only serve the purpose of motivating the constructive proof and are not logical necessities for the proof.

The idea involved can be definitely traced back to the function theory of one complex variable and more or less imitates the construction of a Riemann surface of a meromorphic function. One may say that HLP together with unique path lifting (UPL) property capture all the homotopic properties of a covering projection, though they fall a little short of characterizing a covering projection. Thus, a covering space can be thought of as a suitable space of classes of paths in the given space. The lemma below justifies this. We begin with a definition:

Definition 3.4.19 Let X be a locally path connected and path connected space, $x_0 \in X$ be any point. Then the set $\mathbf{P}(\mathbf{X}, \mathbf{x_0})$ of all paths in X beginning at x_0, with the compact-open-topology is called the *path space* over X. There is an obvious map $e : \mathbf{P}(\mathbf{X}, \mathbf{x_0}) \to X$ defined by $e(\omega) = \omega(1)$, which is surjective. This is called the *evaluation map*.

Lemma 3.4.20 The path space $\mathbf{P} := P(X, x_0)$ is contractible and the evaluation map $e : \mathbf{P} \longrightarrow X$ is an open map.

Proof: Check that the map $\Lambda : \mathbf{P} \times \mathbb{I} \times \mathbb{I} \longrightarrow X$ given by $\Lambda(\omega, t, s) = \omega(ts)$ defines a homotopy of the constant map with the identity of $\mathbf{P}$. Therefore $\mathbf{P}$ is contractible. To prove that e is an open mapping, recall that the collection of sets of the form:

$$< K, U >= \{\gamma \in \mathbf{P} : \gamma(K) \subseteq U\}$$

is a subbase for the compact-open-topology. So it suffices to prove that, the image under e of the intersection of finitely many such sets is open in X. So let $\omega \in < K_i, U_i >$, $i = 1, \ldots, n-1$, and $\omega(1) = x$. If $1 \in K_i$, then note that, $x \in U_i$. So, let U_n be a path connected neighbourhood of x contained in those U_i for which $1 \in K_i$. Let $0 < \epsilon < 1$ be such that $\omega([\epsilon, 1]) \subseteq U_n$ and such that $[\epsilon, 1] \cap K_j = \emptyset$ whenever $1 \notin K_j$. Put $K_n = [\epsilon, 1]$ so that, $\omega(K_n) \subseteq U_n$. Let

$$W = \bigcap_{1 \leq i \leq n} < K_i, U_i >$$

We claim that, $e(W) = U_n$. Clearly $e(W) \subseteq U_n$. Given any point $y \in U_n$ choose a path $\tau : [\epsilon, 1] \to U_n$ from $\omega(\epsilon)$ to y. If γ is defined to be equal to ω on $[0, \epsilon]$ and equal to τ on $[\epsilon, 1]$, then one observes that, γ is in W and $e(\gamma) = y$. This completes the proof of the claim that $e(W) = U$. Hence the map e is open. ♠

Lemma 3.4.21 Let X be a locally path connected and connected space. Suppose that, $p : \overline{X} \to X$ is a covering projection with $\overline{X}$ path connected, $x_0 \in X, \bar{x}_0 \in \overline{X}$ and $p(\bar{x}_0) = x_0$. Then the induced map $p_* : P(\overline{X}, \bar{x}_0) \to P(X, x_0)$ given by $\omega \mapsto p \circ \omega$ is a homeomorphism.

Proof: Clearly, p_* is continuous. Given $\omega \in \mathbf{P}(X, x_0)$ let $\bar{\omega}$ be the unique lift of ω at $\bar{x}_0$. Define $\phi(\omega) = \bar{\omega}$. By ULP, it follows that ϕ is the inverse p_*. It remains to prove that ϕ is continuous.

Since $\mathbb{I}$ is a compact Hausdorff space, it follows that $\phi : P(X, x_0) \to P(\overline{X}, \bar{x}_0)$ is continuous iff the associated map $\Phi : P(X, x_0) \times \mathbb{I} \to \overline{X}$ given by $\Phi(\omega, t) = \bar{\omega}(t)$ is continuous. So, let us prove the continuity at an arbitrary point $(\omega_0, s_0) \in P(X, x_0) \times \mathbb{I}$. We shall consider the case $0 < s_0 < 1$ and leave the two extreme cases to you as an exercise. Let U be any open set of $\overline{X}$ such that $\Phi(\omega_0, s_0) = \bar{\omega}_0(s_0) \in U$. Without loss of generality, we may assume that $p(U)$ is evenly covered by p.

Choose a partition $0 = t_0 < t_1 < \cdots < t_n = 1$ of $\mathbb{I}$ such that $s_0 \in (t_j, t_{j+1})$ for some j and such that there are evenly covered connected open sets V_i in X with $\omega_0([t_i, t_{i+1}]) \subset V_i$. We may as well assume that $p(U) = V_j$. Let $U_i \subset \overline{X}$ be open sets mapped homeomorphically to V_i and such that $\bar{\omega}_0([t_i, t_{i+1}]) \subset U_i$ for each i. Also note that $U_j = U$.

Put $\Omega = \cap_{i=1}^n < [t_i, t_{i+1}], V_i >$. Then $\Omega \times (t_j, t_{j+1})$ is an open neighbourhood of (ω_0, s_0). It suffices to show that $\Phi(\Omega \times (t_j, t_{j+1})) \subset U$.

Let $(\omega, s) \in \Omega \times (t_j, t_{j+1})$. Let $\bar{\omega}$ be the lift of ω at $\bar{x}_0$. If $p^{-1} : V_1 \to U_1$ is the local inverse of p, then, by ULP, we must have $\bar{\omega}(t) = p^{-1}(\omega(t))$, for all $t \in [0, t_1]$. Then $\bar{\omega}[0, t_1] \subset U_1$. Since $\omega(t_1) \in V_2$, it follows that $\bar{\omega}(t) = p^{-1}\omega(t)$ for all $t \in [t_1, t_2]$ where $p^{-1} : V_2 \to U_2$ is the local inverse of p. Therefore $\bar{\omega}[t_1, t_2] \subset U_2$ and so on. Inductively, it follows that $\bar{\omega}([t_i, t_{i+1}]) \subset U_i$ for all i and hence, in particular, $\bar{\omega}((t_j, t_{j+1})) \subset U_j = U$, i.e., $\Phi(\omega, s) \in U$. This completes the proof. ♠

Remark 3.4.22 We have already proved that the evaluation map $e : \mathbf{P}(X, x) \to X$ is an open mapping. For a path connected space X this is clearly a surjection as well. Therefore X can be thought of as a quotient space of $\mathbf{P}(X, x)$. From the above lemma it follows that every covering space of X is also a quotient space of $\mathbf{P}(X, x)$. Thus, if at all, a simply connected covering space exists for X, then it has to be a certain quotient of $\mathbf{P}(X, x)$. This is the main idea in the following proof.

Proof of Theorem 3.4.17

Let $x_0 \in X$ be a fixed base point and consider the space $\mathbf{P}$ of all paths in X with x_0 as the initial point. This space is given the compact-open-topology. Define an equivalence relation in $\mathbf{P}$ by saying $\omega \sim \gamma$ iff the two paths are path-homotopic. Let $\overline{X}$ be the quotient space of equivalence classes and let $\phi : \mathbf{P} \longrightarrow \overline{X}$ be the quotient map. Since $\omega \sim \gamma$ implies that $\omega(1) = \gamma(1)$, it follows that the evaluation map $e : \mathbf{P} \longrightarrow X$ factors through ϕ to define a map $p : \overline{X} \to X$. We shall claim that this is the object that we are seeking.

To show that p is a covering projection, let V be any path connected open set in X such that $i_{\#} : \pi_1(V) \longrightarrow \pi_1(X)$ is the trivial homomorphism. Let us temporarily call such an open set 'ambiently 1-connected'. Then we claim that V is evenly covered by p. Since X can be covered by ambiently 1-connected open sets, this would prove that p is a covering projection.

Given $\omega \in \mathbf{P}$ such that $\omega(1) \in V$ consider the set

$$V_{[\omega]} = \{[\omega * \omega'] \in \overline{X} \ : \ \omega' \text{ is a path in } V\}.$$

Then
(i) $p^{-1}(V) = \cup\{V_{[\omega]} \ : \ \omega(1) \in V\}$. (For, $[\omega] \in V_{[\omega]}$.)
(ii) $[\tau] \in V_{[\omega]} \Longrightarrow V_{[\omega]} = V_{[\tau]}$. Therefore $V_{[\omega]} \cap V_{[\tau]} \neq \emptyset \Longrightarrow V_{[\omega]} = V_{[\tau]}$.
Thus we have proved that $p^{-1}(V)$ is a disjoint union

$$p^{-1}(V) = \coprod V_{[\omega]}.$$

Since V is path connected, it follows that $p : V_{[\omega]} \longrightarrow V$ is surjective. To show that it is injective, suppose $p[\tau_1] = p[\tau_2]$ where $[\tau_i] \in V_{[\omega]}$. Then $\tau_i \simeq \omega * \omega_i$ for some $\omega_i \subset V$. Also $\omega_1(1) = \omega_2(1)$. Since $i_{\#} : \pi_1(V) \longrightarrow \pi_1(X)$ is trivial, we know that $[\omega_1 * \omega_2^{-1}] = 1$. Therefore,

$$[\tau_1] = [\omega * \omega_1] = [\omega * \omega_1 * \omega_2^{-1} * \omega_2] = [\omega * \omega_2] = [\tau_2].$$

This proves that $p : V_{[\omega]} \longrightarrow V$ is injective. Observe that the openness of e implies that p is also open.

To show that each $V_{[\omega]}$ is open, we actually show that $\phi^{-1}(V_{[\omega]})$ is open in $\mathbf{P}$. Let $\lambda \in \phi^{-1}(V_{[\omega]})$. Cover λ with finitely many ambiently 1-connected open subsets, $V_1, \ldots, V_n = V$, and get a partition $0 = t_0 < \cdots < t_n = 1$, such that, $\lambda([t_i, t_{i+1}]) \subset V_i$. If W denotes the set of all the paths λ' at x_0, such that, $\lambda'([t_i, t_{i+1}]) \subset V_i, \ \forall i = 1, \ldots, n$, then by the definition of the compact open topology, W is open in $\mathbf{P}$ and $\lambda \in W$. Using the fact that each V_i is ambiently 1-connected, it can be seen that each element of W is path homotopic to $\lambda * \omega'$ with $\omega' \subset V$. Therefore, $W \subset \phi^{-1}(V_{[\lambda]}) \subset \phi^{-1}(V_{[\omega]})$. This proves that $V_{[\omega]}$ is open.

Finally, it remains to prove that, $\overline{X}$ is connected and simply connected. Again the connectivity follows from that of $\mathbf{P}$. To show the simple connectivity of $\overline{X}$, let $\Lambda : \mathbb{I} \longrightarrow \overline{X}$ be a loop at $[C(x_0)]$ where $[C(x_0)]$ denotes the homotopy class of the constant loop at x_0. In order to show that this loop is null-homotopic in $\overline{X}$, by the injectivity of $p_{\#}$ it suffices to show that $\omega := p \circ \Lambda$ is null homotopic in X.

Now, the path $\Omega : \mathbb{I} \longrightarrow \mathbf{P}$ defined by $\Omega(t)(s) = \omega(ts)$ is such that $e \circ \Omega = \omega$. Therefore $p \circ \phi \circ \Omega = \omega$. Thus we have two lifts, Λ and $\phi \circ \Omega$ of ω in $\overline{X}$. Moreover, $(\phi \circ \Omega)(0) = [C(x_0)] = \Lambda(0)$. Therefore, by unique path lifting property of p it follows that $\Lambda = \phi \circ \Omega$. Therefore $\phi \circ \Omega(1) = \Lambda(1) = [C(x_0)]$ which means that $\omega = \Omega(1)$ is null-homotopic.

This completes the proof of Theorem 3.4.17. ♠

Remark 3.4.23

(1) It may happen that a space may not have any simply connected covering space. It may even happen that a space may have a universal covering projection over it, but no simply

connected covering projections. An interested reader may look into the exercises below for such examples.

(2) Besides local properties, a covering space shares a lot of global properties of the base space such as being a topological group, etc. We end this section with a sample result about CW-structures which has several applications. See the last section for some of them.

Theorem 3.4.24 *Let $X' \to X$ be a connected covering projection where X is a CW-complex. Then X' admits a CW-structure in such a way that*
(i) *p maps each open n-cell of X' onto an open n-cell of X for each n.*
(ii) *Each covering transformation f of p permutes the set of n-cells of X' for each n.*
(iii) *Let $F = p^{-1}(x_0), x_0 \in X$ be any fibre and C_n, C'_n denote the sets indexing the n-cells in X, X', respectively. Then there is bijection $C'_n \approx C_n \times F$.*
In particular, if X is a finite CW-complex and $X' \to X$ is k-sheeted then $\chi(X') = k\chi(X)$.
(iv) *If the CW-structure of X is coming from a simplicial complex structure then the CW-structure on X' is also a simplicial complex and p is simplicial with respect to these simplicial structures.*

Proof: Let the n-cells of X be indexed by the sets C_n with characteristic maps $\phi_{\alpha,n} : \mathbb{D}^n \to X^{(n-1)}$. We put $(X')^{(n)} = p^{-1}(X^{(n)})$ and inductively prove that $X'^{(n)} \setminus X'^{(n-1)}$ is the disjoint union of n-cells index by $C_n \times F$.

Since $X^{(0)}$ is a discrete subset of X, it follows that $X'^{(0)} = p^{-1}(X^{(0)})$ is a discrete space. Now suppose we have described $X'^{(n-1)}$. Fix a base point $* \in \mathbb{D}^n$. Since $\mathbb{D}^n$ is simply connected, by lifting criterion, each map $\phi_{\alpha,n} : \mathbb{D}^n \to X$ has a unique lift $\phi_{\alpha,q,n}$ at each of the points $q \in p^{-1}(\phi_{\alpha,n}(*)) \approx F$. Clearly, each $\phi_{\alpha,q,n}$ is a homeomorphism of the interior of $\mathbb{D}^n$ onto its image. So, we can declare that these are the characteristic functions for n-cells of X'. Clearly $p \circ \phi_{\alpha,q,n} = \phi_{\alpha,n}$. Since for each $\alpha \in C_n$, $\{\phi_{\alpha,q,n}\}$ gives all possible disjoint lifts of $\phi_{\alpha,n}$, it follows that the union of these n-cells covers $X'^{(n)}$.

It remains to verify why the topology of X' is the weak topology with respect to these cells. There are different ways of seeing this. We shall leave this as an exercise. (See Exercise (vii) below.) The rest of the conclusions are all obvious. ♠

Exercise 3.4.25 In all the exercises below, it is assumed that the spaces are locally contractible and connected.

(i) Show that, a simply connected space has no nontrivial covering projection over it, i.e., if $p : \overline{X} \to X$ is a covering projection with X simply connected and $\overline{X}$ connected, then p is a homeomorphism.

(ii) **The real projective space $\mathbb{P}^n$:**
Recall from Remark 1.3.5 (j), the $(n+1)$-dimensional real vector space $\mathbb{R}^{n+1}$. Introduce an equivalence relation amongst the non zero vectors in it as follows: $v_1 \sim v_2$ iff there exists a scalar $\lambda \neq 0$ such that, $v_1 = \lambda v_2$. Verify that, this is indeed an equivalence relation on $\mathbb{R}^{n+1} \setminus \{0\}$. Denote the quotient space by $\mathbb{P}^n$.
(a) If $p : \mathbb{S}^n \to \mathbb{P}^n$ denotes the restriction of the quotient map, to the unit sphere $\mathbb{S}^n$ in $\mathbb{R}^{n+1}$, then show that, p itself is a quotient map. Indeed, p merely identifies a unit vector v with its antipodal $-v$. Deduce that, $\mathbb{P}^n$ is compact and Hausdorff.
(b) Verify that, p is a covering projection with the number of sheets equal to 2.
(c) Use this to show that the fundamental group of $\mathbb{P}^n$ is $\mathbf{Z}_2$ for $n \geq 2$.
(d) Show that, $\mathbb{P}^1$ is homeomorphic to $\mathbb{S}^1$, even though the map $p : \mathbb{S}^1 \to \mathbb{P}^1$ is not a homeomorphism.
(e) Finally, for $n \geq 2$, prove that any map $f : \mathbb{P}^n \to \mathbb{S}^1$ is null homotopic.

(iii) Show that $z \mapsto z^n$ defines a covering projection of $\mathbb{S}^1$ onto itself, for all positive integers. What is the number of sheets of this cover? Are there any other covering projections over $\mathbb{S}^1$. How about determining all of them? What happens if we change $\mathbb{S}^1$ to $\mathbb{C} \setminus \{0\}$? What is the universal covering space of $\mathbb{C} \setminus \{0\}$?

(iv) If $p : \overline{X} \to X$ and $q : \overline{Y} \to Y$ are covering projections, show that their Cartesian product is a covering projection. Thus finite product of covering projections is a covering projection. What do you expect about an arbitrary product of covering projections?

(v) Show that if X is locally path connected/locally contractible then so is its suspension SX. Also, further, show that, if X is connected, then SX is simply connected.

(vi) We have already seen that the fundamental group of the bouquet of two circles is non abelian. (See Example 3.3.10.) Here is yet another method. Let Y be the subspace of $\mathbb{R}^2$, consisting of all horizontal lines with integral y coordinate and all vertical lines with integral x-coordinate.

    (a) Show that, any of the unit squares contained in Y is a retract of Y. In particular, deduce that, Y is not simply connected.

    (b) Restrict the map $\exp \times \exp : \mathbb{R}^2 \to \mathbb{S}^1 \times \mathbb{S}^1$ to the subspace Y and denote it by p. Show that, p is a covering projection onto its image $X \subset \mathbb{S}^1 \times \mathbb{S}^1$.

    (c) Show that, $X = \mathbb{S}^1 \times \{1\} \cup \{1\} \times \mathbb{S}^1$.

    (d) Let ω and γ denote the loops at $(1,1)$ which go around once, the first and the second circles, respectively. Lift the loops $\omega * \gamma$ and $\gamma * \omega$ at $(0,0)$ in Y. Conclude that, in the fundamental group of X, $[\omega]$ and $[\gamma]$ do not commute. Thus the bouquet of two circles provides an example of a space with its fundamental group non commutative.

(vii) Let X' be a covering space of a Hausdorff space X. Show that X' is compactly generated iff X is.

(viii) If X is a finite CW-complex and $p : \bar{X} \to X$ is a (finite) r-sheeted covering, show that $\chi(\bar{X}) = r\chi(X)$.

(ix) Let $p : \bar{X} \to X$ be a covering projection of path connected spaces. Show that $p_{\#} : \pi_n(\bar{X}, \bar{x}_0) \to \pi_n(X, x_0)$ are isomorphisms for $n \geq 2$.

3.5   Group Action

    Group actions occur in a natural way in all branches of mathematics. It is an essential part of modern geometry. It may be used as a technical tool in the study of certain symmetries of mathematical objects. In Theorem 3.4.12, we have come across the 'action' of the group of covering transformation on the total space of the covering while constructing other covering projections. Many of the examples that we have discussed also arise out of group actions. Here, we are not pursuing the theory of actions in general, but would like to relate it to the covering space theory.

Definition 3.5.1 Let X be a set, and G be a group. By a left-action of G on X, we mean a function

$$\mu : G \times X \to X$$

written as

$$\mu(g, x) = gx$$

(for simplicity) such that, the following two properties hold:

(i) <u>Associativity</u> : $\forall\ g, h \in G$ and $x \in X$, $h(gx) = (hg)x$.

(ii) <u>Identity</u> : If e denotes the identity of G, then $ex = x$ for all $x \in X$.

Given $x \in X$, we introduce the notation,

$$G^x := \{g \in G\ :\ gx = x\}, \quad G_x = \{gx\ :\ g \in G\}.$$

G^x is called the *isotropy subgroup of G at x*. (Check that it is indeed a subgroup of G.) G_x is called the orbit of x.

We introduce an equivalence relation in X as follows: $x \sim x'$ iff there is $g \in G$ such that $x' = gx$. Then the orbits are nothing but the equivalence classes under this equivalence relation. We denote the set of orbits by $_G \backslash X$. Let us denote by $q : X \to_G \backslash X$, the quotient map $x \mapsto [x]$.

Remark 3.5.2

(1) Similarly one can define a right action also. In fact, given any left-action as above, if we define $xg = g^{-1}x$, then it is easily seen that we obtain a right action and vice versa. Thus, it suffices to study one of them. The notation for the set of orbits for a right action is $X/_G$.

(2) It follows easily that for each $g \in G$, the assignment $x \mapsto gx$ is a bijection which we shall denote by L_g. Then the assignment $g \mapsto L_g$ itself defines a group homomorphism of G into the group $\sum X$, of permutations of X. Conversely, given a group homomorphism $L : G \to \sum X$, one can define

$$\mu(g, x) = L_g(x)$$

to get an action of G on X.

Definition 3.5.3 An action of G on X is said to be *transitive*, if for each pair $x, y \in X$, there exists $g \in G$ such that, $gx = y$. It is called *effective*, (or *faithful*) if $gx = x$, $\forall\ x \in X$ implies that, $g = e$. This is equivalent to say that, the corresponding homomorphism L is injective. We say the action is *fixed point free* (sometimes merely *free*) if $gx = x$ for some x implies $g = e$.

Definition 3.5.4 Restrictions and extension of group actions Let a group G act on a set X. Given a homomorphism $\rho : G' \to G$ of groups, we make G' act on X by the formula:

$$g'x := \rho(g)x.$$

This is called the restriction action of G to G'. (This name is borrowed from the special case when ρ is the inclusion homomorphism of a subgroup.) On the other hand, given a homomorphism $\alpha : G \to H$, we construct the set $X[\alpha]$ on which H acts as follows: Consider $H \times X$ and the action of G on it by the formula:

$$g(h', x) := (h'\alpha(g^{-1}), gx).$$

Let $X[\alpha]$ denote the orbit space of $H \times X$ under this action and let $[h, x]$ denote the orbit of the point (h, x). The action of H on $H[\alpha]$ is defined by the formula

$$h[h', x] = [hh', x].$$

It is a matter of straightforward verification to see that this indeed defines a left action of H on $X[\alpha]$. We refer to this as the extension of G action to H.

Definition 3.5.5 Now suppose that, X is a topological space. Then by an action of G on X, we shall always mean an action as above such that, with the discrete topology on G, the map $\mu : G \times X \to X$ is continuous.[2]

In other words, here the homomorphism $g \mapsto L_g$ takes values inside the subgroup Homeo$(X) \subset \Sigma X$, of self-homeomorphisms of X.

We give the quotient topology to $_G\backslash X$; a subset U of $_G\backslash X$ is open iff $q^{-1}(U)$ is open in X. Thus, the set of orbits becomes a topological space called the orbit space.

In group theory, group actions are most useful to study the properties of the groups themselves. Here we shall use them to study the properties of the quotient space $_G\backslash X$ that arise.

Definition 3.5.6 We say the action is *even* (classically, *properly discontinuous*[3]) if given any $x \in X$ there exists an open neighbourhood U of x in X, such that $gU \bigcap U = \emptyset$ for all $g \neq e$. Here gU denotes the set $\{gx : x \in U\}$. We shall call such a neighbourhood U of x an *even neighbourhood*. It is easy to see that an even action is fixed point free. The converse is not true in general but holds if we assume G is finite and X is Hausdorff. (See Exercise 3.5.23.)

Remark 3.5.7

(i) It is easily checked that the restriction and extension of an even action are even.

(ii) Given a topological space, clearly the entire group Homeo(X) of all homeomorphisms acts on the space X. We get more interesting actions by taking subgroups of Homeo(X).

(iii) Given a covering projection $p : \overline{X} \to X$, we know that the deck transformation group $G(p)$ is a subgroup of Homeo$(\overline{X})$ and hence we have a topological action of $G(p)$ on $\overline{X}$. Using an even covering for X, it is easily checked that this action is even. If the covering is normal, then Theorem 3.4.7 yields an action of $\pi_1(X, x)$ on $\overline{X}$, such that the orbit space is precisely X. The following theorem describes the converse situation.

Theorem 3.5.8 *Let a group G act evenly on a topological space E. Let B denote the quotient space consisting of all orbits $\{[e] : e \in E\}$ and let $q : E \to B$ be the quotient map. Then q is a normal covering projection. If E is connected, then the group of covering transformations of q is isomorphic to G. Further, there is an exact sequence of groups and homomorphisms:*

$$(1) \longrightarrow q_\#(\pi_1(E)) \longrightarrow \pi_1(B) \xrightarrow{\psi} G \longrightarrow (1).$$

Proof: Given any $e \in E$, let U be an even open neighbourhood as in the Definition 3.5.3. Then we claim $[U] = \{[e'] \in B : e' \in U\}$ is an evenly covered open neighbourhood of $[e]$: For

$$q^{-1}([U]) = \{e' : [e'] = [e] \text{ for some } e' \in U\} = \bigcup_{g \in G} gU$$

and each gU is open. Therefore $[U]$ is open. Further, by the evenness of U any two distinct translations of U are disjoint and each translation gU of U is mapped homeomorphically onto $[U]$. This proves that the quotient map q is a covering projection.

---

[2]In the standard literature, this kind of action is called a 'discontinuous action'. The justification for this rather confusing terminology is perhaps in the fact that classically, we were interested in the study of a 'topological group action on a topological space' and the present one is only a special case when the topology on the topological group involved is discrete.

[3]Read [Fulton, 1995] page no. 159, for a very good reason to discontinue this age-old terminology.

The proof that q is a normal covering is similar to the argument used in Lemma 3.4.6: Choose any base point $b \in B$ and $e \in E$ such that $[e] = b$. Let ω be a loop at b in B and $\bar{\omega}$ be a lift of ω in E at e. Then for any $e' \in [e]$ there is a $g \in G$ such that $ge = e'$. It follows that $g\bar{\omega}$ is the lift of ω at e'. Clearly, $g\bar{\omega}$ is a loop iff $\bar{\omega}$ is. This proves that q is a normal covering.

Now assume that E is connected. Clearly for every $g \in G$, $L_g : E \to E$ is a covering transformation of q. Conversely given any $\phi \in \mathbf{G}(q)$, let $\phi(e) = e'$. Then $[e'] = [e]$ and hence there is $g \in G$ such that $ge = e'$. But then L_g and ϕ are two covering transformations with $L_g(e) = \phi(e)$. Therefore $L_g = \phi$.

In view of Theorem 3.4.7, the rest of the claim follows. ♠

Example 3.5.9 Most of the situations of covering projections arise out of even actions of a group. This is certainly the situation when we have a discrete subgroup H of a Lie group G and we take the homogeneous space G/H. The quotient map $G \to G/H$ is a covering projection. Indeed, the all too familiar example that we started with, viz., $\exp : \mathbb{R} \to \mathbb{S}^1$ is one such, with $G = \mathbb{R}$ and $H = \mathbb{Z}$. Here are a few other examples.

(i) **The torus** Let u, v be a vector space basis for the 2-dimensional real vector space, $\mathbb{R}^2$. Let $\mathbb{Z}^2 = \mathbb{Z}(u, v)$ be the additive subgroup generated by u and v. Let $\mathbf{T}$ denote the quotient group $\mathbb{R}^2/\mathbb{Z}^2$ and p the quotient map. With the usual topology on $\mathbb{R}^2$ and the quotient topology on $\mathbf{T}$, show that p is a covering projection. What is the fundamental group of $\mathbf{T}$? Treating $\mathbb{R}^2$ as the complex plane, we get a complex manifold structure on $\mathbf{T}$. These spaces are called *elliptic curves*. They are complex 1-dimensional manifolds which are compact and connected, i.e., *Riemann surfaces*. (See [Shastri, 2009] for more.) Indeed, each $\mathbf{T}$ is homeomorphic to $\mathbb{S}^1 \times \mathbb{S}^1$. Clearly, this generalizes to any finite dimension and we have $\mathbb{R}^n$ as the universal covering space of $\mathbb{S}^1 \times \cdots \times \mathbb{S}^1$.

(ii) **The real projective space** Consider the antipodal action of $\mathbb{Z}_2$ on $\mathbb{S}^n$ with the quotient space the real projective space $\mathbb{P}^n$. Since $\mathbb{S}^n$ is simply connected (see Corollary 3.4.8 and Exercise 3.4.25.(v)), it follows that $\pi_1(\mathbb{P}^n) = \mathbb{Z}_2$, $n \geq 2$.

(iii) **The lens space** Let p, q be coprime numbers. Let ζ be a primitive p^{th} root of unity and treat it as the generator of the group $\mathbb{Z}_p$. Represent $\mathbb{S}^3$ as a closed subspace of unit vectors in $\mathbb{C} \times \mathbb{C}$. Define an action of $\mathbb{Z}_p$ on $\mathbb{S}^3$ by the formula $\zeta \mapsto \phi_q$, where the diffeomorphism $\phi_q : \mathbb{S}^3 \to \mathbb{S}^3$ is given by

$$\phi_q : (z_0, z_1) \mapsto (\zeta z_0, \zeta^q z_1).$$

Since q is coprime to p, ϕ_q is a diffeomorphism of order p. Notice that the orbit of any point is independent of the choice of the primitive root that we have chosen. The quotient space is denoted by $L(p, q)$ and is called a *lens space*. Verify that the action is fixed point free and hence even. Note that $L(p, q)$ is a closed connected 3-manifold. Obviously, its fundamental group is isomorphic to $\mathbb{Z}_p$.

More generally, given a p and a sequence $q_1, \ldots, q_r$ of numbers which are coprime to p, we define an action of $\mathbb{Z}_p$ on $\mathbb{S}^{2r+1}$ by

$$(\zeta, (z_0, z_1, \ldots, z_r)) \mapsto (\zeta z_0, \zeta^{q_1} z_1, \ldots, \zeta^{q_r} z_r).$$

Verify that this is a fixed point free action. Let us denote by $L(p, q_1, \ldots, q_r) =: L$ the quotient space and the quotient map $\mathbb{S}^{2r+1} \to L$, which is a p-fold covering and $\pi_1(L) \approx \mathbb{Z}_p$.

Even more generally, we can define the infinite dimensional lens spaces as follows:

Given an infinite sequence $q_1, q_2, \ldots$, of numbers coprime to p, define an action of $\mathbb{Z}_p$ on the infinite dimensional sphere $\mathbb{S}^\infty$ by treating it as space of unit vectors in the vector space $\mathbb{C}^\infty$:

$$(\zeta, (z_0, z_1, \ldots)) \mapsto (\zeta z_0, \zeta^{q_1} z_1, \ldots)$$

The verification that the action is fixed point free is the same in the finite dimensional case and we obtain a covering projection. $\mathbb{S}^\infty \to L(p, q_1, q_2, \ldots)$. Note that in all these cases, only the mod p values of $q_1, q_2, \ldots$, etc., really matter. Also, the quotient spaces do not depend on the choice of the primitive p^{th} root of unity. These are called infinite lens spaces.

(a) $L(p, q)$ is homeomorphic to $L(p, -q)$. To see this, consider the homeomorphism $f : \mathbb{S}^3 \to \mathbb{S}^3$ given by $f(z_0, z_1) = (z_0, z_0^t z_1)$ for some integer t. Choosing $t = -2q$ we see that $f \circ \phi_q = \phi_{-q} \circ f$. This shows that f is an equivariant homeomorphism of the two actions. Therefore it induces a homeomorphism of the two quotient spaces.

(b) Similarly, for $qq' \equiv 1(p)$, consider $T(z_0, z_1) = (z_1, z_0)$. Then $T \circ \phi_q = (\phi_q)^{q'} \circ T$. This means T is an equivariant diffeomorphism of the ϕ_q-action to $\phi_q^{q'}$-action, which is equivalent to $\phi_{q'}$ with the choice of the primitive root being ζ^q. Hence $L(p, q)$ is homeomorphic to $L(p, q')$. Combining with (a) it follows $L(p, -q')$ is also homeomorphic to $L(p, q)$.

It is a classical result due to Reidemeister that the converse is also true:

Theorem 3.5.10 (Reidemeister) *$L(p, q)$ is homeomorphic to $L(p, q')$ iff $q \pm q' \equiv 0$ or $qq' \equiv \pm 1$ modulo p.*

As seen above, combining (a) and (b) we get the proof of the 'if' part. The proof of the 'only if' part and its generalization are beyond the scope of this exposition. Also, the following theorem of Whitehead, which, we state here without proof, completely answers the homotopy classification of the 3-dimensional lens spaces. An interested reader may refer to [Cohen, 1973].

Theorem 3.5.11 (Whitehead) *$L(p, q)$ is homotopy equivalent to $L(p, q')$ iff qq' is a square modulo p.*

Remark 3.5.12

(1) The name 'lens space' refers to the following description of these spaces. A lens is a region L bounded by two 2-spheres in $\mathbb{R}^3$ intersecting along a circle, the sharp edge of the lens. All lens spaces can be thought of as quotients of such regions by certain identifications along the boundary. Working inside $\mathbb{S}^3 \subset \mathbb{C} \times \mathbb{C}$, let us denote by

$$D = \{x \in \mathbb{S}^3 \; : \; d(1, x) \le d(1, \zeta)\} \cap \{x \in \mathbb{S}^2 \; : \; d(1, x) \le d(1, \zeta^{p-1})\}.$$

Check that this lens-like subspace is a fundamental domain for the action of ϕ_q. The boundary of D itself is divided into two caps by the circle S which is the intersection of the two hyperplanes with the sphere. Thus, the quotient space $L(p, q)$ can be viewed as the quotient space of D by a single identification along the boundary, viz., the map ϕ_q itself which can be thought of as the reflection in the plane through S followed by a rotation through an angle $2q\pi/p$. Note that if $q = 0$ then it is easy to see that the resulting quotient space is homeomorphic to $\mathbb{S}^3$. (Indeed, in this case, the quotient map $q : \mathbb{S}^3 \to L(p, 0)$ is a 'ramified' covering ramified along a circle.

(2) We can determine the fundamental group of $L(p, q)$ from this latter description, since it provides us with a structure of a CW-complex on the lens space as follows. Note that all the p points on the edge get identified to a single vertex in $L(p, q)$. Likewise, all the p edges are mapped onto a single loop ω. This is the 1-skeleton. Thus the fundamental group

is generated by this loop ω. Now the loop represented by the entire circular edge of the lens itself winds around the loop ω p-times. The two caps become one single 2-cell which bounds the edge and hence kills $p[\omega]$. Therefore the fundamental group of the 2-skeleton is isomorphic to $\mathbb{Z}/p\mathbb{Z}$. Finally, the interior of the lens L is the 3-cell in $L(p,q)$ which has no effect on the fundamental group. We conclude that $\pi_1(L(p,q)) \approx \mathbb{Z}_p$. This argument will be completely justified by Corollary 3.8.12, in a later section.

What happens when E is not connected? This question turns out to be not just an idle curiosity. It is an idea due to Grothendieck which ultimately yields an elegant proof of all versions of the van Kampen theorem, albeit for spaces which admit simply connected coverings. Let us introduce a convenient terminology.

Definition 3.5.13 By a G-*covering* $\xi = (E, P, B)$ we mean a covering projection $p : E \to B$ obtained as the quotient map corresponding to an even action of the group G on E.

All examples discussed above are G-coverings for some group G.

Definition 3.5.14 Given two G-coverings over the same base space B, a G-*map* $\alpha : \xi \to \xi'$ is a continuous map $\alpha : E \to E'$ such that $\alpha(gz) = g\alpha(z)$, for all $g \in G$ and $z \in E$. (In that case we clearly have $p' \circ \alpha = p$.) There is then a category of G-coverings and G-maps over a given base space B. Two G-coverings are said to be G-*equivalent* if there is a G-map between them which is a homeomorphism.

A somewhat unusual but important fact is:

Lemma 3.5.15 Every G-map of G-coverings over a base space is a G-isomorphism.

Proof: If $\alpha : E \to E'$ is such a map then, first of all, it maps fibres of p to the fibres of p' over the same point, i.e., $\alpha(p^{-1}(b)) \subset p'^{-1}(b)$. Since every fibre is a G-orbit, we have $\alpha : Ge \to G\alpha(e)$. Since $\alpha(ge) = g\alpha(e)$, it follows that α is surjective. By the evenness of the action it also follows that α is injective. Finally, from the evenness of the action itself, it follows that α is an open mapping also. Therefore α is a homeomorphism. $\spadesuit$

Clearly a G-covering is a special type of covering projection. Also a G-equivalence $f : E_1 \to E_2$ of two G-coverings obviously defines a usual equivalence of coverings. The question now is how far the converse is true. To understand this properly, start with an action of G on a space E, the associated quotient map $p : E \to B$ and an automorphism $\varphi : G \to G$. Define a new action of G, $\circ : G \times E \to E$ by the formula

$$g \circ e = \varphi(g)e.$$

It is clear that the resulting quotient map is nothing but the same as $p : E \to B$ and so the two coverings are equivalent. However, the identity map $Id : E \to E$ is not a G-map unless $\varphi : G \to G$ is the identity. It is not clear why there should be any G-map $E \to E$ at all. Here is a simple example to the contrary. Consider the subgroup G of $\mathbb{S}^1$ consisting of cube roots of unity and let $\varphi(\xi) = \xi^2$. Then φ is an automorphism and there is no homeomorphism $f : \mathbb{S}^1 \to \mathbb{S}^1$ such that $f(\xi z) = \xi^2 f(z), \xi G, z \in \mathbb{S}^1$. (See Exercise 3.5.23.(iv).) So, we may either modify the definition of G-equivalence to accommodate this situation or learn to live with this reality that the G-equivalence is a bit stronger equivalence than the ordinary equivalence of covering projections. In any case, the following theorem gives the true picture.

Theorem 3.5.16 *Let B be a connected space and E_1, E_2 be any two connected G-coverings over it. They are equivalent as covering projections iff there exists an automorphism $\varphi : G \to G$ and a homeomorphism $f : E_1 \to E_2$ such that $f(gz) = \varphi(g)f(z), g \in G, z \in E_1$.*

Proof: We need to prove only the "if" part. Given a homeomorphism $f : E_1 \to E_2$ such that $p_2 \circ f = p_1$, we must produce an automorphism $\varphi : G \to G$ with the above property. Fix a base point $b_0 \in B$ and $e_0 \in E_1$ such that $p_1(e_0) = b_0$. It follows that for each $g \in G$, there is a unique $\varphi(g) \in G$ such that $f(ge_0) = \varphi(g)f(e_0)$. Now for a fixed $g \in G$, consider the two maps $E_1 \to E_2$ given by

$$e \mapsto f(ge); \quad e \mapsto \varphi(g)f(e).$$

Both are lifts of p_1 and agree at the point e_0 and hence agree everywhere. This just means $f(gz) = \varphi(g)f(z), g \in G, z \in E$. It remains to show that φ is an automorphism. We leave this to the reader as an entertaining exercise. ♠

Remark 3.5.17 We may choose a certain base point $b_0 \in B$ and consider G-coverings over B also to have base points sitting over b_0. Then G-maps are required to preserve the base points. In what follows we shall identify $\pi_1(B, b_0)$ with the group J of covering transformations of a simply connected covering $p : E \to B$ with a base point $e_0 \in E$ as in Corollary 3.4.8. The following theorem is what we were aiming at.

Theorem 3.5.18 *Let G be any group. Let B be a connected, locally path connected and semi-locally simply connected space with a base point b_0. Then there is a canonical bijection between the set $\mathcal{G}(B)$ of equivalence classes of pointed G-coverings over B and the set of all group homomorphisms $\mathrm{Hom}(\pi_1(B, b_0); G)$.*

Proof: Let $\xi = (E, p, B)$ be the universal covering space with a base point $e_0 \in E$. Given a homomorphism $\alpha : \pi_1(B, b_0) =: J \to G$, let $E[\alpha]$ denote the extension of J-action to a G-action and let $p_\alpha : E[\alpha] \to B$ be the corresponding quotient map. Then $\xi[\alpha] = (E[\alpha], p_\alpha, B)$ is a G-covering and we take the equivalence class of this to be $\mu(\alpha)$. We claim that $\mu : \mathrm{Hom}(J, G) \to \mathcal{G}(B)$ is a bijection.

Consider a G-covering $p' : E' \to B$ with a base point z_0 representing a class ζ. Let $\bar{p} : E \to E'$ be the lift of p through p' such that $\bar{p}(e_0) = z_0$. Given $\phi = [\omega] \in J$, recall that $\phi(e_0)$ is defined to be the end-point of the lift $\tilde{\omega}$ of ω at e_0. It follows that $p' \circ \bar{p}(\phi(e_0)) = p(e_0) = b_0$ and hence there exists a unique $g \in G$ such that $gz_0 = \bar{p}(\phi(e_0))$. We define $\alpha(\phi) = g$. That is $\alpha(\phi)$ is defined by the equation

$$\alpha(\phi)(z_0) = \bar{p} \circ \phi(e_0). \tag{3.1}$$

Since both $\alpha(\phi) \circ \bar{p}$ and $\bar{p} \circ \phi$ are lifts of p through p', which agree at the point e_0, it follows that

$$\alpha(\phi) \circ \bar{p} = \bar{p} \circ \phi.$$

Therefore, for all $\phi, \psi \in J$, we have

$$\alpha(\phi) \circ \bar{p} \circ \psi = \bar{p} \circ \phi \circ \psi.$$

This implies, in particular,

$$\alpha(\phi) \circ \bar{p} \circ \psi(e_0) = \bar{p} \circ \phi \circ \psi(e_0).$$

The LHS is equal to $\alpha(\phi) \circ \alpha(\psi) \circ \bar{p}(e_0)$ whereas the RHS is equal to $\bar{p} \circ (\phi \circ \psi)(e_0)$. Therefore by (3.1), it follows that $\alpha(\phi \circ \psi) = \alpha(\phi) \circ \alpha(\psi)$, i.e., α is a homomorphism. We will leave it to the reader to verify that the homomorphism α depends only on the isomorphism class ζ. Thus we get the definition of $\nu : \mathcal{G}(B) \to \mathrm{Hom}(J, G)$ viz., $\nu[\zeta] = \alpha$, which, as we shall soon see, is the inverse of μ.

(Incidentally, check that this also verifies that $\bar{p}$ is a J-map where we treat $p' : E' \to B$ also as a J-space via α. Therefore, the map $(g, e) \mapsto g\bar{p}(e)$ defines a map of G-coverings $\tilde{p} : E[\alpha] \to E'$. By Lemma 3.5.15, it is an isomorphism. This implies that $\mu \circ \nu(\zeta) = \zeta$.

On the other hand, given a homomorphism $\alpha : J \to B, \mu(\alpha)$ is represented by the G-covering $E[\alpha]$. It follows that the lift $\bar{p} : E \to E[\alpha]$ of $p : E \to B$ such that $\bar{p}(e_0) = [1, e_0]$, is given by $\bar{p}(e) = [1, e]$. Therefore the homomorphism $\nu(\mu[\alpha]) : J \to G$ is determined by the formula

$$\nu(\mu[\alpha])(\phi)(e_0) = \bar{p}(\phi(e_0)) = [1, \phi(e_0)] = [\alpha(\phi), e_0] = \alpha(\phi)[1, e_0].$$

That means $\nu \circ \mu[\alpha] = \alpha$. Therefore ν is the inverse of μ.

Finally, we come to the canonical property of μ. Given a map $f : B' \to B$ and a G-covering $q : Z \to B$, we know that the pullback f^*q is a G-covering over B'. On the other hand, homomorphism $f_\# : \pi_1(B', b'_0) \to \pi_1(B, b_0)$ induces a map $f_\#^* : \mathrm{Hom}(\pi_1(B, b_0), G) \to \mathrm{Hom}(\pi_1(B', b'_0), G)$. We want to show that the following diagram is commutative:

$$
\begin{array}{ccc}
\mathcal{G}(B) & \xrightarrow{\quad f^* \quad} & \mathcal{G}(B') \\
{\scriptstyle \nu}\downarrow & & \downarrow{\scriptstyle \nu} \\
\mathrm{Hom}(\pi_1(B, b_0), G) & \xrightarrow{\quad f_\#^* \quad} & \mathrm{Hom}(\pi_1(B', b'_0), G)
\end{array}
$$

We have identified π_1 with the group of covering transformation of the universal covering. Under this identification, we must figure out what the homomorphism $f_\# : \pi_1(B', b'_0) \to \pi_1(B, b_0)$ corresponds to. Note that there is a unique map $\bar{f} : E' \to E$ such that $f \circ p' = p \circ \bar{f}$ and $\bar{f}(e'_0) = e_0$. A covering transformation $\psi : E' \to E'$ corresponds to an element $[\omega'] \in \pi_1(B', b'_0)$ by the rule

$$\psi(e'_0) = \tilde{\omega}'(1),$$

where $\tilde{\omega}'$ is the lift of ω' in E' at the point e'_0. It follows that $\bar{f} \circ \tilde{\omega}'$ is the lift of $f \circ \omega'$ in E at the point $\bar{f}(e'_0) = e_0$. Likewise, the covering transformation $f_\#(\psi)$ is defined by the rule, we have

$$f_\#(\psi)(e_0) = \bar{f} \circ \tilde{\omega}'(1) = \bar{f}\psi(e'_0).$$

Since both $f_\#(\psi) \circ \bar{f}, \psi \circ \bar{f}$ are lifts of $f \circ p'$ and agree at the point e'_0, we get the following commutative diagram:

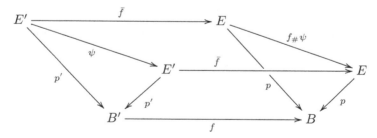

Also, for $\zeta = (Z, q, B) \in \mathcal{G}(B)$, with base point $z_0 \in Z$, the total space of $f^*(\zeta)$ is defined by

$$Z' := \{(b', z) \in B' \times Z : f(b') = q(z)\}$$

with the base point $z'_0 = (b'_0, z_0)$. Let $\bar{p} : E \to Z$ be the lift of the universal covering $p : E \to B$ such that $\bar{p}(e_0) = z_0$. Then it follows that $\bar{p}' : E' \to Z'$ defined by

$$\bar{p}'(e') = (p'(e'), \bar{p} \circ \bar{f}(e'))$$

is the lift of the universal covering $p' : E' \to B'$.

We have to show that $\nu(f^*(\zeta)) = f_\#^* \circ \nu(\zeta) = \nu(\zeta) \circ f_\#$. That means for every covering transformation $\psi : E' \to E'$ where $p' : E' \to B'$ is the universal covering, we have to show that

$$\nu \circ f^*(\zeta)(\psi) = \nu(\zeta)(f_\#\psi)$$

in G.

The LHS satisfies the formula

$$\nu \circ f^*(\zeta)(\psi)(z_0') = \bar{p}'(\psi(e_0')) = (b_0', \bar{p} \circ \bar{f}(\psi(e_0'))).$$

Note that the action of G on Z' is via the second factor and hence $gz_0' = g(b_0', z_0) = (b_0', gz_0)$ for any $g \in G$. Therefore, we have $\nu \circ f^*(\zeta)(\psi)(z_0') = (b_0', \nu \circ f^*(\zeta))(\psi)z_0$ and

$$\nu \circ f^*(\zeta)(\psi)z_0 = \bar{p} \circ \bar{f} \circ (\psi) \circ (e_0') = \bar{p} \circ f_\#(\psi)(e_0) = nu(\zeta)(f_\#\psi)z_0.$$

This proves the claim. ♠

As promised, we shall use these results in Section 3.7 in establishing the van Kampen theorem. We end this section with another useful and standard concept in covering space theory.

Definition 3.5.19 Let G be acting on a connected topological space X. A connected subset $D \subset X$ of X is called a fundamental domain for the action of G on X, if
(i) $X = \cup_{g \in G} gD$ and
(ii) for any $x \in \text{int } D$, $gx \in D$ iff $g = e$.

Remark 3.5.20 Condition (i) of the above definition tells us that the quotient map $q : X \to X/G$ restricted to D is also a quotient map onto X/G. Condition (ii) tells us that $q : D \to X/G$ is injective in the interior of D, i.e., the identifications are taking place only on the boundary of D in X. Thus fundamental domains help us to get a better picture of the quotient space under a group action. Note that if D is a fundamental domain then so are all of its translates $gD, g \in G$.

Example 3.5.21 Any interval of length 1 in $\mathbb{R}$ is a fundamental domain for the action of $\mathbb{Z}$ on $\mathbb{R}$. The quotient space is easily obtained by identifying the two end-points of the interval. Similarly, any square with sides of unit length and parallel to the axes is a fundamental domain for the action of $\mathbb{Z}^2$ on $\mathbb{R}^2$. The quotient space is obtained by identifying the two pairs of opposite sides in an orientation preserving fashion. For the action of $\mathbb{Z}_n$ on $\mathbb{C}^*$ given by $z \mapsto \zeta z$, where ζ is a primitive n^{th} root of unity, we can take any closed sector of angle precisely $2\pi/n$ as a fundamental domain. For the antipodal action of $\mathbb{Z}_2$ on $\mathbb{S}^n$, we can take any closed half-sphere as a fundamental domain. This fact is used very nicely in getting the CW-structure on projective spaces.

Example 3.5.22 Fundamental domain for the Poincaré homology sphere Similar to the description for lens spaces as above, it is possible to carry out fundamental domain description for other finite group actions on $\mathbb{S}^3$. Here we shall give a description of the Poincaré homology 3-sphere P_{120} using fundamental domain. We have seen that the group of symmetries of the dodecahedron is a subgroup I_{60} of $SO(3)$ of order 60. Indeed, consider $\mathbb{S}^3$ as the space of unit quaternions. Then the conjugation action of $\mathbb{S}^3$ on itself is orthogonal and the subspace of all purely imaginary unit quaternions is invariant under this action. Hence we obtain a homomorphism $\mathbb{S}^3 \to SO(3)$ which is easily checked to be a submersion and hence surjective. [See [Shastri, 2011] for more details.] The kernel of this homomorphism is precisely the centre $\{\pm 1\}$ of the group $\mathbb{S}^3$. Thus, the inverse image $\hat{I}$ of I_{60} is a discrete subgroup of $\mathbb{S}^3$ which is of order 120. (This is called the binary icosahedral group.) The quotient map

$$q : \mathbb{S}^3 \to \mathbb{S}^3/\hat{I}$$

is therefore a covering projection. From covering space theory, it follows that $\pi_1(\mathbb{S}^3/\hat{I}) = \hat{I}$. Let

$$D = \{x \in \mathbb{S}^3 \; : \; d(x,1) \le d(x,g), g \in \hat{I}\}.$$

A standard argument shows that D is a fundamental domain for the action of $\hat{I}$. In order to get any useful information out of this, we need to describe D in a better way. It is clear that D is the intersection of all these half-spaces with $\mathbb{S}^3$. Let A be the set of barycentres of the twelve faces of the unit dodecahedron in $\mathbb{S}^2$, where we treat $\mathbb{S}^2$ as the space of purely imaginary quaternions. Let $B = \{\mathbf{v}/\|\mathbf{v}\| \; : \; \mathbf{v} \in A\}$. Put

$$C = \{\sin\theta + \cos\theta\,\mathbf{u} \; : \; \mathbf{u} \in B, \theta = \pm 2\pi/5\}.$$

It is easily checked that $C \subset \hat{I}$ and defines the set of points in $\hat{I}$ closest to 1, each at a distance $\sqrt{2 - 2\sin 2\pi/5}$. Therefore

$$D = \{x \in \mathbb{S}^3 \; : \; d(x,1) \le d(x,g) \; : \; g \in C\}.$$

We can now check that under the stereographic projection from the centre of $\mathbb{R}^4$ onto the subspace $1 \times \mathbb{R}^3$ where $0 \times \mathbb{R}^3$ is the subspace of purely imaginary quaternions, the image of D is mapped onto a copy of dodecahedron, defined by the twelve hyperplanes as above. We may think of D as a 'spherical dodecahedron'. In all, there are 120 translates of D which make up the $\mathbb{S}^3$. A point $x \in D$ is in the boundary of D iff $d(1,x) = d(x,g)$ for some $g \in C$. Clearly, translations by $g^{-1}, g \in C$ defines a 'curvilinear' homeomorphism of the face

$$\{x \in \mathbb{S}^2 \; : \; d(1,x) = d(g,x)\} \mapsto \{x \in \mathbb{S}^2 \; : \; d(1,y) = d(g^{-1},y)\}$$

Therefore, the quotient space $\mathbb{S}^3/\hat{I}$ is homeomorphic to the space obtained from the spherical dodecahedron D by identifying its opposite faces via a Euclidean translation followed by a rotation through an angle $2\pi/5$. This is precisely how Poincaré defined P_{120}. From the covering space theory, it follows that P_{120} is a closed 3-manifold.

Incidentally, the set of points $\hat{I} \subset \mathbb{S}^3$ define a vertex set of a triangulation of $\mathbb{S}^3$. We have to consider the boundary complex of the convex polyhedron which is the convex hull of $\hat{I}$ in $\mathbb{R}^4$, and radially project it over $\mathbb{S}^3$ to get this triangulation.

Exercise 3.5.23

(i) Consider the 3-fold covering $p : \overline{X} \to X$ of Example 3.3.10 and determine $\mathbf{G}(p)$.

(ii) Let $\mathbb{Z} = (t)$ be the infinite cyclic group generated by t, multiplicatively and let λ be an irrational number. Consider the action of $\mathbb{Z}$ on $\mathbb{S}^1$ given by $tz = e^{\lambda\pi i}z$. Show that this is a fixed point free action but not an even action.

(iii) Suppose $s : B \to E$ is a section for the projection map $p : E \to B$ of a G-covering, i.e., $p \circ s = Id_B$. Show that $p : E \to B$ is a trivial G-covering.

(iv) Consider the 3-fold covering $p : \mathbb{S}^1 \to \mathbb{S}^1$ given by $z \mapsto z^3$. Determine the Galois group $\mathbf{G}(p)$.

3.6 Pushouts and Free Products

In this section, we shall deal with some group theoretic background needed to describe the fundamental group of a union of subspaces in terms of the fundamental groups of these subspaces. These results go under the name Seifert–van Kampen theorems.

Definition 3.6.1 Let Λ be an indexing set. By a *diagram* of groups and homomorphisms we mean

(a) a collection $G_i, i \in \Lambda$ of groups,

(b) a collection of homomorphisms $\alpha_i : G_i \to G$ and

(c) a collection of homomorphism $\eta_{ij} : G_{ij} \to G_i$ such that $\alpha_i \circ \alpha_{ij} = \alpha_j \alpha_{ji}$, for all $i \neq j \in \Lambda$.

Such a diagram $(G, G_i, \alpha_i, G_{ij}, \eta_{ij})$ is called a *pushout diagram* if for each diagram $(G', G_i, G_{ij}, \alpha'_i, \eta_{ij})$, there exists a unique homomorphism $\gamma : G \to G'$ such that $\gamma \circ \alpha_i = \alpha'_i$, for all $i \in \Lambda$.

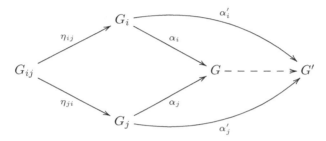

When $G_{ij} = (e)$, the trivial group for all $i \neq j$, the pushout diagram is called a free product of the family $\{G_i \ : \ i \in \Lambda\}$ and we express this by writing

$$G = *_{i \in \Lambda} G_i.$$

Remark 3.6.2 In other words, a pushout diagram is an initial object in an appropriate category of diagrams. Given a collection of groups and homomorphisms $\eta_{ij} : G_{ij} \to G_i$ the problem is to determine the existence and uniqueness of a pushout. The uniqueness is easy to determine: Suppose we have two pushout diagrams $(G, G_i, G_{ij}, \alpha_i, \eta_{ij}), (G', G_i, G_{ij}, \alpha'_i, \eta_{ij})$. Applying the existence part of the definition of pushout for the two systems in either direction, we get $\gamma : G \to G'$ and $\gamma' : G' \to G$ such that $\gamma \circ \alpha_i = \alpha'_i$ and $\gamma' \circ \alpha'_i = \alpha_i$. Therefore $(\gamma \circ \gamma') \circ \alpha'_i = \alpha'_i$. But $Id_{G'} : G' \to G'$ also satisfies the same property. Therefore, the uniqueness part of the definition of pushout for G' gives $\gamma \circ \gamma' = Id_G$. Similarly, we also get $\gamma' \circ \gamma = Id_G$.

It is for the existence part that we have to work harder. The beauty of this categorical definition via universal property is that many results will follow from applying the universal property rather than the actual description of the groups and homomorphisms involved. However, it is hard to claim that we know 'everything' about the pushout diagrams from its definition, the trouble being in unravelling the definition properly.

The proof of the existence of the pushouts will be presented in two stages. Granting the existence of the free product, we shall first show the existence of the pushout. We shall then come to the proof of the existence of free products.

Theorem 3.6.3 *Let $\eta_{ij} : G_{ij} \to G_i$ be a collection of homomorphisms of groups. Let $H = *_{i \in I} G_i$ be the free product of the collection $\{G_i\}$ together with the homomorphisms $\alpha_i : G_i \to H$. Let N be the normal subgroup in H generated by the set*

$$\{\alpha_i \eta_{ij}(h_{ij}) \alpha_j \eta_{ji}(h_{ij}^{-1}) \ : \ h_{ij} \in G_{ij}, \quad i \neq j \in I\}.$$

Let $q : H \to H/N =: G$ be the quotient homomorphism and $\beta_i = q \circ \alpha_i$. Then the collection $(G, G_i, G_{ij}, \beta_i, \eta_{ij})$ is a pushout diagram.

Proof: It follows easily that $\beta_i \circ \eta_{ij} = \beta_j \circ \eta_{ji}, i \neq j$. Thus the collection $(G, G_i, G_{ij}, \beta_i, \eta_{ij})$ is a diagram.

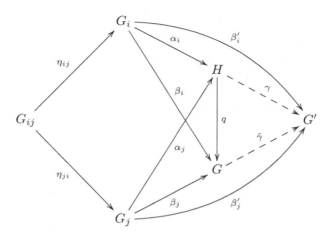

If $(G', G_i, \beta_i', \eta_{ij})$ is another diagram, then by the definition of the free product, there is a unique homomorphism $\gamma : H \to G'$ such that $\gamma \circ \alpha_i = \beta_i'$. Since $\eta_{ij} \circ \beta_i' = \eta_{ji} \circ \beta_j'$, it follows that $\gamma \circ \alpha_i \circ \eta_{ij} = \gamma \circ \alpha_j \circ \eta_{ji}$ and hence $\gamma(N) = (e)$. Therefore, γ factors down to define a homomorphism $\tilde{\gamma} : G \to G'$, i.e., $\tilde{\gamma} \circ q = \gamma$. Clearly, $\gamma' \circ \beta_i = \beta_i'$. The uniqueness of γ' follows from that of γ. ♠

The rest of this section will be essentially devoted to a constructive proof of existence of the free product.

Given a family $\{G_i\}_{i \in \Lambda}$ of groups, consider the set $\hat{W} := \hat{W}(\cup_i G_i)$ of all finite sequences $(x_1, x_2, \ldots, x_k)$ where each x_j comes from some G_{i_j}. We include the empty sequence also in this set and denote it by $\square$. Given two such sequences $s = (x_1, x_2, \ldots, x_k), s' = (x_1', x_2', \ldots, x_l')$ we define their product ss' to be the sequence obtained by juxtaposing s with s' :

$$ss' = (x_1, x_2, \ldots, x_k, x_1', x_2', \ldots, x_l').$$

Of course, by definition, we take

$$s\square = \square s = s, \quad \forall \ s.$$

This makes $\hat{W}$ into a semigroup, the associativity of this operation being completely obvious. We shall write s_j for the jth entry in the sequence s. Elements of $\hat{W}$ are called words in $\sqcup_i G_i$.

We now define a relation $\sim$ in $\hat{W}$: We say s' is obtained by s by an elementary collapsing and write $s \searrow s'$ if s' is obtained by one of the following two operations:
(a) There is some j such that $s_j = 1$ and we delete this entry from s to obtain s';
(b) There is some j such that s_j, s_{j+1} belong to the same group G_i and we combine these two entries to a single entry $s_j s_{j+1}$ to obtain s'.

We say $s \sim s'$ if there exists a finite sequence s^i of members of $\hat{W}$, such that $s = s^1, s' = s^n$ and for each i either $s^j \searrow s^{j+1}$ or $s^{j+1} \searrow s^j$.

It follows easily that $\sim$ is an equivalence relation. Here are some illustrations:
(1) $\searrow \square$, $(g,) \searrow (gg')$, if $g, g \in G_i$. Also $(1, 1, \ldots, 1) \sim \square$, $(a, b, b^{-1}, a^{-1}) \sim \square$ etc.

We shall denote the equivalence class of s by $[s]$ with a simplification that the equivalence class of $\square$ will be denoted by $\square$ itself rather than $[\square]$. The set of these equivalence classes of words will be denoted by $[W]$.

It is also clear that if $s \searrow s'$ then for any sequence t we have

$$st \searrow s't, \quad ts \searrow ts'.$$

From this it follows that the multiplication on $\hat{W}$ factors down to a multiplication on $[W]$:

$$[s][s'] = [ss'], \quad \square[s] = s\square = [s].$$

Given any word $s = (x_1, \ldots, x_k)$ we define the inverse word

$$s^{-1} := (x_k^{-1}, \ldots, x_1^{-1}).$$

It is clear that $ss^{-1} \sim \square \sim s^{-1}s$. Therefore, $[W]$ together with the above multiplication is a group.

The construction of the free product is actually over. However, it takes some more time and effort to see that this is the creature that we have been looking for.

We say $s \in \hat{W}$ is a reduced word if either $s = \square$ or the following conditions hold:
(a) No entry s_j is equal to the unit of a group.
(b) Every consecutive pair of elements s_j, s_{j+1} come from different groups G_i.

Since every elementary collapsing reduces the length of a word by 1, it is clear that every word is collapsible in finitely many steps to a reduced word. This is slightly stronger than saying that the equivalence class of every word contains a reduced word.

What is of fundamental importance to us is that **the equivalence class of every word contains a unique reduced word.**[4] Let us see why this is so.

We shall denote the subset of $\hat{W}$ consisting of reduced words by W. Let $P(W)$ denote the group of all permutations of W. Fix a group G_i. We shall define a subgroup

$$L(G_i) = \{L_g \; : \; g \in G_i\} \subset P(W)$$

isomorphic to G_i. We take $L_g = Id$ if $g = 1$. To each $1 \neq g \in G_i$, we define $L_g : W \to W$ as follows:

$$L_g(\square) = \begin{cases} (g), & \text{the singleton sequence consisting of } g, \; g \neq 1; \\ \square, & \text{if } g = 1; \end{cases}$$

$$L_g(s_1, \ldots, s_k) = \begin{cases} (g, s_1, s_2, \ldots, s_k), & \text{if } s_1 \notin G_i; \\ (gs_1, s_2, \ldots, s_k), & \text{if } s_1 \in G_i \text{ but } s_1 \neq g^{-1}; \\ (s_2, \ldots, s_k), & \text{if } s_1 = g^{-1} \text{ and } k \geq 2; \\ \square, & \text{if } s_1 = g^{-1} \text{ and } k = 1. \end{cases}$$

One readily checks that L_g is injective. Let us check that it is surjective. Given any $s = (s_1, \ldots, s_k) \in W$ if $s_1 \notin G_i$ then $(g^{-1})s \in W$ and $L_g((g^{-1})s) = s$. Suppose now $s_1 \in G_i$. If $s_1 \neq g$ then clearly $s' = (g^{-1}s_1, s_2, \ldots s_k) \in W$ and $L_g(s') = s$. Finally, if $s_1 = g$, then $L_g(s_2, \ldots, s_k) = s$. This proves that L_g is surjective. Therefore, $L_g \in P(W)$.

It is also clear that $L_g = L_h$ iff $g = h$ by simply taking the values of L_g and L_h on the empty sequence $\square$. Finally, we need to verify that L is a homomorphism, i.e., $L_{gh} = L_g \circ L_h$, for $g, h \in G_i$. This is also not difficult but routine and lengthy. Just for completeness we shall write this down in full detail.

---

[4] As a student and as a teacher, I have found this, one of the most subtle points. I was not satisfied with most of the expositions I have read.

First we observe that $L_g(s_1, s_2 \ldots, s_k) = (L_g(s_1))(s_2, \ldots s_k)$ for every reduced word s of length $k \geq 2$. Therefore it is enough to verify that

$$L_{gh}(s) = L_g \circ L_h(s) \tag{3.2}$$

for reduced words of length 1 since the general case will simply follow from juxtaposing the rest of the sequence everywhere. Next, in case any of g, h, gh is equal to 1, the verification of (3.2) is easy. So, we assume that none of g, h, gh is equal to 1.

Now let $s = (s_1)$ with $s_1 \notin G_i$. Then both sides of (3.2) are equal to (gh, s_1) and so we are done. Now suppose $s_1 \in G_i$. Then (3.2) is a consequence of the associativity in the group G_i.

This establishes a monomorphism $L : G_i \to P(W)$ which we shall denote more specifically by L_i. We identify G_i with the subgroup $L_i(G_i)$ of $P(W)$. Let G denote the group generated by $L_i(G_i)'s, i \in \Lambda$. We claim that G is isomorphic to the group $[W]$. Indeed, given any $s = (s_1, \ldots, s_k) \in \hat{W}$ with $s_j \in G_{i_j}$, we define

$$\psi(s) = L_{i_1}(s_1) \circ \cdots \circ L_{i_k}(s_k).$$

Then clearly $\psi(s) \in G$ and $\psi : W \to G$ is a homomorphism of semigroups. It is also clear that if $s \searrow s'$, then $\psi(s) = \psi(s')$. Therefore the homomorphism ψ takes the same value on members of any equivalence class and hence defines a homomorphism $\bar{\psi} : [W] \to G$.

Given $g \in G$, by definition, there exist finitely many $g_i \in G_i$ such that $g = L_1(g_1) \circ \cdots L_k(g_k)$. Take $s = (g_1, \ldots, g_k) \in \hat{W}$. Then clearly, $\psi(s) = g$ which implies $\bar{\psi}[s] = g$. Now suppose, $\bar{\psi}[s] = \bar{\psi}[s']$. Choose reduced words s, s' to represent their class. Then $\psi(s) = \psi(s')$ and therefore, $\psi(s)(\square) = \psi(s')(\square)$. Now for any reduced word $s = (s_1, \ldots, s_k)$ we have $\psi(s)(\square) = (s_1, \ldots, s_k)$. It follows that $s = s'$ and hence $[s] = [s']$. This establishes that $\bar{\psi} : [W] \to G$ is an isomorphism.

Incidentally, we have also established that in every equivalence class $[s]$ there is a unique reduced word, viz., $\bar{\psi}[s](\square)$.

Therefore, from now on, we can drop the notation $[W]$ and G and identify both with W and treat it as a group. We shall denote the monomorphisms $g \mapsto (g)$ by $\eta_i : G_i \to W$. It remains to prove that $(W, \{\eta_i\})$ is a free product. Given any homomorphisms $f_i : G_i \to H, i \in I$, we define $f : \hat{W} \to H$ by the formula

$$f(s_1, \ldots, s_k) = f_{i_1}(s_1) \circ \cdots \circ f_{i_k}(s_k), s_j \in G_{i_j}, 1 \leq j \leq k,$$

and verify that f is a homomorphism which takes the same value on each equivalence class. Therefore, there is a well-defined homomorphism $f : W \to H$ which clearly has the property $f \circ \eta_i(g) = f(g)$, for all $g \in G_i$ and for all $i \in I$. The uniqueness of such a f follows from the fact $W (= G)$ is generated by $\cup_i \eta_i(G_i)$.

This completes the construction of the free product. We shall denote it by

$$*_{i \in I} G_i.$$

For the sake of future reference, we shall summarise this in the following:

Theorem 3.6.4 *Let $\{G_i\}$ be a collection of groups. Then the set W of all reduced words in the set $\sqcup_i(G_i \setminus \{1\})$ forms the free product $*_i G_i$ under the usual law of composition: concatenation and reduction.*

We shall leave the following useful observations as exercises to you:

Theorem 3.6.5 *The free product is functorial in the following sense. If $\alpha_i : G_i \to H_i$ are homomorphisms then there is a unique homomorphism $\alpha : *_{i \in I} G_i \to *_{i \in I} H_i$ which equals α_i restricted to G_i.*

Definition 3.6.6 Let S be a non empty set. By a free group $F := F(S)$ with a basis S we mean a group F which contains the set S and which has the following universal property: Given any group H and a function $f : S \to H$ there is a unique homomorphism $\hat{f} : F \to H$ such that $\hat{f}(s) = f(s), s \in S$.

Remark 3.6.7 It is a simple and enjoyable exercise to convert this into a categorical definition, viz., as an initial object of some appropriate category. It then follows easily that a free group on a given set is indeed unique. One can go through the construction of the free product and make a few appropriate changes and get the construction of the free group. Instead we take the following route. Note that if $S = \{s\}$ is a singleton, then $F(S)$ is the infinite cyclic group generated by s. More generally, we have,

Theorem 3.6.8 *Let G_s be the infinite cyclic group generated by s. Then the free product $*_{s \in S} G_s$ is the free group $F(S)$.*

Proof: Clearly, $S \subset *_s G_s$. Given $f : S \to H$ there is a unique homomorphism $f_s : G_s \to H$ such that $f_s(s) = s$. Now by the universal property of $*G_s$ there is a unique $\hat{f} : *G_s \to H$ such that $\hat{f}|_{G_s} = f_s$. Of course this implies $\hat{f}(s) = f(s)$. ♠

Exercise 3.6.9 Let X be any set and $W(X)$ denote the set of all 'free' words in X together with the empty word $\square$. Show that $W(X)$ is a monoid.

3.7 Seifert–van Kampen Theorem

Having developed the group theoretic background needed, we can now take up the study of van Kampen theorems via covering space techniques. Let us begin with an alternative proof of a weaker version of theorem 1.2.31.

Corollary 3.7.1 (A simple version of Seifert–van Kampen Theorem) Let U and V be any two open, path connected, and simply connected subspaces of a space X, such that $U \cap V$ is path connected and $X = U \bigcup V$. Suppose X admits a simply connected covering space. Then X is simply connected.

Proof: Let $p : \overline{X} \to X$ be the universal covering space of X. Let $x \in X$ and $\bar{x} \in \overline{X}$ be such that, $p(\bar{x}) = x$. Since U and V are simply connected, the inclusion maps can be lifted to say, $f : U \to \overline{X}$, $g : V \to \overline{X}$ such that, $f(x) = \bar{x} = g(x)$. Since both $f|_{U \cap V}$ and $g|_{U \cap V}$ are the lifts of the inclusion map, the unique lifting property implies that, $f|_{U \cap V} = g|_{U \cap V}$. Thus f and g patch up to define a continuous map $h : X \to \overline{X}$ such that $p \circ h = Id_X$. This means p is a homeomorphism (see Exercise 3.1.7.(iv)). Since $\overline{X}$ is simply connected, we are through. ♠

Remark 3.7.2

(i) The assumption that U, V are both open subsets can be replaced by the assumption that they are closed subsets.

(ii) Notice that the proof of Theorem 1.2.31 is completely elementary based on a familiar technique that we use in analysis. This technique can be adopted to prove all forms of the van Kampen theorem. Moreover, unlike in the above corollary, it does not assume that X admits simply connected covering and hence it is applicable in more general situations.

(iii) What is then the point of introducing the covering space technique for computing the fundamental group? To begin with, this approach gives a different point of view which was exploited by several mathematicians to give many different results especially in the topology of 3-manifolds. (See [Serre, 1980] for instance.) Currently, this idea has grown into a big branch of mathematics called geometric group theory. We shall therefore give a little bit of exposure to this method by presenting a proof of one more result which is a sort of prototype of many of these results. For simplicity of the notation, we have temporarily dropped writing down the base points at which the fundamental groups are taken; however, base point should be obvious from the context.

We begin with the following basic result the proof of which is a matter of routine check-up.

Lemma 3.7.3 Patching-up covering spaces Let $X = U \bigcup V$ be the union of two path connected spaces, $p : \tilde{U} \to U, q : \tilde{V} \to V$ be two connected coverings. Suppose $W = U \bigcap V$ is path connected and there is a covering isomorphism $\phi : (p^{-1}(W), p, W) \to (q^{-1}(W), q, W)$, i.e., a homeomorphism $\phi : p^{-1}(W) \to q^{-1}(W)$ such that $q \circ \phi = p$. Then there is a covering $\tau : \tilde{X} \to X$ such that $\tilde{X} = \tilde{U} \bigcup \tilde{V}$ and $\tau|_{\tilde{U}} = p, \tau|_{\tilde{V}} = q$.

Proof: We simply define $\tilde{X}$ to be the quotient space of the disjoint union $\tilde{U} \sqcup \tilde{V}$ by the identification $z \sim \phi(z)$ for all $z \in p^{-1}(W)$. Because ϕ is a homeomorphism of open subsets, the quotient map $\lambda : \tilde{U} \sqcup \tilde{V} \to \tilde{X}$ restricted to $\tilde{U}$ (and $\tilde{V}$, respectively) is a homeomorphism onto an open subset of the quotient space $\tilde{X}$, with which we identify $\tilde{U}$ (and respectively, $\tilde{V}$. Also, the map $p \sqcup q$ factors down to define a map $\tau : \tilde{X} \to X$ which is an extension of p as well as q. It is easily checked that if $\{U_i\}$ and $\{V_j\}$ are even coverings for p and q, respectively, then $\{U_i\} \bigcup \{V_j\}$ is an even covering for λ. ♠

Theorem 3.7.4 Pushout *Let $X = U_1 \bigcup U_2$ where U_i are open and path connected semi-locally simply connected such that $U_1 \bigcap U_2 = W$ is also path connected. Let $x_0 \in W$ be the base point for all the fundamental groups involved and let i_1, i_2, j_1, j_2, etc. be the inclusion induced homomorphisms on the fundamental groups:*

$$i_r : \pi_1(W) \to \pi_1(U_r); \quad j_r : \pi_1(U_r) \to \pi_1(X), \quad r = 1, 2.$$

Assume that $U_r, r = 1, 2$ are semi-locally simply connected. Then for every group G and every pair of homomorphisms $\alpha_j : \pi_1(U_j) \to G, j = 1, 2$ such that $\alpha_1 \circ i_1 = \alpha_2 \circ i_2$, there exists a unique homomorphism $\gamma : \pi_1(X) \to G$ such that $\gamma \circ j_1 = \alpha_1$ and $\gamma \circ j_2 = \alpha_2$.

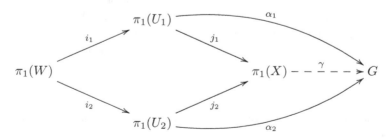

Proof: Let $p_j : \tilde{U}_j \to U_j, j = 1, 2$ be the G-coverings over U_j corresponding to the homomorphisms $\alpha_j, j = 1, 2$. Note that p_j are unique up to isomorphism. It then follows that the restrictions of p_1 and p_2 on W are both G-coverings isomorphic to the G-covering corresponding to the homomorphism $\alpha_1 \circ i_1 = \alpha_2 \circ i_2$ and hence isomorphic to each other. Therefore, from the above lemma, we can patch-up the two G-coverings p_1, p_2 to obtain a G-covering $p : \tilde{X} \to X$. Once again note that the isomorphism class of p is determined by the isomorphism classes of its restrictions over U_1 and U_2. If $\gamma : \pi_1(X) \to G$ is the corresponding homomorphism, it follows that $j_r \circ \gamma$ should correspond to the coverings obtained by restriction over U_r and hence equals $\alpha_r, r = 1, 2$. The uniqueness of γ follows from the uniqueness of the covering p. ♠

As an immediate corollary we obtain:

Theorem 3.7.5 (Seifert–van Kampen) *Let $X = U \bigcup V$ be the union of two open connected spaces such that, $W = U \bigcap V$ is connected and simply connected. Assume that, X is locally contractible also. Then the inclusion induced homomorphisms $i_\# : \pi_1(U) \to \pi_1(X)$, $j_\# : \pi_1(V) \to \pi_1(X)$ are injective and $\pi_1(X)$ is the free product of their images.*

Remark 3.7.6 This theorem has tremendous potential for application. For the proof of the general form of this theorem, viz., without the local contractibility assumption but under some other minor constraints, the reader is referred to the excellent book [Massey, 1977], or the TIFR Lecture notes [de Rham, 1969]. There are other versions of this theorem, such as when the space is written as a union of several open subspaces, and when any two of them do not necessarily intersect in the same subspace. For instance, we may have $X = U \bigcup V$, both U, V being open simply connected, but $U \bigcap V$ being not connected, say, $U \bigcap V = A_1 \sqcup A_2$ with both A_i being simply connected. A typical case of this is the most familiar situation to us, viz., the circle $\mathbb{S}^1$ written as the union of $\mathbb{S}^1 \setminus \{N\}$ and $\mathbb{S}^1 \setminus \{S\}$, where the points N, S denote the north and south poles, respectively. Indeed, the above construction suitably modified, would yield a modified proof that, the fundamental group of $\mathbb{S}^1$ is infinite cyclic.

As an immediate corollary, we can deduce the following result, the details are left to the reader as a simple exercise.

Theorem 3.7.7 *Let X be the one-point union of copies of $\mathbb{S}^1$ indexed over a set J. Then $\pi_1(X)$ is a free group over the set J. In particular, its rank $= \#(J)$.*

Exercise 3.7.8 Prove the above theorem.

3.8 Applications

We shall begin with a direct application of the last result in the previous section to calculate the fundamental group of any pseudo-graph, i.e., a connected 1-dimensional CW-complex. This computation itself will then be applied to deduce other results.

Definition 3.8.1 By a graph we mean a 1-dimensional simplicial complex. By a pseudo-graph we mean a 1-dimensional CW-complex. The 0-cells are then called vertices and the 1-cells are called edges. A connected pseudo-graph is called a tree if it is contractible. By a subtree of G, we mean a subcomplex T of G which is a tree.

Remark 3.8.2 The difference between a graph and a pseudo-graph is that in a graph loops and multi-edges between two vertices are not allowed. In the literature, you may find some of the following results treated for graphs but they hold equally well for pseudo-graphs also.

Proposition 3.8.3 Let T be a subtree of a pseudo-graph G. Then the quotient map $G \to G/T$ is a homotopy equivalence.

Proof: This follows from the fact $T \subset G$ is a cofibration. (See Theorem 1.6.4.)

Theorem 3.8.4 *Let G be a connected (non empty) pseudo-graph, and T_0 be a subtree in it. Then there exists a subtree T in G containing T_0, and such that T contains all vertices of G.*

Proof: Let T_1 be the subcomplex of G which is the union of T_0 and all those edges e_α in G which have two distinct end-points of which precisely one end-point is in T_0. Since it is easy to see that for each such e_α, $T_0 \subset T_0 \bigcup e_\alpha$ is a SDR from which it follows that $T_0 \subset T_1$ is a SDR. In particular, this implies that T_1 is a tree. Now repeat this procedure with T_0 replaced by T_1 etc. to get a sequence of subtrees

$$\cdots \subset T_i \subset T_{i+1} \subset \cdots$$

so that each T_i is SDR of T_{i+1}. As in the proof of Lemma 2.4.2, it now follows that T_0 is a SDR of the subcomplex $T = \bigcup_i T_i$. In particular, T is a subtree containing T_0.

Now suppose there is a vertex $v \in G$. By connectivity, there is a path ω from a point of T_0 to the point v. Since the vertex set of G is a discrete space, this ω passes through only finitely many of them. This just means that $v \in T_k \subset T$ for some k. Thus T contains all the vertices of G. ♠

Theorem 3.8.5 *Every connected pseudo-graph is homotopy equivalent to a bouquet of circles.*

Proof: Starting from $T_0 = \{\star\}$, where $\star$ is a vertex of G, take a tree T in G containing all vertices of G. Then the quotient map $G \to G/T$ is a homotopy equivalence, G/T is a CW-complex with just one vertex and hence is a bouquet of circles. ♠

Remark 3.8.6 The number of circles in G/T is precisely equal to the number of edges in $G \setminus T$.

Corollary 3.8.7 Given a connected pseudo-graph G, $\pi_1(G)$ is a free group of rank equal to the number of edges outside any maximal tree T in G.

Proof: Indeed, fixing a maximal tree $T \subset G$, let $\{e_j\}$ be the set of edges of G not in T. Choose any vertex $v_0 \in T$ as the base point. Join the end-points of e_i to v_0 by the unique edge-path inside T to complete them to loops E_j bases at v_0. In view of the above theorem, it is clear that the homotopy class $[E_J]$ of these loops gives a basis for $\pi_1(G)$. ♠

Corollary 3.8.8 For a finite connected pseudo-graph G, the rank of $\pi_1(G)$ is equal to $1 - \chi(G)$.

Proof: The number of edges in any maximal tree is equal to $f_0(G) - 1$. The rank of $\pi_1(G)$ is equal to $f_1(G) - (f_0(G) - 1) = 1 - f_0(G) + f_1(G) = 1 - \chi(G)$. ♠

Theorem 3.8.9 (Nielson–Schreier) *Every subgroup of a free group is free. Indeed, if F is a free group of finite rank r and F' is a subgroup of finite index k then the rank r' of F' is given by $r' = 1 - k + kr$.*

Proof: Let F be a free group with S as basis. Then $F \approx \pi_1(X)$ where $X = \bigvee_{s \in S} \mathbb{S}^1_s$, a bouquet of circles indexed over S. Let $p : X' \to X$ be the connected covering corresponding to the subgroup F'. Since X' is also a 1-dimensional CW-complex, (see Theorem 3.4.24), it follows that F' is free.

For the more elaborate part, we take $S = \{1, 2, \ldots, r\}$. Since F' is of index k, it follows that the number of sheets of p is k. Therefore, $\chi(X') = k\chi(X)$. (See Exercise 3.4.25.(viii).) But then $r' = 1 - \chi(X') = 1 - k\chi(X) = 1 - k(1 - r) = 1 - k + kr$. ♠

The following theorem is yet another example how covering space techniques can be used in computation of the fundamental group.

Theorem 3.8.10 *Let A and Y be connected and locally contractible spaces $f : A \longrightarrow Y$ be any map. Then the inclusion induced homomorphism $\eta : \pi_1(Y) \to \pi_1(C_f)$ is surjective and the kernel is the normal subgroup N generated by the image of $f_\sharp : \pi_1(A) \to \pi_1(Y)$.*

Proof: Observe that it easily follows that $N \subset \operatorname{Ker}\eta$. We have to prove the equality $N = \operatorname{Ker}\eta$ and the surjectivity of η.

Once again, the idea is to describe the universal covering space of C_f explicitly so that we can identify the group of covering transformations, which is anyway, isomorphic to $\pi_1(C_f)$.

Let X denote the cone over A and $\star$ be the vertex of the cone over A. Note that A is a strong deformation retract of $X \setminus \star$. This yields that Y is a strong deformation retract of the open set $U := C_f \setminus \star$. Also the image V of $X \setminus A$ is an open subset of C_f, which is star-shaped at $\star$ and hence is simply connected. We can now write $C_f = U \bigcup V$, where $U = C_f \setminus \star$. Then $U \bigcap V =: B$, is homeomorphic to $A \times (0,1)$.

Let $p : \overline{U} \to U$ be the covering of U corresponding to the normal subgroup N generated by the image of $(i \circ f)_\sharp$. By the lifting criterion, it follows that, B is evenly covered by p. We take a copy of V for each copy of B in $\overline{U}$ and glue them up to $\overline{U}$ along these copies of B to obtain the space W. (In practice, this can also be viewed as adding just one extra point, a copy of $\star$, at each of the lift of B at the end $A \times 0$.) Clearly, the map $p : \overline{U} \to U$ extends to a map $\bar{p} : W \to C_f$ by sending all the extra points to the point $\star$. It is easy to see that, q is actually a covering projection. Figure 3.4 shows just a double cover of C_f. The bullet indicates the apex point of C_f, V is the region shaded by slanted lines and U is the region shaded by vertical lines.

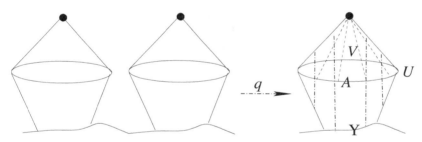

FIGURE 3.4. Universal covering of the mapping cone

We want to show that q is the universal cover. So, let $q : Z \to C_f$ be an universal covering projection.

First of all notice that, $N \subset \operatorname{Ker}\eta$ implies that the map $i \circ p : \overline{U} \to C_f$ has a lift $q_1 : \overline{U} \to Z$. Now the simply connectivity of V implies that, V is evenly covered by q. It follows that, the map q_1 restricted to any one copy of B in $\overline{U}$ can be extended uniquely over the corresponding copy of V, homeomorphically into a copy of V inside Z. The map q' is obtained by piecing all these extensions together. Since Z is the universal covering space, it follows that q' is a homeomorphism.

Notice that, the group of covering transformations of $\bar{p}$ is the same as that of p. Indeed, given a covering transformation ψ of $\bar{p}$, the restriction of ψ to $\bar{U}$ defines a covering transformation of p. This restriction is injective because $\bar{U}$ is connected. It is surjective also, i.e., given a covering transformation ϕ of p, the copies of A are permuted by ψ. Hence ψ extends uniquely to a covering transformation of $\bar{p}$, by the corresponding permutation of the 'extra points' filled in the copies of A.

Thus, we have proved that the group of covering transformations of the universal covering of C_f is equal to $\pi_1(U)/N$ which is the same as $\pi_1(Y)/N$. ♠

Remark 3.8.11 We have put some conditions on Y such that it has a universal covering space. This is necessary in the proof given. However, the conclusion of the theorem holds without this extra assumption and a direct proof can also be given. Once again, the messy part will be in showing that $\mathrm{Ker}\,\eta \subset N$.

Corollary 3.8.12 Let X be a space obtained by attaching 2-cells E_α^2 to a path connected space A via maps $f_\alpha : \mathbb{S}^1 \to A$. Then the fundamental group $\pi_1(X, y_0)$ is isomorphic to the quotient of $\pi_1(Y, y_0)$ by the normal subgroup generated by elements $[f_\alpha] \in \pi_1(A, y_0)$.

Proof: For the sake of simplicity, assume first that all maps f_α preserve the base points, i.e., $f_\alpha(1) = y_0$. If the number of the cells attached to A is one then we can appeal to the above theorem. The case of finitely many cells follows by induction. The general case follows by taking direct limits. ♠

Remark 3.8.13 This is the result that was needed to justify the argument that we used to describe the fundamental group of a lens space in Remark 3.5.12. We shall have many opportunities to apply this result. Other topics of computation of π_1 and its relation with other topological notions will be given at an appropriate time. We shall end this section, with a further discussion on of the famous *Poincaré homology 3-sphere*.

Example 3.8.14 The Poincaré Manifold P_{120}: Here we shall compute $\pi_1(P_{120})$ directly using Poincaré's description of P_{120} rather than the covering projection $\mathbb{S}^3 \to P_{120}$ as done earlier in Example 3.5.22. We begin with the dodecahedron in $\mathbb{R}^3$. It has 12 faces, each of which is a regular pentagon. There is an obvious affine linear isomorphism obtained by the composite of the parallel translation of any face to its opposite face followed by rotation through an angle $\pi/5$. We use these homeomorphisms to identify each face with its opposite face. The resulting space is the Poincaré 3-fold P_{120}. The notation is justified by the following fact:

There are infinitely many spaces of this type, viz., closed 3-manifolds which have the homology of the 3-sphere but with non trivial fundamental group. Among these, P_{120} is the only one which has finite fundamental group and the order of this group is 120.

As in the case of Lens spaces in the above example, we immediately get a CW-structure on the quotient space from the polyhedral structure of the dodecahedron. Just as before, the interior of the polyhedron is going to give one 3-cell which has no effect on the fundamental group. So, we concentrate on the 2-skeleton of P_{120}; the schematic diagram of which is depicted in Figure 3.5, in which, instead of one single 3-dimensional picture, we have drawn two 2-dimensional pictures of the dodecahedron.

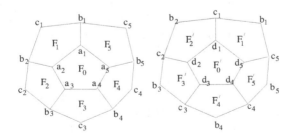

FIGURE 3.5. A 2-dimensional picture of the dodecahedron

The pairs of opposite faces are $\{F_0, F_0'\}$ and $\{F_i, F_{i+3}'\}, i = 1, 2, 3, 4, 5$. Thus, the 2-skeleton of the quotient can now be obtained merely as a quotient of the 2-dimensional polyhedron represented by the first diagram. We need the second diagram only to determine the identifications taking place within the vertices and edges of the first diagram. The identification of these faces as indicated above, results in the identification of the vertices

$$d_i \leftrightarrow a_{i+1}; b_i \leftrightarrow a_{i+2}; c_i \leftrightarrow a_{i+4}, i = 1, 2, 3, 4, 5$$

and so there are precisely 5 distinct equivalence classes of vertices, all of which can be represented by $a_i, i = 1, \ldots, 5$. We also get the following oriented edge identifications:

$$[b_i, c_i] \leftrightarrow [a_{i+2}, a_{i+4}]; [c_i, b_{i+1}] \leftrightarrow [a_{i+4}, a_{i+3}], [a_i, b_i] \leftrightarrow [a_i, a_{i+2}]$$

which leaves precisely 10 distinct edge classes. Thus it is easily seen that the 1-skeleton of the quotient is the complete graph K_5 on five vertices, which we can continue to represent by $a_1, \ldots, a_5$. (See Figure 3.6.) Now the six pentagons in the quotient space are:
$\bar{F}_0 = (a_1 a_2 a_3 a_4 a_5) = \bar{F}'_0$; $\bar{F}_1 = (a_1 a_3 a_5 a_4 a_2) = \bar{F}'_4$; $\bar{F}_2 = (a_1 a_5 a_3 a_2 a_4) = \bar{F}'_5$,
$\bar{F}_3 = (a_1 a_4 a_3 a_5 a_2); = \bar{F}'_1$ $\bar{F}_4 = (a_1 a_3 a_2 a_5 a_4) = \bar{F}'_2$; $\bar{F}_5 = (a_1 a_3 a_4 a_2 a_5) = \bar{F}'_3$;
obtained by merely replacing the vertices of the original pentagons F_i with the corresponding a_i in their equivalence class.

We may choose any point of K_5 as the base point and let us choose a_1. We can then choose a maximal tree containing a_1 in K_5 and let us choose the edges $a_1 a_2, a_1 a_3, a_1 a_4$ and $a_1 a_5$ to form this maximal tree. Each of the remaining six edges forms a loop at a_1 along with the obvious auxiliary edges in the maximal tree and we shall denote them as indicated in the figure. With arbitrarily chosen orientation, the loop classes of them form a basis for the fundamental group of the 1-skeleton K_5, a free group of rank 6.

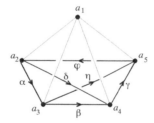

FIGURE 3.6. The 1-skeleton of the Poincaré manifold

Now the six relations from the six faces can be written down by correctly running around their boundary in any one of the directions. We obtain:

$$
\begin{aligned}
\alpha\beta\gamma &= 1; & \eta\gamma^{-1}\delta^{-1} &= 1; & \eta^{-1}\alpha^{-1}\delta &= 1; \\
\beta^{-1}\eta\varphi &= 1; & \alpha^{-1}\varphi^{-1}\gamma^{-1} &= 1; & \beta\delta^{-1}\varphi^{-1} &= 1.
\end{aligned}
$$

The idea is to eliminate as many generators as possible and rewrite the generators and relations in a more comprehensive way. There are several ways to do this, most of them leading to nowhere. Here is one such which leads us to some meaningful conclusion:

In the first round, use second and fifth relations, respectively, to eliminate $\delta = \eta\gamma^{-1}$ and $\varphi = \gamma^{-1}\alpha^{-1}$. The remaining relations now become

$$
\begin{aligned}
\alpha\beta\gamma &= 1; & \eta^{-1}\alpha^{-1}\eta\gamma^{-1} &= 1; \\
\beta^{-1}\eta\gamma^{-1}\alpha^{-1} &= 1; & \beta\gamma\eta^{-1}\alpha\gamma &= 1.
\end{aligned}
$$

In II round, use first relation to eliminate $\gamma = \beta^{-1}\alpha^{-1}$. This leaves us with the following three relations:

$$
\begin{aligned}
\eta^{-1}\alpha^{-1}\eta\alpha\beta &= 1; \\
\beta^{-1}\eta\alpha\beta\alpha^{-1} &= 1; \\
\alpha^{-1}\eta^{-1}\alpha\beta^{-1}\alpha^{-1} &= 1.
\end{aligned}
$$

The last relation can be rewritten as $\eta = \alpha\beta^{-1}\alpha^{-2}$. Use this to eliminate η, to obtain:

$$\alpha^2\beta\alpha^{-1}\alpha^{-1}\alpha\beta^{-1}\alpha^{-2}\alpha\beta = 1; \quad \beta^{-1}\alpha\beta^{-1}\alpha^{-2}\alpha\beta\alpha^{-1} = 1.$$

These are same as:

$$\alpha^2\beta\alpha^{-1}\beta^{-1}\alpha^{-1}\beta = 1: \quad \beta^{-1}\alpha\beta^{-1}\alpha^{-1}\beta\alpha^{-1} = 1.$$

Now if we take the abelianization of this group, viz., by declaring that the two generators α, β commute also, then the above two relations immediately imply that the group is trivial. This implies that the first homology group $H_1(P_{120}; \mathbb{Z}) = (0)$. (See Section 4.5). Is $\pi_1(P_{120}) = 1$? To settle this, we need to work a little harder.

We introduce another generator τ and put it equal to $\alpha\beta^{-1}$ and use it to eliminate $\beta = \tau^{-1}\alpha$:

$$\alpha^2\tau^{-1}\alpha^{-1}\tau\alpha^{-1}\tau^{-1}\alpha = 1; \quad \alpha^{-1}\tau^2\alpha^{-1}\tau^{-1} = 1.$$

These two are equivalent to

$$\alpha^3 = (\tau\alpha)\tau^{-1}\alpha\tau; \quad \tau\alpha = \alpha^{-1}\tau^2.$$

If we substitute in the first relation for $\tau\alpha$ from the second, we get, $\alpha^3 = \alpha^{-1}\tau\alpha\tau$. Therefore the two relations together are equivalent to:

$$\alpha^4 = \tau\alpha\tau; \quad \alpha\tau\alpha = \tau^2.$$

This can be rewritten as:

$$\alpha^5 = (\tau\alpha)^2 = \tau^3.$$

Why is this group non trivial? The geometry of the dodecahedron has the answer. There are two rotations of the dodecahedron which we may call A and T. The rotation A is through an angle $2\pi/5$ about the line joining the centres of any two of the opposite faces. Rotation T is through an angle $2\pi/3$ about the line joining any two opposite vertices. Clearly $A^5 = T^3 = 1$. The non trivial thing which we have to check is the fact that $(TA)^2 = 1$. Thus it follows that there is a non trivial homomorphism from the group

$$\pi_1(P_{120}) = \langle \alpha, \tau \; : \; \alpha^5 = (\tau\alpha)^2 = \tau^3 \rangle$$

to the group of rotations of the dodecahedron given by $\tau \mapsto T$ and $\alpha \mapsto A$.

If you do not like this geometric way, you are welcome to check that there is a non trivial homomorphism of $\pi_1(P_{120})$ into the alternating group on 5 letters given by

$$A \mapsto (12345); T \mapsto (253).$$

Indeed, both these homomorphisms are actually surjective (take this as an exercise) and the kernel is precisely a group of order two generated by

$$\alpha^5 = (\tau\alpha)^2 = \tau^3$$

which happens to be the only non trivial element in the centre of the group. Therefore P_{120} is of order 120 and is called the binary-dodecahedral group (or the binary-icosahedral group, this last name being justified by the fact that due to duality, the dodecahedron and the icosahedron have the same group of symmetries).

3.9 Miscellaneous Exercises to Chapter 3

1. Give an example of a surjective local homeomorphism which is not a covering projection. Give an example of a covering projection $p : \overline{X} \longrightarrow X$ and a subspace A of $\overline{X}$ such that $p|_A$ is surjective but not a covering projection.

2. For a locally path connected and connected covering $p : \overline{X} \longrightarrow X$, show that the Galois group $\mathbf{G}(p)$ is isomorphic to the quotient group $N(K)/K$, where $K = p_{\#}(\pi_1(\overline{X}, \bar{x}))$ and $N(K)$ is the normalizer of K in $\pi_1(X, x)$.

3. Compute $\pi_1(\mathbb{P}^2)$ using Theorem 3.8.10.

4. Given a topological n-manifold X (II-countable, and Hausdorff), choose a countable open cover $\{U_j\}$ for X where each U_j is homeomorphic to $\mathbb{R}^n$. For any two i, j the number of connected components of $U_i \bigcap U_j$ is countable. Choose one point $x_{i,j,n}$ in each of these components and let $P_{i(j,n,m)}$ be a path joining $x_{i,j,n}$ to $x_{i,j,m}$ in U_i. Let x_0 be one of these points. Show that any loop in X based at x_0 is homotopic to a composite of finitely many paths $P_{i(j,m,n)}$ and their inverses. Conclude that $\pi_1(X, x_0)$ is countable.

5. Show that the Cartesian product of infinitely many copies of $\mathbb{S}^1$ does not admit a universal covering.

6. **Hawaiian rings** Here is another popular example of a connected locally path connected space which is not locally contractible. Let $C(p, r)$ be the circle in $\mathbb{R}^2$ with centre p and radius r. Let $Y_n := \bigcup_{m \geq n} C((0, 1/n), 1/n), n \geq 1$. Show that,
   (a) Each $C((0, 1/n), 1/n)$ is a retract of Y.
   (b) Every connected open neighbourhood of $(0, 0)$ in Y contains some Y_n as a retract.
   (c) Y_n is not simply connected.
   (d) Let $p : Z \to Y$ be a covering projection, $z \in Z$ be such that, $p(z) = (0, 0)$ and U be an open neighbourhood of z, sitting over an evenly covered open neighbourhood of $(0, 0)$. Then every connected open neighbourhood U' of z is a retract of Z and contains a subspace V as a retract and homeomorphic to Y_n, for some large n.
   (e) Z is not simply connected, and thus Y has no simply connected covering space.
   (f) Show that Y does not admit any universal covering.
   (g) Put W equal to the union of the x-axis $\mathbb{R} \times 0$ and the circles $C((4m, 1/n), 1/n)$ for all $m\mathbb{Z}, n \geq 2$. Construct an infinite sheeted covering projection $p : W \to Y$.
   (h) Let $\hat{W}$ be the union of W, the line $y = 2$, the circles $C((4m, 1), 1), C((4m, 2), 1/n), n \neq |m| + 2$. Construct a 2-sheeted covering $q : \hat{W} \to W$ in such a way that $p \circ q$ is not a covering projection. Figure 3.7 represents $\hat{W}$.

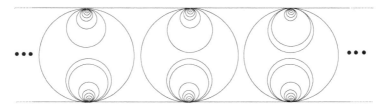

FIGURE 3.7. Hawaiian covering space

7. Let $Y \subset \mathbb{R}^2 \times \{0\}$ be as in the above example and $p = (0,0,1) \in \mathbb{R}^3$. Let cY be the subspace of $\mathbb{R}^3$ which is the union of all line segments $tp + (1-t)y$, $0 \le t \le 1, y \in Y$. Let $-cY$ be the reflection of cY in the origin, viz., the set of all $-z$ where $z \in cY$. Consider $D = cY \bigcup -cY \subset \mathbb{R}^3$. (Abstractly, D is the one-point union of two copies of cY but the point at which the one-point union formed is a 'bad' point.)

   (a) Show that D is not simply connected. (Hint: Look at the loop which winds around C_n and then around $-C_n$ successively for each n.)

   (b) Show that any connected covering of D is homeomorphic to D. (Hint: See Remark 3.4.25.)

   [Clearly (a), (b) together imply that D has no simply connected covering and is the universal cover of itself.]

Topological Groups

Let G be a group. Let the underlying set of G have a topological structure also. Further suppose that the group operations (multiplication, viz., $G \times G \to G$ and taking inverse $G \to G$) are continuous with respect to this topology. Then, we call G a *topological group*. Examples of topological groups are a dime a dozen in mathematics. Any group can be considered as a topological group with the discrete topology on the underlying space. Such topological groups are called *discrete groups*.

8. Show that a subgroup H of G as a subspace, has a discrete topology iff there exists an open neighbourhood U of the identity element $e \in G$ such that $U \bigcap H = \{e\}$.

9. By left multiplication, any subgroup H can be made to act on the whole group. Then the orbit space is nothing but the space of right cosets of H. If H is a discrete subgroup, as seen above it means that, H acts evenly on G. In particular, $G \to G/H$ is a covering projection. Supply full details of all these claims.

10. Verify that any finite dimensional vector space V over $\mathbb{R}$ or $(\mathbb{C})$ is a topological group. If $\{u_1, \ldots, u_n\}$ is a basis for V, then show that, the additive subgroup $\mathbb{Z}\{u_1, \ldots, u_n\}$ generated by $\{u_1, \ldots, u_n\}$ is a discrete subgroup of V.

11. Let G be any topological group.

   (a) Show that a subgroup H of G is open iff there exists an open neighbourhood V of e such that $V \subset H$.

   (b) Show that every open subgroup of G is closed also.

   (c) Show that the connected (path connected) component of G containing the identity $e \in G$ is a closed subgroup.

   (d) If G is a connected group and V is a neighbourhood of e, then show that V is a set of generators of G.

   (e) If H is a normal discrete subgroup of G show that the centraliser of any element in H is an open subgroup. [Hint: Given $x \in H$, choose a neighbourhood N of x such that $N \bigcap H = \{x\}$. Now given $y \in C(x)$, choose a neighbourhood V of y such that $VxV^{-1} \subset N$.]

   (f) Suppose G is connected. Show that any discrete normal subgroup of G is central.

   (g) Let G be a locally path connected, connected and simply connected topological group and let H be a discrete subgroup of G. Then show that $\pi_1(G/H)$ is isomorphic to H. Apply this to compute the fundamental group of several examples that we have already considered.

(h) Suppose G is connected, locally path connected and having a universal covering space, $p : \overline{G} \to G$. Show that, $\overline{G}$ is also a topological group in such a way that, $p : \overline{G} \to G$ is a homomorphism. Also show that, $Ker\ p$ is a central subgroup, isomorphic to $\pi_1(G, e)$. (Hint: Fix $\overline{e} \in p^{-1}(e)$. If $\mu : G \times G \to G$ denotes the group law, take the lift of $\mu \circ (p \times p)$ so that $(\overline{e}, \overline{e})$ is mapped to $\overline{e}$. Use the unique lifting property repeatedly to show the group laws. Later use the above exercise.)

(i) Suppose G is connected, locally path connected and having a universal covering space. Prove that the fundamental group of G is abelian. (Compare this with Exercise 1.9.39.)

12. Consider the quaternion algebra $\mathbb{H}$ which is an algebra over $\mathbb{R}$ generated by the symbols $\mathbf{i}, \mathbf{j}, \mathbf{k}$ with the multiplication defined by $\mathbf{i}^2 = \mathbf{j}^2 = \mathbf{k}^2 = -1$ and

$$\mathbf{ij} = \mathbf{k} = -\mathbf{ji}; \mathbf{jk} = -\mathbf{kj}; \mathbf{ki} = \mathbf{j} = -\mathbf{ik}.$$

Consider the unit quaternions as the 3-dimensional sphere $\mathbb{S}^3$. Show that the conjugation by a unit quaternion defines an orthogonal transformation of the 3-dimensional vector space of all quaternions of the form $b\mathbf{i} + c\mathbf{j} + d\mathbf{k}$ and hence defines a homomorphism $\Theta : \mathbb{S}^3 \to SO(3)$. Show that Θ is surjective. Compute the kernel of Θ. Now, can you conclude anything about $\pi_1(SO(3))$?

13. Consider the space X obtained from the cube $\mathbb{I}^3$ by deleting two line segments AB and CD where $AB = \{1/3\} \times \{1/2\} \times \mathbb{I}$; $CD = \{2/3\} \times \mathbb{I} \times \{1/2\}$. (See Figure 3.8 (I).) Label the vertices of the cube as indicated. Choose 1 as the base point for the fundamental group of X. Let us take paths along the edges of the cube and indicate them by merely writing the vertices through which they pass in that sequence. For instance, the loops $12341 =: \alpha$ and $12781 =: \beta$ represent the two generators $\pi_1(X)$.
(a) Show that 1876581 also represents α. Likewise 1436541 represents β.
(b) Show that the loop 123456781 represents $\alpha\beta^{-1}\alpha^{-1}\beta$.

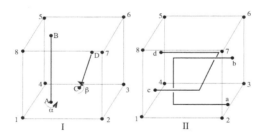

FIGURE 3.8. A 3-disc with two holes

Now consider the space Y obtained from $\mathbb{I}^3$ by deleting two polygonal arcs ab and cd. (See Figure 3.8 (II).) For instance, the arc ab consists of the three segments

$$(1/3, 1) \times 1/2 \times 1/3; 1/3 \times 1/2 \times (1/3, 2/3), \quad (1/3, 1) \times 1/2 \times 2/3.$$

(c) Show that there is homeomorphism $f : \mathbb{I}^3 \to \mathbb{I}^3$ such that $f(X) = Y$.
(d) Now let 13 and 68 denote the diagonals segments. Determine the element in $\pi_1(Y)$ represented by 13681.

14. Let now X be a (compactly generated Hausdorff) topological space with a base point e. Put $J(X) = W(X \setminus \{e\})$. Put a topology on $J(X)$ as follows. Let $J^m(X)$ be the set of all those elements of $W(X)$ which are of length $\leq m$. There is a natural surjection from $\phi_m : X^m \to J^m(X)$ where $\phi_m(x_1, \ldots, x_m)$ is the reduced word obtained after cutting down all occurrences of e in the word $x_1 \cdots x_m$. (Note that $\pi_1(e) = \square$ the empty word.) Give $J^m(X)$ the quotient topology (equivalently, declare ϕ_m a proclusion). Take the topology on $J(X)$ to be the one coinduced by the family $\{J^m(X)\}_{m \geq 1}$. The space $J(X)$ is called *James' reduced product* of X.

(a) Show that each $J^m(X)$ is Hausdorff.

(b) Let X_*^m denote the subspace of all $(x_1, \ldots, x_m)$ in which $x_i = e$ for at least one i. Then $\pi_m : (X^m, X_*^m) \to (J^m(X), J^{m-1}(X))$ is a relative homeomorphism.

(c) The inclusion map $J^{m-1}(X) \to J^m(X)$ is a cofibration.

(d) $J(X)$ is compactly generated Hausdorff space.

(e) The multiplication in $J(X)$ is continuous and hence $J(X)$ is a topological monoid (also called strictly associative H-space.)

(f) J defines a covariant functor from the category of compactly generated topological spaces to the category of topological monoids.

(g) $J(X)$ is the 'free topological semigroup' in the following sense. Given any continuous function $f : X \to M$ where M is a topological semigroup there exists a unique continuous homomorphism $J(f) : J(X) \to M$ which is an extension of f.

(Hint: See [Whitehead, 1978] page no. 326.)

15. **Local Systems** Let X be a topological space. By a *local system* on X we mean a covariant functor from the fundamental groupoid $\mathcal{P}_X$ of X to any category $\mathcal{C}$. (See Example 1.8.3.9.)

(a) Given a category $\mathcal{C}$, show that there is a category $LS(X; \mathcal{C})$ whose objects are local systems on X with values in $\mathcal{C}$.

(b) Given a map $f : X \to Y$ there is a covariant functor $f^* : LS(Y; \mathcal{C}) \to LS(X; \mathcal{C})$. Also if $\phi : \mathcal{C} \to \mathcal{C}'$ is a covariant functor, then there is a covariant functor $\phi_* : LS(X; \mathcal{C}) \to LS(X; \mathcal{C}')$. Thus $LS(-; -)$ is itself a functor which is contravariant in the first slot and covariant in the second.

(c) Given a local system Γ on X, show that there is homomorphism

$$\gamma(x_0) : \pi_1(X, x_0) \to \operatorname{Aut} \Gamma(x_0).$$

(See Exercise 1.8.20.(iii).)

(d) Two local systems on X with values in $\mathcal{C}$ are said to be equivalent, if they are equivalent as objects in $LS(X; \mathcal{C})$. Show that if X is path connected, then two local systems Γ_1, Γ_2 on X are equivalent iff there is an equivalence $\alpha : \Gamma_1(x_0) \to \Gamma_2(x_0)$ such that $\hat{\alpha} \circ \gamma_1(x_0)$ is conjugate to $\gamma_2(x_0)$ in $\operatorname{Aut} \Gamma_2(x_0)$, where $\hat{\alpha}$ is as in Exercise 1.8.20.(iii).

(e) Let X be path connected, $A \in \mathcal{C}$ be any object. Given any homomorphism $\alpha : \pi_1(X, x_0) \to \operatorname{Aut} A$ show that there is a local system Γ on X with values in $\mathcal{C}$ such that $\Gamma(x_0) = A$ and $\gamma(x_0) = \alpha$.

(f) Conclude that given an object $A \in \mathcal{C}$, a path connected space X, and a base point x_0, equivalence classes of local systems Γ on X with $\Gamma(x_0) = A \in \mathcal{C}$ are in one-one correspondence with the conjugacy classes of homomorphisms $\pi_1(X, x_0) \to \operatorname{Aut} A$. (For some examples, the reader may like to see Chapter 10.)

Chapter 4

Homology Groups

Recall the homology version of Cauchy's theorem in complex analysis of 1-variable ([Shastri, 2009] p.214):

Theorem 4.0.1 *Let Ω be an open subset in $\mathbb{C}$ and γ be a cycle in Ω. Then the following conditions on γ are equivalent:*
(a) $\int_\gamma f \, dz = 0$ *for all holomorphic functions f on Ω.*
(b) *The winding number of γ around a, $\eta(\gamma, a) = 0$ for all $a \in \mathbb{C} \setminus \Omega$.*

A cycle γ satisfying the condition (a) or equivalently, (b), was called null-homologous.

As indicated in the introduction to Chapter 1, the homology and cohomology theories have their roots in complex analysis. In the integral calculus of several variables, especially in Stokes' theorem, you may witness a neat interaction between homology and cohomology theories. However, it was Poincaré who formalized the idea of homology theory in his seminal paper 'Analysis Situs' and it took many more years for the arrival of cohomology theory and finally both of them were properly established in 1952 by Eilenberg and Steenrod [Eilenberg–Steenrod, 1952].

In order to bring out the full force of these results, as well as save time, it is good to temporarily isolate these two concepts and study each of them in a more abstract set-up. Naturally, there is a lot of algebra to be worked out. Among these, cohomology is more narutal and some of you may have come across with it while studying differential forms. On the other hand, being more geometric, homology is easier to comprehend. So, we shall deal with homology in this chapter and take up the study of cohomology in the next chapter.

In Section 4.1, we shall introduce some minimal amount of homological algebra needed to launch singular homology. In Section 4.2, we shall construct the singular homology and study its basic properties. In section 4.3, we shall construct some other homology groups and state results which relate them to singular homology. In section 4.4, we shall give some wonderful applications of these results. Section 4.5 will introduce the reader to the beautiful relation between homotopy and homology by studying the same between the fundamental group and the first singular homology group. More about this relation will be taken up in a later chapter. This chapter will end with a discussion of all the postponed proofs. The reader may take her own time to read this last section.

4.1 Basic Homological Algebra

In this section, we shall begin with a minimum introduction to 'homological algebra' necessary to understand the homology groups. The most important result here is the Snake lemma with two ready-to-use corollaries, the Four lemma and the Five lemma. More and more homological algebra will be introduced as and when required. Throughout this chapter, R will denote a general commutative ring with a unit, though often you may assume that this ring is nothing but the ring of integers $\mathbb{Z}$.

Definition 4.1.1 Consider a direct sum

$$C. := C_* := \oplus_{n \in \mathbb{Z}} C_n$$

of R-modules. Often we call C_*, a graded R-module with its nth graded component C_n. Members of C_n are also called homogeneous elements of C_* of degree n. Both these notations for a graded module are common in the literature and hence we would like you to get familiar with both of them right from the beginning.

A R-module homomorphism $f : C_* \longrightarrow C'_*$ is called a graded homomorphism if there exists d such that $f(C_r) \subset C'_{r+d}$ for all r. We then call d the (homogeneous) degree of f. We shall denote $f|_{C_r}$ by f_r if necessary–and often we may simply write f itself for f_r provided there is no confusion.

By a *chain complex* (C_*, ∂) of R-modules, we mean a graded R-module C_*, together with an endomorphism $\partial := \partial_* : C_* \longrightarrow C_*$ of degree -1 with the property $\partial \circ \partial = 0$. Observe that ∂ consists of a sequence $\{\partial_n : C_n \longrightarrow C_{n-1}\}$ of R-module homomorphisms such that $\partial_n \circ \partial_{n-1} = 0$ for all n. The endomorphism ∂ is called the *differential* or the boundary map of the chain complex. Often we shall not mention the map ∂ at all and merely say C_* is a chain complex.

If C_* and C'_* are two chain complexes then by a chain map $f = f_* : C_* \longrightarrow C'_*$ we mean a graded module homomorphism of degree 0 that commutes with the corresponding differentials, i.e., a sequence $f_n : C_n \longrightarrow C'_n$ of R-module homomorphisms such that $\partial'_n \circ f_n = f_{n-1} \circ \partial_n$ for all n. It is straightforward to check that there is a category of chain complexes of R-modules and chain maps. We shall denote this category by $\mathbf{Ch_R}$ with a further simplification $\mathbf{Ch} := \mathbf{Ch_{\mathbb{Z}}}$.

There is an obvious way to define the direct sum of a family of chain complexes $\{(C^\alpha_., \partial^\alpha)\}_{\alpha \in \Lambda}$: the graded module is taken to be the direct sum $\oplus_\alpha C^\alpha_.$ and the boundary map $\partial = \oplus_\alpha \partial^\alpha$. For instance, if $(C^1_., \partial^1), (C^2_., \partial^2)$ are two chain complexes then their direct sum (C, ∂) is defined by

$$C_n = C^1_n \oplus C^2_n, \ \forall \ n; \quad \partial(c^1 \oplus c^2) = \partial^1(c^1) \oplus \partial^2(c^2).$$

It is straightforward to check that $\partial \circ \partial = 0$.

Definition 4.1.2 A sequence of R-modules

$$M' \xrightarrow{\alpha} M \xrightarrow{\beta} M''$$

is said to be *exact* at M if Ker β = Im α. A sequence

$$\cdots \longrightarrow M_{n-1} \longrightarrow M_n \longrightarrow M_{n+1} \longrightarrow \cdots$$

is said to be exact if it is exact at each of the modules M_n. By a short exact sequence we mean an exact sequence of the form

$$0 \longrightarrow M' \longrightarrow M \longrightarrow M'' \longrightarrow 0.$$

Definition 4.1.3 A sequence of chain complexes and chain maps

$$0 \longrightarrow C'_. \xrightarrow{f_.} C_. \xrightarrow{g_.} C''_. \longrightarrow 0 \tag{4.1}$$

is said to be exact if for each n the corresponding sequence of modules

$$0 \longrightarrow C'_n \xrightarrow{f_n} C_n \xrightarrow{g_n} C''_n \longrightarrow 0$$

is exact.

Remark 4.1.4 There is a category of short exact sequences of chain complexes of R-modules, defined in an obvious way. These will be used to 'split-up' and study longer sequences of modules. We shall now introduce the homology groups to measure the deviation of a chain complex from being exact.

Definition 4.1.5 Given a chain complex C_*, define the *homology group* of C_* to be the graded R-module

$$H_*(C_*) := \oplus_{n \in \mathbb{Z}} H_n(C_*)$$

by taking

$$H_n(C_*) := \operatorname{Ker} \partial_n / \operatorname{Im} \partial_{n+1}, \quad \forall \; n \in \mathbb{Z}.$$

Observe that, if $f : C_* \longrightarrow C'_*$ is a chain map then f induces a graded homomorphism $H_*(f) : H_*(C_*) \longrightarrow H_*(C'_*)$ in the obvious way. In addition, this has the *naturality* property, viz., $H_*(Id) = Id$ and if g is another chain map such that $f \circ g$ is defined, then $H_*(f \circ g) = H_*(f) \circ H_*(g)$. Thus, H_* is a *covariant functor* from the category of chain complexes to the category of graded modules. One of the easiest algebraic results is the following, the proof of which is left to the reader as a warm-up exercise.

Theorem 4.1.6 *The homology of a direct sum of chain complexes is isomorphic to the direct sum of the homology of chain complexes.*

Remark 4.1.7 The above property may be named 'additivity of the homology'. However, we need to work in a more general set-up of (4.1) wherein the module C in the middle may not be the direct sum of the two end-modules. The question is: *do we have $H_*(C_.)$ isomorphic to the direct sum of $H_*(C')$ and $H_*(C'')$; or at least, do we have a corresponding short exact sequence of the homology modules?* That is the kind of additivity we will be interested in. However, the answer, in general, is NO. So, we are forced to refer to the property of homology groups as in Theorem 4.1.6 'weak additivity'. The best thing that we can say about this additivity of homology, viz., Theorem 4.1.10 below is one of the very basic results in homological algebra and the following lemma prepares us for it.

Lemma 4.1.8 (Snake lemma) Given a commutative diagram of R-module homomorphisms: where the two horizontal sequences are exact, there exists a R-module homomor-

$$
\begin{array}{ccccccccc}
& & M' & \xrightarrow{\alpha} & M & \xrightarrow{\beta} & M'' & \longrightarrow & 0 \\
& & \Big\downarrow{f'} & & \Big\downarrow{f} & & \Big\downarrow{f''} & & \\
0 & \longrightarrow & N' & \xrightarrow{\alpha'} & N & \xrightarrow{\beta'} & N'' & &
\end{array}
$$

FIGURE 4.1. Snake lemma

phism
$\delta : \operatorname{Ker} f'' \longrightarrow \operatorname{Coker} f'$, called the *connecting homomorphism* such that the sequence

$$\operatorname{Ker} f' \xrightarrow{\alpha} \operatorname{Ker} f \xrightarrow{\beta} \operatorname{Ker} f'' \xrightarrow{\delta} \operatorname{Coker} f' \xrightarrow{\alpha'} \operatorname{Coker} f \longrightarrow \operatorname{Coker} f''$$

is exact. Moreover, the connecting homomorphism δ has the naturality properties, so that the above assignment of a 'snake' to the corresponding 'six-term' exact sequence of modules defines a covariant functor.

Proof: The homomorphism δ is defined as follows: Given $x'' \in \operatorname{Ker} f''$, pick $x \in M$ such that $\beta(x) = x''$. Then it follows that $\beta' \circ f(x) = 0$ and hence $f(x) \in \operatorname{Im} \alpha'$. Pick $y' \in N'$ such that $\alpha'(y') = f(x)$. Take $\delta(x'') = y' + \operatorname{Im} f'$, the class represented by y' in $\operatorname{Coker} f'$. Observe that, if $x'' = 0$ then we could have picked up anything in $\operatorname{Im} \alpha$ for x. But then $f(x) = \alpha' \circ f'(x')$ for some $x' \in M'$. This would imply that $\delta(x'') = 0$ in $\operatorname{Coker} f'$. This in fact shows that δ is well defined. That it is a R-module homomorphism and fits the bill in the lemma can be verified in a straightforward way in a similar manner. We shall verify, say the exactness at $\operatorname{Ker} f''$ and leave the rest of it as an exercise to the reader.

First, we have to show that $\delta \circ \beta = 0$. So let $x'' = \beta x$ where $x \in \operatorname{Ker} f$. In defining δ, we can use this particular x. But then $f(x) = 0$ and hence we are forced to pick up $0 \in N'$ such that $\alpha'(0) = f(x)$. Therefore, $\delta(x'') = 0$. Conversely, let $x'' \in \operatorname{Ker} f''$ be such that $\delta(x'') = 0$. This means that we have $x \in M$ such that $f(x) = \alpha'(y')$ and $y' \in \operatorname{Im} f'$. Say, $y' = f'(x')$ for some $x' \in M'$. Then take $x_1 = x - \alpha(x')$. It follows that $f(x_1) = 0$ and $\beta(x_1) = x''$. Therefore $x'' \in \beta(\operatorname{Ker} f)$. This proves the exactness at $\operatorname{Ker} f''$.

We shall call a commutative diagram of R-modules as in Figure 4.1, with the two horizontal sequences being exact, a 'snake'. A morphism from one snake to another snake is yet another commutative diagram of R-module homomorphisms as shown in the next figure.

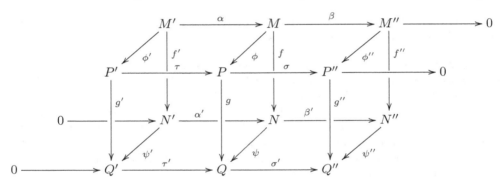

Clearly we then have a diagram of R-module homomorphisms wherein the two horizontal sequences are exact.

$$\operatorname{Ker} f' \xrightarrow{\alpha} \operatorname{Ker} f \xrightarrow{\beta} \operatorname{Ker} f'' \xrightarrow{\delta} \operatorname{Coker} f' \xrightarrow{\alpha'} \operatorname{Coker} f \xrightarrow{\beta'} \operatorname{Coker} f''$$
$$\downarrow{\phi'} \quad\quad \downarrow{\phi} \quad\quad \downarrow{\phi''} \quad\quad \downarrow{\psi'} \quad\quad \downarrow{\psi} \quad\quad \downarrow{\psi''}$$
$$\operatorname{Ker} g' \xrightarrow{\tau} \operatorname{Ker} g \xrightarrow{\sigma} \operatorname{Ker} g'' \xrightarrow{\delta} \operatorname{Coker} g' \xrightarrow{\tau'} \operatorname{Coker} g \xrightarrow{\sigma'} \operatorname{Coker} g''$$

It is straightforward to check that this assignment defines a functor from the category of 'snakes' to category of 'six-terms'. ♠

Remark 4.1.9 The student is advised to go through the definition of connecting homomorphism in the above lemma carefully and memorise it. For, despite all the theories that we are going to develop, often while dealing with the connecting homomorphism, quoting theorems and lemmas does not help—we need to go through this construction of δ itself.

Theorem 4.1.10 *Given a short exact sequence of chain complexes*

$$0 \longrightarrow C'_* \xrightarrow{\alpha} C_* \xrightarrow{\beta} C''_* \longrightarrow 0$$

there is a functorial long exact sequence of homology groups

$$\longrightarrow H_n(C'_*) \xrightarrow{H_n(\alpha)} H_n(C_*) \xrightarrow{H_n(\beta)} H_n(C''_*) \xrightarrow{\delta_n} H_{n-1}(C'_*) \xrightarrow{H_{n-1}(\alpha)} H_{n-1}(C_*) \longrightarrow \quad (4.2)$$

Proof: The chain maps α and β, respectively, induce chain maps of quotient modules:

$$\bar{\alpha} : C'_*/\mathrm{Im}\,\partial' \longrightarrow C_*/\mathrm{Im}\,\partial; \ \bar{\beta} : C_*/\mathrm{Im}\,\partial \longrightarrow C''_*/\mathrm{Im}\,\partial''.$$

Also upon restriction to $\mathrm{Ker}\,\partial'$ and $\mathrm{Ker}\,\partial$, respectively, they define chain maps:

$$\alpha' : \mathrm{Ker}\,\partial' \longrightarrow \mathrm{Ker}\,\partial; \ \beta' : \mathrm{Ker}\,\partial \longrightarrow \mathrm{Ker}\,\partial''.$$

For each n, we then have a snake as follows.

$$
\begin{array}{ccccccc}
C'_n/\mathrm{Im}\,\partial'_{n+1} & \xrightarrow{\bar{\alpha}_n} & C_n/\mathrm{Im}\,\partial_{n+1} & \xrightarrow{\bar{\beta}_n} & C''_n/\mathrm{Im}\,\partial''_{n+1} & \longrightarrow & 0 \\
\downarrow{\scriptstyle \partial'_n} & & \downarrow{\scriptstyle \partial_n} & & \downarrow{\scriptstyle \partial''_{n+1}} & & \\
0 \longrightarrow \mathrm{Ker}\,\partial'_{n-1} & \xrightarrow{\alpha'_{n-1}} & \mathrm{Ker}\,\partial_{n-1} & \xrightarrow{\beta'_{n-1}} & \mathrm{Ker}\,\partial''_{n-1} & &
\end{array}
$$

Now notice that the kernel of $[\partial'_n : C'_n/\mathrm{Im}\,\partial'_{n+1} \to \mathrm{Ker}\,\partial'_{n-1}]$ is isomorphic to $\mathrm{Ker}\,\partial'_n/\mathrm{Im}\,\partial_{n+1} = H_n(C')$. Likewise the cokernel of this homomorphism is isomorphic to $H_{n-1}(C')$. The same is true of ∂_n and ∂''_n. Therefore the associated 'six-term' exact sequence is nothing but

$$H_n(C'_*) \xrightarrow{H_n(\alpha)} H_n(C_*) \xrightarrow{H_n(\beta)} H_n(C''_*) \xrightarrow{\delta_n} H_{n-1}(C'_*) \xrightarrow{H_{n-1}(\alpha)} H_{n-1}(C_*) \xrightarrow{H_{n-1}(\beta)} H_{n-1}(C''_*).$$

Since this is true for all n we obtain (4.2). ♠

We shall now state two ready-to-use corollaries to the snake lemma, which go a long way in exploiting the long exact sequences of homology that occur in topology. You are welcome to write down a direct proof of each of them, which will give you some familiarity with the technique of 'diagram chasing' used in the snake lemma.

Corollary 4.1.11 Consider the following commutative diagram of R-modules and R-linear maps in which the two rows are exact. If f_1 and f_3 are isomorphisms then so is f_2.

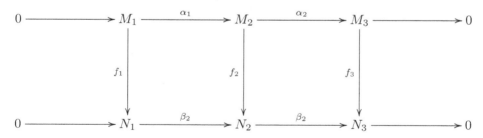

Proof: Exercise.

Corollary 4.1.12 (Four lemma) Consider the following commutative diagram of R modules and R-linear maps in which the two rows are exact. Suppose that f_1 is surjective and f_4 is injective. Then
(i) f_2 is injective $\implies$ f_3 is injective.
(ii) f_3 is surjective $\implies$ f_2 is surjective.

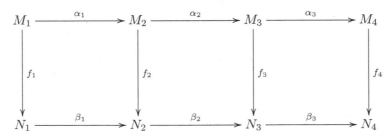

Proof: Hint: Reduce the given diagram into two diagrams as in the snake lemma.

Corollary 4.1.13 (Five lemma) In the following diagram of R-modules, the two rows are given to be exact. If f_1, f_2, f_4 and f_5 are isomorphisms then f_3 is also an isomorphism:

$$
\begin{array}{ccccccccc}
M_1 & \longrightarrow & M_2 & \longrightarrow & M_3 & \longrightarrow & M_4 & \longrightarrow & M_5 \\
\downarrow f_1 & & \downarrow f_2 & & \downarrow f_3 & & \downarrow f_4 & & \downarrow f_5 \\
M_1' & \longrightarrow & M_2' & \longrightarrow & M_3' & \longrightarrow & M_4' & \longrightarrow & M_5'
\end{array}
$$

Proof: Apply four lemma twice, first to f_1, f_2, f_3, f_4 and then to f_2, f_3, f_4, f_5. ♠

We shall now initiate a discussion on a famous invariant known as Euler-characteristic.

Definition 4.1.14 By an *additive function* on a category of R-modules we mean an integral valued set theoretic function ℓ on the isomorphism classes of R-modules such that whenever we have a short exact sequence

$$ 0 \longrightarrow M' \longrightarrow M \longrightarrow M'' \longrightarrow 0 $$

of modules, $\ell(M) = \ell(M') + \ell(M'')$. By a simple induction, one can easily verify that if

$$ 0 \longrightarrow M_1 \longrightarrow \cdots \longrightarrow M_n \longrightarrow 0 $$

is exact, then $\sum_i (-1)^i \ell(M_i) = 0$ for any additive function (Exercise). On the category of finitely generated R modules over a ring R, which is a principal ideal domain (PID), the rank function is one of the most important additive functions.

Definition 4.1.15 Let ℓ be an additive function on some category $\mathcal{F}$ of R-modules. A chain complex $C_.$ of R-modules is said to be of *finite type with respect to* ℓ if all C_n are objects in $\mathcal{F}$ and $\ell(C_n) = 0$ for almost all n. In case, R is PID, and ℓ is the rank function, this is the same as saying that all the C_n are of finite rank and most of them have rank equal to zero. We then merely refer to $C_.$ to be of *finite type*.

Remark 4.1.16 Thus, for example, (when $R = \mathbb{Z}$), a finite type chain complex of abelian groups need not be of *finite type with respect to* some additive function other than the rank function.

Theorem 4.1.17 *Let $C_.$ be a chain complex of R-modules of finite type with respect to an additive function ℓ. Then*

$$ \sum_n (-1)^n \ell(C_n) = \sum_n (-1)^n \ell(H_n(C_*)). $$

Proof: For each n, we have the short exact sequence

$$0 \longrightarrow \text{Ker } \partial_n \longrightarrow C_n \longrightarrow \text{Im } \partial_n \longrightarrow 0$$

and hence $\ell(C_n) = \ell(\text{Ker } \partial_n) + \ell(\text{Im } \partial_n)$. On the other hand, by definition, $H_n(C_*) = \text{Ker } \partial_n/\text{Im } \partial_{n+1}$. Hence, $\ell(\text{Ker } \partial_n) = \ell(H_n(C_*)) + \ell(\text{Im } \partial_{n+1})$. Substituting these on the left hand side of the equation and observing that

$$\sum_n (-1)^n \ell(\text{Im } \partial_n) = -\sum_n (-1)^n \ell(\text{Im } \partial_{n+1}),$$

establishes the required equality. ♠

Definition 4.1.18 Let R be a PID and M graded module of finite type. (This implies that the rank of M_n is finite for all n and is zero except for finitely many of n.) We define the *Euler characteristic*

$$\chi(M) = \sum_i (-1)^i rk(M_i),$$

where rk denotes the rank function.

The following theorem is an immediate consequence of Theorem 4.1.17.

Theorem 4.1.19 *Let $C_.$ be a finite type chain complex of R-modules over a PID R. Then*

$$\chi(H(C_.)) = \chi(C_.) \tag{4.3}$$

Remark 4.1.20 The above theorem tells us that the Euler characteristic of a finitely generated chain complex is the same as the Euler characteristic of its homology groups. In our context, R is either the ring of integers or a field. (So, it should not cause you any worry even if you are not familiar with PIDs at this stage.) The notion of the Euler characteristic plays a very important role in the development of algebraic topology and differential geometry on the whole. It manifests itself in a variety of ways from the simple formula $v - e + f = 2$ of Euler to the Atiyah–Singer index theorem for elliptic differential operators. (See [Shastri, 2011] for several equivalent differential topological definitions of Euler characteristic.)

We shall now introduce the concept of 'chain-homotopy' which is the algebraic analogue of homotopy.

Definition 4.1.21 Two chain maps $f, g : C_* \longrightarrow C'_*$ of degree 0, are said to be *chain homotopic* if there exists a graded homomorphism $D_* : C_* \longrightarrow C'_*$ of degree 1 such that $D \circ \partial + \partial' \circ D = f - g$. It is easily verified (Exercise) that *chain homotopy* is an equivalence relation. The idea behind this definition will be clear when we consider the topological situation, from which it has emerged. The important property of chain homotopy that we are interested in is:

Lemma 4.1.22 Chain homotopic maps induce the same homomorphism of homology groups.

Proof: Exercise. ♠

Remark 4.1.23 By definition a *co-chain complex* is a graded R-module with an endomorphism δ of degree 1 such that $\delta^2 = 0$. The Hom functor converts a chain complex into a cochain complex, viz., given a chain complex

$$\cdots \xrightarrow{\partial} C_{n+1} \xrightarrow{\partial} C_n \xrightarrow{\partial} C_{n-1} \longrightarrow \cdots$$

denoted by (C_*, ∂), the following is a co-chain complex:

$$\cdots \xrightarrow{\delta^*} \operatorname{Hom}(C_{n-1}, R) \xrightarrow{\delta^*} \operatorname{Hom}(C_n, R) \xrightarrow{\delta^*} \operatorname{Hom}(C_{n+1}, R) \longrightarrow \cdots$$

denoted by C^* or (C^*, δ^*) where $\delta^*(f) := f \circ \partial$ for $f \in \operatorname{Hom}(C_*, R)$. Cohomology of a chain complex (C_*, ∂) is defined to be $H_*(C^*)$ denoted by $H^*(C_*; R)$. There exists a canonical R−module homomorphism

$$h : H^*(C_*; R) \longrightarrow \operatorname{Hom}(H_*(C_*; R), R)$$

defined as follows: let $\alpha \in H^n(C_*; R)$, i.e., $\alpha \in \operatorname{Ker} \delta^* / \operatorname{Im} \delta^*$. By definition of δ^*, $\alpha \in \operatorname{Ker} \delta^*$ means that $\alpha \circ \partial : C_{n+1} \longrightarrow R$ is the 0-map. Now consider the restriction $Res(\alpha)$ of α to $\operatorname{Ker} \delta^*$. Clearly this gives rise to a well-defined element in $\operatorname{Hom}(H_*(C_*, R), R)$. One can verify that h is an isomorphism if $R = k$ is a field for the simple reason that every linear subspace is a direct summand. The discussion in the general case will be taken in a later chapter.

Exercise 4.1.24

(i) Let R be a PID. Recall that if M is a finitely generated module over R, and $\operatorname{Tor} M$ denotes the submodule of all torsion elements in M, then the quotient module $M/\operatorname{Tor} M$ is a free module of finite rank which is equal to the rank of M itself. For any endomorphism of $f : M \to M$, there is an induced endomorphism $\bar{f} : M/\operatorname{Tor} M \to M/\operatorname{Tor} M$ of the free R-module. Fixing a basis for this module, we can then consider the matrix of $\bar{f}$. We define $tr(f)$ to be the trace of this matrix, which is independent of the choice of the matrix and hence can be called the trace of f.

(a) Let

$$
\begin{array}{ccccccccc}
0 & \longrightarrow & A & \longrightarrow & B & \longrightarrow & C & \longrightarrow & 0 \\
& & \downarrow{\scriptstyle f} & & \downarrow{\scriptstyle g} & & \downarrow{\scriptstyle h} & & \\
0 & \longrightarrow & A & \longrightarrow & B & \longrightarrow & C & \longrightarrow & 0
\end{array}
$$

be a commutative diagram of finitely generated R-modules and homomorphisms. Show that $tr(f) + tr(h) = tr(g)$.

(b) Let $C_.$ be a chain complex of finite type over R. For any chain endomorphism $f : C_. \to C_.$ we define the *Lefschetz number*

$$L(f) = \sum_n (-1)^n tr(f_n). \qquad (4.4)$$

If $f_* : H_*(C_.) \to H_*(C_.)$ is the induced homomorphism on the homology, show that $L(f_*) = L(f)$. We shall meet this again in Section 4.6.

(ii) Prove the Five lemma and the Four lemma directly by diagram chasing, i.e., without using the snake lemma.

4.2 Singular Homology Groups

We shall construct the singular chain complex of a space, discuss basic properties of it such as functoriality, dimension axiom, additivity, excision and homotopy invariance, some of these without proof. The proofs themselves will be given in a separate section. At this stage, it may be a better idea to learn how to use these results than learning the proofs.

We shall also compute the singular homology groups in the simplest cases, viz., for $\mathbb{S}^{n-1}$ and $(\mathbb{D}^n, \mathbb{S}^{n-1})$.

The Construction

Definition 4.2.1 Let X be any topological space and $n \geq 0$ be an integer. By a *singular n-simplex* in X, we mean a continuous map $\sigma : |\Delta_n| \to X$. By a *singular n-chain* in X, we mean a finite sum $\sum_i n_i \sigma_i$, where $n_i \in \mathbb{Z}$ and σ_i are singular n-simplices in X. We can add any two chains in an obvious way to obtain another chain by simply following the rule

$$n\sigma + m\sigma = (n + m)\sigma \tag{4.5}$$

Then the collection of all singular n-chains in X becomes a free abelian group, with the empty sum playing the role of the zero-element. The collection of all singular n-simplices is then a basis for this group. We denote this group by $S_n(X)$.

If $f : X \to Y$ is a map of topological spaces, then $\sigma \mapsto f \circ \sigma$ extends to define a graded group homomorphism $f_. : S_.(X) \to S_.(Y)$. Since $(g \circ f) \circ \sigma = g \circ (f \circ \sigma)$, it follows that the association $X \rightsquigarrow S_.(X)$ defines a functor.

Now if A is a subspace of X and σ is a singular simplex in A then it can be considered as a singular simplex in X via the inclusion map $A \hookrightarrow X$. In this way, $S_n(A)$ becomes a subgroup of $S_n(X)$. We denote the quotient group $S_n(X)/S_n(A)$ by $S_n(X, A)$. Note that $S_n(X, A)$ is a free abelian group with the collection of all singular n-simplexes σ in X whose image is not contained in A as a basis. If $A = \emptyset$, then clearly $S_n(X, A) = S_n(X)$.

We shall now make the graded group $S_.(X, A) = \oplus_n S_n(X, A)$ into a chain complex.

Definition 4.2.2 For each integer $r \geq 0$, the *face* map or the *face operator* $F^r : \mathbb{R}^\infty \to \mathbb{R}^\infty$ is defined by

$$F^r(\mathbf{e}_s) = \left\{ \begin{array}{ll} \mathbf{e}_s, & \text{if } s < r, \\ \mathbf{e}_{s+1}, & \text{if } s \geq r \end{array} \right.$$

and extended linearly.

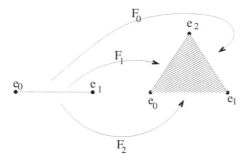

FIGURE 4.2. The face operators

Remark 4.2.3 Note that for each $n \geq r$, F^r carries Δ_{n-1} onto the $(n-1)$-face of Δ_n opposite to the vertex $\mathbf{e}_r$ in an order preserving manner. So, in order to save on cumbersome notations, we shall denote each F^r restricted to any $\Delta_n, n \geq r$ also by F^r, the exact meaning being understood from the context. Figure 4.2 depicts the face operators F_0, F_1, F_2 restricted to Δ_1.

Definition 4.2.4 For $n \geq 1$ and for any singular n-simplex σ in X, we define

$$\partial_n(\sigma) = \sum_{r=0}^{n} (-1)^r \sigma \circ F^r$$

and extend it linearly to obtain homomorphisms $\partial_n : S_n(X, A) \to S_{n-1}(X, A)$. We define $S_n(X, A) = (0)$ for $n \leq -1$ and $\partial_n = 0$ for $n \geq 0$.

Proposition 4.2.5 $(S.(X, A), \partial.) = \{S_n(X, A), \partial_n\}$ is a chain complex and the association $(X, A) \rightsquigarrow (S.(X, A), \partial)$ is a functor.

We need to show that $\partial \circ \partial = 0$. For this we need the following lemma, the proof of which is a straightforward exercise.

Lemma 4.2.6 $F^r \circ F^s = F^{s-1} \circ F^r$ if $r < s$.

To prove the proposition, we shall first show that $\partial^2(\sigma) = 0$ for any singular n-simplex, $n \geq 2$:

$$
\begin{aligned}
\partial^2(\sigma) &= \partial(\sum_r (-1)^r \sigma \circ F^r) \\
&= \sum (-1)^{r+s} \sigma_{r,s} \circ F^r \circ F^s \\
&= \sum_{r \geq s} (-1)^{r+s} \sigma \circ F^r \circ F^s + \sum_{r < s} (-1)^{r+s} \sigma \circ F^r \circ F^s \\
&= \sum_{r \geq s} (-1)^{r+s} \sigma \circ F^r \circ F^s + \sum_{r < s} (-1)^{r+s} \sigma \circ F^{s-1} \circ F^r \quad (\Diamond) \\
&= \sum_{r \geq s} (-1)^{r+s} \sigma \circ F^r \circ F^s - \sum_{i \geq j} (-1)^{i+j} \sigma \circ F^i \circ F^j \quad (\heartsuit) \\
&= 0
\end{aligned}
$$

($\Diamond$ = by the lemma above; $\heartsuit$ = by putting $s - 1 = i, r = j$ in the second sum).

Given a map $f : (X, A) \to (Y, B)$ the fact that $f.$ defines a chain map follows from the fact that in defining ∂ we take composition with the face operators F^r on the right, whereas, the composition with f is taken on the left and we have $f \circ (\sigma \circ F^r) = (f \circ \sigma) \circ F^r$. This is the essential step in the proof of the functorial property; the rest of the details are left to the reader. ♠

Definition 4.2.7 The *relative singular homology groups* of a topological pair (X, A) are the homology groups $H_*(S.(X, A))$ and are denoted by $H_*(X, A)$. Taking $A = \phi$, we get the *homology groups* $H_*(X)$.

Remark 4.2.8 Given any commutative ring R, we could have formed the free module $S_n(X; R)$ of all the singular n-chains in X. Exactly as before, we would have got a chain complex $S.(X; R)$, subchain complex $S.(A; R)$ and the quotient complex $S.(X, A; R)$, etc. The corresponding homology groups $H_*(X; R)$, etc., are then called the homology groups X with coefficients in R. Clearly they form a graded R-module by themselves. It is customary to drop the notation R when $R = \mathbb{Z}$.

Basic Properties of Singular Homology

It is easily seen that $\{S_\cdot(X,A), \partial_\cdot\}$ is a functor from the category of topological pairs to the category of chain complexes; thus $H_*(X,A)$ and $H_*(X)$ are also functors. Given $f : (X,A) \to (Y,B)$, we shall denote by $f_\cdot$, the map induced by f on $S_\cdot(X,A)$ and by f_* the map induced by $f_\cdot$ on $H_*(X,A)$. The functoriality is summed-up in the following facts:
(i) Given continuous functions $f : X \longrightarrow Y$ and $g : Y \longrightarrow Z$, we have, $(g \circ f)_* = g_* \circ f_*$;
(ii) If Id denotes the identity map on any space, then $(Id)_*$ is again the identity map on the homology.

One of the most fundamental topological properties of $H_*(X,A)$ is the homotopy invariance.

Theorem 4.2.9 (Homotopy invariance) *Let $f, g : (X,A) \to (Y,B)$ be homotopic to each other. Then $f_* = g_*$ on $H_*(X,A)$.*

Proof: Let $G : (X,A) \times \mathbb{I} \to (Y,B)$ be a homotopy between f and g (i.e., $G : X \times \mathbb{I} \to Y$ is a map such that $G(A \times \mathbb{I}) \subset B$, $H(x,0) = f(x)$, $G(x,1) = g(x)$, $\forall x \in X$). Consider the inclusion maps $\eta_t : (X,A) \to (X,A) \times \mathbb{I} = (X \times \mathbb{I}, A \times \mathbb{I})$ given by

$$\eta_t(x) = (x,t), \quad t \in \mathbb{I}.$$

Then, $G \circ \eta_0 = f$ and $G \circ \eta_1 = g$. Passing onto homology, this implies that $G_* \circ (\eta_0)_* = G_* \circ (\eta_1)_*$. Therefore it suffices to prove that $(\eta_0)_* = (\eta_1)_* : H_*(X,A) \to H_*((X,A) \times \mathbb{I})$. For this purpose, we construct a chain homotopy $h : S_\cdot(X,A) \to S_\cdot(X \times \mathbb{I}, A \times \mathbb{I})$ between $(\eta_0)_*$ and $(\eta_1)_*$ (at the chain group level). This chain homotopy is called *prism operator*. For future reference we shall state this as a separate lemma:

Lemma 4.2.10 There exist functorial chain homotopies

$$h : S_q(X,A) \to S_{q+1}((X,A) \times \mathbb{I}), \ q \geq 0,$$

between the two inclusion induced maps $\eta_0, \eta_1 : S_\cdot(X,A) \to S_\cdot((X,A) \times \mathbb{I})$.

We shall postpone the proof of this lemma to the last section, and take for granted the Theorem 4.2.9 as proved.

As an immediate consequence, we obtain:

Corollary 4.2.11 Homotopy-equivalent topological spaces have isomorphic homology groups; in particular, a contractible space has the homology groups of a point-space.

Example 4.2.12

(i) **Homology of a point space**

Let n be any positive integer. There is a unique singular n-simplex $\sigma : |\Delta_n| \to \{\star\}$ and hence $S_n(\star) \approx \mathbb{Z}$ for each $n \geq 0$. Observe that $\partial : S_{2n+1}(\star) \to S_{2n}(\star)$ is the zero-map because Δ_{2n+1} has an even number of $(2n)$-faces and the terms in $\partial(\sigma)$ cancel in pairs. However, similar reasoning tells us that $\partial : S_{2n}(\star) \to S_{2n-1}(\star)$ is an isomorphism. Thus we have a chain complex

$$\cdots \xrightarrow{0} \mathbb{Z} \xrightarrow{\cong} \mathbb{Z} \xrightarrow{0} \mathbb{Z} \xrightarrow{\cong} \mathbb{Z} \longrightarrow 0$$

Thus it follows that $H_0(\star) \simeq \mathbb{Z}$ and $H_i(\star) = (0)$ for all $i > 0$.

(ii) **Homology of path components**

Let $\{X_j \; : \; j \in J\}$ be the set of path connected components of X so that $X = \sqcup_{j \in J} X_j$. Since $|\Delta_n|$ is a path connected space, it follows that any singular n-simplex in X is contained completely in one of the path components X_j. Therefore, it follows that the entire chain complex $S_.(X)$ is the direct sum of the subchain complexes $S_.(X_j)$. Hence the homology is also a direct sum of $H_*(X_j), j \in J$. Thus in practice, we can often assume that the space itself is path connected, while dealing with its homology groups.

(iii) **Homology long exact sequence of the pair** (X, A)

By definition of $S_.(X, A)$, we have an exact sequence of chain complexes

$$0 \to S_.(A) \to S_.(X) \to S_.(X, A) \to 0.$$

By Theorem 4.1.10, this yields the long exact sequence of homology groups

$$\left.\begin{aligned} \cdots \to H_i(A) \to H_i(X) \to H_i(X, A) \xrightarrow{\delta} H_{i-1}(A) \to \cdots \\ \cdots \to H_1(X, A) \xrightarrow{\delta} H_0(A) \to H_0(X) \to H_0(X, A), \end{aligned}\right\} \quad (4.6)$$

which is functorial. This sequence comes handy in many computations as we shall see soon.

(iv) **Reduced homology and H_0 of a path connected space**

Let us compute $H_0(X)$ for any path connected space X. Note that $S_0(X)$ is a free abelian group on the underlying set of the space X. For any singular 1-simplex σ in X, we have $\partial_1(\sigma) = \sigma(\mathbf{e}_1) - \sigma(\mathbf{e}_0)$. We define $\varepsilon : S_0(X) \to \mathbb{Z}$ by $\varepsilon(\sum n_i x_i) = \sum n_i$. Then ε is a surjective homomorphism. Since $\epsilon \circ \partial_1(\sigma) = 0$ for all 1-simplexes σ, it follows that $\epsilon \circ \partial_1 = 0$. Finally to show that $\mathrm{Ker}\, \epsilon \subset \mathrm{Im}\, \partial_1$, consider an element $\sum_i n_i x_i$ with $\sum_i n_i = 0$. Choose any point $z_0 \in X$. We can then join z_0 to each x_i by a path σ_i which can be treated as a singular 1-simplex. Then $\partial(\sum_i n_i \sigma_i) = \sum_i n_i x_i - (\sum_i n_i) z_0 = \sum_i n_i x_i$. This proves $\partial(S_1(X)) = \mathrm{Ker}\, \epsilon$. Hence

$$H_0(X) = S_0(X)/\partial(S_1(X)) = S_0(X)/\mathrm{Ker}\; \varepsilon \; \approx \mathbb{Z}.$$

The homomorphism $\varepsilon : S_0(X) \to \mathbb{Z}$ is called the *augmentation* homomorphism. We define the extended chain complex $\tilde{S}(X)$ by

$$\tilde{S}_n(X) = \begin{cases} S_n(X), & \text{if } n \geq 0, \\ \mathbb{Z}, & \text{if } n = -1, \\ (0), & \text{if } n < -1; \end{cases} \quad \text{and} \quad \tilde{\partial}_n = \begin{cases} \partial_n, & n \geq 1, \\ \varepsilon, & n = 0, \\ 0, & n < 0. \end{cases}$$

The homology groups of this complex are denoted by $\widetilde{H}_*(X)$ and are called *reduced homology groups*. Clearly, $\widetilde{H}_i(X) = (0)$ for $i < 0$. Also $\widetilde{H}_i(X) \approx H_i(X)$ for $i > 1$. However in dimension zero, we have $\widetilde{H}_0(X) \oplus \mathbb{Z} \approx H_0(X)$. In particular, for a path connected space $\widetilde{H}_0(X)$ is (0).

Remark 4.2.13

(a) Recall that a contractible space has the homotopy type of a point space and hence by homotopy invariance, it follows that every contractible space has homology isomorphic to that of a point space. Now the only non zero term in the homology of a point space is H_0 which is infinite cyclic. The motivation for introducing the augmentation and thereby the extended chain complex may be attributed to a desire to have homology groups completely vanishing for contractible spaces.

(b) Note that for any non empty subspace $A \subset X$, we have $\tilde{H}_*(X, A) = H_*(X, A)$ and (4.6) is valid if we replace H_i by $\tilde{H}_i$ everywhere. You have to do some checking only at the last three groups.

At this stage, we still do not have any powerful tool to compute the homology. We shall now discuss one of the most important tools, viz., *Mayer–Vietoris* in the computation of singular homology.

Given subsets X_1 and X_2 of a topological space X, let us denote the inclusion maps of X_i in to X by $\eta_i, i = 1, 2$. They induce inclusion maps of the chain complexes which we shall denote again by $\eta_i : S_.(X_i) \to S_.(X), i = 1, 2$. Now consider the chain map

$$(\eta_1, -\eta_2) : S_.(X_1) \oplus S_.(X_2) \to S_.(X).$$

What is the kernel of this map? Remember that $S_.(X)$ is a free module. Thus, if $\alpha_i \in S_.(X_i), i = 1, 2$ are such that $\alpha_1 - \alpha_2 = 0$, then it follows that $\alpha_1 = \alpha_2 \in S_.(X_1 \cap X_2)$. Therefore the kernel of the above chain map is $S_.(X_1 \cap X_2)$. Clearly, the image of this chain map is the submodule generated by $S_.(X_1)$ and $S_.(X_2)$ in $S_.(X)$ which we shall denote by $S_.(X_1) + S_.(X_2)$ so that we have a short exact sequence of chain complexes

$$0 \to S_.(X_1 \cap X_2) \to S_.(X_1) \oplus S_.(X_2) \to S_.(X_1) + S_.(X_2) \to 0 \qquad (4.7)$$

This will then give a long exact sequence of homology modules from which it would be possible to get a lot of information on $H_*(S_.(X_1) + S_.(X_2))$ from $H_*(X_1), H_*(X_2)$ and $H_*(X_1 \cap X_2)$. The crucial question is the following:

Question: *Can we replace the modules $H_*(S_.(X_1) + S_.(X_2))$ with $H_*(X)$; if not always, at least under some suitable conditions?*

Toward an affirmative answer to this question, we proceed, beginning with the following technical lemma.

Lemma 4.2.14 Let $X = A \cup B$. Then the following statements are equivalent.
(a) $S_.(A) + S_.(B) \to S_.(X)$ induces isomorphisms in homology.
(b) $[S_.(A) + S_.(B)]/S_.(B) \to S_.(X)/S_.(B)$ induces isomorphisms in homology.
(c) $[S_.(A) + S_.(B)]/S_.(A) \to S_.(X)/S_.(A)$ induces isomorphisms in homology.
(d) $S_.(A)/S_.(A \cap B) \to S_.(X)/S_.(B)$ induces isomorphisms in homology.
(e) $S_.(B)/S_.(A \cap B) \to S_.(X)/S_.(A)$ induces isomorphisms in homology.

Proof: The equivalence of (a) and (b) is a consequence of the Five lemma applied to the ladder of homology long exact sequences given by the ladder of short exact sequences

$$
\begin{array}{ccccccccc}
0 & \longrightarrow & S_.(B) & \longrightarrow & S_.(A) + S_.(B) & \longrightarrow & [S_.(A) + S_.(B)]/S_.(B) & \longrightarrow & 0 \\
& & \downarrow & & \downarrow & & \downarrow & & \\
0 & \longrightarrow & S_.(B) & \longrightarrow & S_.(X) & \longrightarrow & S_.(X)/S_.(B) & \longrightarrow & 0.
\end{array}
$$

The equivalence of (b) and (d) follows from the commutative diagram below in which the horizontal arrow is the isomorphism given by the Nöether's isomorphism theorem.

$$
\begin{array}{ccc}
S_.(A)/S_.(A \cap B) & \longrightarrow & [S_.(A) + S_.(B)]/S_.(B) \\
& \searrow \qquad \swarrow & \\
& S_.(X)/S_.(B) &
\end{array}
$$

Interchanging A, B gives the equivalence of (a), (c) and (e).   ♠

Definition 4.2.15 Let X be a topological space, A, B be two subspaces such that $X = A \cup B$. We then say that the inclusion map $(A, A \cap B) \hookrightarrow (X, B)$ an excision map. A pair $\{A, B\}$ of subspaces of a space X is called an *excisive couple* for singular homology, if the inclusion map

$$S_\cdot(A) + S_\cdot(B) \hookrightarrow S_\cdot(A \cup B)$$

induces isomorphisms in homology or if any of the other four equivalent conditions in the above lemma is satisfied.

Condition (d) of this lemma immediately gives the following theorem.

Theorem 4.2.16 (Homology excision) *The inclusion map* $(A, A \cap B) \hookrightarrow (A \cup B, B)$ *of an excisive couple* $\{A, B\}$ *induces an isomorphism of homology groups* $H_*(A, A \cap B) \to H_*(A \cup B, B)$.

Note that whether a pair $\{A, B\}$ is excisive or not, the inclusion map $(A, A \cap B) \hookrightarrow (A \cup B, B)$ is an excision map. This terminology refers to the set-theoretic fact, viz., $A \cup B \setminus B = A \setminus A \cap B$. Not all pairs are excisive couples. The basic examples of excisive couples are provided by the following theorem.

Theorem 4.2.17 *If* $X = X_1 \cup X_2 = \mathrm{int}_X(X_1) \cup \mathrm{int}_X(X_2)$ *then,* $\{X_1, X_2\}$ *is an excisive couple for the singular homology.*

We shall postpone the proof of this theorem to a later section. The reader is advised to skip the proof of this theorem in the first reading, and come back to it at leisure. The knowledge of the proof is not all that essential in understanding most of the material that follows. However, any serious student of algebraic topology has to read at least one proof of this theorem and at least once. For the present, the following remark suffices: Given a singular n-simplex σ in X, one subdivides Δ_n in such a way that, the image of each simplex of this subdivision is contained in X_1 or X_2. One thinks of the original singular simplex σ as an appropriate sum of these little pieces. Strictly speaking though, they are different chains and most of the effort is to show that they represent the same element in the homology groups. Of course one has to do all these in a canonical fashion.

Example 4.2.18 Here are a few examples of excisive couples which occur in practice.
(a) $\{\mathbb{D}^n, \mathbb{R}^n \setminus \{0\}\}$ is an excisive couple and hence the inclusion map $(\mathbb{D}^n, \mathbb{D}^n \setminus \{0\}) \hookrightarrow (\mathbb{R}^n, \mathbb{R}^n \setminus \{0\})$ will induce isomorphisms in all singular homology groups.
(b) Let N denote the north pole on $\mathbb{S}^n$ and $\bar{U}$ denote the closed upper hemisphere. Then $\{\mathbb{S}^n \setminus \{N\}, \bar{U}\}$ is an excisive couple, which gives isomorphisms of $H_*(\bar{U}, \bar{U} \setminus \{N\}) \to H_*(\mathbb{S}^n, \mathbb{S}^n \setminus \{N\})$.

In practical situations, often X_1, X_2 are closed subspaces and hence the above theorem cannot be applied immediately. Then the following result comes to the rescue:

Theorem 4.2.19 *Let* $X = A \cup B$ *where* A, B *are closed subspaces. If* $A \hookrightarrow X$ *or* $B \hookrightarrow X$ *is a cofibration, then* $\{A, B\}$ *is an excisive couple.*

Proof: Suppose $A \hookrightarrow X$ is a cofibration. Then $CA \hookrightarrow X \cup CA$ is a cofibration, where CA denotes the cone over A. (See Exercise 1.6.15.(v).) By Theorem 1.6.4, it follows that the quotient map $q : X \cup CA \to X \cup CA / CA$ is a homotopy equivalence. Consider the following

diagram

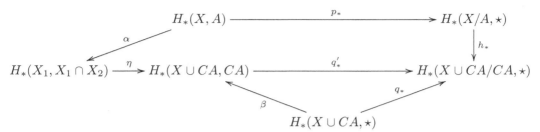

in which we want to prove that p_* is an isomorphism, We have:

(i) h_* is an isomorphism because h is a homeomorphism;

(ii) $X_1 = X \cup A \times [0, 1/2], X_2 = CA$, so that the pair (X_1, X_2) deformation retracts onto the pair (X, A) and hence we have the inclusion induced isomorphism α.

(iii) The pair $\{X_1, X_2\}$ satisfies the condition in the excision theorem above and hence is an excisive couple which implies that η is an isomorphism.

(iv) CA is contractible and hence from the homology exact sequence of the pair $(X \cup CA, CA)$, β is an isomorphism.

(v) q is a homotopy equivalence and therefore q_* is an isomorphism.

It follows that q'_* is an isomorphism and therefore p_* is an isomorphism. ♠

Corollary 4.2.20 A pair $\{X_1, X_2\}$ of closed subspaces of a space is an excisive couple for singular homology under the following conditions:

(a) $(X_i, X_1 \cap X_2)$ is a relative of CW-complex for $i = 1$ or $i = 2$.

(b) $X_i = |K_i|$ where K_i are simplicial subcomplexes of a simplicial complex K with $|K| = X$.

Proof: For (a), use Theorems 2.4.5 and 4.2.19. Statement (b) is a special case of (a). ♠

We shall now concentrate on putting the excision theorem to good use, by first deriving an important ready-to-use result:

If $\{X_1, X_2\}$ is an excisive couple then, in the long exact sequence of homology groups given by (4.7), we can replace the groups $H_*(S_.(X_1) + S_.(X_2))$ by $H_*(X_1 \cup X_2)$ to obtain the **Mayer–Vietoris** sequence:

$$\left. \begin{aligned} \cdots &\to H_{i+1}(X_1 \cup X_2) \xrightarrow{\partial} H_i(X_1 \cap X_2) \xrightarrow{i_*} H_i(X_1) \oplus H_i(X_2) \xrightarrow{j_*} H_i(X_1 \cup X_2) \to \cdots \\ \cdots &\to H_1(X_1 \cup X_2) \xrightarrow{\partial} H_0(X_1 \cap X_2) \to H_0(X_1) \oplus H_0(X_2) \to H_0(X_1 \cup X_2) \end{aligned} \right\} (4.8)$$

with

$$i_*(z) = (i_{1*}z, -i_{2*}z); \quad j_*(z_1, z_2) = j_{1*}z_1 + j_{2*}z_2 \tag{4.9}$$

where,

$$i_1 : X_1 \cap X_2 \hookrightarrow X_1, \quad i_2 : X_1 \cap X_2 \hookrightarrow X_2$$
$$j_1 : X_1 \hookrightarrow X_1 \cup X_2, \quad j_2 : X_2 \hookrightarrow X_1 \cup X_2$$

are inclusion maps. The Mayer–Vietoris sequence is one of the major ready-to-use-tools in computing homology.

As a simple consequence of Mayer–Vietoris, we shall prove:

Theorem 4.2.21 (Homology suspension theorem) *Let X be a topological space. Then there is a canonical isomorphism*

$$S : \tilde{H}_n(X) \to H_{n+1}(SX), \quad n \geq 0,$$

where SX denotes the suspension of X ($SX = \mathbb{S}^0 \star X$).

Proof: Note that the suspension SX is always path connected and hence $H_*(SX) \approx \tilde{H}_*(SX)$ in positive dimensions. So, we shall establish isomorphisms $S : \tilde{H}_n(X) \to \tilde{H}_{n+1}(SX), n \geq 0$. We have:

$$\mathbb{S}^0 \star X = \{-1, 1\} \star X$$

is the union of two cones over X, viz., $\{N\} \star X$ and $\{S\} \star X$ identified along their common base X. Also for any cone $\{v\} \star X$, we have $\{v\} \star X \setminus \{v\}$ is homeomorphic to $(0, 1] \times X$, which can be deformed to $1 \times X$. It follows that both $\mathbb{S}^0 \star X \setminus \{S\} = A_{-1}$ and $\mathbb{S}^0 \star X \setminus \{N\} = A_1$ are contractible and $X \hookrightarrow A_{-1} \cap A_1$ is a deformation retract. Since A_1 and A_{-1} are both open in $\mathbb{S}^0 \star X$, we have a Mayer–Vietoris sequence

$$\tilde{H}_{i+1}(A_1) \oplus \tilde{H}_{i+1}(A_{-1}) \to \tilde{H}_{i+1}(\mathbb{S}^0 \star X) \xrightarrow{\partial} \tilde{H}_i(A_1 \cap A_{-1}) \to \tilde{H}_i(A_1) \oplus \tilde{H}_i(A_{-1})$$

which yields an exact sequence

$$0 \to \tilde{H}_{i+1}(SX) \xrightarrow{\partial} \tilde{H}_i(A_1 \cap A_{-1}) \to 0$$

for each i and hence the connecting homomorphism ∂_i is an isomorphism. Being a deformation retract, $\eta : X \hookrightarrow A_1 \cap A_{-1}$ also induces isomorphisms in homology. The composite

$$S : \tilde{H}_i(X) \xrightarrow{\quad \eta_* \quad} \tilde{H}_i(A_1 \cap A_{-1}) \xrightarrow{\quad \partial_i^{-1} \quad} \tilde{H}_{i+1}(SX)$$

is therefore, an isomorphism.   ♠

Remark 4.2.22

1. Whenever a pair of closed subspaces $\{X_1, X_2\}$ is given, introducing intermediate pairs of open sets $\{U_1, U_2\}$ such that X_i are deformation retract of U_i is a typical way excision Theorem 4.2.17 is used in practice. This is the role played by subsets $A_\pm$ in the proof of the above theorem. Alternatively, we could have used Theorem 4.2.19 as follows: Note that the inclusion $X \hookrightarrow CX$ is a cofibration and hence by Exercise 1.6.15.(v), the inclusion $N * X \hookrightarrow SX$ is a cofibration. Therefore the pair $\{\{N\} * X, \{S\} * X\}$ is an excisive couple.

2. Here is an explicit description of the homology suspension homomorphism S. For any singular n-simplex σ in X, let $[v]\sigma$ denote the $n+1$ simplex in the cone $\{v\} \star X$ which sends $\mathbf{e}_0$ to v and $\mathbf{e}_s$ to $\sigma(\mathbf{e}_{s-1})$ for $s \geq 1$. Let $S(\sigma) = [N]\sigma - [S]\sigma$. Then S can be extended linearly to a homomorphism $S : S_n(X) \to S_{n+1}(SX)$. It is a matter of straightforward verification that S is a chain map and hence gives a homomorphism of the homologies. Indeed, directly going through the definition of ∂ we see that $\partial Sc = c$ for any cycle c in X and therefore, S is equal to $\partial^{-1} \circ \eta_*$.

Example 4.2.23 Computation of $H_*(\mathbb{S}^n)$ We know that $\mathbb{S}^n$ is homeomorphic to the suspension of $\mathbb{S}^{n-1}$. Therefore, starting with the homology of $\mathbb{S}^0$, one can directly apply the homology suspension theorem above inductively, and compute the homology of all spheres. At each step the missing information is about $\tilde{H}_0(\mathbb{S}^n), n \geq 1$ which clearly vanishes since $\mathbb{S}^n, n \geq 1$ are path connected. Therefore

$$\tilde{H}_k(\mathbb{S}^n) = \begin{cases} 0, & k \neq n; \\ \mathbb{Z}, & k = n. \end{cases} \tag{4.10}$$

We would like to know explicitly representative cycles of elements of these simplest homology groups $H_n(\mathbb{S}^n) \approx \mathbb{Z}$. For this we shall use the S-triangulation of the spheres introduced in Example 2.7.10.(iii). We need to introduce some elaborate notation here. Let us begin with denoting the two point space $\mathbb{S}^0$ by $\{u_0, v_0\}$. However, when we are taking suspension of this space, we need another copy of $\mathbb{S}^0$ which we shall denote by $\{u_1, v_1\}$. Thus the S-triangulation of $\mathbb{S}^1$ consists of four edges $\{u_0, u_1\}, \{v_0, u_1\}, \{u_0, v_1\}, \{v_0, v_1\}$. Inductively, we shall use $\{u_n, v_n\}$ to denote the nth-copy of $\mathbb{S}^0$ and then the n-simplexes of the S-triangulation of $\mathbb{S}^n$ will be of the form $\sigma \cup \{u_n\}$ or $\sigma \cup \{v_n\}$ where σ runs over all the $(n-1)$-simplexes of $\mathbb{S}^{n-1}$.

Now if σ is any singular k-simplex in $\mathbb{S}^{n-1}$, we shall denote by $[\sigma, u_n]$ the $(k+1)$-simplex in $\mathbb{S}^n$ defined by

$$[\sigma, u_n](t\alpha + (1-t)\mathbf{e}_{k+1}) = t\sigma(\alpha) + (1-t)u_n, \quad \alpha \in |\Delta_k|, \ 0 \le t \le 1.$$

Similarly, the singular $(k+1)$-simplex $[\sigma, v_n]$ is defined. Now if $c = \sum_i n_i \sigma_i$ is any singular k-chain in $\mathbb{S}^{n-1}$, we define $S(c) = \sum_i n_i [\sigma_i, v_n] - \sum_i n_i [\sigma_i, u_n]$. Finally, we define a n-chain g_n in $\mathbb{S}^n$ inductively by the formula,

$$g_0 = v_0 - u_0; \qquad g_n = S(g_{n-1}), n \ge 1.$$

We claim that g_n is a n-cycle which represents a generator of $\tilde{H}_n(\mathbb{S}^n)$. For $n = 0$, we have seen this in Example 4.2.12.(iv). For $n \ge 1$, observe that $\partial(g_n) = g_{n-1}$ where ∂ is the connecting homomorphism of the Mayer–Vietoris sequence as in the previous theorem. We need to merely go through the initial steps in the proof of the snake lemma in defining this connecting homomorphism. This implies $\eta_*^{-1}\partial(g_n) = (g_{n-1})$ and since $\eta_*^{-1}\partial$ is an isomorphism, we are through.

Example 4.2.24 Homology of the pair $(\mathbb{D}^n, \mathbb{S}^{n-1})$
Since $\mathbb{D}^n$ is contractible, its reduced homology vanishes. On the other hand, from the previous example we know $H_*(\mathbb{S}^{n-1})$. Piecing them together via the long exact sequence

$$\cdots \longrightarrow H_i(\mathbb{D}^n) \longrightarrow H_i(\mathbb{D}^n, \mathbb{S}^{n-1}) \longrightarrow H_i(\mathbb{S}^{n-1}) \longrightarrow \cdots$$

we get

$$H_i(\mathbb{D}^n, \mathbb{S}^{n-1}) = \begin{cases} 0, & i \ne n; \\ \mathbb{Z}, & i = n. \end{cases} \tag{4.11}$$

Example 4.2.25 Homology of the torus
As another illustration of application of Mayer–Vietoris sequence, we shall compute the singular homology groups of $\mathbb{S}^1 \times \mathbb{S}^1$. Take $U_\pm = \mathbb{S}^1 \setminus \{\mp 1\}$. Then $\mathbb{S}^1 \times U_\pm$ are open in $\mathbb{S}^1 \times \mathbb{S}^1$, and

$$\mathbb{S}^1 \times \mathbb{S}^1 = \mathbb{S}^1 \times U_+ \cup \mathbb{S}^1 \times U_-.$$

Therefore we can apply Mayer–Vietoris (4.8) to get an exact sequence. Since $S^1 \times \mathbb{S}^0 \hookrightarrow (\mathbb{S}^1 \times U_+) \cap (\mathbb{S}^1 \times U_-)$ is a SDR, we can replace $H_*((\mathbb{S}^1 \times U_+) \cap (\mathbb{S}^1 \times U_-))$by $H_*(\mathbb{S}^1 \times \mathbb{S}^0)$. Therefore we have,

$$0 \longrightarrow H_2(\mathbb{S}^1 \times \mathbb{S}^1) \xrightarrow{\ \partial\ } H_1(\mathbb{S}^1 \times \mathbb{S}^0) \xrightarrow{\ i_*\ }$$

$$H_1(\mathbb{S}^1 \times U_+) \oplus H_1(\mathbb{S}^1 \times U_-) \xrightarrow{\ j_*\ } H_1(\mathbb{S}^1 \times \mathbb{S}^1) \xrightarrow{\ \partial\ }$$

$$H_0(\mathbb{S}^1 \times \mathbb{S}^0) \xrightarrow{\ i_*\ } H_0(\mathbb{S}^1 \times U_+) \oplus H_0(\mathbb{S}^1 \times U_-) \xrightarrow{\ j_*\ } H_0(\mathbb{S}^1 \times \mathbb{S}^1).$$

The first term in the above sequence actually corresponds to $H_2(\mathbb{S}^1 \times U_+) \oplus H_2(\mathbb{S}^1 \times U_-)$. Since $U_\pm$ are contractible, we have $H_*(\mathbb{S}^1 \times U_\pm) \approx H_*(\mathbb{S}^1)$. We can therefore use (4.10). In particular, the first entry is 0 as we have indicated above. For the same reason, the third, fourth, sixth and seventh terms are isomorphic to $\mathbb{Z} \oplus \mathbb{Z}$. All the higher homology terms, which we have not written, vanish. Since $\mathbb{S}^1 \times \mathbb{S}^1$ is connected the last term is $\mathbb{Z}$. Clearly j_* is surjective on H_0 level and hence its kernel is isomorphic to $\mathbb{Z}$. Therefore, the kernel of i_* is also infinite cyclic and hence $\partial : H_1(\mathbb{S}^1 \times \mathbb{S}^1) \to H_0(\mathbb{S}^1 \times \mathbb{S}^0)$ has an infinite cyclic image. From (4.9) it follows that $j_*(1,0)$ is non zero and $j_*(1,-1) = 0$. Therefore both the image and kernel of $j_* : H_1(\mathbb{S}^1 \times U_+) \oplus H_1(\mathbb{S}^1 \times U_-) \to H_1(\mathbb{S}^1 \times \mathbb{S}^1)$ are infinite cyclic. Therefore, it follows that $H_1(\mathbb{S}^1 \times \mathbb{S}^1) \approx \text{Im} j_* \oplus \text{Im} \partial = \mathbb{Z} \oplus \mathbb{Z}$. It also follows that the kernel of $i_* : H_1(\mathbb{S}^1 \times \mathbb{S}^0)$ is infinite cyclic and hence $H_2(\mathbb{S}^1 \times \mathbb{S}^1) \approx \mathbb{Z}$. And $H_i(\mathbb{S}^1 \times \mathbb{S}^1) = 0$ for $i > 2$. In conclusion, we have,

$$H_i(\mathbb{S}^1 \times \mathbb{S}^1) = \begin{cases} \mathbb{Z}, & i = 0, 2; \\ \mathbb{Z} \oplus \mathbb{Z}, & i = 1; \\ 0, & \text{otherwise.} \end{cases}$$

Remark 4.2.26

1. The example is a typical way Mayer–Vietoris can be employed in specific situations. Later, we shall have more elegant proof of the fact proved in the above example. The reader can now try the exercise below to get more familiar with Mayer–Vietoris.

2. Incidentally, the discussion in Example 4.2.23 provides a proof of the fact that in Theorem 4.2.21, the isomorphism $H_i(X) \to H_{i+1}(SX)$ is actually obtained by 'suspending' each cycle and is therefore canonical.

3. The functoriality, the homotopy invariance and the excision property of the singular homology are so important that they have been raised to the status of 'axioms for homology'. Most often, in deriving a certain result concerned with singular homology, we need not appeal to the actual construction of singular homology but only to these axioms. Indeed, along with one more property called 'dimension axiom' discussed in Example 4.2.12.(i), Eilenberg and Steenrod ([Eilenberg–Steenrod, 1952]) proved that all homology theories on the category of compact polyhedrons coincide. At that time genuine examples of 'homology theories' which may not satisfy the dimension axiom (such as K-theory) were not known. (Of course, one can artificially 'shift' the dimension even for the singular homology by merely defining $H_i^\dagger(X) = H_{i+k}(X)$ for a fixed k.) However, notice that property discussed in Example 4.2.12.(ii), viz., being a direct sum of homologies of path components, is not a consequence of these axioms.

4. With all this, at this stage, we still do not have any 'good way' of computing singular homology. Even in the simple case of a map $f : \mathbb{S}^n \to \mathbb{S}^n$, though the homology modules are known, we do not know how to compute $f_* : H_n(\mathbb{S}^n) \to H_n(\mathbb{S}^n)$. In the next section, we shall introduce a number of alternative homology modules as tools which help us in this task.

Exercise 4.2.27

(i) Compute the homology of $\mathbb{S}^p \times \mathbb{S}^q$ using Mayer–Vietoris.

(ii) Compute $H_*(\vee_k \mathbb{S}^{n_k})$ where $\vee_k \mathbb{S}^{n_k}$ denotes the bouquet of spheres of varying dimensions n_k. In particular, compute $H_*(\vee_k \mathbb{S}^n)$ where each $n_k = n$.

(iii) Use the generator of $H_n(\mathbb{S}^n; \mathbb{Z})$ as given in Example 4.2.23 to compute the induced homomorphism $(\alpha_n)_* : H_n(\mathbb{S}^n) \longrightarrow H_n(\mathbb{S}^n)$, where $\alpha_n : \mathbb{S}^n \to \mathbb{S}^n$ is the antipodal map. [Since this group is isomorphic to $\mathbb{Z}$ and any homomorphism $\phi : \mathbb{Z} \longrightarrow \mathbb{Z}$ is given by multiplication by an integer, given any continuous map $f : \mathbb{S}^n \longrightarrow \mathbb{S}^n$, we define $\deg f$ to be the integer m such that $f_*(c) = mc$ on $H_n(\mathbb{S}^n)$. We call this the *degree* of f.

Now, computing $(\alpha_n)_*$ simply means that we have to determine $\deg \alpha_n$.]

(iv) Compute the degree of maps $\mathbb{S}^1 \longrightarrow \mathbb{S}^1$ given by $z \mapsto z^n$ and $z \mapsto \bar{z}$ and compare it with Exercise 1.9.21. What do you conclude?

(v) Compute the degree of any reflection in a plane $r : \mathbb{S}^n \to \mathbb{S}^n$ say,

$$r : (x_0, x_1, \ldots, x_n) \mapsto (-x_0, x_1, \ldots, x_n).$$

More generally compute the degree of $f_k : \mathbb{S}^n \to \mathbb{S}^n$ given by

$$f_k(x_0, x_1, \ldots, x_n) = (-x_0, \ldots, -x_k, x_{k+1}, \ldots, x_n).$$

(vi) (Try this exercise only if you know the definition of degree of a smooth map between two compact, oriented, n-manifolds.) There is a notion of geometric degree of a map $f : M \longrightarrow N$ where M and N are orientable manifolds, which you may have come across in differential topology. The two notions coincide: Start with a connected compact orientable manifold M without boundary and of dimension m. Assume that it is triangulable and fix a triangulation. Then it is possible to choose orientations on n-simplexes of M in such a way that whenever two of them meet along a $(n-1)$-face, they induce opposite orientation on the face. If you take the sum c of all these oriented simplexes, it follows that $c \in S_m(X)$ and $\partial(c) = 0$. Moreover, $\bar{c} \in H_m(M)$ generates the infinite cyclic group $H_m(M)$. Thus if M, N are two such manifolds, then for any smooth map $f : M \to N$, we can talk about $\deg f$ in two different ways. The two notions coincide.

Prove this for any smooth map $f : \mathbb{S}^n \longrightarrow \mathbb{S}^n$.

(vii) **Homology exact sequence of a triple**

Given topological spaces $B \subset A \subset X$, show that there is a long exact sequence of homology groups

$$\cdots \to H_n(A, B) \to H_n(X, B) \to H_n(X, A) \to H_{n-1}(A, B) \to \cdots$$

which is functorial in the triples (X, A, B).

(viii) Suppose $q : X \to Y$ is a quotient map $A \hookrightarrow X$ is a cofibration and $q : (X, A) \to (Y, q(A))$ is a relative homeomorphism, i.e., $q : X \setminus A \to Y \setminus q(A)$ is a homeomorphism. Then show that q induces isomorphisms in homology $H_*(X, A) \to H_*(Y, q(A))$.

4.3 Construction of Some Other Homology Groups

In this section we discuss variants of singular homology. These include
(A) smooth singular homology for smooth manifolds,
(B) simplicial and singular simplicial homology for simplicial complexes,
(C) CW-homology and cellular CW-homology for CW-complexes.

We shall state how each of them is related with the singular homology of the underlying topological space. The proofs are all postponed to the last section. Each of these homology groups enhances our knowledge of singular homology and helps in computation in special cases.

Smooth Singular Homology of Smooth Manifolds

For this paragraph, we presume that the reader is familiar with the basic theory of smooth manifolds. Consider the category of $\mathcal{C}^1$-manifolds. We must expand this category a little bit by allowing objects such as $M \times \mathbb{I}$ where M is a manifold with or without boundary. Perhaps the easiest way to do this is to consider subspaces of Euclidean spaces with the induced smooth structure on them. By Whitney's embedding theorem, it follows that this will take care of all smooth manifolds as well as objects such as $M \times \mathbb{I}$. In particular, note that the standard n-simplexes with the induced smooth structure on them are objects in this category. Obviously, morphisms in this category are smooth functions on these objects, i.e., those which are restrictions of some smooth functions on some open neighbourhood of the object in the Euclidean space.

Let M be a smooth object. We denote by $S_n^{sm}(M)$ the free abelian subgroup of $S_n(M)$ generated by smooth maps $\sigma : \Delta_n \to M$. It is easily verified that $S_{\cdot}^{sm}(M) = \oplus_{n \geq 0} S_n^{sm}(M)$ is a chain subcomplex of $S_{\cdot}(M)$. The homology of this chain complex is called the *smooth singular homology of M* and is denoted by $H_{\cdot}^{sm}(M)$.

It is easily checked that the assignment $M \rightsquigarrow S_{\cdot}^{sm}(M)$ is a covariant functor. The smooth singular homology functor satisfies the standard properties similar to the singular homology. It satisfies the homotopy invariance as in Theorem 4.2.9. We shall prove this later along with the proof of Theorem 4.2.9. In particular it would follow that if M is a smoothly contractible smooth object then its smooth singular homology is the same as that of a point space. Moreover, it is also easy to see the following.

Corollary 4.3.1 $H_{\cdot}^{sm}(*) = H_{\cdot}(*)$.

Clearly there is a homology long exact sequence of a pair of smooth objects (X, A) and the inclusion map $\eta : S_{\cdot}^{sm} \to S_{\cdot}$ induces a ladder of homomorphisms of the two exact sequences.

$$\cdots \longrightarrow H_n^{sm}(A) \longrightarrow H_n^{sm}(X) \longrightarrow H_n^{sm}(X,A) \xrightarrow{\partial} H_{n-1}^{sm}(A) \longrightarrow \cdots \quad (4.12)$$
$$\downarrow{\eta_*} \qquad \downarrow{\eta_*} \qquad \downarrow{\eta_*} \qquad \downarrow{\eta_*}$$
$$\cdots \longrightarrow H_n(A) \longrightarrow H_n(X) \longrightarrow H_n(X,A) \longrightarrow H_{n-1}(A) \longrightarrow \cdots$$

Note that the discussion preceding Lemma 4.2.14 and the conclusion of the lemma 4.2.14 holds verbatim for $S_{\cdot}^{sm}$ of smooth manifolds. In a similar fashion to that of Definition 4.2.15, one can then define excisive couples for sm-homology. What is somewhat non trivial is that the analogue of Theorem 4.2.17 is also true:

Theorem 4.3.2 *If $X = X_1 \cup X_2 = \text{int}_X(X_1) \cup \text{int}_X(X_2)$ then, $\{X_1, X_2\}$ is an excisive couple for the smooth singular homology.*

We shall prove this later along with the proof of Theorem 4.2.9. Taking this for granted, you are tempted to repeat verbatim the homology suspension theorem, etc., as in the case of singular homology, but hold on: we need to explain what we mean by the suspension in the smooth category... Instead, it is time to have:

Theorem 4.3.3 *The inclusion map* $\eta : S^{sm}_{.} \to S_{.}$ *is a functorial chain equivalence and hence defines a natural equivalence of the two homology modules.*

Here again, you will have to wait for the proof of this until the next chapter.

The Simplicial Homology and the Singular Simplicial Homology

The singular chain complex of a topological space is too huge to carry out certain types of computations of homology groups. At least in the case of polyhedron $|K|$, we can remedy this situation by introducing a subcomplex of $S_{.}(|K|)$ which yields the same homology groups as $S_{.}(|K|)$. We have seen earlier, via the simplicial approximation theorem, that for a polyhedron, the set of simplicial maps is capable of capturing the topological features of the space, at least up to homotopy type, in a certain loose sense. This being so, it should be possible to have similar treatment while dealing with homology groups of a polyhedron. To be a little more precise, we can think of taking only the simplicial maps $\sigma : |\Delta_n| \to |K|$, instead of taking all continuous functions, while forming the singular homology groups. If this turns out to be 'meaningful', then it would have the same kind of advantages over the singular chain complex, as a simplicial map over a continuous map, which we have witnessed earlier. For instance, the chain groups become combinatorial objects and will be extremely handy compared to the ordinary singular chain groups. This is what we would like to study now.

Definition 4.3.4 Let (K, L) be a simplicial pair. Let $\mathcal{S}_n(K)$ be the subgroup of $S_n(|K|)$ generated by all simplicial maps $\lambda : |\Delta_n| \to |K|$. Let $\mathcal{S}_n(K, L) = \mathcal{S}_n(K)/\mathcal{S}_n(L)$. Note that $\mathcal{S}_n(K, L)$ is a free abelian group and can be regarded as a subgroup of $S_n(|K|, |L|) = S_n(|K|)/S_n(|L|)$ with a basis consisting of those simplicial maps $\lambda : \Delta_n \to K$ such that $\lambda(\Delta_n)$ is not contained in L. If ∂ denotes the boundary map of the singular chain complex $S_{.}(|K|)$, then clearly $\partial(\mathcal{S}_n(K)) \subset \mathcal{S}_{n-1}(K)$. Therefore, $\mathcal{S}_{.}(K) = \oplus_{n \geq 0}\mathcal{S}_n(K)$ and $\mathcal{S}_{.}(K, L) = \oplus_{n \geq 0}\mathcal{S}_n(K, L)$ form subchain complexes of $S_{.}(|K|)$ and $S_{.}(|K|, |L|)$, respectively.

Example 4.3.5 $H_0(\mathcal{S}_{.}(K))$ **of a connected complex** K

Recall that if K is connected (i.e., $|K|$ is connected), then any two vertices u, v in K can be joined by a sequence of edges (see Exercises 2.11.2). This then yields a 1-chain σ such that $\partial(\sigma) = v - u$. Proceeding as in Example 4.2.12.4, we conclude that $H_0(\mathcal{S}(K)) \approx \mathbb{Z}$.

Definition 4.3.6 A simplicial map $\lambda : \Delta_n \longrightarrow K$ can be displayed by simply writing down the images of the vertices $\mathbf{e}_i$ of Δ_n under λ in that order. Thus an ordered $(n + 1)$-tuple of vertices of K defines a simplicial map iff all those vertices belong to a simplex of K. Just in order to distinguish them from mere $(n + 1)$-tuples and think of them as simplicial simplexes, we shall display them inside square brackets as:

$$[v_0, v_1, \ldots, v_n]$$

where $v_i \in F$ for a k-simplex F in K. (Note that here the vertices v_i need not be distinct.)

In particular, consider now the case when $K = \Delta_n$ is the standard n-simplex. Then the face maps $F^i : \Delta_{n-1} \to \Delta_n$ are simplicial and can be represented in the form

$$F^i = [\mathbf{e}_0, \mathbf{e}_1, \ldots, \mathbf{e}_{i-1}, \mathbf{e}_{i+1}, \ldots, \mathbf{e}_n].$$

We introduce some convenient notation here:

$$F^i := [\mathbf{e}_0, \mathbf{e}_1, \ldots, \mathbf{e}_{i-1}, \mathbf{e}_{i+1}, \ldots, \mathbf{e}_n] =: [\mathbf{e}_0 \ldots, \widehat{\mathbf{e}}_i, \ldots, \mathbf{e}_n].$$

Here the hat $\widehat{}$ indicates that the corresponding entry is deleted from the sequence. In this notation, it follows that

$$\partial[v_0, \ldots, v_n] = \sum_i (-1)^i [v_0, \ldots, \widehat{v_i}, \ldots, v_n] \qquad (4.13)$$

Example 4.3.7 The chain complex $\mathscr{S}_.(\Delta_n)$

Clearly, there are precisely $(n+1)^{q+1}$ singular simplicial q-simplices in Δ_n because they are in one-to-one correspondence with sequences of length $(q+1)$ in the letters $\{e_0, \ldots, e_n\}$. Thus the chain complex $\mathscr{S}_.(\Delta_n)$ looks like

$$\cdots \mathbb{Z}^{(n+1)^{q+1}} \longrightarrow \mathbb{Z}^{(n+1)^q} \longrightarrow \cdots \longrightarrow \mathbb{Z}^{(n+1)^2} \longrightarrow \mathbb{Z}^{(n+1)}.$$

Given a singular n-simplex σ and any vertex $v \in \Delta_n$, we shall use the notation $v\sigma$ to denote the singular $(n+1)$- simplex defined by

$$v\sigma(e_i) = \begin{cases} v, & i = 0; \\ \sigma(v_{i-1}), & i \geq 1. \end{cases}$$

We can then extend this notation when the n-simplex σ is replaced by any n-chain in Δ_n :

$$v\sigma = v\left(\sum_i n_i \sigma_i\right) = \sum_i n_i v\sigma_i.$$

It follows that

$$\partial(v\sigma) = \sigma - v\partial\sigma. \qquad (4.14)$$

This allows us a simple way to construct a chain homotopy of the $Id : \mathscr{S}_.(\Delta_n) \to \mathscr{S}_.(\Delta_n)$ with the chain map $\zeta : \mathscr{S}_.(\Delta_n) \to \mathscr{S}_.(\Delta_n)$ defined as follows. Fix any vertex v_0 of Δ_n. Take $\zeta_0(v) = v_0$, and for all vertices v in Δ_n and take $\zeta_k \equiv 0$ for $k > 0$. Check that ζ is indeed a chain map. Now for each $k \geq 0$, take $P : \mathscr{S}_k(\Delta_n) \to \mathscr{S}_{k+1}(\Delta_n)$ to be the map

$$P(\sigma) = v_0\sigma.$$

From (4.14), it easily follows that

$$P \circ \partial + \partial \circ P = Id - \zeta.$$

The immediate consequence of this is that on homology the two induced homomorphisms are the same $Id = Id_* = \zeta_* : H(\mathscr{S}_.(\Delta_n)) \to H(\mathscr{S}_.(\Delta_n))$. Since it is easily seen that $\zeta_* = 0$ on H_q for $q > 0$ we conclude that $H_q(\mathscr{S}_.(\Delta_n)) = 0$ for $q > 0$. Also since any simplex is connected we know $H_0(\mathscr{S}_.(\Delta_n)) \approx \mathbb{Z}$.

Let Σ_{n+1} denote the permutation group on $(n+1)$-letters. Then to each $\alpha \in \Sigma_{n+1}$, we have a simplicial homeomorphism $G_\alpha : |\Delta_n| \to |\Delta_n|$. This defines a right-action of Σ_{n+1} on $\mathscr{S}_n(\Delta_n)$ as follows: First define the right action of α on the basis elements of $\mathscr{S}_n(\Delta_n)$ by the formula,

$$\lambda^\alpha = \lambda \circ G_\alpha$$

where $\lambda : |\Delta_n| \to |\Delta_n|$ is a simplicial map. Finally define

$$\left(\sum_j n_j \lambda_j\right)^\alpha = \sum_j n_j (\lambda \circ G_\alpha).$$

Definition 4.3.8 A simplicial map $\lambda : |\Delta_n| \to |K|$ is called a *degenerate n-simplex* if the dimension of the simplex $\lambda(\Delta_n)$ is strictly less than n. Let $\mathbb{S}_n^0(K)$ denote the subgroup generated by degenerate simplicial maps λ, and elements of the form

$$\lambda - sgn(\alpha)\lambda^\alpha,$$

where λ runs over all simplicial maps $\lambda : \Delta_n \longrightarrow K$ and α runs over Σ_{n+1}.

Lemma 4.3.9 $\partial(\mathbb{S}_n^0(K)) \subset \mathbb{S}_{n-1}^0(K)$.

Proof: We shall first check that the boundary of a degenerate simplex is in $\mathbb{S}_{n-1}^0(K)$. So, let $\lambda : \Delta_n \longrightarrow K$ be a simplicial map with $\dim \lambda(\Delta_n) = m < n$.

First suppose $m < n - 1$. This means that λ maps either three vertices to the same vertex or two pairs of vertices to the same vertex. Since $\partial(\lambda)$ is a sum of simplicial maps each obtained by deleting only one vertex of Δ_n, it follows that $\partial(\lambda)$ is a sum of degenerate $(n-1)$-simplexes.

Next, consider the case when $m = n - 1$, say, λ maps i^{th} and j^{th} vertices to the same vertex, for some $0 \le i < j \le n$. In this case, $\partial(\lambda)$ consists of a sum of a number of degenerate $(n-1)$-simplices and the two terms

$$(-1)^i[v_0, v_1, \ldots, v_{i-1}, \hat{v}_i, v_{i+1}, \ldots, v_j, \ldots, v_n] \\ (-1)^j[v_0, v_1, \ldots, v_{i-1}, v_i, v_{i+1}, \ldots, v_{j-1}, \hat{v}_j, v_{j+1} \ldots, v_n]. \tag{4.15}$$

But observe that since $v_i = v_j$, $[v_0, v_1, \ldots, v_{i-1}, v_i, v_{i+1}, \ldots v_{j-1}, \hat{v}_j, v_{j+1} \ldots, v_n]$ is obtained from $[v_0, v_1, \ldots, v_{i-1}, \hat{v}_i, v_{i+1}, \ldots v_j, \ldots, v_n]$ by performing $i+j-1$ transpositions. Therefore, the sum in (4.15) is of the form $\tau - sgn(\alpha)\tau^\alpha$. This proves that $\partial(\lambda) \in \mathbb{S}_{n-1}^0(K)$.

To show that the boundary of elements of the form $\lambda - (sgn\alpha)\lambda^\alpha$ is in $\mathbb{S}_{n-1}^0(K)$, it suffices to do so under the assumption that α is a transposition. But then $\partial(\lambda + \lambda^\alpha)$ is again a sum of such expressions. ♠

Definition 4.3.10 Let $C_n(K) = \mathbb{S}_n(K)/\mathbb{S}_n^0(K)$. It follows that $\partial : \mathbb{S}_n(K) \to \mathbb{S}_{n-1}(K)$ induces a homomorphism on these quotient groups, which we shall again denote by $\partial : C_n(K) \to C_{n-1}(K)$. Clearly $\partial^2 = 0$ and hence $\{C_.(K), \partial\}$ is a chain complex. This chain complex is called the *simplicial chain complex* of K. The homology groups $H_*(C_.(K))$ are called the *simplicial homology groups* of K.

Example 4.3.11 Let us take a closer look at the simplicial chain complex $C(\Delta_n)$ of the n-simplex Δ_n. To begin with, there is no confusion about 0-chains; they are merely linear combinations of vertices of Δ_n and so $C_0(K) = \mathbb{Z}^{n+1}$. Now let us look at 1-chains. We know $\mathbb{S}_1(\Delta_n) = \mathbb{Z}^{(n+1)^2}$. But any simplex given by a non-injective map is a degenerate simplex and so goes into $\mathbb{S}_1^0(\Delta_n)$. Therefore, we need to take only pairs $[u, v]$ with $u \ne v$. But then the action of the permutation group takes $[u, v]$ to $[v, u]$ and hence chains of the form $[u, v] + [v, u]$ belong to $\mathbb{S}_1^0(\Delta_n)$. This just means that we need to count an edge of Δ_n just once, in whichever order we may choose. Since there are precisely $\binom{n+1}{2}$ edges in Δ_n, it follows that $C_1(\Delta_n) = \mathbb{Z}^{\binom{n+1}{2}}$. The boundary operator is clearly given by

$$\partial[u, v] = [v] - [u].$$

Likewise it follows that $C_q(\Delta_n) = \mathbb{Z}^{\binom{n+1}{q+1}}$ and $\partial : C_q(\Delta_n) \to C_{q-1}(\Delta_n)$ is given by

$$\partial[u_0, \ldots, u_q] = \sum_i (-1)^i[u_0, \ldots, \widehat{u_i}, \ldots, u_q]$$

Thus the entire chain complex $C(\Delta_n)$ looks like

$$0 \cdots 0 \longrightarrow \mathbb{Z}^{\binom{n+1}{n+1}} \longrightarrow \mathbb{Z}^{\binom{n+1}{n}} \longrightarrow \cdots \longrightarrow \mathbb{Z}^{\binom{n+1}{2}} \longrightarrow \mathbb{Z}^{n+1}.$$

Write down a full description of this for the case $n = 3$ along with the boundary operators. Can you show that this chain complex is exact?

Remark 4.3.12 Note that $C_n(K)$ is purely an algebraic object, which can be described as the free abelian group on the n-simplices of K. A priori, this has very little say on the topology on $|K|$. Obviously, the boundary maps themselves depend on the incidence relations within K and hence the homology groups depend 'merely' on the combinatorial nature of K. Thus, the following results come as a surprise.

Theorem 4.3.13 (Simplicial versus singular) *The inclusion map*

$$\iota : \mathcal{S}_{\cdot}(K, L) \to S_{\cdot}(|K|, |L|)$$

is a chain homotopy equivalence.

Theorem 4.3.14 (Singular-simplicial versus simplicial) *The quotient map*

$$\varphi : \mathcal{S}_{\cdot}(K, L) \to C_{\cdot}(K, L)$$

is chain homotopy equivalence.

As an easy consequence, we have:

Theorem 4.3.15 *For pairs of simplicial complexes (K, L), there are canonical isomorphisms of the singular homology groups $H(S_{\cdot}(|K|, |L|))$, the singular-simplicial homology groups $H(\mathcal{S}_{\cdot}(K, L))$ and the simplicial homology groups $H(C_{\cdot}(K, L))$.*

Remark 4.3.16
(1) We are going to postpone the proofs of these theorems also. The homology groups of $C_{\cdot}(K, L)$ will be referred to as the *simplicial homology of the simplicial pair* (K, L).
(2) A priori, this depends upon the actual simplicial complex. However, once we have proved the above theorem, since singular homology is a homeomorphism invariant, it follows that so is simplicial homology. Thus, while computing the homology groups, we are free to choose suitable triangulations of a polyhedron.
(3) We note that, even though $C_{\cdot}(K, L)$ is defined as a quotient complex of a free chain complex, it is also a free complex. One begins with a choice of an arbitrary total order on the vertices of K. Then to each n-simplex in K, one can choose the non degenerate singular n-simplex given by the order preserving map. This set will form a free basis for $C_n(K)$. In case K is a finite complex, each $C_n(K)$ will be of finite rank also. Moreover, the boundary homomorphisms are given by incidence matrices, i.e., matrices whose entries are 0 or ± 1, the data being completely determined by the face relations and the chosen orientations on each simplex. Thus, simplicial homology is quite simple from the computational point of view.
(3) We also note that, $\mathcal{S}_{\cdot}(K, L)$ plays an intermediary role in connecting $C_{\cdot}(K)$ and $S_{\cdot}(|K|)$.
(4) Suppose K_1 and K_2 are subcomplexes of a simplicial complex K such that $K = K_1 \cup K_2$. Since any simplex (singular or not) of K is a simplex in K_1 or in K_2, it follows that

$$\mathcal{S}(K_1) + \mathcal{S}(K_2) = \mathcal{S}(K); \quad C(K_1) + C(K_2) = C(K).$$

This just means that (K_1, K_2) is an excisive couple for the singular simplicial homology as well as simplicial homology and hence we have the Mayer–Vietoris type exact sequence for both these homologies, which comes very handy in computing the homology of the union from that of the pieces.

An easy but a striking consequence of equivalence of singular and simplicial homology is the following:

Corollary 4.3.17 Let (K, L) be a polyhedral pair. If $\dim K \leq n$, then $H_k(|K|, |L|) = 0$, $\forall\, k > n$ and $H_n(|K|, |L|)$ is a free abelian group. If K is a compact polyhedron, then $H_*(|K|, |L|)$ is finitely generated.

Proof: Easy.   ♠

Remark 4.3.18 Euler characteristic revisited
   Recall that in Exercise 2.11 (iii), we have defined

$$\chi(K) := (-1)^i f_i(K)$$

as the Euler characteristic of the simplicial complex K. For the simplicial chain complex $C.(K)$, we have another definition of $\chi(C.(K))$ in 4.1.18. Since each $C_i(K)$ is a free abelian group over the set of i-simplices of K, it immediately follows that

$$\chi(K) = \chi(C.(K)).$$

Since we have also proved that $\chi(C.(K)) = \chi(H(C.(K))$ and since $H(C.(K))$ is canonically isomorphic to $H_*(|K|)$, we conclude that the Euler characteristic of K is a topological invariant. Indeed, for any topological space X with finitely generated singular homology, we define

$$\beta_i(X) := \operatorname{rank} H_i(X; \mathbb{Z})$$

and call it the ith-Betti number of X. We then have,

$$\chi(X) = \chi(H_*(X)) = \sum_i (-1)^i \beta_i(X).$$

Because of the topological invariance of the homology groups, Betti numbers can be computed using any triangulation of a space and the corresponding simplicial homology.

Exercise 4.3.19

(i) Let K be a finite simplicial complex. Describe the simplicial chain complex of cK, the cone over K in terms of the simplicial chain complex of K. Deduce that $C.(cK)$ is exact. [Hint: Fix a total order on the vertices of K. Extend this order on the vertices of cK by taking the apex of cK to be the least element.]

(ii) Write down explicitly the chain complex $C.(\Delta_n)$. Using the above exercise, compute $H_*(\mathbb{S}^{n-1})$. (Remember that $|\dot{\Delta}_n|$ is homeomorphic to $\mathbb{S}^{n-1}$.)

(iii) Define the subdivision chain map $Sd : C.(K) \to C.(sd\,K)$, inductively, as follows. On C_0, it is defined to be linear extension of the inclusion map. Suppose we have defined it on C_{n-1}. For any n-simplex σ in K, define $Sd(\sigma) = \beta(\sigma)Sd(\partial(\sigma))$ where $\beta(\sigma)$ denotes the barycentre of σ.
(a) Verify that Sd is a chain map and show that it is functorial.
(b) If $\tau : sd\,K \to K$ is a simplicial approximation to Id, then show that, $C(\tau) \circ Sd = Id$.
(c) Show that $\tau_* : H_*(|sdK|) \to H_*(|K|)$ is the identity isomorphism.
(d) Show that Sd induces the identity isomorphism on homology.

CW-Homology and Cellular Singular Homology

We begin with:

Theorem 4.3.20 *Let X be obtained by attaching n-cells $\{e^n{}_j\}$ to Y via the attaching maps $\alpha_j : \mathbb{S}^{n-1} \to Y$. Let $\alpha : \oplus_{j \in J}\mathbb{Z} \to H_{n-1}(Y)$ be the homomorphism which when restricted to the jth summand is equal to $(\alpha_j)_* : \mathbb{Z} = H_{n-1}(\mathbb{S}^{n-1}) \to H_{n-1}(Y)$. Then*
(a)

$$H_k(X, Y) = \begin{cases} 0, & k \neq n, \\ \oplus_{j \in J}\mathbb{Z}, & k = n. \end{cases}$$

(b) *The inclusion $\iota : Y \to X$ induces isomorphisms $H_k(Y) \to H_k(X)$ for all $i \neq n - 1, n$.*
(c) *Moreover, $\iota_* : H_{n-1}(Y) \to H_{n-1}(X)$ is a surjection with its kernel equal to $\text{Im}\, \alpha$*
(d) *and $\iota_* : H_n(Y) \to H_n(X)$ is injective with $H_n(X) \approx \iota_*(H_n(Y)) \oplus \text{Ker}\, \alpha$.*

Proof: (a) Choose a point $p_j \in \text{int}\, e_j^n$ for each j and put $U = X \setminus \{p_j : j \in J\}$. Then observe that Y is a deformation retract of U. Therefore from the long homology exact sequence of the triple (X, U, Y), it follows that the inclusion map $\eta_1 : (X, Y) \longrightarrow (X, U)$ induces an isomorphism. Put $V = \cup_j \text{int}\, e_j^n$. Since U is open in X, it follows that $\{U, V\}$ is an excisive couple. Therefore, the inclusion map $\eta_2 : (V, V \cap U) \longrightarrow (X, U)$ induces isomorphism of homology groups. Now the pair $(V, V \cap U)$ is the disjoint union of the pairs $(\text{int}\, e_j^n, \text{int}\, e_j^n \setminus \{p_j\})$. Hence the homology is the direct sum of the homology of these pairs. Now appeal to Example (4.2.24).
Now (b) (c) and (d) are easy consequences of the long homology exact sequence of the pair (X, Y). ♠

Remark 4.3.21 We have made a very good beginning in this lemma of understanding the homology of a CW-complex. We shall now launch a systematic study of the homology of a CW-complex. Our theme is 'from parts to the whole' and the main tools are long homology exact sequences and excision.

Lemma 4.3.22 Let X be a CW-complex. Then for each integer $n \geq 0$, we have
(a) $H_k(X^{(n)}, X^{(n-1)})$ vanishes for $k \neq n$ and is isomorphic to the free abelian group of rank equal to the number of n-cells in X.
(b) $H_k(X^{(n)}) = (0), k > n$.
(c) The inclusion map $\eta_n : X^{(n)} \longrightarrow X$ induces isomorphism in homology for $k < n$ and surjection for $k = n$.

Proof: (a) This is a direct consequence of Lemma 4.3.20 (a).
(b) Since $X^{(0)}$ is a discrete space, we know $H_k(X^{(0)}) = (0)$ for all $k > 0$. Inductively let us assume that the statement (b) holds for $n - 1$. Let $k > n$. In the long exact sequence of $(X^{(n)}, X^{(n-1)})$:

$$H_k(X^{(n-1)}) \longrightarrow H_k(X^{(n)}) \longrightarrow H_k(X^{(n)}, X^{(n-1)})$$

the first term is zero by induction and the third term is zero by (a). Hence the middle term is also zero.
(c) First, we assume that X is finite dimensional say $\dim X = m$. If $n \geq m$, then $X^{(n)} = X$ and so there is nothing to prove. So we may as well assume $n < m$. Then for each $i \geq 0$, and $k < n$, we have, the inclusion induced map

$$H_k(X^{(n+i)}) \longrightarrow H_k(X^{(n+i+1)})$$

is an isomorphism. (Use the homology exact sequence and (a). Composite of finitely many such isomorphisms yields the isomorphism

$$H_k(X^{(n)}) \longrightarrow H_k(X^{(m)}) = H_k(X).$$

For $k = n$, this fails only at the very first instance, viz., $H_k(X^{(n)}) \longrightarrow H_k(X^{(n+1)})$. However, even here, the map is surjective, since the next group in the sequence is $H_n(X^{(n+1)}, X^{(n)}) = (0)$.

Now suppose X is infinite dimensional. Suppose the inclusion map $\eta_n : X^{(n)} \longrightarrow X$ induces a non-injective homomorphism $H_k(X^{(n)}) \longrightarrow H_k(X)$ for some $k < n$. Let c be a k-cycle in $X^{(n)}$ which is a boundary in X say $c = \partial y$. Since y is a finite combination of $(k+1)$-singular simplices, its support is compact. Therefore y is a $(k+1)$-chain in a finite skeleton $X^{(m)}$ of X. This means that $H_k(X^{(n)}) \longrightarrow H_k(X^{(m)})$ is not injective, which is a contradiction to what we have proved above, in the finite dimensional case. This proves injectivity. The proof of surjectivity is similar and easier. ♠

Remark 4.3.23 Motivated by the construction of the chain complex $C_.(K)$ for a simplicial complex K and the results obtained above, we now make the following definition of a chain complex associated to a CW-complex.

Definition 4.3.24 Let X be a CW-complex. For each n, let

$$i_n : X^{(n)} \longrightarrow X^{(n+1)} \text{ and } j_n : X^{(n)} \longrightarrow (X^{(n)}, X^{(n-1)})$$

denote the inclusion maps. Put

$$C_n^{CW}(X) := H_n(X^{(n)}, X^{(n-1)}); \quad d_n := j_{n-1} \circ \partial_n, n \geq 1$$

where $\partial_n : H_n(X^{(n)}, X^{(n-1)}) \longrightarrow H_{n-1}(X^{(n-1)})$ is the connecting homomorphism in the long exact sequence of the pair $(X^{(n)}, X^{(n-1)})$.

The following lemma shows that $d_n \circ d_{n+1} = 0$ for all n and hence $C_*^{CW}(X)$ is a chain complex. This is called the *cellular chain associated to X*. The homology of this chain complex will be called the *cellular homology of X*.

Lemma 4.3.25 We have, $d_n \circ d_{n+1} = 0$.

Proof: In the commutative diagram (Figure 4.3), the two vertical sequences are respectively part of the exact sequences of the pairs $(X^{(n+1)}, X^{(n)})$ and $(X^{(n-1)}, X^{(n-2)})$, whereas the horizontal sequence is that of $(X^{(n)}, X^{(n-1)})$. In particular, we have $\partial_n \circ j_n = 0$ and hence $d_n \circ d_{n+1} = j_{n-1} \circ \partial_n \circ j_n \circ \partial_{n+1} = 0$. ♠

Theorem 4.3.26 $H_n^{CW}(X) \approx H_n(X)$.

Proof: Once again we refer to Figure 4.3. Observe that the appearance of the three zero groups is justified by Lemma 4.3.22. Since j_{n-1} is injective, it follows that $\operatorname{Ker} d_n = \operatorname{Ker} \partial_n = \operatorname{Im} j_n \approx H_n(X^{(n)})$ since j_n is injective. Therefore, $H_n^{CW}(X) = \operatorname{Ker} d_n / \operatorname{Im} d_{n+1} \approx H_n(X^{(n)})/\operatorname{Im} \partial_{n+1} \approx H_n(X^{(n+1)}) \approx H_n(X)$. ♠

Example 4.3.27 Homology of $\mathbb{CP}^n$

An immediate consequence of this theorem is that if a CW-complex X does not have any cells of dimension k, then $H_k(X) = (0)$. This fact fits very neatly in computing the homology of complex projective spaces. Recall that $\mathbb{CP}^n$ has cell decomposition consisting

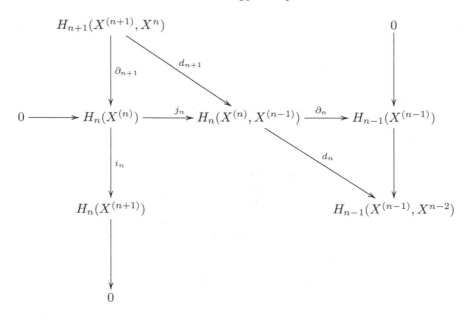

FIGURE 4.3. Construction of cellular homology

of one cell of dimension $2k$ for $0 \leq k \leq n$. Therefore the cellular chain complex $C^{CW}(\mathbb{CP}^n)$ looks like

$$0 \longrightarrow \mathbb{Z} \longrightarrow 0 \longrightarrow \mathbb{Z} \longrightarrow 0 \longrightarrow \ldots \longrightarrow \mathbb{Z} \longrightarrow 0 \longrightarrow \mathbb{Z}$$

with 0 and $\mathbb{Z}$ occurring alternatively. Therefore

$$H_i(\mathbb{CP}^n) \approx \begin{cases} \mathbb{Z}, & i = 2k, \ 0 \leq k \leq n; \\ (0), & \text{otherwise.} \end{cases}$$

Remark 4.3.28 In order to exploit the cellular homology further, we should try to understand the boundary homomorphisms d_n of this chain complex. Here is a description of d_n in terms of the attaching maps of the CW-structure. Let $\phi_\alpha : (\mathbb{D}^n, \mathbb{S}^{n-1}) \to (X^{(n)}, X^{(n-1)})$ denote the collection of characteristic maps of n-cells in X and let $f_\alpha : \mathbb{S}^{n-1} \to X^{(n-1)}$ denote the corresponding attaching maps. We know that $C_n = H_n(X^{(n)}, X^{(n-1)})$ is freely generated over the classes $[\phi_\alpha]$. Therefore, in order to determine d_n we have only to find expressions for $d_n([\phi_\alpha])$ in $H_{n-1}(X^{(n-1)}, X^{(n-2)})$. So, let us denote the collection of characteristic maps of $(n-1)$-cells by ψ_β. Observe that the quotient space $X^{(n-1)}/X^{(n-2)}$ is homeomorphic to the bouquet of spheres $\vee_\beta \mathbb{S}_\beta^{n-1}$. Indeed the indexing of these spheres is such that the $(n-1)$-cell ψ_β is mapped onto the sphere $\mathbb{S}_\beta^{n-1}$ under the quotient map $q_{n-1} : X^{n-1} \to X^{(n-1)}/X^{(n-2)}$. Also q_{n-1} induces an isomorphism of C_{n-1} and $H_{n-1}(\vee_\beta \mathbb{S}_\beta^{n-1})$. Let $p_\beta : \vee_\beta \mathbb{S}_\beta^{n-1} \to \mathbb{S}_\beta^{n-1}$ be the projection onto the βth component. Put $d_n[\phi_\alpha] = \sum_\beta n_{\alpha,\beta}[\psi_\beta]$. Note that $\partial_n[\phi_\alpha] = [f_\alpha]$. It follows that $n_{\alpha,\beta}$ is nothing but the degree of the map $p_\beta \circ q_{n-1} \circ f_\alpha : \mathbb{S}^{n-1} \to \mathbb{S}_\beta^{n-1}$.

Example 4.3.29 Homology groups of lens spaces

Recall that we have constructed a CW-structure for each lens space in Example (iii). You are welcome to use them together with the above remark to compute the corresponding CW-chain complex of the lens spaces and thereby, their homology groups. Here, we shall give yet another (simpler) CW-structure to each lens space $X = L_p(q_1, \ldots, q_r)$ $(p \geq 3)$ with exactly one cell for each dimension $k, 0 \leq k \leq 2r + 1$. The chain complex will look like

$$0 \cdots 0 \longrightarrow \mathbb{Z} \xrightarrow{0} \mathbb{Z} \xrightarrow{p} \mathbb{Z} \xrightarrow{0} \mathbb{Z} \xrightarrow{p} \cdots \qquad \cdot \longrightarrow \mathbb{Z} \xrightarrow{0} \mathbb{Z}. \qquad (4.16)$$

We shall first describe a CW-structure on $\mathbb{S}^{2s+1}$ inductively, so that the action of $\mathbb{Z}_p$ on it will be cellular and then take the quotient structure. So, let $\zeta = e^{2\pi \iota / p}$. We declare all points ζ^j, $0 \leq j < p$, as 0-cells and the arcs in $\mathbb{S}^1$ from ζ^k to ζ^{k+1} as the 1-cells. For 2-cells, e_k^2, we take the half-spheres in the $\mathbb{R}$-linear span of $\mathbb{S}^1 \times 0$ and $(0, \zeta^k)$ containing point $(0, \zeta^k)$ and with boundary $\mathbb{S}^1 \times 0$. Note that each of these discs have $\mathbb{S}^1 \times 0$ as their boundary which also constitutes their points of intersection. Therefore, the union of any two of them bounds a 3-disc in $\mathbb{S}^3$. For the 3-cells, we take these 3-discs e_k^3 bounded by e_{k-1}^2 and e_k^2, $0 \leq k \leq p-1$. Verification that this describes a cell structure on $\mathbb{S}^3$ and that the cell structure is $\mathbb{Z}_p$-action invariant is straightforward.

Now, inductively, suppose we have described the cell-structure on $\mathbb{S}^{2s-1}$. Put $\tau_k = (0, \ldots, 0, \zeta^k) \in \mathbb{C}^{s+1}$ and let e_k^{2s} be the $2s$-dimensional disc which is contained in the intersection of $\mathbb{S}^{2s}$ with the $\mathbb{R}$-linear span of $\mathbb{R}^{2s}$ and τ_k, and which contains the point τ_k and has boundary $\mathbb{S}^{2s-1}$. And for $(2s+1)$-discs, we take the portions of $\mathbb{S}^{2s+1}$ bounded by e_{k-1}^{2s}, e_k^{2s}. Once again, the verification that this extends the cell structure on $\mathbb{S}^{2s-1}$ to a cell structure of $\mathbb{S}^{2s+1}$ and that it is invariant under the $\mathbb{Z}_p$-action is straightforward. Since the action is clearly transitive on the set of r-cells, it follows that the quotient space has a cell structure with precisely one cell in each dimension, $0 \leq s \leq 2s + 1$.

Thus, in particular, it follows that the CW-chain complex of this CW-structure on X has the property $C_k = \mathbb{Z}$, $0 \leq k \leq 2r + 1$ and (0) otherwise. It remains to describe $\partial_k : C_k \to C_{k-1}$. Clearly $\partial_1 = 0$ since the 1-cell is attached to a single vertex. The boundary of any of the oriented 2-cells in $\mathbb{S}^{2r+1}$ is the oriented circle $\mathbb{S}^1$ which is clearly wrapped onto the 2-cell p-times. That is to say that the quotient map $q : \mathbb{S}^{2r+1} \to X$ restricted to $\mathbb{S}^1$ is of degree p. Therefore it follows that ∂_2 is the multiplication by p. On the other hand, each oriented 3-cell e_k^3 in $\mathbb{S}^3 \subset \mathbb{S}^{2r+1}$ is bounded by precisely two oriented discs e_{k-1}^2, e_{k+1}^2 and hence $\partial(e_k^3) = e_k^2 - e_{k-1}^2$. Since the quotient map restricted to the interior of each of these 2-cells is orientation preserving, it follows that $\partial_3 = 0$. The description of $\partial_{2k}, \partial_{2k+1}, k \geq 2$ is identical to those of ∂_2, ∂_3, respectively, thereby establishing (4.16).

We conclude that

$$H_i(L_p(q_1, \ldots, q_r); \mathbb{Z}) = \begin{cases} \mathbb{Z}, & i = 0, 2r + 1; \\ \mathbb{Z}_p, & i = 2k - 1, 1 \leq k \leq r; \\ (0), & \text{otherwise.} \end{cases} \qquad (4.17)$$

Finally, we now come to yet another homology for CW-complexes which lies 'between' CW-homology and singular homology and which we call *cellular singular homology*. The case is similar to the simplicial homology and singular simplicial homology of a simplicial complex.

Definition 4.3.30 Let (X, A) be a CW-complex. Let $C_n^{cell}(X)$ be the free abelian group generated by the set of all continuous maps $\sigma : \Delta_n \to X$ which are cellular, in the standard cell structure of Δ_n. Clearly, this forms a subgroup of $S_n(X)$ and one can easily verify that $\partial(C_n^{cell}) \subset C_{n-1}^{cell}$. In other words, $C_{\cdot}^{cell}(X) = \oplus_n C_n^{cell}(X)$ forms a subchain complex of $S_{\cdot}(X)$.

Theorem 4.3.31 *The inclusion map $C_{\cdot}^{cell}(X) \subset S_{\cdot}(X)$ is a chain equivalence.*

We shall postpone the proof of this theorem to the end of the chapter. However, notice that the CW-chain complex itself can be thought of as a subcomplex of this complex via the characteristic maps. Moreover, by the naturality, it easily extends to the case of relative CW-complexes as well. This theorem will be of theoretical importance to us. (see, e.g., in the construction of Leray–Serre spectral sequence in Chapter 13.)

Exercise 4.3.32

(i) Use cellular homology to compute $H_*(\mathbb{S}^p \times \mathbb{S}^q)$.

(ii) For a finite CW-complex X, show that $\chi(X) = \sum_i(-1)^i f_i(X)$, where $f_i(X)$ is the number of i-cells in X.

(iii) If X and Y are finite CW-complexes, show that $\chi(X \times Y) = \chi(X)\chi(Y)$.

(iv) Suppose X is obtained by attaching a n-cell to a space Y, $n \geq 2$, via an attaching map $\phi : \mathbb{S}^{n-1} \to Y$. Let $\alpha \in H_{n-1}(Y)$ denote the image of a generator of $H_{n-1}(\mathbb{S}^{n-1};\mathbb{Z}) \approx \mathbb{Z}$. Show that

$$H_q(X) \approx \begin{cases} H_q(Y), & q \neq n-1, n; \\ H_{n-1}(Y)/(\alpha), & q = n-1; \\ H_n(Y), & q = n \text{ and } \alpha \text{ is of infinite order.} \\ H_n(Y) \oplus \mathbb{Z}, & q = n \text{ and } \alpha \text{ is of finite order.} \end{cases}$$

Examine the case when $n = 1$. Write down the corresponding statement and prove it.

4.4   Some Applications of Homology

In this section, we shall give some of the popular applications of homology theory. Many of these results can be arrived at by different ad hoc methods. We begin with Brouwer's fixed point theorem, and then go on to improve this to get Lefschetz's fixed point theorem. We then prove the hairy ball theorem, Jordan–Brouwer's separation theorem and Brouwer's theorem on invariance of domain. The section will end with a brief discussion of embeddings of spheres in spheres and the construction of Alexander's horned sphere, which is an embedded 2-sphere in $\mathbb{S}^3$ such that one of the components of the complement is a 3-disc, whereas the other component is not simply connected.

Lemma 4.4.1 Let A be a retract of X. Then $H_*(A)$ is a retract of $H_*(X)$, i.e., $H_*(A)$ is a direct summand of $H_*(X)$.

Proof: Let $r : X \to A$ be a retraction. Then $r_* : H_*(X) \to H_*(A)$ is a retraction, by functoriality, viz., $r_* \circ \eta_* = (r \circ \eta)_* = (Id_A)_*$. Here $\eta : A \to X$ denotes the inclusion map.♠

Corollary 4.4.2 For any $n \geq 0, \mathbb{S}^n$ is not a retract of $\mathbb{D}^{n+1}$.

Proof: We have seen that, $H_n(\mathbb{S}^n)$ is 'non trivial' and hence cannot be a subgroup of the 'trivial' group $H_n(\mathbb{D}^{n+1})$.   ♠

Remark 4.4.3 Recall that we have proved the above result in a different way using simplicial approximation and the Sperner lemma. We have also seen that the above result is actually equivalent to Brouwer's fixed point theorem. (See Theorem 2.9.13.) The present proof is obviously shorter though it uses the big machine of homology. Our next application is a generalization of this result on any compact polyhedron. Of course, now we shall have some hypothesis on the map itself. Recall that, given a chain map $\tau : C \to C$ on a chain complex, which is finitely generated we define the Lefschetz number $L(\tau)$ to be the alternate sum of the traces.

Also recall that, if τ_* denotes the homomorphism induced on the homology groups, then $L(\tau) = L(\tau_*)$. For any continuous map $f : X \to X$ on a space X that has finitely generated homology groups, we now define $L(f) = L(f_*)$. The well-known Lefschetz fixed point theorem reads as follows:

Theorem 4.4.4 (Lefschetz fixed point theorem) *Let X be a compact polyhedron, $f : X \to X$ be a continuous map. If $L(f) \neq 0$, then f must have a fixed point.*

Remark 4.4.5 Notice that, since X is a compact polyhedron, its homology is finitely generated and so $L(f) = L(f_*)$ makes sense. Secondly, if $X = \mathbb{D}^n$ then f is homotopic to the identity map of $\mathbb{D}^n$. Since $L(f)$ is a homotopy invariant, it follows that $L(f) = L(Id)$. But $L(Id)$ is nothing but the Euler characteristic of the space $\mathbb{D}^n$. Therefore, $L(Id) = 1$ since $\mathbb{D}^{n+1}$ is contractible. Thus the requirement of the theorem is satisfied. So, in conclusion, we can say that f has a fixed point. This shows that Lefschetz fixed point theorem is a generalization of Brouwer's fixed point theorem. Indeed, we have just derived a stronger version of BFT, viz., any self map of a compact contractible space has a fixed point.

Proof: Given a compact polyhedron X and a continuous map, $f : X \to X$ such that $L(f) \neq 0$, we have to show that, f has a fixed point.

We shall assume that f has no fixed point and arrive at a contradiction. Fixing some linear metric on X, we can find $\epsilon > 0$ such that, $d(x, f(x)) > \epsilon$, $\forall\, x \in X$. (Why?) Let K be a simplicial complex that triangulates X and such that

$$mesh\, K = \max\{\text{diam}\, F\ :\ F \in K\} < \epsilon/3.$$

[For instance we can start with any simplicial complex that triangulates X and then take K to be the (sufficiently often) iterated barycentric subdivision of it to make its mesh as small as we please (Lemma 2.8.8).

Let $\phi : L = sd^k K \to K$ be a simplicial approximation to f.

Then $|\phi|$ is homotopic to f and hence the two maps have the same Lefschetz number.

Now, let ρ be any simplex of $L = sd^k K$, such that, $|\rho| \subset |\sigma|$ for some simplex σ of K. We claim that

$$(|\phi|(|\rho|)) \cap |\sigma| = \emptyset. \tag{4.18}$$

For otherwise, say $x \in |\rho|$, and $|\phi|(x) \in |\sigma|$. Then, both $x, |\phi|(x)$ belong to $|\sigma|$ and hence $d(x, |\phi|(x)) \leq \text{diam.}(|\sigma|) < \epsilon/3$. On the other hand, $d(|\phi|(x), f(x)) < \epsilon/3$ and hence $d(x, f(x)) < \epsilon$, which is a contradiction.

Our task has been complicated by the fact that, while obtaining the simplicial approximation ϕ, we had to subdivide only the domain, and so ϕ almost never has the same domain and range. This difficulty is overcome at least in two different ways:

Method I: We can use the CW-chain complex associated with K. Since ϕ is a simplicial map $L \to K$, it follows that $|\phi|$ is a cellular map of the CW-complex K. Let $\alpha_n : H_n(K^{(n)}, K^{(n-1)}) \to H_n(K^{(n)}, K^{(n-1)})$ be the homomorphism induced by $|\phi|$. If $\tau \in K$ is an oriented n-simplex, it follows from (4.18), that $|\phi|(|\tau|) \cap |\tau| = \emptyset$. Therefore, the coefficient of τ in $\alpha_n(\tau)$ will be zero. It follows that the trace of the matrix representing α_n is zero. Therefore $L(|\phi|) = 0$.

Method II Consider the subdivision chain map $Sd : C_.(K) \to C_.(sd\, K)$, as defined in Exercise 4.3.19.(iii). We have seen that Sd induces identity isomorphism on homology. By repeated application of this we know that $Sd^k = Sd \circ \cdots \circ Sd$ (k-copies) induces isomorphism on homology. Therefore, $L(f) = L(|\phi|) = L(|Sd^k| \circ |\phi|)$. We can compute the Lefschetz number of $|Sd^k| \circ |\phi| : |sd^k K| \to |sd^k K|$ at the chain level using the chain complex $C_.(sd^k K)$ with the basis elements from the simplexes of $sd^k K$. From (4.18) it follows that in the expression for $C(Sd^k \circ \phi) \circ Sd^k(\rho)$, the simplex ρ does not occur at all. Hence, the matrix of $C(Sd^k \circ \phi)$ on $C_n(sd^k K)$ has all its diagonal entries 0. Since this is true for all n, we conclude that, $L(Sd^k \circ \phi) = 0$.

In either method, this contradiction completes the proof of the theorem. ♠

Remark 4.4.6 We are now in a position to prove the famous hairy ball theorem. Loosely speaking, this theorem says that one cannot comb one's hair without parting at least at one point. The precise statement, however, is the following:

Theorem 4.4.7 (Hairy ball theorem) *There is no continuous map* $f : \mathbb{S}^{2n} \to \mathbb{S}^{2n}$ *such that, for each* $x \in \mathbb{S}^{2n}$, $f(x)$ *is orthogonal to* x.

Proof: We shall actually prove that for any continuous map $f : \mathbb{S}^{2n} \to \mathbb{S}^{2n}$ there is a point $x \in \mathbb{S}^{2n}$ such that $f(x) = x$ or $f(x) = -x$. Supposing on the contrary, it follows that the line joining x and $f(x)$ is well defined and does not pass through the origin. Therefore, the map

$$H(x,t) := \frac{(1-t)f(x) + tx}{\|(1-t)f(x) + tx\|}$$

is well defined and continuous. Clearly H is a homotopy from f to the identity map Id of $\mathbb{S}^{2n}$. Hence, f induces identity homomorphism on the homology. In particular, it follows that, $L(f) = \chi(\mathbb{S}^{2n}) = 2$. Hence, by LFT, it follows that, f must have a fixed point contradicting our assumption.   ♠

Remark 4.4.8 The above theorem has its origin in differential topology, wherein it is stated as follows: *There is no nowhere vanishing tangent vector field on* $\mathbb{S}^{2n}$. Observe that all odd dimensional spheres possess such vector fields. For example: $(x_1, x_2, \ldots, x_{2n}) \to (x_1, -x_2, \ldots, x_{2n-1}, x_{2n})$ is one such.

Let us now give another application of LFT. Recall that for $n \geq 1$, $H_n(\mathbb{S}^n) \approx \mathbb{Z}$. Therefore for any continuous map $f : \mathbb{S}^n \longrightarrow \mathbb{S}^n$ the homomorphism $f_* : H_n(\mathbb{S}^n) \longrightarrow H_n(\mathbb{S}^n)$ is given by multiplication by an integer. This integer is called the degree of f. (See Exercise 4.2.27.)

Also $H_0(\mathbb{S}^n) \approx \mathbb{Z}$ and $f_* : H_0(\mathbb{S}^n) \longrightarrow H_0(\mathbb{S}^n)$ is always the identity homomorphism. Since all other homology groups vanish it follows that

$$L(f) = 1 + (-1)^n \deg f.$$

As a special case, consider the antipodal map $\alpha : \mathbb{S}^n \longrightarrow \mathbb{S}^n$ given by $x \mapsto -x$. Since this has no fixed points, it follows that $L(\alpha) = 0$. Therefore we have,

Theorem 4.4.9 *The degree of the antipodal map on* $\mathbb{S}^n$ *is equal to* $(-1)^{n+1}$ *for all* $n \geq 1$.

Remark 4.4.10 More generally, if $f : \mathbb{S}^{2n} \longrightarrow \mathbb{S}^{2n}$ is a homeomorphism, then f_* is an isomorphism and hence $\deg f = \pm 1$ according as f preserves or reverses orientation. If in addition, f has no fixed points then $L(f) = 0$ and hence $\deg f = -1$. which means that f is orientation reversing. In particular, this implies that

Corollary 4.4.11 Let G be a group of odd order acting on $\mathbb{S}^{2n}$ through homeomorphisms. Then for each $g \in G$, there exists $v \in \mathbb{S}^{2n}$ such that $gv = v$.

We end this section with two more celebrated results. The 1-dimensional version of the first result is known as the Jordan curve theorem. The proof that Camille Jordan gave in 1905 was not accepted by many mathematicians.[1] Since then, several authors have given various proofs of of this result and the degree of accuray of these proofs also varies. The general result was proved by L. E. J. Brouwer in 1911, more or less as presented below.

---

[1]See [Gamelin–Greene, 1997] for an elegant proof.

Theorem 4.4.12 (Jordan–Brouwer separation theorem) *An $(n-1)$-sphere embedded in an n-sphere separates it into exactly two components. These two components have their common boundary as the embedded $(n-1)$-sphere.*

Theorem 4.4.13 (Brouwer's invariance of domain) *Let U and V be any two subspaces of $\mathbb{R}^n$ homeomorphic to each other. If U is open in $\mathbb{R}^n$ then so is V.*

We need the following lemmas:

Lemma 4.4.14 Let A be a subset of $\mathbb{S}^n$, homeomorphic to $\mathbb{I}^k$ for some $0 \leq k \leq n$. Then the reduced homology groups of $\mathbb{S}^n \setminus A$ all vanish.

Lemma 4.4.15 Let B be a subset of $\mathbb{S}^n$ homeomorphic to $\mathbb{S}^k$, for some $0 \leq k \leq n-1$. Then the reduced homology groups of $\mathbb{S}^n \setminus B$ are all zero except that, $\tilde{H}_{n-k-1}(\mathbb{S}^n \setminus B) = \mathbb{Z}$.

Granting these two lemmas for a while, let us proceed to prove the Theorems:

Proof of Theorem 4.4.12: Let B be a subset of $\mathbb{S}^n$ homeomorphic to $\mathbb{S}^{n-1}$. From the above lemma, we have $\tilde{H}_0(\mathbb{S}^n \setminus B) = \mathbb{Z}$ and this implies that, $\mathbb{S}^n \setminus B$ has precisely two components. Let us call them U and V. Note that, both U and V are open subsets of $\mathbb{S}^n$. By definition, the boundary set $\partial(U) := \bar{U} \setminus \text{int}(U) = \bar{U} \setminus U$. Since V is open and disjoint from U, no point of V is in the closure of U. Thus it follows that, $\partial(U) \subset B$. By symmetry, it remains to show that, every point b of B is a closure point of U. This is the same as showing that, any arbitrary small neighbourhood N of b intersects $\bar{U}$. Inside N, we can find a neighborhood N_1 of b in B such that $B = N_1 \cup N_2$ and both N_i are homeomorphic to $\mathbb{I}^{n-1}$. Hence by Lemma 4.4.14, it follows that $\mathbb{S}^n \setminus N_2$ is path connected. We choose points $p \in U$ and $q \in V$ and join them by a path ω lying in $\mathbb{S}^n \setminus N_2$. Since U and V are different components of $\mathbb{S}^n \setminus B$, it follows that, ω intersects N_1. If we now follow the path ω from $p \in U$ until we hit N_1, the point we get in N_1 is definitely in the closure of U, hence $N \cap \bar{U} \neq \emptyset$. Hence $b \in \bar{U}$. This proves that, $B \subset \partial(U)$. ♠

Proof of Theorem 4.4.13: By taking one-point compactifications we can assume that both U and V are subspaces of $\mathbb{S}^n$. Let $\phi : U \to V$ be a homeomorphism and let U be open in $\mathbb{S}^n$. Let $x \in U$ and $y = \phi(x)$. We should produce an open subset of $\mathbb{S}^n$ containing y and contained in V. Let A be a neighborhood of x in $\mathbb{S}^n$ homeomorphic to a disc and contained in U. Then its boundary $\partial(A)$ is homeomorphic to $\mathbb{S}^{n-1}$, and so is $B = \phi(\partial(A))$. Hence by Theorem 4.4.12, $\mathbb{S}^n \setminus B$ has precisely two connected components. Clearly, $A \setminus \partial A$ is connected and hence $\phi(A \setminus \partial A) = \phi(A) \setminus B$ is connected. By Lemma 4.4.14, $\mathbb{S}^n \setminus \phi(A)$ is connected. On the other hand, $\mathbb{S}^n \setminus B = (\mathbb{S}^n \setminus \phi(A)) \amalg (\phi(A) \setminus B)$. Hence, these two sets must be the components of $\mathbb{S}^n \setminus B$. In particular, it follows that, $\phi(A) \setminus B$ is an open subset of $\mathbb{S}^n$. Also it contains the point y and is a subset of V. Thus we have succeeded in producing an open neighbourhood of x contained in V. This proves that V is open. ♠

It remains to prove the two lemmas. Since the proof of Lemma 4.4.15 is easier, we shall present that one first, though we need to use Lemma 4.4.14 for it.

Proof of Lemma 4.4.15: We shall use Mayer–Vietoris sequence and induction on k. For $k = 0$, $\mathbb{S}^0$ has two points and we know that $\mathbb{S}^n \setminus B$ is homotopy equivalent to $\mathbb{S}^{n-1}$ and we have already computed the homology groups of $\mathbb{S}^{n-1}$ (see (4.10). So, the statement is true for $k = 0$.

Assume the statement of the lemma for $k - 1$. Write $B = A_1 \cup A_2$ so that both A_i are homeomorphic to $\mathbb{I}^k$ and $A_1 \cap A_2$ is homeomorphic $\mathbb{S}^{k-1}$. Both $\mathbb{S}^n \setminus A_i$ are open in $\mathbb{S}^n$. Hence, we can apply the Mayer–Vietoris sequence:

$$\ldots \tilde{H}_{q+1}(\mathbb{S}^n \setminus A_1) \oplus \tilde{H}_{q+1}(\mathbb{S}^n \setminus A_2) \to \tilde{H}_{q+1}(\mathbb{S}^n \setminus (A_1 \cap A_2)) \to$$

$$\tilde{H}_q(\mathbb{S}^n \setminus B) \to \tilde{H}_q(\mathbb{S}^n \setminus A_1) \oplus \tilde{H}_q(\mathbb{S}^n \setminus A_2) \to \dots$$

Now by Lemma 4.4.14, we have $\tilde{H}(\mathbb{S}^n \setminus A_i) = 0$. Thus the middle arrow is an isomorphism. Now by the inductive assumption the result follows.   ♠

Proof of Lemma 4.4.14: Here also, we use induction on k. If $k = 0$, then A is a single point and we know that, $\mathbb{S}^n \setminus A$ is contractible and hence, its reduced homology groups vanish. Assume the result for $k - 1$. Also suppose $\tilde{H}_q(\mathbb{S}^n \setminus A) \neq 0$ for some q and let c be a cycle representing a non zero element $[c] \in \tilde{H}_q(\mathbb{S}^n \setminus A)$. Let $h : \mathbb{I}^k \to A$ be a homeomorphism. Write $A = A_1 \cup A_2$, where $A_1 = h(\mathbb{I}^{k-1} \times [0, 1/2])$ and $A_2 = (\mathbb{I}^{k-1} \times [1/2, 1])$. Apply Mayer–Vietoris sequence to the open sets $\mathbb{S}^n \setminus A_1$, $\mathbb{S}^n \setminus A_2$, to obtain the exact sequence:

$$\tilde{H}_{q+1}(\mathbb{S}^n \setminus A_1 \cap A_2) \to \tilde{H}_q(\mathbb{S}^n \setminus A) \to \tilde{H}_q(\mathbb{S}^n \setminus A_1) \oplus \tilde{H}_q(\mathbb{S}^n \setminus A_2) \to H_q(\mathbb{S}^n \setminus A_1 \cap A_2).$$

Since $A_1 \cap A_2$ is homeomorphic to $\mathbb{I}^{k-1}$, the two end-groups in the above sequence are zero by induction hypothesis. Therefore the middle arrow represents an isomorphism. Recall that if $\eta_i : \mathbb{S}^n \setminus A \longrightarrow \mathbb{S}^n \setminus A_i$ are the inclusion maps, then the middle arrow is nothing but $((\eta_1)_*, (\eta_2)_*)$. It follows that, $(\eta_i)_*[c]$ is non zero at least for one of $i = 1, 2$, say $i = 1$.

We now iterate this construction on A_1 in place of A and so on, to obtain a sequence of closed sets:

$$A \supset A_1 \supset \dots \supset A_r \supset A_{r+1} \dots$$

such that $\cap_r A_r := T$ is homeomorphic to $\mathbb{I}^{k-1}$ and the image of $[c]$ in $\tilde{H}_q(\mathbb{S}^n \setminus A_r)$ is non zero. But by induction hypothesis $\tilde{H}_q(\mathbb{S}^n \setminus T) = 0$. Thus there exists a $(q+1)$-chain τ in $\mathbb{S}^n \setminus T$ such that, $\partial(\tau) = c$. Now every chain is supported on a compact set and every compact subset of $\mathbb{S}^n \setminus T$ is contained in $\mathbb{S}^n \setminus A_r$, for some r. This means that the image of $[c]$ is zero in $\tilde{H}_q(\mathbb{S}^n \setminus A_r)$, which is a contradiction.   ♠

Example 4.4.16 Alexander's horned sphere

Let us consider embeddings $f : \mathbb{S}^k \hookrightarrow \mathbb{S}^n$ for any $0 \leq k \leq n$. Two such embeddings f, g are said to be equivalent if there is a map $H : \mathbb{S}^k \times \mathbb{I} \to \mathbb{S}^n$ such that each $H(-, t)$ is an embedding and $H(-, 0) = f, H(-, 1) = g$. Such a map H is called an isotopy. Clearly this defines an equivalence relation on the set of all embeddings $\mathbb{S}^k \hookrightarrow \mathbb{S}^n$. The two extreme cases, viz., $k = 0, n$ are the easiest to handle: for $k = 0$, there is only one equivalence class and for $k = n$, there are two! Lemma 4.4.15 tells us that the complement of these embedded spheres are homologically trivial except in dimension $n - k - 1$ and so will not help us in detecting different isotopy classes. A deep result in PL-topology tells us that (with mild restrictions on the embeddings) if $n - k \geq 3$ then any two embeddings are isotopic. Therefore, the interesting cases left out are $n - k = 2, 1$. The case $k = n - 2$, viz., the study of codimension 2 embeddings goes under the name *knot theory* which is a fully grown branch of algebraic topology with applications in several fields.

We now come to the case $k = n - 1$. For $n = 2$, there is the so called Jordan–Schoenflies theorem (see Theorem 5.3.2) which says that any two embeddings of a circle in the plane are equivalent. With additional mild restrictions, the same holds for all n. Here we shall discuss the classical example due to J. W. Alexander [Alexander, 1924] of a 'wild' embedding (which does not satisfy the so-called mild restriction) of $\mathbb{S}^2$ in $\mathbb{S}^3$ such that the complement consists of two components one of which is a 3-disc and the other is not simply connected. This example also serves to illustrate the fact that you cannot apply the van Kampen theorem here, for the intersection of the closure of the two components is a 'wild' 2-sphere.

By deleting the point at infinity, we can carry out all our constructions inside $\mathbb{R}^3$. We shall define a sequence of closed subsets

$$F_1 \supset F_2 \supset \dots$$

inductively and take $B = \cap_k F_k$. We begin with F_1 as the image of a standard embedding of $\mathbb{D}^2 \times \mathbb{S}^1 \hookrightarrow \mathbb{R}^3$, say for example, a tubular neighbourhood of $\mathbb{S}^1 \times 0 \subset \mathbb{R}^3$. The core of F_1 is a circle of length 2π. Look at Figure 4.4 (A) which represents two embeddings of $\mathbb{D}^2 \times \mathbb{I}$ inside another $\mathbb{D}^2 \times \mathbb{I}$ in a particular way.

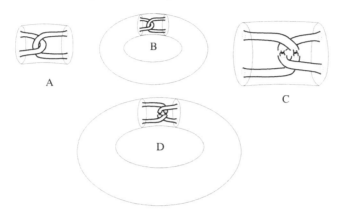

FIGURE 4.4. Alexander's horned sphere

Cut out a copy of $\mathbb{D}^2 \times \mathbb{I}$ of core length $\pi/5$, say, and replace it with a copy of A of the same core length as shown in Figure 4.4 (B). Call this F_2. The first stage construction is over. Now inside F_2, we repeat the above process on both the handles H_1, H_2 in the copy of A, viz., cut out a copy of $\mathbb{D}^2 \times \mathbb{I}$ of core length $\pi/5^2$ and fit a copy of A in its place to obtain F_3. The II stage construction is over. Repeat this process to get the sequence $\{F_k\}$ and take $B = \cap F_k$. This is the so-called Alexander's horned disc and its boundary is the horned sphere.

In order to prove that B is homeomorphic to $\mathbb{D}^3$, we reconstruct B as an increasing union $B_1 \subset B_2 \subset \cdots$ where each B_i is homeomorphic to $\mathbb{D}^3$. Take B_1 to be the closure of F_1 minus the first copy of A. Clearly B_1 is homeomorphic to $\mathbb{D}^3$. Take B_2 to be the closure of the complement in F_2 of the two copies of A that are removed from F_2. Note that there is a homeomorphism $h_1 : B_2 \to B_1$ which is identity restricted to a smaller disc $B_1' \subset B_1$. Repeat this process to see that there are discs $B_k' \subset B_k$ and homeomorphisms $h_k : B_{k+1} \to B_k$ such that $h_k|_{B_k'} = id$ and $\cup_k B_k' = B$. Now define $h = h_1 \circ h_2 \circ h_3 \circ \cdots$ and check that h is a continuous bijection and hence a homeomorphism.

Clearly the boundary of B, viz., $h^{-1}(\partial B_1)$ is an embedded sphere in $\mathbb{R}^3$. We shall now show that the complement $W = \mathbb{R}^3 \backslash B$ is not simply connected. However, as observed earlier, we know that $\tilde{H}_i(W) = (0), i \geq 0$. In the next section, we shall see a general result which implies that $H_1(W)$ is actually the abelianization of $\pi_1(W)$. Therefore, as a consequence we shall know that $\pi_1(W)$ is equal to its own commutator. Indeed, this is precisely what we shall directly prove.

Put $G_k = \mathbb{R}^3 \setminus F_k$. Then $\cdots \subset G_k \subset G_{k+1} \subset \cdots$ and $W = \cup_k G_k$. Note that $\pi_1(G)$ is generated by the loop γ represented by $\partial \mathbb{D}^2 \times \{p\}$ contained in a copy of A and which goes around the core of F_1 exactly once. G_2 is the union of G_1 and a copy of $A \setminus H_1 \cup H_2$. The intersection is $\partial \mathbb{D}^2 \times \mathbb{I}$. Note that $\pi_1(A \setminus H_1 \cup H_2)$ is a free group on two generators, say α_1, β_1 represented by loops that go around each of these handles H_1 and H_2. Use Exercise 3.9.13 to see that γ is equal to $[\alpha_1, \beta_1^{-1}]$. Therefore, by Van Kampen's theorem, $\pi_1(G_2)$ is the quotient of a free group on three generators $\{\alpha_1, \beta_1, \gamma\}$ by the relation $\gamma = [\alpha_1, \beta_1^{-1}]$. This is clearly isomorphic to the free group generated by α_1, β_1. Repeating this argument, it follows that $\pi_1(G_3)$ is a free group on four new generators, the two generators coming from $\pi_1(G_2)$ being identified with some commutators in these new generators. Thus each $\pi_i(G_k)$ is a free group on 2^{k-1} generators and there are monomorphisms $f_k : \pi_1(G_k) \to \pi_1(G_{k+1})$ with their image contained in the commutator subgroup. Now $\pi_1(W)$ is the direct limit of this system which clearly contains a copy of a free group on any finite number of generators and which has the property that every element in it is a product of commutators. In particular, the abelianization of $\pi_1(W)$ is trivial.

Exercise 4.4.17 Let X be a compact polyhedron.

   (i) If $f : X \longrightarrow X$ is null homotopic then show that f has a fixed point.

  (ii) Suppose X has a topological group structure, such that the path connected component of e is not $\{e\}$. Then show that $\chi(X) = 0$.

 (iii) Show that $\mathbb{S}^{2n}$ cannot have any topological group structure.

  (iv) Let G be a path connected topological group which acts continuously on X. Suppose $\chi(X) \neq 0$. Show that for each $g \in G$, there exists $x \in X$ such that $gx = x$.

4.5   Relation between π_1 and H_1

We shall assume that our topological space X is path connected. Let $x_0 \in X$ be the base point. Given an element of $\pi_1(X, x_0)$ represented by a loop $\omega : \mathbb{I} \to X$, we can view ω as a 1-cycle in X, by thinking of $\mathbb{I}$ as the geometric realization of the standard 1-simplex Δ_1. This 1-cycle then represents an element (ω) in $H_1(X)$. The aim of this section is to see that the simple assignment

$$[\omega] \mapsto (\omega)$$

called *Hurewicz homomorphism* can be used to give a precise relation between $\pi_1(X, x_0)$ and $H_1(X)$.

Lemma 4.5.1 Let $\omega_i : \mathbb{I} \longrightarrow X$ be any two paths so that $\omega_1 * \omega_2$ is defined. Then as a singular 1-chain, $\omega_1 * \omega_2 - \omega_1 - \omega_2$ is null-homologous.

Proof: The proof is obvious from Figure 4.5 which defines a singular 2-simplex $\sigma : \Delta_2 \longrightarrow X$ such that

$$\partial \sigma = \omega_1 * \omega_2 - \omega_1 - \omega_2.$$

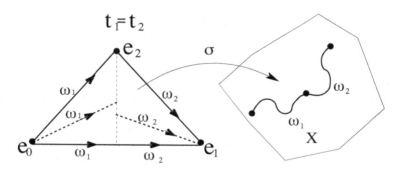

FIGURE 4.5. Chain equivalence under subdivision

Indeed, with the convention that $\sum_i t_i = 1, 0 \leq t_0, t_1, t_2 \leq 1$, we have:

$$\sigma(t_0 \mathbf{e}_0 + t_1 \mathbf{e}_1 + t_2 \mathbf{e}_2) = \begin{cases} \omega_1 \left(\frac{2 - 2t_0}{1 + t_2} \right), & t_0 \geq t_1; \\ \omega_2 \left(\frac{2t_1 + t_2 - 1}{1 + t_2} \right), & t_0 \leq t_1. \end{cases}$$

The lemma follows. ♠

Theorem 4.5.2 *The assignment $[\omega] \mapsto (\omega)$ defines a functorial surjection*

$$\varphi : \pi_1(X, x_0) \to H_1(X); \quad \varphi[\omega] = (\omega).$$

whose kernel is precisely the commutator subgroup of $\pi_1(X, x_0)$.

(Thus φ defines $H_1(X)$ as the abelianization of $\pi_1(X, x_0)$.)

Proof: The first thing to do is to show that φ is well-defined. (See Figure 4.6.) Let $H : \mathbb{I} \times \mathbb{I} \to X$ be a path homotopy of ω_1 with ω_2. Consider the following triangulation of $\mathbb{I} \times \mathbb{I}$. The vertex set of the simplicial complex K consists of points, $v_0 = (0,0), v_1 = (1,0), v_2 = (1,1)$ and $v_3 = (0,1)$.

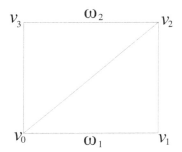

FIGURE 4.6. Hurewicz map is well-defined

The edges are $\{v_0, v_1\}, \{v_1, v_2\}, \{v_0, v_2\}, \{v_0, v_3\}$ and $\{v_2, v_3\}$. The 2-simplexes in $|K|$ are $\{v_0, v_1, v_2\}$ and $\{v_0, v_2, v_3\}$. Let $\alpha, \beta : \Delta_2 \longrightarrow K$ be defined by $\alpha = [v_0, v_1, v_2], \beta = [v_0, v_2, v_3]$. Consider the 2-chain in K given by $\tau = H \circ \alpha + H \circ \beta$. We claim that $\partial(\tau) = \omega_2 - \omega_1$ and that will show that $(\omega_2) = (\omega_1)$ in $H_1(X)$.

First observe that $H/0 \times \mathbb{I}$ and $H/1 \times \mathbb{I}$ are constant maps at x_0. Thus $H \circ [v_0, v_3] = H \circ [v_1, v_2]$. Now

$$\partial(\tau) = H \circ ([v_1, v_2] - [v_0, v_2] + [v_0, v_1]) + H \circ ([v_2, v_3] - [v_0, v_3] + [v_0, v_2])$$

which is easily seen to be equal to

$$H \circ [v_0, v_1] + H \circ [v_2, v_3] = \omega_1 - \omega_2.$$

Therefore φ is well-defined.

Now the Lemma 4.5.1 immediately yields that φ is a group homomorphism.

The functoriality of φ is left to the student as a straightforward exercise.

We shall denote $\pi_1(X, x_0)$ simply by π_1 and $H_1(X)$ by H_1, for the rest of the proof.

Let $\psi : \pi_1 \longrightarrow \pi_1^{ab}$ denote the canonical quotient homomorphism of π_1 onto the abelianization of π_1. Since H_1 is an abelian group, it follows that, $\varphi : \pi_1 \longrightarrow H_1$ factors through the abelianization to define a homomorphism

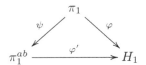

such that $\varphi' \circ \psi = \varphi$. It remains to show that, φ' is an isomorphism. We shall construct an explicit inverse to φ'.

For each $x \in X$, choose an arbitrary path ω_x from x_0 to x in X except that, ω_{x_0} should be chosen to be the constant loop at x_0. To each singular 1-simplex σ in X, define $\theta(\sigma)$ to be the element in π_1^{ab} represented by the loop $\omega_a * \sigma * (\omega_b)^{-1}$, where $a = \sigma(0)$ and $b = \sigma(1)$. Then θ extends to a homomorphism $\theta : S_1(X) \longrightarrow \pi_1^{ab}$. We will show that $\theta \circ \partial(S_2(X)) = (0)$. Then it would follow that θ defines a homomorphism, denoted by $\tilde{\theta} : H_1(X) \to \pi_1^{ab}$. This homomorphism is then easily seen to be the inverse of φ'.

For any singular 2-simplex $\gamma : \Delta_2 \longrightarrow X$, we shall prove that $\theta(\partial(\gamma)) = 0$. Let $\gamma(e_0) = a$, $\gamma(e_1) = b$, and $\gamma(e_2) = c$. Let $F^i : \Delta_1 \longrightarrow \Delta_2$ be the face maps $F^i : \mathbb{R}^\infty \to \mathbb{R}^\infty$ as given in 4.2.2. Then

$$\theta(\partial(\gamma)) = \psi(\omega_a * (\gamma \circ F^2) * (\gamma \circ F^0) * (\gamma \circ F^1) * (\omega_a)^{-1})$$

Since $(\gamma \circ F^2) * (\gamma \circ F^0) * (\gamma \circ F^1)$ is a null homotopic loop, it follows that the term on the RHS above is zero. Thus $\tilde{\theta}$ is well defined.

Given $[\tau] \in \pi_1(X, x_0)$, $\varphi([\tau])$ is represented by the 1-cycle τ itself. Therefore, $\tilde{\theta}(\varphi[\tau]) = \psi([\tau])$ in $\pi_1^{ab}(X)$.

On the other hand, given any element $\lambda \in H_1$ represented by a cycle $\sigma = \sum_i \lambda_i$, Lemma 4.5.1 shows that $\varphi(\sum_i \theta(\lambda_i))$ represents the same element λ, since all the extra edges introduced along paths $\omega'_x s$ cancel out in pairs. This means that $\varphi' \circ \tilde{\theta}(\lambda) = \lambda$ and so, $\tilde{\theta} = (\varphi')^{-1}$. ♠

Example 4.5.3 As a simple application, we can now compute the CW-chain complex of $\mathbb{S}^1 \times \mathbb{S}^1$. Recall (see Example 2.3.3) that treating $\mathbb{S}^1$ as a CW-complex with a single 0-cell and a single 1-cell, the product CW-complex will then have a single 0-cell and two 1-cells and a single 2-cell. It follows that $C_0 = \mathbb{Z}, C_1 = \mathbb{Z} \oplus \mathbb{Z}, C_2 = \mathbb{Z}$ and $C_q = 0$ for $q > 2$. It is also clear that $\partial_1 : C_1 \to C_0$ is the zero map. Now $\partial_2 : C_2 \to C_1$ is given by the degree of two maps obtained by composing the attaching map $\mathbb{S}^1 = \partial(\mathbb{I} \times \mathbb{I}) \to \mathbb{S}^1 \vee \mathbb{S}^1$ with the two projections $\mathbb{S}^1 \vee \mathbb{S}^1 \to \mathbb{S}^1$. We have seen that the attaching map which is actually the boundary of the product of the two characteristic maps of the 1-cells, represents the element $xyx^{-1}y^{-1}$ in $\pi_1(\mathbb{S}^1 \vee \mathbb{S}^1)$. Passing onto the homology via the Hurewicz map this represents the trivial element in $H_1(\mathbb{S}^1 \vee \mathbb{S}^1)$. It follows that the two degrees are zero and hence $\partial_2 = 0$.

4.6   All Postponed Proofs

Homotopy Invariance Theorem for $S.$ and S^{sm}

Here we shall present a proof of Theorem 4.2.9. [Those of you who have studied Poincaré's lemma in differential topology may notice some similarity in the proof there (see [Shastri, 2011] p. 116) and the proof of this theorem given below.]

We shall first concentrate on singular homology. As seen before, we need to construct the prism operators $h : S_q(X, A) \to S_{q+1}((X, A) \times \mathbb{I})$ (4.2.10), which define a chain homotopy between $(\eta_0)_*$ and $(\eta_1)_*$ (at the chain group level). This h will be functorial, in the sense that if $\alpha : (X, A) \to (Y, B)$ is any map, then we have a commutative diagram

$$
\begin{array}{ccc}
S.(X, A) & \overset{h}{\to} & S.((X, A) \times \mathbb{I}) \\
\alpha_* \downarrow & & \downarrow (\alpha \times id)_* \\
S.(Y, B) & \overset{h}{\to} & S.((Y, B) \times \mathbb{I}).
\end{array}
$$

But then, if σ is a singular n-simplex, we would have

$$h(\sigma) = h \circ \sigma_*(\xi_n) = (\sigma \times id)_* h(\xi_n)$$

where $\xi_n : |\Delta_n| \to |\Delta_n|$ is the identity map. Thus it suffices to define $h(\xi_n)$ for each n. We demand that h satisfies the equation

$$(\eta_1)_*(\sigma) - (\eta_0)_*(\sigma) = h \circ \partial(\sigma) + \partial \circ h(\sigma) \tag{4.19}$$

for all singular n simplexes σ. By functoriality, this will be the case, if we manage to have,

$$(\eta_1)_*(\xi_n) - (\eta_0)_*(\xi_n) = h \circ \partial(\xi_n) + \partial \circ h(\xi_n). \tag{4.20}$$

For we can simply apply $(\sigma \times Id)_*$ to both sides of (4.20) to get (4.19).

We now need to introduce some notation. In $\Delta_n \times \mathbb{I}$, let

$$b^i = (\mathbf{e}_i, 0), \quad c^i = (\mathbf{e}_i, 1) \text{ and}$$
$$\gamma^n = (\beta(\Delta_n), 1/2) \quad (\text{where } \beta(\Delta_n) \text{ is the barycentre of } \Delta_n).$$

(See Example 2.7.12 and Figure 2.19.) For any simplicial complex K, if $\sigma : \Delta_n \to K$ is a simplicial map with $\sigma(\mathbf{e}_i) = v_i$, then we shall denote σ by $[v_0, ..., v_n]$. If $x \in |s|$, where $|s| \supset \sigma(\Delta_n)$, then $x\sigma := x[v_0, ..., v_n]$ denotes the linear singular $(n+1)$-simplex defined by

$$\left. \begin{array}{ll} (x\sigma)(\mathbf{e}_0) = x & \text{and} \\ (x\sigma)(\mathbf{e}_i) = v_{i-1}, & i \geq 1 \end{array} \right\} \tag{4.21}$$

and extended linearly. If $\rho = \sum_j n_j \sigma_j$ is a singular n-chain with *supp* $\rho \subset |s|$ then for any $x \in |s|$, it makes sense to talk of $x\rho = \sum_j n_j x \sigma_j$, which is a singular $(n+1)$-chain.

Now the definition of h is completed by induction on n: Define

$$h(\xi_0) = \gamma^0([c^0] - [b^0]).$$

Then h is defined on $S_0(X, A)$ for all (X, A). Note that if $h(\xi_r)$ is defined for all $r \leq n-1$, then h is defined on $S_r(X, A)$ for all $r \leq n-1$ and for all (X, A). So inductively define

$$h(\xi_n) = \gamma^n([c^0, ..., c^n] - [b^0, ..., b^n] - h \circ \partial(\xi_n)).$$

By induction again we have,

$$\begin{aligned} \partial \circ h(\partial \xi_n) &= (\eta_1)_*(\partial \xi_n) - (\eta_0)_*(\partial \xi_n) - h \circ \partial^2 \xi_n \\ &= \partial((\eta_1)_*(\xi_1) - (\eta_0)_*(\xi_n)) \\ &= \partial([c^0, ..., c^n] - [b^0, ..., b^n]). \end{aligned}$$

Hence

$$\begin{aligned} \partial \circ h(\xi_n) &= [c^0, ..., c^n] - [b^0, ..., b^n] - h \circ \partial(\xi_n) \\ &\quad - \gamma^n(\partial[c^0, ..., c^n], -\partial[b^0, ..., b^n] - \partial \circ h \circ \partial(\xi_n)) \\ &= (\eta_1)_*(\xi_n) - (\eta_0)_*(\xi_n) - h \circ \partial(\xi_n) \end{aligned}$$

as required.

Now consider the case of S^{sm} on the smooth category. If we replace the word 'map' by 'smooth map' everywhere in the above proof, the entire thing goes through without any trouble, until we come to the notation (4.21). However, we need this only for the case when K is the 'prism' simplicial complex underlying $\Delta_n \times \mathbb{I}$ as constructed in Example 2.7.12. All simplexes here will be automatically smooth and the inductive formula for h takes values inside smooth chains. This gives the proof of the homotopy invariance of smooth singular homology, H^{sm}. ♠

Excision Theorems

Here we shall present proofs of Theorems 4.2.17 and 4.3.2.
We begin with a definition:

Definition 4.6.1 Subdivision chain map We define, respectively, a functorial chain map Sd and a chain homotopy D of Sd with the identity map

$$Sd : S_.(X) \to S_.(X); \quad D : S_.(X) \to S_.(X)$$

of Sd inductively as follows:
(1) If τ is a 0-chain, define $Sd(\tau) = \tau$ and $D(\tau) = 0$.
(2) Having defined Sd and D on $(n-1)$-chains, we shall now define them on n-chains. First, let ξ_n be the identity singular n-simplex on $|\Delta_n|$. Take

$$Sd(\xi_n) = \beta(\Delta_n)(Sd(\partial(\xi_n))); \quad D(\xi_n) = \beta(\Delta_n)Sd(\partial(\xi_n)) - \xi_n - D\partial(\xi_n)).$$

Now for any singular n-simplex σ, take

$$Sd(\sigma) = \sigma_*(Sd(\xi_n)); \quad D(\sigma) = \sigma_*(D(\xi_n)).$$

(Here σ_* denotes the homomorphisms induced by σ on the chain complexes.) Then extend them linearly over all the n-chains. Notice that, once we define these maps on the 'universal' singular simplexes ξ_n, then the rest of the definition is forced on us by the functoriality. (Just like in the proof of homotopy invariance, this is a typical example of what one generally does in such situations and worth noting down.)

To see that, Sd is a chain map, we prove $\partial \circ Sd(\tau) = Sd(\partial(\tau))$ by induction on the (homogeneous) degree n of the chain τ. If $n = 0$, there is nothing to prove. Having proved this for $n - 1$, we note that it is enough to consider the case $\tau = \xi_n$. Then we have.

$$
\begin{aligned}
\partial \circ Sd(\xi_n) &= Sd(\partial(\xi_n)) - \beta(\Delta_n)(\partial(Sd(\partial(\xi_n)))) \\
&= Sd(\partial(\xi_n)) - \beta(\Delta_n)(Sd \circ \partial(\partial(\xi_n))) \\
&= Sd(\partial(\xi_n)),
\end{aligned}
$$

since $\partial \circ \partial = 0$. Now to show that $\partial \circ D + D \circ \partial = Sd - Id$, we again induct on the degree of τ. At degree 0, there is no statement at all. At degree 1, this follows from the fact that $D_0 = 0$. Suppose we have proved the statement for $(n - 2)$-chains. As a consequence we have,

$$
\begin{aligned}
\partial \circ D(\partial(\xi_n)) &= (\partial \circ D + D \circ \partial)(\partial(\xi_n)) \\
&= (Sd - Id)(\partial(\xi_n)) \\
&= \partial(Sd(\xi_n)) - \partial(\xi_n) \\
&= \partial(Sd(\xi_n) - \xi_n)
\end{aligned}
$$

Thus, $\partial(Sd(\xi_n) - \xi_n - D(\partial(\xi_n))) = 0$. Now we have,

$$
\begin{aligned}
(D \circ \partial + \partial \circ D)(\xi_n) &= D \circ \partial(\xi_n) + \partial(\beta(\Delta_n)(Sd(\xi_n) - \xi_n - D \circ \partial(\xi_n))) \\
&= D \circ \partial(\xi_n) + Sd(\xi_n) - \xi_n - D \circ \partial(\xi_n) \\
&\quad -\beta(\Delta_n)(\partial(Sd(\xi_n) - \xi_n - D(\partial(\xi_n)))) \\
&= Sd(\xi_n) - \xi_n,
\end{aligned}
$$

as required.
For the sake of future use, we make two observations which we state as theorems.

Theorem 4.6.2 *The subdivision chain map and the chain homotopy D have the property that*
(a) *for any smooth object M, $Sd(S^{sm}_.(M) \subset S^{sm}_.(M)$, $D(S^{sm}_.(M)) \subset S^{sm}_.(M)$ and define a chain map and chain homotopy of the subcomplex $S^{sm}_.$ and*
(b) *for any simplicial complex K,*

$$Sd(\mathbf{S}_.(K)) \subset \mathbf{S}_.(sd\,K), D(\mathbf{S}_.(K)) \subset \mathbf{S}_.(sd\,K)$$

and respectively define a chain map and a chain homotopy at the subcomplex levels also. Moreover,
(c) *this further induces chain map and chain homotopy at the quotient complexes*

$$Sd : C(K) \longrightarrow C(sd\,K); \ D : C(K) \longrightarrow C(sd\,K).$$

$$
\begin{array}{ccc}
S_.(|K|) & \xrightarrow{Sd} & S_.(|K|) = S_.(|sd\,K|) \\
\uparrow & & \uparrow \\
\mathbf{S}_.(K) & \xrightarrow{Sd} & \mathbf{S}_.(sd\,K) \\
\downarrow & & \downarrow \\
C(K) & \xrightarrow{Sd} & C(sd\,K)
\end{array}
$$

Theorem 4.6.3 *If $\tau : sd\,K \longrightarrow K$ is a simplicial approximation to identity map then on the chain complex $C(K)$ we have $\tau_* \circ Sd = Id$.*

The proofs of the above theorems are straightforward.

Coming to the proof of excision for the singular homology:

Let $\phi : H(S_.(X_1) + S_.(X_2)) \to H(X)$ be the map induced by the inclusion. Now given a singular n-simplex σ in X, we get two open sets $(\sigma)^{-1}(\mathrm{int}\,X_1)$ and $(\sigma)^{-1}(\mathrm{int}\,X_2)$ which cover $|\Delta_n|$. Clearly, there exists k such that $sd^k \Delta_n$ is finer than this covering. If $Sd^k = Sd \circ Sd \ldots \circ Sd$ (k-copies), then it follows that the singular n-chain $Sd^k(\sigma)$ belongs to the subgroup $S_.(X_1) + S_.(X_2)$. Thus if τ is a singular n-chain, then there exists a k, for which $Sd^k(\tau) \in S_.(X_1) + S_.(X_2)$. Note that, $Sd^k(\tau) \in S_.(X_1) + S_.(X_2)$ implies its boundary is also there. Further, note that both Sd and D map the chain subgroup $S_.(X_1) + S_.(X_2)$ to itself and so Sd induces identity homomorphism on the homology of $S_.(X_1) + S_.(X_2)$. In particular, for any cycle τ in $S_.(X_1) + S_.(X_2)$, $Sd^k(\tau)$ and τ represent the same element in $H(S_.(X_1) + S_.(X_2))$. Now, given a n-cycle τ, for some k, we have $Sd^k(\tau) \in S_.(X_1) + S_.(X_2)$ and hence the surjectivity of ϕ follows. To prove the injectivity, let τ_1 and τ_2 be any two n-cycles in $S_.(X_1) + S_.(X_2)$ and let α be a $(n+1)$-chain in $S_.(X)$, such that $\partial(\alpha) = \tau_1 - \tau_2$. Then for large k, $Sd^k(\alpha) \in S_.(X_1) + S_.(X_2)$ and we have $\partial Sd^k(\alpha) = Sd^k(\tau_1) - Sd^k(\tau_2)$. On the other hand, from our earlier observation, $Sd^k(\tau)$ and τ represent the same element in $H(S_.(X_1) + S_.(X_2))$. This proves the injectivity of ϕ, thereby completing the proof of the excision Theorem 4.2.17.

Now, in view of Theorem 4.6.2(a) the above proof works verbatim for the smooth singular chain complex as well, thereby proving Theorem 4.3.2. ♠

Proof of Singular-Simplicial versus Singular Theorem

We shall directly prove that induced map $\iota_* : H(\mathbf{S}_.(K)) \to H(|K|)$ is an isomorphism in homology. Since both chain complexes are free, it would then follow from general algebraic considerations that ι is a chain equivalence. (See Theorem 6.1.6.)

We first note that it is enough to prove this when K is finite. For arbitrary K, we can then take direct limit over finite subcomplexes of K.

Let K be a finite complex. We induct on the number r of simplexes in K. If $r = 1$ then the statement is obviously true. Assume the result for all simplicial complexes with number of simplexes $< r$ and $r > 1$. Let s be a simplex in K of maximal dimension. Let L be the subcomplex of K consisting of all simplexes in K other than s. Then by induction hypothesis $\iota_* : H_.(\mathcal{S}_.(L)) \to H_.(S_.(|L|))$ is an isomorphism. which restricts to an isomorphism $\iota_* : H_.(\mathcal{S}(\dot{s})) \to H_.(|\dot{s}|)$. Here $\dot{s}$ denotes the boundary simplicial complex of the simplex s. We also know that both $H_.(\mathcal{S}_.(s))$ and $H_.(S_.(|s|))$ vanish in positive dimensions (see Examples 4.3.7) and hence $\iota_* : H_.(\mathcal{S}(s)) \to H_.(|s|)$ is an isomorphism. Since any pair $\{K_1, K_2\}$ of subcomplexes of a simplicial complex is an excisive couple for both singular and singular simplicial complexes we get the two corresponding Mayer–Vietoris sequences. Denoting temporarily $H_.(\mathcal{S}(K))$ by $\hat{H}_.(K)$ we then have a commutative diagram:

$$
\begin{array}{ccccccccc}
\hat{H}_n(\dot{s}) & \longrightarrow & \hat{H}_n(s) \oplus \hat{H}_n(L) & \longrightarrow & \hat{H}_n(K) & \longrightarrow & \hat{H}_{n-1}(\dot{s}) & \longrightarrow & \hat{H}_{n-1}(s)) \\
\downarrow & & \downarrow & & \downarrow & & \downarrow & & \downarrow \\
H_n(|\dot{s}|) & \longrightarrow & H_n(|s|) \oplus H_n(|L| & \longrightarrow & H_n(|K| & \longrightarrow & H_{n-1}(|\dot{s}|) & \longrightarrow & H_{n-1}(|s|)
\end{array}
$$

in which the two rows are Mayer–Vietoris exact sequences and the vertical arrows are all induced by ι. By the Five lemma, it follows that $\iota_* : \hat{H}_n(K) \to H_n(|K|)$ is also an isomorphism for all n. This completes the proof of Theorem 4.3.13. ♠

Equivalence of Singular-Simplicial and Simplicial Homologies

Here we shall present a proof Theorem 4.3.14.

Fix a total order on the set of vertices of K. Then for each n, each n-simplex σ in K can be displayed uniquely as a strictly monotonically increasing sequence $(v_{i0}, \ldots, v_{in})$ and hence defines a unique element of $\mathcal{S}_n(K)$. This assignment extended linearly defines a splitting $\alpha_n : C_n(K) \to \mathcal{S}_n(K)$ of the quotient map $\varphi : \mathcal{S}_n(K) \to C_n(K)$. It can be easily checked that the $\alpha = \{\alpha_n\}$ is a chain map $C(K) \to \mathcal{S}(K)$ such that $\varphi \circ \alpha = Id_{C(K)}$. It remains to define a chain homotopy $h : \alpha \circ \varphi \approx Id_{\mathcal{S}(K)}$.

Observe that both φ and α preserve subcomplexes of K, viz., for any subcomplex $L \subset K$ we have $\varphi(\mathcal{S}(L)) \subset C(L)$ and $\alpha(C(L)) \subset \mathcal{S}(L)$. The chain homotopy that we are going to construct will also have this property. Hence we can easily pass onto relative chain complexes as well.

We have to define $h : \mathcal{S}_n(K) \to \mathcal{S}_{n+1}(K)$ so that

$$\partial \circ h + h \circ \partial = \alpha \circ \varphi - Id. \tag{4.22}$$

The construction of h is carried out by induction on n. We note that $\alpha_0 : C_0(K) \to \mathcal{S}_0(K)$ is the identity map. Therefore we can start with $h(\sigma) = 0$ for all 0-simplexes. Now suppose we have defined $h : \mathcal{S}_{n-1}(K) \to \mathcal{S}_n(K)$ so as to satisfy

$$h \circ \partial + \partial \circ h = \alpha \circ \varphi - Id \tag{4.23}$$

and such that for each $(n-1)$ singular simplex σ, the support of $h(\sigma)$ is contained in the support of σ. Now let τ be any singular n-simplex in K and $s = \operatorname{supp} \tau \in K$. We need to find a $(n+1)$-chain $h(\tau)$ supported in s and such that

$$\partial \circ h(\tau) = \alpha \circ \varphi(\tau) - \tau - h \circ \partial(\tau).$$

Obviously, it is necessary that the chain on the RHS of the above equation must be a cycle, i.e.,

$$0 = \partial(\alpha \circ \varphi(\tau) - \tau - h \circ \partial(\tau)) = \alpha \circ \varphi(\partial\tau) - \partial(\tau) - \partial \circ h(\partial(\tau)).$$

This follows from the inductive step (4.23). It is also clear that support of this cycle is contained in s. Since the homology of $H_n(\mathbb{S}(\Delta_q)) = 0, n > 0$ (see Example 4.3.7), it follows that this cycle is a boundary and hence there is a chain $h(\tau) \in \mathbb{S}_{n+1}(s) \subset \mathbb{S}_{n+1}(K)$ satisfying our requirement. This completes our hunt for a choice of $h(\tau)$ and thereby completes the construction of the chain homotopy h.   ♠

Equivalence of CW-homology and Cellular Singular Homology

Here we shall present a proof of Theorem 4.3.31. The basic idea here is the one we have met in the proof of the homotopy invariance.

Lemma 4.6.4 (Retraction operator) Let C. be a subcomplex of $S(X, A)$ (freely) generated by some singular simplexes. Assume that to each singular simplex $\sigma : \Delta_q \to X$ there exists a *singular prism* (i.e., a continuous map) $P\sigma : \mathbb{I} \times \Delta_q \to X$ with the following properties:
(a) $P\sigma(0, z) = \sigma(z)$;
(b) $P\sigma|_{1 \times \Delta_q} \in C.$;
(c) $P(\sigma \circ F^i) = P\sigma \circ (1 \times F^i)$, for each face operator F^i;
(d) if $\sigma \in C.$, then $P\sigma(t, z) = \sigma(z)$; and
(e) if $\sigma \in S(A)$ then $P\sigma(\mathbb{I} \times \Delta_q) \subset A$.
Then there is a chain deformation retraction $\tau : (S(X), S(A)) \to (C., S.(A) \cap C)$. In particular, the inclusion $C. \to S.(X)$ induces isomorphism in the homology.

Proof: Taking $\tau(\sigma) = P\sigma|_{1 \times \Delta_q}$, and extending linearly over $S(X, A)$, we obtain the map $\tau : S.(X, A) \to C.$. It follows that τ is a chain retraction (use (c)). Now a chain homotopy D from Id to τ is defined by taking $D(\sigma) = h(\sigma)$, where h is the prism operator defined in Lemma 4.2.10.   ♠

Now the proof of Theorem 4.3.31 is completed by appealing to the cellular approximation theorem. Given any simplex $\sigma : \Delta_n \to X$, all we do is choose a cellular approximation to it and a homotopy of the original map with the approximation, which gives the prism $P\sigma$. Of course, if σ is already a cellular map then we take this homotopy itself to be the identity, viz., $P\sigma(z, t) = \sigma(z)$.   ♠

Remark 4.6.5 The above lemma plays a crucial role in the proof of the Hurewicz theorem (see Chapter 10.)

4.7   Miscellaneous Exercises to Chapter 4

1. (a) Suppose $f_\alpha : A_\alpha \to B_\alpha$ is a directed system of homomorphisms of abelian groups with its direct limit $f : A \to B$. Show that $\{\text{Ker } f_\alpha\}$ and $\{\text{Im } f_\alpha\}$ form directed systems and
$$\text{Ker } f = \varinjlim_\alpha \text{ Ker } f_\alpha; \quad \text{Im } f = \varinjlim_\alpha \text{ Im } f_\alpha.$$

   (b) Show that direct limit of a family of long exact sequences of abelian groups is a long exact sequence.

   (c) Show that the singular chain complex of a topological space is the direct limit of the singular chain complexes of its compact subsets.

(d) Show that homology commutes with direct limits, i.e., given a directed system $(C.)_\alpha$ of chain complexes

$$H(\varinjlim_\alpha (C.)_\alpha) = \varinjlim_\alpha H((C.)_\alpha).$$

(e) Conclude from the above two exercises that the singular homology of a space is the direct limit of the singular homology of its compact subsets.

2. Given an open covering $\mathcal{U} = \{U_i\}$ of a space X, we define the chain complex $S_*^{\mathcal{U}}$ of X to be the subchain complex generated by those singular simplices σ whose image is contained in one of the members of $\mathcal{U}$. Show that the inclusion map $S_*^{\mathcal{U}}(X) \to S_*(X)$ induces isomorphism on homology. The chain complexes $S_*^{\mathcal{U}}$ are called *complexes of small chains*. [Hint: Use iterated barycentric subdivisions as in the proof of the excision theorem.]

3. Let X be the union of two closed sets X_1, X_2 such that $\mathrm{cl}\,(X_1 \setminus X_2) \cap \mathrm{cl}\,(X_2 \setminus X_1) = \emptyset$. Show that $\{X_1, X_2\}$ is an excisive couple. Put $A = X_1 \cap X_2$. Deduce that $H(X, A) = H(X_1, A) \oplus H(X_2, A)$. Generalize this to the case when X is the union of finitely many closed sets.

4. Show that if K_1, K_2 are two triangulations of a compact space then $\dim K_1 = \dim K_2$.

5. Compute the integral homology groups of the double combspace given in Exercise 1.9.17 and show that it is trivial. (See Exercise (ii) in 10.4.17.)

6. The graph is called *bipartite* if the vertex set V can be written as a disjoint union $V = V_1 \sqcup V_2$ so that there are no edges within V_1 or V_2 but every vertex of V_1 is jointed to every vertex of V_2. If $\#(V_i) = r_i$, then this graph is denoted by K_{r_1, r_2}. A finite graph on n vertices in which every pair of vertices is joined is called a complete graph and is denoted by K_n. A graph whose underlying topological space can be embedded in $\mathbb{R}^2$ (or in $\mathbb{S}^2$) is called a planar graph.

   Show that K_5 and $K_{3,3}$ are non planar. [Hint: Use the Jordan curve theorem.] (In fact a classical theorem due to Kuratowski states that a finite graph is planar iff it contains no subgraphs homeomorphic to K_5 or $K_{3,3}$.]

7. Consider the subspace Y of $\mathbb{R}^2 = \mathbb{C}$ which is the union of the three line segments $[0, 1], [0, \omega], [0, \omega^2]$, where ω is a primitive cube root of 1. Prove or disprove that there is an embedding (piecewise linear) of $Y \times Y$ inside $\mathbb{R}^3$.

Chapter 5

Topology of Manifolds

Manifolds are central objects of study in topology. Earlier we introduced CW-complexes and simplicial complexes. These are in some sense a slight generalization of manifolds and help us in understanding manifolds better.

In Section 5.1, let us get familiar with the general concept of a topological manifold and some of its point-set topological properties. The topological invariance of domain is a crucial result which helps us to 'define' manifolds with boundary in an unambiguous way.

In Section 5.2, we shall study triangulability of manifolds. For a triangulated manifold X, Poincaré introduced the concept of 'a dual cell decomposition' using which he showed that the ith Betti number $\beta_i(X)$ is equal to $\beta_{n-i}(X)$. In this section, we shall sketch this result closely following Poincaré's arguments (see [Diedonne, 1989] for a vivid account of this). A rigorous formulation and proof of the so-called Poincaré duality will be taken up in a later chapter after we introduce cohomology and cup product, etc.

The classification of 1-dimensional manifolds is easy: every connected 1-dimensional manifold is homeomorphic (diffeomorphic) to precisely one of the following four:
(1) $\mathbb{R}$,
(2) the ray $[0, \infty)$,
(3) the closed interval $[0, 1]$ or
(4) the unit circle $\mathbb{S}^1$.
For an easy proof of this fact, we refer the reader to [Shastri, 2011]. In Section 5.3, we shall show that all 2-dimensional manifolds are triangulable and use this to obtain a topological classification of compact surfaces. In Section 5.4, we shall introduce the basic theory of vector bundles. Vector bundles are one of the nicest tools for studying smooth manifolds.

5.1 Set Topological Aspects

Definition 5.1.1 Let $n \in \mathbb{N}$. Let $X \neq \emptyset$ be a topological space. By a (n-dimensional) *chart* for X we mean a pair (U, ψ) consisting of an open neighbourhood U of x and a homeomorphism $\psi : U \longrightarrow \mathbb{R}^n$ onto an open subset of $\mathbb{R}^n$. By an *atlas* $\{(U_j, \psi_j)\}$ for X, we mean a collection of charts for X such that $X = \cup_j U_j$. If there is an atlas for X, we say X is *locally Euclidean.*

A chart (U, ψ) is called a *chart at* $x_0 \in X$ if $\psi(x_0) = 0$. Writing $\psi = (\psi_1, \ldots, \psi_n)$, these n component functions ψ_i are called *local coordinate functions* for X at x.

Let $n \geq 1$ be an integer and X be a topological space. We say X is a *topological manifold of dimension n* if X is :
(i) locally Euclidean, i.e., there is an atlas consisting of n-dimensional charts,
(ii) a Hausdorff space and
(iii) II-countable, i.e., it has a countable base for its topology.

Any countable discrete space is called a 0-*dimensional manifold.*

Remark 5.1.2

1. We would like to include the empty space also as a topological manifold. However, there is no good way of assigning a dimension to it. Some authors prefer it to be of dimension -1, and some others $-\infty$. Indeed, the best way would be to treat it as a manifold of any dimension as and when required. In what follows a manifold is always assumed to be nonempty unless it obviously follows from the context that a particular one is empty.

2. Observe that once a chart (U, ψ) exists around a point $x_0 \in X$, then we can choose a chart (V, ϕ) at x_0 such that $\phi(V) = \mathbb{R}^n$. For, by composing with a translation, we can assume that $\psi(x_0) = 0$ and then we can choose $r > 0$ such that the open ball $B_r(0) \subset \psi(U)$ and put $V = \psi^{-1}(B_r(0))$, and $\phi = f \circ \psi$ where $f : B_r(0) \to \mathbb{R}^n$ is the homeomorphism (actually a diffeomorphism) given by $\mathbf{x} \mapsto \frac{\mathbf{x}}{r^2 - \|\mathbf{x}\|^2}$.

3. For an atlas, it is necessary to assume that the integer n is the same for all the charts. Of course, if X is connected, it is a consequence of the *topological invariance of domain* (see Theorem 4.4.13) that the integer n is the same for all charts.

4. For a topological space that is locally Euclidean, the II-countability condition is equivalent to many others, such as metrizability or paracompactness. We find II-countability the most suitable for our purpose and at times, without hesitation, we shall assume that a manifold is metrizable as well.

Example 5.1.3

(i) Clearly, differential manifolds inside Euclidean spaces that you may have studied in your calculus course are topological manifolds in the above sense.

(ii) The boundary of the unit square $\mathbb{I} \times \mathbb{I}$ as a subspace of $\mathbb{R}^2$ is a topological 1-manifold. You may have learnt that this is not a smooth submanifold of $\mathbb{R}^2$ because of those corner points.

(iii) Let X be the union of the two axes in $\mathbb{R}^2$. If U is any connected neighbourhood of $(0,0)$ in X then $U \setminus \{(0,0)\}$ has four components. It follows that X cannot have any chart covering $(0,0)$ and hence fails to be a topological manifold.

(iv) Let X be the set of all real numbers together with one extra point that we shall denote by $\tilde{0}$. We shall make X into a topological space as follows: Let $\mathcal{T}$ be the collection of all subsets A of X of the form $A = B \cup C$ where
(a) B is either empty or an open subset of $\mathbb{R}$ in the usual topology and
(b) C is either empty or is such that $(C \cap \mathbb{R}) \cup \{0\}$ is a neighbourhood of 0 in $\mathbb{R}$.
We leave it to you to verify that $\mathcal{T}$ forms a topology on X in which $\mathbb{R}$ is a subspace. Since $\tilde{0}$ also has neighbourhoods that are homeomorphic to an interval, it follows that X has an atlas. It is easily seen that X has a countable base also. But however, observe that X fails to be a Hausdorff space, since neighbourhoods of 0 and $\tilde{0}$ cannot be disjoint. This space can also be thought of as the quotient space obtained by taking two copies of $\mathbb{R}$ and identifying every non zero real number in one copy with the corresponding number in the other copy.

(v) Likewise, one can also give examples of spaces that are Hausdorff and locally Euclidean but are not II-countable. The typical example is the so-called *long line:* Consider an uncountable set Ω_0 which is well ordered in such a way that every element has only countably many predecessors. Put $X = \Omega_0 \times [0,1)$. Define a total order $\ll$ on X as follows:

$$(\alpha, t) \ll (\beta, s) \quad \text{if } \alpha < \beta \text{ or } \alpha = \beta \text{ and } t < s.$$

With the order topology induced by this order, X is a connected, Hausdorff space, having a smooth structure, locally diffeomorphic to $\mathbb{R}$. But X does not have a countable base. (For more details, See [Joshi, 1983].)

(vi) Another type of non-example is obtained by taking the disjoint union of manifolds of different dimensions. Thus, the subspace of $\mathbb{R}^2$, consisting of the x-axis together with the point $(0,1)$, is not a manifold.

Let us now consider some examples of manifolds that do not occur **naturally** as subspaces of any Euclidean space.

(vii) **The real projective spaces** The foremost one is the n-dimensional real projective space $\mathbb{P}^n$. This is the quotient space of the unit sphere $\mathbb{S}^n$ by the antipodal action, viz., each element x of $\mathbb{S}^n$ is identified with its antipode $-x$. For more details see Example 2.2.11(v), wherein we have proved that $\mathbb{P}^n$ is compact.

Given $x \in \mathbb{S}^n$ consider V to be the set of all points in $\mathbb{S}^n$, that are at a distance less than $\sqrt{2}$ from x. Then check that $U = q(V)$ is a neighbourhood of $[x]$ in $\mathbb{P}^n$ and q itself restricts to a homeomorphism from V to U. Since V is anyway homeomorphic to an open subset of $\mathbb{R}^n$, this proves the existence of an n-dimensional atlas for $\mathbb{P}^n$.

To see that $\mathbb{P}^n$ is Hausdorff, let $[x] \neq [y] \in \mathbb{P}^n$ be two points. Clearly, in $\mathbb{S}^n$, we can choose $\epsilon > 0$ such that $B_\epsilon(\pm x) \cap B_\epsilon(\pm y) = \emptyset$. It then follows that $q(B_\epsilon(x))$ and $q(B_\epsilon(y))$ are disjoint neighbourhoods of $[x]$ and $[y]$ in $\mathbb{P}^n$.

(viii) One can merely take a connected topological manifold X, and take a connected covering space $p : \tilde{X} \to X$. Since p is a local homeomorphism, it follows that $\tilde{X}$ is locally Euclidean. It is not very hard to prove that $\tilde{X}$ is Hausdorff (see Exercise 3.1.7(i)). However, why should $\tilde{X}$ be II-countable or metrizable? If we know that the cardinality of the fibre is countable (which is necessary if $\tilde{X}$ has to be II-countable), then of course it easily follows that $\tilde{X}$ is II-countable: We can take a countable base $\{V_i\}$ for X consisting of evenly covered open sets and then take the collection of all open sets in $\tilde{X}$ homeomorphic to V_i's under p. Thus we are led to the question: Is the fundamental group of any topological manifold countable? The answer is yes (see Corollary 5.1.18) and hence we conclude that $\tilde{X}$ is a manifold of the same dimension as X. Moreover, it is not hard to see that additional structures such as a C^r-structure or a complex analytic structure, etc., on X can also be pulled back onto $\tilde{X}$.

Let us introduce the notation

$$\boldsymbol{H}^n = \{(x_1, \ldots, x_n) \ : \ x_n \geq 0\}$$

for the closed upper half space in $\mathbb{R}^n$.

Definition 5.1.4 A topological space X is called a *manifold with boundary* if it is a II-countable, Hausdorff space, such that each point x of X has an open neighbourhood U_x and a homeomorphism $\phi : U_x \longrightarrow \boldsymbol{H}^n$ onto an open subset of $\boldsymbol{H}^n$.

Denote by int X, the set of all those points in x having a neighbourhood U_x homeomorphic to an open subset of

$$\text{int } \boldsymbol{H}^n = \{(x_1, \ldots, x_n) \in \mathbb{R}^n \ : \ x_n > 0\}.$$

Clearly this forms an open subset of X and is a topological n-manifold in the old sense. Can you see why int X is non empty if X is non empty? The complement of int X in X is denoted by ∂X and is called the *boundary* of X. Clearly it is a closed subset of X.

Remark 5.1.5

1. It may happen that ∂X is empty which means precisely that X is a manifold. The points of ∂X are characterized by the following property. There is a neighbourhood U_x of x and a homeomorphism $\phi : U_x \longrightarrow \boldsymbol{H}^n$ such that the nth-coordinate of $\phi(x)$ vanishes, i.e., $\phi_n(x) = 0$. This is again a simple consequence of the topological invariance of domain (Theorem 4.4.13).

2. It follows that $\hat{U} = \phi^{-1}(\mathbb{R}^{n-1} \times 0)$ is a neighbourhood of x in ∂X if we take ϕ as above. Also, then ϕ itself restricts to a homeomorphism $\phi : \hat{U} \longrightarrow (\mathbb{R}^{n-1} \times \{0\}) \cap \phi(U)$. As a consequence, it follows that ∂X, if non empty, is itself a topological $(n-1)$-dimensional manifold (without boundary).

3. We shall most often use the word 'manifold' to mean a manifold without boundary. Often the results that we state for them are valid for manifolds with boundary as well, though we cannot take them for granted. Indeed, whenever, special attention is needed for manifolds with boundary, we shall take care to mention them.

One of the most important tools in the study of manifolds is partition of unity. If you have studied this earlier for differential manifolds, you may skip reading the proofs of Lemma 5.1.6 and Theorem 5.1.7 given below.

Lemma 5.1.6 Let X be a topological manifold. Then there exists a nested sequence of open subsets $\{W_i\}$ in X such that
(i) $\overline{W}_i$ is compact for each i;
(ii) $\overline{W}_i \subset W_{i+1}$, for each i;
(iii) $X = \cup_i W_i$.

Proof: Using II-countability, it follows that there is a countable family of diffeomorphisms $\phi_i : U_i \to \mathbb{R}^n$, where U_i are open in X and $X = \cup_i \phi_i^{-1}(\mathbb{D}^n)$. Put $V_i = \phi_i^{-1}(\mathbb{D}^n)$. Then each V_i is an open subset of X. The closure of V_i is compact being homeomorphic to the closure of the open ball $\mathbb{D}^n$.

Put $W_1 = V_1$. Inductively having defined W_k, satisfying (i) and (ii), we can select finitely many members of $\{V_i\}$, that cover $\overline{W}_k$. Let W_{k+1} be the union of these members and V_{k+1}. Check Property (iii).   ♠

Theorem 5.1.7 Partition of unity on abstract manifolds: *Let X be any subspace of a manifold Y and $\{U_\alpha\}_{\alpha \in \Lambda}$ be an open covering of X. Then there exists a countable family $\{\theta_j\}$ of continuous real valued functions on Y with compact support such that*
(i) *$0 \leq \theta_j(x) \leq 1$, for all j and $x \in X$;*
(ii) *for each $x \in X$ there exists a neighbourhood N_x of x in X, such that only finitely many of θ_j are nonzero on N_x;*
(iii) *for each j, $(\text{supp}\, \theta_j) \cap X \subset U_{\alpha_j}$ for some α_j; and*
(iv) *$\sum_j \theta_j(x) = 1$, for all $x \in X$.*

Proof: As before, we may replace X by a neighbourhood of X in Y and assume that X itself is a smooth n-manifold. By the previous lemma, X is the increasing union of a countable family of open sets $\{K_i\}_{i \geq 1}$ with the closure of each K_i being compact. We set $K_0 = \emptyset$.

We shall first construct a countable family $\{B_{ij}\}$ of open sets in X with homeomorphisms

$\psi_{ij} : B_{ij} \to \mathbb{D}^n$ such that if we put $\frac{1}{2} B_{ij} := \psi_{ij}^{-1}(\mathbb{D}^n_{1/2})$, then $\{\frac{1}{2} B_{ij}\}$ is a covering of X, and is a locally finite open refinement of $\{U_\alpha\}$.

Inductively, suppose $\{B_{ij}\}$ have been constructed for $i \leq k$ so that $\{\frac{1}{2} B_{ij}\}$ covers $\overline{K}_k$. For each point $x \in \overline{K}_{k+1} \setminus K_k$, we can choose a neighbourhood W_x contained in some member of $\{U_\alpha\}$ and not intersecting $\overline{K}_{k-1}$. We can further assume that there is a homeomorphism $\psi_x : W_x \to \mathbb{D}^n$ such that $\psi_x(x) = 0$. Since $\{\psi_x^{-1}(\mathbb{D}^n_{1/2})\}$ is an open cover of the compact space $\overline{K}_{k+1} \setminus K_k$, there exist finitely many $x_1, \ldots, x_r$ such that

$$\overline{K}_{k+1} \setminus K_k \subset \cup_{1 \leq i \leq r} \psi_{x_j}^{-1}(\mathbb{D}^n_{1/2}).$$

Put $B_{(k+1),j} := \psi_{x_j}^{-1}(\mathbb{D}^n)$. It then follows that

$$\overline{K}_{k+1} \subset \cup_{ij} \{B_{ij} \ : \ i \leq k+1\}.$$

Inductively, the construction of the family $\{B_{ij}\}$ is over. Clearly, it is an open refinement of the family $\{U_\alpha\}$ and covers X. To see that the family $\{B_{ij}\}$ is *locally finite*, given $x \in X$, suppose $x \in K_k$. Then, take $N_x = K_k$. Clearly, N_x does not meet any of the B_{ij} for $i \geq k+2$ and the family $\{B_{ij} \ : \ i \leq k+2\}$ is finite.

Now, let $\beta : \mathbb{D}^n \to \mathbb{I}$ be the function

$$\beta(x) := 2 \min \left\{ d(x, \mathbb{S}^{n-1}), \frac{1}{2} \right\}.$$

Put $\eta_{i,j} = \beta \circ \psi_{i,j}$, where $\psi_{i,j} : B_{i,j} \to \mathbb{D}^n$ are the homeomorphisms chosen earlier. Extend $\eta_{i,j}$ by zero over all of Y. Now re-index this family by single integers and call it $\{\eta_j\}$. It is easily verified that this family satisfies properties (i),(ii) and (iii). Now define $\eta = \sum_i \eta_i$ which makes sense and is continuous. Moreover, $\eta(x) \geq 1$ for all $x \in X$ since each x belongs to some $\frac{1}{2} B_j$ and then $\eta_j(x) = 1$. Take $\theta_j := \eta_j / \eta$ and verify that this family of functions is as required. ♠

Remark 5.1.8 If we begin with a 'smooth' manifold Y, the maps $\{\theta_j\}$ can be chosen to be smooth. The only difference is replace above β by a smooth 'bump function with similar properties (see [Shastri, 2011]).

As an immediate consequence of existence of partition of unity, we shall now obtain the collar neighbourhood theorem.

Definition 5.1.9 Let X be a manifold with boundary $\partial X \neq \emptyset$. By a collar of ∂X in X we mean an open subset U of X with a homeomorphism $\varphi : U \to \partial X \times [0, \infty)$ such that for all $x \in \partial X$, $\varphi(x) = (x, 0)$.

Notice that there is nothing special in the choice of the interval $[0, \infty)$ in the above definition. We can as well take $[0, \epsilon)$ for any $\epsilon > 0$.

Theorem 5.1.10 *In every manifold X, ∂X has a collar neighbourhood. Moreover for any collar neighbourhood U of ∂X, $X \setminus U$ is homeomorphic to X.*

Proof: Let Y be the space obtained by 'attaching an external collar to X', viz., Y is the quotient of the disjoint union of X and $\partial X \times [-1, 0]$ by identifying $x \in \partial X$ with $(x, 0) \in \partial X \times [-1, 0]$. Observe that Y is also a n-manifold with its boundary homeomorphic to $\partial X \times \{-1\}$. The idea is to define a homeomorphism $f : X \to Y$ and then take $U = f^{-1}(\partial X \times [-1, 0))$.

Begin with a (countable) partition of unity $\{\theta_i\}$ on ∂X so that supp θ_i is contained in a coordinate open set U_i of ∂X together with a homeomorphism ϕ_i from $U_i \times [0,1)$ onto an open subset V_i of X. Put $\eta_0 = 0, \eta_k = \sum_{i=1}^{k} \theta_i$; and

$$Z_k := \{(x,t) \ : \ x \in U_k, -\eta_{k-1}(x) \leq t \leq 1\}; \ Z'_k := \{(x,t) \ : \ x \in U_k, \ -\eta_k(x) \leq t \leq 1\}.$$

Let $\alpha_k : Z_k \to Z'_k$ be the homeomorphism which linearly stretches the segment $[-\eta_{k-1}(x), 1]$ homeomorphically onto the segment $[-\eta_k(x), 1]$, for each x.

Put $Y_k = X \cup \{(x,t) \ : \ -\eta_k(x) \leq t \leq 0\}$ and let $\beta_k : Z'_k \to Y_k$ be the embeddings given by

$$\beta_k(x,t) = \phi_k(x,t), \ t \geq 0; \text{ and } \beta_k(x,t) = (x,t), \ t \leq 0.$$

We define homeomorphisms $f_k : Y_{k-1} \to Y_k$ as follows:

$$f_k(z) = \begin{cases} z, & z \in X \setminus \phi_k(U_k \times [0,1)); \\ (x,t), & z = (x,t), \ x \notin U_k; \\ \beta_k \circ \alpha_k(x,t), & x \in U_k. \end{cases}$$

Finally put $f = \cdots \circ f_k \circ f_{k-1} \circ \cdots \circ f_1$.

First of all note that on the complement of $V = \cup_i \phi_i(U_i \times [0,1))$, f is identity. On V itself, f makes sense, since given any point $x \in \partial X$, there are only finitely many i for which $x \in U_i$ and $f_k(x,t) = (x,t)$ if $x \notin U_k$. Indeed in a neighbourhood of x, all f_k are identity except those k for which $\theta_k(x) \neq 0$. For this reason, f is also a proper mapping. Since $\sum_k \theta_k(x) = 1$, it follows that f is surjective. Since each f_k is an embedding f is injective. Therefore f is a homeomorphism. ♠

Remark 5.1.11 Naturally, we now have many other consequences of the existence of countable partition of unity such as Urysohn's lemma, the Tietze's extension theorem, etc. For example, if $\{\theta_i\}$ is a countable partition of unity on X then $f : X \to [0, \infty)$ defined by $f(x) = \sum_k k\theta_k(x)$ is a proper mapping.

Exercise 5.1.12

(i) Show that a manifold is connected iff it is path connected iff its interior is connected.

(ii) (This exercise is intended as some kind of an explanation for our obsession with Hausdorffness and locally Euclidean spaces.) Consider $\mathbb{R}$ with the cofinite topology. Certainly it is not Hausdorff. But is it locally Euclidean? More generally, take the affine space $\mathbb{C}^n$ with the Zariski topology. Is it locally Euclidean? Is it contractible?

The following result tells us that after all, we could have just stuck to the study of subsets of Euclidean spaces for studying manifolds. As we shall see, this single result has several implications on topological, homotopical and homological properties of a manifold.

Theorem 5.1.13 *Every n-manifold is homeomorphic to a closed subset of $\mathbb{R}^{2n+1}$.*

Theorem 5.1.14 *For every topological n-manifold X, the set of embeddings of X in $\mathbb{I}^{2n+1}$ is dense in the space $(\mathbb{I}^{2n+1})^X$ of all continuous maps (with topology of uniform convergence on compact sets, which is the same as the compact-open-topology).*

Remark 5.1.15 The proofs of these theorems are somewhat lengthy and hard. The smooth version of Theorem 5.1.13 goes under the name *easy Whitney embedding theorems* which you may read from many books such as [Shastri, 2011]. However, for the topological case, there are not many references available. You are welcome to see this in the excellent old book by Hurewicz and Wallman ([Hurewicz–Wallman, 1948], Theorem V-3). Or you may choose to work through a sequence of exercises (a dozen of them) from Chapter 4 of [Spanier, 1966]. You may choose to read a nice proof of the embedding Theorem 5.1.13 from [Munkres, 1984(1)]. Here, we shall be satisfied with an easy proof of the following weaker version:

Theorem 5.1.16 *Every compact manifold is homeomorphic to a closed subset of some Euclidean space.*

Proof: Cover X by finitely many open sets $\{U_i\}_{1 \le i \le k}$ such that there is a homeomorphism $f_i : U_i \to \mathbb{R}^n$. Let $\eta : \mathbb{R}^n \to \mathbb{S}^n$ be the inverse of the stereographic projection and $g_i : X \to \mathbb{S}^n$ be the extension of $\eta \circ f_i$ which sends $X \setminus U_i$ to the north pole. Put $g = g_1 \times g_2 \cdots \times g_k :$ $X \to (\mathbb{S}^n)^k \subset \mathbb{R}^{nk+k}$. Verify that g is a one-one mapping. Since X is compact, g is a homeomorphism onto a closed subset. ♠

In order to derive some homotopical and homological properties of a manifold, we need the following:

Lemma 5.1.17 Every locally contractible closed subset X of $\mathbb{R}^N$ is a retract of some neighbourhood of X in $\mathbb{R}^N$.

Proof: For $\delta > 0$, let P_δ denote the partition of $\mathbb{R}^N$ into cubical boxes of side-length δ by hyper-planes parallel to the coordinate planes. Let W_1 be the union of all cubes in P_1 which do not meet X. Inductively, let W_k be the union of all those cubes in $P_{1/2^{k-1}}$ which do not meet X and which are not contained in W_{k-1}. Then

$$\mathbb{R}^N \setminus X := W = \cup_{k=1}^\infty W_k$$

and W has a CW-structure in which all these cubes of various sizes form the N-dimensional cells. (Do not confuse W_k with the k-skeleton of W.) Note that each W_k is a closed subset of W.

We shall construct a subcomplex V of W and a function $r : X \cup V \to X$ such that $r(x) = x$, $x \in X$. We shall then show that $X \cup V$ contains a neighbourhood of X and r is continuous on this neighbourhood, which will complete the proof of the lemma.

The constructions of V and r will be done simultaneously and inductively. Of course, we start with $r(x) = x$, $x \in X$. Take the 0-skeleton of V to be $W^{(0)}$, i.e., all vertices of W. Given $\sigma \in V^{(0)}$, choose any point $r(\sigma) \in X$ such that

$$d(\sigma, r(\sigma)) < 2d(\sigma, X).$$

(Here $d(A, B)$ denotes the infimum of the Euclidean distances $d(a, b)$ where $a \in A, b \in B$.) This completes the constructions of $V^{(0)}$ and $r : X \cup V^{(0)} \to X$.

Having defined $V^{(k-1)}$ and $r : X \cup V^{(k-1)} \to X$, let us take σ to be any k-cell of W such that $\partial \sigma \subset V^{(k-1)}$. If $r : \partial \sigma \to X$ has an extension then let us put this σ inside $V^{(k)}$; not otherwise. Having put this σ inside $V^{(k)}$ take $r : \sigma \to X$ to be any extension which satisfies the property

$$d(\sigma, r(\sigma)) < 2d(\sigma, r'(\sigma)) \tag{5.1}$$

for all extensions $r' : \sigma \to X$ of $r : \partial \sigma \to X$. This completes the constructions of $V^{(k)}$ and $r : X \cup V^{(k)} \to X$. By induction the constructions of V and r are complete.

Note that by the very choice, r is continuous on V. So we have to verify continuity at points $x \in X$ only.

Given $x \in X$ and $\epsilon > 0$ put $\epsilon_{2N} = \epsilon$ and choose inductively $\epsilon_N > \cdots > \epsilon_0$ such that

$$3\epsilon_i < \epsilon_{i+1}, i = 0, \ldots, N - 1$$

so that if $B_i := X \cap B_{\epsilon_i}(x)$ then the inclusion map $B_i \subset B_{i+1}$ is null homotopic. (This is where local contractibility is used.) Let $U_t = B_t(x)$, where $0 < t < \epsilon_0/4$.

We first claim that if σ is any cell of W contained in U_t then $\sigma \in V$ and $d(x, r(\sigma)) < \epsilon$. Suppose a is a 0-cell and $a \in U_t$. Then

$$d(x, r(a)) \leq d(x, a) + d(a, r(a)) < t + 2d(a, x) < 3t < \epsilon_0.$$

Now suppose for all $(k-1)$-cells τ of V contained in U_t we have $d(x, r(\tau)) < \epsilon_{2k-2}$. Let σ be a k-cell of W contained in U_t. Then $r(\partial\sigma) \subset B_{\epsilon_{2k-2}}$ and the inclusion $B_{\epsilon_{2k-2}} \subset B_{\epsilon_{2k-1}}$ is null homotopic. From (5.1), it follows that $d(\sigma, r(\sigma)) < 2\epsilon_{2k-1}$ and therefore,

$$d(x, r(\sigma)) \leq d(x, \sigma) + d(\sigma, r(\sigma)) < t + 2\epsilon_{2k-1} < \epsilon_{2k}.$$

Therefore, by induction, we have, $r(U_t \cap V) \subset U_\epsilon$.

Note that this however, does not complete the proof of continuity of r at x; nor does this prove that V contains a neighbourhood of $x \in X$. For we have not proved $U_t \subset V$ since U_t will necessarily have many points belonging to cells σ in V where σ itself is not completely in U_t.

So, we claim that there exist $s > 0$ such that if any cell σ of W intersects U_s then the entire cell σ is contained in U_t. Combined with the earlier claim this will then complete both the proofs.

If this is not the case, there exists a sequence $n_i \longrightarrow \infty$ and cells σ_i in W such that $\sigma_i \cap U_{1/n_i} \neq \emptyset$ and $\sigma \cap U_t^c \neq \emptyset$. In particular, this implies that the diameter of σ_i is bigger than $t/2$ for all large n_i. If m is an integer such that $t/2 > 1/m$, this means that all σ_i are inside W_m which has a positive distance from x, say, δ. If i is such that $1/n_i < \delta$, this will contradict the choice that $\sigma_i \cap U_{1/n_i} \neq \emptyset$. This completes the proof of the lemma. ♠

Corollary 5.1.18 The fundamental group, the homology and cohomology groups of a compact manifold are finitely generated.

Proof: Embed a given compact manifold X in some Euclidean space $\mathbb{R}^N$ and take a neighbourhood U of X and a retraction $r : U \to X$. Choose $\delta > 0$ to be less than $1/3$ of $d(X, \mathbb{R}^N \setminus U)$ and let P_δ denote the partition of $\mathbb{R}^N$ into cubical boxes of size δ by planes parallel to the coordinate planes. Let Y be the union of all those boxes of P_δ which meet X. Then Y is a finite CW-complex $X \subset Y \subset U$. The retraction r restricts to a retraction $r : Y \to X$. Now $r \circ i = Id_X$ implies that $r_\# \circ i_\# = Id_\#$ on the fundamental groups, $r_* \circ i_* = Id_*$ on homology (and $i^* \circ r^* = Id^*$ on cohomology) groups. Since Y has fundamental group, homology (cohomology) modules finitely generated, the same must hold for X as well. ♠

Corollary 5.1.19 The fundamental group and the homology modules of a manifold are countable.

Proof: By Lemma 5.1.6, we can write $X = \cup_i W_i = \cup_i \bar{W}_i$ where each $\bar{W}_i$ is compact. Therefore $\pi_1(X) = \varinjlim_i \pi_1(\bar{W}_i)$ and each $\pi_1(\bar{W}_i)$ is countable being finitely generated. Therefore $\pi_1(X)$ is countable. The same argument holds for homology modules. ♠

Exercise 5.1.20

(i) Give examples of Hausdorff, II-countable spaces that are not manifolds.

(ii) Can you think of some manifolds, other than projective spaces, that do not `naturally` occur as subspaces of Euclidean spaces?

(iii) Let X be a manifold, $U \subset X$ be a connected open subset. Given $p, q \in U$ show that there exists a homeomorphism $f : X \to X$ such that $f(x) = x$ for all $x \in X \setminus U$ and $f(p) = q$. (It is in this strong sense that the group of homeomorphisms of X acts 'transitively' on X for any connected manifold. Compare Exercise (iv) below. [Hint: See Exercise 1.5.19.(i)]

(iv) Let X be a connected manifold of dimension $n \geq 2$ and k any positive integer. Given any two k-subsets $\{a_1, \ldots, a_k\}, \{b_1, \ldots, b_k\} \subset X$, show that there exists a homeomorphism $f : X \to X$ such that $f(a_i) = b_i, i = 1, 2, \ldots, k$. What is the best that we can say here if $\dim X = 1$?

(v) Given a connected n-manifold and any finite subset $F \subset X$, show that there is an n-cell $E \subset X$ such that F is contained in the interior of E.

(vi) Prove (without using embeddability results) that every locally Euclidean, II-countable, Hausdorff space is paracompact or read it from some book, say, e.g. [Dugundji, 1990].

(vii) Show that every subspace of a manifold is paracompact.

5.2   Triangulation of Manifolds

In this section, we shall begin a discussion on triangulation of manifolds. The three fundamental classical questions here are:
(A) Can every topological (smooth) manifold be triangulated?
(B) Given two triangulations K_1, K_2 of a topological (smooth) manifold, is K_1 combinatorially equivalent to K_2?
(C) Does every triangulated manifold carry a smooth structure?
The classification of 1-dimensional manifolds, stated that:

Theorem 5.2.1 *Every connected 1-dimensional manifold is homeomorphic to one of the following four manifolds:*
(1) $\mathbb{R}$;   (2) $\mathbb{S}^1$;   (3) $\mathbb{I}$;   (4) $[0, 1)$.

See [Shastri, 2011] for an easy proof. It follows easily, that every 1-dimensional manifold is triangulable. In this section, we shall present a proof of the result that all (compact) 2-dimensional manifolds are triangulable (a result due to Rado, proved in 1924). The triangulability of all 3-dimensional manifolds is a deep result due to Edwin Moise [Moise, 1977], who also proved that any two triangulations K_1, K_2 of the same 3-manifold are combinatorially equivalent, i.e., there are subdivisions of K_i' of K_i such that K_1' is isomorphic to K_2'.

A theorem of Cairns (1935) says that every smooth manifold is triangulable. (See [Whitehead, 1940] for a proof and more.]

And there are triangulable manifolds which do not admit any smooth structure. The first example came in 1960 due to Kervaire in dimension 10 which was soon improved to dimension 8 by Eells and Kuiper. Siebenmann has constructed an example of a 5-manifold which cannot be triangulated. With these classical results, we may safely say that the above three questions have been answered satisfactorily.

A triangulation of a manifold has necessarily more combinatorial structure. We make a beginning of the study of these properties. Besides theoretical importance, it provides a very effective tool in the study of topological properties of manifolds. We shall then introduce the notion of 'combinatorial triangulation' and use it to sketch a proof of Poincaré duality as envisaged by Poincaré himself. More technical versions and rigorous proofs of Poincaré duality will be discussed in a subsequent chapter. Finally, we shall describe a result due to Munkres about a certain local homological condition. This result has found applications in the algebra of face rings.

We begin with a general lemma about 'local' homology of a simplicial complex. In what follows, we have suppressed the coefficient group in the homology groups. One can take it to be any field or the group of integers according to one's requirements.

Lemma 5.2.2 Let $x \in\ <F>$, i.e., a point in the interior of $|F|$ for some simplex $F \in K$. Then
$$H_i(|K|, |K| \setminus \{x\}) \approx \widetilde{H}_{i-\dim F-1}(Lk(F)).$$

Proof: For any simplex $F \in K$, the open star, $st(F)$ is an open subset of $|K| = X$ and hence by Theorem 4.2.17, $\{St(F), X \setminus \{x\}\}$ is an excisive couple in $X = | St(F) | \cup X \setminus \{x\}$. By Theorem 4.2.16,
$$H_i(| St(F) |, | St(F) | \setminus\{x\}) \approx H_i(X, X \setminus \{x\}).$$

Recall that $| St(F) | \setminus\{x\}$ contains $| Lk(F) * \mathcal{B}(F) |$ as a deformation retract (see Corollary 2.10.5). Therefore
$$H_i(| St(F) |, | Lk(F) * \mathcal{B}(F) |) \approx H_i(| St(F) |, | St(F) | \setminus\{x\}).$$

Since $| St(F) |$ is contractible, we have
$$H_i(| St(F) |, | Lk(F) * \mathcal{B}(F) |) \overset{\partial}{\underset{\approx}{}} \widetilde{H}_{i-1}(| Lk(F) * \mathcal{B}(F) |)$$
$$\approx \widetilde{H}_{i-1-\dim F}(Lk(F))$$

(by the homology suspension Theorem 4.2.21).   ♠

Definition 5.2.3 Let K be a finite simplicial complex. We say K is *pure* (of dim n) if each simplex in K is a face of a n-simplex.

For a topological manifold X without boundary, given any point $x \in X$ we can choose a neighbourhood U of x homeomorphic to $\mathbb{R}^n$. By excision, it follows that
$$H_i(X, X \setminus \{x\}) = H_i(U, U \setminus \{x\}) \approx H_i(\mathbb{R}^n, \mathbb{R}^n \setminus \{0\}).$$

Combined with the above lemma, we immediately have the following.

Theorem 5.2.4 *Let X be a connected, compact topological manifold (without boundary) and K be a simplicial complex such that $| K | = X$. Then the following holds:*
(i) For all non empty faces F of K, we have, $\widetilde{H}_i(Lk(F)) = (0)$ for $i < \dim Lk(F)$ and $\approx \mathbb{Z}$ for $i = \dim Lk(F)$, i.e., $Lk(F)$ is a homology sphere of dimension equal to $\dim Lk(F)$.
(ii) K is pure of dimension n.
(iii) Every $(n-1)$-simplex of K occurs as the face of exactly two n-simplices.
(iv) Given any two n-simplexes σ and τ in K, there is a chain of n-simplexes connecting σ and τ, i.e., there exist n-simplices $s_1, \ldots, s_k$ in K such that $s_i \cap s_{i+1}$ is an $(n-1)$-face for $i = 1, \ldots, k-1$ and $s_1 = \sigma, s_k = \tau$.

Proof: We shall first prove (ii). If F is a maximal simplex then $<F>$ is an open set in $|K| = X$ and hence for any $x \in\ <F>$, $\{X \setminus \{x\}, |F|)$ is an excisive couple. Also $\partial|F|$ is a SDR of $|F| \setminus \{x\}$ and therefore we have,
$$H_i(X, X \setminus \{x\}) \approx H_i(|F|, |F| \setminus \{x\}) \approx H_{i-1}(|F| \setminus \{x\}) \approx H_{i-1}(\partial|F|).$$

This means $\partial |F|$ is a homology $(n-1)$-sphere, which means $\dim F = n$.
(i) is immediate from the Lemma 5.2.2, since $\dim Lk(F) = n - \dim F - 1$.
(iii) If F is any $(n-1)$ simplex, then from (i) we get

$$\mathbb{Z} \approx H_n(X, X \setminus \{x\}) \approx \tilde{H}_0(Lk(F)).$$

Since $Lk(F)$ is just a set of vertices one for each n-simplex of which F is a face, we are done.

(iv) Clearly, on the set of all n-simplexes in K, having a chain of n-simplexes connecting them is an equivalence relation. We need to show that there is just one equivalence class. Assuming on the contrary, let A be the subcomplex spanned by all n-simplexes in one of the equivalence classes and B the subcomplex spanned by the rest of the n-simplexes. Then clearly $|A|, |B|$ are closed subspaces of X. Since X is connected, it follows that $A \cap B \neq \emptyset$. Let F be a maximal simplex in $A \cap B$. It follows that $\dim F < n - 1$. This implies $\dim Lk_K(F) > 0$. By part (i), we conclude that $\tilde{H}_0(Lk_K(F)) = (0)$ and hence $Lk_K(F)$ is connected. Clearly $Lk_K(F) \cap A \neq \emptyset \neq Lk_K(F) \cap B$. It follows that there is an edge $e \in Lk_K(F)$ with one vertex $u \in A \setminus B$ and another vertex $v \in B \setminus A$. If s is a n-simplex such that $F \cup e \subset s$, then either $s \in A$ or $s \in B$, which implies $v \in A$ or $u \in B$, which is a contradiction. ♠

Definition 5.2.5 A simplicial complex which satisfies the properties (ii), (iii) and (iv) of the above theorem is called a *pseudo-manifold* of dimension n. If we weaken the condition (iii) by replacing the phrase 'exactly two' by 'at most two' we get the definition of a pseudo-manifold with boundary.

As an immediate corollary of this definition we get the following important result:

Theorem 5.2.6 *Given any connected, n-dimensional triangulated pseudo-manifold K with boundary, there exists a triangulated convex polyhedron P in $\mathbb{R}^n$ and affine linear isomorphisms $\phi_i : F_i \to F_i'$ where $F_1, \ldots, F_k, F_1', \ldots, F_k'$ are some (all, if $\partial |K| = \emptyset$) distinct facets of P such that K is the quotient of P by the identifications*

$$x \sim \phi_i(x), \quad x \in F_i, \quad i = 1, 2, \ldots, k. \tag{5.2}$$

Proof: Label the n-simplexes of K so that for $i \geq 2$, any n-simplex σ_i has at least one $(n-1)$-face (i.e., a facet) common with some $\sigma_j, j < i$. [This is possible because of condition (iv) in Theorem 5.2.4.] Let K_j be the subcomplex spanned by $\cup_{i \leq j} \sigma_i$. Inductively, we shall construct triangulated convex polyhedrons $P_1 \subset \cdots \subset P_j \subset P_{j+1} \subset \cdots \subset \mathbb{R}^n$ and surjective simplicial maps $\Theta_j : P_j \to K_j$ such that for each j
(a) Θ_j is a bijection on each simplex and defines a homeomorphism restricted to the interior of P_j;
(b) $\Theta_j|_{P_{j-1}} = \Theta_{j-1}$.

For $j = 1$, we can take P_1 to be any geometric n-simplex in $\mathbb{R}^n$ and $\Theta_1 : P_1 \to K_1$ to be a bijection of vertices. Suppose we have arrived at the stage j. Now σ_{j+1} shares one $(n-1)$-face F with one of the n-simplexes in K_j say, τ, i.e., $\sigma_{j+1} = F \cup \{u\}$ and $\tau = F \cup \{u'\}$. Let $F' \subset P_j$ be such that $\Theta_j(F') = F$. It follows that $F' \subset \partial P_j$. Consider the convex region bounded by the hyperplanes spanned by F' and other facets of P_j intersecting F. If v' is any point in the interior of this convex region, and τ_{j+1} is the convex hull of v' and F', we put $P_{j+1} = P_j \cup \tau_{j+1}$. It is easy to check that P_{j+1} is again a triangulated convex polyhedron. We define Θ_{j+1} to be equal to Θ_j on P_j and $\Theta_{j+1}(v') = v$. Clearly Θ extends to a simplicial map which surjects onto P_{j+1}. Injectivity of Θ_{j+1} in the interior of P_{j+1} is easily verified. By induction, this completes the construction of $P_k = P$ and $\Theta_k = \Theta$. The only points where Θ is not injective are in ∂P. There again, on each facet, Θ is injective. It may happen that two distinct faces are mapped on to the same facet in K. Label them in pairs as F_i, F_i'.

Put $\psi_j = \Theta|_{F_i}, \psi'_j = \Theta_{F'_i}$. Define $f_i = (\psi'_i)^{-1} \circ \psi_i$. It follows that K is then isomorphic as a simplicial complex to the quotient space of P by the relations (5.2), as required. ♠

Remark 5.2.7

(a) The method of proof in the above theorem is the beginning of a technique known as 'cut-and-paste-technique', in low-dimensional topology. We have illustrated this a little bit right in Section 1.1. In the next section, we shall use this technique to classify surfaces.

(b) A partial converse of the above theorem is also true with some minor modifications. First of all, you will have to take the second barycentric subdivision, viz., given a triangulated convex polyhedron P and an equivalence relation on the boundary faces as above, the quotient of the second barycentric subdivision of P is a pseudo-manifold. We leave the details to the reader as an exercise. This result raises the question as to when the quotient is actually a manifold. Indeed, it is not even true that condition (i) of Theorem 5.2.4 holds for such a quotient space in general. However, in dimension ≤ 2, there is no need to put any extra condition and the result is true. In dimension 3, there is a nice criterion in terms of the Euler characteristic due to Poincaré. It uses two important facts, viz.,

(i) A closed surface S is homeomorphic to $\mathbb{S}^2$ iff $\chi(S) = 2$. (See Section 5.3)

(ii) If L is a compact odd dimensional manifold with non empty boundary, then $2\chi(L) = \chi(\partial L)$ (see Remark 5.2.17 (b) below). Let us assume these and now state and see how to prove the above result of Poincaré.

Theorem 5.2.8 *Let K be a pseudo-manifold of dimension 3 (without boundary) obtained by identification of pairs of facets of a convex polyhedron P in $\mathbb{R}^3$. If $\chi(X) = 0$ then $X := |K|$ is a closed 3-manifold.*

Proof: We need to prove the local Euclideanness of X. For points which are images of points of interior of P, there is no problem. Similarly, those points which are in the image of an interior of a facet also, there is no problem, since two 3-simplexes will come together at such a facet and the neighbourhood of the point in the quotient space is the union of two half discs along their base and hence is a full disc. Next we consider a point in the interior of an edge e. The topology of the neighbourhood of such a point depends on the topology of the link of e because $St(e) = Lk(e) \star e$ and hence $|St(e)|$ is homeomorphic to the iterated cone $C(C(|Lk(e)|))$. Clearly, the $Lk(e)$ is 1-dimensional pseudo-manifold. We claim that it must be connected. For if not, we can write $Lk(e) = A \sqcup B$ as a disjoint union of two subcomplexes. We can then partition the set of identifications on the boundary of P into two sets and obtain a quotient space X' in which the points lying over the interior of the edge e will be partitioned to define two different edges e_1, e_2 and X is obtained by a further identification of e_1 with e_2. Since such extra identifications are not allowed, this is a contradiction. Therefore $Lk(e)$ is a circle and hence $St(e)$ is a 3-disc.

It remains to check the local Euclideanness at the image v_i of vertices of ∂P. By passing to a subdivision of K, we may assume that $St(v_i)$ are disjoint in $|K| = X$. We need to show that each $Lk(v_i)$ is a topological 2-sphere, which is the same as showing that $\chi(Lk(v_i)) = 2$ for all i. Let $S = \cup_i St(v_i)$, and let L be the closure of $K \setminus S$ so that $K = L \cup S$ and $L \cap S = \sqcup Lk(v_i)$. Then L is a topological 3-manifold with boundary $\partial L = L \cap S$. We know from 5.2.17 (b) that $\chi(S \cap L) = \chi(\partial L) = 2\chi(L)$. Also,

$$0 = \chi(X) = \chi(K) = \chi(S) + \chi(L) - \chi(S \cap L) = k - \frac{1}{2}\chi(S \cap L).$$

This means that $\sum_{i=1}^{k} \chi(Lk(v_i)) = 2k$. Notice that each $Lk(v_i)$ is a surface and hence $\chi(Lk(v_i)) \leq 2$ with equality holding iff $Lk(v_i) = 2$. It follows that each $Lk(v_i)$ is a 2-sphere for each i. ♠

Definition 5.2.9 Let X be a topological n-manifold. We say a triangulation $f : |K| \to X$ is combinatorial, if the link of every k-simplex in K is combinatorially equivalent to the boundary complex of the standard $(n-k)$-simplex.

Remark 5.2.10 Obviously, the requirement in the above definition is more stringent than the conclusion in the above theorem which holds for any triangulation of a manifold. J. H. C. Whitehead (1940) strengthened the theorem of Cairns by proving that every smooth manifold has a combinatorial triangulation ([Whitehead, 1940]). Kirby and Siebenmann and independently Lashof and Rosenberg came up with a unique obstruction class in $H^4(M, \mathbb{Z}_2)$ for any given topological manifold M of dimension ≥ 5 to admit a combinatorial triangulation. And soon, Siebenmann actually showed that for each $n \geq 5$ there are closed manifolds M of dimension n, which have no combinatorial triangulations.

Definition 5.2.11 Let K be a simplicial complex which triangulates a given n-manifold and $sd\,K$ be its first barycentric subdivision. (See Section 2.8 for details.) Given any k-simplex $\sigma \in K$ consider the subcomplex σ^* of $sd\,K$ defined as follows:

$$\sigma^* = \{(\hat{\sigma}_1, \hat{\sigma}_2, \ldots, \hat{\sigma}_k) \ : \ \sigma \subseteq \sigma_1 \subset \sigma_2 \subset \cdots \subset \sigma_k, \ \ \sigma_i \in K\}.$$

Here $\hat{\tau}$ denotes the barycentre of τ. We call σ^* the *dual cell* of σ.

We need a notion which is a minor modification of the link.

Definition 5.2.12 Let $\sigma \in K$. We shall define $Lk'(\sigma)$ to be the subcomplex of $sd\,K$ given by

$$Lk'(\sigma) := \{(\hat{\sigma}_1, \hat{\sigma}_2, \ldots, \hat{\sigma}_k) \ : \ \sigma \subset \sigma_1 \subset \sigma_2 \subset \cdots \subset \sigma_k, \ \ \sigma_i \in K, \sigma \neq \sigma_1\}$$

and call it the *link of σ in $sd\,K$*.

Remark 5.2.13 Notice the difference in the usual definition of the $Lk_L(\sigma)$, as introduced in Definition 2.10.3 and $Lk'(\sigma)$ one. There, σ is taken be a simplex in a subcomplex L of K. Here there is no such requirement nor $sd\,K$ is a subcomplex of K.

Clearly $Lk'\sigma \subset \sigma^*$. The following lemma is easy to verify.

Lemma 5.2.14 The inclusion of vertices extends linearly to define a piecewise linear homeomorphism $Lk'(\sigma) \star \{\hat{\sigma}\} \to \sigma^*$. Also there is a canonical isomorphism $\eta : Lk(\sigma) \to Lk'(\sigma)$ given by the vertex map η which maps v to the barycentre of the simplex $\sigma \cup \{v\}$.

Theorem 5.2.15 *If $|K|$ is a combinatorial triangulation of a n-manifold, then for each k-simplex σ in K, $|\sigma^*|$ is a $(n-k)$-cell.*

Proof: Since $Lk(\sigma)$ is combinatorially equivalent to the boundary complex $\mathcal{B}(\Delta_{n-k})$ and hence $|Lk(\sigma)|$ is a topological $(n-k-1)$-sphere. Now use Lemma 5.2.14. ♠

Theorem 5.2.16 *Let K be a combinatorial triangulation of a n-manifold. Define $X_K^{(i)}$ to be the union of all $|\sigma^*|$, where σ ranges over all $(n-i)$-simplices of K. Then $X_K = \cup_i X_K^{(i)}$ defines a CW-structure on $|K|$.*

Proof: By the previous theorem each $X_K^{(i)}$ is a union of cells. The boundary of the i-cell σ^* is clearly contained in $X_K^{(i-1)}$. So the characteristic maps are simply the inclusions. Finally, for any $\tau \in sd\,K$, since $\tau = (\hat{\sigma}_1, \ldots, \hat{\sigma}_k)$ for some chain $\sigma_1 \subset \cdots \subset \sigma_k$ of simplexes in K, it is clear that the open simplex $\langle \tau \rangle$ is contained in the unique dual cell σ_1^*. From this it follows that X_K defines a cell decomposition of $|sd\,K| = |K|$. ♠

Remark 5.2.17

(a) Note that the assignment $\sigma \mapsto \sigma^*$ defines a bijection of i-simplexes of K with the $(n-i)$-cells of X_K. Already this implies that for an odd dimensional closed combinatorial manifold, the Euler characteristic is 0 since we can compute it in two different ways: $\chi(K) = \chi(X) = \chi(X_K)$ and we have

$$\chi(K) = \sum_i (-1)^i f_i(K) = \sum_i (-1)^i f_{n-i}(X_K) = -\sum_i (-1)^{n-i} f_{n-i}(X_k) = -\chi(X_K).$$

where f_i is the number of i-faces of K, and $f_i(X_K)$ are number of i-cells in the CW-complex X_K.

(b) As a simple consequence of this, we can derive the formula $2\chi(X) = \chi(\partial X)$ where X is a compact, odd dimensional combinatorial manifold with boundary ∂X. For then, we can glue two copies of X along their boundary via the identity map to obtain the double $2X$ which becomes a closed combinatorial manifold of the same dimension. But then $0 = \chi(2X) = 2\chi(X) - \chi(\partial X)$ (see Exercise 2.11. (iii)).

(c) Poincaré of course did not stop here. We shall describe his great result here, which says that for a n-manifold X, the ith Betti number is equal to the $(n-i)^{th}$ Betti number, $\beta_i(X) = \beta_{n-i}(X)$. (The modern versions of this great theorem will be discussed in a latter chapter.) So, we make an artificial and temporary device here. Let us work with field coefficients R. Given a finitely generated chain complex (A_*, ∂) of R-modules, consider the 'inverted dual' complex $(B_*, \hat{\partial})$ defined as follows: $B_i = \text{Hom}(A_{n-i}, R)$, $\hat{\partial} : B_i \to B_{i-1}$ is given by $\hat{\partial}(\phi) = \phi \circ \partial$. Verification that $(B_*, \hat{\partial})$ is a chain complex is straightforward. We denote the homology modules $H^i(K; R) := H_i(B_*)$. The important thing to note is that the rank of the homology modules of this chain complex are the Betti numbers of X labeled in the reverse order:

$$\text{rank } H_i(A) = \text{rank } H_{n-i}(B) = \text{rank } H^{n-i}(K; R).$$

Theorem 5.2.18 (A simple version of Poincaré duality) *Let M be a closed, triangulated, and orientable manifold of dimension n. Then its Betti numbers satisfy the relation $\beta_i(M) = \beta_{n-i}(M)$.*

We shall first prove a mod 2 version of this statement, just to understand the basic idea clearly, and so that the notion of orientability does not trouble us.

Proposition 5.2.19 Let K be a combinatorial triangulation of a compact manifold. Let $W_*(K)$ denote the CW-chain complex of $\mathbb{Z}_2$-vector spaces with basis over the dual cells of $sd\, K$, as in the previous theorem. Let $C(K)$ denote the simplicial chain complex (over $\mathbb{Z}_2$) of K. Let $\phi : C_k(K) \to B_k = \text{Hom}(W_{n-k}(K), \mathbb{Z}_2)$ denote the linear map given by

$$\phi(\sigma)(\tau^*) = \begin{cases} 1, & \text{if } \tau = \sigma, \\ 0, & \text{otherwise,} \end{cases}$$

and extended linearly. Then ϕ is an isomorphism of the chain complexes $C_* \to B_*$. In particular ϕ induces isomorphisms $H_k(K; \mathbb{Z}_2) \to H^{n-k}(K; \mathbb{Z}_2)$.

Proof: The only thing that needs to be checked is the fact that ϕ is a chain map, i.e., for each k we must show that the following diagram is commutative.

$$
\begin{array}{ccc}
C_k & \xrightarrow{\;\;\partial\;\;} & C_{k-1} \\
{\scriptstyle\phi}\downarrow & & \downarrow{\scriptstyle\phi} \\
\text{Hom}(W_{n-k}; \mathbb{Z}_2) & \xrightarrow{\;\;\hat{\partial}\;\;} & \text{Hom}(W_{n-k+1}; \mathbb{Z}_2)
\end{array}
$$

So, for every k-simplex $F \in K$ we must verify that $\phi \circ \partial(F) = \hat{\partial} \circ \phi(F)$. But by definition of $\hat{\partial}$, we have $\hat{\partial} \circ \phi(F) = \phi(F) \circ \partial$. Therefore, we must show that for each $(k-1)$-simplex $G \in K$,

$$\phi \circ \left(\sum_i F_i \right)(G^*) = \phi(F) \circ \partial(G^*)$$

where G^* denotes the dual $(n-k+1)$-cell in $sd\,K$ and where F_i denote the $(k-1)$-faces of F. The left hand side is 1 iff $G = F_i$ for some i. The right hand side equals 1 iff in the sum ∂G^*, the dual cell F^* occurs. This is the same as saying that the boundary of the dual cell G^* contains the dual cell F^* which is the same as saying that G is facet of F. This set theoretic fact is easily verified. ♠

We shall now sketch the proof of Theorem 5.2.18 in the orientable case.

Definition 5.2.20 By an orientation on a n-simplex for $n \geq 1$, we mean an equivalence class of labelling the vertices, two such labellings being treated equivalent if one is obtained from the other via an even permutation.

Remark 5.2.21 It follows easily that there are precisely two orientations on every n-simplex, $n \geq 1$. We call them orientations opposite of each other. Often an oriented n-simplex is displayed as $[v_0, \ldots, v_n]$. Sometimes when one of the orientations is preferred for some reason and referred to as positive orientation, then the other one is called the negative orientation. For instance on the standard simplex $\Delta_n = \{e_0, e_1, \ldots, e_n\}$, $[e_0, \ldots, e_n]$ is called the positive orientation. Also, we have $[v_0, v_1] = -[v_1, v_0]$.

For the sake of logical consistency, we define two orientations on every 0-simplex v as well, and denote them by $[v]$ and $-[v]$.

Definition 5.2.22 Let $\sigma = [v_0, \ldots, v_n]$ be an oriented simplex. Then $(-1)^i [v_0, \ldots, \hat{v}_i, \ldots, v_n]$ are the $(n-1)$-faces of σ with the induced orientations.

For instance, the oriented vertices of the oriented edge $[v_0, v_1]$ are $-[v_0], [v_1]$; the oriented edges of the oriented triangle $[v_0, v_1, v_2]$ are $[v_0, v_1], [v_1, v_2]$ and $[v_2, v_0] = -[v_0, v_2]$.

Definition 5.2.23 Let K be a triangulation of a n-manifold. By an orientation on K, we mean a choice of orientation on each n-simplex so that the orientations induced on a common $(n-1)$-face from any two n-simplices are opposite.

Remark 5.2.24 Not all combinatorial manifolds are orientable. It turns out that given a manifold X whether a combinatorial triangulation of X is orientable or not depends just on X itself and not on the particular choice of the triangulation.

Now, let K be a combinatorial triangulation of a n-manifold which is orientable. Fix an orientation on K. Given a n-simplex F, the barycentric subdivision $F' \subset sd\,K$ automatically acquires an orientation from that of F. On the other hand, if G is a k-face of F with an arbitrary orientation, then each $(n-k)$-simplex H in $Lk_{F'}(G)$ can be given an orientation so that the orientation of a k-simplex of G' followed by the orientation of H gives the orientation of the n-simplexes of F'. One has to verify that with these orientations the dual cell G^* gets oriented. This is where the (global) orientability condition has to be used. We can now take the CW-chain complex of X_K with these oriented cells as generators. The gist is that now we can work with integer coefficients and the entire discussion above that we had for the $\mathbb{Z}_2$-coefficients goes through and yields the following duality theorem, from which theorem 5.2.18 follows. We leave the details to the reader.

Theorem 5.2.25 Oriented Poincaré Duality *Let K be an oriented combinatorial triangulation of a compact manifold. Then there is a chain isomorphism $\phi : C_k(K) \to \mathrm{Hom}_{\mathbb{Z}}(W_{n-k}(K); \mathbb{Z})$. In particular, $H_k(K; \mathbb{Z}) \approx H^{n-k}(K; \mathbb{Z})$.*

Remark 5.2.26

1. All these results can be extended to arbitrary triangulations of manifolds (not necessarily combinatorial ones) and manifolds with boundary as well, which we shall not discuss here, since we are going to treat this in more generality in a later chapter.

2. Given a triangulable n-manifold, we would like to minimize either the number of vertices or the number of n-simplexes required. Many related questions together form an active branch of combinatorial topology called minimal triangulations. See [Datta, 2007] for more details.

3. There is then a generalization of the notion of a simplicial complex to what are known as polyhedrons, in which your building blocks are not necessarily simplexes but more general convex polytopes in the Euclidean spaces. They form the objects of a category called **PL**-category which lies between the categories **Diff** and **Top**.

4. We shall end this section with a digressional note if only to indicate how a very important result due to Munkres can be derived effortlessly from Lemma 5.2.2. In the following theorem statement (b) is Reisner's condition for the face ring of a simplicial complex K to be *Cohen–Macaulay*. These results were used in a non trivial way by Stanley in the solution of the upper bound conjecture.

Theorem 5.2.27 (Munkres) *With $X = |K|$ where K is a finite simplicial complex of dimension n, the following two statements are equivalent:*
(a) $\widetilde{H}_i(X) = \widetilde{H}_i(X, X \setminus \{x\}) = (0)$ *for $i < n$ and for every $x \in X$.*
(b) $\widetilde{H}_i(Lk(F)) = (0)$ *for $i < \dim Lk(F)$ and for every face $F \in K$.*

Observe that since the empty set is also a face in K, according to our convention, and $Lk_K(\emptyset) = K$, condition (b) implies that $\widetilde{H}_i(K) = (0)$ for $i < n$.

We shall first prove two lemmas.

Lemma 5.2.28 Either of the conditions (a) and (b) implies that K is pure.

Proof: Suppose (a) holds. Let F be a maximal simplex in K and x be a point in the open simplex $< F >$. Then

$$
\begin{aligned}
H_i(X, X \setminus \{x\}) &\approx H_i(| St(F) |, | St(F) | \setminus \{x\}) \\
&= H_i(| F |, | F | \setminus \{x\}) \\
&\approx H_i(| F |, | \mathcal{B}(F) |) \approx H_i(\mathbb{D}^k, \mathbb{S}^{k-1})
\end{aligned}
$$

where $k = \dim F$. By (a), $k \geq n$ and hence $k = n$. Now suppose (b) holds. Suppose K is not pure. Let L_1 be the collection of all n-simplices and their faces. Let L_2 be the collection of all simplices F which are not a face of any n-simplex, and all the faces of such simplices. Then L_1 and L_2 are subcomplexes of K, $K = L_1 \cup L_2$ and $L_1 \neq \emptyset \neq L_2$. Since X is connected $L_1 \cap L_2 \neq \emptyset$. Let σ be a maximal simplex in $L_1 \cap L_2$. Then $\sigma \neq \emptyset$ and is a proper face of a simplex F in L_2 and hence

$$\dim \sigma < \dim F < n.$$

Also σ is a proper face of a n-simplex $F' \in L_1$. Hence $Lk(\sigma)$ intersects both $L_1 \setminus \sigma$ and $L_2 \setminus \sigma$. Since σ is maximal in $L_1 \cap L_2$, it follows that $Lk(\sigma)$ is disconnected. Hence $\widetilde{H}_0(Lk(\sigma)) \neq (0)$. But, $\dim \sigma \leq n - 2$ and $\sigma \subset s' \in L_1$ and so $\dim (Lk(\sigma)) \geq 1$. This is contradicting (b). ♠

Remark 5.2.29 Observe that we have used only the second part of condition (a) and condition (b), only for non empty faces of K in proving the purity. It is worth watching our steps in this light, in the proof of Theorem 5.2.27 that follows.

Proof of the Theorem 5.2.27 By Lemma 5.2.28 either of the conditions implies K is pure. Hence for each $F \in K, |St(F)|$ has dimension n. Hence

$$\dim Lk(F) + \dim F + 1 = n.$$

Now,

condition (a)

$\Longleftrightarrow$ $\widetilde{H}_i(X) = (0)$ for $i < n$ and

$\widetilde{H}_{j-\dim \ F-1}(Lk(F)) = (0) \ \forall \ \emptyset \neq \ F \in K,$ and $j < n$ (by Lemma 5.2.2).

$\Longleftrightarrow$ $\widetilde{H}_j(X) = (0)$ and

$\widetilde{H}_{j-n+\dim \ Lk(F)}(Lk(F)) = (0)$ for $j < n$ and for every $F \in K$.

$\Longleftrightarrow$ condition (b).

This completes the proof of Munkres' theorem. ♠

5.3 Classification of Surfaces

In this section, we shall study the classification problem for surfaces. Originally, this is attributed to August Möbius (1790–1868) and Camille Jordan (1838–1922) who considered surfaces embedded in the 3-dimensional Euclidean space. Our treatment is somewhat general. We shall first of all show that every compact surface is triangulable and then use this triangulation for getting a classification up to homeomorphism. For a classification of smooth surfaces up to diffeomorphism using Morse theory, see [Shastri, 2011]. It turns out that both these classifications coincide, in the sense that every homeomorphism class of a compact surface contains a unique diffeomorphism class.

Definition 5.3.1 By a *surface* we shall mean a compact 2-dimensional connected topological manifold (without boundary).

In this section we shall first show that every connected compact 2-dimensional manifold without boundary is *triangulable* and then use this to classify them. The proof of the classification is combinatorial in nature. It turns out that just orientability and Euler characteristic are enough invariants to classify them. The connectivity of the surface is not a logical necessity but only a convenience as we immediately see that we can first write the given compact surface as a union of its connected components.

The proof of the existence of triangulations on a surface that we are going to present now is due to Tibor Rado (1920), and we shall follow the presentation given in Ahlfors and Sario [Ahlfors–Sario, 1960], with minor modifications. We observe that we need to use the famous Jordan curve theorem(JCT). This is not a problem since we have already proved the celebrated Jordan–Brouwer separation theorem. Indeed, we shall actually need the stronger form of it which goes under the name Jordan–Schönflies theorem(JST) which was proved by Schönflies around 1892. A proof of this can be found in [Moise, 1977]. More recently, another proof of this appeared in the American Mathematical Monthly (see [Thomassen, 1992]). According to the author himself, this proof is not much simpler than the earlier ones but somewhat shorter. The novelty in [Thomassen, 1992] is that it uses graph theory: The entire thing hinges on the Kuratowski's criterion for planarity of graphs. The point of the above remarks is to indicate to what extent we can keep these results elementary, for example without bringing in the heavy machinery of homology theory. However, as we shall see, for the uniqueness part of the classification, we need to bring in homology or the fundamental group.

Theorem 5.3.2 (Jordan–Schönflies theorem) *Every Jordan curve γ in $\mathbb{C}$ can be mapped onto the unit circle in $\mathbb{C}$ by a homeomorphism $\mathbb{C} \longrightarrow \mathbb{C}$.*

Here, by a *Jordan curve* we mean a simple closed curve. Analogously, we will use the word *Jordan arc* to mean a simple arc, i.e., the image of a one-to-one continuous mapping defined on a closed interval. The bounded component of the complement of a Jordan curve in $\mathbb{C}$ is called a *Jordan region*. We shall use the term *Jordan path* to mean either a Jordan curve or a Jordan arc. The two important immediate consequences of the JST that we are going to use are:

Lemma 5.3.3 A Jordan path in $\mathbb{C}$ has no interior.

Lemma 5.3.4 The closure of each Jordan region is a 2-cell, i.e., homeomorphic to the closed disc.

These two consequences can also be derived directly from JCT and the Riemann mapping theorem if you like. For Lemma 5.3.3, we can also use the theorem of invariance of domain. So, to that extent, we need not really have to depend upon JST. In any case, using JST, observe that Lemmas 5.3.3 and 5.3.4 follow from the corresponding statements about the unit circle in place of an arbitrary Jordan curve. For the circle itself these statements are obvious. The proof of Lemmas 5.3.5, 5.3.6 and 5.3.7 are left to the reader just to catch her attention to what is going on.

Lemma 5.3.5 Let γ be a Jordan arc in a Jordan region A such that $\gamma \cap \partial A = \partial \gamma$. Then $A \setminus \gamma$ consists of two Jordan regions.

Lemma 5.3.6 Let A be a Jordan region and B be a Jordan subregion. Let the Jordan curve $\gamma = \partial B$ be such that $\gamma \cap \partial \overline{A}$ is a singleton set. Then $A \setminus \overline{B}$ is an open 2-cell.

Lemma 5.3.7 Let Γ be a finite connected pseudo-graph embedded in $\mathbb{C}$. (See Definition 3.8.1.) Then $\mathbb{C} \setminus \Gamma$ consists of an unbounded component and a finite number of Jordan regions and open 2-cells.

Definition 5.3.8 Let $\{U_i\}$ be a family of open sets in a surface S. We say $\{U_i\}$ is of *finite character* if the following conditions are satisfied:
(i) The family $\{U_i\}$ is locally finite, i.e., each point of S has a neighbourhood which intersects only finitely many U_i's.
(ii) The closure $\overline{U_i}$ of U_i in S is a closed 2-cell.
(iii) Each $J_i := \partial \overline{U_i}$ meets at most finitely many other J_j's.
(iv) $J_i \cap J_j$ has finitely many connected components (which may be either arcs or points) for each i, j.

We shall first prove:

Proposition 5.3.9 A surface is triangulable if it has an open covering of finite character.

(Observe that as stated above, the result is true even for non-compact surfaces. We shall prove the theorem only for the compact case. The idea of the proof is the same for the non-compact case also, though the details would be different.)

Proof: Let $\{U_i\}$ be a finite open covering of S that is of finite character. By discarding any set U_i contained inside another U_j, we shall assume that no U_i is contained in any other U_j. Clearly the number of U_i is at least two.

Case 1. Suppose that for some $i \neq j$, we have $J_i \subset U_j$. In this case we will actually show that S is homeomorphic to the sphere $\mathbb{S}^2$. Since the sphere is triangulable, this will do. We claim that $U_i \cup U_j = S$. This will follow if we show that $U_i \cup U_j$ is closed in S, because this set is already open and by assumption S is connected. Since J_i is a Jordan curve in U_j, it follows that $U_j \setminus J_i$ consists of a Jordan region A and another region, say T. Also J_i is the common boundary of these two regions. Since U_i is connected and not contained in U_j, it follows that $U_i \cap A = \emptyset$ and $T \cap U_i \neq \emptyset$. But then, since T is connected and does not intersect J_i, it follows that $T \subset U_i$. Hence $\overline{T} \subseteq \overline{U_i}$ and in particular, $J_j \subset U_i$. Therefore,

$$\overline{U_i \cup U_j} \subseteq \overline{U_i} \cup \overline{U_j} \subseteq U_i \cup J_i \cup U_j \cup J_j \subseteq U_i \cup U_j.$$

Therefore, $U_i \cup U_j$ is closed as required. This also shows that $S = A \cup \overline{U_i}$. If $h : U_i \longrightarrow \mathbb{D}^2_+$ is a homeomorphism, where $\mathbb{D}^2_+$ denotes the upper hemisphere in $\mathbb{R}^3$, then we can extend this homeomorphism to a homeomorphism of $S \longrightarrow \mathbb{S}^2$, by mapping $\overline{A}$ homeomorphically onto the lower hemisphere. (Any homeomorphism of $\mathbb{S}^1$ to itself extends to a homeomorphism of $\mathbb{D}^2$ to itself.)

Case 2. We can now assume that no J_i is contained in any U_j. This means that for each i, j such that $U_i \cap U_j \neq \emptyset$, we have $J_i \cap J_j \neq \emptyset$. Also, $J_i \cap \overline{U_j}$ consists of finitely many Jordan curves and Jordan arcs each with its boundary on J_j. (Observe that it may happen that $J_i \subset \overline{U_j}$ with $J_i \cap J_j$ being a singleton set.) It follows that $\Gamma := \cup_i J_i$ is connected and $S \setminus \Gamma$ is the finite union of open 2-cells. In other words, Γ is an embedding of a connected pseudo-graph. We first subdivide the pseudo-graph Γ into a graph by introducing extra vertices in the interior of each of the Jordan arcs (and curves) that make up Γ. By choosing a point in the interior of each 2-cell in $S \setminus \Gamma$, we then perform the star-construction to obtain a triangulation of S. ♠

Before proceeding with the proof of the existence of a covering with finite character we make a technical definition.

Definition 5.3.10 Let Γ be a set of Jordan paths on a surface S. We say Γ is *discrete* in S if every point of S has a neighbourhood U that meets finitely many arcs or curves in Γ.

Observe that any finite set of Jordan paths is discrete. The following lemma is crucial, in the construction of a covering of finite character.

Lemma 5.3.11 Let Γ be discrete in S. Then for any region Ω in S, $\Omega \cap \Gamma$ is discrete in Ω.

Proof: Let $x \in \Omega$. Choose a neighbourhood U of x such that $\overline{U} \subset \Omega$ and U meets only finitely many paths in Γ. If infinitely many paths in $\Gamma \cap \Omega$ meet U then it follows that there is at least one path, say γ, in Γ such that infinitely many components of $\gamma \cap \Omega$ meet U. Starting at some point on γ and moving along γ in one or the other direction, we can obtain a sequence of points on the (Jordan) path γ say $x_1, y_1, x_2, y_2 \ldots$ such that $x_i \in U$ and $y_i \notin \Omega$. Passing to a subsequence if necessary, we may assume that this sequence converges to a point z. Since this point is the limit of $\{x_i\}$ it is in $\overline{U}$. Since it is also the limit of $\{y_i\}$ it cannot be in the open set Ω. This contradicts the fact $\overline{U} \subset \Omega$. Hence U meets finitely many arcs of $\Gamma \cap \Omega$. ♠

Lemma 5.3.12 Let Γ be a set of Jordan paths in the closed unit disc $\mathbb{D}^2$ in $\mathbb{R}^2$, so that $\Gamma \cap \operatorname{int} \mathbb{D}^2$ is discrete in $\operatorname{int} \mathbb{D}^2$. Given any two points $z_1, z_2 \in A := \mathbb{D}^2 \setminus \Gamma \cap \partial \mathbb{D}^2$, there exists a Jordan path γ in $\mathbb{D}^2$, joining z_1, z_2 and intersecting each component of Γ in finitely many arcs and points and lying in $\operatorname{int} \mathbb{D}^2$ except perhaps for its end-points.

Proof: Observe that if we can join z_1 to z_2 and also z_2 to z_3 inside $\mathbb{D}^2$ in the manner described in the lemma, then it is clear that we can join z_1 to z_3 in the same manner. Thus it is enough to prove the lemma for a fixed $z_1 \in A$. So, let X be the set of all points of A that can be joined to z_1 in this manner. We have to show that $X = A$. (We use a typical topological argument here: first showing that X is a non empty open and closed subset of A and then appealing to the connectivity of A.) To see that X is open, let $z \in X$. Choose a neighbourhood U of z inside $\mathbb{D}^2$ such that U intersects only those arcs of Γ to which z belongs. If z is on the boundary of $\mathbb{D}^2$, then this is possible since z is not a point of Γ. If it is in $\operatorname{int} \mathbb{D}^2$ then this is possible since $\Gamma \cap \operatorname{int} \mathbb{D}^2$ is discrete. (Observe that in case $z \notin \Gamma$, this means that U does not meet Γ.) Now in order to prove that $U \subseteq X$, it suffices to show that each point in U can be joined to z in the above manner, inside $\mathbb{D}^2$. So, given any $x \in U$ take any Jordan arc λ in U joining x to z. If λ has the required property, then well and good! In any case, take the arc λ_1 on λ that starts at x and has its other end-point y_1, the point at which λ meets $\Gamma \cap U$ for the first time. Let λ_2 be the arc in Γ from y_1 to z. Then we see that $\lambda_1 \cup \lambda_2$ is an arc joining x to z inside $\mathbb{D}^2$ and having the required property. The argument to show that $A \setminus X$ is open is exactly similar.   ♠

The next proposition will complete the proof of the existence of triangulation.

Proposition 5.3.13 For any surface S, there exists an open covering of S of finite character.

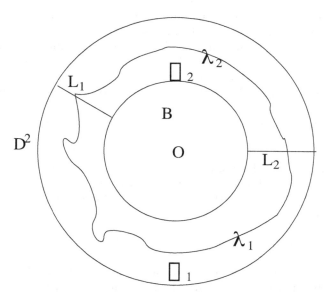

FIGURE 5.1. Construction of an open covering of finite character

Proof: Here again we shall prove the theorem for compact surfaces only. Let $h_i : \mathbb{D}^2 \longrightarrow S$ be homeomorphisms of the closed disc onto closed subsets of S such that the images of the half-discs form a finite open cover of S, i.e., $S = \cup_{i=1}^{n} h_i(B)$, where B denotes the open disc of radius $1/2$ around 0. Put $U_i = h_i(B)$, $V_i = \operatorname{int} h_i(\mathbb{D}^2)$. We shall find open 2-cells W_i such that $U_i \subseteq W_i \subset V_i$, $\forall\, i$ and such that the covering W_i is of finite character.

Take $W_1 = U_1$. Assume that we have already found W_i, $1 \leq i \leq k-1$, such that $\{W_1, W_2, \ldots, W_{k-1}\}$ is of finite character. We should find W_k such that $U_k \subseteq W_k \subset V_k$ and such that $\gamma_k := \partial W_k$ intersects $\gamma_1 \cup \gamma_2 \cup \cdots \cup \gamma_{k-1}$ in finitely many arcs and points. (If ∂U_k has this property then we can choose $W_k = U_k$, but this may not be the case). Let $\Gamma = \gamma_1 \cup \gamma_2 \cdots \cup \gamma_{k-1}$. Then, each curve in Γ is a Jordan curve and Γ has finitely many such curves. Thus Γ has no interior. Let us put $\Gamma' = h_k^{-1}(\Gamma)$. Then Γ' has no interior in $\mathbb{D}^2$. So we can choose points z_1 and z_2 in the interior of the annulus $\mathbb{D}^2 \setminus B$ such that z_i do not lie on Γ and also they are not on the same line through 0. Let L_i be the radial segments through z_i to the boundary of the annulus. Then the annulus $\mathbb{D}^2 \setminus B$ is cut into two Jordan regions Ω_1 and Ω_2, respectively, by L_1 and L_2. (See Figure 5.1.) By Lemma 5.3.11, it follows that $\Omega_i \cap \Gamma'$ is discrete in Ω_i, for $i = 1, 2$, respectively. Join z_1 and z_2 by two Jordan arcs λ_i with their interior lying inside Ω_i, respectively, and intersecting Γ in finitely many arcs and points (Lemma 5.3.12). Then $\lambda_1 \cup \lambda_2$ is a Jordan path lying in the annulus, bounding an open 2-cell R inside $\mathbb{D}^2$. Take $W_k = h_k(R)$. This completes the proof of the proposition. ♠

Toward the classification of surfaces, we fix a triangulation on a connected compact surface X. We then appeal to Theorem 5.2.6 which provides us a representation of X as a quotient of convex polygon P of $2n$ sides whose sides have to be identified pairwise by homeomorphisms which are linear on each side. This process can be completely described purely combinatorially as follows:

Let us agree once and for all, that we shall trace the boundary any convex polygon in the anticlockwise direction. We are free to start from any vertex. We then label the edges by letters $\mathbf{a}, \mathbf{b}, \mathbf{c}$, etc. As soon as we meet an edge which is being identified with an edge which has already been labeled, we shall use the same letter to label this new edge also. However, we have to take care of another aspect, viz., whether the identification is orientation preserving or orientation reversing. To indicate the orientation reversing we shall use labellings $\mathbf{a}^{-1}, \mathbf{b}^{-1}, \mathbf{c}^{-1}$, etc. Of course, we stop as soon as we have arrived back where we started. Since the starting point is arbitrary, what this process yields is that the surface X is completely determined by the cyclic sequence

$$\mathbf{a}_1^{\epsilon_1} \mathbf{a}_2^{\epsilon_2} \ldots, \mathbf{a}_n^{\epsilon_n} \tag{5.3}$$

in which each letter occurs precisely twice. For the simplicity of the notation, for an edge $\mathbf{a}^{+1}$ occurring with $+1$ sign, we drop this sign and simply write it as a. A sequence such as (5.3) is called a *a canonical polygon*. We shall use bold face capitals $\mathbf{A}, \mathbf{B}$, etc., to denote any sequence such as (5.3). For $n \geq 4$, we can represent a canonical polygon by a regular convex polygon P in $\mathbb{R}^2$ with n sides with its sides appropriately labeled. Observe that we allow the exceptional case when $n = 2$ also. In this case, we do not get a convex polygon in $\mathbb{R}^2$. However, in this case, we take P to be the unit disc with its boundary being divided into two edges, the sequence itself being $\mathbf{aa}^{-1}$ or $\mathbf{aa}$. From Exercise 1.3.11.(iii) (a),(b), it follows that we get $\mathbb{S}^2$ in the first case, and $\mathbb{P}^2$ in the second case. Therefore, in what follows we can concentrate only on the case $n \geq 4$.

If both $\mathbf{a}$ and $\mathbf{a}^{-1}$ occur in this sequence for some edge $\mathbf{a}$, we call that pair of edges $\{\mathbf{a}, \mathbf{a}^{-1}\}$ *type I pair*. Otherwise, the pair is of *type II*. While representing a canonical polygon by a picture, instead of the exponents ± 1 over the letters, we shall use arrows to indicate direction.

Several canonical polygons may define the same surface up to homeomorphism. Our next step is to make a list which should include all possible topological types as well as have no redundancy. Next, we must identify some simple transformations on the canonical polygons which do not change the topological type and then keep applying these transformations so as to bring any given canonical polygon to one in the list. Let us illustrate this with an example.

Example 5.3.14 Consider the surface given by the sequence $\mathbf{abab^{-1}}$. We have seen that this represents the Klein bottle. Mark the diagonal of the square piece of the paper with a (thick) arrow and letter $\mathbf{c}$ and cut the square along this arrow as shown in Figure 5.2. Bring the upper triangular piece down, flip it and identify the two triangles along the edge $\mathbf{a}$, taking care to preserve the orientation of the edge while identifying. The new figure is not a rectangle but we treat it as a rectangle with the sides marked $\mathbf{bbcc}$ which is a simpler canonical polygon which represents connected sum of two copies of $\mathbb{P}^2$.

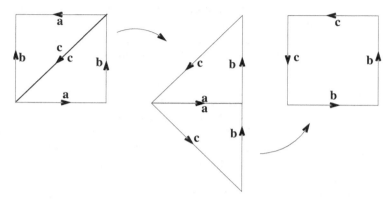

FIGURE 5.2. Transformations of canonical polygons

Theorem 5.3.15
(A) *Every compact connected surface without boundary is homeomorphic to the surface defined by one of the following canonical polygons:*
(i) $\mathbf{aa^{-1}}$.
(ii) $\mathbf{a_1 b_1 a_1^{-1} b_1^{-1} \cdots a_g b_g a_g^{-1} b_g^{-1}}$, $g \geq 1$.
(iii) $\mathbf{a_1 a_1 \cdots a_n a_n}$, $n \geq 1$.
(B) *Moreover, any two distinct members of this list give surfaces which are non homeomorphic.*

The canonical polygons listed above are said to be in the *normal form*. The rest of this section will be occupied by the proof of part (A) of the theorem which will be achieved in five steps, and then giving three different proofs of the uniqueness part (B).

Beginning with an arbitrary canonical polygon, we would like to *transform* it to one in the normal form without changing the homeomorphism type of the surface defined by it.

First observe that in (ii) and (iii), all the vertices are identified to a single point. This may not be true for an arbitrary canonical polygon. So, our first aim would be find a reduction process which will ensure that all vertices are identified to a single point. One of the simplest thing to do is to cancel out a pair of edges $\cdots \mathbf{aa^{-1}} \cdots$

Step 1 Elimination of adjacent edges of type I:
We will now show that if D has at least four edges, then an adjacent pair of edges of type I can be eliminated, until we end up with case (i) or there are no adjacent pairs of edges of type I. Figure 5.3 illustrates this process. And Exercise 1.3.11(iii) provides justification. The relation may be coded as:

$$\mathbf{Aaa^{-1}} \sim \mathbf{A}. \tag{5.4}$$

We can assume that the new P is again convex. By repeated application of this, we eliminate all adjacent pairs of type I.

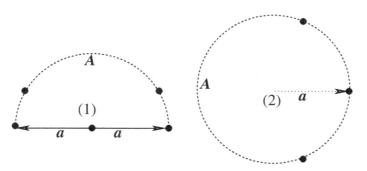

FIGURE 5.3. Elimination of adjacent edges of type I

Step 2 Reduction to the single-vertex-class case:

We consider the equivalence classes of vertices under the quotient map $P \to X$. By step 1, we assume that there are no adjacent edges of type I. Suppose there are at least two equivalence classes of vertices and consider a class [P] with the least number of elements in it. We can pick a vertex P in this class so that the next vertex Q on ∂P is not in this class (see Figure 5.4). Let R be the other adjacent vertex.

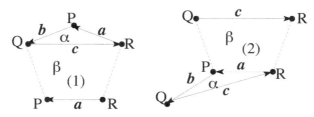

FIGURE 5.4. Reduction to single vertex class

Make a cut along the line labeled **c** from Q to the other vertex of the edge **a**. Glue the two edges labelled **a** together. In the new polygon, there is one less vertex in the equivalence class of P and one more vertex in the equivalence class of Q. The corresponding relation reads as follows:

$$\mathbf{abAa^{-1}B} \sim \mathbf{c^{-1}AbcB}. \tag{5.5}$$

If needed, we now carry out Step 1 again. Note that each time Step 1 is performed, the number of edges as well as the number of vertices go down without increasing the number of equivalence classes of vertices. Step 2, on the other hand, keeps all these numbers the same. By repeated applications of these two steps, we keep reducing the number of vertices from the classes with the least number of vertices each time and hence, at some stage one of these classes has to disappear. (Indeed, if a class of vertices has just one vertex in it, this implies the two edges incident at this vertex must of the type I and hence we can perform Step 1 to get rid of them.) This way we keep reducing the number of classes themselves until there is only one class left.

Thus from now on we shall also assume that all vertices are identified to a single point. It is worth noting that none of the reduction operations that we are going to perform from now onward will disturb this property. (In fact, the only way to create another equivalence class of vertices is to perform the reverse operation of Step 1.)

Step 3 To make any pair of edges of type II adjacent:

Suppose there is a non-adjacent pair of edges of type II labeled **a**, as depicted in Figure 5.5.

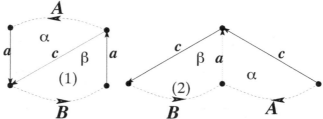

FIGURE 5.5. Bringing type-II edges together

Cut along the line labeled **c** and paste along **a**. By repeated application of this, we make all pairs of edges of type II adjacent. The corresponding relation may be coded as:

$$\mathbf{aAaB} \sim \mathbf{ccBA}^{-1}. \tag{5.6}$$

Note that such an operation does not disturb some other pair **bb** which is already adjacent. Therefore, repeated application of this will make all the pairs of type II adjacent. At this stage if there is no pair of edges of type I left, we have found the canonical polygon has become $\mathbf{a_1a_1a_2a_2\cdots a_na_n}$ as in (iii) of Theorem 5.3.15.

Step 4 Handling a pair of edges of type I: Suppose there is at least one pair of edges of type I, say labeled **a**, (which is necessarily not adjacent). Then P has the form as shown in Figure 5.6. where both A, B are non empty.

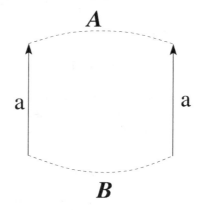

FIGURE 5.6. Handling the interlocked edges

If no edge in **A** is identified with any edge in **B**, then there will be at least two equivalence classes of vertices, viz., the endpoints of **a**. This contradicts the assumption that there is only one equivalence class of vertices. Therefore at least one edge in **A** is identified with one edge in **B**. This pair cannot be of type II, since all pairs of type II are adjacent. Therefore the polygon P has the form as depicted in Figure 5.7(1).

We will transform the polygon so that the two interlocked pairs of edges become consecutive.

First cut Figure 5.7(1) along **c** and paste together along **b** to obtain 5.7(2). The polygon 5.7(3) is nothing but 5.7(2) redrawn so as to be represented by a convex polygon. Cut 5.7(3)

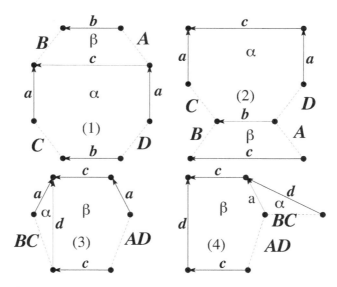

FIGURE 5.7. Inter-locked edges brought together

along $\mathbf{d}$ and paste together along $\mathbf{a}$ to obtain 5.7(4), as desired. The corresponding relation may be coded as:

$$\mathbf{aAbBa^{-1}Cb^{-1}D} \sim \mathbf{aca^{-1}CBc^{-1}AD} \sim \mathbf{dcd^{-1}c^{-1}ADCB}. \tag{5.7}$$

By repeated application of this, we get the canonical polygon with edges marked

$$\mathbf{a_1 b_1 a_1^{-1} b_1^{-1} \cdots a_m b_m a_m^{-1} b_m^{-1} c_1 c_1 c_2 c_2 \cdots c_n c_n}.$$

If either $m = 0$ or $n = 0$ then we are done. It remains to consider the case when both m and n are positive. See Figure 5.8.

Step 5 Handling the case of mixed types:
Combining results of the two cut-and-pastes done in Figure 5.8, we obtain the following relation:

$$\mathbf{Aaba^{-1}b^{-1}cc} \sim \mathbf{Axyxzyz}. \tag{5.8}$$

We can now apply step 3 successively to make this into $\mathbf{Axxyyzz}$.

By repeated application of this step, all interlocking pairs of type I can be replaced by two pairs of type II, so that the canonical polygon

$$\mathbf{a_1 b_1 a_1^{-1} b_1^{-1} \cdots a_m b_m a_m^{-1} b_m^{-1} c_1 c_1 c_2 c_2 \cdots c_n c_n}$$

is transformed into $\mathbf{x_1 x_1 \cdots x_k x_k}$, where $k = 2m + n$. This completes the proof of Theorem 5.3.15.

We shall now turn to the proof of part (B), using cellular homology, simplicial homology, or the fundamental group.

Proof I Cellular homology of the canonical polygon:
The canonical polygon describes a CW-structure K on the surface that is obtained by the identification of the edges on the boundary of the polygon. From the lists in Theorem 5.3.15, we need to consider the cases (ii) and (iii) only, since case (i) clearly gives the sphere $\mathbb{S}^2$.

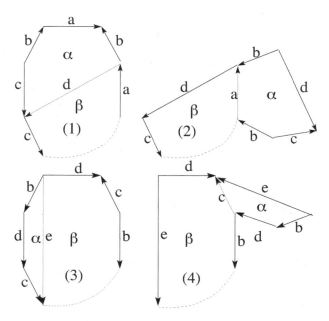

FIGURE 5.8. Mixed type

Denoting the number of edges by $2m$ ($m = 2g$ or $m = n$), the number 0-cells, 1-cells and 2-cells is, respectively, equal to $1, m, 1$. Therefore the cellular chain complex of K looks like:

$$0 \longrightarrow \mathbb{Z} \xrightarrow{\partial_2} \mathbb{Z}^m \xrightarrow{\partial_1} \mathbb{Z} \longrightarrow 0$$

Since each 1-cell is attached to the same single vertex, the boundary operator $\partial_1 \equiv 0$. On the other hand, ∂_2 being the homomorphism $H_1(\partial D) \longrightarrow H_1(K^{(1)})$, depends upon the actual canonical polygon: in case (ii), $m = 2g$, and $\partial_2 = 0$. Therefore, $H_2(K) = \mathbb{Z}$ and $H_1(K) = \mathbb{Z}^{2g}$.

In case (iii), ∂_2 sends the generator to twice the sum of the generators of $H_1(K^{(1)})$. Therefore, $H_2(K) = 0$ and $H_1(K) = \mathbb{Z}^{n-1} \oplus \mathbb{Z}/2\mathbb{Z}$.

Case (i) anyway gives $X = \mathbb{S}^2$ and hence $H_1(K) = 0$ and $H_2(K) = \mathbb{Z}$.

Since cellular homology is a topological invariant, this completes the proof of Theorem 5.3.15.

Incidentally, we have proved that $\chi(T_{2g}) = 2 - 2g$ and $\chi(P_n) = 2 - n$.

The theorem below just sums it all.

Theorem 5.3.16 *Let S_1, S_2 be compact, connected 2-dimensional topological manifolds without boundary. Then S_1 and S_2 are homeomorphic if and only if their Euler characteristics are equal and both are orientable or both non-orientable.*

Proof: The only thing that we need to say here is that compact connected surface S without boundary is orientable iff $H_2(S, \mathbb{Z}) = \mathbb{Z}$. This is the case only for $S = \mathbb{S}^2$ or $= T_{2g}$ in the list. (A rigorous proof of this will be done in a later chapter.)   ♠

Proof II Canonical triangulation of a normal form:

Figure 5.9 depicts the triangulation of the canonical polygon aa which represents the projective space $\mathbb{P}^2$. We will give the general construction below.

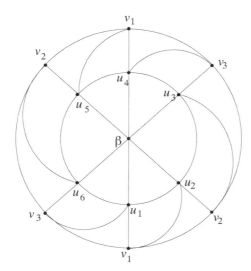

FIGURE 5.9. Canonical triangulation of a canonical polygon

Let the canonical polygon have k edges and k vertices (without regard to the identifications).

(i) Introduce two additional vertices in the interior of each edge, making $3k$ vertices and $3k$ edges.

(ii) Join the barycentre β of the polygon to all the $3k$ vertices.

(iii) Denoting by $\nu_1, \cdots, \nu_{3k}$ the vertices on the boundary arranged cyclically, introduce a new vertex u_i in the interior of the edge $[\beta, \nu_i]$. Cut each triangle $\{\beta, \nu_i, \nu_{i+1}\}$ into three triangles by introducing the edges $[u_i, u_{i+1}]$ and $[\nu_i, u_{i+1}]$.

After the identifications on the boundary of the canonical polygon, this gives a triangulation K of S.

Let us now consider the canonical triangulation of a polygon D in the normal form (ii) or (iii). Then observe that, in the triangulation K induced on the surface, all the original vertices are identified to a single point. Also, each new vertex on the boundary of D is identified with precisely one other such vertex. None of the new vertices introduced in the interior of D is identified with any other vertex. Therefore, the number of vertices of K equals $1 + k + 1 + 3k + 2 = 4k + 2$. Likewise one sees that the number of edges of K equals $3k/2 + 12k = (27/2)k$. Of course the number of triangles is the easiest to count and is equal to $9k$. Therefore, the Euler number is given by $\chi(K) = 2 - k/2$. Observe that $k = 4g$ or $2n$ according to cases (ii) and (iii). Hence, it follows that no two surfaces within (ii) have the same Euler characteristic. Also they all have different Euler characteristic than the case (i). Similarly no two surfaces within the list (iii) have the same Euler number.

Proof III The fundamental group: Clearly the fundamental group of a surface can be easily determined by the CW-structure K given by the canonical polygon. (See Corollary 3.8.12.)

The 1-skeleton $K^{(1)}$ of the CW-structure is a bouquet of r circles, the number r being equal to either $2g$ or n for T_g and for P_n, respectively. The convex polygon gives just one 2-cell. The attaching map of this 2-cell represents a certain element t in $\pi_1(K^{(1)})$ which is easily determined by the sequence representing the canonical polygon. The fundamental group $\pi_1(K)$ is the quotient of the free group $\pi_1(K^{(1)})$ by the normal subgroup generated by t. Therefore we have:

Theorem 5.3.17 *The fundamental group of a surface is given by the generators and relation:*

$$\pi_1(T_g) = \langle x_1, y_1, \ldots, x_g, y_g \mid \prod_{i=1}^{g} [x_i, y_i] \rangle;$$
$$\pi_1(P_n) = \langle x_1, \ldots, x_n \mid x_1^2 \cdots x_n^2 \rangle.$$

Remark 5.3.18

1. It is not hard to see that each of the groups listed in the theorem belongs to a distinct isomorphism class. Indeed, their abelianizations themselves will be non isomorphic. As a consequence we can conclude that the topological type of a surface is completely determined by the fundamental group which comes as a small surprise. (There is no such surprise in a similar result for compact 1-dimensional manifold though.) This may be attributed to a deeper property of surfaces: viz., barring two exceptional cases of $\mathbb{S}^2$ and $\mathbb{P}^2$, all connected closed surfaces have the universal covering space either $\mathbb{C}$ or the upper half plane, both of which are contractible. (This result is not at all obvious without the use of function theory of one complex variable.)

   One can now ask whether such is the case in higher dimensional manifolds as well, viz., suppose we have a n-manifold X covered by a Euclidean space. Is the homeomorphism type of X determined by the fundamental group of X? This problem goes under the name Borel's conjecture which has been verified in every known case. A complete solution is yet to come. See [Farrell, et al., 2002] or

   http://publications.ictp.it/lns/vol. 9.html

   for more details.

2. It is not hard to obtain a classification of all connected compact 2-manifolds with boundary. By capping-off all the boundary components we obtain a surface. Thus any compact 2-manifold with boundary is obtainable by removing finitely many disjoint discs—it can be shown that it does not matter from where you remove these discs.

3. The passage from the compact to the non compact case is quite hard as compared to the 1-dimensional case. For instance, simple problems such as characterization of all open subsets of $\mathbb{R}^2$ does not seem to have a satisfactory solution.

Exercise 5.3.19

(i) Determine the surfaces given by the sequences $\mathbf{aba^{-1}bcc, abcabc, abc^{-1}abc}$.

(ii) Show that every orientable closed surface is the zero set of a real polynomial in three real variables.

5.4   Basics of Vector Bundles

The elementary algebraic topology of smooth manifolds differs essentially from that of general topological manifolds, because of the role played by the tangent bundle. The notion of vector bundles over arbitrary topological spaces is an easy generalization of the notion of the tangent bundle of a smooth manifold. In this section, we shall present some basic facts of vector bundles on arbitrary topological space, though our main objects of study will be those on manifolds. You will see that with very mild restrictions such as paracompactness on the topological spaces, the theory becomes a powerful tool in the study of algebraic topological aspects of spaces.

Definition 5.4.1 Let B be a topological space. By a real vector bundle of rank k over B we mean an ordered pair $\xi = (E, p)$, where E is a topological space $p : E \to B$ is a continuous function such that for each $b \in B$, the fibre $p^{-1}(b) =: \xi_b$ is a k-dimensional $\mathbb{R}$-vector space satisfying the following local triviality condition:
(LTC) To each point $b \in B$ there is an open neighbourhood U of b and a homeomorphism $\phi : p^{-1}(U) \to U \times \mathbb{R}^k$ such that:
(i) $\pi_1 \circ \phi = p$ and
(ii) $\pi_2 \circ \phi : p^{-1}(b') \to \mathbb{R}^k$ is an isomorphism of vector spaces for all $b' \in U$.
    Here $\pi_1 : U \times \mathbb{R}^k \to U$ and $\pi_2 : U \times \mathbb{R}^k \to \mathbb{R}^k$ are projection maps.
    E is called the total space of ξ and B is called the base. Often a vector bundle of rank k is also called a *k-plane bundle*. When $k = 1$, we also call it a *line bundle*.
    If $\xi = (E_i, p_i, B_i)$, $i = 1, 2$ are two vector bundles, a morphism $\xi_1 \to \xi_2$ of vector bundles consists of a pair $(f, \bar{f})$ of continuous maps such that the diagram

$$
\begin{array}{ccc}
E_1 & \xrightarrow{\ \bar{f}\ } & E_2 \\
\downarrow{\scriptstyle p_1} & & \downarrow{\scriptstyle p_2} \\
B_1 & \xrightarrow{\ f\ } & B_2
\end{array}
$$

is commutative and such that $\bar{f}|_{p_1^{-1}(b)}$ is $\mathbb{R}$-linear. If both f and $\bar{f}$ are homeomorphisms also, then we say $(f, \bar{f})$ is a vector bundle isomorphism. In this situation we say that the two bundles are isomorphic.
    Often while dealing with vector bundles over a fixed base space B, we require a bundle morphism $(f, \bar{f}) : (E_1, p_1, B) \to (E_2, p_2, B)$ to be such that $f = Id_B$.
    The simplest example of a vector bundle of rank k over B is $B \times \mathbb{R}^k$. These are called *trivial vector bundles.* In fact any vector bundle isomorphic to a product bundle is called a trivial vector bundle. We shall denote this by $\Theta^k := B \times \mathbb{R}^k$, the base space of the bundle being understood by the context.
    Given a subspace $E' \subset E$ where $\xi = (E, p, B)$ is a vector bundle, consider the restriction map $p' = p|_{E'}$. We say $\xi' = (E', p', B)$ is a subbundle of ξ iff
(i) $\xi' = (E', p', B)$ is a vector bundle on its own (in particular, p' is surjective).
(ii) The inclusion map $p'^{-1}(b) \subset p^{-1}(b)$ is a linear map for each b.

Remark 5.4.2 It is easy to construct ξ' satisfying (ii) without satisfying (i). Also, if ξ is trivial, it does not mean a subbundle ξ' is also trivial.

Example 5.4.3

(i) A simple example of a non trivial vector bundle is the infinite Möbius band M: Consider the quotient space of $\mathbb{R} \times \mathbb{R}$ by the equivalence relation $(t, s) \sim (t + 1, -s)$. The first projection gives rise to a map $p : M \to \mathbb{S}^1$ which we claim is a non trivial real vector bundle of rank 1 over $\mathbb{S}^1$. It is easy to see that the complement of the 0-section in the total space of this bundle is connected. Therefore, the bundle cannot be the trivial bundle $S^1 \times \mathbb{R}$. Indeed the total space of this bundle is not even homeomorphic to $S^1 \times \mathbb{R}$ but to see that needs a little bit more topological arguments.

(ii) The tangent bundle $\tau(X) := (TX, p, X)$ of any smooth submanifold $X \in \mathbb{R}^N$ is a typical example of a vector bundle of rank n, where $n = \dim X$. It satisfies the additional smoothness condition, viz., both total and base spaces are smooth manifolds, the projection map p is smooth and the homeomorphisms $\phi : p^{-1}(U) \to U \times \mathbb{R}^n$ are actually diffeomorphisms. For a smooth manifold B, a vector bundle which satisfies this additional smoothness condition will be called a smooth vector bundle. On a manifold X embedded in $\mathbb{R}^N$, we get another vector bundle, viz., the normal bundle, $\nu(X)$ which is also a smooth vector bundle.

(iii) Let $B = \mathbb{P}^n$ be the n-dimensional real projective space. The canonical line bundle $\gamma_n^1 = (E, p, \mathbb{P}^n)$ is defined as follows: Recall that $\mathbb{P}^n$ can be defined as the quotient space of $\mathbb{S}^n$ by the antipodal action.

$$E = \{([x], \mathbf{v}) \in \mathbb{P}^n \times \mathbb{R}^{n+1} \; : \; \mathbf{v} = \lambda x\}$$

That is, over each point $[x] \in \mathbb{P}^n$ we are taking the entire line spanned by the vector $x \in \mathbb{S}^n$ in $\mathbb{R}^{n+1}$. Let $p : E \to \mathbb{P}^n$ be the projection to the first factor. The verification that this data forms a line bundle is easy. The case $n = 1$ is an interesting one. The base space $\mathbb{P}^1$ is then diffeomorphic to $\mathbb{S}^1$. However, the bundle γ_1^1 is the infinite Möbius band we considered above.

We begin with the following fundamental criterion to construct/detect isomorphisms between vector bundles.

Lemma 5.4.4 Let $f : \xi \to \zeta$ be a bundle map from one vector bundle over B to another. Then f is an isomorphism of vector bundles iff f restricted to each fibre is an isomorphism of vector spaces.

Proof: The only thing that we have to verify is the continuity of f^{-1}. This then can be done locally and hence the problem reduces to the case when ξ, ζ are trivial. In this case, a bundle map $f : B \times \mathbb{R}^k \to B \times \mathbb{R}^k$ is determined by

$$f(b, \mathbf{v}) = (b, A(b)\mathbf{v})$$

where $b \mapsto A(b)$ is a continuous map $A : B \to M(n; \mathbb{R})$ the space of real $n \times n$ matrices. The hypothesis that f restricts to an isomorphism on each fibre is the same as saying that each $A(b)$ is invertible and hence we have a continuous map $A : U \to GL(k, \mathbb{R})$. This then means that $b \mapsto A^{-1}(b)$ is also continuous. Therefore, correspondingly, the map given by

$$(b, \mathbf{v}) \mapsto (b, A^{-1}(b)(\mathbf{v}))$$

is continuous which is nothing but f^{-1}.   ♠

Definition 5.4.5 Let $\xi = (E, p, B)$ be a vector bundle. By a *section* of ξ we mean a continuous (smooth) map $\sigma : B \to E$ such that $p \circ \sigma = Id_B$. A section σ is said to be nowhere zero, if $\sigma(b) \neq 0$ for each $b \in B$.

Remark 5.4.6
(i) A simple example of a section is the zero-section which assigns to each $b \in B$ the 0-vector in $p^{-1}(b)$. (Use (LTC) to see that the zero-section is continuous.)
(ii) It is easy to see that the trivial bundle has lots of sections. Indeed if $\sigma : B \to B \times \mathbb{R}^k$ is a section then it is for the form,

$$\sigma(b) = (b, f(b))$$

where $f : B \to \mathbb{R}^k$ is continuous and conversely. Thus the set of sections of Θ^k is equal to

$\mathcal{C}(B, \mathbb{R}^k)$.

(iii) More generally, the space of sections $\Gamma(\xi)$ can be given a module structure on the ring of continuous functions $\mathcal{C}(B; \mathbb{R})$ and the study of this module is essentially all about the study of vector bundles over B.

Theorem 5.4.7 *A vector bundle ξ of rank k is trivial iff there exist sections $\{\sigma_1, \ldots, \sigma_k\}$ which are linearly independent at every point of B.*

Definition 5.4.8 By a continuous/smooth *vector field* on a smooth manifold X we mean a continuous/smooth section of the tangent bundle. By a *parallelizable manifold,* we mean a smooth manifold X whose tangent bundle is trivial.

Remark 5.4.9 Alternatively, a manifold is parallelizable iff there exists n smooth vector fields $\{\sigma_1, \ldots, \sigma_n\}$ such that for each $p \in X$, we have

$$\{\sigma_1(p), \ldots, \sigma_n(p)\}$$

is linearly independent in $T_p(X)$.

Operations on Vector Bundles

Pullback bundle Given a triple $\xi = (E, p, B)$ (of topological spaces and continuous map) and a continuous function $f : B' \to B$ the pullback $f^*\xi = (E', p', B')$ is defined by

$$E' = \{(b', e) \in B' \times E \ : \ f(b') = p(e)\}; \ \ p'(b', e) = b'.$$

If map p satisfies a certain topological condition (such as homeomorphism, local homeomorphism, proper mapping, covering projection, etc.) often it is the case that the same condition is satisfied by the map p'. Thus, if the triple ξ is a vector bundle of rank k, it follows that so is the triple $f^*\xi$. Moreover, we have a continuous map $\bar{f} : E' \to E$ such that the diagram is commutative:

$$\begin{array}{ccc} E' & \xrightarrow{\bar{f}} & E \\ \downarrow{p'} & & \downarrow{p} \\ B & \xrightarrow{f} & B \end{array}$$

Notice that (LTC) for $f^*\xi$ follows from the observation that if ξ is the trivial bundle then so is $f^*\xi$. The pullback bundle has the following universal property. Given any vector bundle $\xi'' = (E'', p'', B')$ over B' and a continuous map $\bar{g} : E'' \to E$ such that $p \circ \bar{g} = f \circ p''$ there exists a unique bundle map $g' : E'' \to E$ over B', i.e., such that $p' \circ g' = p''$ with the property that $\bar{g} = \bar{f} \circ g'$.

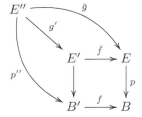

A special case of the pullback construction is obtained when B' is a subspace of B and $f = \eta : B' \hookrightarrow B$ is the inclusion. We then denote $\eta^*(\xi)$ by $\xi|_{B'}$.

Now suppose $\xi' = (E', p', B')$ and $\xi = (E, p, B)$ are two vector bundles and $(f, \bar{f}) : \xi' \to \xi$ is a bundle map

$$
\begin{array}{ccc}
E' & \xrightarrow{\ \bar{f}\ } & E \\
{\scriptstyle p'}\downarrow & & \downarrow{\scriptstyle p} \\
B' & \xrightarrow{\ f\ } & B
\end{array}
$$

This gives us unique bundle map $f' : \xi' \to f^*\xi$ which covers f. The proof of the following theorem is a warm-up exercise for the reader.

Theorem 5.4.10 *The bundle map $f' : \xi' \to f^*\xi$ is an isomorphism iff $\bar{f}$ restricts to an isomorphism on each fibre.*

Remark 5.4.11 Note that f itself is just a continuous function and need not be a homeomorphism. However, if f is a homeomorphism, then f is covered by a homeomorphism $\bar{f}$ iff the two bundles ξ' and $f^*(\xi)$ over B' are isomorphic. More generally, if f is a homeomorphism, then there is 1-1 correspondence between bundle maps $(f, \bar{f}) : \xi' \to \xi$ and bundle maps $(Id_B, g) : \xi' \to f^*(\xi)$. This is the reason why we assume that a bundle map $\xi \to \zeta$ of two vector bundles over the same base space B is of the form (Id_B, g).

Cartesian product Given ξ, ξ' we can take $\xi \times \xi' = (E \times E', p \times p', B \times B')$ in the usual way, as a vector bundle of rank $= rk(\xi) + rk(\xi')$. For this bundle, the fibre over a point (b, b') clearly equals $\xi_b \times \xi'_{b'}$. Of particular interest is the special case when ξ' is of rank 0, i.e., $p : E' \to B'$ is a homeomorphism. We denote the product in this case simply by $\xi \times B'$.

Whitney sum Let ξ, ξ' be bundles over the same base B. Consider the diagonal map $\Delta : B \to B \times B$. The Whitney sum of ξ and ξ' is defined by

$$\xi \oplus \xi' = \Delta^*(\xi \times \xi')$$

the pullback of the Cartesian product via the diagonal map. Put $\xi \oplus \xi' := (E(\xi \oplus \xi'), q, B)$. We have a commutative diagram

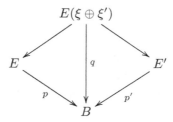

The Whitney sum indeed corresponds to taking sums of subbundles in the following sense.

Lemma 5.4.12 Let ξ_1, ξ_2 be subbundles of ζ such that for each $b \in B$, ζ_b is equal to the direct sum of the subspaces $(\xi_1)_b$ and $(\xi_2)_b$. Then ζ is isomorphic to $\xi_1 \oplus \xi_2$.

Proof: Define $\phi : E(\xi_1 \oplus \xi_2) \to E(\zeta)$ by $\phi(\mathbf{v}_1, \mathbf{v}_2) = \mathbf{v}_1 + \mathbf{v}_2$. ♠

Definition 5.4.13 A bundle ξ is said to be *stably trivial* if $\xi \oplus \Theta^k$ is trivial for some $k \geq 1$.

Remark 5.4.14 Of course, trivial bundle is stably trivial. However, there are many stably trivial bundles which are not trivial. The simplest examples are tangent bundles of spheres of dimension $\neq 1, 3, 7$. For $n = 1, 3, 7$ the tangent bundle of $\mathbb{S}^n$ is actually trivial. In general, the tangent bundle of $\mathbb{R}^{n+1}$ is trivial and hence restricts to a trivial bundle over $\mathbb{S}^n$ and can be written as the Whitney sum of the tangent bundle with the normal bundle. Now the normal bundle to the $\mathbb{S}^n$ in $\mathbb{R}^{n+1}$ is easily seen to be trivial. This means that the tangent bundle is stably trivial. Finally, by the hairy ball theorem, it follows that there are no non vanishing vector fields on any even dimensional spheres. Therefore the tangent bundle of any $\mathbb{S}^{2n}$ is not trivial. To see this result for odd dimensional spheres other than $n = 1, 3, 7$, it requires a little more effort.

Riemannian Metric Structure

Let ξ be a real vector bundle of rant k. A Riemannian metric on ξ is a continuous function $\beta : E(\xi \oplus \xi) \to \mathbb{R}$ such that restricted to each fibre, β is an inner product.

It is easy to see that on any trivial bundle we can give the standard inner product of $\mathbb{R}^k$ itself on each fibre. More generally, given a continuous map $\hat{\beta} : B \to M(k, \mathbb{R})$ taking values inside symmetric positive definite matrices, we can associate a Riemannian metric on $\Theta^k = B \times \mathbb{R}^k$ by the rule:

$$\beta(\mathbf{v}_1, \mathbf{v}_2)_b = \mathbf{v}_1^t \hat{\beta}(b) \mathbf{v}_2,$$

where $\mathbf{v}_1, \mathbf{v}_2$, etc. are treated as column vectors. And conversely, every Riemannian metric on Θ^k corresponds to a continuous map from B to the space of symmetric positive definite real $k \times k$ matrices.

Given two bundles with Riemannian metrics one can seek bundle maps which respect the inner products. We can then talk about 'isometries' of such bundles. The simplest question one can ask is: 'What are all isometrically inequivalent metrics on a trivial bundle?' The answer is:

Theorem 5.4.15 *Any two Riemannian metrics on Θ^k are isometrically equivalent.*

Proof: Gram–Schmidt process. ♠

Orthogonal complement

Given a Riemannian bundle ξ and a subbundle ξ', the orthogonal complement $(\xi')^\perp$ of ξ' in ξ is defined by

$$E((\xi')^\perp) = \{\mathbf{v} \in \xi_b \ : \ \mathbf{v} \perp \xi_b'\}$$

together with the projection $p : E((\xi')^\perp) \to B$. The non trivial thing to verify is the (LTC) which follows once again, from Gram–Schmidt's process.

Remark 5.4.16 It is not true that every vector bundle can be given a Riemannian structure. The following result is the 'most' general in this respect in a certain sense.

Theorem 5.4.17 *Let B be a paracompact space. Then every vector bundle over B has a Riemannian structure on it.*

Proof: Partition of unity. ♠

Transition functions

Given an open covering $\{U_i\}$ of B and local trivializations $\phi_i : p^{-1}(U_i) \to U_i \times \mathbb{R}^k$, of a bundle $\xi = (E, p, B)$, for each pair (i, j) of indices, consider the isomorphisms of the trivial bundles:

$$\phi_i \circ \phi_j^{-1} : (U_i \cap U_j) \times \mathbb{R}^k \to (U_i \cap U_j) \times \mathbb{R}^k.$$

They are of the form

$$(b, \mathbf{v}) \mapsto (b, \lambda_{ij}(b)(\mathbf{v}))$$

for some continuous maps $\lambda_{ij} : U_i \cap U_j \to GL(k; \mathbb{R})$. These are called the **transition functions** of the bundle ξ. They satisfy the following two 'cocycle conditions':
(CI) $\lambda_{ii}(b) = Id$ for all i;
(CII) For $b \in U_i \cap U_j \cap U_t$ we have

$$\lambda_{jt}(b) \circ \lambda_{ij}(b) = \lambda_{it}(b).$$

We would like to reverse the picture: Starting with an open covering $\{U_i\}$ of B and a family $\lambda = \{\lambda_{ij}\}$ of continuous functions $\lambda_{ij} : U_i \cap U_j :\to GL(k; \mathbb{R})$, we define a vector bundle $\xi_\lambda = (E_\lambda, p_\lambda, B)$ of rank k as follows: On the disjoint union $\tilde{E} = \sqcup_i U_i \times \mathbb{R}^k$ define an equivalence relation by saying that

$$(b, \mathbf{v}) \sim (b, \lambda_{ij}(b)(\mathbf{v}))$$

for each pair (i, j) such that $b \in U_i \cap U_j$ and for all $\mathbf{v} \in \mathbb{R}^k$.

The two cocycle conditions ensure that the identifications are compatible and define an equivalence relation. Denote the quotient space by E_λ.

Observe that the projection maps $\pi_1 : U_i \times \mathbb{R}^k \to U_i$ all patch up to define a continuous map $p_\lambda : E \to B$. Indeed verify that the inclusion $U_i \times \mathbb{R}^k \to \tilde{E}$ followed by the quotient map $\tilde{E} \to E_\lambda$ is a homeomorphism onto $p_\lambda^{-1}(U_i)$ and so we obtain homeomorphisms $\psi_i : U_i \times \mathbb{R}^k :\to p^{-1}(U_i)$. Since each identification map $\lambda_{ij}(b) : \{b\} \times \mathbb{R}^k \to \{b\} \times \mathbb{R}^l$ is an isomorphism of vector bundles we get a unique vector bundle structure on each fibre $p^{-1}(b)$. Taking $\phi_i = \psi_i^{-1} : p_\lambda^{-1}(U_i) \to U_i \times \mathbb{R}^k$, we get (LTC) for the bundle ξ_λ. For these local trivializations, one can easily verify that

$$\phi_i \circ \phi_j^{-1}(b, \mathbf{v}) = (b, \lambda_{ij}(b)(\mathbf{v}))$$

getting back to where we started.

It is obvious that the topology of the total space as well as the bundle will heavily depend upon the nature of the transition functions. Indeed, if we start off with a bundle ξ and a local trivialization, the union of all local trivializations defines a map $\Phi : \tilde{E} \to E$ which in turn defines a bundle isomorphism $\xi_\lambda \to \xi$.

The transition function description allows us a sure way of carrying out vector space operations on vector bundles. For example, if ξ and η are two bundles over B, get a common open covering on which we have local trivializations for both the bundles. Let $\lambda_\xi, \lambda_\eta$ be the corresponding families of transition functions. Define the family $\lambda_\xi \oplus \lambda_\eta$ by the formula

$$(b, \mathbf{v}, \mathbf{u}) \mapsto (b, (\lambda_\xi)_{ij}(b)(\mathbf{v}), (\lambda_\eta)_{ij}(b)(\mathbf{u})).$$

It is a matter of straightforward verification to see that the resulting vector bundle is isomorphic to the Whitney sum $\xi \oplus \eta$. If you want to construct the bundle $\text{Hom}(\xi, \eta)$ all that you have to do is to consider the transition functions

$$\text{Hom}(\lambda_\xi, \lambda_\eta) : (U_i \cap U_j) \times \text{End}(\mathbb{R}^k, \mathbb{R}^l) \to (U_i \cap U_j) \times \text{End}(\mathbb{R}^k, \mathbb{R}^l)$$

defined by

$$(b, \alpha) \mapsto (b, (\lambda_\eta)_{ij}^{-1} \circ \alpha \circ (\lambda_\xi)_{ij}).$$

Likewise, the exterior powers $\wedge^i \xi$ are constructed out of the transition functions which are fibre-wise i^{th} exterior power of the transition functions of ξ.

Exercise 5.4.18 Show that

$$\wedge^2(\xi \oplus \eta) \cong \wedge^2(\xi) \oplus \wedge^2(\eta) \oplus \xi \otimes \eta.$$

Remark 5.4.19 A simplistic point of view of the entire theory of vector bundles is that it is nothing but the continuous/smooth version of linear and multilinear algebra. A simple illustration of this occurs in the construction of the normal bundle: local triviality of the normal bundle is a consequence of carrying out the Gram–Schmidt process, on a set of continuous/smooth vector valued functions which are independent everywhere. Another simple example is that the polar decomposition is a continuous smooth process and hence yields the following: If μ, μ' are two Riemannian metrics on a given vector bundle ξ then there exists a fibre preserving homeomorphism $f : E(\xi) \to E(\xi)$ such that $\mu \circ (f, f) = \mu'$. (Compare Theorem 5.4.15.)

Example 5.4.20 Recall that a finite dimensional vector space V and its dual V^* are isomorphic to each other. However, given a vector bundle ξ its dual bundle, in general, may not be isomorphic to ξ. The reason is that the isomorphism between V and V^* is **not** canonical. On the other hand, it follows easily that $(\xi)^{**}$ is isomorphic to ξ. However, if ξ carries a Riemannian metric, then fixing one such, we get an isomorphism $\xi \cong \xi^*$.

Example 5.4.21 Consider the tangent bundle $\tau := \tau(\mathbb{P}^n)$. Using the double covering map $\phi : \mathbb{S}^n \to \mathbb{P}^n$, we can describe the total space of τ by

$$E(\tau) = \{[\pm x, \pm \mathbf{v}] \ : \ x \in \mathbb{S}^n, \mathbf{v} \perp x, \mathbf{v} \in \mathbb{R}^{n+1}\}.$$

Observe that $D\phi : E(\tau(\mathbb{S}^n)) \to E(\tau)$ has the property $D(\phi)(x, \mathbf{v}) = D(\phi)(y, \mathbf{u})$ iff $(y, \mathbf{u}) = \pm(x, \mathbf{v})$. Therefore, $D\phi$ is actually the quotient map.

On the other hand a pair $(x, \mathbf{v}) \in \mathbb{S}^n \times \mathbb{R}^{n+1}$ such that $\mathbf{v} \perp x$ also determines a linear map on the 1-dimensional subspace $[x]$ spanned by x to its orthogonal complement. Note that the pair $(-x, -\mathbf{v})$ also determines the same linear map. Therefore, we can identify the quotient space with the space of linear maps from 1-dimensional subspaces to their complements in $\mathbb{R}^{n+1}$. This then also describes the vector bundle $\mathrm{Hom}(\gamma_n^1, (\gamma_n^1)^{\perp})$ over $\mathbb{P}^n$. We have established:

Theorem 5.4.22 $\mathrm{Hom}(\gamma_n^1, (\gamma_n^1)^{\perp}) \cong \tau(\mathbb{P}^n)$.

Exercise 5.4.23 If ξ_j are all vector bundles over the same base space B, prove that

$$\mathrm{Hom}(\xi_1, \xi_2 \oplus \xi_3) \cong \mathrm{Hom}(\xi_1, \xi_2) \oplus \mathrm{Hom}(\xi_1, \xi_3).$$

Exercise 5.4.24 If η is a line bundle show that $\mathrm{Hom}(\eta, \eta) \cong \Theta^1$, the trivial line bundle.

Theorem 5.4.25 *Let τ denote the tangent bundle of $\mathbb{P}^n$. Then*

$$\tau \oplus \Theta^1 \cong \gamma_n^1 \oplus \cdots \oplus \gamma_n^1$$

the $(n+1)$-fold Whitney sum of the canonical line bundle.

Proof:

$$\begin{aligned}
\tau \oplus \Theta^1 &\cong \mathrm{Hom}(\gamma_n^1, \gamma_n^{\perp}) \oplus \mathrm{Hom}(\gamma_n^1, \gamma_n^1) \\
&\cong \mathrm{Hom}(\gamma_n^1, \gamma^{\perp} \oplus \gamma_n^1) \\
&\cong \mathrm{Hom}(\gamma_n^1, \Theta^{n+1}) \\
&\cong \mathrm{Hom}(\gamma_n^1, \Theta^1) \oplus \cdots \oplus \mathrm{Hom}(\gamma_n^1, \Theta^1).
\end{aligned}$$

Since $\mathbb{P}^n$ is compact, every vector bundle over it admits a Riemannian metric. Therefore, every vector bundle is isomorphic to its dual over $\mathbb{P}^n$. The theorem follows. ♠

Homotopical Aspect

Lemma 5.4.26 Let $B = X \times [a, c], a < b < c$. Suppose ξ is a vector bundle over $B \times [a, c]$ such that $\xi|_{B \times [a,b]}$ and $\xi|_{B \times [b,c]}$ are trivial bundles. Then ξ itself is trivial.

Proof: Let $\phi_1 : \xi|_{X \times [a,b]} :\to (X \times [a, b]) \times \mathbb{R}^k$ and $\phi_2 : \xi|_{X \times [b,c]} :\to (X \times [b, c]) \times \mathbb{R}^k$ be some trivializations. Consider the isomorphism $\phi_1 \circ \phi_2^{-1} : X \times \{b\} \times \mathbb{R}^k \to X \times \{b\} \times \mathbb{R}^k$ which can be written in the form

$$(x, b, \mathbf{v}) \mapsto (x, b, \lambda_x(\mathbf{v})).$$

It follows that if $\lambda(x, t, \mathbf{v}) = (x, t, \lambda_x(\mathbf{v}))$, then λ is an automorphism of the trivial bundle $B \times [b, c] \times \mathbb{R}^k$. Now define $\phi : E(\xi) \to X \times [a, c] \times \mathbb{R}^k$ by

$$\phi(e) = \left\{ \begin{array}{ll} \phi_1(e), & \text{if } \pi(e) \in B \times [a, b]; \\ \lambda \circ \phi_2(e), & \text{if } \pi(e) \in B \times [b, c]. \end{array} \right.$$

Verify ϕ defines a trivialization of ξ. ♠

Lemma 5.4.27 Let ξ be a vector bundle over $X \times [a, b]$. Then there is an open covering U_i of X such that $\xi|_{U_i \times [a,b]}$ is trivial for each i.

Proof: Easy.

Theorem 5.4.28 *Let ξ be a vector bundle over $X \times \mathbb{I}$ where X is paracompact. Then $\xi, (\xi|_{X \times 1}) \times I$ and $(\xi|_{X \times 0}) \times I$ are all isomorphic to each other.*

Corollary 5.4.29 Let $f, g : X \to Y$ be two homotopic maps. Then for any vector bundle ξ' over Y, we have $f^*\xi' \cong g^*\xi'$.

Proof of the corollary If $H : X \times I \to Y$ is a homotopy from f to g consider the bundle $\xi = H^*(\xi')$ over $X \times I$. By the above theorem, $\xi|_{X \times 0}$ and $\xi|_{X \times 1}$ are isomorphic. But they are respectively equal to $f^*(\xi')$ and $g^*(\xi')$.

The proof of the theorem itself is obtained easily via the following proposition.

Proposition 5.4.30 Let X be a paracompact Hausdorff space and ξ be a vector bundle over $X \times \mathbb{I}$. Then there is a bundle map $(r, \bar{r}) : \xi \to \xi$, where $r(x, t) = (x, 1)$ and $\bar{r}$ is an isomorphism on each fibre.

We shall prove this proposition for the case when X is compact and Hausdorff. The general case does not involve any deeper ideas but only technically more difficult.

Since a compact Hausdorff space is normal, we can get a finite open covering $\{U_1, \ldots, U_n\}$ of X such that
(i) $\xi|_{U_i \times \mathbb{I}}$ is trivial for each i;
(ii) there is a continuous map $\alpha_i : X \to \mathbb{I}$ such that $\overline{\alpha_i^{-1}(0, 1]} \subset U_i$, for each i; and
(iii) for every $x \in X$, $\max\{\alpha_1(x), \ldots, \alpha_n(x)\} = 1$.

For each i, choose trivializations $h_i : U_i \times \mathbb{I} \times \mathbb{R}^k \to p^{-1}(U_i \times \mathbb{I})$ over $U_i \times \mathbb{I}$ and define bundle maps $(r_i, \bar{r}_i) : \xi \to \xi$ as follows: $r_i(x, t) = (x, \max\{\alpha_i(x), t\})$; whereas,

$$\bar{r}_i(e) = \left\{ \begin{array}{ll} h_i(x, \max\{\alpha_i(x), t\}, \mathbf{v}), & \text{if } e = h_i^{-1}(x, t, \mathbf{v}) \in p^{-1}(U_i \times \mathbb{I}) \\ e, & \text{if } e \notin p^{-1}(U_i \times \mathbb{I}). \end{array} \right.$$

Then clearly r_i is continuous. Since $\bar{r}_i$ is identity outside the support of α_i and is continuous over $U_i \times \mathbb{I}$, it is continuous all over. Moreover, restricted to each fibre, it is a linear isomorphism also. Now consider the composition

$$(r, \bar{r}) := (r_1, \bar{r}_1) \circ \cdots (r_n, \bar{r}_n).$$

All that you have to do is to check that $r(b, t) = (b, 1)$. ♠

The Grassmann Manifolds and the Gauss Map

Fix integers $1 \leq k \leq n$. Let $G_{n,k}$ denote the set of all k-dimensional subspaces of $\mathbb{R}^n$. Let $V_{n,k}$ denote the subspace of $\mathbb{S}^{n-1} \times \cdots \times \mathbb{S}^{n-1}$ (k factors) consisting of ordered k-tuples $(\mathbf{v}_1, \ldots, \mathbf{v}_k)$ such that $\langle \mathbf{v}_i, \mathbf{v}_j \rangle = \delta_{ij}$. There is a surjective map $\eta : V_{n,k} \to G_{n,k}$ and we declare this as a quotient map so as to topologise $G_{n,k}$. This is called the Grassmann manifold of type (n, k). Also, $V_{n,k}$ is called the Steifel manifold of type (n, k).

Exercise 5.4.31 Show that there is a diffeomorphism of the homogeneous space

$$\frac{O(n)}{O(k,) \times O(, n-k)} \to G_{n,k}$$

Here $O(k,) \subset O(n), O(, n-k) \subset O(n)$ denote, respectively, the subgroups on $O(n)$ which keep the last $n - k$ basic vectors $\{\mathbf{e}_{k+1}, \ldots, \mathbf{e}_n\}$ (respectively, the first k basic vectors $\{\mathbf{e}_1, \ldots, \mathbf{e}_k\}$) fixed.

Consider the triple $\gamma_n^k = (E, \pi, B)$ where, $B = G_{n,k}$,

$$E = E(\gamma_n^k) = \{(V, \mathbf{v}) \in G_{n,k} \times \mathbb{R}^n \; : \; \mathbf{v} \in V\}$$

and $\pi = \pi_1$ the restriction of the projection to the first factor. One can show as in the case $k = 1$ that this defines a k-plane bundle over $G_{n,k}$.

Now consider $\mathbb{R}^n$ as the subspace $\mathbb{R}^n \times 0$ of $\mathbb{R}^{n+1}$. This then induces an inclusion of $G_{n,k} \xrightarrow{\iota} G_{n+1,k}$. Moreover there is a bundle inclusion:

$$
\begin{array}{ccc}
E(\gamma_n^k) & \longrightarrow & E(\gamma_{n+1}^k) \\
\downarrow & & \downarrow \\
G_{n,k} & \xrightarrow{\iota} & G_{n+1,k}
\end{array}
$$

Now consider the spaces

$$G_k = \cup_{n \geq k} G_{n,k}, \quad E(\gamma^k) = \cup_{n \geq k} E(\gamma_n^k)$$

with the weak topology, i.e., $F \subset G_k$ (respectively, $E(\gamma^k)$) is closed iff $F \cap G_{n,k}$ (respectively, $F \cap E(\gamma_n^k)$) is closed in $G_{n,k}$ (respectively, in $E(\gamma_n^k)$). It is not difficult to see that the corresponding projection maps patch up to define a projection map $\pi : E(\gamma^k) \to G_k$ giving a vector bundle γ^k of rank k over G_k.

G_k is called the infinite Grassmann. Indeed, this is nothing but the space of all k-dimensional subspaces of the infinite direct sum $\mathbb{R}^\infty = \mathbb{R} \oplus \mathbb{R} \oplus \cdots$. Also, γ^k is called the tautological (canonical) vector bundle over G_k.

Definition 5.4.32 Let ξ be a k-plane bundle over B. A map $g : E(\xi) \to \mathbb{R}^n$, $k \leq n \leq \infty$ is called a Gauss map on ξ, if $g|_{\xi_b}$ is a linear monomorphism for all $b \in B$.

Example 5.4.33 The second projection $\pi_2 : G_{n,k} \times \mathbb{R}^n \to \mathbb{R}^n$ restricted to $E(\gamma_n^k)$ is a Gauss map on γ_n^k for all $k \leq n \leq \infty$. Indeed, these Gauss maps give rise to all other Gauss maps as elaborated in the following lemma.

Remark 5.4.34 Recall the Gauss map you have come across in your calculus course. Given a smooth closed surface $S \subset \mathbb{R}^3$, the Gauss map $g : S \to \mathbb{S}^2$ was defined by the rule $g(x) =$ the unit outward normal to the surface at the point x. This map occurs in the proof of the

Gauss–Bonnet theorem. The Gauss map that we have defined above is a direct generalization of this concept. If you ignore the direction of the normals and merely take the normal line then you get map $\hat{g} : S \to \mathbb{P}^2 = G_{3,1}$. Note that $G_{3,1}$ is canonically diffeomorphic to $G_{3,2}$ via the orthogonal complement and under this dual map $\hat{g}$ corresponds to the map $S \to G_{3,2}$ which assigns to each point $x \in S$ the tangent plane to S at x. In other words, this is the Gauss map on the tangent bundle of S according to the definition given above.

Lemma 5.4.35 Let $(f, \bar{f}) : \xi \to \gamma_n^k$ be a bundle map which is an isomorphism on each fibre. Then $\pi_2 \circ \bar{f}$ is a Gauss map for ξ. Conversely, given a Gauss map $g : E(\xi) \to \mathbb{R}^n$ there exists a bundle map $(f, \bar{f}) : \xi \to \gamma_n^k$ which is an isomorphism on each fibre such that $\pi_2 \circ \bar{f} = g$.

Proof: The first part is clear. To prove the converse, we define $f(b) = g(\xi_b) \in G_{n,k}$ and $\bar{f}(e) = (f(p(e)), g(e))$. Use (LTC) to see that f is continuous and therefore $\bar{f}$ is continuous. Other requirements are straightforward. ♠

Proposition 5.4.36 Any k-plane bundle ξ over a paracompact Hausdorff space admits a Gauss map into $\mathbb{R}^\infty$.

Proof: Since B is paracompact, there exists a countable open covering U_i of B, a partition of unity α_i subordinate to the cover $\{U_i\}$ and trivializations $h_i : U_i \times \mathbb{R}^k \to p^{-1}(U_i)$. Define $g(e) = \sum_i g_i(e)$ where $g_i : E(\xi) \to \mathbb{R}^k$ is zero outside $p^{-1}(U_i)$ and on $p^{-1}(U_i)$, we have

$$g_i(e) = \alpha_i(p(e))\pi_2(h_i^{-1}(e)).$$

Theorem 5.4.37 *Let B be a paracompact space. Given a k-dimensional vector bundle ξ over B, there exists a continuous map $f : B \to G_k$ such that $\xi \cong f^*(\gamma^k)$. Moreover, if $f' : B \to G_k$ is another such continuous map then f is homotopic to f'.*

Proof: Let $g : \xi \to \mathbb{R}^\infty$ be a Gauss map as in the previous proposition. Then by the above lemma, we get a bundle map $(f, \bar{f}) : \xi \to \gamma^k$ such that $\pi_2 \circ \bar{f} = g$ and $\bar{f}$ is an isomorphism on each fibre. This in turn induces a bundle map $(Id, \eta) : \xi \to f^*\gamma_k$ which is again an isomorphism on each fibre and hence is bundle isomorphism. This proves the first part.

To prove the second part, we note that an isomorphism $\xi \cong f'^*(\gamma^k)$ induces a bundle map $(f', \bar{f}') : \xi \to \gamma^k$ which in turn corresponds to a Gauss map $g' : E(\xi) \to \mathbb{R}^\infty$. Likewise to get a homotopy between f and f' it is enough to produce a homotopy $g_t : E(\xi) \to \mathbb{R}^\infty$ of g and g' through Gauss maps.

Let $\mathbb{R}^{ev}, \mathbb{R}^{odd}$ be subspaces of $\mathbb{R}^\infty$ consisting of elements whose odd-place coordinates (respectively, even-place coordinates) are zero. Let $ev : \mathbb{R}^\infty \to \mathbb{R}^{ev}$ and $odd : \mathbb{R}^\infty \to \mathbb{R}^{odd}$ be the maps defined by

$$ev(x_1, \ldots, x_n, 0, \ldots) \mapsto (0, x_1, 0, x_2, \ldots); \ odd(x_1, x_2, \ldots, x_n, 0, \ldots) \mapsto (x_1, 0, x_2, 0, \ldots).$$

Then ev and odd are monomorphisms and are homotopic through monomorphisms to the identity map:

$$tx + (1-t)ev\,(x); \ \ tx + (1-t)odd\,(x).$$

Therefore it follows that g is homotopic to $ev \circ g$ and g' is homotopic to $odd \circ g'$. Now consider the homotopy

$$g_t(e) = (1-t)(ev \circ g)(e) + t(odd \circ g')(e)$$

between $ev \circ g$ and $odd \circ g'$. Injectivity of g_t follows from the fact that the line joining $ev \circ g(e)$ and $odd \circ g'(e)$ does not pass through the origin in the vector space $\mathbb{R}^\infty$, since $ev \circ g(e)$ and $odd \circ g'(e)$ are linearly independent for all e. ♠

Remark 5.4.38 We have come to a junction in the study of isomorphism class of vector bundles over a fixed base space B. We can proceed now in different directions. One such direction is K-theory which is beyond the scope of this book. Another direction is the study of characteristic classes. In a subsequent chapter we take up the study a little bit of characteristic classes of vector bundles.

Exercise 5.4.39 We have proved Theorem 5.4.28 only in the case B is compact. Complete its proof in the general case, by completing the proof of Proposition 5.4.30 in the general case. [Hint: See Exercise 2.2.23.(ii).]

5.5   Miscellaneous Exercises to Chapter 5

(Several exercises here may require familiarity with some differential topology(see [Shastri, 2011].)

1. Show that if $X \subset Y$ where X and Y are manifolds then $\dim X \leq \dim Y$. Further, if $\dim X < \dim Y$ then X has empty relative interior in Y.

2. Let X be a CW-complex of dimension $\leq n$, and ξ be a vector bundle of rank $n+1$ on X. If ξ is stably trivial then show that ξ is trivial.

3. Let $X = \mathbb{S}^{k_1} \times \cdots \times \mathbb{S}^{k_r}$ be a finite product of spheres with each $k_i \geq 1$.
   (a) Show that there is a smooth embedding $X \subset \mathbb{R}^n$ where $n = (\sum_i k_i) + 1$.
   (b) Show that X is s-parallelizable, i.e., its tangent bundle is stably trivial.
   (c) If $r \geq 2$ and one of the k_i is odd, show that X is parallelizable.

4. Let X be a connected, $(n-1)$-dimensional smooth submanifold of $\mathbb{R}^n$.
   (a) Show that $\mathbb{R}^n \setminus X$ has at most two connected components.
   (b) For any subset A of X homeomorphic to $\mathbb{D}^{n-1}$ show that $\mathbb{R}^n \setminus (X \setminus A)$ is connected.
   (c) Deduce that $\mathbb{R}^n \setminus X$ has precisely two connected components.
   (d) Show that the normal bundle of X in $\mathbb{R}^n$ is trivial. In particular, conclude that X is orientable and s-parallelizable.

5. Show that the mapping cylinder of the map $f : \mathbb{S}^1 \to \mathbb{S}^1$ given by $z \mapsto z^n$ can be embedded in the solid torus $\mathbb{D}^2 \times \mathbb{S}^1$ and the mapping cone of the same map can be embedded in $\mathbb{D}^2 \times \mathbb{D}^2$.

6. Let X be a connected finite CW-complex of dimension 2 of the form $(\vee_{i=1}^k \mathbb{S}_i^1) \cup \{e_j^2\}_{j=1}^l$. Show that X can be embedded in $\mathbb{R}^4$.

7. Let X be a connected smooth manifold and Y be a smooth submanifold of codimension ≥ 2. Show that $X \setminus Y$ is path connected and the inclusion induced homomorphism $\pi_1(X \setminus Y) \to \pi_1(X)$ is surjective.

8. Let X be a n-manifold with boundary, and $f : \mathbb{S}^{k-1} \times \mathbb{D}^{n-k} \to \partial X$ be an embedding, $1 \leq m$. By attaching a k-handle to X via f we mean forming the quotient space $X[f]$ of $X \sqcup \mathbb{D}^k \times \mathbb{D}^{n-k}$ by the identification $x \sim f(x)$ for all $x \in \mathbb{S}^{k-1} \times \mathbb{D}^{n-k}$. Show that
   (a) For $k = 1$, $\pi_1(X[f]) \approx \mathbb{Z} * \pi_1(X)$.
   (b) For $k = 2$, $\pi_1(X[f]) \approx \pi_1(X)/N$ where N is the normal subgroup generated by the image of $f_\# : \mathbb{S}^{k-1} \times 0 \to \pi_1(X)$.
   (c) For $k \geq 3$, $\pi_1(X[f]) \approx \pi_1(X)$.

9. Let X_1 and X_2 be two connected n-dimensional manifolds with boundary and $f_i :$ $\mathbb{D}^{n-1} \to \partial X_i$ be any two embeddings. Then the boundary connected sum $X_1 \#_b X_2$ is defined to be the quotient of $X_1 \sqcup X_2$ via the identification $f_1(x) \sim f_2(x)$. Show that for $n \geq 3$,

$$\pi_1(X_1 \#_b X_2) \approx \pi_1(X_1) * \pi_1(X_2).$$

10. Let X be manifold of dimension $n \geq 3$. Show that every element of $\pi_1(M)$ can be represented by an embedded loop.

11. Show that every finitely presented group is the fundamental group of a closed orientable smooth manifold of dimension 4.

12. Let $n \geq 1$ be an integer and G be group which is abelian if $n \geq 2$. By a *Moore space*[1] $M = M(G, n)$ of type (G, n), we mean a connected CW-complex M such that $\pi_i(M) = 0$ for $i < n$ and $\pi_n(M) = G$ and $H_i(M, \mathbb{Z}) = (0)$ for $i > n$. Construct such a space when $n \geq 2$. However, even when G is an abelian group, there may not exist a Moore space of type $(G, 1)$. See [Varadarajan, 1966] for more details.) Also see Exercise 10.4.17.(i).

13. Compute the homology of a totally disconnected space.

14. Show that the total space of the tangent bundle for $V_{n,k}$ is given by

$$\{((v_1, \ldots, v_k), u) \in V_{n,k} \times \mathbb{R}^n \mid u \perp v_i, \ i = 1, \ldots, k\}.$$

15. Show that $\tau(G_{n,k}) \approx \operatorname{Hom}(\gamma_n^k, (\gamma_n^k)^\perp)$.

---

[1] Watch out! Some authors have different definitions and the objects are really not the same either.

Chapter 6

Universal Coefficient Theorem for Homology

In this short chapter, we introduce homology groups with coefficients in an arbitrary module G over a given ring R and establish some algebraic relations between these groups and the homology groups with coefficients in R. On the way, we introduce the powerful and almost only known method to establish natural equivalences of various functors taking values in the category of chain complexes, viz., the method of acyclic models. As an easy consequence we get a proof of Theorem 4.3.3.

As a natural generalization, we then obtain the so-called Künneth formula for the homology of the tensor product of two chain complexes. With the help of Eilenberg–Zilber map this is then converted to a formula relating the singular homology of product of two spaces with those of the factors.

Throughout this chapter, R will denote a commutative ring with a unit. However, all important results will be established for the case when R is a PID. We also restrict ourselves to chain complexes which are non negatively graded and do away with the lower suffixes $C_{\cdot}$ or C_* and use the simplified notation C.

6.1   Method of Acyclic Models

We begin with a modest result which happens to be the precursor of the result to come later. Understanding the proof of this result helps to a large extent in understanding the later ones. Next, introduce an important notion of algebraic mapping cone. We then go on to formulate the method of acyclic models. As a simple application, we obtain a proof of Theorem 4.3.3.

Lemma 6.1.1 Let C, C' be non negative chain complexes, C be free and C' be acyclic in positive dimensions, i.e., $H_q(C') = 0$ for all $q > 0$. Then to every homomorphism $\phi : H_0(C) \to H_0(C')$, there is a chain map $\tau : C \to C'$ such that $H_0(\tau) = \phi$. Moreover, any two chain maps $\tau, \tau' : C \to C'$ such that $H_0(\tau) = H_0(\tau')$ are chain homotopic.

Proof: The construction of the chain map τ (and the chain homotopy D) is done inductively on q. We fix a basis $\{c_{qj}\}$ for each C_q and observe that it is enough to define $\tau_q(c_{qj})$ (respectively $D_q(c_{qj})$ for each q such that

(a1) $\partial' \tau_q(c_{qj}) = \tau_{q-1}(\partial(c_{qj}))$ and

(b1) $\partial' D_q(c_{qj}) = \tau_q(c_{qj}) - \tau'_q(c_{qj}) - D_{q-1}(\partial(c_{qj}))$

and extend them all over C_q linearly.

We choose $\tau_0(c_{0j}) \in C'_0$ to be such that $\phi[c_{0j}] = [\tau_0(c_{0j})]$ and extend it linearly over C_0. Verify that $H_0(\tau) = \phi$ so that the definition of τ_0 is complete.

Indeed, we observe that $[\tau_0(\partial(c_{1j})] = \phi[\partial c_{1j}] = 0$, and hence by the definition of $H_0(C')$, these are boundary elements in C'_0. So, we can choose $\tau_1(c_{1j}) \in C'_1$ so that $\partial'(\tau_1(c_{1j})) = \tau_0(\partial c_{1j})$. Extending linearly over C_1 completes the definition of τ_1.

Having defined τ_{q-1} appropriately, we observe that $\partial' \circ \tau_{q-1} \circ \partial(c_{qj}) = \tau_{q-2} \circ \partial^2(c_{qj}) = 0$. By the acyclicity of C' in dimension $q > 0$, it follows that we can choose $\tau_q(c_{qj}) \in C'_q$ so that $\partial'(\tau_q(c_{qj})) = \tau_{q-1}(\partial(c_{qj}))$ and extend τ_q linearly all over C_q. This completes the construction of τ.

The construction of the chain homotopy D is left to the reader as an exercise. ♠

Before going further, we would like to convert this lemma into a useful tool for studying chain equivalences. For this purpose we introduce the algebraic version of the mapping cone.

Definition 6.1.2 For any chain map $\tau : C \to C'$ the mapping cone $C(\tau) = \{\bar{C}_q, \bar{\partial}_q\}$ of τ is the chain complex

$$\bar{C}_q = C_{q-1} \oplus C'_q; \quad \bar{\partial}_q(c, c') = (-\partial_{q-1}(c), \tau(c) + \partial'_q(c'))$$

Verification that $C(\tau)$ is indeed a chain complex is easy. Also note that if C, C' are free then $C(\tau)$ is free.

Theorem 6.1.3 *A chain map $\tau : C \to C'$ is a chain equivalence iff its mapping cone is chain contractible.*

Proof: Suppose $\tau' : C' \to C$ is a chain map and $D : C \to C$, $D' : C' \to C'$ are chain homotopies

$$D : \tau' \circ \tau \approx 1_C; \quad D' : \tau \circ \tau' \approx 1_{C'}.$$

Put $\bar{D}(c, c') = (a, b)$ where

$$a = D(c) + \tau' D' \tau(c) - \tau' \tau D(c) + \tau(c');$$

$$b = D' \tau D(c) - D' D \tau(c) - D'(c').$$

Verify that $\bar{D}$ is a chain contraction of $C(\tau)$. Conversely, assume that we have a chain contraction $\bar{D}$ of $C(\tau)$. Define $\tau' : C' \to C$, $D : C \to C$ and $D' : C' \to C'$ by the equations:

$$(\tau'(c'), -D'(c')) = \bar{D}(0, c'); \quad (D(c), -) = \bar{D}(c, 0).$$

Verify that D and D' are chain homotopies $\tau' \circ \tau \approx Id_C$ and $\tau \circ \tau' \approx Id_{C'}$ respectively. ♠

Corollary 6.1.4 A chain map between two free chain complexes is a chain equivalence iff its mapping cone is acyclic.

Lemma 6.1.5 Given a chain map $\tau : C \to C'$, there is a short exact sequence of chain complexes

$$0 \longrightarrow C' \longrightarrow \bar{C} \longrightarrow \hat{C} \longrightarrow 0$$

where $\bar{C}$ is the mapping cone of $\tau : C \to C'$ and $\hat{C}$ is the chain complex defined by $\hat{C}_q = C_{q-1}$ and $\hat{\partial}_q = -\partial_{q-1}$. In particular, the mapping cone is acyclic iff $\tau_* : H(C) \to H(C')$ is an isomorphism.

Proof: The first part follows by direct verification. The second part follows from the long homology exact sequence associated with this short exact sequence. ♠

Combining this lemma with the above corollary we obtain:

Theorem 6.1.6 *A chain map $\tau : C \to C'$ of two free chain complexes is a chain equivalence iff it induces isomorphism on homology.*

Remark 6.1.7 We are now ready for a far-reaching generalization of Lemma 6.1.1 to functors taking values in chain complexes. The freeness of C reduces the task of defining a homomorphism to the task of defining it on a set of basic elements. The generalization comes in the direction of the category on which the functors are taken. You may see the analogy of this in the generalization of the notion of uniform convergence to the notion of equicontinuity.

Definition 6.1.8 By a *category $\mathcal{C}$ with models $\mathcal{M}$* we mean a category $\mathcal{C}$ together with a set $\mathcal{M} = \{M_j : j \in J\}$ of objects in $\mathcal{C}$ called *models*. Let $\mathcal{F} : \mathcal{C} \rightsquigarrow \mathcal{A}b$ be a covariant functor. A set $\{f_j\}_{j \in J}$ such that $f_j \in \mathcal{F}(M_j)$, where $M_j \in \mathcal{M}$ for each j, is called a *basis for $\mathcal{F}$* if for every object X in $\mathcal{C}$, the set $\{\mathcal{F}(s)(f_j) : s \in hom(M_j, X), j \in J\}$ is a basis for the abelian group $\mathcal{F}(X)$. Any functor $\mathcal{F}$ with a basis as above is called a *free functor on $\mathcal{C}$ with models $\mathcal{M}$*.

Remark 6.1.9 It follows that $\mathcal{F}$ is the composite of two functors: the first one assigns to each object $X \in \mathcal{C}$ the set $\{\mathcal{F}(s)(f_j) : s \in hom(M_j, X), j \in J\}$; the second one is the free abelian group functor $\mathbf{Ens} \rightsquigarrow \mathcal{A}b$ which assigns to each set the free abelian group generated by the set. Note that if a functor $\mathcal{F}$ is free with models $\mathcal{M}$ and some basis and if $\mathcal{M}'$ is a set of objects which contains $\mathcal{M}$, then $\mathcal{F}$ is free with models $\mathcal{M}'$ also.

Definition 6.1.10 A graded abelian group $A = \oplus_q A_q$ is said to be free if each A_q is free. In this situation, a graded set $\{\alpha_j^q\}$ is called a basis for A if for each fixed q, $\{\alpha_j^q\} \subset A_q$ is a basis for A_q. A covariant functor $\mathcal{F}$ on a category $\mathcal{C}$ with models to the category of chain complexes is said to be *free* if $\mathcal{F}_q$ is free for all $q \in \mathbb{Z}$.

Example 6.1.11

(a) Let $\mathcal{C} = \mathbf{Top}$. For each $q \geq 0$ the functor $X \rightsquigarrow S_q(X)$, the free group of q-chains in X, is free with model $\mathcal{M}_q = \{\Delta_q\}$ and basis $\{\xi_q\}$. The chain complex $S.(X)$ is free with models $\{\Delta_q : q \geq 0\}$ and basis $\{\xi_q : q \geq 0\}$.

(b) Let K be a simplicial complex and $\mathcal{K}$ be the category of all subcomplexes with $hom(L_1, L_2) = \{\iota\}$ (where ι is the inclusion map) iff L_1 is a subcomplex of L_2 and $hom(L_1, L_2) = \emptyset$ otherwise. Fix a total order on the vertices of K and consider the oriented simplicial chain complex functor $C(L)$ on this category. Choose $\mathcal{M} = \{s : s \in K\}$ as the model, where we treat each $s \in K$ as a subcomplex of K. Then C is free with models $\mathcal{M}$ and basis $\{\sigma(s) : s \in K\}$ where $\sigma(s)$ denotes the oriented simplex with its support on s.

(c) Consider the category of simplicial complexes and simplicial maps with models $\{\Delta_q\}_{q \geq 0}$. On this category, consider the simplicial singular chain complex functor $\mathcal{S}$. The collection $\{\xi_q\}$ forms a basis for this functor.

(d) Consider the category $\mathbf{Diff}$ of smooth objects and smooth maps with models $\{\Delta_q\}_{q \geq 0}$. The smooth singular chain complex is a free functor with basis $\{\xi_q\}_{q \geq 0}$.

(e) Consider the point-category $\mathcal{P}$ with a single object denoted by $\star$ and with a single morphism denoted by $Id : \star \to \star$. Any free abelian group A can be thought of as a functor on this category with model $\star$ and basis being any basis of A. Also, any free chain complex $C.$ can be thought of as a free functor with basis as a graded basis $\{\alpha_j^q\}$.

Definition 6.1.12 A functor $\mathcal{F}$ on a category with models $\mathcal{M}$ taking values in the category of non negative chain complexes is called *acyclic in positive dimensions* if $H_q(\mathcal{F}(M)) = 0$ for all $q > 0$ and for all $M \in \mathcal{M}$.

Remark 6.1.13 The examples cited above are all acyclic in positive dimensions.

Theorem 6.1.14 *Let $\mathcal{C}$ be a category with models $\mathcal{M}$. Suppose $\mathcal{F}, \mathcal{F}'$ are two covariant functors from $\mathcal{C}$ to the category of chain complexes such that $\mathcal{F}$ is free and non negative and $\mathcal{F}'$ is acyclic in positive dimensions. Then*

(a) *Any natural transformation $H_0(\mathcal{F}) \to H_0(\mathcal{F}')$ is induced by a natural chain map $\tau : \mathcal{F} \to \mathcal{F}'$.*

(b) *Two natural chain maps $\tau, \tau' : \mathcal{F} \to \mathcal{F}'$ inducing the same natural transformation on $H_0(\mathcal{F}) \to H_0(\mathcal{F}')$ are naturally chain homotopic.*

Proof: In order to prove (a) and (b), respectively, for each object $X \in \mathcal{C}$, we must define

(a1) a chain map $\tau(X) : \mathcal{F}(X) \to \mathcal{F}'(X)$ and

(b1) a chain homotopy $D(X) : \tau(X) \approx \tau'(X)$

such that if $h : X \to Y$ is a morphism in $\mathcal{C}$ then

(a2) $\tau(Y) \circ \mathcal{F}(h) = \mathcal{F}'(h) \circ \tau(X)$ and

(b2) $D(Y) \circ \mathcal{F}(h) = \mathcal{F}'(h) \circ D(X)$.

$$
\begin{array}{ccc}
\mathcal{F}(X) \xrightarrow{\mathcal{F}(h)} \mathcal{F}(Y) & \qquad & \mathcal{F}(X) \xrightarrow{\mathcal{F}(h)} \mathcal{F}(Y) \\
\tau(X) \downarrow \qquad \downarrow \tau(Y) & & D(X) \downarrow \qquad \downarrow D(Y) \\
\mathcal{F}'(X) \xrightarrow{\mathcal{F}'(h)} \mathcal{F}'(Y) & & \mathcal{F}'(X) \xrightarrow{\mathcal{F}'(h)} \mathcal{F}'(Y)
\end{array}
$$

We shall define $\tau_q(X)$ (and $D_q(X)$) inductively so that

(a3) $\partial' \circ \tau_q(X) = \tau_{q-1}(X) \circ \partial$ and

(b3) $\partial' \circ D_q(X) = \tau_q(X) - \tau'_q(X) - D_{q-1}(X) \circ \partial$.

Let $\{f_j \in \mathcal{F}_q(M_j) \ : \ j \in J_q\}$ be a basis for $\mathcal{F}_q$ for each $q \geq 0$. Then by definition,

$$\{\mathcal{F}_q(s)(f_j) \ : \ s \in hom(M_j, X), j \in J_q\}$$

is a basis $\mathcal{F}_q(X)$. It follows that $\tau_q(X)$ (respectively ($D_q(X)$) is determined by the collection

(a4) $\{\tau_q(M_j)(f_j) \ : \ j \in J_q\}$ (respectively)

(b4) $\{D_q(M_j)(f_j) \ : \ j \in J_q\}$

and by the linearity property:

(a5) $\tau_q(X)(\sum_i n_{ij} \mathcal{F}_q(s_{ij})(f_j)) = \sum_i n_{ij} \mathcal{F}'_q(s_{ij})(\tau_q(M_j)(f_j))$ and

(b5) $D_q(X)(\sum_i n_{ij} \mathcal{F}_q(s_{ij})(f_j)) = \sum_i n_{ij} \mathcal{F}'_{q+1}(s_{ij}) D_q(M_j)(f_j)$, respectively.

Thus having defined τ_i (and D_i) for $i < q$, we need to define

(a6) $\tau_q(M_j)(f_j)$ so that $\partial' \tau_q(M_j)(f_j) = \tau_{q-1}(M_j)(\partial f_j)$ and

(b6) $D_q(M_j)(f_j)$ such that

$$\partial' D_q(M_j)(f_j) = \tau_q(M_j)(f_j) - \tau'_q(M_j)(f_j) - D_{q-1}(M_j)\partial(f_j)$$

for all $j \in J_q$, respectively.

(a7) Given a natural transformation $\phi : H_0(\mathcal{F}) \to H_0(\mathcal{F}')$, we define $\tau_0(M_j)(f_j)$ to be any element of $\mathcal{F}'(M_j)$ such that the homology class

$$[\tau_0(M_j)(f_j)] = \phi(M_j)([f_j])$$

for all $j \in J_0$. By (a5), $\tau_0(X)$ gets defined for all X and we have for $f \in \mathcal{F}_0(X)$, $[\tau_0(X)(f)] = \phi(X)[g]$. In particular, for any $j \in J_1$, $\tau_0(M_j)(\partial f_j)$ is a boundary in $\mathcal{F}'_0(M_j)$. Hence, we can choose $\tau_1(M_j)(f_j) \in \mathcal{F}'(M_j)$ so that $\partial \tau_1(M_j)(f_j) = \tau_0(M_j)(\partial f_j)$. Equation (a5) then takes care of the definition of $\tau_1(X)$ for all X. Now for some $q > 1$, assuming that we have defined τ_i for $i < q$, we observe that the RHS of (a6) is a cycle in $\mathcal{F}'_{q-1}(M_j)$ (because (a3) is satisfied for $q - 1$). Since $q > 1$, $H_{q-1}(\mathcal{F}'(M_j)) = 0$ so that we can choose $\tau_q(M_j)(f_j)$ so as to satisfy (a3). This completes the definition of τ.

(b7) We leave the details of the rest of the proof of (b) to the reader as an exercise. ♠

Remark 6.1.15 From this theorem, we can recover Lemma 6.1.1 via the singleton category $\mathcal{P}$ (as described in Example 6.1.11.(e)), and thinking of any free chain complex as a free functor on this category with a basis. Then the statement of the theorem reduces to the statement of the lemma. Similarly, we can prove:

Corollary 6.1.16 A non negative, acyclic free chain complex C is contractible.

Proof: Once again we take the category with models as in the above remark. It follows that C is a free functor on this category which is acyclic in positive dimensions. Since $H_0(C) = 0$ it follows that the chain maps $Id_C, 0_C$ induce the same homomorphism on $H_0(C) \to H_0(C)$. Therefore $Id_C \approx 0_C$ which means, by definition, that C is contractible. ♠
   More generally we have:

Theorem 6.1.17 *Let $\mathcal{C}$ be a category with models $\mathcal{M}$. Let $\mathcal{F}, \mathcal{F}'$ be any two functors from $\mathcal{C}$ to the category of non negative chain complexes, both being free and acyclic in positive dimensions with models $\mathcal{M}$. Then any natural transformation $\Gamma : \mathcal{F} \to \mathcal{F}'$ which induces a natural equivalence on $H_0(\mathcal{F}) \to H_0(\mathcal{F}')$ is a natural equivalence.*

Proof: Let $\theta : H_0(\mathcal{F}') \to H_0(\mathcal{F})$ be the inverse natural transformation to the equivalence $\Gamma_0 : H_0(\mathcal{F}) \to H_0(\mathcal{F}')$. Let $\Theta : \mathcal{F}' \to \mathcal{F}$ be the natural transformation which induces θ on H_0. Then the composites $\Theta \circ \Gamma : \mathcal{F} \to \mathcal{F}$ and $\Gamma \circ \Theta : \mathcal{F}' \to \mathcal{F}'$ induce identity equivalence on H_0's. Since $\mathcal{F}, \mathcal{F}'$ are free and acyclic in positive dimensions, it follows that $\Theta \circ \Gamma$ and $\Gamma \circ \Theta$ are chain homotopic to respective identity transformations. That just means that Γ is a natural equivalence with its inverse Θ.
   We can now fulfill the promise of a proof of Theorem 4.3.3 which we restate as:

Corollary 6.1.18 The inclusion map $\eta : S^{sm}_{\cdot} \to S_{\cdot}$ of the smooth singular chain complex into the singular chain complex is a natural equivalence. In particular, $\eta_* : H^{sm}_*(M) \to H_*(M)$ is an isomorphism for any smooth manifold M.

Proof: Both chain functors are free with models $\{\Delta_q\}_{q \geq 0}$ and acyclic in positive dimensions, since they satisfy homotopy axiom. ♠
   This result will be used in Chapter 8 in establishing de Rham's theorem.

6.2   Homology with Coefficients: The Tor Functor

   We first introduce homology with coefficients in a module and state a result, which involves the torsion functor. We then introduce the torsion functor as a measure of deviation of tensor product from being an exact functor. When R is a PID, this becomes quite simple and yet a very useful tool in many topological situations. We first prove a simpler version of the above result based on which the general result will be proved using yet another notion, viz., free approximations to chain complexes.

Definition 6.2.1 Given a chain complex C of R-modules and a R-module G, we define $C \otimes G$ to be the chain complex $\{C_n \otimes G, \partial_n \otimes Id_G\}$. The homology modules $H_{\cdot}(C \otimes G)$ are called the homology of C with coefficient group G and is denoted by $H_{\cdot}(C; G)$.

Remark 6.2.2 For each fixed chain complex C, the assignment $G \rightsquigarrow H(C; G)$ defines a functor. So does the assignment $C \rightsquigarrow H(C; G)$, for each fixed G. Given $c \in Z_q(C)$ and $g \in G$, check that $c \otimes g$ is a cycle and hence we get a bilinear map

$$H_q(C) \times G \to H_q(C; G); \quad ([c], g) \mapsto [c \otimes g]$$

which in turn yields a homomorphism

$$\mu : H_q(C) \otimes G \to H_q(C; G); \quad [c] \otimes g \mapsto [c \otimes g].$$

Notice that if C is a chain complex of abelian groups and G is a R-module then $C \otimes_{\mathbb{Z}} R$ is a chain complex of R-modules and we have the canonical isomorphism

$$C \otimes_{\mathbb{Z}} R \otimes_R G \approx C \otimes_{\mathbb{Z}} G; \quad c \otimes r \otimes g \mapsto c \otimes rg.$$

In other words, the homology modules of $C \otimes R$ over R with coefficients G are isomorphic to the homology modules of C over $\mathbb{Z}$ with coefficients in $R \otimes_R G \approx G$.

Remark 6.2.3 Since taking tensor product commutes with taking direct sum it follows that if

$$0 \longrightarrow C' \longrightarrow C \longrightarrow C'' \longrightarrow 0$$

is a split short exact sequence of chain complexes, then

$$0 \longrightarrow C' \otimes G \longrightarrow C \otimes G \longrightarrow C'' \otimes G \longrightarrow 0$$

is a short exact sequence and hence there is a long exact sequence of homology modules. Such is the case with singular (cellular, simplicial) chain complexes of pairs and triples and hence we get various long homology exact sequences with coefficients in a module as well. The big difference in homology with coefficients in a module is due to the fact that the homomorphism μ is **not**, in general, an ismorphism, i.e., *taking homology does not commute with taking tensor products.* The precise relationship is the content of the following theorem.

Theorem 6.2.4 General universal coefficient theorem for homology *Let R be a principal ideal domain. On the subcategory of the product category of chain complexes C and modules G over R such that $C \star G$ is acyclic there is a functorial short exact sequence*

$$0 \longrightarrow H_q(C) \otimes G \longrightarrow H_q(C; G) \longrightarrow H_{q-1}(C) \star G \longrightarrow 0$$

and this sequence is split.

Here $\star$ denotes the torsion product. Proving the above theorem just amounts to establishing definition and certain algebraic properties of this product. This is what we plan to do in the next section. First, we shall prove a slightly weaker version of the above theorem which suffices for the study of singular homology. However, for applications in cohomology later, we shall need the full force of the above theorem. The rest of this section will be devoted to the proof of this theorem.

Definition 6.2.5 Let A be an R-module. By a resolution of A (over R) we mean an exact sequence

$$\ldots \longrightarrow C_n \xrightarrow{\partial_n} \cdots \cdots \longrightarrow C_0 \xrightarrow{\epsilon} A \longrightarrow 0$$

of R-modules. If each C_n is a free module, then we call it a free resolution of A.

Remark 6.2.6 Recall from Example 4.2.12(d) that by an augmentation of a non negative chain complex C, we mean a surjective map $\epsilon : C_0 \to R$ and extend the chain complex C by defining $\partial_0 = \epsilon : C_0 \to R = C_1$ and $C_q = 0$ for all $q < -1$. The same thing can be done with any R-module A in place of R. Thus a resolution of A consists of a chain complex augmented over A and such that the augmented chain complex is acyclic. Starting with an R-module A, we take $C_0 = F(A)$ the free module generated by the set A and $\partial_0 := \epsilon : C_0 \to A$ to be the map which sends the basic elements to themselves. Inductively, we repeat this process with A replaced by $\operatorname{Ker} \partial_k$ to get a surjective homomorphism $C_{k+1} \to \operatorname{Ker} \partial_k$ and compose this with the inclusion $\operatorname{Ker} \partial_k \subset C_k$ to obtain $\partial_{k+1} : C_{k+1} \to C_k$. This yields a free resolution of A called the *canonical free resolution of A.*

Theorem 6.2.7 *Let C be a free non negative chain complex augmented over A and let C' be a resolution of A'. Then any homomorphism $\phi : A \to A'$ extends to a chain map $\hat{\phi} : C \to C'$ compatible with the augmentations. Moreover, any two such chain maps are chain homotopy equivalent.*

Proof: Note that $A \approx H_0(C), A' \approx H_0(C')$ and hence giving a homomorphism $\phi : A \to A'$ corresponds to giving one $H_0(C) \to H_0(C')$. Now the theorem follows from Lemma 6.1.1.♠

Corollary 6.2.8 Any two free resolutions of a given module are canonically chain equivalent chain complexes.

Definition 6.2.9 Fix a free resolution C, say the canonical one for a given module A. Then for any module B, it follows that the modules $H_q(C, B)$ are well defined independent of the choice of the resolution C and we use the notation:

$$\text{Tor}_q(A, B) := H_q(C, B); \quad \text{Tor.} = \oplus_{q \geq 0} \text{Tor}_q.$$

Remark 6.2.10 Clearly Tor. defines a covariant functor in both the slots. Since the tensor product commutes with direct sums and direct limits, so do all the Tor_q's. By definition $\text{Tor}_0(A, B) = (C_0 \otimes B)/\text{Im}(\partial_1 \otimes 1) = \text{Ker}\,(\epsilon \otimes 1)$. By the exactness of

$$C_0 \otimes B \longrightarrow A \otimes B \longrightarrow 0$$

it follows that $\text{Tor}_0(A; B) = A \otimes B$.

Definition 6.2.11 We put $A \star B := \text{Tor}_1(A, B)$.

So far we have not used the fact that R is a PID and so all results are valid for any commutative ring with a unit. Now assume that R is a PID. Then every submodule of a free module is free and hence any module A admits a free resolution

$$0 \longrightarrow C_1 \longrightarrow C_0 \longrightarrow A \longrightarrow 0$$

and hence $H_q(A; B) = 0$ for all $q \geq 2$. There is a short exact sequence

$$0 \longrightarrow A \star B \longrightarrow C_1 \otimes B \longrightarrow C_0 \otimes B \longrightarrow A \otimes B \longrightarrow 0 \ .$$

In fact, $A \star B = H_1(C \otimes B) = \text{Ker}\,(C_1 \otimes B \to C_0 \otimes B)$, since $C_2 = 0$.

Example 6.2.12 The following properties of the torsion product are easily verified. (Here modules are taken over a PID R.)

(a) If A is free then $A \star B = 0$.

(b) $A \star R = 0$.

(c) $A \star B = 0$ for all free R modules B. In particular, If R is a field then $A \star B = 0$ for all R modules A, B.

(d) If A or B is torsion free then $A \star B = 0$.

(e) $(R/tR) \star B = \{b \in B \ : \ tb = 0\}$ for any non zero $t \in R$.

(f) If A, B are any two finitely generated modules, determine $A \star B$.

We are now ready to prove a simpler version of Theorem 6.2.4.

Theorem 6.2.13 (Universal coefficient theorem for homology) *Let R be a PID and C be a free chain complex of R-modules and let G be any module over R. Then there is a short functorial exact sequence*

$$0 \longrightarrow H_q(C) \otimes G \stackrel{\mu}{\longrightarrow} H_q(C;G) \longrightarrow H_{q-1}(C) \star G \longrightarrow 0 \qquad (6.1)$$

and this sequence splits. Here $\mu([c] \otimes g) = [c \otimes g]$.

Proof: We define two chain complexes Z and B as follows:
$Z_q = Z_q(C); B_q = B_{q-1}(C)$, with trivial boundary operators on both of them. It follows that we have a short exact sequence

$$0 \longrightarrow Z \stackrel{\eta}{\longrightarrow} C \stackrel{\lambda}{\longrightarrow} B \longrightarrow 0$$

of free modules. Here $\eta(z) = z$ and $\lambda(c) = \partial(c)$, where ∂ is the boundary operator of C. Because B_q are free, the sequence splits also. Therefore we can apply Remark 6.2.3 to obtain a long homology exact sequence

$$\dots \longrightarrow H_q(Z;G) \stackrel{\eta_*}{\longrightarrow} H_q(C;G) \stackrel{\lambda_*}{\longrightarrow} H_q(B;G) \stackrel{\partial}{\longrightarrow} H_{q-1}(Z;G) \longrightarrow \dots$$

Since Z and B have trivial boundary operators, the same is true of $Z \otimes G$ and $B \otimes G$. Therefore, $H_q(Z;G) = Z_q \otimes G; H_q(B;G) = B_{q-1}(C) \otimes G$. Moreover, under these identifications the connecting homomorphism $\partial : H_q(B;G) \to H_{q-1}(Z,G)$ becomes $\iota_{q-1} \otimes 1$, where $\iota_{q-1} : B_{q-1}(C) \to Z_{q-1}$ is the inclusion. Thus the long homology sequence becomes

$$\dots B_q(C) \otimes G \stackrel{\iota_q \otimes 1}{\longrightarrow} Z_q(C) \otimes G \stackrel{\eta_*}{\longrightarrow} H_q(C;G) \stackrel{\lambda_*}{\longrightarrow} B_{q-1}(C) \otimes G \stackrel{\iota_{q-1} \otimes 1}{\longrightarrow} Z_{q-1}(C) \otimes G \dots$$

which in turn yields a sequence of short exact sequences

$$0 \longrightarrow \mathrm{Coker}\,(\iota_q \otimes 1) \longrightarrow H_q(C,G) \longrightarrow \mathrm{Ker}\,(\iota_{q-1} \otimes 1) \longrightarrow 0. \qquad (6.2)$$

What are the two modules on either side of $H_q(C;G)$? Since R is a PID, and C is free, the short exact sequence

$$0 \longrightarrow B_q(C) \longrightarrow Z_q(C) \longrightarrow H_q(C) \longrightarrow 0$$

is a free presentation of $H_q(C)$. Therefore upon taking tensor product with G we get an exact sequence

$$0 \longrightarrow H_q(C) \star G \longrightarrow B_q(C) \otimes G \stackrel{\iota_q \otimes 1}{\longrightarrow} Z_q(C) \otimes G \longrightarrow H_q(C) \otimes G \longrightarrow 0.$$

This means that $H_q(C) \star G = \mathrm{Ker}\,(\iota_q \otimes 1)$ and $H_q(C) \otimes G = \mathrm{Coker}\,\iota_q \otimes 1$. Substituting these into (6.2) yields the exact sequence 6.1. Unravelling through all these notations, one also checks that the homomorphism $H_q(C) \otimes G \to H_q(C;G)$ is nothing but μ.

It remains to see that the sequence splits. (Caution: The splitting is not functorial.) Since B_{q-1} is a free module we can choose a splitting h_q for ∂_q, i.e., $\partial_q \circ h_q = Id$. This means that $(\partial_q \otimes 1) \circ (h_q \otimes 1)$ factors through $\iota_{q-1} \otimes 1$. Therefore, $h_q \otimes 1$ maps $\mathrm{Ker}\,(\iota_{q-1} \otimes 1)$ inside $\mathrm{Ker}\,(\partial_q \otimes 1)$ and hence induces a homomorphism $\mathrm{Ker}\,(\iota_{q-1} \otimes 1) = H_{q-1}(C) \star G \to H_q(C;G)$ which is a right inverse to $\lambda_* : H_q(C;G) \to H_{q-1}(C) \star G \subset H_q(B;G)$. ♠

Remark 6.2.14 As an immediate consequence, we have the universal coefficient theorem for singular, simplicial and cellular homology theories, since in each case, we have a free chain complex to begin with.

Definition 6.2.15 For any topological pair (X, A) and any abelian group G, we define

$$H_*(X, A; G) := H_*(S_.(X, A) \otimes G).$$

Corollary 6.2.16 Let $f : X \to Y$ be a continuous map inducing isomorphism $f_* : H_*(X; \mathbb{Z}) \to H_*(Y; \mathbb{Z})$. Then $f_* : H_*(X; G) \to H_*(Y; G)$ is an isomorphism for all coefficient groups G.

We shall need one more technical result before we take up the proof of Theorem 6.2.4. All modules here are over a PID R. Chain complexes are assumed to be non negative for simplicity.

Definition 6.2.17 By a *free approximation* $\tau : \bar{C} \to C$ of a chain complex C we mean a free chain complex $\bar{C}$ and a chain map τ such that
(a) τ is surjective and
(b) the induced homomorphism $\tau_* : H_*(\bar{C}) \to H_*(C)$ is an isomorphism.

Lemma 6.2.18 Free approximation exists and is unique up to a chain homotopy equivalence.

Proof: Given a chain complex C, let F_q be a free module and $\alpha_q : F_q \to Z_q(C)$ an epimorphism. Put $F'_q = \alpha_q^{-1}(B_q(C))$. Since F'_q is also free there exist $\beta_q : F'_q \to C_{q+1}$ such that $\partial_{q+1} \circ \beta_q = \alpha_q|_{F'_q}$. Take $\bar{C}_q = F_q \oplus F'_{q-1}$ and $\bar{\partial}_q(x, y) = (y, 0)$. Clearly, $(\bar{C}, \bar{\partial})$ is then a free chain complex. Define $\tau(x, y) = \alpha_q(x) + \beta_{q-1}(y)$. Verify that $\tau : \bar{C} \to C$ is a chain map. To see that τ_q is surjective, let $c \in C_q$. Then there exist $x \in F'_{q-1}$ such that $\alpha_{q-1}(y) = \partial_q(c) \in B_{q-1}(C)$. Now check that $\tau(0, y) - c \in Z_q(C)$. Therefore we can choose $x \in F_q$ such that $\alpha_q(x) = \tau(0, y) - c$. Verify that $\tau(x, y) = c$. Finally, $\operatorname{Ker} \tau_q = F_q = \alpha_q^{-1}(Z_q(C))$ and $\operatorname{im} \tau_{q+1} = F'_q \subset \alpha_q^{-1}(B_q(C))$. Since $\tau_q|_{F_q} = \alpha_q$, it follows that $\tau_* : H_q(\bar{C}) \to H_q(C)$ is the isomorphism induced by α_q (by the second isomorphism theorem). The uniqueness part follows from a result which is somewhat more general. We state and prove this as a separate lemma. ♠

Lemma 6.2.19 Let $\tau : \bar{C} \to C$ be a free approximation. Given a free chain complex C' and a chain map $\tau' : C' \to C$, there exists a chain map $\bar{\tau} : C' \to \bar{C}$ such that $\tau \circ \bar{\tau} = \tau'$. Moreover any two such chain maps are chain homotopic.

Proof: Consider the short exact sequence

$$0 \longrightarrow \operatorname{Ker} \tau \longrightarrow \bar{C} \overset{\tau}{\longrightarrow} C \longrightarrow 0.$$

Since $\bar{C}$ is a free chain complex, so is $\operatorname{Ker} \tau$ (R is a PID). Since τ_* is an isomorphism on homology, from the homology long exact sequence it follows that $\operatorname{Ker} \tau$ is acyclic. From Corollary 6.1.16, it follows that $\operatorname{Ker} \tau$ is contractible. Let $D = \{D_q : \operatorname{Ker} \tau_q \to \operatorname{Ker} \tau_{q+1}\}$ be a chain contraction of $\operatorname{Ker} \tau$.

Since C' is free and τ is surjective, there exist homomorphisms $\phi_q : C'_q \to \bar{C}_q$ such that $\tau_q \circ \phi_q = \tau'_q$. Put $h_q = \bar{\partial}_q \phi_q - \phi_{q-1} \partial'_q$. Check that $\tau_{q-1} \circ h_q = 0$. Therefore $h_q : C'_q \to \operatorname{Ker} \tau_{q-1}$. Put $\bar{\tau}_q = \phi_q - D_{q-1} h_q$. Verify that $\bar{\tau}$ is a chain map and $\tau \bar{\tau} = \tau'$.

Finally, if $\bar{\tau}, \bar{\tau}' : C' \to \bar{C}$ are two chain maps such that $\tau\bar{\tau} = \tau\bar{\tau}'$, clearly $\bar{\tau} - \bar{\tau}'$ is a chain map which takes values in $\operatorname{Ker}\tau$. Verify that $\{D_q \circ (\bar{\tau} - \bar{\tau}')\}$ defines a chain homotopy from $\bar{\tau}$ to $\bar{\tau}'$. ♠

Proof of Theorem 6.2.4: Let $\tau : \bar{C} \to C$ be a free approximation. Then we have an exact sequence

$$0 \longrightarrow \operatorname{Ker}\tau \longrightarrow \bar{C} \overset{\tau}{\longrightarrow} C \longrightarrow 0$$

in which $\operatorname{Ker}\tau$ is chain contractible. Since the above sequence is a free presentation of C we have an exact sequence

$$0 \longrightarrow C \star G \longrightarrow (\operatorname{Ker}\tau) \otimes G \overset{\iota \otimes 1}{\longrightarrow} \bar{C} \otimes G \overset{\tau \otimes 1}{\longrightarrow} C \otimes G \longrightarrow 0$$

which gives two exact sequences

$$0 \longrightarrow C \star G \longrightarrow (\operatorname{Ker}\tau) \otimes G \longrightarrow \operatorname{Im}(\iota \otimes 1) \longrightarrow 0,$$

$$0 \longrightarrow \operatorname{Im}(\iota \otimes 1) \longrightarrow \bar{C} \otimes G \overset{\tau \otimes 1}{\longrightarrow} C \otimes G \longrightarrow 0.$$

In the first one, $C \star G$ is acyclic by hypothesis. Since $\operatorname{Ker}\tau$ is contractible, so is $(\operatorname{Ker}\tau) \otimes G$. By the long homology exact sequence it follows that $\operatorname{Im}(\iota \otimes 1)$ is acyclic. Therefore, the second exact sequence yields an isomorphism

$$(\tau \otimes 1)_* : H(\bar{C} \otimes G) \to H(C \otimes G).$$

We now have a commutative diagram

$$
\begin{array}{ccccccccc}
0 & \longrightarrow & H_q(\bar{C}) \otimes G & \overset{\mu}{\longrightarrow} & H_q(\bar{C} \otimes G) & \longrightarrow & H_{q-1} \star G & \longrightarrow & 0 \\
& & \downarrow{\scriptstyle \tau_* \otimes 1} & & \downarrow{\scriptstyle (\tau \otimes 1)_*} & & \downarrow{\scriptstyle \tau_* \star 1} & & \\
0 & \longrightarrow & H_q(C) \otimes G & \overset{\mu}{\longrightarrow} & H_q(C \otimes G) & \dashrightarrow & H_{q-1} \star G & \longrightarrow & 0
\end{array}
$$

in which the top row is exact by Theorem 6.2.13 and the vertical arrows are isomorphisms. Therefore there is a unique way that the dotted arrow in the bottom sequence can be defined so as to make the entire diagram commutative. It then follows that the bottom row is also exact. Since the top one splits, the bottom one also splits. The functoriality of the bottom row and its independence from the choice of the free approximation $\bar{C}$ follow from Lemma 6.2.19. ♠

Exercise 6.2.20 In all these exercises, we assume that R is a PID and the modules are over R.

(i) Compute $\mathbb{Z}_m \star \mathbb{Z}_n$ for any two positive integers m, n.

(ii) Given a short exact sequence of modules

$$0 \longrightarrow A' \longrightarrow A \longrightarrow A'' \longrightarrow 0 \tag{6.3}$$

and given a module B such that A'' or B is torsion free, show that there is a short exact sequence

$$0 \longrightarrow A' \otimes B \longrightarrow A \otimes B \longrightarrow A'' \otimes B \longrightarrow 0.$$

(iii) Let C be torsion free chain complex. Given a short exact sequence

$$0 \longrightarrow G' \longrightarrow G \longrightarrow G'' \longrightarrow 0$$

of modules there is a functorial connecting homomorphism $\beta : H(C; G'') \to H(C; G')$ of degree -1 and a functorial exact sequence

$$\cdots \longrightarrow H_q(C; G') \longrightarrow H_q(C, G) \longrightarrow H_q(C, G'') \overset{\beta}{\longrightarrow} H_{q-1}(C; G') \longrightarrow \cdots$$

This homomorphism β is called *the Bockstein homology homomorphism* corresponding to the coefficient sequence (6.3).

(iv) Given a module M and short exact sequence

$$0 \longrightarrow G' \longrightarrow G \longrightarrow G'' \longrightarrow 0$$

of R-modules there is a six-term exact sequence

$$0 \longrightarrow M \star G' \longrightarrow M \star G \longrightarrow M \star G''$$

$$M \otimes G' \longrightarrow M \otimes G \longrightarrow M \otimes G'' \longrightarrow 0$$

(v) Show that there is a functorial isomorphism

$$A \star B \approx B \star A.$$

(vi) For any R module A, it is also common practice to denote the torsion submodule

$$\{a \in A \ : \ \lambda a = 0, \text{ for some } \lambda \in R\}$$

by the symbol $\text{Tor}A$. Let $\iota : \text{Tor}\, A \to A$ be the canonical inclusion map. Show that $\iota_A \star \iota_B : \text{Tor}\, A \star \text{Tor}\, B \approx A \star B$.

(vii) Let $\tau : C \to C'$ be a chain map between two torsion free chain complexes. Suppose $\tau_* : H_*(C) \to H_*(C')$ is an isomorphism. Then show that for any R-module G,

$$(\tau \otimes 1)_* : H_*(C \otimes G) \approx H_*(C' \otimes G).$$

(viii) Show that for any chain complex C of abelian groups of finite type and for any field $\mathbb{K}$, we have

$$\chi(C) = \sum_i (-1)^i \dim_{\mathbb{K}} H_i(C; \mathbb{K}).$$

6.3   Künneth Formula

As a direct generalization of the results proved in the previous section, we obtain a formula which expresses the homology of the tensor product of two chain complexes in terms of the homology of the two factor complexes. Through the Eilenberg–Zilber map, we then obtain a formula for the homology of the product of two spaces in terms of the homology of the factors.

Definition 6.3.1 The tensor product $C \otimes C'$ of two chain complexes $\{C_q, \partial_q\}$ and $\{C'_q, \partial'_q\}$ is defined to be $\{C''_n, \partial''_n\}$ with

$$C''_n = \oplus_{p+q=n} C_p \otimes C'_q; \quad \partial''_n(c \otimes c') = \partial_p(c) \otimes c' + (-1)^p c \otimes \partial'_q(c'), \quad c \in C_p, c' \in C'_q. \quad (6.4)$$

Replacing $\otimes$ by $\star$, we get the definition of the chain complex $C \star C'$.

Remark 6.3.2 Observe that if $C'_q = 0$ except for $q = 0$ then $C \otimes C'$ is nothing but $C \otimes C'_0$ and hence the tensor product of two chain complexes is a direct generalization of the tensor product of a chain complex and a module. We shall see that the corresponding universal coefficient theorem expresses the homology $H(C \otimes C')$ in terms of $H(C)$ and $H(C')$. The obvious generalization of the homomorphism μ in this case is a degree 0 homomorphism

$$\mu : H(C) \otimes H(C') \to H(C \otimes C')$$

given by

$$[c] \otimes [c'] \mapsto [c \otimes c']$$

for $c \in Z_p(C)$ and $c' \in Z_q(C')$.

We begin with a one-step generalization of Theorem 6.2.13.

Theorem 6.3.3 *Let R be a PID, C, C' be chain complexes of R-modules with C' free over R. Then there is a functorial short exact sequence*

$$0 \longrightarrow [H(C) \otimes H(C')]_n \overset{\mu}{\longrightarrow} H_n(C \otimes C') \longrightarrow [H(C) \star H(C')]_{n-1} \longrightarrow 0.$$

If C is also free then this sequence splits.

Proof: As in the proof of Theorem 6.2.13, we begin with the short exact sequence

$$0 \longrightarrow Z' \longrightarrow C' \longrightarrow B \longrightarrow 0$$

of free chain complexes (because C' is free.) Taking tensor product with C this yields a short exact sequence

$$0 \longrightarrow C \otimes Z' \longrightarrow C \otimes C' \longrightarrow C \otimes B' \longrightarrow 0$$

which, in turn, yields the long homology exact sequence

$$\dots \longrightarrow H_q(C \otimes Z') \longrightarrow H_q(C \otimes C') \longrightarrow H_q(C \otimes B') \overset{\partial_*}{\longrightarrow} H_{q-1}(C \otimes Z') \longrightarrow \dots.$$

For each fixed integer p, consider the chain complex M^p defined as follows:

$$(M^p)_n = C_{n-p} \otimes Z'_p; \quad \hat{\partial}_n = \partial_{n-p} \otimes 1.$$

Since Z' has trivial boundary operators it follows that $C \otimes Z'$ is the direct sum of the chain complexes: $C \otimes Z' = \oplus_p M^p$. Since Z' is free it follows that

$$H_n(C \otimes Z') = \oplus_p H_n(M^p) = \oplus_p H_{n-p}(C) \otimes Z'_p \approx \oplus_{p+q=n} H_q(C) \otimes Z_p(C').$$

Likewise

$$H_n(C \otimes B') \approx \oplus_{p+q=n-1} H_q(C) \otimes B_p(C').$$

Under these isomorphisms, the homomorphism ∂_* corresponds to the sum $\oplus_{q+p=n}(-1)^q \otimes \eta_p$ where $\eta_p : B_p(C') \subset Z_p(C')$ is the inclusion. Therefore there are short exact sequences

$$0 \longrightarrow \oplus_{q+p=n}[\text{Coker } [(-1)^q \otimes \eta_p] \longrightarrow H_n(C \otimes C') \longrightarrow \oplus_{q+p=n-1}[\text{Ker } [(-1)^q \otimes \eta_p] \longrightarrow 0 \quad (6.5)$$

Now, we have to identify the two terms in (6.5). Consider the free resolution

$$0 \longrightarrow B_p(C') \xrightarrow{(-1)^q \eta_p} Z'_p \longrightarrow H_p(C') \longrightarrow 0 .$$

Take tensor product with $H_q(C)$ to get an exact sequence

$$0 \longrightarrow H_q(C) \star H_p(C') \longrightarrow H_q(C) \otimes B_p(C')$$
$$\downarrow{\scriptstyle (-1)^q \otimes \eta_p}$$
$$H_q(C) \otimes Z_p(C') \longrightarrow H_q(C) \otimes H_p(C') \longrightarrow 0.$$

Taking direct sum over $q + p = n$ yields the required expressions for the two end groups in the short sequence.

It is routine to check that the first homomorphism is indeed equal to μ. We shall leave the proof of the splitting of this sequence under the assumption that C is free to the reader. [Hint: Construct a left inverse for μ.] ♠

Remark 6.3.4 We get a similar short exact sequence if we assume that C is free instead of C' is free. The two short exact sequences are the same if both C and C' are free. More generally, we can merely assume that $C \star C'$ is acyclic and obtain the same result. Further, we can then bring in arbitrary coefficients as well with mild restrictions. These are the next stage generalizations of Theorem 6.2.13.

Theorem 6.3.5 *On the category of ordered pairs of chain complexes C, C' such that $C \star C'$ is acyclic, there is a functorial short exact sequence*

$$0 \longrightarrow [H(C) \otimes H(C')]_q \xrightarrow{\mu} H_q(C \otimes C') \longrightarrow [H(C) \star H(C')]_{q-1} \longrightarrow 0$$

and this sequence splits.

Proof: The idea is to replace C and C' by their free approximations and use the previous result. So, let $\tau : \bar{C} \to C, \tau' : \bar{C}' \to C'$ be free approximations and let

$$0 \longrightarrow \tilde{C}' \xrightarrow{i'} \bar{C}' \xrightarrow{\tau} C' \longrightarrow 0$$

be the exact sequence given by τ'. Since $\bar{C}'$ is (torsion) free, the six-term exact sequence (see Exercise (iv) in 6.2.17) gives an exact sequence

$$0 \longrightarrow C \star C' \longrightarrow C \otimes \tilde{C}' \longrightarrow C \otimes \bar{C}' \xrightarrow{1 \otimes \bar{\tau}} C \otimes C' \longrightarrow 0$$

We know that $\tilde{C}'$ is contractible. Since $C \star C'$ is acyclic by hypothesis, as in the proof of the previous theorem, we have an isomorphism

$$(1 \otimes \bar{\tau})_* : H_*(C \otimes \bar{C}') \to H_*(C \otimes C').$$

Next we consider the exact sequence defined by τ:

$$0 \longrightarrow \tilde{C} \xrightarrow{\iota} \bar{C} \xrightarrow{\tau} C \longrightarrow 0.$$

Tensoring this with the free chain complex $\bar{C}'$, we get

$$0 \longrightarrow \tilde{C} \otimes \bar{C}' \longrightarrow \bar{C} \otimes \bar{C}' \xrightarrow{\tau \otimes 1} C \otimes \bar{C}' \longrightarrow 0.$$

Since $\tilde{C}$ is contractible, it follows that

$$(\tau \otimes 1)_* : H_*(\bar{C} \otimes \bar{C}') \approx H_*(C \otimes \bar{C}').$$

Now take the composite isomorphism $(\tau \otimes \tau')_* = (1 \otimes \tau')_* \circ (\tau \otimes 1)_*$ and argue as in the previous theorem to complete the proof.   ♠

Finally, given R-modules G, G' and chain complexes C, C' the canonical isomorphism

$$(C \otimes G) \otimes (C' \otimes G') \approx (C \otimes C') \otimes (G \otimes G')$$

induces an isomorphism in homology. Composing this with the functorial homomorphism μ we get the canonical homomorphism

$$\mu' : H(C; G) \otimes H(C'; G') \to H(C \otimes C'; G \otimes G').$$

This is called the homology cross product which gives the final Künneth formula:

Theorem 6.3.6 *Given torsion free chain complexes C and C' and modules G, G' such that $G \star G' = 0$ there is a functorial short exact sequence*

$$0 \to [H(C; G) \otimes H(C', G')]_q \xrightarrow{\mu'} H_q(C \otimes C'; G \otimes G') \to [H(C; G) \star H(C'; G')]_{q-1} \to 0$$

and this sequence splits.

Proof: By taking the two chain complexes to be $C \otimes G$ and $C' \otimes G'$ in the previous theorem, it suffices to check that $(C \otimes G) \star (C' \otimes G')$ is acyclic. Indeed, we shall prove that this chain complex is actually the trivial one. Let

$$0 \longrightarrow F_1 \longrightarrow F_0 \xrightarrow{\eta} G \longrightarrow 0$$

be a free presentation of G. Since $G \star G' = 0$, we get an exact sequence

$$0 \longrightarrow F_1 \otimes G' \xrightarrow{\eta \otimes Id_{G'}} F_0 \otimes G' \longrightarrow G \otimes G' \longrightarrow 0.$$

Since $C \otimes C'$ is torsion free, the six-term exact sequence gives an exact sequence

$$0 \longrightarrow (C \otimes C') \otimes (F_1 \otimes G') \xrightarrow{Id_{C \otimes C'} \otimes (\eta \otimes Id_{G'})} (C \otimes C') \otimes (F_0 \otimes G') \qquad (6.6)$$

$$\downarrow$$

$$(C \otimes C') \otimes (G \otimes G') \longrightarrow 0.$$

On the other hand, we also have, the short exact sequence

$$0 \longrightarrow C \otimes F_1 \xrightarrow{Id_C \otimes \eta} C \otimes F_0 \longrightarrow C \otimes G \longrightarrow 0$$

with $C \otimes F_0$ torsion free. Therefore, tensoring with $(C' \otimes G')$, it gives an exact sequence

$$0 \to (C \otimes G) \star (C' \otimes G') \longrightarrow (C \otimes F_1) \otimes (C' \otimes G') \qquad (6.7)$$

$$\downarrow{(Id_C \otimes \eta) \otimes Id_{C' \otimes G'}}$$

$$(C \otimes F_0) \otimes (C' \otimes G').$$

Under the usual isomorphism, $(C \otimes F_1) \otimes (C' \otimes G') \approx (C \otimes C') \otimes (F' \otimes G')$, etc., the homomorphism $(Id_C \otimes \eta) \otimes Id_{C' \otimes G'}$ corresponds to the homomorphism $Id_{C \otimes C'} \otimes (\eta \otimes Id_{G'})$. From (6.6), we have $Id_{C \otimes C'} \otimes (\eta \otimes Id_{G'})$ is injective. Hence from (6.7), it follows that $(C \otimes G) \star (C' \otimes G') = 0$. ♠

Remark 6.3.7 Taking μ' as in Theorem 6.3.6, we define the *cross product*

$$u \times v := \mu'(u \otimes v). \tag{6.8}$$

Apart from being bilinear, the cross product satisfies the following commutativity relation with the connecting homomorphism of long exact sequences.

Theorem 6.3.8 *Given short exact sequences* $0 \to \bar{\bar{C}} \to C \to \bar{C} \to 0$ *of chain complexes and elements* $u \in H(C; G), v \in H(V'; G')$ *we have*

$$\partial_*(u \times v) = \partial_* u \times v; \quad \partial_*(v \times u) = (-1)^{deg\,v} v \times \partial_* u. \tag{6.9}$$

Proof: Fix a cycle $c' \in Z(C' \otimes G')$ representing v. Now taking tensor product on the right with c' defines chain maps $\tau_M : M \otimes G \to (M \otimes G) \otimes (C' \otimes G')$ for each of the chain complexes $M = \bar{\bar{C}}, C, \bar{C}$. Moreover, these three chain maps fit together to define a morphism of the corresponding short exact sequences. Because the connecting homomorphism is functorial, we obtain a commutative diagram:

$$
\begin{array}{ccccc}
H(C \otimes G) & \xrightarrow{\tau_*} & H((C \otimes G) \otimes C' \otimes G') & \xrightarrow{\approx} & H((C \otimes C') \otimes (G \otimes G')) \\
\downarrow{\scriptstyle \partial_*} & & \downarrow{\scriptstyle \partial_*} & & \downarrow{\scriptstyle \partial_*} \\
H(\bar{C} \otimes G) & \xrightarrow{\bar{\tau}} & H((\bar{C} \otimes G) \otimes C' \otimes G') & \xrightarrow{\approx} & H((\bar{C} \otimes C') \otimes (G \otimes G')).
\end{array}
$$

This gives the first part of the statement. The proof of the second part is similar except that taking tensor product on the left with c' becomes a chain map only after introducing the multiplicative factor $(-1)^{\deg c'}$. ♠

The link between algebra of tensor products and the topology of product spaces is the so-called Eilenberg–Zilber map.

Theorem 6.3.9 (Eilenberg–Zilber) *On the category of ordered pairs of topological spaces* X *and* Y, *there is a natural chain equivalence* ζ *of the functor* $S_.(X \times Y)$ *to the functor* $S_.(X) \otimes S_.(Y)$.

Proof: We choose models for this category as $\mathcal{M} = \{(\Delta_p, \Delta_q)\}_{p,q \geq 0}$.

Let $d_n : \Delta_n \to \Delta_n \times \Delta_n$ be the diagonal map. Given any $\sigma : \Delta_n \to X \times Y$ we have $\sigma = (p_1 \circ \sigma, p_2 \circ \sigma) \circ d_n$. Conversely, given any two $\sigma' : \Delta_n \to X$ and $\sigma'' : \Delta_n \to Y$ we take $(\sigma', \sigma'') \circ d_n$ to get a singular n-simplex in $X \times Y$. This means that $\{d_n\}$ is a basis for $S_n(X \times Y)$ and hence $S_.(X \times Y)$ is free with models $\{(\Delta_n, \Delta_n)\}$. Therefore it is free with models $\mathcal{M}$ as well. Moreover, since $|\Delta_p \times \Delta_q|$ is contractible, it follows that $\tilde{S}(\Delta_p \times \Delta_q)$ is acyclic. Thus $\tilde{S}(X \times Y)$ is acyclic with models $\mathcal{M}$.

Next, we also observe that $S_p(X)$ is free with basis $\xi_p \in S_p(\Delta_p)$. Therefore it follows that $S_p(X) \times S_q(Y)$ is free with basis

$$\{\xi_p \otimes \xi_q \in S_p(\Delta_p) \otimes S_q(\Delta_q)\}.$$

Therefore $[S_.(X) \otimes S_.(Y)]_n$ is free with basis $\{\xi_p \otimes \xi_q\}_{p+q=n}$. This means that $S_.(X) \otimes S_.(Y)$ is free with models $\mathcal{M}$.

Finally, since $\tilde{S}(\Delta_p)$ is contractible, the augmentation map $\epsilon : S(\Delta_p) \to \mathbb{Z}$ is a chain equivalence. It follows that

$$\epsilon \otimes \epsilon : S_.(\Delta_p) \otimes S_.(\Delta_q) \to \mathbb{Z} \otimes \mathbb{Z} = \mathbb{Z}$$

is also a chain equivalence. This just means that $S_.(\Delta_p) \otimes S_.(\Delta_q)$ is acyclic in positive dimensions. Thus we have shown that $S_.(X) \otimes S_.(Y)$ is free and acyclic with models $\mathcal{M}$.

The conclusion of the Theorem now follows from the method of acyclic models Theorem 6.1.14. ♠

The map $\zeta : S_.(X \times Y) \to S_.(X) \otimes S_.(Y)$ is called the *Eilenberg–Zilber map*.

Exactly similar to the above theorem, we can prove:

Theorem 6.3.10 *Given topological spaces X, Y, Z there is a chain homotopy commutative diagram where the vertical maps are natural equivalences.*

$$
\begin{array}{ccc}
S_.(X \times (Y \times Z)) & \xrightarrow{\approx} & S_.((X \times Y) \times Z) \\
\zeta_{X,Y \times Z} \downarrow & & \downarrow \zeta_{X \times Y, Z} \\
S_.(X) \otimes S_.(Y \times Z) & & S_.(X \times Y) \otimes S_.(Z) \\
1 \otimes \zeta_{Y,Z} \downarrow \approx & & \approx \downarrow \zeta_{X,Y} \otimes 1 \\
S_.(X) \otimes (S_.(Y) \otimes S_.(Z)) & \xrightarrow{\approx} & (S_.(X) \otimes S_.(Y)) \otimes S_.(Z)
\end{array}
$$

Theorem 6.3.11 *For any two topological spaces X, Y, there is a chain homotopy commutative diagram*

$$
\begin{array}{ccc}
S_.(X \times Y) & \xrightarrow{\approx} & S_.(Y \times X) \\
\zeta_{X,Y} \downarrow \approx & & \approx \downarrow \zeta_{Y,X} \\
S_.(X) \otimes S_.(Y) & \xrightarrow{\approx} & S_.(Y) \otimes S_.(X)
\end{array}
$$

where the bottom map sends

$$x \otimes y \mapsto (-1)^{\deg x \deg y}(y \otimes x)$$

and the vertical maps are the natural chain equivalences.

We need to strengthen Theorem 6.3.9 to include the relative case.

Theorem 6.3.12 *On the category of ordered pairs of topological pairs $(X, A), (Y, B)$ satisfying the condition that $\{X \times B, A \times Y\}$ is an excisive couple in $X \times Y$, there is a natural equivalence of the functors*

$$\hat{\zeta} : [S_.(X)/S_.(A)] \otimes S_.(Y)/S_.(B)] \rightsquigarrow S_.(X \times Y)/S_.(X \times B \cup A \times Y) := S_.((X, A) \times (Y, B)).$$

Proof: That $\{X \times B, A \times Y\}$ is an excisive couple in $X \times Y$ is the same as saying that the inclusion induced map

$$\eta : S_.(X \times Y)/[S_.(X \times B) + S_.(A \times Y)] \to S_.(X \times Y)/S_.(X \times B \cup A \times Y)$$

is a chain equivalence. The Eilenberg–Zilber map $\zeta : S_.(X \times Y) \to S_.(X) \otimes S_.(Y)$ takes $S_.(X \times B)$ and $S_.(A \times Y)$ into $S_.(X) \otimes S_.(B)$ and $S_.(A) \otimes S_.(Y)$, respectively. Therefore ζ induces a chain equivalence

$$\bar{\zeta} : S_.(X \times Y)/[S_.(X \times B) + S_.(A \times Y)] \to S_.(X) \otimes S_.(Y)/[S_.(X) \otimes S_.(B) + S_.(A) \otimes S_.(Y)]$$

For purely algebraic reasons, there is a functorial isomorphism

$$[S_.(X)/S_.(A)] \otimes [S_.(Y)/S_.(B)] \to S_.(X) \otimes S_.(Y)/[S_.(X) \otimes S_.(B) + S_.(A) \otimes S_.(Y)]$$

This isomorphism followed by $\bar{\zeta}^{-1}$ followed by η is the required equivalence $\hat{\zeta}$. ♠

Definition 6.3.13 Let

$$\mu : H_*(X, A; G) \otimes H_*(Y, B; G) \to H_*([S_.(X) \otimes S_.(Y)/S_.(A) \otimes S_.(Y) + S_.(X) \otimes S_.(B)] \otimes G \otimes G')$$

be as in Theorem 6.3.6. Let $\hat{\zeta}_*$ be the isomorphism induced by the chain equivalence given by the above theorem. Put $\mu' = \hat{\zeta}_* \circ \mu$. For $u \in H_p(X, A; G)$ and $v \in H_q(Y, B; G')$ we define the homology cross product

$$u \times v := \mu'(u \otimes v) = \hat{\zeta}_* \mu(u \otimes v). \tag{6.10}$$

Since $S_.(X)/S_.(A)$ and $S_.(Y)/S_.(B)$ are free chain complexes, Theorem 6.3.6 is applicable. Combining this with the above theorem we get the Künneth formula for singular homology, which obviously involves the homology cross product:

Theorem 6.3.14 (Künneth formula) *Let R be a PID and G, G' be R-modules such that $G \star G' = 0$. Then for any excisive couple $\{X \times B, A \times Y\}$ in $X \times Y$ we have the functorial short exact sequence*

$$0 \longrightarrow [H_*(X, A; G) \otimes H_*(Y, B; G')]_q$$
$$\downarrow \nu$$
$$H_q((X, A) \times (Y, B); G \otimes G') \longrightarrow [H(X, A; G) \star H(Y, B; G')]_{q-1} \longrightarrow 0$$

and this sequence splits.

Remark 6.3.15 This theorem is extremely useful especially when $R = G = G'$ is a field, since the torsion product identically vanishes and hence the cross product is an isomorphism. This is also useful when R is arbitrary but the homology groups involved are torsion free.

We shall now merely list a number of properties of the cross product and leave their verification to the reader.

Theorem 6.3.16
(a) *The cross product is functorial in both slots, i.e., given the maps $f : (X, A) \to (X', A')$, and $g : (Y, B) \to (Y', B')$, we have*

$$(f \times g)_*(u \times v) = f_*(u) \times g_*(v).$$

(b) *The cross product is associative.*
(c) *The cross product is skew-commutative, i.e., if $T : X \times Y \to Y \times X$ is the map that interchanges the factors, then*

$$T_*(u \times v) = (-1)^{(\deg u)(\deg v)} v \times u.$$

(d) *Let ∂ and ∂' denote the connecting homomorphism of the Mayer–Vietoris sequence of an excisive couple $\{(X_1, A_1), (X_2, A_2)\}$ in X and the exact homology sequence of a pair (Y, B), respectively. Then for $u \in H_p(X_1 \cup X_2, A_1 \cup A_2; G)$ and $v \in H_q(Y, B; G')$, we have:*

$$\partial(u \times v) = \partial(u) \times v; \quad \partial'(v \times u) = (-1)^q v \times u.$$

(e) *Let $\pi : X \times Y \to X$ be the projection map and $\epsilon : H(Y; G') \to G'$ be the homomorphism induced by the augmentation map. Then*

$$\pi_*(u \times v) = \mu'(u \otimes \epsilon(v)).$$

In particular, if $\deg v > 0$, then $\pi_*(u \times v) = 0$.

We shall end this section with another important natural transformation which will be used in a later chapter.

Theorem 6.3.17 *On the category of topological spaces, there exists a functorial chain map $\tau(X) : S(X) \to S(X) \otimes S(X)$ which preserves augmentations and any two such chain maps are chain homotopy equivalent.*

Proof: The existence is a direct consequence of part (a) of the method of acyclic models in Theorem 6.1.14, since $S(X)$ is free with models $\{\Delta_q\}_{q \geq 0}$ and $S(X) \otimes S(X)$ is acyclic in positive dimensions with models $\{\Delta_q\}_{q \geq 0}$. The uniqueness up to chain homotopy follows from part (b) since chain maps preserving augmentations induce the same homomorphism on H_0. ♠

Example 6.3.18 Alexander–Whitney diagonal approximation

Consider the Eilenberg–Zilber natural equivalence $\zeta : S_.(X \times X) \to S_.(X) \otimes S_.(X)$ given by Theorem 6.3.9. Let $d(x) = (x, x)$ and consider $S(d) : S_.(X) \to S_.(X \times X)$. Then we get a natural transformation $\zeta \circ S(d) : S_.(X) \to S_.(X) \otimes S_.(X)$ which preserves augmentations. Therefore this is a specific representative of the class of natural transformations given by the above theorem. This is the reason we call every member in this class *a diagonal approximation*.

While dealing with cellular chain complexes or simplicial chain complexes, this representative is not particularly a good one since d is **not** a cellular map (nor a simplicial one). Here is another diagonal approximation which has better geometric behaviour. For $0 \leq i \leq q$ the *front i-face* is the orientated i-simplex ${}_i\xi^q = [\mathbf{e}_0, \dots, \mathbf{e}_i] \subset \Delta_q$. Similarly *the back i-face* is the i-dimensional oriented simplex $\xi_i^q = [\mathbf{e}_{q-i}, \dots, \mathbf{e}_q]$. Given a singular simplex $\sigma : \Delta_q \to X$ its front and back i faces are defined to be ${}_i\sigma = \sigma|_{{}_i\xi^q}$ and $\sigma_i = \sigma|_{\xi_i^q}$. Now the *Alexander–Whitney diagonal approximation* is given by:

$$\alpha(\sigma) = \sum_{i+j=q} {}_i\sigma \otimes \sigma_j \tag{6.11}$$

for any singular q simplex σ in X. It is easily checked that this is a natural transformation of functors preserving the augmentations. Therefore, it is indeed a diagonal approximation. We shall exploit this while dealing with products in cohomology.

Remark 6.3.19 Why have we singled-out the singular chain complex $S_.(X)$ in the discussion of the Eilenberg–Zilber theorem and the Whitney diagonal approximation? Why not consider $\mathscr{S}_.(K)$ for simplicial complexes and $C^{CW}(X)$ for CW-complexes? For $\mathscr{S}_.(X)$ the method of acyclic model fails— we do not know any model with basis with respect to which both the functors $\mathscr{S}_.(X) \otimes \mathscr{S}_.(Y)$ and $\mathscr{S}_.(X \times Y)$ are free and acyclic. Perhaps the simple reason is that the diagonal map $d : \Delta_n \to \Delta_n \times \Delta_n$ is not 'simplicial'. However, the Alexander–Whitney diagonal approximation makes perfect sense here.

With the CW-chain complex the story is somewhat different. The standard product CW-structure on $X \times Y$ seems to give a module theoretic isomorphism between $C^{CW}(X \times Y)$ and $C^{CW}(X) \otimes C^{CW}(Y)$. Whether this isomorphism is a homomorphism of chain complexes is another matter. However, the diagonal approximation utterly fails. One can give plenty of examples to illustrate the point that the CW-chain complex loses a lot of information which is present in the singular chain complex. We shall be able to offer more explanation of this point in the next chapter.

6.4 Miscellaneous Exercises to Chapter 6

1. Let R be a PID and A be a non negatively graded module over R which is finitely generated and free. If B, C are any two (non negatively) graded modules of finite type such that $A \otimes B \approx A \otimes C$ then show that $B \approx C$.

2. Let X_1, X_2 be any two subspaces of a topological space X. Show that the following conditions are equivalent:
   (a) $\{X_1, X_2\}$ is an excisive couple.
   (b) For every subspace $A \subset X_1 \cap X_2$, there is an exact Mayer–Vietoris sequence

   $$\cdots \longrightarrow H_q(X_1 \cap X_2, A) \longrightarrow H_q(X_1, A) \oplus H_q(X_2, A)$$

   $$H_q(X_1 \cup X_2, A) \longrightarrow H_{q-1}(X_1 \cap X_2, A) \longrightarrow \cdots$$

   (c) For every subspace Y such that $X_1 \cup X_2 \subset Y \subset X$, there is an exact Mayer–Vietoris sequence

   $$\cdots \longrightarrow H_{q+1}(Y, X_1 \cup X_2) \longrightarrow H_q(Y, X_1 \cap X_2)$$

   $$H_q(Y, X_1) \oplus H_q(Y, X_2) \longrightarrow H_q(Y, X_1 \cup X_2) \longrightarrow \cdots$$

3. Use Theorem 6.3.14 to compute the integral homology of a finite product of spheres. Conclude that if $X = \mathbb{S}^{p_1} \times \cdots \times \mathbb{S}^{p_r}, Y = \mathbb{S}^{q_1} \times \cdots \times \mathbb{S}^{q_s}$ with $p_1 \leq p_2 \leq \cdots \leq p_r$ and $q_1 \leq q_2 \leq \cdots \leq q_s$ then X is homeomorphic to Y iff $r = s$ and $(p_1, \ldots, p_r) = (q_1, \ldots, q_r)$.

4. Let $x_0 \in X$ be a non degenerate base point in a connected space X. Identify the one point union $X \vee X$ with the subspace $X \times x_0 \cup x_0 \times X$ of $X \times X$. For $t = 1, 2$, let $p_t : X \vee X \to X$ be the restrictions of projections and $j_t : X \to X \vee X, i_t : X \to X \times X$ be the inclusions, given by $i_1(x) = j_1(x) = (x, x_0), i_2(x) = j_2(x) = (x_0, x)$. Let $k : X \vee X \to X \times X$ and $\ell : X \times X \to (X \times X, X \vee X)$ be the inclusions. Let $\Delta : X \to X \times X$ be the diagonal map. Prove the following:
   (a) $j_{t*} : H_*(X) \to H_*(X \vee X)$ represent $H_*(X \vee X)$ as a direct sum, whereas $p_{i*} : H_*(X \vee X) \to H_*(X)$ represent it as a direct product.
   (b) $k_* : H_q(X \vee X) \to H_q(X \times X)$ is a split monomorphism for all q.
   (c) $\ell_* : \operatorname{Ker} p_{1*} \cap \operatorname{Ker} p_{2*} \to H_q(X \times X, X \vee X)$ is an isomorphism for all q.
   (d) The homomorphisms $i_{1*}, i_{2*}, \beta = (\ell)_*^{-1} : H_q(X \times X, X \vee X)$ represent $H_*(X \times X)$ as a direct sum, whereas $(p_1)_*, (p_2)^*, \ell_*$ represent it as a direct product.

5. We continue with the notation of the previous two exercises. An element $u \in H_q(X)$ is called a *primitive* if $\Delta_*(u) = i_{1*}u + i_{2*}u$.
   (a) Show that the set $P_q(X)$ of all primitive elements in $H_q(X)$ is a submodule. [Hint: It is equal to $\operatorname{Ker} \ell_* \circ \Delta_*$.]
   (b) If $f : X \to Y$ is a map of spaces with non degenerate base points, then $f_*(P_*(X)) \subset P_*(Y)$.
   (c) If $\tilde{H}_q(X) = 0$ for $q < n$, then $P_q(X) = H_q(X)$ for $q < 2n$.
   (d) Every element in $H_n(\mathbb{S}^n)$ is a primitive, $(n > 0)$.
   (e) An element $u \in H_q(X)$ is called *spherical* if there is a map $f : \mathbb{S}^q \to X$ such that $u \in \operatorname{Im} f_*$. Prove that every spherical element is a primitive.

Chapter 7

Cohomology

Cohomology in its simplest form is related to homology just the way a vector space V is related to its dual V^*. Its traces can be found in the integral calculus of several variables, for instance, in line integrals in complex analysis. It started appearing in topology in the works of Poincaré in his duality theorem. It appeared in Alexander's duality theorem, in the intersection theory due to Alexander and Lefschetz, de Rham's result on smooth differential forms on manifolds and in Pontrjagin's duality. It took several more authors and many more years to get properly established until in 1952, when the now classical book by Eilenberg and Steenrod on *Foundations of Algebraic Topology* appeared. The word 'cohomology' was invented by H. Whitney.

In Section 7.1, we shall quickly go through the algebra of cochain complexes and introduce the singular, simplicial and CW-cohomology of topological pairs. In Section 7.2, we shall establish universal coefficient theorem for cohomology. In Section 7.3, we shall introduce the cup and cap products and their basic properties. We shall carry out some interesting computation in Section 7.4. In the last section we shall introduce cohomology operations and study Steenrod squares to some extent.

7.1 Cochain Complexes

Definition 7.1.1 A cochain complex $C^{\cdot}$ is a $\mathbb{Z}$-graded R-module with a graded homogeneous homomorphism of degree $= 1$. It is customary to denote the components of a cochain complex by superscripts C^q and call the graded homomorphisms $d^q : C^q \to C^{q+1}$, the *differentials* or the *boundary homomorphisms*. The homology modules $\operatorname{Ker} d^q / \operatorname{Im} d^{q-1}$ are called the *cohomology modules* of $C^{\cdot}$. These are denoted by $H^q(C^{\cdot})$.

Remark 7.1.2
Given a cochain complex $(C^{\cdot}, d)$, define $(C_{\cdot}, \partial)$ by the formula $C_n = C^{-n}$ and $\partial_n = d^{-n}$. Then it can be easily checked that $(C_{\cdot}, \partial)$ is a chain complex, from which we can recover the original cochain complex completely by the same process. Moreover, $H_n(C_{\cdot}) = H^{-n}(C^{\cdot})$. Thus, correctly interpreted, all the results that we have proved for homology of chain complexes hold ditto for cochain complexes. Thus the cohomology module is a covariant functor from the category of cochain complexes to the category of graded modules. Given an exact sequence of cochain complexes

$$0 \longrightarrow C'^{\cdot} \overset{\phi}{\longrightarrow} C^{\cdot} \overset{\psi}{\longrightarrow} C''^{\cdot} \longrightarrow 0$$

there is a functorial connecting homomorphism $\delta^{\cdot}$ and a long-exact sequence of cohomology modules:

$$\cdots \longrightarrow H^q(C''^{\cdot}) \overset{\delta}{\longrightarrow} H^{q+1}(C'^{\cdot}) \overset{\phi^*}{\longrightarrow} H^q(C^{\cdot}) \overset{\psi^*}{\longrightarrow} H^q(C''^{\cdot}) \longrightarrow \cdots$$

As usual, let R denote a principal ideal domain. Analogous to Theorems 6.2.13 and 6.3.3 we have, respectively, the following two results.

Theorem 7.1.3 *Given a free cochain complex $C^\cdot$, an R-module G, and $q \geq 0$, there is a functorial short exact sequence*

$$0 \longrightarrow H^q(C^\cdot) \otimes G \xrightarrow{\mu} H^q(C^\cdot \otimes G) \longrightarrow H^{q+1}(C^*) \star G \longrightarrow 0$$

and this sequence splits.

Theorem 7.1.4 *Let G, G' be R-modules such that $G \star G' = 0$. Then for each pair (C, C') of torsion free cochain complexes C and C', there is a functorial short exact sequence*

$$0 \longrightarrow [H(C,G) \otimes H(C',G')]^q \xrightarrow{\mu'} H^q(C \otimes C'; G \otimes G') \longrightarrow [H(C;G) \star H(C';G')]^{q+1} \longrightarrow 0$$

which splits, for each $q \geq 0$.

Remark 7.1.5 Let $C^\cdot$ be a non negative cochain complex, i.e., $C^q = 0$ for $q < 0$. An augmentation of $C^\cdot$ over a module G is a monomorphism $\eta : G \to C^0$ such that $\delta^0 \circ \eta = 0$. Let $G^\cdot$ be the atomic cochain complex with $G^0 = G$ and $G^q = 0$ for all other q. Then η defines an exact sequence of cochain complexes

$$0 \longrightarrow G^\cdot \xrightarrow{\eta} C^\cdot \longrightarrow \tilde{C}^\cdot \longrightarrow 0$$

and $\tilde{C}^\cdot$ is called the associated reduced cochain complex. Observe that $\tilde{C}^0 = \operatorname{Coker} \eta$ and $\tilde{C}^q = C^q, \quad q \neq 0$. It follows immediately that the reduced cohomology modules

$$\tilde{H}(C^\cdot) := H(\tilde{C}^\cdot)$$

are given by $\tilde{H}^q(C^\cdot) = H^q(C^\cdot)$ for $q \neq 0$ and the exact sequence

$$0 \longrightarrow G \xrightarrow{\eta} H^0(C^\cdot) \longrightarrow \tilde{H}^0(C^\cdot) \longrightarrow 0 \,.$$

Remark 7.1.6 Given a smooth manifold M, let $\Omega^p(M)$ denote the vector space over $\mathbb{R}$ of all smooth p-forms on M. Then the graded vector space $\Omega^\cdot(M)$ together with the external derivation d defines a natural cochain complex called de Rham complex. The homology modules of this complex are called the de Rham cohomology modules of M.

That $\partial^2 = 0$ is the same as saying that $Ker\,[d : \Omega^p(X) \to \Omega^{p+1}(X)]$ contains the image of $d : \Omega^{p-1}(X) \to \Omega^p(X)$. For $X = \mathbb{R}^3$, this is the same as saying curl $\circ$ grad $= 0$ and div $\circ$ curl $= 0$. Classically, if you had a differential equation such as

$$pdx + qdy = 0, \tag{7.1}$$

solving this equation means that we have to find a function f such that $df = pdx + qdy$. In the terminology of the de Rham complex, this just means that the 1-form $\omega = pdx + qdy$ is in the image of $d : \Omega^0 \to \Omega^1$. Of course, in that case, we also know that it is necessary to assume that

$$\frac{\partial p}{\partial y} = \frac{\partial q}{\partial x}. \tag{7.2}$$

and in that case, we also know how to solve this equation in $\mathbb{R}^2$. Classically, the equation (7.1) where p, q satisfy (7.2) is called an *exact equation*. Notice how this terminology has morphed to become 'exact sequences' in modern times.

Given a smooth map $\sigma : \Delta_p \to M$, and $\omega \in \Omega^p(M)$, there is the *pullback p-form* $\sigma^*(\omega)$ on Δ_p. The integral $\int_{\Delta_p} \sigma^*(\omega) \in \mathbb{R}$ is well defined. This way, we can think of ω as defining a linear map on the $\mathbb{R}$-vector space of all smooth singular chains in M. The following definition is a sweeping generalization of this classical phenomenon.

Definition 7.1.7 Given a topological pair (X, A), and an abelian group G, the singular cochain complex $S^{\cdot}(X, A; G)$ with coefficients G is defined by

$$S^q(X, A; G) = \text{Hom}\,(S_q(X)/S_q(A), G)$$

and $\delta^q : S^{q-1} \to S^q$ is given by

$$\delta^q(\omega) = \partial_q \circ \omega, \text{ for } \quad \omega : \text{Hom}\,(S_{q-1}(X)/S_{q-1}(A), G).$$

Likewise, we define the simplicial cochain complex and the singular simplicial cochain complex of a pair (K, L) of simplicial complex to be

$$\mathbb{S}^{\cdot}(K, L; G) = \text{Hom}\,(\mathbb{S}_{\cdot}(K)/\mathbb{S}_{\cdot}(L); G); \quad C^{\cdot}(K, L; G) = \text{Hom}\,(C_{\cdot}(K)/C_{\cdot}(L); G)$$

with the differentials defined appropriately. The CW-cochain complex of a CW-pair (X, A) is also defined similarly:

$$C^{\cdot}_{CW}(X, A; G) := \text{Hom}\,(C^{CW}_{\cdot}(X, A), G).$$

The homology modules of these cochain complexes are then defined to be $H^{\cdot}(X, A; G)$; $H^{\cdot}(\mathbb{S}^{\cdot}(X, A); G), \ldots,$ etc.

More generally, given a chain complex $C_{\cdot}$ and a module G over a ring R, we define $H^{\cdot}(C, G)$ to be the homology $H_{\cdot}(\text{Hom}\,(C_{\cdot}, G))$ modules of the cochain complex $\text{Hom}\,(C_{\cdot}; G)$.

Alas! Even though $S_{\cdot}(X), \mathbb{S}_{\cdot}(X)$, etc., are all free chain complexes, $\text{Hom}(S_{\cdot}(X); R)$, etc., are not free chain complexes in general. This renders Theorems 7.1.3 and 7.1.4 somewhat useless, while dealing with these cochain complexes.

In the next section, we shall take up the study of cohomology with coefficients, in a somewhat general algebraic set-up.

7.2   Universal Coefficient Theorem for Cohomology

In this section, apart from a few basic properties of cohomology, we shall first establish a relation between homology and cohomology and also a universal coefficient theorem for cohomology which relates cohomology with coefficients in a module G over a ring R to cohomology with coefficients in R.

Lemma 7.2.1 Given an exact sequence

$$M' \xrightarrow{\alpha} M \xrightarrow{\beta} M'' \longrightarrow 0$$

of R-modules and an R-module G, there is an exact sequence

$$0 \longrightarrow \text{Hom}\,(M'', G) \xrightarrow{\beta^*} \text{Hom}\,(M, G) \xrightarrow{\alpha^*} \text{Hom}\,(M', G).$$

Lemma 7.2.2 Given a split-short exact sequence

$$0 \longrightarrow M' \xrightarrow{\alpha} M \underset{\longleftarrow}{\overset{\beta}{\longrightarrow}} M'' \longrightarrow 0$$

of R-modules and a R-module G, there is a split-exact sequence

$$0 \longrightarrow \text{Hom}\,(M'', G) \xrightarrow{\beta^*} \text{Hom}\,(M, G) \underset{\longleftarrow}{\overset{\alpha^*}{\longrightarrow}} \text{Hom}\,(M', G) \longrightarrow 0.$$

Both the lemmas are proved in a straightforward way. There is a chain complex version of the second lemma above. We need a definition.

Definition 7.2.3 A chain short exact sequence of chain complexes

$$0 \to C' \to C \to C'' \to 0$$

is called a *split-short exact sequence* if for each q, we have the short exact sequences that split:

$$0 \longrightarrow C'_q \longrightarrow C_q \mathrel{\substack{\longrightarrow \\ \longleftarrow}} C''_q \longrightarrow 0.$$

Caution: Note that it is **not** required that the splittings collectively define a chain map.

Theorem 7.2.4 *Given a split-short exact sequence of chain complexes*

$$0 \longrightarrow C' \xrightarrow{\alpha} C \xrightarrow{\beta} C'' \longrightarrow 0$$

and any R-module G, there is a functorial long exact sequence

$$\cdots \longrightarrow H^q(C''; G) \longrightarrow H^q(C; G) \longrightarrow H^q(C'; G) \xrightarrow{\delta^*} H^{q+1}(C''; G) \longrightarrow \cdots$$

This in turn yields the Mayer–Vietoris sequence exactly as in the case of homology:

Corollary 7.2.5 For any excisive couple of topological pairs $\{(X_1, A_1), X_2, A_2)\}$ and any R-module G, there is a functorial exact sequence of cohomology modules:

$$\cdots \xrightarrow{\delta^*} H^q(X_1 \cup X_1, A_1 \cup A_2; G) \xrightarrow{j^*} H^q(X_1, A_1; G) \oplus H^q(X_2, A_2; G)$$

$$\downarrow{i^*}$$

$$H^q(X_1 \cap X_2, A_1 \cap A_2; G) \xrightarrow{\delta^*} \cdots,$$

$$j^*(c) := (i_1^*(c), i_2^*(c)); \quad i^*(c_1 + c_2) := i_1^*(c_1) - i_2^*(c_2),$$

where $i_1 : (X_1, A_1) \hookrightarrow (X_1 \cup X_2, A_1 \cup A_2), i_2 : (X_2, A_2) \hookrightarrow (X_1 \cup X_2, A_1 \cup A_2)$ are inclusion maps.

Remark 7.2.6 Note that the Hom functor converts direct sum into direct product, i.e., $\mathrm{Hom}(\oplus_j M_j; G) \approx \times_j \mathrm{Hom}(M_j, G)$. Therefore we get the following result.

Theorem 7.2.7 *The singular cohomology module of a space is the direct product of the singular cohomology modules of its path-connected components.*

Remark 7.2.8 Observe that Hom functor does not commute with direct limits. Therefore, there is no analogue of Exercise 4.7.1.((i)c) for cohomology of a direct limit. In particular, it is not true that the singular cohomology of a space is the inverse limit of the singular cohomology of its compact subspaces.

Remark 7.2.9 There is a result in cohomology identical to the Bockstein homomorphism in homology (see Exercise 6.2.20(c)). We shall leave it to the reader to come up with the correct statement and then figure out the proof as well.

Definition 7.2.10 Given modules A, B over R, let $C_. \to A \to 0$ be the canonical free presentation of A. We define the *extension modules* $\mathrm{Ext}^q(A, B)$ to be equal to the homology modules $H^q(\mathrm{Hom}\,(C, B))$.

Now suppose R is a PID. Then we get a free presentation

$$0 \longrightarrow C_1 \longrightarrow C_0 \longrightarrow A \longrightarrow 0$$

and hence $\mathrm{Ext}^q(A, B) = 0$ for $q > 1$ and it follows that there is a short exact sequence

$$0 \longrightarrow \mathrm{Hom}\,(A, B) \longrightarrow \mathrm{Hom}\,(C_0, B) \longrightarrow \mathrm{Hom}\,(C_1, B) \longrightarrow \mathrm{Ext}^1(A, B) \longrightarrow 0.$$

Check that $\mathrm{Hom}\,(A, B) = H^0(C_., B) = \mathrm{Ext}^0(A, B)$ and

$$\mathrm{Ext}(A, B) := \mathrm{Ext}^1(A, B) = H^1(C_., B) = \mathrm{Hom}\,(C_1, B)/\mathrm{Im}[\mathrm{Hom}\,(\partial_1, 1)] = \mathrm{Coker}\,[\mathrm{Hom}\,(\partial_1, 1)].$$

It follows that $\mathrm{Ext}(A, B)$ is contravariant in A and covariant in B.

Example 7.2.11 Let R be a PID.

1. If A is free, we have a free presentation $0 \to A \to A \to 0$. Hence $\mathrm{Ext}(A, B) = 0$.

2. For the cyclic module $A = R/tR$, for some $0 \neq t \in R$, we have the free presentation:

$$0 \longrightarrow R \xrightarrow{\ h\ } R \longrightarrow A \longrightarrow 0$$

where $h(s) = ts$ is the multiplication by t. Under the isomorphism $\mathrm{Hom}\,(R, B) \approx B$, $\mathrm{Hom}\,(h, 1)$ corresponds to the multiplication by t on B and hence we get $\mathrm{Ext}^1(R/tR, B) = \mathrm{Coker}\,\mathrm{Hom}\,(h, 1)] \approx B/tB = (R/tR) \otimes B$.

3. Since Hom commutes with finite direct sums it follows that for any finitely generated torsion module A, we have $\mathrm{Ext}(A, B) \approx A \otimes B$. (However, note that this isomorphism is not canonical.) In particular, for any finite abelian group G, $\mathrm{Ext}^1(G, \mathbb{Z}) \approx G$. Also, this shows that $\mathrm{Ext}^1(A, B), \mathrm{Ext}^1(B, A)$ are not isomorphic, in general.

Exercise 7.2.12 Let A, B be any two modules over R. By an extension of B by A we mean a short exact sequence of R modules

$$0 \longrightarrow B \longrightarrow E \longrightarrow A \longrightarrow 0.$$

Two such extensions are said to be equivalent if there is a commutative diagram

$$
\begin{array}{ccccccccc}
0 & \longrightarrow & B & \longrightarrow & E & \longrightarrow & A & \longrightarrow & 0 \\
& & \downarrow & & \downarrow & & \downarrow & & \\
0 & \longrightarrow & B & \longrightarrow & E' & \longrightarrow & A & \longrightarrow & 0
\end{array}
$$

where the vertical arrows are isomorphisms. Clearly this defines an equivalence relation on the set of all extensions of B by A. Let us temporarily denote these equivalence classes by $[E]$.

Now fix a free presentation $C_. \to A \to 0$. Given an extension of B by A, we get a commutative diagram of R linear maps as follows.

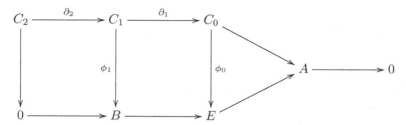

The map ϕ_0 exists because C_0 is free and $E \to A$ is surjective. Since $\phi_0 \circ \partial_1(C_1) \subset B$, it defines the map $\phi_1 : C_1 \to B$. Observe that $\phi_1 \circ \partial_2 = 0$ and hence $\phi_1 \in \mathrm{Ker}[\mathrm{Hom}\,(\partial_2, 1)] \subset \mathrm{Hom}\,(C_1, B)$, defines a unique element $[\phi_1] \in H^1(C_.; B) = \mathrm{Ext}(A, B)$.

(a) Show that the cohomology class $[\phi_1]$ depends only on the equivalence class $[E]$ of the extension and the function so defined is a bijection of the set of equivalence classes of extensions of B by A with $\mathrm{Ext}(A, B)$. [Hint: to show surjectivity, use push-out diagrams.]

(b) Use this bijection to interpret the meaning of the sum of two extensions and multiplication of an extension by an element of R.

Because of functoriality of Ext, given a chain complex $C_.$ of R-modules and a R module G, we can define the cochain complex $\mathrm{Ext}(C_., B)$ in an obvious way. Moreover, for any R module G', there is the obvious R-linear map

$$h : H^q(C; G) \to \mathrm{Hom}\,(H_q(C; G'), G \otimes G')$$

satisfying

$$h([f])(c \otimes g') = f(c) \otimes g'$$

for $[f] \in H^q(C, G), c \in H_q(C), g' \in G'$. [Of course, you need to verify that the RHS is independent of the choices made in LHS.] We can now state the result that connects homology and cohomology of a chain complex.

Theorem 7.2.13 (Universal coefficient theorem for cohomology) *Let R be a PID. Given a free chain complex $C_.$ and a R-module G, there is a functorial exact sequence*

$$0 \longrightarrow \mathrm{Ext}(H_{q-1}(C), G) \longrightarrow H^q(C; G) \longrightarrow \mathrm{Hom}\,(H_q(C), G) \longrightarrow 0$$

and this sequence splits, for all $q \geq 1$.

Proof: This is similar to the proof of Theorem 6.2.13 and details are left to the reader. ♠

Corollary 7.2.14 *Let $f : X \to Y$ be a map which induces isomorphism in integral homology $f_* : H_*(X) \to H_*(Y)$. Then with any abelian group G we have isomorphisms $f^* : H^*(Y; G) \to H^*(X; G)$.*

Proof: By functoriality of the exact sequence in the above theorem, we have a commutative diagram

$$
\begin{array}{ccccccccc}
0 & \longrightarrow & \mathrm{Ext}(H_{q-1}(Y), G) & \longrightarrow & H^q(Y; G) & \longrightarrow & \mathrm{Hom}\,(H_q(Y), G) & \longrightarrow & 0 \\
& & \downarrow{\scriptstyle \mathrm{Ext}(f_*, 1)} & & \downarrow{\scriptstyle f^*} & & \downarrow{\scriptstyle \mathrm{Hom}(f_*, 1)} & & \\
0 & \longrightarrow & \mathrm{Ext}(H_{q-1}(X), G) & \longrightarrow & H^q(X; G) & \longrightarrow & \mathrm{Hom}\,(H_q(X), G) & \longrightarrow & 0.
\end{array}
$$

Since f_* is an isomorphism, the two extreme vertical arrows are isomorphisms. By Five lemma, the central vertical arrow is also an isomorphism. ♠

Example 7.2.15 It readily follows from the above theorem that $H^0(X, G)$ is isomorphic to the direct product of copies of G indexed over the set of path connected components of X. Of course, you can derive this directly from the definition of cohomology. On the other hand consider the following result:

If $H_q(X)$ is finitely generated for all q, then the torsion submodule of $H^q(X)$ is isomorphic to the torsion submodule of $H_{q-1}(X)$; and the free part of $H^q(X)$ is isomorphic to the free part of $H_q(X)$.

This is an easy consequence of the above theorem (by putting $G = \mathbb{Z}$) and not at all obvious from the definitions.

Remark 7.2.16 One may interpret Corollary 7.2.14 as follows: If two topological spaces cannot be 'distinguished' through their homology groups, then they cannot be distinguished by their cohomology either. In other words, we seem to have all the information about cohomology already 'available' in homology. Maybe so! However, in practice, we shall soon see that cohomology readily gives more information about a topological space than the homology through its richer algebraic structure. The catch is that in Corollary 7.2.14 the isomorphism of homology is not just a module isomorphism but has to be induced by a continuous map defined between the two spaces.

As a typical example consider $\mathbb{CP}^2$, the complex projective space of complex dimension 2, and the wedge-sum, $\mathbb{S}^2 \vee \mathbb{S}^4$. It is not difficult to prove that if X denotes either of them then $H_0(X), H_2(X), H_4(X)$ are isomorphic to $\mathbb{Z}$ and all other homology groups are trivial. So is the case with cohomology. However, in the next section, we shall see that the two spaces are not homotopy equivalent. Indeed, there is no map from one space to the other which produces these isomorphisms in homology modules. We shall prove this by studying an additional algebraic structure called the cup product on the cohomology modules.

Before we begin talking about product structures, it is useful to have a 'universal coefficient theorem' for cohomology just in terms of cohomology, i.e., one which relates $H^*(X, G)$ with $H^*(X, R)$. We shall end this section with such a result.

Lemma 7.2.17 Let M, G, G' be any R-modules. Consider the homomorphism

$$\mu : \mathrm{Hom}\,(M, G) \otimes G' \to \mathrm{Hom}\,(M, G \otimes G')$$

given by $\mu(f \otimes g')(a) = f(a) \otimes g'$. The homomorphism is functorial and if M is free and G' is finitely generated, then μ is an isomorphism.

Proof: We leave it to you to verify the functoriality. To prove the second part, first we observe that the result holds for $G = R$. Since the two functors Hom and $\otimes$, commute with finite direct sums, the result follows when G' is finitely generated and free. Finally, let G' be given by a short exact sequence

$$0 \longrightarrow F_1 \longrightarrow F_0 \longrightarrow G' \longrightarrow 0$$

where F_0, F_1 are finitely generated and free. (Here we are using the fact that R is a PID.) Upon tensoring with $\mathrm{Hom}\,(M, G)$ and also first tensoring with G and then taking $\mathrm{Hom}\,(M, -)$ we get commutative diagram

$$
\begin{array}{ccccccc}
\mathrm{Hom}\,(M, G) \otimes F_1 & \longrightarrow & \mathrm{Hom}\,(M, G) \otimes F_0 & \longrightarrow & \mathrm{Hom}\,(M, G) \otimes G' & \longrightarrow & 0 \\
\downarrow{\scriptstyle \mu_1} & & \downarrow{\scriptstyle \mu_0} & & \downarrow{\scriptstyle \mu} & & \downarrow \\
\mathrm{Hom}\,(M, G \otimes F_1) & \longrightarrow & \mathrm{Hom}\,(M, G \otimes F_0) & \longrightarrow & \mathrm{Hom}\,(M, G \otimes G') & \longrightarrow & 0.
\end{array}
$$

The first row is exact because $\otimes$ is right-exact. The second one is exact because $\text{Hom}\,(M, -)$ is exact where M is free. Since the first, second and fourth vertical arrows are isomorphisms, by Four lemma, the third vertical arrow μ is also an isomorphism.

There is some sort of a generalization of μ as follows: Given any modules M, M', G, G' consider

$$\mu : \text{Hom}\,(M, G) \otimes \text{Hom}\,(M', G') \to \text{Hom}\,(M \otimes M', G \otimes G')$$

given by

$$\mu(f \otimes f')(g \otimes g') = f(g) \otimes f'(g'). \tag{7.3}$$

Observe that if $M' = R$ then under the canonical isomorphism $\text{Hom}\,(M', G) \approx G'$, this μ corresponds to the μ of the above lemma.

Lemma 7.2.18 Let M, M' be free modules and
(i) M and M' be finitely generated OR
(ii) M and G' be finitely generated. Then μ as in (7.3) is an isomorphism.

Proof: Since M, M' are both free, so is $M \otimes M'$.
(i) In this case $M \otimes M'$ is finitely generated also. Since Hom and $\otimes$ commute with finite direct sum on either side, and since the result is trivially true when $M = M' = R$, we are through.
(ii) We first consider the case when $G' = R$, viz., we want to show that

$$\mu : \text{Hom}\,(M, G) \otimes \text{Hom}\,(M', R) \to \text{Hom}\,(M \otimes M', G)$$

is an isomorphism. This is trivial if $M' = R$. In the general case, we can appeal to the fact that Hom and $\otimes$ commute with finite direct sums. Now let G' be any finitely generated module. We then have a commutative diagram

$$
\begin{CD}
\text{Hom}\,(M, G) \otimes \text{Hom}\,(M', R) \otimes G' @>{1 \otimes \mu}>> \text{Hom}\,(M, G) \otimes \text{Hom}\,(M', G') \\
@V{\mu \otimes Id_{G'}}VV @VV{\mu}V \\
\text{Hom}\,(M \otimes M', G) \otimes G' @>{\mu}>> \text{Hom}\,(M \otimes M', G \otimes G').
\end{CD}
$$

We have seen that the horizontal arrows are isomorphisms in the previous lemma. The left-side vertical arrow is an isomorphism as seen above. Therefore the right-side vertical arrow is also an isomorphism. ♠

Definition 7.2.19 We say a graded module C is of finite type if each C_q is a finitely generated module.

Thus a finite type C is finitely generated iff $C_q = 0$ for almost all q. The following lemma allows us to transfer the hypothesis of 'finite type' on the homology to that on the chain complex.

Lemma 7.2.20 Let C be a free chain complex such that $\text{H}(C)$ is of finite type. Then there is a free chain complex of finite type which is chain equivalent to C.

Proof: For each q, pick up a finitely generated (free) submodule F_q of $Z_q(C) \subset C_q$ which surjects onto $H_q(C)$. Let F_q' be the kernel of the epimorphism $F_q \to H_q(C)$. Put $C_q' = F_q \oplus F_{q-1}'$ and $\partial'(c, c') = (c', 0)$. Clearly (C', ∂') is a free chain complex of finite type. It is also clear that $H_q(C') = F_q/F_q' = H_q(C)$.

To get a chain equivalence from C' to C, since F_q' is a free submodule of $Z_q(C)$, there exists a homomorphism $\phi_q : F_q' \to C_{q+1}$ such that $\partial_{q+1} \circ \phi_q = Id$. Put $\tau(c, c') = c + \phi_{q-1}(c')$. Verify that τ is a chain map $\tau : C' \to C$ and induces isomorphism in homology. Since both C and C' are free, it follows that τ is a chain equivalence (see Theorem 6.1.6). ♠

Theorem 7.2.21 *Let C be a chain complex of free R-modules and G be a R-module. Assume that* (i) *G is finitely generated OR* (ii) *$H(C)$ is finite type. Then there is a functorial short exact sequence*

$$0 \longrightarrow H^q(C; R) \otimes G \xrightarrow{\ \mu\ } H^q(C; G) \longrightarrow H^{q+1}(C; R) \star G \longrightarrow 0$$

and the sequence splits (but not functorially), for each $q \geq 0$.

Proof: (i) Since G is finitely generated, we have $\mu : \mathrm{Hom}\,(C; R) \otimes G \approx \mathrm{Hom}\,(C, G)$. Since $\mathrm{Hom}\,(C, R)$ is torsion free, $\mathrm{Hom}\,(C; R) \star G = 0$. Hence the result follows from Theorem 6.2.4, by the standard method of converting a cochain complex to a chain complex.
(ii) If $H(C)$ is of finite type, by the lemma, we can replace C by a chain complex C' which is of finite type and argue with C'. Now by Lemma 7.2.18, $\mathrm{Hom}\,(C', R) \otimes G \approx \mathrm{Hom}\,(C', G)$ for all modules G. From the left-exactness property of Hom, this implies that $C' \star G = 0$. Therefore, we can again apply Theorem 6.2.4. ♠

7.3 Products in Cohomology

As in the case of homology, there is the cross product in cohomology from the tensor product of the cohomologies of two spaces X and Y to the cohomology of their product $X \times Y$. This cross product has similar properties as listed in Theorem 6.3.16 for homology cross product and enters in the Künneth formula for cohomology of the product.

When $X = Y$, the diagonal map $d : X \to X \times X$ can be used to pull-back this product from $H^*(X \times X)$ to $H^*(X)$ which provides a multiplicative structure on the graded module $H^*(X)$ called the cup product. This is the extra feature of cohomology which makes it more informative than the homology modules with far-reaching applications. The genesis of this goes back to the product of two differential forms on a manifold and indeed very closely related to it.

Definition 7.3.1 Given an excisive couple $\{X \times B, A \times Y\}$, consider the composite of the functorial homomorphism and the Eilenberg–Zilber equivalence

$$\mathrm{Hom}\,(S_.(X)/S_.(A), G) \otimes \mathrm{Hom}\,(S_.(Y)/S_.(B), G')$$

$$\mu \downarrow$$

$$\mathrm{Hom}\,([S_.(X)/S_.(A)] \otimes [S_.(Y)/S_.(B)], G \otimes G')$$

$$\downarrow \zeta$$

$$\mathrm{Hom}\,(S_.(X \times Y)/S_.(X \times B \cup A \times Y), G \otimes G').$$

Let μ' denote the homomorphism induced by this composite in homology, i.e.,

$$H_*(\mathrm{Hom}\,(S_.(X)/S_.(A), G) \otimes \mathrm{Hom}\,(S_.(Y)/S_.(B); G'))$$

$$\downarrow \mu'$$

$$H_*(\mathrm{Hom}\,[S_.(X \times Y)/S_.(X \times B \cup A \times Y), G \otimes G']).$$

Given $u \in H^p(X, A; G), v \in H^q(Y, B; G')$, we define their cross product

$$u \times v := \mu'(u \otimes v) \in H^{p+q}((X, A) \times (Y, B); G \otimes G').$$

As a special case of Theorem 7.1.4, we get the Künneth formula for cohomology:

Theorem 7.3.2 *Let $\{X \times B, A \times Y\}$ be an excisive couple in $X \times Y$. Let R be a PID and G, G' be modules over R such that $G \star G' = 0$. Then there is a functorial exact sequence*

$$0 \longrightarrow [H(C;G) \otimes H(C';G')]^q \overset{\mu'}{\longrightarrow} H^q(C \otimes C'; G \otimes G') \longrightarrow [H(C;G) \star H(C';G')]^{q+1} \longrightarrow 0$$

and this sequence splits, for all $q \geq 0$.

Definition 7.3.3 Consider R-modules G, G', G'' and a fixed pairing $\varphi : G \otimes G' \to G''$ (i.e., a R-linear map,) and two subspaces $A_i \subset X$ such that $\{X \times A_2, A_1 \times X\}$ is excisive in $X \times X$. Then for $u \in H^p(X, A_1; G)$ and $v \in H^q(X, A_2; G')$ we define

$$u \smile v = \varphi_* d^*(u \times v) \in H^{p+q}(X, A_1 \cup A_2; G'').$$

Of course, we can take any topological space X and $A_1 = A_2 = \emptyset$, and $G = G' = R$ with the standard pairing $R \otimes R \to R$ given by the multiplication in R to get a cup product

$$\smile : H^p(X; R) \otimes H^q(X; R) \to H^{p+q}(X; R).$$

From the corresponding properties of the cross product, it follows easily that the cup product satisfies the properties listed in the theorem below.

Theorem 7.3.4
(a) *Given excisive couples $\{X \times A_2, A_1 \times X\}$ and $\{Y \times B_2, B_1 \times Y\}$ and a map $f : X \to Y$ such that $f(A_1) \subset B_1, f(A_2) \subset B_2$, we have*

$$f^*(u \smile v) = f^*(u) \smile f^*(v).$$

(b) *With respect to the standard pairings $R \otimes G \approx G \approx G \otimes R$, and for $1 \in H^0(X, A_1; R), u \in H^q(X, A_2; R)$, we have,*

$$1 \smile u = u = u \smile 1.$$

(c) *Given a commutative diagram of pairings*

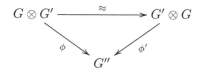

and for any $u \in H^p(X, A_1; G), v \in H^q(X, A_2; G')$, we have,

$$u \smile v = (-1)^{pq} v \smile u.$$

(d) *Similarly, with respect to as associative data of pairings,*

$$G_1 \otimes (G_2 \otimes G_3) \overset{\approx}{\longrightarrow} (G_1 \otimes G_2) \otimes G_3 \overset{\phi_{12} \otimes 1}{\longrightarrow} G_{12} \otimes G_3 \qquad (7.4)$$

$$\downarrow^{1 \times \phi_{23}} \qquad\qquad\qquad\qquad\qquad\qquad\qquad \downarrow^{\psi}$$

$$G_1 \otimes G_{23} \overset{\psi'}{\longrightarrow} G_{123}$$

we have the following associativity of the cup product:

$$u_1 \smile (u_2 \smile u_3) = (u_1 \smile u_2) \smile u_3.$$

(e) *Under the connecting homomorphisms of appropriate Mayer–Vietoris sequences we have:*

$$\delta^*(u \smile \iota^* v) = \delta^*(u) \smile v; \quad \delta^*(\iota^* u \smile v) = (-1)^q v \smile \delta^* u.$$

(f) *The cup product is skew-mutually distributive with the cross product. With the appropriate excisive couples and pairings we have:*

$$(u_1 \times u_2) \smile (v_1 \times v_2) = (-1)^{\deg u_2 \deg v_1}(u_1 \smile v_1) \times (u_2 \smile v_2).$$

Before, we take up the discussion of examples and computations, etc., we shall introduce one more 'product' which is closely related to the cup product:

It is convenient to introduce the following notation: Given $f \in \mathrm{Hom}(C'_q, G)$ and $c = \oplus_i c_i \in C$, we set

$$\langle f, c \rangle := f(c_q).$$

Note that in this notation we have,

$$\langle \delta f, c \rangle = \langle f, \partial_{q+1}(c_{q+1}) \rangle.$$

Given a pairing $\phi : G \otimes G' \to G''$ and chain complexes C, C', consider the functorial homomorphism

$$h : \mathrm{Hom}\,(C', G) \otimes (C \otimes C' \otimes G') \to C \otimes G''$$

given by

$$h(f \otimes c \otimes c' \otimes g') = c \otimes \varphi(f(c') \otimes g'). \tag{7.5}$$

Note that if $f \in \mathrm{Hom}(C'_q, G), \lambda \in (C \otimes C')_{n+1} \otimes G'$ then

$$\partial h(f \otimes \lambda) = (-1)^{n-q} h(\delta f \otimes \lambda) + h(f \otimes \partial \lambda). \tag{7.6}$$

In particular, if f is a cocycle and λ is a cycle then it follows that $h(f \otimes \lambda)$ is a cycle; further if $f = \delta g$ is a coboundary or $\lambda = \partial c$ is a boundary, it follows that $h(f \otimes \lambda) = h(\delta g \otimes \lambda) = (-1)^{n-q} \partial h(g \otimes \lambda)$ or $h(f \otimes \lambda) = h(f \otimes \partial c) = \partial h(f \otimes c)$ and hence in either case, $h(f \otimes \lambda)$ is boundary.

Now given a topological space X, let $\tau_X : S_.(X) \to S_.(X) \otimes S_.(X)$ be any diagonal approximation (say, as in (6.11)), then we get a functorial homomorphism

$$\tilde{\tau} := \tilde{\tau}_X : \mathrm{Hom}\,(S_.(X), G) \otimes (S_.(X) \otimes G') \to S_.(X) \otimes G''$$

given by

$$\tilde{\tau}(f \otimes c \otimes g') = h(f \otimes \tau_X(c) \otimes g').$$

Indeed, if we use Alexander–Whitney diagonal approximation, then for $\lambda = \sum_c c \otimes g'_c$, we have,

$$h(f \otimes \lambda) = \sum_c {}_{n-q} c \otimes \varphi(\langle f, c_q \rangle \otimes g'_c). \tag{7.7}$$

Let now A_1, A_2 be subspaces of X. If $f \in S^q(X; G)$ vanishes on A_1 then for any $c \in S(A_1) \otimes G'$, we have $\tilde{\tau}(f \otimes c) = 0$. Therefore, if $f \in Z^q(X, A_1; G)$ and $c \in S(X) \otimes G'$ is such that $\partial c \in (S(A_1) + S(A_2)) \otimes G'$, then

$$\partial(\tilde{\tau}(f \otimes c)) = \tilde{\tau}(f \otimes \partial c) \tag{7.8}$$

belongs to $S(A_2) \otimes G''$. Furthermore, if f is the coboundary of a cochain which vanishes on A_1 OR if c is a boundary modulo $(S(A_1) + S(A_2)) \otimes G'$, then $\tilde{\tau}(f \otimes c)$ is a boundary modulo $S(A_2) \otimes G''$. Therefore, the assignment

$$\{f\} \otimes \{c\} \mapsto \{\tilde{\tau}(f \otimes c)\}$$

defines a homomorphism

$$H^q(X, A_1; G) \otimes H_n(S(X)/S(A_1) + S(A_2), G') \to H_{n-q}(X, A_2; G'').$$

It follows that if $\{A_1, A_2\}$ is an excisive couple in $A_1 \cup A_2$ then $\tilde{\tau}$ induces a functorial homomorphism:

$$H^q(X, A_1; G) \otimes H_n(X, A_1 \cup A_2; G') \to H_{n-q}(X, A_2; G'')$$

called the *cap product*. For $u \in H^q(X, A_1; G)$ and $z \in H_n(X, A_1 \cup A_2; G')$, we write $u \frown z$ to denote the image of the element $u \otimes z$ under this homomorphism.

The following properties of the cap product are all easy to prove similar to the corresponding properties of the cup product.

Theorem 7.3.5 (a) *With the obvious pairing $R \otimes G \approx G$, we have,*

$$1 \frown z = z.$$

(b) *Given an associative data of pairing as in (7.4) we have*

$$u \frown (v \frown z) = (u \smile v) \frown z.$$

(c) *Under the augmentation $\epsilon : H_0(X; G \otimes G') \to G \otimes G'$, we have*

$$\epsilon(u \frown z) = \langle u, z \rangle.$$

(d) **Projection formula** *Given a map $f : X \to Y$, we have*

$$f_*(f^*(u) \frown z) = u \frown f_*(z).$$

(e) *If $i : A_1 \cup A_2 \to X, j : A_2 \to A_1 \cup A_2$ are the inclusions, and $\partial : H_n(X, A_1 \cup A_2) \to H_{n-1}(A_1 \cup A_2), \partial' : H_{n-q}(X, A_2) \to H_{n-q-1}(A_2)$ are connecting homomorphisms of homology exact sequences of appropriate pairs, then for $u \in H^q(X, A_1), z \in H_n(X, A_1 \cup A_2)$, we have*

$$j_* \partial'(u \frown z) = i^*(u) \frown \partial z.$$

Similar formulae hold under connecting homomorphisms of exact sequences of triples, excisive couples, etc. (The proof is directly from (7.8).)

(f) *The cap product* **mutually distributes with cross products***:*

$$(u_1 \times u_2) \frown (z_1 \times z_2) = (-1)^{mq}(u_1 \frown z_1) \times (u_2 \frown z_2)$$

where

$$u_1 \in H^p(X, A_1; G_1), u_2 \in H^q(Y, B_1; G_2), z_1 \in H_m(X, A_1 \cup A_2; G_1'), z_2 \in H_n(Y, B_1 \cup B_2; G_2').$$

(g) *Let $\phi : X \times Y \to Y$ be the projection map. Given $z_1 \in H_p(X), z_2 \in H_q(Y), u_1 \in H^q(Y)$ we have*

$$\phi^*(u) \frown (z_1 \times z_2) = \langle u, z_2 \rangle z_1.$$

(h) *Given $f \in Z^p(X, A; R)$, consider the chain maps*

$$\alpha : C_*(X, A; R) \to C_*(X; R), \quad \beta : C^*(X; G) \to C^*(X, A; G)$$

given by

$$\alpha(c) = c \frown f; \quad \beta(\sigma) = \sigma \smile f.$$

Then $\operatorname{Hom}(\alpha, G) = \beta$ and hence the same holds at the cohomology level also.

Part (g) follows directly from (7.5), whereas the last part (h) follows from (7.7), with a little effort. Indeed, for any singular q-simplex λ in X, we have,

$$
\begin{aligned}
\operatorname{Hom}(\alpha, 1_G)(\sigma)(\lambda) &= (\sigma \circ \alpha)(\lambda) \\
&= \sigma(\lambda \frown f) = \sigma(\,_{q-p}\lambda \otimes f(\lambda_p) \\
&= \sigma(_{n-q}\lambda)f(\lambda_p) \\
&= (\sigma \smile f)(\lambda) \\
&= \beta(\sigma)(\lambda).
\end{aligned}
$$

7.4 Some Computations

In this section, we shall compute the cohomology algebras of some standard spaces such as product of spheres, one point unions and connected sums, Riemann surfaces, projective spaces, etc. The emphasis is on the methods rather than the particular examples chosen. We shall also give one application of the computation of cohomology algebra of real projective space.

Example 7.4.1 Cohomology of Cartesian product

We shall work with coefficients in $\mathbb{Z}$ unless specified otherwise. Let X, Y be two CW-complexes. Let $\sigma : \Delta_p \to X, \rho : \Delta_q \to Y$ denote a p-cell in X and q-cell in Y. Then we have $\sigma \times \rho : \Delta_p \times \Delta_q :\to X \times Y$ representing a $(p+q)$-cell. If u, v are some cochains in X and Y, respectively, then it follows that

$$\langle u \times v, \sigma \times \rho \rangle = (-1)^{pq} \langle u, \sigma \rangle \langle v, \rho \rangle.$$

In particular, let us consider $X = \mathbb{S}^p, Y = \mathbb{S}^q, X \times Y = \mathbb{S}^p \times \mathbb{S}^q$. Direct application of the Künneth formula yields that

$$H_*(X \times Y) \approx H_*(S^p) \otimes H_*(S^q); \quad H^*(X \times Y) \approx H^*(S^p) \otimes H^*(S^q).$$

Indeed if z_1, z_2 denote the homology classes represented by the p-cell and the q-cell, it follows that $z_1 \times z_2$ is a generator of $H_{p+q}(X)$; the similar statement holds in cohomology as well. Moreover, if $u_1 \in \operatorname{Hom}(H_p(X); \mathbb{Z})$ and $u_2 \in \operatorname{Hom}(H_q(Y); \mathbb{Z})$ are dual to z_1, z_2 then

$$\langle u_1 \times u_2, z_1 \times z_2 \rangle = (-1)^{pq} \langle u_1, z_1 \rangle \langle u_2, z_2 \rangle = (-1)^{pq}.$$

Let ϕ_X, ϕ_Y denote the respective projections on $X \times Y$. Then we have

$$
\begin{aligned}
(\phi_X^*(u_1) \smile \phi_Y^*(u_2)) \frown (z_1 \times z_2) &= \phi_X^*(u_1) \frown (\phi_Y^*(u_2) \frown (z_1 \times z_2)) \\
&= (-1)^{pq} \phi_X^*(u_1) \frown (\langle u_2, z_2 \rangle z_1) \\
&= (-1)^{pq} \langle u_1, z_1 \rangle = (-1)^{pq}.
\end{aligned}
$$

Therefore, it follows that $\phi_X^*(u_1) \smile \phi_Y^*(u_2)$ is a generator of $H^{p+q}(X \times Y)$. Also note that $\phi_X^*(u_1) \smile \phi_X^*(u_1) = \phi_X^*(u_1 \smile u_1) = 0$. Similarly $\phi_Y^*(u_2)^2 = 0$.

Furthermore, if we have $p = q$ then $H^p(\mathbb{S}^p \times \mathbb{S}^p) \approx \mathbb{Z}^2$ generated by the $v_i = \phi_i^*(u_i), i = 1, 2$ and any $v \in H^p(\mathbb{S}^p \times \mathbb{S}^p)$ is uniquely expressible as $m_1 v_1 + m_2 v_2, m_i \in \mathbb{Z}$. It follows that

$$v^2 := v \smile v = \begin{cases} 2m_1 m_2 (u_1 \times u_2), & \text{if } p \text{ is even;} \\ 0, & \text{otherwise.} \end{cases}$$

The arguments here can be easily adopted to prove a similar result in the relative situation

$$H^i(\mathbb{D}^i, \partial \mathbb{D}^i) \otimes H^j(\mathbb{D}^j, \partial \mathbb{D}^j) \to H^{i+j}(\mathbb{D}^{i+j}, \partial \mathbb{D}^{i+j}).$$

If $\phi_i : \mathbb{D}^{i+j} \to \mathbb{D}^i$ and $\phi_j : \mathbb{D}^{i+j} \to \mathbb{D}^j$ denote the projection maps to the first i (respectively last j coordinates, and $u \in H^i(\mathbb{D}^i, \partial \mathbb{D}^i)$, $v \in H^j(\mathbb{D}^j, \partial \mathbb{D}^j)$ denote generators, then

$$\phi_i^*(u) \smile \phi_j^*(v) = u \times v \tag{7.9}$$

is a generator.

Example 7.4.2 Cohomology of one-point-union

Let us assume that the base points occurring here are all non degenerate. Let $Y = Z_1 \vee Z_2$ be the one point union of two connected spaces, let $\eta_i : Z_i \to Y$ be the inclusion map and $q_i : Y \to Z_i$ be the retractions which pinch Z_j to a single point $j \neq i$, respectively. Using Mayer–Vietoris, it follows easily that $H^*(Y)$ is a direct sum of $q_i^*(H^*((Z_i)), i = 1, 2$ in positive dimensions and of course, $H^0(Y) = \mathbb{Z}$. We claim that

$$q_1^*(u_1) \smile q_2^*(u_2) = 0$$

for any $u_i \in H^{p_i}(Z_i), p_1, p_2 > 0$. Note that an element $u \in H^*(Y)$ is zero iff $\eta_i^*(u) = 0$ for both $i = 1, 2$. Now

$$\eta_1^*(q_1^*(u_1) \smile q_2^*(u_2)) = (q_1 \circ \eta_1)^*(u_1) \smile (q_2 \circ \eta_1)^*(u_2) = (q_1 \circ \eta_1)^*(u_1) \smile 0 = 0$$

because $q_2 \circ \eta_1$ is a point map. Likewise, it follows that $\eta_2^*(q_1^*(u_1) \smile q_2^*(u_2)) = 0$. This result gets generalized easily to one-point union of an arbitrary family of pointed connected spaces: There is a graded ring isomorphism:

$$H^*(\vee_\alpha X_\alpha) \approx \prod_\alpha H^*(X_\alpha).$$

Here, on the RHS the product of graded rings has to be defined correctly: As a graded module its 0th-component is taken to be R, whereas its qth component for $q > 0$ is taken to be the direct product of all q^{th} components of the rings in the family. The multiplication of two elements of positive degree from two different rings is of course taken to be zero.

Example 7.4.3 Cohomology of connected sum

Let $X = M_1 \# M_2$ be the connected sum to closed n-manifolds. There is a quotient map $q : X \to M_1 \vee M_2$, the onepoint union of M_1 and M_2, which pinches the $(n-1)$-sphere along which the connected sum is performed to a single point. Using Mayer–Vietoris, one can see that $q^* : H^*(M_1 \vee M_2) \to H^*(M_1 \# M_2)$ is an isomorphism in dimensions less than n and is a surjection in dimension n. Let $\tau_i : M_1 \# M_2 \to M_i$ be the composite of q with $q_i : M_1 \vee M_2 \to M_i$. It follows from the previous example that for $u_1 \in H^{p_1}(M_1), u_2 \in H^{p_2}(M_2)$ with $p_1, p_2 > 0$ we have

$$\tau_1^*(u_1) \smile \tau_2^*(u_2) = 0.$$

Example 7.4.4 Relation between cup product and intersection number

In this example, we assume that the reader is familiar with the notion of geometric intersection number of two submanifolds, of complementary dimensions. (See Chapter 7 of [Shastri, 2011].) We illustrate this relation by a discussion of the same on a smooth orientable surface. Let $X = S_g$ be a smooth, connected, orientable, compact surface without boundary. We know that $H_0(X) = \mathbb{Z}, H_1(X) = \mathbb{Z}^{2g}$ and $H_2(X) = \mathbb{Z}$. Therefore by universal coefficient theorem, $H^*(X) \approx \mathrm{Hom}(H_*(X), \mathbb{Z})$. The only non trivial cup product occurs in $H^1(X) \otimes H^1(X) \to H^2(X)$. Thus, there is nothing to discuss if $g = 0$. The case $g = 1$, i.e, $X = \mathbb{S}^1 \times \mathbb{S}^1$ is discussed in the Example 7.4.1, as a part of a more general discussion. In particular, we see that if u_i are cohomology elements dual to z_i where $z_i, i = 1, 2$ are elements of $H_1(\mathbb{S}^1 \times \mathbb{S}^1)$ given by the two coordinate inclusions $\mathbb{S}^1 \to \mathbb{S}^1 \times \mathbb{S}^1$, then $u_1 \smile u_2$ is a generator of $H^2(X)$ and with appropriate orientations, we have $\langle u_1 \smile u_2, z_1 \times z_2 \rangle = (u_1 \smile u_2) \frown (z_1 \times z_2) = 1$. Note that the intersection number of the two submanifold $\mathbb{S}^1 \to \mathbb{S}^1 \times \mathbb{S}^1$ given by the coordinate inclusions is also equal to 1.

Now we consider the case $g \geq 2$. Let $\eta_i : \mathbb{S}^1 \to X$ be any two embeddings which meet transversely at a single point. Let $Y \subset \mathbb{S}^1 \times \mathbb{S}^1$ denote the union $\mathbb{S}^1 \times 1 \smile 1 \times \mathbb{S}^1$ in $\mathbb{S}^1 \times \mathbb{S}^1$. Clearly η_i together define an embedding of $\eta : Y \to X$. It is not difficult to see that this embedding extends to an embedding of tubular neighbourhoods $\eta : U(Y) \to V(\eta(Y))$ in the respective spaces. Note that the complement of $U(Y)$ in $\mathbb{S}^1 \times \mathbb{S}^1$ is a disc. Putting $W = X \setminus V(Y)$, it follows that X is the connected sum of $\mathbb{S}^1 \times \mathbb{S}^1$ and another surface $\hat{W}$, where $\hat{W}$ is the closed orientable surface obtained by capping-off the boundary component of W. Also, the factor $\mathbb{S}^1 \times \mathbb{S}^1$ corresponds to the surface obtained by capping off $\bar{U}(Y)$. Let $\phi : X \to \mathbb{S}^1 \times \mathbb{S}^1$ be the map which pinches W to a single point. It is easily checked that ϕ is a degree one map, and (hence or otherwise) $\phi_* : H_*(X) \to H_*(\mathbb{S}^1 \times \mathbb{S}^1)$ is surjective and $\phi^* : H^*(\mathbb{S}^1 \times \mathbb{S}^1) \to H^*(X)$ is injective. Note that $\phi \circ \eta_i$ are coordinate inclusions $\mathbb{S}^1 \to \mathbb{S}^1 \times \mathbb{S}^1$. Put $u_i = \phi^*(v_i)$ where $v_i \in H^1(\mathbb{S}^1 \times \mathbb{S}^1), i = 1, 2$ are the elements corresponding to the two inclusion maps η_i. It follows that $u_1 \smile u_2 = \phi^*(v_1 \smile v_2)$ and hence is a generator of $H^2(X)$. Proceeding this way, using a handle decomposition of S_g, we get generators $u_1, v_1, \dots, u_g, v_g$ for $H^1(X)$ such that

$$u_i \smile u_j = 0 = v_i \smile v_j, u_i \smile v_j = \delta_{ij}[\hat{X}]$$

where $[\hat{X}]$ denotes a generator of $H^2(X)$. Such a basis is called a symplectic basis for the cup product form which is a skew symmetric form in this case. The reader is urged to come back to this example after studying Poincaré duality, etc., from Chapter 8.

Example 7.4.5 Cup product in $\mathbb{S}^1 \times \mathbb{S}^1$ using a triangulation

Let us compute the cup product in $H^*(\mathbb{S}^1 \times \mathbb{S}^1)$ in another way. For this we shall use a triangulation of the $X = \mathbb{S}^1 \times \mathbb{S}^1$, say the standard one, as given in Figure 7.1

Since we have labeled the vertices by integers, this gives a natural order on each edge and each 2-simplex σ. Fix an orientation on the surface X which will then induce orientation on each 2-simplex. Define $\epsilon_\sigma = \pm 1$ according as the two orientations on σ coincide or not. The 2-cycle $\sum \epsilon_\sigma \sigma$ where the summation is taken over all 2-simplexes represents the generator $[X]$ of $H_2(X)$ corresponding to the orientation.

We shall now define two simplicial 1-cocycles φ, ψ which represent a set of generators for H^1. These are indicated by the dotted lines in the picture, one horizontal and the other vertical. Note that the intersection of each of these lines with an edge is either empty or a single point of transversal intersection. We follow the convention that $\varphi(\sigma)$ is zero if the horizontal line does not meet the edge σ and is equal to ± 1 if it meets the edge. Here again, the sign has to be determined depending on whether the orientations defined by the segment of φ and the edge coincides with the orientation on the surface. Similarly we define

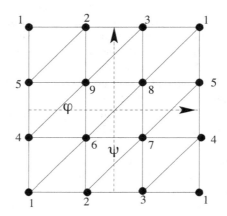

FIGURE 7.1. Computing the cup product in $\mathbb{S}^1 \times \mathbb{S}^1$

ψ with respect to the vertical dotted line. It is easily verified that φ and ψ are co-cycles and respectively take the value 1 (or 0) on the 1-cycles which represent the two standard generators in H_1.

Now we use the Alexander–Whitney diagonal approximation to compute $\varphi \smile \psi$ at the cochain level itself. Note that the sum of all the ordered 2-simplexes is a generator of H_2. Even if one of the two dotted lines does not meet a 2-simplex σ, it follows that

$$(\varphi \smile \psi)(\sigma) = \varphi(_1\sigma)\psi(\sigma_1) = 0.$$

Therefore, we have to concentrate only on the two central 2-simplexes. Finally, we see that
$(\varphi \smile \psi)[678] = \varphi[67]\psi[78] = 0$;
$(\varphi \smile \psi)[689] = \varphi[68]\psi[89] = 1.1 = 1$.
Therefore $(\varphi \smile \psi)[X] = 1$. Since we know beforehand that $H^2(X) \approx \mathbb{Z}$, this proves that $\varphi \smile \psi$ must be a generator of $H^2(X)$. It is also interesting to check that

$$(\psi \smile \varphi)[X] = (\psi \smile \varphi)[678] = (-1)(1) = 1.$$

Example 7.4.6 Cup product in $\mathbb{P}^2$.

As in the previous example, we begin with a triangulation of $\mathbb{P}^2$, label the vertices and take the induced orientations on each simplex. Since $\mathbb{P}^2$ is a non orientable surface, we shall now work with $\mathbb{Z}_2$ coefficients. The sum $\sum \sigma$ of all the 2-simplexes is a 2-cycle mod 2 which represents the generator of $H_2(\mathbb{P}^2; \mathbb{Z}_2)$. The dotted line in Figure 7.2 is chosen so that it intersects every edge transversely in at most one point and this point is not a vertex and defines a 2-cochain φ by the same rule as before.

It is easily checked that

$$(\varphi \smile \varphi)[\sum \sigma] = (\varphi \smile \varphi)[450] = \varphi[45]\varphi[50] = 1.$$

Therefore $\varphi \smile \varphi$ is the generator of $H^2(\mathbb{P}^2; \mathbb{Z}_2)$.

It is interesting to examine what happens if you just change the labeling of the vertices. For instance, if you interchange 6 and 0 then

$$(\varphi \smile \varphi)[\sum \sigma] = (\varphi \smile \varphi)([678] + [560] + [456]) = 1 + 1 + 1 = 1.$$

The method above is not effective in determining the cup product in $\mathbb{P}^n, n \geq 3$, which we shall discuss below with a different method.

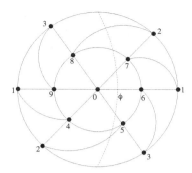

FIGURE 7.2. Computing the cup product in the projective plane

Remark 7.4.7 We have seen in Example 4.5.3 that the CW-chain complex of $X = \mathbb{S}^1 \times \mathbb{S}^1$ is given by

$$\cdots 0 \longrightarrow \mathbb{Z} \xrightarrow{\partial_2} \mathbb{Z}^2 \xrightarrow{\partial_1} \mathbb{Z}$$

where $\partial_2 = 0 = \partial_1$. It is also easily checked that this is the CW-chain complex of $Y = \mathbb{S}^1 \vee \mathbb{S}^1 \vee \mathbb{S}^2$. However, we have now seen that the cup product in $H^*(X)$ is non trivial whereas the same is completely trivial in $H^*(Y)$. Thus, it is clear that the CW-chain complex loses some vital information which is present in the singular chain complex. It should be noted that even though the chain complexes are isomorphic, there is no map $X \to Y$ or $(Y \to X)$ which induces an isomorphism of the two CW-chain complexes.

Example 7.4.8 Cohomology algebra of real projective spaces
The cellular chain complex of $\mathbb{P}^n$ with integer coefficients is given by

$$\mathbb{Z} \longrightarrow \cdots \xrightarrow{0} \mathbb{Z} \xrightarrow{.2} \mathbb{Z} \xrightarrow{0} \mathbb{Z} \xrightarrow{.2} \mathbb{Z} \xrightarrow{0} \mathbb{Z}$$

from which it follows that $H^i(\mathbb{P}^n; \mathbb{Z}_2) \approx \mathbb{Z}_2$ for $0 \le i \le n$ and 0 for $i > n$. We claim that

$$H^*(\mathbb{P}^2; \mathbb{Z}_2) \approx \mathbb{Z}_2[\alpha]/(\alpha^{n+1}), \tag{7.10}$$

the *truncated polynomial algebra* over $\mathbb{Z}_2$ generated by one variable α of degree 1. For $n = 1$ and 2, we have seen the proof of this statement.

We shall now use $\mathbb{Z}_2$ coefficients for cohomology groups and drop mentioning it temporarily.

Note that the inclusion $\mathbb{P}^{n-1} \subset \mathbb{P}^n$ induces an isomorphism of the cohomology algebras in dimensions $\le n-1$. Thus, inductively, assuming that α^{n-1} is the generator of $H^{n-1}(\mathbb{P}^{n-1}) \approx H^{n-1}(\mathbb{P}^n)$, it remains to prove that α^n is the generator of $H^n(\mathbb{P}^n)$.

We shall indeed prove a seemingly little more general statement that if u_i and u_j denote the generators of H^i and H^j, respectively, then $u_i \smile u_j$ is the generator of $H^{i+j}(\mathbb{P}^{i+j})$.

Put $i + j = n$. Again by cellular homology it follows that $u_i \in H^i(\mathbb{P}^n)$ corresponds, under the inclusion map to the generator $v_i \in H^i(\mathbb{P}^i)$. Next we observe that there are various inclusion maps $\mathbb{P}^i \subset \mathbb{P}^n$ given by putting any j coordinates to zero and any two of these inclusion maps are homotopic to each other. (See Exercise 7.4.13(i).) Therefore we can choose the two inclusions $\eta : \mathbb{P}^i \to \mathbb{P}^n$ and $\psi : \mathbb{P}^j \to \mathbb{P}^n$ given by

$$\eta[x_0, \ldots, x_i] = [x_0, \ldots, x_i, 0, \ldots, 0]; \psi[x_0, \ldots, x_j] = [0, \ldots, 0, x_0, \ldots, x_j].$$

We want to show that $\eta^*(u_i) \smile \psi^*(u_j)$ is the generator of $H^n(\mathbb{P}^n)$ where $u_k \in H^k(\mathbb{P}^k)$ are the generators.

Let U be the open subset of points in $\mathbb{P}^n$ whose i^{th} coordinate is not zero. Then $U = \mathbb{P}^n \setminus L$, where L denotes a copy of $\mathbb{P}^{n-1}$ itself. By cellular cohomology, it follows that $H^n(\mathbb{P}^n) \approx H^n(\mathbb{P}^n, L)$. On the other hand, there is a homeomorphism of U onto $\mathbb{R}^n$ under which $U \cap \eta(\mathbb{P}^i)$ and $U \cap \psi(\mathbb{P}^j)$ correspond to the coordinate subspaces $\mathbb{R}^i \times 0$ and $0 \times \mathbb{R}^j$ with their intersection point $p = [0, \ldots, 0, 1, 0, \ldots, 0]$ (with 1 occurring in the ith place) corresponding to the origin in $\mathbb{R}^n$. Since L is a deformation retract of $\mathbb{P}^n \setminus \{p\}$, we have

$$H^n(\mathbb{P}^n, L) \approx H^n(\mathbb{P}^n, \mathbb{P}^n \setminus \{p\}) \approx H^n(U, U \setminus \{p\}) \approx H^n(\mathbb{R}^n, \mathbb{R}^n \setminus \{0\}).$$

By naturality of the cup product, the problem is reduced to the problem of computing the cup product of the generators of $H^i(\mathbb{R}^i, \mathbb{R}^i \setminus 0)$ and $H^j(\mathbb{R}^j, \mathbb{R}^j \setminus 0)$ in $H^n(\mathbb{R}^n, \mathbb{R}^n \setminus \{0\})$. By excision again, this problem is the same as determining the cup product of generators of $H^i(\mathbb{D}^i, \partial \mathbb{D}^i)$ and $H^i(\mathbb{D}^j, \partial \mathbb{D}^j)$ in $H^n(\mathbb{D}^n, \partial \mathbb{D}^n)$. This is precisely what we have seen at the end of Example 7.4.1.

Note that it follows immediately that the cohomology algebra $H^*(\mathbb{P}^\infty; \mathbb{Z}_2)$ is isomorphic to the polynomial algebra $\mathbb{Z}_2[\alpha]$ without any truncation.

Remark 7.4.9 Similar arguments with complex coordinates and quaternion coordinates yield that

$$H^*(\mathbb{CP}^\infty; \mathbb{Z}) \approx \mathbb{Z}[\beta_2]; \qquad H^*(\mathbb{HP}^\infty; \mathbb{Z}) \approx \mathbb{Z}[\gamma_4]$$

where β_2 and γ_4 have degree 2 and 4, respectively. Of course for finite dimensional projective spaces, you have to truncate them at the appropriate dimension. For yet another method of computing $H^*(\mathbb{P}^\infty)$ see the exercises below.

As an application of the computation of $H^*(\mathbb{P}^n; \mathbb{Z}_2)$ we can now give a proof of the Borsuk–Ulam theorem (see Exercise 1.9.24) in all dimensions. (For a proof using differential topology see [Shastri, 2011].) We begin with:

Corollary 7.4.10 Let $n > m \geq 1$, $p : \mathbb{S}^m \to \mathbb{P}^m$ be the double cover. Then for any map $f : \mathbb{P}^n \to \mathbb{P}^m$ there exists $f' : \mathbb{P}^n \to \mathbb{S}^m$ such that $p \circ f' = f$.

Proof: By the lifting criterion, the conclusion follows if we show that $f_\# : \mathbb{Z}_2 = \pi_1(\mathbb{P}^n) \to \pi_1(\mathbb{P}^m)$ is the trivial map. If $m = 1$, $\pi_1(\mathbb{P}^m) = \mathbb{Z}$ and hence $f_\# = 0$. Now consider the case when $m > 1$ and hence $\pi_1(\mathbb{P}^m) = \mathbb{Z}_2$. So, if $f_\#$ is not the trivial map then it is an isomorphism. Passing onto homology, this implies that $f_* : H_1(\mathbb{P}^n) \to H_1(\mathbb{P}^m)$ is an isomorphism. This in turn implies that f^* is also an isomorphism:

$$f^* : H^1(\mathbb{P}^m; \mathbb{Z}_2) = \mathrm{Hom}(H_1(\mathbb{P}^m), \mathbb{Z}_2) \approx \mathrm{Hom}(H_1(\mathbb{P}^n); \mathbb{Z}_2) = H^1(\mathbb{P}^n; \mathbb{Z}_2).$$

If α denotes the generator of $H^1(\mathbb{P}^m; \mathbb{Z}_2)$ then $\alpha' := f^*(\alpha) \in H^1(\mathbb{P}^n; \mathbb{Z}_2)$ is also a generator. But this in turn implies that $(\alpha')^{m+1} = f^*(\alpha^{m+1}) = 0$, and hence $\alpha'^n = 0$. This contradicts (7.10). ♠

Using this result, proving the following two results is similar to that of Exercise 1.9.24 and left to the reader.

Corollary 7.4.11 For $n > m \geq 1$, there is no map $\phi : \mathbb{S}^n \to \mathbb{S}^m$ such that $\phi(-x) = -\phi(x)$ for all $x \in \mathbb{S}^n$.

Theorem 7.4.12 (Borsuk–Ulam theorem) *Given any map* $f : \mathbb{S}^n \to \mathbb{R}^n, n \geq 1$, *there exist* $x \in \mathbb{S}^n$ *such that* $f(x) = f(-x)$.

Exercise 7.4.13

(i) Show that any two coordinate inclusions $\mathbb{P}^m \hookrightarrow \mathbb{P}^n, m < n$, are homotopic (isotopic) to each other.

(ii) **Lusternik–Schnirelmann theorem** Suppose $\mathbb{S}^n$ has been written as a union of $n+1$ closed sets. Then show that at least one of these sets will contain a pair of antipodal points. (See Exercise 1.9.25.)

(iii) Let X be the quotient space of $\mathbb{S}^n \times \mathbb{S}^n$ modulo the relation $(x,p) \sim (p,x), x \in \mathbb{S}^n$ where $p = (1,0,\ldots,0)$ is the base point. Show that X has a CW-structure with one cell for dimensions $0, n, 2n$, respectively. Compute the cohomology algebra $H^*(X;\mathbb{Z})$.

(iv) **Another approach to $H^*(\mathbb{P}^\infty)$**
(a) Verify that the infinite dimensional sphere $\mathbb{S}^\infty$ has a CW-structure with two cells in each dimension and this CW-structure is invariant under the antipodal action $T(x) = -x$.
(b) Denote the two n-cells by g_n and Tg_n. Let W_* denote the CW-chain complex of the above CW-structure on $\mathbb{S}^\infty$. Verify that $W_n = \mathbb{Z}^2 = (\{g_n, Tg_n\})$, i.e., the free abelian group generated by the two generators. Also verify that the boundary operator is given by

$$\partial(g_n) = g_{n-1} + (-1)^n Tg_{n-1} \tag{7.11}$$

and extended linearly.
(c) Check that W_* is acyclic.
(d) Note that $T \circ \partial = \partial \circ T$ and $T^2 = Id$. Conclude that there is $\mathbb{Z}_2$ action on W_*. Also, check that the induced CW-structure on $\mathbb{P}^\infty = \mathbb{S}^\infty/T$ is nothing but the familiar one discussed in Example 2.2.11.(vi) and the associated CW-chain complex of $\mathbb{P}^\infty$ is nothing but W/T.
(e) Let e_n denote the image of g_n in $\mathbb{P}^\infty$. Check that $\partial(e_{2n}) = 2e_{2n-1}$ and $\partial(e_{2n-1}) = 0$.
(f) Conclude that $\tilde{H}_i(\mathbb{P}^\infty;\mathbb{Z}) = 0$ if i is even and $\mathbb{Z}_2$ if i is odd.
(g) Use UCT to conclude that $H_i(\mathbb{P}^\infty;\mathbb{Z}_2) \approx \mathbb{Z}_2$, and $H^i(\mathbb{P}^\infty;\mathbb{Z}_2) \approx \mathbb{Z}_2$ for all i.
(h) Define the diagonal map $r : W \to W \otimes W$ by

$$r(g_n) = \sum_{0 \leq j \leq n} (-1)^{j(n-j)} g_j \otimes T^j g_{n-j}; \quad r(Tg_n) = T(r(g_n)),$$

(where $T^j = T$ if j is odd and $= Id$ if j is even) and extend linearly. Verify that r is a chain map, where the boundary operator in $W \otimes W$ is the standard one, viz., $\partial(u \otimes v) = \partial u \otimes v + (-1)^{\deg u} u \otimes \partial v$.
(i) Note that the group $\mathbb{Z}_2$ acts on W via T. Verify that under the diagonal action of $\mathbb{Z}_2$ on $W \otimes W$, r is equivariant and hence induces a chain map $s : W/T \to W/T \otimes W/T$ (via $(W \otimes W)T$) given by

$$s(e_n) = \sum_{0 \leq j \leq n} (-1)^{j(n-j)} e_j \otimes e_{n-j}.$$

Check that s is a chain approximation to the diagonal map $d : \mathbb{P}^\infty \to \mathbb{P}^\infty \times \mathbb{P}^\infty$.
(j) Thus, you can define the cup product in $\mathbb{P}^\infty$ using s and conclude that $[u_n] \smile [u_m] = u_{n+m}$ where $u_k \in H^k(\mathbb{P}^\infty;\mathbb{Z}_2)$ is the generator, for each $k \geq 1$.
(k) Conclude that $H^*(\mathbb{P}^\infty;\mathbb{Z}_2)$ is the polynomial algebra $\mathbb{Z}_2[u_1]$ over a single generator of degree 1.

7.5   Cohomology Operations; Steenrod Squares

The advantage of cohomology over homology lies in the fact that it has more algebraic structure. In the previous section, we have seen just a little bit of this. We would like to take this a step further by introducing the notion of cohomology operations, discuss a little bit of one important example, viz., 'Steenrod squares'. Interested readers may then look up in other sources such as [Whitehead, 1978] or [Mosher–Tangora, 1968] for more. Cohomology operations can easily detect whether a particular homomorphism between cohomology modules of topological spaces is induced by a continuous map or not. As we shall see, this leads to several interesting applications.

Definition 7.5.1 Let p, q be integers and M, M' be R-modules. By a cohomology operation of type (p, q, M, M') we mean a natural transformation of the functor $H^p(-; M)$ to $H^q(-; M')$. A cohomology operation Θ is said to be additive if each $\Theta_{(X,A)}$ is a homomorphism of abelian groups.

Example 7.5.2
(1) Any coefficient homomorphism $\phi : M \to M'$ induces a cohomology operation of type (p, p, M, M') for all p, viz., the induced homomorphism $\phi_* : H^p(X, A; M) \to H^p(X, A; M')$. Note that the commutativity of the following diagram

$$
\begin{array}{ccc}
H^p(X, A; M) & \xrightarrow{\ \phi_*\ } & H^p(X, A; M') \\
{\scriptstyle f^*}\downarrow & & \downarrow{\scriptstyle f^*} \\
H^p(Y, B; M) & \xrightarrow{\ \phi_*\ } & H^p(Y, B; M')
\end{array}
$$

is what makes ϕ_* a natural transformation. This is an additive operation.
(2) Given a short exact sequence of R-modules

$$0 \to G' \to G \to G'' \to 0$$

for every p, there is the Bockstein homomorphism $\beta^* : H^p(X, A; G'') \to H^{p+1}(X, A; G')$, viz., the connecting homomorphism of the long cohomology exact sequence which is a cohomology operation of type $(p, p+1, G'', G')$ for every p. This is also additive.
(3) For each p, q, there is a cohomology operation Θ_p of type (q, pq, R, R) called the pth power operation, viz., $\Theta_p(x) = x^p$. In general these are not additive. The $(m+1)^{\text{th}}$ power operations of type $(1, m+1, \mathbb{Z}_2, \mathbb{Z}_2)$ were used in the proof of Corollary 7.4.10. In what follows we shall concentrate on one special power Θ_2 and some peculiar variants of it called 'reduced squares', because they decrease the degree as compared to the square operation Θ_2.

The Construction of Cup-i Products

For any topological pair (X, A), we now consider the singular chain complex $S_{\cdot}(X, A) \otimes W$, where W is the CW-chain complex of $\mathbb{S}^\infty$ introduced in Exercise 7.4.13.(iv).

Lemma 7.5.3 There is a natural chain map

$$\tau : S_{\cdot}(X, A) \otimes W \to S_{\cdot}(X, A) \otimes S_{\cdot}(X, A)$$

preserving the augmentation and unique up to a chain homotopy.

Proof: We want to prove that there is a chain map $\tau : S_.(X, A) \otimes W \to S_.(X, A) \otimes S_.(X, A)$ which preserves augmentation and is functorial in the pair (X, A). Naturally the best method is to apply the method of acyclic models. We consider the ring R to be the integral group ring of the group $\mathbb{Z}_2$; indeed R is isomorphic to the quotient of the polynomial ring $\mathbb{Z}[T]$ by the ideal $(1 - T^2)$. Note that the $\mathbb{Z}_2$-action on $\mathbb{S}^\infty$ via the antipodal map induces an action on W and makes it a free R-module. Then $S_.(X) \otimes_\mathbb{Z} W$ becomes a free chain complex of R-modules via the action of R on W and with models $\{\Delta^q\}$ and basis $\{\xi_q \otimes g_j\}$. On the other hand, we take an action of $\mathbb{Z}_2$ on $S_.(X) \otimes_\mathbb{Z} S_.(X)$ via $T(a \otimes b) = (-1)^{\deg a \deg b} b \otimes a$. Clearly, $S_.(X) \otimes_\mathbb{Z} S_.(X)$ is acyclic with models $\{\Delta^q\}$. It follows that we can apply Theorem 6.1.14. ♠

We now need to unravel τ.

Lemma 7.5.4 The degree 0 chain map $\tau : S_.(X) \otimes W \to S_.(X) \otimes S_.(X)$ corresponds to a sequence of morphisms $\tau_j : S_.(X) \to S_.(X) \otimes S_.(X)$ given by

$$\tau_j(c) = \tau(c \otimes g_j), j \geq 0. \tag{7.12}$$

(Just for notational completeness we put $\tau_j = 0$ for $j < 0$.) They satisfy the property

$$(\partial \tau_j - \tau_j \partial)(c) = (-1)^{\deg c}(1 + (-1)^j T)\tau_{j-1}(c) \tag{7.13}$$

because τ is a chain map and $\mathbb{Z}_2$-equivariant (use (7.11) and (6.4)).

Definition 7.5.5 For each integer $i \geq 0$, define a 'cup-i product'

$$S^p(X) \otimes S^q(X) \to S^{p+q-i}(X); \quad (u, v) \mapsto u \smallsmile_i v$$

by the formula

$$u \smallsmile_i v(c) = \langle (u \otimes v), \tau(c \otimes g_i) \rangle = (u \otimes v)(\tau_i(c)), \quad c \in S_{p+q-i}(X).$$

It is useful to know that the cup-i products are defined on integral chains. Using (7.13) we can express $\delta(u \smallsmile_i v)$ in terms of cup-i and cup-(i-1) products as in (7.14) below. This means keeping track of a lot of cumbersome $\pm$ signs. The reader may first of all ignore these signs and get the mod 2 version which suffices most of the time.

Lemma 7.5.6 For $u \in S^p(X), v \in S^q(X)$, the coboundary formula (7.13) can be rewritten as follows:

$$\delta(u \smallsmile_i v) = \delta u \smallsmile_i v + (-1)^p u \smallsmile_i \delta v - (-1)^{p+q-i} u \smallsmile_{i-1} v - (-1)^{pq+p+q} v \smallsmile_{i-1} u. \tag{7.14}$$

Proof: Let $c \in S_r(X)$ where $r = p + q - i$. We have

$$
\begin{aligned}
\delta(u \smallsmile_i v)(c) &= (u \smallsmile_i v)\partial(c) = (u \otimes v)(\tau_i(\partial c)) \\
&= (u \otimes v)[\partial \circ \tau_i - (-1)^r(1 + (-1)^i T)\tau_{i-1}](c) \\
&= \delta(u \otimes v)(\tau_i(c)) - (-1)^r[(u \otimes v) + (-1)^{pq+i}(v \otimes u)](\tau_{i-1}(c)) \\
&= (\delta u \otimes v)(\tau_i(c)) + (-1)^p(u \otimes \delta v)(\tau_i(c)) - (-1)^r[(u \smallsmile_{i-1} v) + (-1)^{pq+i}(v \smallsmile_{i-1})](c)) \\
&= [\delta u \smallsmile_i v + (-1)^p u \smallsmile_i \delta v - (-1)^{p+q-i} u \smallsmile_{i-1} v - (-1)^{pq+p+q} v \smallsmile_{i-1} u)](c).
\end{aligned}
$$

Note that we have used (7.13) at the second step above and Theorem 6.3.11 in the third step. ♠

Construction of Squaring Operations

Let now $\tau_i^* : \mathrm{Hom}\,(S_.(X) \otimes S_.(X); \mathbb{Z}_2) \to \mathrm{Hom}(S_.(X), \mathbb{Z}_2)$ be homomorphisms of degree $-i$ given by $(\tau_i^* f)(\sigma) = f(\tau_i \sigma)$, for $\sigma \in S_q(X), f \in \mathrm{Hom}(S_.(X) \otimes S_.(X); \mathbb{Z}_2)$. For a mod 2 q-cochain c^* on X, define

$$Sq^i c^* = \left\{ \begin{array}{ll} 0, & i > q, \\ \tau_{q-i}^*(c^* \otimes c^*), & i \le q. \end{array} \right.$$

(With our convention that $\tau_j = 0$ for $j < 0$, we simply have $Sq^i = (\Delta \circ \tau_{q-i})^*$, where Δ is the diagonal map.)

Lemma 7.5.7 The operators $\{Sq^i\}$ pass on to the cohomology level and define additive homomorphisms (which we shall denote by the same symbols) $Sq^i : H^q(X, A) \to H^{q+i}(X, A)$ defined by $Sq^i[u^*] = [Sq^i u^*]$.

Proof: (1) If u^* is a cocycle so is $Sq^i u^*$.

$$\begin{array}{lll} \delta(Sq^i u^*)(\sigma) & = & \tau_{q-i}^*(u^* \otimes u^*)(\partial \sigma) = (u^* \otimes u^*)(\tau_{q-i}\partial \sigma) \\ & = & (u^* \otimes u^*)(\partial \tau_{q-i} + (1 + T)\tau_{q-i-1})(\sigma) \\ & = & (u^* \otimes u^*)(\partial \tau_{q-i}(\sigma)) + 0 \\ & = & \delta(u^* \otimes u^*)(\tau_{q-i}\sigma). \end{array}$$

The second step is valid because of (7.13) and the third step is valid because $(u^* \otimes u^*)T(\lambda) = u^* \otimes u^*(\lambda)$ for all λ. The conclusion follows.
(2) If u^* is a boundary, so is $Sq^i u^*$.
Let $u^* = \delta v^*$. Then

$$\begin{array}{lll} (Sq^i u^*)(\sigma) & = & \tau_{q-i}^*(\delta v^* \otimes \delta v^*)(\sigma) = \delta(v^* \otimes \delta v^*)(\tau_{q-i}\sigma) \\ & = & (v^* \otimes \delta v^*)(\tau_{q-i}\partial + (1 + T)(\tau_{q-i-1})(\sigma) \\ & = & \tau_{q-i}^*(v^* \otimes u^*)\partial(\sigma) + \delta(v^* \otimes v^*)\tau_{q-i-1}(\sigma) \\ & = & \tau_{q-i}^*(v^* \otimes u^*)\partial(\sigma) + (v^* \otimes v^*)(\tau_{q-i-1}\partial\sigma) + (v^* \otimes v^*)(1 + t)(\tau_{q-i-2}\sigma) \\ & = & \tau_{q-i}^*(v^* \otimes u^*)\partial(\sigma) + (v^* \otimes v^*)(\tau_{q-i-1}\partial\sigma) + 0 \\ & = & \delta[\tau_{q-i}^*(v^* \otimes u^*) + \tau_{q-i-1}^*(v^* \otimes v^*)] \end{array}$$

(3) For any two cocycles u_1^*, u_2^*, we have,

$$Sq^i(u_1^* + u_2^*) = Sq^i u_1^* + Sq^i u_2^* + \delta\tau_{q-i-1}^*(u_1^* \otimes u_2^*).$$

We have,

$$\begin{array}{lll} Sq^i(u_1^* + u_2^*)(\sigma) & = & [(u_1^* + u_2^*) \otimes (u_1^* + u_2^*)]\tau_{q-i}(\sigma) \\ & = & (u_1^* \otimes u_1^* + u_2^* \otimes u_2^* + (u_1^* \otimes u_2^*)(1 + T)(\tau_{q-i}\sigma) \\ & = & (Sq^i u_1^* + Sq^i u_2^*)(\sigma) + (u_1^* \otimes u_2^*)(\tau_{q-i-1}\partial + \partial\tau_{q-i-1})(\sigma) \\ & = & (Sq^i u_1^* + Sq^i u_2^*)(\sigma) + \delta\tau_{q-i-1}^*(u_1^* \otimes u_2^*)(\sigma) + 0, \end{array}$$

the last term being zero because $\delta(u_1^* \otimes u_2^*) = 0$. ♠

Remark 7.5.8 Note that the homomorphisms Sq^i are independent of the choice of τ because any two such chain transformations are chain homotopic. These operators are called the *Steenrod squares*. They are completely characterized by the properties listed below. We shall prove that the properties are satisfied but will not include the proof of uniqueness. Interested readers may consult [Whitehead, 1978] or [Mosher–Tangora, 1968].

Theorem 7.5.9 *The Steenrod square operators $Sq^i : H^q(X, A; \mathbb{Z}_2) \to H^{q+i}(X, A; \mathbb{Z}_2)$ satisfy the following properties:*
(a) $Sq^0 = Id$.
(b) *If* $\deg u = i$ *then* $Sq^i u = u \smile u$.
(c) *If* $\deg u < i$ *then* $Sq^i u = 0$.
(d) **Cartan product formula** *For any excisive couple $\{X \times B, A \times Y\}$ in $X \times Y$, we have*

$$Sq^q(u \times v) = \sum_{i+j=q} Sq^i u \times Sq^j v \qquad (7.15)$$

in $H^*((X, A) \times (Y, B); \mathbb{Z}_2)$.

Proof: The oriented simplicial chain complex $C_.(\Delta_q)$ can be thought of as a subcomplex of the singular chain complex $S_.(X)$. (Working with $\mathbb{Z}_2$ coefficients, there is a unique orientation on each simplex and hence there is no ambiguity whatsoever.) If $F^j : \Delta^n \to \Delta^m$ is an n-face map then the induced map $F^j_* : S_.(\Delta^n) \to S_.(\Delta^m)$ carries the subchain complex $C_.(\Delta^n)$ inside $C_.(\Delta^m)$. Since $C_.(\Delta_q)$ is acyclic for each q, it follows that we can choose $\tau'_j s$ in such a way that $\tau_j(\xi_q) \in C_.(\Delta_q) \otimes C_.(\Delta_q) \subset S_.(\Delta_q) \otimes S_.(\Delta_q)$. Since $[C_.(\Delta^q) \times C_.(\Delta^q)]_s = 0$ if $s > 2q$, we conclude that $\tau_j(\xi_q) = 0$ for $j > q$. Once again by canonical property this implies that $\tau_j(\sigma) = 0$ for any $\sigma \in S_q(X)$ with $j > q$.

Next we shall prove that $\tau_q(\xi_q) = \xi_q \otimes \xi_q$. We induct on q. For $q = 0$, $\tau_0(\xi_0)$ should have non zero augmentation and there is just one element, viz., $\xi_0 \otimes \xi_0 \in C_.(\Delta^0) \otimes C_.(\Delta^0)$, which has non zero augmentation. Therefore $\tau_0(\xi_0) = \xi_0 \otimes \xi_0$. Assume $q > 0$ and $\tau_{q-1}(\xi_{q-1}) = \xi_{q-1} \otimes \xi_{q-1}$. Once again there is only one non zero element in $[C_.(\Delta^q) \otimes C_.(\Delta^q)]_{2q}$ and hence either $\tau_q(\xi_q) = 0$ or $\tau_q(\xi_q) = \xi_q \otimes \xi_q$.

Suppose $\tau_q(\xi_q) = 0$. Since $\tau_q(\partial \xi_q) = 0$, (7.13) yields $(1 + T)\tau_{q-1}(\xi_q) = 0$. Putting $\xi^{(i)}_{q-1} = F^i \circ \xi_{q-1}$, and $\tau_{q-1}(\xi_q) = \sum_i (a_i \xi^{(i)}_{q-1} \otimes \xi_q + b_i \xi_q \otimes \xi^{(i)}_{q-1})$, $(1+T)\tau_{q-1}(\xi_q) = 0$ implies that $a_i = b_i$. We now have

$$(1 + T)\tau_{q-2}(\xi_q) = (\partial \tau_{q-1} + \tau_{q-1}\partial)(\xi_q).$$

Let us compute the coefficient of $\xi_{q-1} \otimes \xi_{q-1}$ on either side of this equation. This is clearly 0 on the LHS. On the RHS the first term contributes $a_0 + b_0 = 0$. The second term is equal to

$$\begin{aligned} \tau_{q-1}(\partial \xi_q) &= \tau_{q-1}\left(\sum_i F^i \circ \xi_{q-1}\right) = \sum_i F^i \tau_{q-1}(\xi_{q-1}) \\ &= \sum_i F^i(\xi_{q-1} \otimes \xi_{q-1}) = \sum_i \xi^{(i)}_{q-1} \otimes \xi^{(i)}_{q-1} \end{aligned}$$

and therefore contributes 1. This absurdity proves that the first choice is wrong and we have the second choice $\tau_q(\xi_q) = \xi_q \otimes \xi_q$.

Therefore, it follows that for all $\sigma \in S_q(X)$, $\tau_q(\sigma) = \sigma \otimes \sigma$.

We can now dispose of (a), (b) and (c). Property (c) is obvious from the definition. For (a), we have

$$(Sq^0 u^*)(\sigma) = (u^* \otimes u^*)\tau_q(\sigma) = (u^* \otimes u^*)(\sigma \otimes \sigma) = (u^*(\sigma))^2 = u^*(\sigma)$$

(the last equality being valid, since we are working in $\mathbb{Z}_2$) which proves that $Sq^0 = Id$. Since τ_0 is a diagonal approximation, it follows that $\tau_0^*(u^* \otimes u^*) = [u^*] \smile [u^*]$, for any cocycle u^*. This implies $Sq^q u = u \smile u$ for any $u \in H^q(X)$. This proves (c).

It remains to prove the Cartan formula (7.15). First of all, since $u \smile v = d^*(u \times v)$ where $d : X \to X \times X$ is the diagonal map, the Cartan formula is easily seen to be equivalent to the following: For $u, v \in H^*(X, A)$, we have

$$Sq^k(u \smile v) = \sum_{i+j=k} Sq^i u \smile Sq^j v. \qquad (7.16)$$

Let us denote τ on a space X by τ^X. Consider the category of products $X \times Y$ of topological spaces and look at the chain transformations $\tau^{X \times Y}$ which leads to the definition of $Sq^q(u \times v)$ on the LHS of (7.16). We claim that the RHS is also given by a chain transformation satisfying (7.13) on this category. Of course, we should also check that both these transformations are augmentation preserving. Then by the method of acyclic models, it follows that these two transformations are chain homotopic and hence on the cohomology level, they are equal thereby proving (7.16).

So, let $\bar{T} : (S_.(X) \otimes S_.(X)) \otimes (S_.(Y) \otimes S_.(Y)) \to (S_.(X) \otimes S_.(Y)) \otimes (S_.(X) \otimes S_.(Y))$ be the map which interchanges the second and third factors. The sequence of natural transformations $\{\hat{\tau}_k := \bar{T} \sum_{i+j=k} t^k \tau_i^X \otimes \tau_j^Y\}$ is such a sequence satisfying (7.13) and such that $\hat{\tau}$ is a chain transformation which preserves augmentation. Let $u_1^* \in S^p(X), u_2^* \in S^q(Y)$ and σ_1 be a singular r-simplex in X and σ_2 be a singular s-simplex in Y. We need to consider the case $p \leq r \leq 2p, \ q \leq s \leq 2q$ with $r + s = p + q + k$. Then

$$\hat{\tau}^*_{p+q-k}(u_1^* \otimes u_2^*) \otimes (u_1^* \otimes u_2^*)(\sigma_1 \otimes \sigma_2)$$
$$= [(u_1^* \otimes u_2^*) \otimes (u_1^* \otimes u_2^*)]\hat{\tau}_{p+q-k}(\sigma_1 \otimes \sigma_2)$$
$$= [(u_1^* \otimes u_1^*) \otimes (u_2^* \otimes u_2^*)] \left(\sum_{i+j=p+q-k} t^{p+q-k} \tau_i^X \sigma_1 \otimes \tau_j^Y \sigma_2 \right)$$
$$= [u_1^* \otimes u_1^*)(\tau_{2p-r}\sigma_1)][u_2^* \otimes u_2^*)(\tau_{2q-s}\sigma_2)] = (Sq^{r-p}u_1^* \otimes Sq^{s-q}u_2^*)(\sigma_1 \otimes \sigma_2).$$

Since this is true for all σ_1, σ_2 as above, putting $r - p = i, q = j$ so that $i + j = k$, we obtain

$$\hat{\tau}^*_{p+q-k}(u_1^* \otimes u_2^*) \otimes (u_1^* \otimes u_2^*) = \sum_{i+j=k} Sq^i u_1^* \otimes Sq^j u_2^*.$$

This completes the proof of the Cartan formula and hence that of the theorem. ♠

Remark 7.5.10 The above four properties may be raised to the status of being called axioms because they completely characterise these cohomology operations, i.e., there is a unique class of such cohomology operations satisfying these four properties. (We shall omit the proof of this which is rather complicated and we shall not use the uniqueness.) Of course, this also means that the following important properties can also be derived from them.

Theorem 7.5.11 *The Steenrod squares commute with coboundary homomorphisms, i.e., the following diagram is commutative:*

$$\begin{array}{ccc} H^n(A; \mathbb{Z}_2) & \xrightarrow{\ \delta\ } & H^{n+1}(X, A; \mathbb{Z}_2) \\ {\scriptstyle Sq^i} \downarrow & & \downarrow {\scriptstyle Sq^i} \\ H^{n+i}(A; \mathbb{Z}_2) & \xrightarrow{\ \delta\ } & H^{n+i+1}(X, A; \mathbb{Z}_2). \end{array}$$

Proof: Consider first the special case, when $A = B \times \dot{\mathbb{I}}$ and $X = B \times \mathbb{I}$. Then every element of $H^*(A; \mathbb{Z}_2)$ is of the form $x \times y$ for $x \in H^*(B; \mathbb{Z}_2)$ and $y \in H^0(\dot{\mathbb{I}}; \mathbb{Z}_2)$. We have $\delta(x \times y) = x \times \delta(y)$. Therefore

$$\begin{aligned} Sq^i(\delta(x \times y)) &= Sq^i(x \times \delta(y)) \\ &= Sq^i(x) \times Sq^0(\delta(y)) \quad \text{by Cartan formula (7.15)} \\ &= Sq^i \times \delta(y) \quad \text{by property (a)} \\ &= \delta(Sq^i(x) \times y) \\ &= \delta(Sq^i(x \times y)) \quad \text{again by (a) and (d).} \end{aligned}$$

The general case can be reduced to this special case as follows: We keep appealing to the fact that both δ and Sq^i are natural transformations. Therefore, we can replace the pair (X, A) by the homotopy equivalent pair $(X \times 0 \cup A \times \mathbb{I}, A \times \mathbb{I})$ first and then by $(X \cup A \times \mathbb{I}, A \times 1)$. Put $Y = X \times 0 \cup A \times \mathbb{I}, Y_1 = A \times \{1\}, Y_2 = X \times 0 \cup A \times [0, 1/2]$. We then have a commutative diagram

$$
\begin{array}{ccc}
H^n(Y_1 \cup Y_2; \mathbb{Z}_2) & \longrightarrow & H^n(Y_1, \mathbb{Z}_2) \\
\downarrow{\delta} & & \downarrow{\delta} \\
H^{n+1}(Y, Y_1 \cup Y_2; \mathbb{Z}_2) & \longrightarrow & H^{n+1}(Y; Y_1; \mathbb{Z}_2)
\end{array}
$$

in which the top arrow is a surjection. Therefore, we can replace (Y, Y_1) by $(Y; Y_1 \cup Y_2)$. By excision, this pair can be replaced by $(A \times [1/2, 1], A \times \{1/2, 1\})$ which is homotopy equivalent to $(A \times \mathbb{I}, A \times \dot{\mathbb{I}})$. ♠

Theorem 7.5.12 Sq^i *commutes with unreduced suspension.*

Proof: This is a direct consequence of the above theorem once you recall that the suspension isomorphism is defined as the composite of

$$
\tilde{H}^n(X) \xrightarrow{\delta} H^{n+1}(CX, X) \xleftarrow{\approx} H^{n+1}(SX).
$$

Remark 7.5.13 Cohomology operations which commute with suspension are called *stable*. They form an important class of cohomology operations. The above simple theorem tells that Steenrod squares belong there.

Theorem 7.5.14 (Adem's relations) *For all integers, i, j, such that $0 < i < 2j$, we have*

$$
R(i, j) := Sq^i Sq^j - \sum_k \binom{j - k - 1}{i - 2k} Sq^{i+j-k} Sq^k \equiv 0 \quad mod \ 2. \tag{7.17}
$$

Here, we follow the convention that $\binom{p}{k} = 0$ if $p < k$. Of course, $Sq^p = 0$ for $p < 0$ also. Hence we need not write the range of the summation explicitly. It is for $0 \leq k \leq \lfloor i/2 \rfloor$.

These relations form a complete set of relations on the Steenrod squares and play a crucial role in employing Steenrod squares in various applications.

We shall neither be able to present a complete proof of these relations nor the proof of the fact that there are no other linear relations. However, at the end of the section, we shall prove that these relations are satisfied over a finite product of infinite projective spaces which is a tiny but important step toward the full proof. Before that, let us study some other interesting properties of Steenrod square, as well as some applications.

Theorem 7.5.15 Sq^1 *is equal to the Bockstein homomorphism corresponding to the coefficient sequence $0 \to \mathbb{Z}_2 \to \mathbb{Z}_4 \to \mathbb{Z}_2 \to 0$.*

Proof: Let $\beta : H^p(X, A; \mathbb{Z}_2) \to H^{p+1}(X, A; \mathbb{Z}_2)$ denote the Bockstein homomorphism. The claim of the theorem is a special case of the following lemma and property (a):

Lemma 7.5.16 $\beta Sq^j = 0$, if j is odd; $\beta Sq^j = Sq^{j+1}$ if j is even.

Start with an integral cocycle $u \in S^p(X, A; \mathbb{Z})$ such that, modulo 2, it represents a certain element $x \in H^p(X, A; \mathbb{Z}_2)$. Then $Sq^j x$ is represented by $(u \smile_{p-j} u)$. Now $\delta(u) = 2a$ for some integral cochain $a \in S^{p+1}(X, A)$. Writing i for $p - j$, by the coboundary formula (7.14), we get

$$
\begin{aligned}
\delta(u \smile_i u) &= 2a \smile_i u + (-1)^p u \smile_i 2a - (-1)^i u \smile_{i-1} u - (-1)^p u \smile_{i-1} u \\
&= 2(a \smile_i u + u \smile_i a) - (-1)^p [u \smile_{i-1} u + (-1)^j u \smile_{i-1} u].
\end{aligned}
$$

Therefore, $\beta Sq^j u$ is represented by the integral cocycle which is half the RHS above. Going mod 2, we get $\beta Sq^j(u)$ is represented by

$$a \smile_i u + u \smile_i a + \eta(u \smile_{i-1} u)$$

where the coefficient η is 1 or 0 according to whether j is even or odd, respectively. But the sum of the first two terms is equal to $\delta(u \smile_{i+1} a) \mod 2$. Therefore $\beta Sq^j(u) = \eta(u \smile_{i-1} u)$ which proves the lemma. ♠

Example 7.5.17 Clearly $Sq^i : H^*(\mathbb{S}^n) \to H^*(\mathbb{S}^n)$ are all zero for $i > 0$. From this we can deduce the same for any bouquet $\vee_k \mathbb{S}^{n_k}$ of spheres also, using the retractions $\vee_k \mathbb{S}^{n_k} \to \mathbb{S}^{n_k}$.

Example 7.5.18 Here is an example of two simply connected CW-complexes having iso-morphic cohomology algebra over integers and yet are of different homotopy type. Let $X = \mathbb{S}^3 \vee \mathbb{S}^5$ and Y be the unreduced suspension of $\mathbb{CP}^2$. That these are simply connected spaces with isomorphic cohomology modules is verified easily. The cup product vanishes in both the cohomology algebras. As seen before, $Sq^i, i > 0$ vanish on X. Let $Z = \mathbb{CP}^2 \times \mathbb{I}$. Then Y is the quotient of Z in which $\mathbb{CP}^2 \times \{0\}$ and $\mathbb{CP}^2 \times \{1\}$ are identified to two distinct points y_0, y_1. We know that if $u \in H^2(\mathbb{CP}^2; \mathbb{Z}) \approx \mathbb{Z}$ is the generator then $u \smile u \in H^4(\mathbb{CP}^2; \mathbb{Z}) \approx \mathbb{Z}$ is the generator. Passing onto $\mathbb{Z}_2$ coefficients, we have $Sq^2 u = u \smile u$ is the generator. If $v \in H^1(\mathbb{I}, \dot{\mathbb{I}})$ is the non zero element, by Cartan formula, we have

$$Sq^2(u \times v) = Sq^2 u \times v.$$

In particular, $Sq^2 : H^3(Z, \partial Z) \to H^5(Z, \partial Z)$ is non trivial where $\partial Z = \mathbb{CP}^2 \times \dot{\mathbb{I}}$. On the other hand, the quotient map $q : (Z, \partial Z) \to (Y, \{y_0, y_1\})$ being a relative homeomorphism induces isomorphism in the relative cohomology modules (see Exercise 4.2.27.(viii)) and hence by naturality we conclude that $Sq^2 : H^3(Y) \to H^5(Y)$ is non trivial. The space Y also serves to illustrate the fact that even when the cup products vanish, the Steenrod squares can be non trivial.

Remark 7.5.19
(a) Note that Sq^i are additive group homomorphisms, on cohomology groups in each di-mension. Indeed, on the entire cohomology ring $H^*(X; \mathbb{Z}_2)) = \oplus_{q \geq 0} H^q(X; \mathbb{Z}_2))$, the total Steenrod square

$$Sq : H^*(X; \mathbb{Z}_2) \to H^*(X; \mathbb{Z}_2); \quad Sq = \sum_i Sq^i$$

makes sense because of property (c), and as an easy consequence of Cartan formula (7.16), Sq is a ring homomorphism of the cohomology ring, i.e.,

$$Sq(a \smile b) = Sq(a) \smile Sq(b).$$

(b) In particular, for any $u \in H^1(X; \mathbb{Z}_2)$, we have $Sq(u) = u + u^2$. Therefore $Sq(u^j) = u^j \sum_k \binom{j}{k} u^k$. Upon comparing the corresponding degree terms, we obtain

$$Sq^i(u^j) = \binom{j}{i} u^{i+j}, \quad u \in H^1(X; \mathbb{Z}_2). \tag{7.18}$$

(c) Two cohomology operations of the same type can be added to get another of the same type. Given $\theta_1 \in Op(p, q; G, G'), \theta_2 \in Op(q, r; G', G'')$, we can compose them to get $\theta_1 \circ \theta_2 \in Op(p, r; G, G'')$. Thus, in the algebra of cohomology operations there is a subalgebra generated by Steenrod squares called the *mod 2 Steenrod algebra*.

(d) As pointed out earlier, Adem's relations are important properties of Steenrod squares. Just to understand what is going on, you may write down Adem's relations for small values of i and j. For instance,

$$Sq^1 Sq^1 = 0; \ Sq^1 Sq^2 = Sq^3; \ Sq^2 Sq^2 = Sq^3 Sq^1, \ Sq^1 Sq^3 = 0, \ Sq^2 Sq^4 = Sq^6 + Sq^5 Sq^1,$$

etc. This leads us to make the following definition.

Definition 7.5.20 Sq^i is said to be *decomposable* if it belongs to the left ideal generated by powers $Sq^j, 1 \leq j < i$. We say Sq^i is *indecomposable* if it is not decomposable.

Note that to say that sq^i is decompasable is the same as saying

$$Sq^i = \sum_{j<i} a_j Sq^j,$$

where each a_j is a product of some squaring operations, whose total order is equal to $i - j$.

Example 7.5.21 Clearly, Sq^1 is indecomposable. Sq^2 is indecomposable because $Sq^1 Sq^1 = 0$ as seen above. On the other hand, Sq^3 is decomposable. Similarly, from what we have seen above, it follows that Sq^4 is indecomposable. One can likewise check that Sq^5, Sq^6 are decomposable. Indeed there is a strong pattern!

Proposition 7.5.22 Sq^i is indecomposable iff i is a power of 2.

Proof: Let $i = 2^n$. To show that Sq^i are indecomposable, we exploit the mod 2 cohomology algebra $H^*(\mathbb{P}^\infty; \mathbb{Z}_2)$. Let $\alpha \in H^1(\mathbb{P}^\infty; \mathbb{Z}_2)$ be the generator. Then

$$Sq(\alpha^i) = (Sq^0 \alpha + Sq^1 \alpha)^i = (\alpha + \alpha^2)^i = \alpha^i + \alpha^{2i}$$

the last equality follows because we are working mod 2. This means that

$$Sq^j \alpha^i = \begin{cases} \alpha^i, & j = 0; \\ \alpha^{2i}, & j = i; \\ 0, & \text{otherwise.} \end{cases}$$

Now if $Sq^i = \sum_{0<j<i} a_j Sq^j$, then it would follow that $\alpha^{2i} = Sq^i(\alpha^i) = 0$ which is absurd.

To prove the converse, let $i = a + b$ where $b = 2^n$ and $0 < a < 2^n$. Using Adem's relations, we get

$$Sq^a Sq^b = \binom{b-1}{a} Sq^{a+b} + \sum_{c>0} \binom{b-c-1}{a-2c} Sq^{a+b-c} Sq^c.$$

Since $b = 2^n$, $\binom{b-1}{a} \equiv 1 \pmod 2$. Therefore, Sq^i is decomposable. ♠

Hopf Invariant

We can now give an application of this to the *Hopf invariant one* problem. For that we need to introduce the Hopf invariant. We presume that the reader has gone through Exercise 1.9.37 by now and knows what higher homotopy groups are. Otherwise, she may consult section 10.2 first and come back here.

Definition 7.5.23 Consider a map $f : \mathbb{S}^{2n-1} \to \mathbb{S}^n$ and the mapping cone $C_f = \mathbb{S}^n \cup_f e^{2n}$, which can be viewed as a CW-complex obtained by attaching a $2n$-cell to $\mathbb{S}^n$ via the map f. Fixing orientations on spheres involved, we obtain generators u, v of $H^n(C_f; \mathbb{Z})$ and $H^{2n}(C_f; \mathbb{Z})$. It follows that $u^2 = H(f)v$ where $H(f)$ is an integer. This integer or rather its modulus $|H(f)|$ is called the 'Hopf invariant' of f.

Remark 7.5.24

(i) Note that the homotopy type C_f is completely determined by the homotopy class of f. Therefore, once we fix orientations on spheres, we obtain a function $H : \pi_{2n-1}(\mathbb{S}^n) \to \mathbb{Z}$. This map may be called the *Hopf map*, which has some nice properties. The idea behind this goes back to the classical study by Hopf of the so-called Hopf fibration $\mathbb{S}^3 \to \mathbb{S}^2$ and his methods were more geometrical. The 'Hopf invariant one' problem is nothing but to determine for what values of n, we have $[f] \in \pi_{2n-1}(\mathbb{S}^n)$ such that $H(f) = \pm 1$.

(ii) Suppose n is odd. Then $H(f) = 0$. This follows easily because of the anti-commutativity of the cup product.

(iii) Recall that $\partial(\mathbb{D}^n \times \mathbb{D}^m) = \mathbb{S}^{n-1} \times \mathbb{D}^m \cup \mathbb{D}^n \times \mathbb{S}^{m-1}$. Given $\alpha : (\mathbb{D}^n, \mathbb{S}^{n-1}) \to (X, x_0)$ and $\beta : (\mathbb{D}^m, \mathbb{S}^{m-1}) \to (X, x_0)$ representing elements $[\alpha] \in \pi_n(X, x_0)$ and $[\beta] \in \pi_m(X, x_0)$, respectively, we define $[\alpha, \beta] \in \pi_{n+m-1}(X, x_0)$ to be the element represented by the map $\gamma : \partial(\mathbb{D}^n \times \mathbb{D}^m) \to X$ defined as follows:

$$\gamma(x, y) = \begin{cases} \alpha(x), & y \in \mathbb{S}^{m-1}; \\ \beta(y), & x \in \mathbb{S}^{n-1}. \end{cases}$$

The element $[\alpha, \beta]$ is called the *Whitehead product* of α with β. It crops up in several places especially when we study product spaces. With the CW-structure for the spheres with one 0-cell and one n-cell, if we take the product CW-structure on $\mathbb{S}^n \times \mathbb{S}^m$, then the attaching map $\mathbb{S}^{m+n-1} \to \mathbb{S}^n \vee \mathbb{S}^m$ represents the Whitehead product $[\iota_n, \iota_m] \in \pi_{n+m-1}(\mathbb{S}^n \vee \mathbb{S}^m)$, where ι_p denote the element represented by the inclusion maps, $p = n, m$.

(iv) Now, let n be an even integer and $[f] = [\iota_n, \iota_n] \in \pi_{2n-1}(\mathbb{S}^n)$ be the Whitehead product. Then there is a map $g : \mathbb{S}^n \times \mathbb{S}^n \to C_f$ which sends each of the n-cells onto the n-cell in C_f homeomorphically and sends the open $2n$ cell onto the open $2n$-cell of C_f homeomorphically. After fixing orientations, let ξ_n and η_{2n} denote the generators of $H^n(C_f; \mathbb{Z})$ and $H^{2n}(C_f; \mathbb{Z})$, and x, y denote the generators of $H^n(\mathbb{S}^n \times \mathbb{S}^n; \mathbb{Z})$. Then $f^*(\xi_n) = x + y$ and $f^*(\eta_{2n}) = x \times y$, the generator of $H^{2n}(\mathbb{S}^n \times \mathbb{S}^n; \mathbb{Z})$. Therefore,

$$f^*(\xi_n \smile \xi_n) = f^*(\xi_n) \smile f^*(\xi_n) = (x + y) \smile (x + y) = 2x \times y = 2f^*(\eta_{2n}).$$

This proves that $\xi_n \smile \xi_n = 2\eta_n$ and hence $|H([f])| = 2$.

(v) For $n = 2, 4$ and 8 one can demonstrate that there exist elements of Hopf invariant one, in a very surprising way, viz., by considering the Hopf fibrations

$$h_2 : \mathbb{S}^3 \to \mathbb{S}^2 = \mathbb{C}\mathbb{P}^1, \quad h_4 : \mathbb{S}^7 \to \mathbb{S}^4 = \mathbb{H}\mathbb{P}^1, \quad h_8 : \mathbb{S}^{15} \to \mathbb{S}^8 = \mathbb{C}\mathbb{A}\mathbb{P}^1,$$

corresponding to the 1-dimensional projective spaces over $\mathbb{C}, \mathbb{H}$ and the Cayley numbers, respectively. We have seen that $\mathbb{C}\mathbb{P}^2 = \mathbb{S}^2 \cup_{h_2} e^4$ and the cup product of the generator of $H^2(\mathbb{C}\mathbb{P}^2; \mathbb{Z})$ with itself is the generator of $H^4(\mathbb{C}\mathbb{P}^2; \mathbb{Z})$. Therefore $H(h_2) = 1$. Similar arguments prove that $H(h_4) = 1$ and $H(h_8) = 1$. The above three fibrations are peculiar in the sense that they were the by-products of certain 'multiplicative structures' on $\mathbb{R}^n$. (See the exercises below.) Therefore one is led to the question whether we have some such algebraic structures on other Euclidean spaces, which was answered by Hurewicz and other authors in the negative. That makes the task of constructing elements of Hopf invariant one a bit difficult of course. Indeed, here is how the Steenrod squares help us to narrow down our search even further.

Proposition 7.5.25 If there is an element in $\pi_{2n-1}(\mathbb{S}^n)$ of Hopf invariant odd then n is a power of 2.

Proof: Consider the space C_f. We know that $u^2 = H(f)v = v$. We also know that $Sq^n(u) = u^2 \neq 0 \mod 2$. Since C_f has trivial cohomology in dimension $0 < i < n$, it follows that Sq^n is not decomposable. Now we appeal to Proposition 7.5.22. ♠

Remark 7.5.26 Indeed, under the hypothesis of the above proposition, n is necessarily equal to 2, 4, 8. This was proved by Frank Adams in a landmark paper [Adams, 1960] using topological K-theory. This is beyond the scope of the present exposition.

In the rest of this section, we shall first reduce the proof of Adem's relations $R(i, j)y = 0$ to the case when $i + j \leq n$, where $y \in H^n(X; \mathbb{Z}_2)$. We shall also verify the relations fully on the products of infinite real projective space $\mathbb{P}^{\infty, n} = \mathbb{P}^\infty \times \cdots \times \mathbb{P}^\infty$. For one more step toward a complete proof, see Section 10.7.

Lemma 7.5.27 Let $R = R(i, j)$ be an Adem's relation. If $Ry = 0$ for every class y of dimension p, then $Rz = 0$ for every class z of dimension $p - 1$.

Proof: Let $u \in H^1(\mathbb{S}^1; \mathbb{Z}_2)$ be the generator. Then we know that $Sq^i u = 0$ for $i > 0$. Therefore, by Cartan's formula, $0 = R(u \times z) = u \times Rz$ in $H^*(\mathbb{S}^1 \times X; \mathbb{Z}_2)$ which implies $Rz = 0$. ♠

Corollary 7.5.28 If $R(i, j)y = 0$ for every class y of degree $> i + j$, then $R(i, j) = 0$.

Lemma 7.5.29 Let y be a cohomology class such that $Ry = 0$ for every Adem relation R. Then

$$Sq^{i-1}Sq^j y + Sq^i Sq^{j-1} y = \sum_k s(k) \left[Sq^{i+j-k-1} Sq^k y + \sum_k Sq^{i+j-k} Sq^{k-1} y \right].$$

Proof: Case 1: Suppose $i = 2j - 2$. Then $i - 2k = 2(j - k - 1)$ and therefore $s(k) = \binom{p}{2p} = 0$ unless $p = j - k - 1 = 0$, i.e., $k = j - 1$. Therefore

$$RHS = Sq^i Sq^{j-1} y + Sq^{i+1} Sq^{j-2} y.$$

Because of $R(i-1, j)$ we have the first term on the LHS equal to $\sum_k \binom{j-k-1}{i-2k-1} Sq^{i+j-k-1} Sq^k y$. Since $\binom{j-k-1}{a-2k-1} = \binom{p}{2p-1} = 0$ unless $p = 1$ i.e., unless $k = j - 2$. Therefore

$$LHS = Sq^{i+1} Sq^{j-2} y + Sq^i Sq^{j-1} y$$

which completes this case.

Case 2: $i = 2j - 1$. Similar argument shows that the two terms on the LHS are equal and all terms on the RHS vanish.

Case 3: $i < 2j - 2$. We then have $R(i, j - 1)$ which gives

$$Sq^i Sq^{j-1} y = \sum_k \binom{j-k-2}{i-2k} Sq^{i+j-k-1} Sq^k y$$

Similarly, by replacing $k - 1$ by k', the second term on the RHS can be rewritten as,

$$\sum_{k'} \binom{j-k'-2}{i-2k'-2} Sq^{i+j-k'-1} Sq^{k'} y.$$

Therefore we are reduced to proving

$$\sum_k (s_1(k) + s_2(k)) Sq^{i+j-k-1} Sq^k y = \sum_k (s_3(k) + s_4(k)) Sq^{i+j-k-1} Sq^k$$

which will follow if we prove

$$s_1(k) + s_2(k) \equiv s_3(k) + s_4(k), \quad \text{modulo } 2 \tag{7.19}$$

where

$$s_1(k) = \binom{i+j-2}{i-2c}, \ s_2(k) = \binom{i+j-1}{i-2c-1}, \ s_3(k) = \binom{i+j-1}{i-2c} \ s_4(k) = \binom{i+j-2}{i-2c-2}.$$

But (7.19) is an elementary consequence of the relation $\binom{p}{q} = \binom{p-1}{q-1} + \binom{p-1}{q}$. ♠

Lemma 7.5.30 Let y be a cohomology class such that $Ry = 0$ for every Adem relation R. Then so is the case with xy where x is any cohomology class of degree 1.

Proof: By Cartan's formula $Sq^j(xy) = xSq^j y + x^2 Sq^{j-1} y$ and hence

$$Sq^i Sq^j (xy) = Sq^i(xSq^j y + x^2 Sq^{j-1}y)$$
$$= xSq^i Sq^j y + x^2 Sq^{i-1} Sq^j y + x^2 Sq^i Sq^{j-1} y + 0 + x^4 Sq^{i-2} Sq^{j-1} y$$

The last term is 0 because $Sq^1(x^2) = \binom{2}{1} x^3$ modulo (2). Similarly, putting $s(k) = \binom{j-k-1}{i-2k}$ we have,

$$\sum_k s(k) Sq^{i+j-k} Sq^k (xy) = x \sum_k Sq^{i+j-k} Sq^k y + x^2 \sum_k s(k) Sq^{i+j-k-1} Sq^k y$$
$$+ x^2 \sum_k Sq^{i+j-k} Sq^{k-1} y + x^4 \sum_k s(k) Sq^{i+j-k-2} Sq^{k-1} y.$$

Note that the first terms on the RHS of the above two formulas are the same. Next, since $i < 2j$, $(i-2) < 2(j-1)$. Since $Ry = 0$ for all R we can take $R = R(i-2, j-1)$ to see that the last terms also match, being the two sides of $x^4 R(i-2, j-1)y$. The middle two terms are equal by the previous lemma. ♠

Lemma 7.5.31 If $H^*(X; \mathbb{Z}_2)$ is generated as an algebra by degree one elements, then all the Adem relations hold in $H^*(X; \mathbb{Z}_2)$.

Proof: $R(1) = 0$ for dimensional reasons for all Adem relations R. Therefore, by the above lemma $R(x) = R(x1) = 0$ for all degree one elements x. Inductively, this implies that $R(z) = 0$ for all monomials in degree one elements. By additivity, this implies $Ry = 0$ for all $y \in H^*(X; \mathbb{Z}_2)$. ♠

The following is then an immediate consequence of Künneth theorem:

Proposition 7.5.32 Adem relations are valid on the cohomology of the product of finitely many copies of the infinite real projective space, $H^*(\mathbb{P}^\infty \times \cdots \times \mathbb{P}^\infty; \mathbb{Z}_2) \approx \mathbb{Z}_2[x_1, \ldots, x_n]$ generated by degree one elements.

Exercise 7.5.33 Establish the following properties of the Whitehead product as defined in Remark 7.5.24.

1. If $m = n = 1$, then $[\alpha, \beta] = \alpha\beta\alpha^{-1}\beta^{-1}$.

2. If $n > 1$ and $m = 1$, then $[\alpha, \beta] = \alpha - h_\beta(\alpha)$.

3. $n, m \geq 2$, then $[,]$ is bilinear and anti-commutative, i.e.,
$$[\alpha, \beta] = (-1)^{mn} [\beta, \alpha].$$

4. For any map $f : (X, x_0) \to (Y, y_0)$, $f_\#[\alpha, \beta] = [f_\#(\alpha), f_\#(\beta)]$.

5. For any path $\omega : \mathbb{I} \to X$ we have $h_{[\omega]}[\alpha, \beta] = [h_{[\omega]}\alpha, h_{[\omega]}\beta]$.

Chapter 8

Homology of Manifolds

We shall discuss the homology and cohomology groups of a manifold. In the first section, we shall discuss the concept of orientability (independently of triangulations). In Section 8.2, we shall prsent various forms of duality theorems. In Section 8.3, some applications of duality are presented. The notions of degree and cobordism are two important topics here. In Section 8.4, we recall some basic facts about differential forms, integration, the Stokes' theorem and the Poincaré lemma, etc., introduce the de Rham cohomology of a smooth oriented manifold and then show that it is canonically isomorphic to singular cohomology with real coefficients.

8.1 Orientability

In this section we shall study the notion of orientability of a topological manifold. We begin with recalling quickly this concept from differential topology. A reader who is not familiar with the concept of orientation on a smooth manifold may take time to look it up in some other source such as [Shastri, 2011].

Definition 8.1.1 Let V be real vector space of dimension $k > 0$. By an orientation on V one means an equivalence class of an ordered basis of V; two bases being equivalent if the transformation matrix taking one to the other is of positive determinant.

Now, let ξ be a vector bundle over B. Then by a pre-orientation on ξ we mean a choice of orientation on each fibre ξ_b. A pre-orientation is called an orientation if it satisfies the following local constancy condition:

There exists an open covering U_i of B on which we have trivializations $h_i : p^{-1}(U_i) \to U_i \times \mathbb{R}^k$ such that the restriction map $h_i : \xi_b \to b \times \mathbb{R}^k$ preserves orientations, where we orient $\mathbb{R}^k$ with the standard orientation.

Remark 8.1.2
(i) Thus on a vector space, there are precisely two orientations.
(ii) Let V, W be two oriented vector spaces. Then we give the orientation on $V \times W$ by first taking the basis for V and then following it up with a basis for W. Thus it is easily seen that $W \times V$ will receive the orientation equal to $(-1)^{kl}$ times that of $V \times W$ where $\dim(V) = k, \dim(W) = l$.
(iii) Let V_0 denote the space $V \setminus \{0\}$. Then one knows that $H_k(V, V_0; \mathbb{Z})$ is isomorphic to $\mathbb{Z}$; so is $H^k(V, V_0; \mathbb{Z})$. Consider the standard k-simplex and its embedding in the linear subspace $L = \{x = (x_1, \ldots, x_{n+1}) : \sum_i x_i = 0\}$ of $\mathbb{R}^{k+1}$ given by

$$x \mapsto x - \beta_k$$

where, $\beta_k = (e_1 + e_2 + \cdots + e_k)/k$ is the barycentre of Δ_k. It is not hard to verify that this embedding defines a singular simplex generating $H_k(L, L_0; \mathbb{Z}) \approx H_k(\mathbb{R}^k, \mathbb{R}_0^k; \mathbb{Z})$. A choice of order on the vertices of Δ_k gives an oriented simplex and hence corresponds to a generator of $H_k(L, L \setminus \{0\}; \mathbb{Z})$. More generally, it follows that for any k-dimensional vector space V, choosing an orientation on V is equivalent to choosing a generator for $H_k(V, V_0; \mathbb{Z})$. Similar statements can be made by using cohomology groups as well.

(iv) Now consider the case of a vector bundle ξ over a manifold B. If U is a trivializing coordinate neighbourhood of a point $b \in B$, then it follows that there is an isomorphism $(j_b)^* : H^k(p^{-1}(U), p^{-1}(U)_0; \mathbb{Z}) \to H^k(\xi_b, (\xi_b)_0; \mathbb{Z})$. Therefore, the local constancy condition can now be described more algebraically as follows:

Lemma 8.1.3 A vector bundle ξ over a manifold B is orientable iff there exists a trivializing open cover $\{U_\alpha\}$ of B and elements $u_\alpha \in H^k(\xi_U, (\xi_U)_0)$ such that for each $b \in U_\alpha$, $(j_b)^*(u_\alpha) \in H^k(\xi_b, (\xi_b)_0; \mathbb{Z})$ is a generator.

For a stronger version of the above lemma see theorem 12.1.1.

Example 8.1.4 Orientation double cover

Given any k-plane bundle $\xi = (E, p, B)$, we shall construct an orientable k-bundle $\tilde{\xi} = (\tilde{E}, \tilde{p}, \tilde{B})$ and a bundle map $(q, \bar{q}) : \tilde{\xi} \to \xi$ as follows: Choose any atlas $\{U_i, h_i\}$ of trivializations of ξ. Fix an orientation on each of $U_i \times \mathbb{R}^k$ orient each $p^{-1}(U_i)$ so that h_i preserve orientations. Now for each i take two copies of $V_i = p^{-1}(U_i)$ and label them by $V_i^\pm$. Let X be the disjoint union of $\{V_i^\pm\}$. For each un-ordered pair of indices $\{i, j\}$, let $W_{ij} = U_i \cap U_j$. On X we define an equivalence relation by the following rule: Given $(b, v) \in W_{ij}^\pm \times \mathbb{R}^k$ identify it with $h_i \circ h_j^{-1}(b, v) \in W_{ij}^\pm \times \mathbb{R}^k$ if $h_i \circ h_j^{-1} : \{b\} \times \mathbb{R}^k \to \{b\} \times \mathbb{R}^k$ is orientation preserving; otherwise identify it with $h_i \circ h_j^{-1}(b, v) \in W_{ij}^\mp \times \mathbb{R}^k$. Let $\tilde{E}$ denote the quotient space of X. Note that X contains two copies of $U_i \times 0$ and under the above identification they define a subspace $\tilde{B}$ of $\tilde{E}$. Then the projection maps p factor through a map $\tilde{p} : \tilde{E} \to \tilde{B}$ defining a k-plane bundle $\tilde{\xi}$. Moreover there is an obvious quotient map which defines a bundle map $(q, \bar{q}) : \tilde{\xi} \to \xi$. Observe that

(i) $\tilde{\xi}$ is always orientable.

(ii) $(q, \bar{q})$ is a double covering.

(iii) $\tilde{\xi}$ is a disjoint union of two copies of ξ iff ξ is orientable.

Remark 8.1.5 In your calculus course, you may have come across the notion of orientability of a smooth manifold X. In the language of vector bundles, this corresponds to an orientation on the tangent bundle of the manifold. The correspondence is given by a diffeomorphism $(\xi)_x, (\xi)_0) \to (U, U \setminus \{x\})$ known as the exponential map, where U is a suitable coordinate neighbourhood of $x \in X$.

Now let X be a general topological manifold. Tangent spaces do not make sense now. However, the algebra takes over from geometry. Orientation at a point $x \in X$ can be defined as a choice of a basis element $\mu_x \in H_n(U, U \setminus \{x\})$, where U is a coordinate neighbourhood of x in X. Following the differential topological insight as observed earlier, we now make the following definition.

Definition 8.1.6 Let $A \subset X$ where X is a topological n-manifold. If the assignment

$$\mu : x \mapsto \mu_x \in H_n(X, X \setminus \{x\}; \mathbb{Z})$$

satisfies the following local compatibility condition (CC) for every $x \in A$, then we say that μ is an orientation on X along A.

Local Compatibility Condition (LCC): Given $x \in A$, there exists an open neighbourhood V of x and $\mu_V \in H_n(X, X \setminus V; \mathbb{Z})$ with the property that if $j_{x,V} : (X, X \setminus V) \hookrightarrow (X, X \setminus x)$ is the inclusion map then $(j_{x,V})_*(\mu_V) = \mu_x$.

Of course if $A = X$, then we merely say that μ is an orientation of X and X is orientable.

Remark 8.1.7

1. By choosing V to be homeomorphic to an open ball, it follows that each $(j_{x,V})_*$ is an isomorphism. Therefore, (CC) is the same as saying that there is a single generator μ_V which maps to each of the generators μ_x. This is the reason why (CC) is also called the *local constancy condition* or *continuity condition*. For brevity, and when it does not cause any confusion, we shall denote the homomorphism $(j_{x,V})_*$ on homology groups also by $j_{x,V}$.

2. We can replace the coefficient group $\mathbb{Z}$ above by any other commutative ring R but then we must use the terminology 'orientation over R'. However, it turns out that we need consider only two important cases, viz., $R = \mathbb{Z}$ or $R = \mathbb{Z}_2$.

3. Since the group $\mathbb{Z}_2$ has a unique generator, it follows that (CC) is satisfied automatically over $\mathbb{Z}_2$. Since $\mathbb{Z}$ has precisely two generators, it also follows that for a connected manifold we can have at most two orientations. Since $\mathbb{Z}$ coefficients can be converted to R-coefficients by merely tensoring with R, it follows that an orientation over $\mathbb{Z}$ yields an orientation over any other ring. These results are summed up in the following proposition:

Proposition 8.1.8

(i) Every manifold is uniquely orientable over $\mathbb{Z}_2$.
(ii) Every connected manifold has at most two orientations over $\mathbb{Z}$.
(iii) If X is orientable over $\mathbb{Z}$ then it is orientable over every commutative ring R.

We now need the topology to play its role of patching up these local choices. In order to put this notion on a proper foundation, we introduce the notion of a general fibre bundle pair and then construct a fibre bundle pair over a topological manifold, analogous to the tangent bundle which serves our purpose.

Definition 8.1.9 By a *fibre bundle pair* over the base space B we mean a total pair $(E, \dot{E})$ and a fibre pair $(F, \dot{F})$ and a bundle projection map $p : E \to B$ which satisfies the following local triviality condition:

There is an open covering $\{U_\alpha\}$ of B and a collection of homeomorphisms $h_\alpha : U_\alpha \times (F, \dot{F}) \to (p^{-1}(U_\alpha), p^{-1}(U_\alpha) \cap \dot{E})$ so that $p \circ h_\alpha$ is the projection to the first factor.

Example 8.1.10

(a) Clearly any product pair $(B \times F, B \times \dot{F})$ with the first projection defines a fibre bundle pair.

(b) Let $p : E \to B$ be a real vector bundle of rank n, and $\zeta : B \to E$ denote the zero section. Then we can take $\dot{E} = E \setminus \zeta(B)$ to get a fibre bundle pair $(E, \dot{E})$.

(c) If $p : E \to B$ is a vector bundle with a Riemannian metric over it, we can take the pair $(U(E), S(E))$ where $U(E)$ and $S(E)$ denote the subspaces of all vectors of length ≤ 1 and $= 1$, respectively.

(d) Given a (locally trivial bundle) $\dot{p} : \dot{E} \to B$, with fibre $\dot{F}$, let E be the mapping cylinder of $\dot{p}$ and $p : E \to B$ be the canonical retraction. Then $(E, \dot{E})$ is a fibre bundle pair with fibre pair $(F, \dot{F})$, where F is the cone over $\dot{F}$.

For any space X, let Δ_X denote the diagonal in $X \times X$:

$$\Delta_X = \{(x, y \in X \times X \; : \; x = y\}.$$

Lemma 8.1.11 Let X be a topological manifold. Then the projection map $p : X \times X \to X$ to the first coordinate defines a fibre bundle pair $(X \times X, X \times X \setminus \Delta_X)$ with fibre pair $(X, X \setminus x)$.

Proof: Given $x \in X$, we must find a neighbourhood U of X and a homeomorphism

$$h : (U \times X, U \times X \setminus \Delta_X) \to U \times (X, X \setminus x)$$

which respects the first coordinates. For simplicity, let us first consider the case when $X = \mathbb{R}^n$ and $x = 0$. We then take $U = B_{1/2}(0)$ the open $\frac{1}{2}$-unit ball, and $\mathbb{D}^n$ the closed ball of radius 1. Define $q : U \times \mathbb{S}^{n-1} \times \mathbb{I} \to U \times \mathbb{D}^n$ by the formula:

$$(a, b, t) \mapsto (a, (1-t)a + tb)).$$

See Figure 8.1. Note that $q(a, b, 0) = (a, a)$ is independent of b. Therefore, if $\eta : \mathbb{S}^{n-1} \times \mathbb{I} \to \mathbb{D}^n$ is the quotient map in the cone construction, then q factors through $Id \times \eta : U \times \mathbb{S}^{n-1} \times \mathbb{I} \to U \times \mathbb{D}^n$ to define a map $h : U \times \mathbb{D}^n \to U \times \mathbb{D}^n$. Verify that h is a homeomorphism which respects the first coordinates. Since $h(a, a) = (a, 0)$, this gives a homeomorphism of the pairs

$$h : (U \times \mathbb{D}^n, U \times \mathbb{D}^n \setminus \Delta_U) \to U \times (\mathbb{D}^n, \mathbb{D}^n \setminus 0).$$

Also, note that $h[a, b, 1] = (a, b)$, i.e., each h_a is identity on $\mathbb{S}^{n-1}$. Therefore, we can extend this homeomorphism by identity to a homeomorphism

$$h : (U \times \mathbb{R}^n, U \times \mathbb{R}^n \setminus \Delta_U) \to U \times (\mathbb{R}^n, \mathbb{R}^n \setminus 0).$$

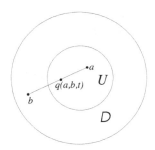

FIGURE 8.1. Existence of orientation bundle

The general case now follows easily: Given any point $x \in X$, we take a homeomorphism $\phi : V \to \mathbb{R}^n$ of some neighbourhood of x such that $\phi(x) = 0$, put $h' = (\phi \times \phi)^{-1} \circ h \circ (\phi \times \phi)$ and extend it by identity to a homeomorphism

$$(U \times X, U \times X \setminus \Delta_X) \to U \times (X, X \setminus x)$$

as required.   ♠

Remark 8.1.12 The orientation bundle

Associated to the fibre bundle pair of Lemma 8.1.11, consider another fibre bundle $q : X_{\mathbb{Z}} \to X$ with fibres equal to $H_n(X, X \setminus x; \mathbb{Z}) \approx \mathbb{Z}$. The total space $X_{\mathbb{Z}}$ is the disjoint union of $X_{\mathbb{Z}} = \coprod_{x \in X} H_n(X, X \setminus x)$ and the projection $q : X_{\mathbb{Z}} \to X$ is the function that takes whole group $H_n(X, X \setminus x)$ to x. We put a topology on $X_{\mathbb{Z}}$ as follows: For any coordinate open subset U of X and an element $\alpha \in H_n(X, X \setminus \bar{U})$, define

$$\langle U, \alpha \rangle = \{j_{x,U}(\alpha) \ : \ x \in U\}.$$

Take $\mathcal{B} = \{\langle U, \alpha \rangle\}$, where U varies over open subsets of X which are homeomorphic to an open disc and $\alpha \in H_n(X, X \setminus \bar{U})$.

Let us verify that this $\mathcal{B}$ is a basis for a topology on $X_{\mathbb{Z}}$. Given $g \in X_{\mathbb{Z}}$, suppose $g \in H_n(X, X \setminus x)$. If U is a neighbourhood of $x \in X$ which is homeomorphic to an open disc then we know that $j_{x,U}$ induces an isomorphism $H_n(X, X \setminus \bar{U}) \approx H_n(X, X \setminus x)$ and so $g \in \langle U, \alpha \rangle$ for some $\alpha \in H_n(X, X \setminus \bar{U})$. Thus $\mathcal{B}$ is a cover of $X_{\mathbb{Z}}$.

Now suppose $g \in \langle U, \alpha \rangle \cap \langle V, \beta \rangle$, i.e., $g = j_{x,U}(\alpha) = j_{x,V}(\beta)$. We can take an open set W homeomorphic to a ball such that $x \in W \subset U \cap V$. Then since $j_{x,W}$ is an isomorphism, we get $\gamma \in H_n(X, X \setminus W)$ such that $j_{x,W}(\gamma) = g$, i.e., $x \in \langle W, \gamma \rangle$. Moreover, we have a commutative diagram of isomorphisms:

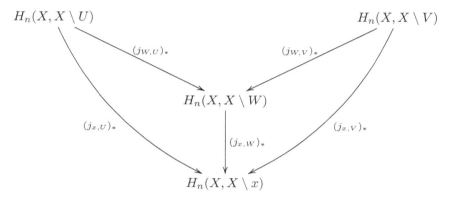

Therefore, $\gamma = j_{W,U}(\alpha) = j_{W,V}(\beta)$. From this it follows that $\langle W, \gamma \rangle \subset \langle U, \alpha \rangle \cap \langle V, \beta \rangle$.

Since basic open sets are mapped onto open sets by q, it follows that q is an open mapping. In fact, for a coordinate open set U in X, we have $q^{-1}(U)$ is the disjoint union of basic open sets $\langle U, \alpha \rangle$ where $\alpha \in H_n(X, X \setminus U) \approx \mathbb{Z}$, and q restricted to each one of them is a bijection. It follows that q is continuous and a covering projection.

Note that each fibre of q has exactly two units. It follows that the subspace $X_u \subset X_{\mathbb{Z}}$ of all units in each fibre of q forms a double cover of X.

Theorem 8.1.13 *Let X be a connected manifold. The following conditions are equivalent:*
(i) X is orientable.
(ii) X is orientable along every loop in X.
(iii) The units X_u in $X_{\mathbb{Z}}$ form a trivial double cover of X.
(iv) $q : X_{\mathbb{Z}} \to X$ is a trivial bundle.
(v) $q : X_{\mathbb{Z}} \to X$ has a continuous section, which is nowhere vanishing. i.e., there is a map $s : X \to X_{\mathbb{Z}}$ such that $s \circ q = Id_X$ and $s(x) \neq 0, x \in X$.

Proof: (i) $\implies$ (ii) is obvious.
Note that X_u is the trivial double cover iff X_u has two components. Therefore, if it were not trivial then X_u is path connected. Given $x \in X$ we can then find a path ω in X_u which joins the two units in the fibre $q^{-1}(x) \approx \mathbb{Z}$. Then $q(\omega)$ is a loop in X along which X will not be orientable. This proves (ii) $\implies$ (iii).

Given a homeomorphism $\phi : X \times \{-1, 1\} \to X_u$ such that $q \circ \phi(x, \epsilon) = x$, let us define $\hat{\phi} : X \times \mathbb{Z} \to X_{\mathbb{Z}}$ by

$$\hat{\phi}(x, n) = |n|\phi(x, n/|n|), n \neq 0, \quad \hat{\phi}(x, 0) = 0.$$

Verify that $\hat{\phi}$ defines a trivialization of $X_{\mathbb{Z}}$. This takes care of (iii) $\implies$ (iv).
(iv) $\implies$ (v) is obvious.
Given a section s as specified take $\mu(x) = s(x)/|s(x)|$. Given $x \in X$, if U is a coordinate neighbourhood of x, then U is evenly covered by q, i.e., $q^{-1}(U)$ is a disjoint union of $\langle U, \alpha \rangle$ where $\alpha \in H_n(X, X \setminus \bar{U})$. By continuity of s this implies that $s(x)$ is a constant for all $x \in U$ and must be a generator. So is $m(x) = s(x)/|s(x)|$ which is then equal to $j_{x,U}(\mu_U)$ for a generator μ_V of $H_n(X, X \setminus U; \mathbb{Z})$. This proves (v) $\implies$ (i).   ♠

Remark 8.1.14 Thus we have established that an orientation μ on a subset A of a manifold X corresponds to a continuous section $s_\mu : A \to X_u$ of the sub-bundle $q : X_u \to X$.

For a subset $A \subset X$, let $\Gamma_c(A)$ denote the collection of all sections of $X_{\mathbb{Z}}$ defined over A and having compact support, i.e., vanishing outside a compact subset. Fibre-wise addition makes sense and using local triviality, one can easily verify the continuity of the sum of two sections. Thus $\Gamma_c(A)$ is an abelian group.
For a closed subset A consider the homomorphism

$$J_A : H_n(X, X \setminus A) \to \Gamma_c(A)$$

given by $J_A(\alpha)(x) = j_{x,A}(\alpha)$. That $J_A(\alpha)$ is a section is easy to verify. Why does it have compact support? Because any element α in the homology is represented by a cycle which has compact support. Why is $J_A(\alpha)$ continuous? If α is represented by a relative cycle c then ∂c is a chain in $X \setminus A$. Therefore for each $x \in A$, we can find a coordinate neighbourhood U of x such that ∂c is contained in $X \setminus \bar{U}$. Thus $[c] \in H_n(X, X \setminus \bar{U})$ and we have for all $x \in U$, $J_A(\alpha)(x) = j_{x,U}([c])$, which is continuous in U. Thus J_A is well defined. Since each $j_{x,A}$ is a homomorphism, so is J_A. In the rest of this section our major concern is in establishing the fact that J_A is an isomorphism when A is compact.

Proposition 8.1.15
(i) J_A is functorial, i.e., if $A \subset B \subset X$ are closed sets then we have the commutative diagram:

$$
\begin{array}{ccc}
H_n(X, X \setminus B) & \longrightarrow & H_n(X, X \setminus A) \\
\downarrow{\scriptstyle J_B} & & \downarrow{\scriptstyle J_A} \\
\Gamma_c(B) & \longrightarrow & \Gamma_c(A)
\end{array}
$$

(ii) For closed subsets A, B of X we have the exact sequence:

$$0 \longrightarrow \Gamma_c(A \cup B) \overset{\sigma}{\longrightarrow} \Gamma_c(A) \oplus \Gamma_c(B) \overset{\tau}{\longrightarrow} \Gamma_c(A \cap B). \tag{8.1}$$

where σ, τ respectively denote the sum and difference of the restrictions.
(iii) If $\{A_i\}$ is a decreasing sequence of compact subsets of X, $A = \cap_i A_i$, then the restrictions $\Gamma(A_i) \to \Gamma(A)$ induce an isomorphism

$$\varinjlim \Gamma(A_i) \overset{\approx}{\longrightarrow} \Gamma(A).$$

Proof: (i) and (ii) are easy. To prove (iii) the important thing to observe is that if s, s' are sections and $x \in X$ is such that $s(x) = s'(x)$ then there is an evenly covered neighbourhood V of x on which $s = s'$. This is an easy consequence of the discreteness of fibres of $X_{\mathbb{Z}}$. Now suppose s, s' are sections over some A_i which agree on A, it follows that there is an open set $U \supset A$ such that $s = s'$ on U. By compactness, there is N such that $A_N \subset U$. This then implies $s = s'$ on A_j for all $j > N$. Therefore $s = s'$ as elements of $\varinjlim \Gamma(A_i)$. To prove the onto-ness, given $s \in \Gamma(A)$, first cover A by finitely many evenly covered neighbourhoods U_i such that $s = s_i$ on U_i. Define

$$V = \{x \in \cup_i U_i \ : \ s_i(x) = s_j(x) \text{ whenever } x \in U_i \cap U_j\}.$$

Then W is open since there are only finitely many pairs (i, j) to be considered. Clearly $A \subset W$. On W, we define $\hat{s}(x) = s_i(x)$ if $x \in U_i \cap W$. Now there is some N such that $A_j \subset W$ for $j > N$ and hence $\hat{s} \in \varinjlim \Gamma(A_i)$. Since $\hat{s} = s$ on A we are through. ♠

Let M be a fixed manifold. Consider the family of closed subsets A of M. Let $P(A)$ be a statement (proposition) defined for all closed sets or for a suitable subclass of closed sets such as compact sets. Consider the following axioms on $P(A)$:
(B1) A is a compact and convex subset of some coordinate neighbourhood in $M \implies P(A)$.
(B2) $P(A), P(B), P(A \cap B) \implies P(A \cup B)$.
(B3) $A_1 \supset A_2 \supset \cdots$ are compact, $P(A_i)$ for all $i \implies P(\cap_i A_i)$.
(B4) $\{A_i\}$ are compact with disjoint neighbourhoods N_i, $P(A_i)$ for all $i \implies P(\cup A_i)$.

Lemma 8.1.16 (Bootstrap Lemma)
(i) Let $P(A) := P(M, A)$ be a statement about compact subsets A of a manifold M which satisfies (B1), (B2) and (B3). Then $P(A)$ is true for all compact subsets A of M.
(ii) Let $P(A) := P(M, A)$ be a statement about closed subsets A of a manifold M which satisfies (B1)-(B4). Then $P(M, A)$ is true for all closed subsets A of M.

Proof:
(i) By (B1), $P(A)$ is true for compact convex subsets of Euclidean subspaces of M. Since intersection of any two such sets in a given Euclidean subspace, is again compact and convex, by (B2), $P(A)$ is true for the union of two such sets. It takes a little more effort to apply induction here. Suppose U is an Euclidean open subset of M and $A_1, \ldots, A_k$ are compact convex subsets of U. Assume inductively that $P(B)$ holds whenever B is the union of $k - 1$ compact convex subsets Then

$$A_k \cap (A_1 \cup \cdots \cup A_{k-1}) = (A_k \cap A_1) \cup \cdots \cup (A_k \cap A_{k-1})$$

and hence $P(A_k \cap (A_1 \cup \cdots \cup A_{k-1}))$ holds by induction. Now (B2) implies $P(\cup_{i=1}^k A_i)$.

We can now remove the convexity hypothesis, i.e., $P(U, A)$ is true for all compact subsets A of a Euclidean open set U. To see this, it suffices to remark that every compact subset A of $\mathbb{R}^n$ is a decreasing intersection of a sequence of compact sets A_i each of which is a finite union of compact convex sets (exercise). For then we can apply (B3). Using (B2) again as above, we get $P(A)$ is true for all A which are finite unions of compact sets each of which is contained in some Euclidean space. Using (B3) again, since every compact set A in M is the intersection of a decreasing sequence of compact sets as above, we obtain $P(A)$ is true for all compact subsets of M.
(ii) Let $f : M \to [0, \infty)$ be a proper mapping. (Such a map exists; see Remark 5.1.11.) Put $A_i = f^{-1}[2i, 2i + 1], B_i = f^{-1}[2i + 1, 2i + 2]$ for $i = 0, 1, 2, \ldots,$. Then each A_i, B_i is compact and $\{A_i\}$ have disjoint neighbourhoods; so have $\{B_i\}$. Put $A = \cup A_i$ and $B = \cup B_i$. Clearly $M = A \cup B$. Therefore given any closed set C of M we can write $C = (A \cap C) \cup (B \cap C)$ and then $A \cap C, B \cap C$ are a union of a family of compact sets which have disjoint neighbourhoods. Moreover, $(A \cap C) \cap (B \cap C)$ is the union of compact sets $C \cap (A_i \cap B_i)$'s and $C \cap (B_i \cap A_{i+1})$. Note that each of these two families of compact sets have disjoint neighbourhoods. Therefore (B4) gives $P(C \cap A), P(C \cap B)$ and $P((C \cap A) \cap (C \cap B))$. Again by (B2) we get $P(C)$. ♠

Theorem 8.1.17 *Let X be a topological n-manifold and A be a closed subset. Then*
(a) $H_i(X, X \setminus A) = 0$ *for* $i > n$; *and*
(b) $J_A : H_n(X, X \setminus A) \to \Gamma_c(A)$ *is an isomorphism.*

Proof: Let us denote the statement of the theorem by $P(A)$. We shall verify that $P(A)$ satisfies the axioms (B1), (B2), (B3) and (B4). From the previous lemma, it then follows that $P(A)$ holds for all closed subsets of X.

Proof of (B1) Given a compact convex subset A of a Euclidean space $\mathbb{R}^n \approx U \subset X$, choose a disc D such that $A \subset \text{int}\, D \subset \bar{D} \subset U$. Then for every $x \in A$, we have the following commutative diagram

$$
\begin{array}{ccc}
H_i(X, X \setminus A) & \longrightarrow & H_i(X, X \setminus x) \\
\big\uparrow{\scriptstyle e_1} & & \big\uparrow{\scriptstyle e_2} \\
H_i(\mathbb{R}^n, \mathbb{R}^n \setminus A) & \longrightarrow & H_i(\mathbb{R}^n, \mathbb{R}^n \setminus x) \\
\big\uparrow{\scriptstyle h_1} & & \big\uparrow{\scriptstyle h_2} \\
H_i(D, \partial D) & \xrightarrow{\;=\;} & H_i(D, \partial D)
\end{array}
$$

where e_1, e_2 are isomorphisms given by excision and h_1, h_2 are isomorphisms induced by the inclusions which are deformation retracts (see Exercise 1.9.2). This gives both (a) and (b) for such A.

Proof of (B2) (a) follows easily from Mayer–Vietoris sequence. For (b), apply Four lemma to the following commutative diagram in which we have used a temporary notation $H_i(X|Y)$ to denote $H_i(X, X \setminus Y)$ (so that the diagram fits within the page).

$$
\begin{array}{ccccccc}
H_{n+1}(X|A \cap B) & \longrightarrow & H_n(X|A \cup B) & \longrightarrow & H_n(X|A) \oplus H_n(X|B) & \longrightarrow & H_n(X|A \cup B) \\
\big\| & & \Big\downarrow{\scriptstyle J_{A \cup B}} & & {\scriptstyle J_A \oplus J_B}\Big\downarrow{\scriptstyle \approx} & & {\scriptstyle J_{A \cap B}}\Big\downarrow{\scriptstyle \approx} \\
0 & \longrightarrow & \Gamma_c(A \cup B) & \longrightarrow & \Gamma_c(A) \oplus \Gamma_c(B) & \longrightarrow & \Gamma_c(A \cap B)
\end{array}
$$

Proof of (B3) Since homology commutes with taking direct limit, (a) follows. For (b), first observe that $\varinjlim \Gamma_c(A_i) = \Gamma_c(\cap_i A)$. Then appeal to the commutative diagram below in which three of the arrows are isomorphisms. Therefore the fourth one is also an isomorphism:

$$
\begin{array}{ccc}
\varinjlim H_n(X, X \setminus A_i) & \xrightarrow{\;\approx\;} & H_n(X, X \setminus \cap_i A_i) \\
{\scriptstyle \approx}\Big\downarrow & & \Big\downarrow \\
\varinjlim \Gamma_c(A_i) & \xrightarrow{\;\approx\;} & \Gamma_c(\cap_i A_i).
\end{array}
$$

Proof of (B4) Put $N = \cup N_i$, $A = \cup A_i$. We have $H_i(X, X \setminus A) \approx H_i(N, N \setminus A)$, by excision. Since homology of a disjoint union is the direct sum of the homologies, (a) follows immediately. For (b) use the following fact: A section on the disjoint union is defined iff it is defined on each part. ♠

Remark 8.1.18 Given a closed subset A, the group $\Gamma_c(A)$ may be trivial. This corresponds to the case when X is non orientable over A. If $\Gamma_c(A)$ has one non zero element, then it will contain an infinite cyclic subgroup, for we can then multiply this nonzero section fibre-wise by any integer. If we assume that A is connected, then an element of $\Gamma(A)$ is completely determined by its value at a single point and therefore, $\Gamma_c(A) \approx \mathbb{Z}$ or (0). For the same reason, if we assume that A is non compact and connected then $\Gamma_c(A) = (0)$. Upon taking $A = X$ we obtain the following:

Corollary 8.1.19 Let X be a connected n-manifold. Then
(a) $H_i(X) = (0)$ for $i > n$.
(b) If X is a compact, then $H_n(X) \approx \mathbb{Z}$ iff X is orientable; otherwise $H_n(X) \approx (0)$.
(c) If X is non compact then $H_n(X) = (0)$.

Remark 8.1.20 The discussion above goes through when we use an arbitrary commutative ring in place of $\mathbb{Z}$ except for a few changes as observed in Remark 8.1.7. Thus for a compact connected manifold X, $H_n(X, \mathbb{Z}_2) \approx \mathbb{Z}_2$ always. Note that for non compact manifold X, $H_n(X; \mathbb{Z}_2)$ is also (0). Thus homology fails to have any say over orientability. We need to rectify this somehow. This will task will be taken up in the next section.

Exercise 8.1.21

(i) Show that a manifold X is orientable if $\pi_1(X)$ has no subgroups of index 2. However, note that this condition is not necessary.

(ii) Show that a manifold is orientable over $\mathbb{Q}, \mathbb{R}$ or $\mathbb{C}$ iff it is orientable.

(iii) Show that every compact subset K of $\mathbb{R}^n$ is the intersection of a decreasing family of compact sets K_i where each K_i is a finite union of compact convex sets.

8.2 Duality Theorems

In this section, we begin with a modification of singular cohomology, which will serve the purpose of detecting orientability of non compact manifolds. This modified cohomology lies somewhere between the so-called Alexander cohomology (which we shall not discuss at all) and the Čech cohomology which we shall discuss in the next chapter. With the help of his we shall be able to present Poincaré duality theorem and some other variants of it.

Definition 8.2.1 Given a topological space X and pairs (A, B) of subspaces of X, a neighbourhood of (A, B) in X is defined to be a pair (U, V) where $A \subset U, B \subset V$ and U, V are open in X. The family of neighbourhoods of (A, B) is treated as a directed (downwards) family via inclusion and we takes the direct limit

$$\bar{H}^*(A, B) := \varinjlim H^*(U, V).$$

Apparently, the group $\bar{H}^*(A, B)$ depends on how the pair (A, B) is embedded in X. However, where X is a manifold and A, B are closed subsets this group is naturally isomorphic to what is called 'the Čech cohomology group', and also to 'Alexander cohomology groups'. We shall not need this here. (See [Dold, 1972] for more details). However, just to feel how this cohomology group is different from the usual cohomology, let us consider the following example.

Example 8.2.2 Consider the compact subset of the $\mathbb{R}^2$ which is called the *topologists sine-loop* as defined in Exercise 1.9.19. Clearly A is path connected. You can easily show that $\pi_1(A, p) = (1)$ where p could be chosen to be any point of A, say $= (1, 1)$. It follows that $H^1(A, \mathbb{Z}) = (0)$. On the other hand, see that there is a fundamental system of neighbourhoods U_i of A in $\mathbb{R}^2$, with $\mathbb{Z} = H^1(U_i)$ and such that the inclusion induced homomorphisms $H^1(U_i) \to H^1(U_{i+1})$ are isomorphisms. Therefore $\bar{H}^1(A) = \mathbb{Z}$.

Now let X be a manifold, (A, B) be a closed pair in X and (U, V) be a neighbourhood of (A, B) in X. Consider the cap product

$$\frown : H^p(U, V) \otimes H_n(U, U \setminus A) \to H_{n-p}(U, V \cup (U \setminus A)).$$

Since $U = V \cup U \setminus B$, it follows that we can write any element $c \in S_n(U, U \setminus A)$ as $c = a + b$ where $a \in S_n(V)$ and $b \in S_n(U \setminus B)$. Since a cocycle $f \in S_p^*(U, V)$ vanishes on $S_n(V)$, we have $f \frown c = f \frown b + f \frown c = f \frown c$. Thus the above cap product takes values inside $H_{n-p}(U \setminus B, U \setminus A)$ and we get the cap product pairing

$$\frown : H^p(U, V) \otimes H_n(U, U \setminus A) \to H_{n-p}(U \setminus B, U \setminus A). \tag{8.2}$$

Now let A be a compact subset of X. Fix an element $\mu_A \in H_n(X, X \setminus A)$. By excision this corresponds to a unique element in $H_n(U, U \setminus A)$, which we shall continue to denote by μ_A. Then (8.2) gives a homomorphism

$$\frown \mu_A : H^p(U, V) \to H_{n-p}(U \setminus B, U \setminus A) \approx H_{n-p}(X \setminus B, X \setminus A). \tag{8.3}$$

Passing to the direct limit this gives a homomorphism:

$$\frown \mu_A : \bar{H}^p(A, B) \to H_{n-p}(X \setminus B, X \setminus A). \tag{8.4}$$

Lemma 8.2.3 Given a compact pair (A, B) in a n-manifold X and an element $\mu_A \in H_n(X, X \setminus A)$, there is a commutative diagram of long exact rows of cohomology and homology modules wherein the vertical arrows represent homomorphisms given by capping with μ_A.

$$
\begin{array}{ccccccccc}
\cdots \longrightarrow & \bar{H}^p(A, B) & \longrightarrow & \bar{H}^p(A) & \longrightarrow & \bar{H}^p(B) & \longrightarrow & \bar{H}^{p+1}(A, B) & \longrightarrow \cdots \\
& \downarrow & & \downarrow & & \downarrow & & \downarrow & \\
\cdots \longrightarrow & H_{n-p}(B^c, A^c) & \longrightarrow & H_{n-p}(X, A^c) & \longrightarrow & H_{n-p}(X, B^c) & \longrightarrow & H_{n-p-1}(B^c, A^c) & \longrightarrow \cdots
\end{array}
$$

Proof: (Here A^c, etc., denote $X \setminus A$, etc.) The exactness of the top row follows from the cohomology exact sequences of the pairs (U, V) upon taking direct limit (see Exercise 4.7.1). The bottom row is the exact homology sequence of the triple $(X, X \setminus B, X \setminus A)$. The only non trivial verification is the commutativity of the last square, wherein the horizontal arrows represent the connecting homomorphisms of the respective exact sequences. This itself follows, upon taking direct limit if we verify the commutativity of the squares:

$$
\begin{array}{ccc}
H^p(V) & \longrightarrow & H^{p+1}(U, V) \\
\downarrow & & \downarrow \\
H_{n-p}(X, B^c) & \longrightarrow & H_{n-p-1}(B^c, A^c)
\end{array}
$$

Now for any $f \in Z^p(V)$, we have $\delta(f \frown \mu_A) = \delta(f) \frown \mu_A \pm f \frown \partial \mu_A$. Since $\mu_A \in H_n(X, A^c)$, $\partial \mu_A = 0 \in H_{n-1}(B^c, A^c)$ and hence $\delta(f \frown \mu_A) = \delta(f) \frown \mu_A$ which verifies the commutativity of the diagram above. ♠

Lemma 8.2.4 Let K_1, K_2 be compact subsets of a n-manifold X with an orientation class $\mu_{K_1 \cup K_2} \in H_n(X, (K_1 \cup K_2)^c)$. Then there is a commutative diagram of long exact columns of cohomology and homology modules, in which the horizontal arrows represent the homo-morphisms given by the cap product with μ :

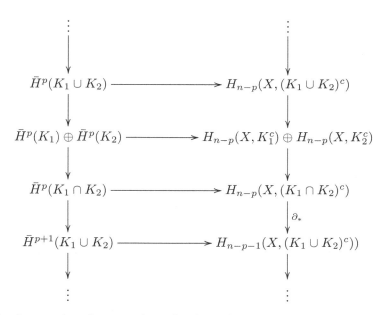

Proof: As in the previous lemma, the only thing that needs to be checked is the commu-tativity of the third square which is a consequence of the commutativity of

$$
\begin{array}{ccc}
H^p(U_1 \cap U_2) & \longrightarrow & H_{n-p}(X,(K_1 \cap K_2)^c) \\
\downarrow{\scriptstyle \delta^*} & & \downarrow{\scriptstyle \partial} \\
H^{p+1}(U_1 \cup U_2) & \longrightarrow & H_{n-p-1}(X,(K_1 \cup K_2)^c)
\end{array}
\tag{8.5}
$$

for every open set $U_i \supset K_i, i = 1, 2$. This is what we have to check then. Now the left-vertical arrow represents the connecting homomorphism δ^* of the short exact sequence of cochain complexes

$$
0 \longrightarrow \operatorname{Hom}(S_*(U_1) + S_*(U_2); R) \longrightarrow \operatorname{Hom}(S_*(U_1); R) \oplus \operatorname{Hom}(S_*(U_2); R)
$$
$$
\downarrow
$$
$$
\operatorname{Hom}(S_*(U_1 \cap U_2); R) \longrightarrow 0
$$

Given $\alpha \in H^p(U_1 \cap U_2)$ represented by a cochain f such that $\delta f = 0$ on $U_1 \cap U_2$, we take some extension of f over $S_*(U_1)$ consider $(f, 0) \in \operatorname{Hom}(S_*(U_1); R) \oplus \operatorname{Hom}(S_*(U_2); R)$ which maps onto f. We then take its coboundary $(\delta f, 0)$ which should come from an element $h \in \operatorname{Hom}(S_*(U_1) + S_*(U_2); R)$. Indeed, check that the element h defined by $h(\sigma_1 + \sigma_2) = (\delta f)(\sigma_1)$ works. By definition $\delta^*(\alpha)$ is represented by this h.

Next, we can represent $\mu_{K_1 \cup K_2}$ by a chain of the form

$$
a = b + c + d + e
$$

where $b \in S_*(U_1 \cap U_2), c \in S_*(U_1 \setminus K_1), d \in S_*(U_2 \setminus K_2)$ and $e \in S_*((K_1 \cup K_2)^c)$. It follows that $\delta^*(\alpha) \frown \mu$ is represented by

$$
h \frown a = h \frown (c + (b + d) + e) = h(c + (b + d)) + h(e) = (\delta f)(c) + h(e) = (\delta f)(c),
$$

the term $h(e)$ being ignored since c is a chain in $(K_1 \cup K_2)^c$.

On the other hand, the right-vertical arrow in (8.5) represents the connecting homomorphism ∂ of the homology long exact sequence associated to the following short exact sequence of chain complexes:

$$0 \longrightarrow \frac{S_*(X)}{S_*((K_1 \cup K_2)^c)} \longrightarrow \frac{S_*(X)}{S_*(K_1^c)} \oplus \frac{S_*(X)}{S_*(K_2^c)} \longrightarrow \frac{S_*(X)}{S_*(K_1^c) + S_*(K_2^c)} \longrightarrow 0$$

The element $\alpha \frown \mu \in H_{n-p}(X, (K_1 \cap K_2)^c)$ is represented by $f \frown a$ modulo $S_*(K_1^c) + S_*(K_2^c)$. We can lift this to $(f \frown a, 0)$ in the direct sum of the modules and take its boundary: $(\partial(f \frown a), 0) = ((\delta f) \frown a \pm f \frown \partial a, 0)$. Now $(\delta f) \frown a = (\delta f)(c) + (\delta f)(d)$. But $(\delta f)(d), f \frown \partial a \in S_*(K_1^c)$ and hence can be ignored. Therefore $\delta^*(\alpha) \frown \mu = \partial(\alpha \frown \mu)$, thereby proving the commutativity of (8.5).   ♠

Given an orientation class μ on a n-manifold X, over a ring R, we may view it as a section s_μ of the unit subbundle of X_R. Then for each compact subset K of X, s_μ restricts to a section over K and hence can be thought of as an element in $\Gamma_c(X_u)$. This in turn yields a unique element $\mu_K \in H_n(X, X \setminus K; R)$ under the isomorphism J_K. Thus we can assign a new meaning to the orientation μ as the collection $\{\mu_K\}$ which automatically satisfies the compatibility condition under inclusion maps. Thus for all compact pairs (K, L) we can denote the homomorphism

$$\frown \mu_K : \bar{H}^p(K, L; G) \to H_{n-p}(X \setminus L, X \setminus K; G)$$

by an unambiguous notation $\frown \mu$.

Theorem 8.2.5 (Alexander–Lefschetz–Poincaré duality theorem) *Let X be a n-manifold with an orientation class μ over a commutative ring R. Then for any R-module G and any compact subsets $L \subset K \subset X$, the cap product*

$$\frown \mu : \bar{H}^p(K, L; G) \to H_{n-p}(X \setminus L, X \setminus K; G) \tag{8.6}$$

is an isomorphism.

Proof: As usual, the coefficient module G is immaterial and hence will be suppressed in the notation that follows. By Lemma 8.2.3 and Five lemma, it suffices to prove (8.6) for the case $L = \emptyset$. Thus we shall prove

$$\frown \mu : \bar{H}^p(K) \to H_{n-p}(X, X \setminus K) \tag{8.7}$$

is an isomorphism.

Let P(K) be the statement that (8.7) holds for K. We shall prove that P(K) satisfies the axioms (B1), (B2), (B3). Then from Bootstrap Lemma 8.1.16(i), the theorem follows. Proof of (B1): For the case $K = \{*\}$ note that both the groups involved in (8.7) are zero for $p \neq 0$; and for $p = 0$ both are isomorphic to the coefficient group and the homomorphism (8.7) is given by $1 \mapsto 1 \frown \mu_x = \mu_x \in H_n(X, X \setminus *)$. Since μ_x is the generator, we are through. Now let K be a compact convex subset of an Euclidean space. We then have the commutative diagram wherein the vertical arrows represent isomorphisms given by homotopies (see Exercise 1.9.2):

$$
\begin{array}{ccc}
\bar{H}^p(K) & \xrightarrow{\frown \mu} & H_{n-p}(X, X \setminus K) \\
\Big\downarrow{\approx} & & \Big\downarrow{\approx} \\
\bar{H}^p(*) & \xrightarrow{\approx} & H_{n-p}(X, X \setminus *).
\end{array}
$$

Proof of (B2): This is a direct consequence of Lemma 8.2.4, and the Five lemma.

Proof of (B3): Let $K_1 \supset K_2 \supset \cdots$ be a decreasing sequence of compact subsets so that $P(K_i)$ holds for each i. This then gives a directed system of isomorphisms, which upon taking the limit, yields the following commutative diagram:

$$
\begin{array}{ccc}
\varinjlim_i \bar{H}^p(K_i) & \xrightarrow{\approx} & \varinjlim_i H_{n-p}(X, X \setminus K_i) \\
\downarrow & & \downarrow{\scriptstyle \approx} \\
\bar{H}^p(K) & \longrightarrow & H_{n-p}(X, X \setminus K)
\end{array}
$$

It remains to see why the first vertical arrow is an isomorphism. This is indeed an easy consequence of a general result about iterated direct limits. However, we shall prove this directly as follows: Using a metric on X or otherwise, we first construct a fundamental system $\{U_{i,j}\}$ of neighbourhoods of K_i such that

$$ U_{1,j} \supset U_{2,j} \cdots $$

holds for each j. We then have

$$ \varinjlim \bar{H}^p(K_i) = \varinjlim_i \varinjlim_j H^p(U_{i,j}) \xrightarrow{\eta} \varinjlim_{i,j} H^p(U_{i,j}) = \bar{H}^p(K). $$

The first equality is obvious. The last equality follows because the directed system $U_{i,j}$ forms a fundamental system of neighbourhoods of K. It remains to see how to obtain the homomorphism η and show that it is an isomorphism. Note that for each pair of indices (t, s), there are canonical homomorphisms

$$ H^p(U_{t,s}) \to \varinjlim_{i,j} H^p(U_{i,j}). $$

For each fixed t, by the universal property of the direct limit (with respect to the index s), these homomorphisms induce a unique homomorphism

$$ \bar{H}^p(K_t) \to \varinjlim_{i,j} H^p(U_{i,j}). $$

Again by the universal property of the direct limit (with respect to t) these homomorphisms in turn induce

$$ \eta : \varinjlim_i \varinjlim_j H^p(U_{i,j}) = \varinjlim \bar{H}^p(K_t) \to \varinjlim_{i,j} H^p(U_{i,j}). $$

Similarly there are canonical homomorphisms

$$ \tau_{i,j} : H^p(U_{i,j}) \to \bar{H}^p(K_i), \quad \lambda_i : \bar{H}^p(K_i) \to \varinjlim \bar{H}^p(K_i). $$

The composites $\lambda_i \circ \tau_{i,j} : H^p(U_{i,j}) \to \varinjlim \bar{H}^p(K_i)$ form a compatible family of homomorphisms and hence define a unique homomorphism

$$ \lambda : \varinjlim_{i,j} H^p(U_{i,j}) \to \varinjlim \bar{H}^p(K_i). $$

Appealing to the universal property of the direct limits, it is easily verified that λ is the inverse of η. This completes the proof of (B3) and thereby the proof of the theorem. ♠

Remark 8.2.6 From this version of the duality theorem, we deduce several other versions. Taking $K = X$ to be a compact manifold, and $L = \emptyset$, we obtain the Poincaré duality theorem. If we do not necessarily take $L = \emptyset$, we get the so-called Lefschetz duality theorem, which is a mild generalization of Poincaré duality theorem. Putting $X = \mathbb{R}^n, L = \emptyset$ and noting that $H_{n-p}(\mathbb{R}^n, \mathbb{R}^n \setminus K) \approx \tilde{H}_{n-p-1}(\mathbb{R}^n \setminus K)$, we get the Alexander duality theorem. There is also another version of Alexander duality. We shall state all this below and leave the details to the reader.

Corollary 8.2.7 (Poincaré duality theorem) Let X be a compact n-manifold (without boundary) oriented over R with an orientation class $\mu_X \in H_n(X; R)$. Then for any R-module G, the cap product

$$\frown \mu_X : H^p(X; R) \to H_{n-p}(X; R)$$

is an isomorphism for all p.

Remark 8.2.8 Note that in this case, an orientation class μ_X is the sum of μ_{X_i}, where each μ_{X_i} is a generator of $H_n(X_i, R) \approx R$, where X_i are path components of X. It is also customary to denote this orientation class for X by $[X]$.

Corollary 8.2.9 (Lefschetz's duality theorem) Let X be a compact n-manifold (without boundary). Suppose $K \subset X$ is a closed subset and $\mu_K \in H_n(X, X \setminus K; R)$ is an orientation class. Then the cap product

$$\frown \mu_K : \bar{H}^p(K; R) \to H_{n-p}(X, X \setminus K; R)$$

is an isomorphism.

Remark 8.2.10 If we closely follow the proofs, you may observe that in the above result, we need not have X to be a manifold all over, i.e., X needs to be locally Euclidean away from the closed set K, i.e., (X, K) is a relative n-manifold.

Corollary 8.2.11 (Alexander duality theorem version 1) Let K be a compact subset of $\mathbb{R}^n$. Then

$$\tilde{H}_{n-p-1}(\mathbb{R}^n \setminus K; R) \approx \bar{H}^p(K; R).$$

Proof: Since $\mathbb{R}^n$ is orientable over $\mathbb{Z}$, we have

$$\bar{H}^p(K; R) \approx H_{n-p}(\mathbb{R}^n, \mathbb{R}^n \setminus K; R) \approx \tilde{H}_{n-p-1}(\mathbb{R}^n \setminus K; R).$$

Note that here the last isomorphism is given by the homology exact sequence of the pair $(\mathbb{R}^n, \mathbb{R}^n \setminus K)$. ♠

Corollary 8.2.12 (Alexander duality theorem version 2) Let K be a non empty closed subset of $\mathbb{S}^n$. Then

$$\tilde{H}_{n-p-1}(\mathbb{S}^n \setminus K; R) \approx \tilde{\bar{H}}^p(K; R).$$

Proof: Again, since $\mathbb{S}^n$ is orientable over $\mathbb{Z}$, we have for $p > 0$

$$\bar{H}^p(K; R) \approx H_{n-p}(\mathbb{S}^n, \mathbb{S}^n \setminus K; R) \approx H_{n-p-1}(\mathbb{S}^n \setminus K; R)$$

and for $p = 0$, we have the following commutative diagram in which the rows are exact:

$$
\begin{array}{ccccccccc}
0 & \longrightarrow & H^0(\mathbb{S}^n) & \longrightarrow & \bar{H}^0(K) & \longrightarrow & \tilde{\bar{H}}^0(K) & \longrightarrow & 0 \\
& & \downarrow{\scriptstyle \approx} & & \downarrow{\scriptstyle \approx} & & \downarrow & & \\
H_n(\mathbb{S}^n \setminus K) & \longrightarrow & H_n(\mathbb{S}^n) & \longrightarrow & H_n(\mathbb{S}^n, \mathbb{S}^n \setminus K) & \longrightarrow & H_{n-1}(\mathbb{S}^n \setminus K) & \longrightarrow & 0
\end{array}
$$

The first homomorphism in the bottom row $H_n(\mathbb{S}^n \setminus K) \to H_n(\mathbb{S}^n)$ vanishes beacuse it factors through $H_n(\mathbb{S}^n \setminus K) \to H_n(\mathbb{S}^n \setminus \{*\}) = 0$. The conclusion of the corollary for $p = 0$ follows. ♠

As an immediate corollary to Lefschetz's duality theorem, we have:

Corollary 8.2.13 Let K be a proper compact subset of an R-orientable, connected n-manifold. Then $\bar{H}^q(K; G) = 0$ for all $q \geq n$ and for all R-modules G.

Remark 8.2.14 In particular, this is so for compact subsets of $\mathbb{R}^n$. You may say that this is something to do with the 'topological dimension' of compact subsets of K. However, the same does not hold for singular homology groups. Barratt and Milnor [Barratt–Milnor, 1962] have shown the existence of compact subsets K of $\mathbb{R}^n, n \geq 3$ which have non vanishing singular homology groups in dimensions $> n$. Indeed one can take $K = \cup_{k \geq 1} K_k \subset \mathbb{R}^3$, where K_k denotes the 2-sphere of radius $1/k$ and centre $(0, 0, 1/k)$.

We shall now consider manifolds M with boundary and denote the boundary of M by ∂M. We know that the interior int $M = M \setminus \partial M$ is then a manifold (without boundary).

Definition 8.2.15 We say M is orientable if int M is orientable.

Theorem 8.2.16 *Let M be a connected, compact orientable n-manifold with boundary with an orientation class $[M]$. Then*
(a) $H_n(M, \partial M) \approx R$.
(b) ∂M is a compact orientable $(n-1)$-manifold without boundary with an orientation class $[\partial M]$, where $[\partial M] = \partial[M]$ under the connecting homomorphism of the pair $(M, \partial M)$.
(c) There is a ladder of exact rows where vertical homomorphisms are duality isomorphisms given by cap products and the squares are commutative up to sign as indicated:

$$H^p(M) \xrightarrow{i^*} H^p(\partial M) \xrightarrow{\delta^*} H^{p+1}(M, \partial M) \xrightarrow{j^*} H^{p+1}(M) \quad (8.8)$$

$$\approx \Big\downarrow \frown[M] \quad (-1)^p \quad \approx \Big\downarrow \cap[\partial M] \quad (-1)^{p+1} \quad \approx \Big\downarrow \frown[M] \quad 1 \quad \approx \Big\downarrow \frown[M]$$

$$H_{n-p}(M, \partial M) \xrightarrow{\partial_*} H_{n-p-1}(\partial M) \xrightarrow{i_*} H_{n-p-1}(M) \xrightarrow{j_*} H_{n-p-1}(M, \partial M).$$

(d) *Suppose $\partial M = K \cup L$ where K, L are $(n-1)$-manifolds with common boundary $K \cap L$ (which may be empty also). Then*

$$\frown[M] : H^p(M, K) \xrightarrow{\approx} H_p(M, L).$$

Proof: (a) Let $\partial M \times [0, 1) \subset M$ be a collar neighbourhood of ∂M in M where in $\partial M \times 0$ is identified with ∂M. Put $K = M \setminus \partial M \times [0, 1)$. Then K is a compact subset of int M, which is orientable. By Theorem 8.2.5, we then have

$$
\begin{aligned}
H_n(M, \partial M) &\approx H_n(M, \partial M \times [0, 1)) && \text{(by homotopy)} \\
&\approx H_n(\text{int } M, \text{int } M \setminus K) && \text{(by excision)} \\
&\approx \bar{H}^0(K) && \text{(by duality)} \\
&\approx \bar{H}^0(M) && \text{(by homotopy)} \\
&\approx H^0(M) \approx R.
\end{aligned}
$$

Therefore, we can identify the orientation class μ of int M with a generator $[M]$ of $H_n(\text{int } M, \text{int } M \setminus K) \approx H_n(M, \partial M) \approx R$.
(b) Let $\partial M = A \sqcup B$, where A is one of the components of ∂M. Then by taking $K = M \setminus B \times (0, 1), L = \partial M$, in the above argument, we obtain

$$H_n(M, B) \approx H_n(M \setminus \partial M, M \setminus K) \approx H^0(K, \partial M) \approx H^0(\partial M, \partial M) = 0.$$

Therefore from the exact sequence of the triple $(M, \partial M, B)$ we have,

$$0 = H_n(M, B) \longrightarrow H_n(M, \partial M) \longrightarrow H_{n-1}(\partial M, B) \xrightarrow{\approx} H_{n-1}(A).$$

This means $H_{n-1}(A)$ contains the ring R and hence by Corollary 8.1.19, A is orientable. Since this is true for each component of ∂M, we get ∂M is orientable. This also shows that $\partial[M]$ is the sum $\sum_A \mu_A$ of generators $\mu_A \in H_{n-1}(A)$, where A runs over components of ∂M. Therefore, we can take $[\partial M] = \partial[M]$ as an orientation class for ∂M.

(c) Consider the following diagram:

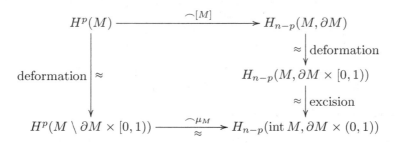

in which all three vertical isomorphisms are induced by inclusion maps, two of them deformations and the third one is excision. Therefore the diagram is commutative. Since the bottom horizontal arrow is the duality isomorphism so is the top one.

Thus, in Diagram (8.8), once we prove the commutativity, the third arrow will also represent an isomorphism by the Five lemma.

The commutativity of the third square is obvious. Let us represent the orientation class $[M, \partial M] \in H_n(M, \partial M)$ by a relative n-cycle c. Then ∂c is a $(n-1)$-cycle in ∂M which represents $[\partial M]$.

To see the commutativity of the first square, let f be a p-cocycle in M. Restriction to ∂M and then taking the cap product with ∂c yields $f|_{\partial M} \frown \partial c = f \frown \partial c = (-1)^p \partial(f \frown c)$. On the other hand, if we first take cap product with c and then take image under the connecting homomorphism $\partial_* : H_{n-p}(M, \partial M) \to H_{n-p-1}(\partial M)$, we get $\partial(f \frown c)$. This proves the commutativity of the first square up to the sign as indicated. The commutativity of the second square is similar and left to the reader as an exercise.

(d) Since $K \cap L$ has collar neighbourhoods both in K and L, it follows that $\{K, L\}$ is an excisive couple in $\partial M = K \cup L$. Therefore, there is the cap product

$$\frown \ : H^p(M, K) \otimes H_n(M, \partial M) \to H_{n-p}(M, L).$$

We have seen that ∂M is orientable with an orientation class $[\partial M] = \partial_*[M]$. Take its image under $H_{n-1}(\partial M) = H_{n-1}(K \cup L) \to H_{n-1}(\partial M, L) \approx H_{n-1}(K, K \cap L)$ which is clearly an orientation class for K (from the local property of the orientation class for ∂M). We then have a ladder of exact sequences in which the squares are commutative up to sign and the first, second, fourth and fifth vertical arrows are isomorphisms:

$$
\begin{array}{ccccccccc}
H^{p-1}(M) & \longrightarrow & H^{p-1}(K) & \longrightarrow & H^p(M, K) & \longrightarrow & H^p(M) & \longrightarrow & H^p(K) \\
{\scriptstyle \frown[M]}\downarrow{\scriptstyle\approx} & & {\scriptstyle\frown[K]}\downarrow{\scriptstyle\approx} & & {\scriptstyle\frown[M]}\downarrow & & {\scriptstyle\frown[M]}\downarrow{\scriptstyle\approx} & & {\scriptstyle\frown[K]}\downarrow{\scriptstyle\approx} \\
& & H_{n-p+1}(K, K \cap L) & & & & & & H_{n-p}(K, K \cap L) \\
& & \downarrow{\scriptstyle\approx} & & & & & & \downarrow{\scriptstyle\approx} \\
H_{n-p+1}(M, \partial M) & \to & H_{n-p}(K \cup L, L) & \to & H_{n-p}(M, L) & \to & H_{n-p}(M, \partial M) & \to & H_{n-p}(K \cup L, L).
\end{array}
$$

By Five lemma, it follows that the third vertical arrow also is an isomorphism. ♠

The following formulation of duality purely in terms of cohomology and cup product is quite useful. It is available over integer coefficients also provided the homology is known to be torsion free. Since we are working with field coefficients, from the universal coefficient theorem it follows that homology and cohomology are related more closely.

Theorem 8.2.17 *Let M be a compact, connected n-manifold with or without boundary and oriented over a field $\mathbb{K}$. Then the cup product*

$$\smile : H^p(M; \mathbb{K}) \otimes H^{n-p}(M, \partial M; \mathbb{K}) \to H^n(M, \partial M; \mathbb{K}) \approx \mathbb{K}$$

given by

$$\alpha \otimes \beta \mapsto \langle \alpha \smile \beta, [M] \rangle$$

defines a non degenerate pairing, viz., $\langle \alpha \smile \beta, [M] \rangle = 0$ for every β iff $\alpha = 0$ (and similarly, $\langle \alpha \smile \beta, [M] \rangle = 0$ for every α iff $\beta = 0$).

Proof: Since we are working over a field $\mathbb{K}$, every module A is free and hence $\text{Ext}(A, B) = 0$ always. Therefore by the universal coefficient theorem,

$$h : H^p(M, \mathbb{K}) \to \text{Hom}(H_p(M, \mathbb{K}); \mathbb{K})$$

is an isomorphism. Now the basic property of the cup and cap product that is employed here is the projection formula 7.3.5(b):

$$(\alpha \smile \beta) \frown [M] = \alpha \frown (\beta \frown [M]).$$

By Poincaré duality, given $\theta \in H_p(M)$, there is a (unique) $\beta \in H^{n-p}(M, \partial M)$ such that $\beta \frown [M] = \theta$. Now suppose $\langle \alpha \smile \beta, [M] \rangle = 0$ for every $\beta \in H^{n-p}(M, \partial M)$. This means $\alpha \frown \theta = 0$ for every $\theta \in H_p(M)$. This means that $h(\alpha) = 0$ as an element of $\text{Hom}(H_p(M; \mathbb{K}); \mathbb{K})$. By universal coefficient theorem this means $\alpha = 0$. Similarly, we see the non degeneracy in the second slot as well. ♠

Remark 8.2.18 Recall that the homology modules of a compact manifold are all finitely generated (Corollary 5.1.18). As an immediate consequence we can say that the assignment $\alpha \mapsto \alpha^*$, where $\alpha^*(\beta) = \langle \alpha \smile \beta, [M] \rangle$ defines an isomorphism

$$H^p(M, \mathbb{K}) \to \text{Hom}(H^{n-p}(M, \partial M; \mathbb{K}); \mathbb{K}).$$

Exercise 8.2.19

(i) Show that removing a finite number of points from a manifold does not affect the orientability of the manifold.

(ii) For a connected closed manifold M, assume that the cardinality of $\pi_1(M, x)$ is odd. Show that $H_{n-1}(M, \mathbb{Z})$ is torsion free.

(iii) Suppose X is a compact connected orientable n-manifold with its first Betti number $\beta_1(X) = 0$. Show that for any proper closed subset A of X, the number of components of $X \setminus A$ is equal to one more than the rank of $\bar{H}^{n-1}(A)$.

(iv) In the above exercise, assume further that $H_1(X) = (0)$. Show that every $(n-1)$-dimensional closed submanifold of X is orientable.

(v) Show that no closed non orientable surface can be embedded in $\mathbb{R}^3$.

(vi) **Homology of the knot complements** Compute $H_{n-1}(\mathbb{S}^n \setminus f(\mathbb{S}^{n-2}); \mathbb{Z})$, where $f : \mathbb{S}^{n-2} \to \mathbb{S}^n$ is an arbitrary smooth embedding.

8.3 Some Applications

This section will contain a few applications of duality.

Theorem 8.3.1 *The cohomology algebras of real, complex and quaternion projective spaces of dimension n are the truncated polynomial algebras over one variable, respectively given by*
(i) $H^*(\mathbb{P}^n; \mathbb{Z}_2) \approx \mathbb{Z}_2[u]/(u^{n+1})$, $\deg u = 1$
(ii) $H^*(\mathbb{CP}^n; \mathbb{Z}) \approx \mathbb{Z}[v]/(v^{n+1})$, $\deg v = 2$;
(iii) $H^*(\mathbb{HP}^n; \mathbb{Z}) \approx \mathbb{Z}[w]/(w^{n+1})$, $\deg w = 4$.

Proof: We shall prove (i) only, the proofs of (ii) and (iii) are similar. Recall that $\mathbb{P}^n$ has a CW-complex structure with a single cell in each dimension $0 \le q \le n$ and the attaching map of the q cell is via the double covering map $\mathbb{S}^{q-1} \to \mathbb{P}^{q-1}$. Therefore the associated chain complex with $\mathbb{Z}_2$ coefficients has $C_q = \mathbb{Z}_2$ for $0 \le q \le n$ and with the boundary maps $\partial_q = 0$ for all $1 \le q \le n$. It follows that the cohomology as a module is as claimed. Moreover, it also follows that the inclusion map $\mathbb{P}^q \to \mathbb{P}^{q+1}$ induces an isomorphism in cohomology in dimensions $\le q$. We shall use this isomorphism to pull back the notations we use for generators of $H^i(\mathbb{P}^q)$ to $H^i(\mathbb{P}^{q+1})$ inductively for $i \le q$. Clearly the result holds for $q = 1$, wherein there is no difference between the module and the algebra. Assume the result to be true for $n \le q$. We then have to show that $u \smile u^q$ is the generator of $H^{q+1}(\mathbb{P}^{q+1})$ which follows from the duality Theorem 8.2.17. We can then justifiably denote this generator by u^{q+1} and proceed.   ♠

Degree of a Map

Definition 8.3.2 Let $f : M \to N$ be any map between connected oriented n-manifolds. Then the degree of f is defined to be the unique integer such that

$$f_*[M] = (\deg f)[N]. \tag{8.9}$$

Remark 8.3.3

1. Note that if M is orientable then for any map $f : M \to M$, the degree is well defined, since after choosing an orientation on $[M]$, we must use the same one for domain as well as codomain of f.

2. If $[\tilde{M}], [\tilde{N}]$, etc., denote the dual elements in the cohomology, it is easily checked that $f^*([\tilde{N}]) = (\deg f)[\tilde{M}]$.

3. **The geometric degree** Temporarily, let us call (8.9) the algebraic degree of f. Now recall that if $f : M \to N$ is a smooth map of oriented smooth compact manifolds, then the geometric degree of f is defined as follows: By Sard's theorem there exists $y \in N$ which is a regular value of f, i.e., for each $x \in M$ such that $f(x) = y$ the tangent map $df_x : T_x M \to T_y N$ is an isomorphism. By inverse function theorem and compactness of M, this implies that there is a coordinate neighbourhood V of y in N (diffeomorphic to a closed disc) such that $\bar{U} := f^{-1}(\bar{V}) = \coprod_j \bar{U}_j$ is a disjoint union of closures of open subsets of M and for each j, $f : \bar{U}_j \to \bar{V}$ is a diffeomorphism. Under the induced orientations, it makes sense whether these diffeomorphisms preserve orientations or not and accordingly, we assign the value $c_j := \pm 1$ to each index j. The sum $\sum_j c_j$ is called the geometric degree of f.

   We claim that the algebraic and geometric degrees of a smooth map are the same.

Consider the following commutative diagram which consists of homomorphisms induced by inclusions or restrictions of f.

$$
\begin{array}{ccc}
H_n(\bar{U}, \partial\bar{U}) & \xrightarrow{f_*} & H_n(\bar{V}, \partial\bar{V}) \\
\eta_* \downarrow \approx & & \downarrow \approx \\
H_n(M, M \setminus U) & \xrightarrow{f_*} & H_n(N, N \setminus V) \\
i_* \uparrow & & \approx \updownarrow \\
H_n(M) & \xrightarrow{f_*} & H_n(N).
\end{array}
$$

Since i_* is injective, the task of determining the bottom arrow is converted into the task of determining the effect of the top arrow on $\eta_*^{-1}(i_*[M]) \in H_n(\bar{U}, \partial\bar{U})$. Now $H_n(\bar{U}, \partial\bar{U}) = \oplus_j H_n(\bar{U}_j, \partial\bar{U}_j)$. Taking the induced orientations $[U_j]$ on each U_j, it follows that $\sum_j \eta_*([U_j]) = i_*[M]$. Since by definition of c_j, we have $f_*([U_j]) = c_j[V]$, the claim follows.

4. Thus from now onward, we can merely speak about the degree of a map, without the qualifier 'algebraic' or 'geometric'.

Corollary 8.3.4 Every map $f : \mathbb{CP}^{2n} \to \mathbb{CP}^{2n}$ is of degree non negative. In particular, $\mathbb{CP}^{2n}$ does not admit any orientation reversing self-homeomorphisms.

Proof: If $f^*[v] = kv$ for some integer k, we have

$$
f^*[\mathbb{CP}^{2n}] = f^*(v^{2n}) = (f^*(v))^{2n} = k^{2n}[\mathbb{CP}^{2n}]
$$

and k^{2n} is non negative for any integer k. ♠

Remark 8.3.5 On the other hand, the map $g : \mathbb{CP}^{2n+1} \to \mathbb{CP}^{2n+1}$ given by $[z_1, \ldots, z_{2n+2}] \mapsto [\bar{z}_1, \ldots, \bar{z}_{2n+2}]$ is an orientation reversing diffeomorphism. For $n = 0$, this is as easy consequence of the fact that $z \mapsto \bar{z}$ is orientation reversing in $\mathbb{C}$. If $v \in H^2(\mathbb{CP}^{2n+2}; \mathbb{Z}) \approx \mathbb{Z}$ denotes a generator, this implies that $g^*(v) = -v$. Therefore, $g^*[\mathbb{CP}^{2n+1}] = -[\mathbb{CP}^{2n+1}]$.

Theorem 8.3.6 *Let $f : M \to N$ be any map between closed connected oriented manifolds of dimension n. Suppose $\deg f = d \neq 0$. Then*
(i) $f_\#\pi_1(M)$ is of index k in $\pi_1(N)$ for some k which divides d.
(ii) $f_* : H_*(M; \mathbb{Q}) \to H_*(N; \mathbb{Q})$ is surjective and $f^* : H^*(N; \mathbb{Q}) \to H^*(M; \mathbb{Q})$ is injective.
(iii) If $\deg f = \pm 1$ then $f_* : H_*(M; \mathbb{Z}) \to H_*(N; \mathbb{Z})$ is split surjective and $f^* : H^*(N; \mathbb{Z}) \to H^*(M; \mathbb{Z})$ is split injective.

Proof: (i) Let $p : \tilde{N} \to N$ be a k-sheeted covering corresponding to the subgroup $f_\#(\pi_1(M)) \subset \pi_1(N)$. Clearly, there is a map $\tilde{f} : M \to \tilde{N}$ such that $p \circ \tilde{f} = f$. If $k = \infty$ then $\tilde{N}$ is non compact and we know $H_n(\tilde{N}) = 0$. It follows that $f_* = 0 : H_n(M) \to H_n(N)$ and $d = 0$ (which is OK as far as k dividing d, but we have assumed that $d \neq 0$). Therefore k is finite and $\tilde{N}$ is an oriented closed manifold and it follows that $d = \deg f = \deg \tilde{f} \cdot \deg p = (\deg \tilde{f})k$.

We shall prove (iii) since a slight modification of it will yield proof of (ii). We shall apply the projection formula (see Theorem 7.3.5)(d): $f_*(f^*(u) \frown z) = u \frown f_*(z)$. For simplicity, by changing the orientation on N we may assume that $f_*[M] = [N]$. Let us write

$$
\phi_i = \frown [M] : H^i(M) \to H_{n-i}(M); \quad \psi_i = \frown [N] : H^i(N) \to H_{n-i}(N).
$$

$$H_i(M) \xrightarrow{\quad f_* \quad} H_i(N)$$

$$\cap[M] \Big\uparrow \qquad\qquad\qquad \Big\uparrow \cap[N]$$

$$H^{n-i}(M) \xleftarrow{\quad f^* \quad} H^{n-i}(N)$$

Put $\alpha_i = \phi_{n-1} \circ f^* \circ \psi_{n-i}^{-1} : H_i(N) \to H_i(M)$. Then by the projection formula above, for any $u \in H^{n-i}(N)$,

$$f_*(f^*(u) \frown [M]) = u \frown f_*[M] = u \frown [N]$$

which is the same as saying $f_* \circ \alpha = Id$. Thus we have constructed an explicit right-inverse of f_*. This also proves that $\beta_i = \psi_i^{-1} \circ f_* \circ \phi_i$ is the left-inverse of f^*. ♠

Remark 8.3.7 The above result is at the starting point of surgery theory. One of the central problems in this theory is as follows: Given a map f as in the above theorem, can we 'modify' M as well as f to $f' : M' \to N$ which is a homotopy equivalence? The kind of modifications that are allowed are called 'spherical modification' or 'surgery' on the map f which try to cut down the kernel and cokernel of f_* so that it becomes an isomorphism. For further reading on this topic, see [Browder, 1965].

Cobordism

We have seen in Remark 5.2.17 how Poincaré duality influences Euler characteristic:

Theorem 8.3.8 *If X is a closed manifold of odd dimension, then $\chi(X) = 0$.*

Let us now see how duality puts certain restrictions on the homology of a manifold $X = \partial M$ which is the boundary of some other manifold. Of course, for simplicity, we shall consider only the compact case. Note that the only connected closed 1-manifold $\mathbb{S}^1$ is the boundary of $\mathbb{D}^2$. All orientable surfaces are boundaries of solids in $\mathbb{R}^3$. Is the simplest non-orientable (closed) surface $\mathbb{P}^2$ a boundary? Wait a minute.

Theorem 8.3.9 *Let M^{2n+1} be a manifold with boundary $\partial M = X^{2n}$. Then $\chi(X)$ is even.*

Proof: Consider the double $2M$ of M, obtained by gluing two copies of M along their common boundary. By Mayer–Vietoris, it follows that $\chi(2M) = 2\chi(M) - \chi(X)$. On the other hand, $2M$ being a closed manifold of odd dimension, we have $\chi(2M) = 0$. ♠

Corollary 8.3.10 None of the manifolds $\mathbb{P}^{2n}, \mathbb{CP}^{2n}, \mathbb{HP}^{2n}$ is a total boundary of a manifold.

Proof: The Euler characteristics of these manifolds are odd. ♠

Example 8.3.11 The above result is not applicable to the Klein bottle $\mathbf{K}$, since its Euler characteristic is 0. A somewhat surprising result is that $\mathbf{K}$ is indeed a boundary. To see this, take $W = \ddot{\mathrm{M}} \times \mathbb{I}$, the product of the Möbius band with an interval. Then $\partial W = \mathbf{K}$, the Klein bottle, being the connected sum of $\mathbb{P}^2$ with itself. (Do not confuse W with a 'thick' Möbius band, because a thickening of an embedding $\mathbf{M} \subset \mathbb{R}^3$ is not homeomorphic to $\mathbf{M} \times \mathbb{I}$.) More generally, let X be any closed n-manifold, $A \subset X$ be homeomorphic to $\mathrm{int}\,\mathbb{D}^n$. Then $M = (X \setminus A) \times \mathbb{I}$ is a $(n+1)$-manifold with boundary homeomorphic to the connected sum of two copies of X. Caution is needed here when X is orientable—then ∂M is the connected sum $X \# (-X)$ where $-X$ denotes X with the opposite orientation.

Following this trick, you can see that a non orientable surface is a boundary iff it has even Euler characteristic. However, in higher dimensions, one has to work harder.

We shall now take up manifolds X of dimension $n = 4m$.

Definition 8.3.12 Let X be an oriented closed manifold of dimension $4m$. Then the cup product pairing $\langle -, - \rangle : H^{2m}(X; \mathbb{R}) \times H^{2m}(X; \mathbb{R}) \to \mathbb{R}$

$$\langle u, v \rangle = (u \smile v) \frown [X]$$

is symmetric and bilinear. From Poincaré duality, it follows that this bilinear form is non degenerate, i.e., the matrix representing it is invertible. Since it is symmetric, all its eigenvalues are real.

The index or the signature of any non degenerate symmetric bilinear form is defined to be the number of positive eigenvalues minus the number of negative eigenvalues. The index or the signature of X is defined to be the index of the above bilinear form.

Theorem 8.3.13 (Thom) *If $X^{4m} = \partial M$, where M is a compact orientable manifold, then the index of X is zero.*

We need the following lemma:

Lemma 8.3.14 Let all coefficients for homology and cohomology be taken over a field R, and let M be an R-oriented manifold of dimension $2k+1$ with $\partial M = X$, $i : X \subset M$. Then

$$\dim \mathrm{Ker}\,[i_* : H_k(X) \to H_k(M)] = \frac{1}{2} \dim H^k(X) = \dim \mathrm{Im}\,[i^* : H^k(M) \to H^k(X)].$$

In particular, $\dim H^k(X)$ is even. Moreover, for any two elements $z_1, z_2 \in \mathrm{Im}\,i^*$, we have $z_1 \smile z_2 = 0$.

Proof: By rank-nullity theorem,

$$\dim \mathrm{Ker}\,i_* + \mathrm{rank}\,i_* = \dim H_k(X).$$

We shall show that $\dim \mathrm{Ker}\,i_* = \mathrm{rank}\,i_*$ from which one equality follows. The other one is similar. Again, since all vector spaces involved are finite dimensional, $H^k(X) = \mathrm{Hom}(H_k(X); R)$, $i^* = \mathrm{Hom}(i_*, 1)$, etc., by elementary linear algebra, we have $\mathrm{rank}\,i_* = \mathrm{rank}\,i^*$. Finally, from the following commutative diagram, wherein the rows are exact and vertical arrows are isomorphisms, it follows that $\frown [X]$ maps $\mathrm{Im}\,i^*$ onto $\mathrm{Ker}\,i_*$ isomorphically and hence $\dim \mathrm{Ker}\,i_* = \mathrm{rank}\,i^*$. The claim follows.

$$
\begin{array}{ccccc}
H^k(M) & \xrightarrow{\;i^*\;} & H^k(X) & \xrightarrow{\;\delta\;} & H^{k+1}(M, X) \\
\scriptstyle\frown[M] \downarrow \scriptstyle\approx & & \scriptstyle\frown[X] \downarrow \scriptstyle\approx & & \scriptstyle\frown[M] \downarrow \scriptstyle\approx \\
H_{k+1}(M, X) & \xrightarrow{\;\partial\;} & H_k(X) & \xrightarrow{\;i_*\;} & H_k(M).
\end{array}
$$

Now, since $\partial[M] = [X]$, given $u, v \in H^k(M)$, we have,

$$i_*[(i^*(u) \smile i^*(v)) \frown [X]] = (\delta \circ i^*(u \smile v)) \frown [M] = 0$$

and since $i_* : H_0(X) \to H_0(M)$ is an isomorphism, it follows that $(i^*(u) \smile i^*(v)) \frown [X] = 0$ which implies $i^*(u) \smile i^*(v) = 0$. ♠

Coming to the proof of Thom's theorem, let p, q denote the number of positive and number of negative eigenvalues of the cup-product bilinear pairing which is non degenerate. Then $d = \dim H^{2m}(X) = p + q$. By the above lemma, it follows that the image of i^* has to be complementary to the subspace on which the cup-product is positive definite and hence $d/2 + p \leq d$ and for similar reason, $d/2 + q \leq d$. Adding these two inequalities we get $d + p + q \leq 2d$ wherein equality holds. Therefore we must have equality everywhere: $d/2 + p = d = d/2 + q$ which means $p = q$. Therefore the index is zero. ♠

Example 8.3.15 It is not hard to see that the cup-product form of a connected sum of two manifolds is the 'orthogonal sum' of the cup-product forms of the two manifolds. Therefore, the index gets added up under connected sum. Likewise, one easily checks that the change of orientation changes the sign of the index. Thus the index of $\mathbb{CP}^2$ is easily seen to be 1. Therefore connected sum of any number of copies of $\mathbb{CP}^2$ is not the boundary of any orientable 5-manifold. On the other hand, if M is any orientable closed manifold by removing an open tubular neighbourhood of $\{*\} \times \mathbb{I}$ from $M \times I$, we have a manifold whose boundary is $M\#(-M)$. In particular, $(\mathbb{CP}^2)\#(-\mathbb{CP}^2)$ is the boundary of a 5-manifold.

Is then $\mathbb{CP}^2\#\mathbb{CP}^2$ the boundary of a non orientable 5-manifold? The answer is yes. To see this, as before we begin with $\mathbb{CP}^2 \times \mathbb{I}$ and then take a connected sum of this with some non orientable closed 5-manifold; e.g., $\mathbb{P}^2 \times \mathbb{S}^3$ will do, i.e., $W = (\mathbb{CP}^2 \times \mathbb{I})\#(\mathbb{P}^2 \times \mathbb{S}^3)$. Now ∂W consists of two nice copies of $\mathbb{CP}^2$. (Note that the operation of taking connected sum with a closed manifold does not disturb the boundary.) Let A be a smoothly embedded arc in W with $\omega \cap \partial W = \partial A$ so that A intersects both the boundary components of W transversely in a single point. Then a tubular neighbourhood B of A is diffeomorphic to $\mathbb{I} \times \mathbb{D}^4$ and we can remove $\mathbb{I} \times \text{int}\,\mathbb{D}^4$ so as to get a manifold W' with boundary equal to $(\mathbb{CP}^2)\#(\mathbb{CP}^2)$ or $(\mathbb{CP}^2)\#(-\mathbb{CP}^2)$ depending on whether the arc A is orientation preserving or orientation reversing. Since W is non orientable, we can always arrange such that the arc A is orientation reversing and then we have the desired W'. Note that there is nothing special about $\mathbb{CP}^2$ in the above discussion and the same holds for any manifold M, i.e., $M\#M$ is the boundary of a (probably non orientable) manifold.

Remark 8.3.16 The above result is the starting point of the cobordism theory due to the works of Pontrjagin and Thom. Two closed n-manifolds M, N are said to be *cobordant* if there is a $(n+1)$-manifold W such that $\partial W = M \sqcup N$. In this terminology a manifold which bounds is said to be *null-cobordant*. On the set of homeomorphism (resp. diffeomorphism) class of closed n-dimensional manifolds, 'cobordism' introduces an equivalence relation. The set Ω^n of these equivalence classes forms a commutative, associative, semigroup under disjoint union, with the class of $\mathbb{S}^n$ as the neutral element. As seen in Example 8.3.16 above, it follows that $M\#M$ is always null-cobordant. There are different versions of this such as 'oriented cobordism', 'framed cobordism', etc., with applications to other areas of mathematics.

Exercise 8.3.17 Consider closed surfaces. Show that all oriented ones are null-cobordant; all non oriented ones of even genus are null cobordant; all non oriented ones of odd genus are cobordant to each other.

8.4 de Rham Cohomology

Let X be a smooth n-dimensional manifold. Consider the functorial graded algebra $\Omega^*(X)$ of smooth differential forms over the ring $\mathcal{C}^\infty(X; \mathbb{R})$. Together with the functorial differential operator $d : \Omega^*(X) \longrightarrow \Omega^*(X)$ this becomes a co-chain complex and is called the *de Rham complex* of X. In Section 4.1, we have indicated how differential forms and the integration theory are at the root of the idea of cohomology theory.

Let us denote the homology modules $H(\Omega^*(X))$ by $H_{DR}^*(X)$. These are called the de Rham cohomology modules of X. It is clear that $X \rightsquigarrow H_{DR}^p(X)$ defines a contravariant functor on the category of smooth manifolds to the category of graded vector spaces over $\mathbb{R}$. Indeed the graded algebra structure on $\Omega^*(X)$ given by the wedge product of smooth forms induces a graded algebra structure on the cohomology modules. In this section, we shall establish that de Rham cohomology of a smooth manifold is naturally isomorphic to the singular cohomology of the manifold. This celebrated result, probably known to Poincaré and explicitly conjectured by Cartan in 1928, was rigorously proved in 1931 by de Rham.

The singular cohomology of smooth singular chains plays an intermediary role here in the proof. The proof we present here may be attributed to Bredon, especially the idea in the bootstrap lemma. The standard proof of this theorem using sheaf theory will be presented in the next chapter. There are other proofs which are slight variations of these themes. For proof based on Weil's idea of double complexes, you may look into www3.nd.edu/ lnicolae/Fanoethesis.pdf.

Apart from some basic properties of differential forms and integration, we need two specific fundamental results—the Stokes theorem for singular chains and the Poincaré lemma. For ready reference, we shall restate them here. For proofs and more details the reader may refer to [Shastri, 2011], [Rudin, 1976].

Definition 8.4.1 By a smooth singular q-simplex in X we mean a map $\sigma : \Delta_q \to X$ which is defined and smooth in a neighbourhood of the standard simplex Δ_q in $\mathbb{R}^{q+1}$. Given a smooth q-form ω on X, consider the integral

$$\Psi(\sigma)(\omega) = \int_\sigma \omega := \int_{\Delta_q} \sigma^*(\omega) = \sum_\sigma n_\sigma \int_\sigma \omega.$$

Here we take Δ_q with its standard orientation. More generally, if $c = \sum_\sigma n_\sigma \sigma$, is a smooth q-chain we can extend this definition linearly:

$$\Psi(c)(\omega) = \int_c \omega.$$

Thus we get a linear map

$$\Psi(\omega) : S_q^{sm}(X) \to \mathbb{R}$$

which in turn defines a linear map

$$\Psi := \Psi_q : \Omega^q(X) \to \mathrm{Hom}(S_q^{sm}(X); \mathbb{R}), \quad 0 \le q \le n.$$

These are called *period maps*.

Theorem 8.4.2 (Stokes) *For any $(q-1)$-form on X and any smooth singular q-chain c in X we have:*

$$\int_c d\lambda = \int_{\partial c} \lambda.$$

[In some expositions, the above theorem may be stated and proved for only open subsets X of $\mathbb{R}^n$. Using functoriality, linearity, etc., it is then quite routine to deduce the general statement from this. It is also possible that you have come across a version that gives treatment for 'cubical' singular chains. Once again, you must be able to adopt the proof easily to the case of simplicial singular chains.]

Thus, the Stokes theorem says that Ψ is a chain map, i.e., $\Psi \circ d = \delta \circ \Psi$. Therefore Ψ defines a homomorphism

$$\Psi^* : H_{DR}^*(X) \to H_{sm}^*(X).$$

It is easily verified that for a point space $\{*\}$, de Rham cohomology groups are trivial, i.e.,

$$H_{DR}^p(*) \approx \left\{ \begin{array}{ll} \mathbb{R}, & p = 0 \\ 0, & p > 0. \end{array} \right.$$

The celebrated Poincaré lemma says that in any convex domain every closed form is exact. In modern terminology, the Poincaré lemma may be described as homotopy invariance of de Rham cohomology:

Theorem 8.4.3 (Poincaré lemma) *Let $H : X \times \mathbb{I} \to Y$ be a smooth homotopy and put $h_t(x) = H(x,t)$. Then the induced homomorphisms $h_t^* : H_{DR}^*(Y) \to H_{DR}^*(X)$ are the same for all $t \in \mathbb{I}$.*

As an immediate consequence, we get

Theorem 8.4.4 (Homotopy invariance of de Rham cohomology) *If $f : X \to Y$ is a smooth homotopy equivalence, then $f^* : H_{DR}^*(Y) \to H_{DR}^*(X)$ is an isomorphism.*

We can now state:

Proposition 8.4.5 Let X be a smooth manifold.
(A) $\Psi^* : H_{DR}^*(X) \to H_{sm}^*(X)$ is an isomorphism.
(B) The inclusion of the chain complexes $\eta : S_*^{sm}(X) \to S_*(X)$ induces an isomorphism of homology modules $\eta_* : H_*^{sm}(X) \to H_*(X)$.

By appealing to universal coefficient theorem, we obtain:

Theorem 8.4.6 (de Rham) *There is a canonical equivalence*

$$(\eta^*)^{-1} \circ \Psi^* : H_{DR}^*(X) \to H^*(X; \mathbb{R})$$

of the two contravariant functors, the de Rham cohomology and the singular cohomology, on the category of smooth manifolds.

The rest of this section will be devoted to the proof of the above proposition, and thereby the proof of the theorem. The key idea is the following lemma which is similar to the bootstrap lemma 8.1.16 which we have seen earlier. Even the proof is more or less the same. However, we shall write down the details to a large extent.

Lemma 8.4.7 Let X be a smooth manifold. Suppose $P(U)$ is a statement about open subsets of X which satisfies the following three axioms.
(i) $P(U)$ is true if U is diffeomorphic to a convex open subset of $\mathbb{R}^n$.
(ii) $P(U), P(V), P(U \cap V) \implies P(U \cup V)$.
(iii) For a disjoint family of open sets $\{U_\alpha\}$, $P(U_\alpha)$ is true for all $\alpha \implies P(\cup_\alpha U_\alpha)$.
Then $P(X)$ is true.

Proof: (i) and (ii) together imply, by induction, that $P(U)$ holds for open subsets of $\mathbb{R}^n$ which are a finite union of convex open subsets, because of the set theoretic identity

$$(U_1 \cup \cdots \cup U_k) \cap U_{k+1} = (U_1 \cap U_{k+1}) \cup \cdots \cup (U_k \cap U_{k+1}).$$

Now given any open set $U \subset \mathbb{R}^n$ choose a proper mapping $f : U \to [0, \infty)$ and put

$$U_i = f^{-1}(i-1, i+1), \quad U_{odd} = \cup_{i=0}^\infty U_{2i+1}, \quad U_{even} = \cup_{i=1}^\infty U_{2i}.$$

Note that the closure of each U_i is compact and hence each of them is a finite union of convex open sets. Therefore $P(U_i)$ is true. The same is true for $U_i \cap U_{i+1}$ as well. Now each of the U_{odd}, U_{even} is a disjoint union of $P(U_i)$ and hence $P(U_{odd}), P(U_{even})$ are true. Moreover, $U_{odd} \cap U_{even}$ is a disjoint union of $U_{2i} \cap U_{2i+1}$ and hence $P(U_{odd} \cap U_{even})$ is true. By (ii) again, since $U = U_{odd} \cup U_{even}$, $P(U)$ is true.

We now have axioms (ii) (iii) and $(i)'$ in place of (i) where
$(i)'$ $P(U)$ is true if U is diffeomorphic to open subsets of $\mathbb{R}^n$. The above arguments can be repeated and yield that $P(X)$ is true. ♠

Proof of Proposition 8.4.5:

Denoting the claims (A), (B) of the proposition by P_A, P_B, respectively, we shall varify that they satisfy the axioms in the above lemma.

(i) This is an easy consequence of homotopy invariance of the three cohomology groups involved.

(ii) This is an easy consequence of excision property satisfied by the three cohomology groups. For singular and smooth singular, we have seen this. For the de Rham complex, we actually have a stronger version: For any open sets U_0, U_1 there is the functorial short exact sequence

$$0 \longrightarrow \Omega^p(U_0 \cup U_1) \xrightarrow{(i_0, -i_1)} \Omega^p(U_0) \oplus \Omega^p(U_1) \xrightarrow{j_0 + j_1} \Omega^p(U_0 \cap U_1) \longrightarrow 0 \qquad (8.10)$$

where i_0, j_0, etc., are appropriate inclusion induced maps. The only non trivial thing that needs to be proved is the surjectivity of $j_0 + j_1$. So let $\omega \in \Omega^p(U_0 \cap U_1)$. Choose a smooth function $f : U_0 \cup U_1 \to \mathbb{I}$ such that $f = 0$ on a neighbourhood of $U_0 \setminus U_1$ and $f = 1$ on a neighbourhood of $U_1 \setminus U_0$. (This is Smooth Urysohn's lemma which can be easily proved by using smooth partition of unity. See for example Section 1.7 of [Shastri, 2011].) Then $f\omega$ is zero in a neighbourhood of $U_0 \setminus U_1$ and hence can be extended to a smooth p-form ω_0 on U_0 by 0. Similarly, $(1 - f)\omega$ can be extended to a p-form ω_1 on U_1. Now on $U \cap V$ it follows that $\omega = f\omega + (1 - f)\omega = j_1(\omega_1) + j_2(\omega_2)$. The exactness of (8.10) follows.

Clearly we have a commutative diagram

$$
\begin{array}{ccccccccc}
0 & \longrightarrow & \Omega^p(U_0 \cup U_1) & \xrightarrow{(i_0, -i_1)} & \Omega^p(U_0) \oplus \Omega^p(U_1) & \xrightarrow{j_0 + j_1} & \Omega^p(U_0 \cap U_1) & \longrightarrow & 0 \\
 & & \downarrow & & \downarrow & & \downarrow & & \\
0 & \longrightarrow & [S_{sm}(U_0) + S_{sm}(U_1)]^* & \longrightarrow & [S_p^{sm}(U_o)]^* \oplus [S_p^{sm}(U_1)]^* & \longrightarrow & [S_p^{sm}(U \cap V)]^* & \longrightarrow & 0
\end{array}
$$

where $[-]^*$ denotes the dual group of linear maps into $\mathbb{R}$ and the vertical arrows present respective Ψ. This in turn yields a ladder of Mayer–Vietoris exact sequences of cohomology modules

$$
\begin{array}{ccccccccc}
\cdots \longrightarrow & H_{DR}^p(U_0 \cup U_1) & \longrightarrow & H_{DR}^p(U_0) \oplus H_{DR}^p(U_1) & \longrightarrow & H_{DR}^p(U_0 \cap U_1) & \longrightarrow & H_{DR}^{p+1}(U_0 \cup U_1) & \longrightarrow \cdots \\
 & \downarrow & & \downarrow & & \downarrow & & \downarrow & \\
\cdots \longrightarrow & H_{sm}^p(U_0 \cup U_1) & \longrightarrow & H_{sm}^p(U_0) \oplus H_{sm}^p(U_1) & \longrightarrow & H_{sm}^p(U_0 \cap U_1) & \longrightarrow & H_{sm}^{p+1}(U_0 \cup U_1) & \longrightarrow \cdots
\end{array}
$$

Now by Five lemma (ii) follows for $P_A(U)$. The same holds for $P_B(U)$ by using the homology Mayer–Vietoris sequences for H_{sm} and H.

(iii) This comes almost freely once you observe that each of the cohomology (homology) modules of a disjoint union of open sets U_α is the direct product (direct sum) of the cohomology (homology) modules of each U_α.

This completes the proof of de Rham's theorem.

8.5 Miscellaneous Exercises to Chapter 8

1. Show that a bundle is orientable iff there exist local trivializations so that the corresponding transition functions have a positive determinant.

2. Show that if ξ is orientable then all the exterior powers $\Lambda^i(\xi)$ are orientable.

3. Let ξ be a line bundle over B. Then ξ is orientable iff it is trivial.

4. Show that a k-plane bundle ξ is orientable iff $\Lambda^k(\xi)$ is orientable iff $\Lambda^k(\xi)$ is trivial. ($\Lambda^k(\xi)$ is called the determinant bundle of ξ.)

5. For every complex bundle ξ, the underlying real bundle $\xi_{\mathbb{R}}$ is orientable.

Chapter 9

Cohomology of Sheaves

The concept of sheaves has its origin in the study of holomorphic functions on a complex manifold. It provides a systematic approach to keep track of local algebraic data on a topological space and aids in arriving at global conclusions. Experience has shown that certain techniques developed in dealing with holomorphic functions are applicable to smooth functions, continuous functions, and so on. Sheaf theory is the outcome of axiomatizing certain basic local properties common to all these functions. In this chapter, we shall present some basic properties of sheaves. For further reading one may look into [Ramanan, 2004].

In Section 9.1, we discuss some generalities about presheaves and sheaves. In Section 9.2, we take up the study of injective modules and injective sheaves and establish the fundamental result that every sheaf has an injective resolution. In Section 9.3, we introduce sheaf cohomology and discuss its relation with ordinary singular cohomology. As an application, we present a proof of de Rham's theorem. Section 9.4 is devoted to the basics of Čech cohomology and importance of having a 'good' covering.

9.1 Sheaves

Here, we shall briefly discuss some generalities of sheaves. We shall begin with the functorial approach to a sheaf and later include the etale space point of view. The reader is advised to consult some other source(s) such as [Godement, 1958] or [Hartshorne, 1977] for more details.

Throughout this section, let $\mathcal{C}$ denote a subcategory of **Ens** with an initial object 0 (which is not necessarily the empty set). Often, we shall be interested in $\mathcal{C} = $ **Ab** or a subcategory of **Ab** such as **R-mod**, etc. Recall from Example 1.8.3.8, that given a topological space, we have a category $\mathcal{U}_X$ of open sets of X and restriction maps.

Definition 9.1.1 By a *presheaf* on X with values in $\mathcal{C}$, we mean a contravariant functor $\mathcal{F} : \mathcal{U}_X \rightsquigarrow \mathcal{C}$ such that $\mathcal{F}(\emptyset) = 0$.

Remark 9.1.2

1. More elaborately, a presheaf $\mathcal{F}$ assigns, to each open set U of X an object $\mathcal{F}(U) \in \mathcal{C}$ and a $\mathcal{C}$-morphism $\mathcal{F}(j_{VU}) =: \rho_{VU} : \mathcal{F}(U) \to \mathcal{F}(V)$ whenever $V \subset U$. This assignment has the properties:
   (i) $\mathcal{F}(\emptyset) = 0$;
   (ii) $\rho_{UU} = Id$;
   (iii) for open sets $W \subset V \subset U$ we have $\rho_{WU} = \rho_{WV} \circ \rho_{VU}$.

2. Since $\mathcal{C}$ is a subcategory of **Ab** it follows that each $\mathcal{F}(U)$ is an abelian group. So, we are free to use set-theoretic terminology such as $s \in \mathcal{F}(U)$, restriction map, etc.

3. Notice the extra notation ρ_{VU} for the morphism $\mathcal{F}(j_{VU})$. This is not only convenient but is meant to remind you that often we are indeed working with 'restriction' maps, even in the codomain category $\mathcal{C}$. (But this is not a logical necessity).

4. A further simplification of the notation is also in vogue: given $s \in \mathcal{F}(U)$, $\rho_{VU}(s)$ is denoted by $s|_V$ and is called the restriction of s to V.

Definition 9.1.3 Given a presheaf $\mathcal{F}$ over a space X and a collection $\mathcal{U}$ of open subsets of X by a *compatible $\mathcal{U}$-family of $\mathcal{F}$* we mean a collection $\{s_U \ : \ s_U \in \mathcal{F}(U)\}_{U \in \mathcal{U}}$ with the property that for any two members $U, V \in \mathcal{U}$ we have

$$s_U|_{U \cap V} = s_V|_{U \cap V}.$$

It is easily checked that the collection of all compatible $\mathcal{U}$-families for $\mathcal{F}$ forms an object in $\mathcal{C}$ which we shall denote by $\mathcal{F}(\mathcal{U})$.

Definition 9.1.4 A presheaf $\mathcal{F}$ on X is called a *sheaf* if whenever $\mathcal{V}$ is an open covering of an open set U in X the following two conditions are satisfied.
(F-I) **Uniqueness:** Suppose $s, t \in \mathcal{F}(U)$ are such that $s|_V = t|_V$ for all $V \in \mathcal{V}$. Then $s = t$.
(F-II) **Existence:** Suppose we are given elements $s_V \in \mathcal{F}(V)$ for each $V \in \mathcal{V}$, such that $s_V|_{V \cap W} = s_W|_{V \cap W}$ for every $V, W \in \mathcal{V}$. Then there is $s \in \mathcal{F}(U)$ such that $s|_V = s_V$ for all $V \in \mathcal{V}$.

In other words, for a presheaf to be a sheaf, for each open cover $\mathcal{V}$ of U and a compatible $\mathcal{V}$-family $\{s_V\}$ of $\mathcal{F}$, we must have a unique element $s \in \mathcal{F}(U)$ such that $s|_V = s_V$ for every $V \in \mathcal{V}$.

Remark 9.1.5 Sheaf theory was developed by Leray and Serre. The popular notation $\mathcal{F}$ for a sheaf is due to the French word *faisceau* (sheaf). Again, as the name indicates, a presheaf is an afterthought, though in our definition a sheaf is a presheaf which satisfies some additional conditions. There is another popular definition of a sheaf which is sometimes more convenient. However, we shall first study a few examples before talking about other things.

Example 9.1.6

(a) **The presheaf of constants** This is one of the most interesting and important presheaves. Given a topological space and an abelian group A, we define $\mathcal{A}(U) = A$ for all open subsets of X except $\mathcal{A}(\emptyset) = (0)$. Of course, each restriction map $\rho_{VU} = Id$ for $U, V \neq \emptyset$. This is called the presheaf of constants in A. Check that this presheaf is not a sheaf in general.

Closely associated to the above presheaf of constants is another one which is actually a sheaf. It is denoted by $\hat{A}$ and defined by $\hat{A}(U) =$ the set of all continuous functions from U to A, where A is given the discrete topology (and with the convention $\hat{A}(\emptyset) = 0$). The morphisms $\hat{A}(j_{VU})$ are taken to be the restriction maps. Notice that if U is connected, then $\hat{A}(U) \approx A$ is the group of all constant functions on U. More generally, for locally connected spaces, $\hat{A}(U)$ is isomorphic to the direct product of copies of A over the set of connected components of U. It is easily checked that this is a presheaf which satisfies the extra conditions (FI) and (FII) for a sheaf. We shall call this one *the sheaf of constants with values in A*. Thus, given an abelian group A, we have a presheaf of constants which we denote by $\mathcal{A}$ and a sheaf of constants with values in A which we denote by $\hat{A}$. The temporary confusion which may arise out of this terminology and notation (which is deliberate) will soon disappear. (See Example 9.1.18.)

(b) **The presheaf of relative singular cochains** Let G be an abelian group. Recall that for any space X, $S^*(X; G) = \text{Hom}(S_*(X), G)$ is called the singular cochain complex X with coefficients in G. Given a topological pair (X, Y), the assignment

$$U \rightsquigarrow S^*(U, U \cap Y; G) = S^*(U; G)/S^*(Y; G)$$

defines a sheaf with the usual restriction morphisms. More generally, if $\mathcal{G}$ is any presheaf of abelian groups, we then consider

$$U \rightsquigarrow S^*(U, U \cap Y; \mathcal{G}(U))$$

to define the so-called presheaf of relative singular cochains with coefficients in $\mathcal{G}$. Here the restriction maps are taken to be the composites of

$$S^*(U, U \cap Y); \mathcal{G}(U)) \xrightarrow{(\rho^{\mathcal{G}}_{VU})_*} S^*((U, U \cap Y); \mathcal{G}(V)) \xrightarrow{(\rho^{S_*}_{VU})^*} S^*((V, V \cap Y); \mathcal{G}(V)).$$

(c) **Structure sheaf of a variety** Let X be an algebraic variety over a field $\mathbb{K}$. For each Zariski open set U of X, let $\mathcal{O}(U)$ denote the ring of regular functions on U. This sheaf is called the structure sheaf of the variety X or the sheaf of regular functions on X and is denoted by $\mathcal{O}_X$.

This leads to a more general concept called 'ringed spaces', viz., a topological space with a presheaf $\mathcal{R}$ of commutative rings. We can then talk about a presheaf $\mathcal{F}$ on X being a module over the ringed space $(X, \mathcal{R})$. This means each abelian group $\mathcal{F}(U)$ is an $\mathcal{R}(U)$-module and all the restriction maps are linear maps of appropriate extension rings. Such a structure is finer than saying $\mathcal{F}$ is a presheaf of R-modules. Indeed, the latter is a special case of the former, when we fix the ring structure on X to be the constant presheaf R. In what follows, when we say a presheaf of modules on an algebraic variety we mean a presheaf of $\mathcal{O}_X$-modules.

Similarly, for a smooth manifold, one can define the structure sheaf by taking the ring of all smooth real valued functions on U; and for a complex manifold, by taking the ring of all holomorphic functions on U and so on. Though this point-of-view is there implicitly in the study of differential topology, often it is not explicitly mentioned.

(d) **The sheaf of sections** Let $p : E \to B$ be a vector bundle. For each open set U in B, take $\Gamma(U)$ to be the vector space of all sections of p over U, i.e., the set of all continuous functions $s : U \to E$ such that $p \circ s = Id$. This sheaf is called the *sheaf of sections* of p and is denoted by Γ. If p is a smooth vector bundle, we usually demand that the sections s be smooth also. Likewise, if p is an analytic bundle over a complex manifold, we may demand that the sections be holomorphic and so on. Classically, this was the prime object of study which led to the present-day sheaf theory. We shall come back to it a little later.

Definition 9.1.7 Given two presheaves $\mathcal{F}, \mathcal{F}'$ over a given topological space and taking values in the same category $\mathcal{C}$, we define a morphism $\alpha : \mathcal{F} \to \mathcal{F}'$ to be a natural transformation of functors. There is then a category whose objects are presheaves on X with values in $\mathcal{C}$ and morphisms being natural transformations of functors. By an isomorphism of two presheaves we mean a morphism with two-sided inverse in this category, i.e., a natural equivalence of the two functors.

Example 9.1.8 Let X be an open subset of $\mathbb{R}^n$ (or a smooth manifold). Consider the sheaves $\mathcal{C}^r, 0 \le r \le \infty$ of smooth real valued functions on X.

(i) The inclusion of the constant functions defines a morphism of a presheaf of rings $\mathbb{R} \to \mathcal{C}^r$.

(ii) There are inclusions $\mathcal{C}^r \to \mathcal{C}^s \to \mathcal{C}^\infty, r < s$; all of these are morphisms of sheaves of rings.

(iii) Taking the total derivative $D : \mathcal{C}^\infty \to \mathcal{C}^\infty$ is an endomorphism if we treat $\mathcal{C}^\infty$ as a sheaf of abelian groups but not as a sheaf of rings. Likewise, taking a partial derivative with respect to a variable also defines an endomorphism $\mathcal{C}^\infty \to \mathcal{C}^\infty$.

Definition 9.1.9 The neighbourhoods of a given point x form a directed set under the reverse inclusion. Given a presheaf $\mathcal{F}$ on X and a point $x \in X$ we define the *stalk* $\mathcal{F}_x$ at x of $\mathcal{F}$ to be the direct limit of the directed system $\{\mathcal{F}(U) : x \in U \text{ open}\}$. For every neighbourhood U of x there is a morphism $\mathcal{F}(U) \to \mathcal{F}_x$ (given by the definition of the direct limit. Given an element $s \in \mathcal{F}(U)$, we denote its image in $\mathcal{F}_x$ by s_x and call it the stalk of s at x.

Notice that given a morphism $\alpha : \mathcal{F} \to \mathcal{F}'$ we get an induced morphism $\alpha_x : \mathcal{F}_x \to \mathcal{F}'_x$.

Example 9.1.10 Let X be a smooth manifold and $\mathcal{C}^\infty$ be the sheaf of real valued smooth functions on X, i.e., for each open set U, $\mathcal{C}^\infty(U)$ is the ring of all smooth real valued functions on U. Then the stalk $\mathcal{C}^\infty_x$ at any point $x \in X$ is nothing but the 'germs' of smooth functions: An element of $\mathcal{C}^\infty_x$ is nothing but an equivalence class of smooth functions $f_U : U \to \mathbb{R}$ where U is a neighbourhood of x; two such functions f_U, g_V are said to be equivalent iff $f_U|_W = g_V|_W$, where $x \in W \subset U \cap V$ and W is an open set. The same comment holds for the sheaf of holomorphic functions on a complex manifold or the sheaf of sections of a vector bundle, etc. More generally, for any sheaf $\mathcal{F}$, it is customary to call elements of $\mathcal{F}_x$ 'germs of sections'. (The reader is advised to spend some time in checking each of these things by herself.)

The importance of the stalks is in the following result:

Theorem 9.1.11 *Let $\alpha : \mathcal{F} \to \mathcal{F}'$ be a morphism of sheaves over X. Then α is an isomorphism iff for each $x \in X$, $\alpha_x : \mathcal{F}_x \to \mathcal{F}'_x$ is an isomorphism.*

Proof: If $\beta : \mathcal{F}' \to \mathcal{F}$ is the inverse of α, it is easily verified that for each $x \in X$, β_x is the inverse of α_x. It is for the converse part that we have to work harder.

Let then $\alpha_x : \mathcal{F}_x \to \mathcal{F}'_x$ be an isomorphism for each $x \in X$. In order to construct the inverse transformation $\beta : \mathcal{F}' \to \mathcal{F}$, it suffices (and is necessary) to show that $\alpha(U) : \mathcal{F}(U) \to \mathcal{F}'(U)$ is an isomorphism for each open set U. For then, we can take $\beta(U) := \alpha(U)^{-1}$ and verify first of all that it is a morphism and then that it is the inverse of α.

To show that $\alpha(U)$ is injective, suppose $\alpha(U)(s) = 0$ for some $s \in \mathcal{F}(U)$. This clearly implies that $\alpha_x(s_x) = 0$ for all $x \in U$. Since α_x is injective, this implies that $s_x = 0$ for all $x \in U$. By the definition of direct limit, it follows that there is a neighbourhood $W_x \subset U$ of x such that $s|_{W_x} = 0$. Since $\{W_x : x \in U\}$ is an open cover, by the uniqueness property of a sheaf it follows that $s = 0$ in $\mathcal{F}(U)$.

To show that $\alpha(U)$ is surjective, start with an element $t \in \mathcal{F}'(U)$. By the surjectivity of α_x there exists an element $s_x \in \mathcal{F}_x$ such that $\alpha_x(s_x) = t_x$ for each $x \in U$. By the definition of direct limit, each s_x is represented by a section $s_{V_x} \in \mathcal{F}(V_x)$, where V_x is a neighbourhood of x in U. Now $\alpha(s_{V_x})$ and $t|_{V_x}$ are two elements of $\mathcal{F}'(V_x)$ such that they represent the same element t_x in $\mathcal{F}'_x$. Therefore, by replacing V_x by a smaller neighbourhood of x if necessary, we may assume that $\alpha(V_x)(s_{V_x}) = t|_{V_x}$ in $\mathcal{F}(V_x)$.

Now U is covered by open sets $\{V_x\}$ and on each V_x we have a section $s_{V_x} \in \mathcal{F}(V_x)$. If y, z are two points of U then $s_{V_y}|_{V_y \cap V_z}$ and $s_{V_z}|_{V_y \cap V_z}$ are two sections which are both mapped onto $t_{V_y \cap V_z}$ by $\alpha(V_y \cap V_z)$. By the injectivity part which is proved above, it follows that

$$s_{V_y}|_{V_y \cap V_z} = s_{V_z}|_{V_y \cap V_z}.$$

Therefore, by the existence property of a sheaf there exists $s \in \mathcal{F}(U)$ such that $s|_{V_x} = s_{V_x}$

for each $x \in X$. We claim that $\alpha(s) = t$. This follows from the uniqueness property of the sheaf $\mathcal{F}'$ since we have $\alpha(s)|_{V_x} = t|_{V_x}$ for each x. ♠

Remark 9.1.12 It pays rich to get familiar with the above proof. You may anticipate that the result is not true for an arbitrary presheaf since the sheaf property is used very heavily in the proof and you are right. There may be many presheaves with the same stalk at every point. The crux of the matter is that one of these presheaves will be a sheaf and it will have a certain universal property. That is our next immediate goal.

Definition 9.1.13 If $\mathcal{U} = \{U_i\}_{i \in I}, \mathcal{V} = \{V_j\}_{j \in J}$ are two families of subsets of X, a function $\alpha : J \to I$ is called an *affinity function* or a *refinement function* if for each $j \in J$, we have $V_j \subset U_{\alpha(j)}$. In this case, we say that the family $\mathcal{V}$ is a refinement of the family $\mathcal{U}$ and write $\mathcal{V} \ll \mathcal{U}$. The collection of all open covers of a space is partially ordered by the refinement relation $\ll$.

Definition 9.1.14 Suppose $\sigma : J \to I$ is a refinement function of open coverings $\{U_i\}, \{V_j\}$. We define $\sigma^* : \mathcal{F}(\mathcal{U}) \to \mathcal{F}(\mathcal{V})$ as follows: Given any compatible $\mathcal{U}$-family $s = \{s_{U_i}\}$ we define the compatible $\mathcal{V}$-family $\sigma^*(s) = t = \{t_{V_j}\}$ by $t_{V_j} = s_{U_{\sigma(j)}}|_{V_j}$. (Note that the definition of $\{t_{V_j}\}$ does not depend on the actual refinement map σ.) This makes the collection $\{\mathcal{F}(\mathcal{U})\}$ into a directed system of modules, where $\mathcal{U}$ ranges over all the open coverings of X.

Theorem 9.1.15 *Given a presheaf $\mathcal{F}$ over X there is a sheaf $\hat{\mathcal{F}}$ over X and a morphism $\Theta : \mathcal{F} \to \hat{\mathcal{F}}$ satisfying the following universal property: Given any morphism $\alpha : \mathcal{F} \to \mathcal{F}'$ where $\mathcal{F}'$ is a sheaf, there is a unique morphism $\hat{\alpha} : \hat{\mathcal{F}} \to \mathcal{F}'$ such that $\alpha = \hat{\alpha} \circ \Theta$. The pair $(\hat{\mathcal{F}}, \Theta)$ itself is unique up to sheaf isomorphisms and is called the completion of $\mathcal{F}$ or the sheaf associated to $\mathcal{F}$. Moreover, at the stalk level, Θ induces an isomorphism $\Theta_x : \mathcal{F}_x \to \hat{\mathcal{F}}_x$ for each x.*

Proof: We take
$$\hat{\mathcal{F}}(U) := \operatorname*{dlim}_{\rightarrow} \{\mathcal{F}(\mathcal{U}) \ : \ \mathcal{U} \text{ is an open cover of } U\}$$

with respect to the partial order given by the refinement relation. Now if $W \subset U$ and $\mathcal{U}$ is an open covering of U then
$$\mathcal{U} \cap W := \{U_i \cap W \ : \ U_i \in \mathcal{U}\}$$

forms an open covering for W which refines $\mathcal{U}$. Therefore, we get a morphism $\mathcal{F}(\mathcal{U}) \to \mathcal{F}(\mathcal{U} \cap W)$. Upon passing to the direct limits, this in turn yields a morphism $\hat{\mathcal{F}}(U) \to \hat{\mathcal{F}}(W)$. One verifies that this way $\hat{\mathcal{F}}$ is a presheaf.

Often it is necessary to appeal to the construction of the direct limit, in understanding $\hat{\mathcal{F}}$. Fix an open set U of X. An element $\hat{s} \in \hat{\mathcal{F}}(U)$ is represented by a compatible family $\{s_{U_\alpha} \in \mathcal{F}(U_\alpha)\}$ for some open cover $\{U_\alpha\}$ of U. Another such family $\hat{t} = \{t_{V_\beta} \in \mathcal{F}(V_\beta)\}$ represents the same element in $\hat{\mathcal{F}}(U)$ iff there is a common open refinement $\{W_\gamma\}$ of the two open covers of U such that
$$s_{U_{\alpha(\gamma)}}|_{W_\gamma} = t_{V_{\beta(\gamma)}}|_{W_\gamma},$$

for all γ where $\gamma \mapsto \alpha(\gamma), \gamma \mapsto \beta(\gamma)$ are some refinement functions.

Notice how the properties F-I and F-II of a sheaf are built into the definition of $\hat{\mathcal{F}}$. All that you need to observe is that if $\{U_\alpha\}_\alpha$ is an open cover of U and for each α, $\{U_{\alpha,j}\}_j$ is an open cover for U_α, then the combined collection $\{U_{\alpha,j}\}_{\alpha,j}$ is an open cover for U. We shall leave the verification that $\hat{\mathcal{F}}$ is a sheaf to the reader.

The morphism $\Theta(U) : \mathcal{F}(U) \to \hat{\mathcal{F}}(U)$ is defined by observing that the singleton $\{U\}$ is a cover of U and hence $\mathcal{F}(U) = \mathcal{F}(\{U\})$ is the module of compatible $\{U\}$-family of elements of $\mathcal{F}$. Therefore, we get a morphism $\mathcal{F}(U) \to \hat{\mathcal{F}}(U)$ in the direct limit, which we take as $\Theta(U)$.

Finally let $\mathcal{F}'$ be any sheaf of modules over X and $\varphi : \mathcal{F} \to \mathcal{F}'$ be any morphism. Let $s \in \hat{\mathcal{F}}(U)$ be an element. Represent this by a compatible $\mathcal{U}$-family $\{s_{U_i} \in \mathcal{F}(U_i)\}$, where $\mathcal{U}$ is an open cover for U. It follows that $\{\varphi(s_{U_i}) \in \mathcal{F}'(U_i)\}$ is a compatible family for $\mathcal{F}'$ and since $\mathcal{F}'$ is a sheaf this family defines a unique element $t \in \mathcal{F}'(U)$. We define $\hat{\varphi}(s) = t$. We have to check that this t is independent of the choice of the compatible family that represents s which is straightforward and left to the reader. It is also easily checked that $\hat{\varphi} \circ \Theta = \varphi$. The uniqueness of $\hat{\varphi}$ is a consequence of the fact that for any $s \in \hat{\mathcal{F}}(U)$ represented by a compatible $\mathcal{U}$-family $\{s_{U_i} \in \mathcal{F}(U_i)\}$ as above, we have $\Theta(s_{U_i}) = s|_{U_i}$. ♠

Remark 9.1.16 Thus, given a presheaf $\mathcal{F}$, in the category of pairs $(\tau, \mathcal{F}')$ where $\tau : \mathcal{F} \to \mathcal{F}'$ is a morphism and $\mathcal{F}'$ is a sheaf, $(\Theta, \hat{\mathcal{F}})$ is an initial object. It is common practice to suppress the morphism Θ and merely write $\mathcal{F} \to \hat{\mathcal{F}}$. We also caution the reader that there are slightly different definitions/constructions of the completion $\hat{\mathcal{F}}$ of a presheaf, which do not coincide on all topological spaces. However, under certain additional topological conditions on the base space X say, that every open subspace of X is paracompact, all these definitions coincide. See for instance, Theorem 9.1.21.

Corollary 9.1.17 $\Theta : \mathcal{F} \to \hat{\mathcal{F}}$ is an isomorphism iff $\mathcal{F}$ is a sheaf.

Proof: Use the universal property of $\hat{\mathcal{F}}$ in the previous theorem, twice, to show that $\hat{\Theta}$ itself is the two-sided inverse of Θ. ♠

Example 9.1.18 Given an abelian group A, consider the constant presheaf $\mathcal{A}$. What is the completion of $\mathcal{A}$? It is the sheaf $\hat{\mathcal{A}}$ of constants with values in A as introduced in Example 9.1.6.(a). To see this, all that you have to note is that a continuous map into A with discrete topology is nothing but a compatible family of locally constant functions taking values in A. With this justification for the name and the notation introduced in the above example, the temporary confusion also should disappear.

Etale Covering

Going back to Remark 9.1.5, we shall now give another description of a sheaf which is actually the origin of sheaves and is due to Oka.

Consider any map $\pi : \bar{X} \to X$. For each open subset U of X, let $\Gamma(U)$ denote the set of all continuous sections $s : U \to \bar{X}$ of π, (i.e., $\pi \circ s(x) = x, x \in U$). Then Γ defines a presheaf on X which is easily verified to be a sheaf. It is called the sheaf of continuous sections of π and temporarily we shall denote it by Γ_π.

To keep the discussion at an elementary level, let us now assume that X is a T_1 space and π is a local homeomorphism. It then follows easily that the stalks of Γ_π are nothing but the fibres of π.

The idea is to reverse this situation. We begin with a presheaf $\mathcal{F}$ over a topological space X and take $\bar{X} = \sqcup \mathcal{F}_x$ and $\pi : \bar{X} \to X$ to be the obvious projection map which sends each member of $\mathcal{F}_x$ to the single point x. We shall topologise $\bar{X}$ in such a way that π is continuous and for every open subset U of X, every member $s \in \mathcal{F}(U)$, the function $\tilde{s} : U \to \bar{X}$ defined by

$$x \mapsto s_x$$

is continuous. Of course, we can do this by merely taking the least topology on $\bar{X}$ such that π is continuous. For then given any open set U of X and $s \in \mathcal{F}(U)$, for any open set V of

X, $\tilde{s}(x) = s_x \in \pi^{-1}(V)$ implies $x \in V$ and therefore $\tilde{s}^{-1}(\pi^{-1}(V)) = U \cap V$ is open in U. This means that $\tilde{s}$ is continuous.

However, we shall put the largest topology on $\bar{X}$ with respect to which all $\tilde{s}$ are continuous. This topological space is called the *etale space associated to $\mathcal{F}$ and is denoted by $Sp\acute{e}(\mathcal{F})$.*

Obviously π is continuous. Indeed, we claim that π is a local homeomorphism. We shall describe the topology of $Sp\acute{e}(\mathcal{F})$ in another way from which the above claim follows easily.

Consider the disjoint union of topological spaces

$$Z = \sqcup U_s$$

where U_s denotes a copy of U for each open set U of X and for each $s \in \mathcal{F}(U)$. Let $q : Z \to \bar{X}$ be the function defined by $q|_{U_s} = \tilde{s}$. Clearly, q is surjective. We claim that $q : Z \to Sp\acute{e}(\mathcal{F})$ is a quotient map. Clearly q is continuous. Indeed, q is an open mapping also. For, suppose W is an open subset of Z. This just means that W is a disjoint union of open subsets $U_s \cap W$, for each $s \in \mathcal{F}(U)$ and each open set U of X. Its image $q(W)$ in $\bar{X}$ is open in the topology of $Sp\acute{e}(\mathcal{F})$ iff $\tilde{s}^{-1}(q(W))$ is open in U for each $s \in \mathcal{F}(U)$ and each open subset U of X. But $\tilde{s}^{-1}(q(W)) = U_s \cap W$ and so, we are done.

Finally, we observe that q is injective on each U_s, since $\pi \circ q = Id$ on each U_s. Therefore, $q(U_s)$ is an open set in $Sp\acute{e}(\mathcal{F})$ on which π is a homeomorphism. This proves that π is a local homeomorphism.

We sum up this discussion in the following theorem.

Theorem 9.1.19 *Let $\pi : \bar{X} \to X$ be a local homeomorphism. For each open subset U of X, let $\Gamma_\pi(U)$ denote the set of all continuous sections $s : U \to \bar{X}$ of π. Then Γ_π defines a sheaf on X called the sheaf of continuous sections of π.*

To obtain a result in the reverse direction we need to assume that every open subset of X is paracompact. We begin with the following purely topological lemma the proof of which is left to the reader as a exercise.

Lemma 9.1.20 *Let $\{U_i\}_{i \in I}$ be a locally finite open cover of a space X and $\{V_i\}$ be a shrinking (i.e., $\{V_i\}_{i \in I}$ is also an open cover of X such that $\bar{V}_i \subset U_i, i \in I$. Then for every $x \in X$ there is an open neighbourhood W_x of x such that*
(i) $I_x = \{i \in I : W_x \cap V_i \neq \emptyset\}$ is finite;
(ii) $i \in I_x$ implies $x \in \bar{V}_i$ and $W_x \subset U_i$.
(iii) If $W_x \cap W_y \neq \emptyset$, then $W_x \cap W_y \subset U_i$ for some $i \in I$.

Theorem 9.1.21 *Let X be a topological space such that every open subset of X is paracompact. Given any presheaf $\mathcal{F}$ on X, the completion $\hat{F}$ of $\mathcal{F}$ is naturally equivalent to the sheaf Γ_π of sections of the etale covering $\pi : Sp\acute{e}(\mathcal{F}) \to X$. Moreover, the assignments $\pi \rightsquigarrow \Gamma_\pi$ and $\mathcal{F} \rightsquigarrow Sp\acute{e}(\mathcal{F})$ define functors which are adjoint of each other.*

Proof: It suffices to construct a natural transformation of functors $\lambda : \Gamma_\pi \to \hat{\mathcal{F}}$ which restricts to the morphism $\Theta : \mathcal{F} \to \hat{\mathcal{F}}$. By the universal property of $\hat{\mathcal{F}}$ the rest of the proof follows. Given an open set V and $\sigma \in \Gamma_\pi(V)$, we shall construct a compatible family $\lambda(\sigma) = \{s_G \in \mathcal{F}(G)\}$, where $\{G\}$ form an open cover V such that if $\sigma = s \in \mathcal{F}(U)$, then $s_G = s|_G$ for each $G \in \{G\}$. The family $\{\sigma^{-1}(q(U_s))\}$ as U varies over all open sets of X and as s varies over $\mathcal{F}(U)$, forms an open cover of V. On each of the members we have $\sigma(x) = s_x$. Take a locally finite open refinement $\{U_i\}_{i \in I}$ of this cover pass onto a shrink $\{V_i\}$ and construct the cover $\{W_x\}$ as in the lemma. We can then say that we have $s_i \in \mathcal{F}(U_i)$ such that $\sigma(x) = (s_i)_x$ for all $x \in U_i$. Now for any two $i, j \in I_x$, $\sigma(x) = (s_i)_x = (s_j)_x$ implies that there is a neighbourhood $W_{ij}(x)$ of x such that $s_i|_{W_{ij}(x)} = s_j|_{W_{ij}(x)}$. Put

$G_x = W_x \cap (\cap_{i,j \in I_x} W_{ij}(x))$. Since $G_x \cap G_y \subset W(x) \cap W(y) \subset U_i$ for some i, it follows that the family

$$\lambda(\sigma) := \{s_{G_x} = s_i|_{G_x}\}$$

is a compatible family. The definition of λ is completed by taking the class represented by $\lambda(\sigma)$ in $\mathcal{F}(V)$. Clearly, λ defines a natural transformation of functors.   ♠

Remark 9.1.22

1. There are several advantages of going to the etale space model especially while dealing with sheaves as compared to presheaves. For instance, even though $\mathcal{F}$ is defined only for open subsets of X, it now makes sense to define $\mathcal{F}(A)$ for any subset A of X, viz., as the module of sections of $\mathrm{Sp\acute{e}}(\mathcal{F})$ defined over A. So, we shall keep going back and forth from one model to another.

2. Of course, if we begin with a sheaf with more algebraic structure then fibres of the etale space will have corresponding algebraic structures. This amounts to considering local homeomorphisms with additional algebraic structures on the fibre.

3. In the above discussion, we 'ignored' any topological structure on the stalks of the presheaf, thereby getting discrete topology on the fibres of the etale space. This was natural because the stalks were certain direct limits taken in the category **Ens**. On the other hand, suppose we start with a presheaf $\mathcal{F}$ which takes values in a sub category of **Top**. Then the stalks $\mathcal{F}_x$ are topological spaces and we could then try to retain this topology in $\mathrm{Sp\acute{e}}(\mathcal{F})$. All that we need to do is take the strongest topology on $\bar{X} = \sqcup_x \mathcal{F}_x$ so that all the sections $s : U \to \bar{X}$ as well as all the inclusion maps $\mathcal{F}_x \to \bar{X}$ are continuous. The etale cover is then a special case of this general construction when all the stalks are discrete spaces. This leads to concepts such as 'locally trivial' fibre spaces and vector bundles, etc. See Exercise 9.5.6.

4. By now, you may have guessed that the conditions (F-I) and (F-II) have something to do with the injectivity of the natural morphism $\Theta : \mathcal{F} \to \hat{\mathcal{F}}$ and then you are right. Indeed, (F-I) is equivalent to saying that for every open set U of X, $\Theta_U : \mathcal{F}(U) \to \hat{\mathcal{F}}(U)$ is injective. (See Exercise 9.5.1.) However, the relationship with surjectivity and condition (F-II) is somewhat subtler. In the absence of F-I, there is no way to get F-II out of surjectivity of Θ. The converse however is true. Because of its importance we shall state this separately as a proposition.

Proposition 9.1.23 If a presheaf $\mathcal{F}$ satisfies (F-II) then $\Theta : \mathcal{F}(U) \to \hat{\mathcal{F}}(U)$ is surjective for all open subsets U of X.

Proof: Given $\sigma \in \mathcal{F}(U)$ represent it by a compatible family $\{s_{U_i} \in \mathcal{F}(U_i)\}$ for some open cover $\{U_i\}$ of U. By F-II, we get an element $s \in \mathcal{F}(U)$ such that $s|_{U_i} = s_{U_i}$, for all i. But then $\Theta(s) = \sigma$.   ♠

Example 9.1.24 The presheaf S^* of singular cochain complexes is not a sheaf. However, it satisfies (F-II) for the simple reason that for each U, $S_*(U; \mathbb{Z})$ is a free abelian group and given an open covering $\{U_i\}$, the subgroup G of $S_*(U; \mathbb{Z})$ generated by $\{S_*(U_i; \mathbb{Z})\}$ is a *pure* subgroup and hence a direct summand. Therefore, given a compatible family $s_{U_i} \in S^*(U_i; A)$ first of all there is a well-defined group homomorphism $s : G \to A$ such that $s|_{S_*(U_i)} = s_{U_i}$. Since G is a direct summand, this s can be extended to a homomorphism $\hat{s} : S_*(U) \to A$. Clearly then $\hat{s}|_{U_i} = s|_{U_i} = s_{U_i}$.

Definition 9.1.25 Sub-presheaf, quotients, kernel, cokernel, etc. By a sub-presheaf $\mathcal{F}'$ of a presheaf $\mathcal{F}$ we mean a presheaf $\mathcal{F}'$ such that for every open set U of X we have $\mathcal{F}'(U)$ is a submodule of $\mathcal{F}(U)$ and the morphisms $\mathcal{F}'(j) : \mathcal{F}'(U) \to \mathcal{F}'(V)$ are obtained by taking restriction of $\mathcal{F}(j)$ for each $j : V \subset U$. If the sub-presheaf happens to be a sheaf on its own then it will be called a subsheaf. We can have a sub-presheaf of a sheaf as well as a subsheaf of a presheaf. The quotient presheaf $\mathcal{F}/\mathcal{F}'$ is defined by taking

$$(\mathcal{F}/\mathcal{F}')(U) = \mathcal{F}(U)/\mathcal{F}'(U).$$

But even if both $\mathcal{F}$ and $\mathcal{F}'$ are sheaves their quotient defined as above may not be a sheaf. Therefore, the common practice is to take the completion of $\mathcal{F}/\mathcal{F}'$ as the quotient sheaf and often use the same notation.

Given a morphism $\phi : \mathcal{F} \to \mathcal{F}'$ of presheaves of modules, we define $\operatorname{Ker} \phi, \operatorname{Coker} \phi, \operatorname{Im} \phi$ to be the presheaves given by

$$U \rightsquigarrow \operatorname{Ker}(\phi(U)), \quad U \rightsquigarrow \operatorname{Coker}(\phi(U)), \quad U \rightsquigarrow \operatorname{Im}(\phi(U)),$$

respectively. The sheaves associated to these presheaves will be denoted by $\ker \phi, \operatorname{coker} \phi, \operatorname{im} \phi$, respectively. Clearly, $\operatorname{Ker} \phi$ and $\operatorname{Im} \phi$ are sub-presheaves of $\mathcal{F}$ and $\mathcal{F}'$, respectively.

Now suppose $\phi : \mathcal{F} \to \mathcal{F}'$ is a morphism of sheaves. Then automatically the presheaf $\operatorname{Ker} \phi$ is a sheaf. Therefore $\operatorname{Ker} \phi = \ker \phi$. This is not the case with Im and Coker in general. However, by the universal property of the completion, there is a morphism $\operatorname{im} \phi \to \mathcal{F}'$ such that the diagram

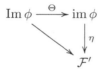

is commutative. It follows easily that at the stalk level, η_x is injective.

We define ϕ to be injective, if $\ker \phi = (0)$. This is equivalent to saying that $\phi(U) : \mathcal{F}(U) \to \mathcal{F}'(U)$ is injective for every open set U. We define ϕ to be surjective if $\operatorname{im} \phi = \mathcal{F}'$. Note that this need not imply that $\phi(U) : \mathcal{F}(U) \to \mathcal{F}'(U)$ is surjective for every open set. However, passing to the stalks, we can say ϕ is surjective iff $\phi_x : \mathcal{F}_x \to \mathcal{F}'_x$ is surjective. In this sense, we can regard $\operatorname{im} \phi$ as a subsheaf of $\mathcal{F}'$ via η as in the previous paragraph.

A sequence of morphisms

$$\mathcal{F}' \xrightarrow{\phi} \mathcal{F} \xrightarrow{\psi} \mathcal{F}''$$

is said to be exact at $\mathcal{F}$ if $\operatorname{im} \phi = \ker \psi$. This is equivalent to say that

$$\mathcal{F}'_x \xrightarrow{\phi_x} \mathcal{F}_x \xrightarrow{\psi_x} \mathcal{F}''_x$$

is exact at $\mathcal{F}_x$.

Example 9.1.26 Let G be any non trivial group and $\hat{G}$ denote the constant sheaf on a totally disconnected space X with at least two points $x_1 \neq x_2$. Let $\mathcal{F}$ be the subsheaf of sections which vanish at x_1 and x_2. The quotient presheaf $\hat{G}/\mathcal{F}$ is not a sheaf. If we denote its completion by $\bar{G}$ then we have an exact sequence

$$0 \to \mathcal{F} \to \hat{G} \to \bar{G} \to 0.$$

However, $\hat{G}(X) \to \bar{G}(X)$ is not surjective, because for any $s \in \hat{G}$, we have $s(x_1) = s(x_2)$ but we can take some $t \in \bar{G}(X)$ such that $t(x_1) \neq t(x_2)$. (In fact $\bar{G}(X) = G \times G$.) This failure of being exact at the section level is precisely what makes the cohomology theory of sheaves interesting and that is what we shall discuss in the next section.

Definition 9.1.27 Restrictions and extensions Given a presheaf $\mathcal{F}$ on X and an open set U in X, we can consider the *restriction of $\mathcal{F}$ to U* as follows: $\mathcal{F}_U(V) = \mathcal{F}(V)$. It is clear how to define restriction maps to make this into a presheaf. It is also clear that if $\mathcal{F}$ were a sheaf then so is $\mathcal{F}_U$.

On the other hand given a presheaf $\mathcal{G}$ on U we define the extension of $\mathcal{G}$ to X as follows:

$$\mathcal{G}^U(V) = \{s \in \mathcal{G}(U \cap V) \ : \ \text{supp}\, s \text{ is a closed subset of } V\}$$

(Here and elsewhere, $\text{supp}\, s$ for $s \in \mathcal{F}(U)$ is by definition the closure in U of the set $\{x \ : \ s(x) \neq 0\}$.) Once again, it is clear how to define restriction maps here so as to make $\mathcal{G}^U$ into a presheaf. It takes a little more effort to see that if $\mathcal{G}$ is a sheaf then so is $\mathcal{G}^U$, viz., in the verification of (F-II). (Notice that for any $s \in \mathcal{G}(U \cap V)$ its support is, by definition, a closed subset of $U \cap V$, whereas to be inside $\mathcal{G}^U(V)$, s should satisfy the extra condition that its support is closed in V.) Thus, given an open covering $\{V_i\}$ of an open subset V of X, and $s \in \mathcal{G}(U \cap V)$, it is enough to show that the support of s is a closed subset of V if for each i, the support of the restriction of s to $U \cap V_i$ is a closed subset of V_i. But this is just an elementary topological fact. It is also clear that for every point $x \notin U$, we have $(\mathcal{G}^U)_x = (0)$.

The following proposition is going to play a key role in the cohomology theory of sheaves.

Proposition 9.1.28 The extension of a restriction of a sheaf is a subsheaf of the given sheaf, i.e., $(\mathcal{F}_U)^U$ is a subsheaf of $\mathcal{F}$. Indeed, the two constructions are 'adjoint' of one another in the following sense: For any sheaf $\mathcal{F}$ on X and any sheaf $\mathcal{G}$ on U, there is a natural isomorphism

$$\eta : \text{Hom}(\mathcal{G}^U; \mathcal{F}) \to \text{Hom}(\mathcal{G}; \mathcal{F}_U)$$

given by the restriction map.

Proof: The first part is obvious.

Given a morphism $\lambda : \mathcal{G}^U \to \mathcal{F}$ of sheaves (defined over X) its restriction $\eta(\lambda)$ to U clearly defines a morphism $\mathcal{G} \to \mathcal{F}_U$ and the assignment itself is a homomorphism. It is easily seen that if $\eta(\lambda) = 0$ then $\lambda = 0$. To prove surjectivity of η, suppose $\tau : \mathcal{G} \to \mathcal{F}_U$ is a homomorphism. We then have, for each open set V of X, a homomorphism $\tau_V : \mathcal{G}(U \cap V) \to \mathcal{F}(U \cap V)$. We wish to define $\hat{\tau} : \mathcal{G}^U(V) \to \mathcal{F}(V)$ which coincides with τ_V if $V \subset U$. Given $s \in \mathcal{G}^U(V)$, i.e., $s \in \mathcal{G}(U \cap V)$ with its support being closed in V, it follows that $\tau(s) \in \mathcal{F}(U \cap V)$ also has its support closed in V. Therefore, we can extend this to a section $\hat{\tau}(s)$ of $\mathcal{F}$ on the whole of V by putting $\hat{\tau}(s)_x = 0$ for points $x \in V \setminus \text{supp}\, s$ and $= \tau(s)_x$, for points $x \in U \cap V$. One can check easily that the assignment $s \mapsto \hat{\tau}(s)$ is a homomorphism $\mathcal{G}^U(V) \to \mathcal{F}(V)$ as required. This proves surjectivity of η.   ♠

Example 9.1.29 For any ring R with a unit 1, consider the constant presheaf of rings R on a connected space X. Then for any presheaf $\mathcal{F}$ of modules over R, what do the groups $\text{Hom}(R(U); \mathcal{F}(U))$ look like? Recall that $\text{Hom}(R, M)$ for any R-module M is canonically isomorphic to M given by $f \mapsto f(1)$. Consider the section $\mathbb{1}_U$ where $\mathbb{1}_U(x) = 1$, for all $x \in U$. Check that $f \mapsto f(\mathbb{1}_U)$ defines an isomorphism $\text{Hom}(R(U), \mathcal{F}(U)) \approx \mathcal{F}(U)$. In particular, $\text{Hom}(R, \mathcal{F})(X) = \mathcal{F}(X)$. Note that the assignment $U \rightsquigarrow \text{Hom}(R(U), \mathcal{F}(U))$ itself defines presheaf which is isomorphic to $\mathcal{F}$. Thus, any section $s \in \mathcal{F}(X)$ can be interpreted as defining a unique homomorphism of the sheaves $R \to \mathcal{F}$. (More generally, this observation is valid for any sheaf which is a module over a presheaf of rings.)

Now fix an open set U of X. What does $\text{Hom}(R_U^U; \mathcal{F})$ look like? By the above proposition, there is a canonical isomorphism of this with $\text{Hom}(R_U; \mathcal{F}_U)$. Combining this with the observation above, we conclude that

$$\text{Hom}(R_U^U; \mathcal{F}) \approx \mathcal{F}(U). \tag{9.1}$$

Definition 9.1.30 Pushout and pullback sheaves Let $f : X \to Y$ be a map of topological spaces and $\mathcal{F}$ be a presheaf on X. Then we define the *direct image presheaf* or the *pushout* sheaf $f_*\mathcal{F}$ on Y by the formula:

$$f_*\mathcal{F}(V) = \mathcal{F}(f^{-1}(V)), V \subset Y \text{ open.}$$

Verify that if $\mathcal{F}$ is a sheaf then so is $f_*\mathcal{F}$.

Now let $\mathcal{F}'$ be a presheaf on Y, then the *pullback* presheaf $f^*(\mathcal{F}')$ is defined by

$$f^*F(U) = \varinjlim \{\mathcal{F}'(V) \ : \ f(U) \subset V, V \text{ open in } Y\}.$$

However, when $\mathcal{F}'$ is a sheaf, it is not necessary that f^*F is a sheaf. For this, we need to pass on to the completion of the above presheaf. Yet, we shall use the notation $f^*\mathcal{F}$ only to denote this sheaf. (See Exercise 9.5. 9.)

Remark 9.1.31 Many expositions use the symbol $f^{-1}(\mathcal{F}')$ for $f^*(\mathcal{F}')$ that we have defined above and call it the inverse of $\mathcal{F}'$ under f. There is a further confusion when X and Y are ringed spaces and $\mathcal{F}'$ is a module over the structure ring $\mathcal{O}_Y$. Then, to begin with $f^*\mathcal{F}'$ defined as above is an $\mathcal{O}_Y$-module, which we need to convert into an $\mathcal{O}_X$-module and call it $f^{-1}(\mathcal{F}')$. We shall **not** be using these notions here and advise the reader to look out carefully for the right concept while reading/using such material.

Many concepts which are available in the category **R-mod**, such as direct sum, tensor product, etc., are also available for presheaves over a given space X with values in **R-mod**. For instance, there is a category of short exact sequences of presheaves

$$0 \to \mathcal{F}' \to \mathcal{F} \to \mathcal{F}'' \to 0. \tag{9.2}$$

Primarily, sheaves of modules play the role of supplying appropriate coefficient groups as one moves around in a topological space. Simultaneously, they store a lot of topological information in them. To begin with, we need to take care of the algebra. As witnessed in Example 9.1.26, an exact sequence of sheaves does not produce, in general, an exact sequence at the section level. However, we can say something positive.

Proposition 9.1.32 Given an exact sequence of sheaves of modules

$$0 \to \mathcal{F} \to \mathcal{G} \to \mathcal{H} \to 0$$

over any topological space X, the induced sequence of modules of sections

$$0 \to \mathcal{F}(X) \to \mathcal{G}(X) \to \mathcal{H}(X)$$

is exact.

Proof: Exercise.

As observed before, due to this failure of right exactness, we are led to look out, first of all, for those sheaves $\mathcal{F}$ which intrinsically satisfy the exactness property at the right-end as well.

This algebraic aspect is precisely what we are going to discuss in the next section. An impatient reader may quickly browse through it and go to the statement of Theorem 9.2.10.

9.2 Injective Sheaves and Resolutions

In this section, we shall study properties of sheaves with respect to extensions of sections. Injective sheaves are the strongest in this sense. We shall establish that every sheaf has an injective resolution which is unique up to homotopy, thereby preparing the ground for the study of cohomology of sheaves.

Definition 9.2.1 A sheaf $\mathcal{F}$ of modules is said to be *flabby* (respectively, *soft*) if for every open (respectively, closed) subset $A \subset X$, every section $s \in \mathcal{F}(A)$ can be extended to a section of the whole space.

Proposition 9.2.2 If $\mathcal{F}$ is flabby then for every exact sequence

$$0 \to \mathcal{F} \to \mathcal{G} \to \mathcal{H} \to 0$$

the induced sequence of modules

$$0 \to \mathcal{F}(X) \to \mathcal{G}(X) \to \mathcal{H}(X) \to 0,$$

is exact.

Proof: We need only to see that every section $s \in \mathcal{H}(X)$ comes from a section $t \in \mathcal{G}(X)$. The exactness $\mathcal{G} \to \mathcal{H} \to 0$ implies that for every $x \in X$, we have the exactness $\mathcal{G}_x \to \mathcal{H}_x \to 0$. This implies that for every $x \in X$, there is an open neighbourhood U_x of x and $t_x \in \mathcal{G}(U_x)$ which is a lift of $s|_{U_x}$. So, we form a collection C of pairs (U,t) where $t \in \mathcal{G}(U)$ is a lift of $s|_U$. We put a partial order on C by saying $(U,t) < (U',t')$ if $U \subset U'$ and $t'|_U = t$. It is easily verified that every chain in C is bounded above. Therefore by Zorn's lemma, there is a maximal element (U,t) in C. All that we need to show is that $U = X$.

If not, suppose $x \in X \setminus U$ and V is a neighbourhood of x such that there is $t' \in \mathcal{G}(V)$ which lifts $s|_V$. This means that $(t - t')|_{V \cap U}$ maps to 0 in $\mathcal{H}$ and hence comes from some $\alpha \in \mathcal{F}(V \cap U)$. Since $\mathcal{F}$ is flabby, there is $\beta \in \mathcal{F}(X)$ which extends α. Now consider $\gamma = t' + \beta \in \mathcal{G}(U)$. Then $\gamma|_{U \cap V} = t_{U \cap V}$. Therefore there is $\tau \in \mathcal{G}(U \cup V)$ such that $\tau|_U = t$ and $\tau|_V = \gamma$. Also this τ maps onto s on $U \cup V$ and $(U \cup V, \tau)$ is a member of C. This contradicts the maximality of (U,t). ♠

In a similar fashion, we can prove the following.

Proposition 9.2.3 If $\mathcal{F}$ is soft over a paracompact Hausdorff space, then for every exact sequence

$$0 \to \mathcal{F} \to \mathcal{G} \to \mathcal{H} \to 0$$

the induced sequence of modules

$$0 \to \mathcal{F}(X) \to \mathcal{G}(X) \to \mathcal{H}(X) \to 0$$

is exact.

Proof: Given $s \in \mathcal{H}(X)$ using paracompactness of X we can find a family of open sets $\{U_i\}$, a shrink $\{W_i\}$ (i.e., open sets W_i such that $W_i \subset \bar{W}_i \subset U_i$ and $\cup_i W_i = X$) and $t_i \in \mathcal{G}(U_i)$ which are lifts of $s|_{U_i}$. Applying Zorn's lemma to the collection of members of $\{W_i\}$ on which we have lifts of s, the proof can be completed as in the previous proposition. The details are left to the reader. ♠

Example 9.2.4

(a) The sheaf of smooth functions on $\mathbb{R}$ is **not** flabby, e.g., take the function tan: $(-\pi/2, \pi/2) \to \mathbb{R}$; nor is the sheaf of holomorphic functions on $\mathbb{C}$ flabby.

(b) However, the sheaf of smooth functions on $\mathbb{R}$ is soft (smooth partition of unity).

(c) The sheaf of holomorphic function is not soft either. The story is the same with the constant sheaf $\mathbb{Z}$.

(d) On a discrete space, every sheaf is flabby as well as soft.

Definition 9.2.5 A sheaf $\mathcal{F}$ is said to be *injective* if given a morphism $\alpha : \mathcal{G}' \to \mathcal{F}$ where $\mathcal{G}'$ is a subsheaf of $\mathcal{G}$, there exists a morphism $\beta : \mathcal{G} \to \mathcal{F}$ which extends α :

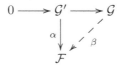

Proposition 9.2.6 Every injective sheaf is flabby.

Proof: Take $\mathcal{G} = R$ the constant sheaf R and $\mathcal{G}' = R_U{}^U$, where U is any given open set of X. Then by (9.1), any section $s \in \mathcal{F}(U)$ corresponds to a homomorphism $\alpha : R_U{}^U \to \mathcal{F}$ and its extension $\beta : R \to \mathcal{F}$ corresponds to an extension of s to a section defined all over X. ♠

Before going further with injective sheaves, let us briefly recall a few facts about injective modules, just for the sake of completeness. For more details, the reader may refer to a book on homological algebra e.g., [Hilton–Stommbach, 1970].

Definition 9.2.7 Let R be a commutative ring with a unit. An R-module D is said to be divisible, if given any $x \in D$ and $\lambda \neq 0$ in R there exist $y \in D$ such that $\lambda y = x$. We say D is injective, if for every pair $A \subset B$ of R-modules and every R-linear map $\alpha : A \to D$, there exists a R-linear map $\beta : B \to D$ which is an extension of α.

A typical example of a divisible abelian group is $\mathbb{Q}/\mathbb{Z}$. The prototype of a projective (free) $\mathbb{Z}$-module is $\mathbb{Z}$ itself. A prototype of an injective $\mathbb{Z}$-module is $\mathbb{Q}/\mathbb{Z}$ which is an easy consequence of the following proposition.

Proposition 9.2.8 Every divisible $\mathbb{Z}$-module is injective.

Proof: Given abelian groups $A \subset B$ and a homomorphism $\alpha : A \to D$ where D is a divisible abelian group, we need to show that there is a homomorphism $\beta : B \to D$ which extends α. The plan is to employ Zorn's lemma. Consider the collection of pairs (A', α') where $A \subset A' \subset B$ and $\alpha' : A' \to D$ is an extension of α. Under the usual partial order, this collection is easily seen to be inductive and hence has a maximal element $(\hat{A}, \beta)$. We claim that $\hat{A} = B$. If not, let $b \in B \setminus \hat{A}$. The set of all $n \in \mathbb{Z}$ such that $nb \in \hat{A}$ is clearly an ideal $= (p)$ say. Now if $p \neq 0$, there exists $y \in D$ such that $py = \alpha(pb)$; otherwise take $y \in D$ to be equal to 0 (or any other element). We can now extend β over the subgroup of B generated by $\hat{A} \cup \{b\}$ as follows. Define $\alpha'(a + \lambda b) := \beta(a) + \lambda y$. If $a_1 + \lambda_1 b = a_2 + \lambda_2 b$, then $(\lambda_1 - \lambda_2)b = a_2 - a_1 \in \hat{A}$ and hence $\lambda_1 - \lambda_2 = \mu p$. Therefore, $\beta(a_2 - a_1) = \beta(\mu p b) = \mu \beta(pb) = \mu p y$ which implies $\alpha'(a_1 + \lambda_1 b) = \alpha'(a_2 + \lambda_2 b)$. Therefore, α' is well defined. Checking that α' is a homomorphism and checking that it agrees with β on $\hat{A}$ is routine and left to the reader. This contradicts the maximality of $(\hat{A}, \beta)$. ♠

Proposition 9.2.9 For any divisible abelian group D and a commutative ring R, $\mathrm{Hom}_{\mathbb{Z}}(R, D)$ is an injective R-module.

Proof: Here we appeal to the general fact that for any R-module A, there is a canonical isomorphism

$$\Phi_A : \mathrm{Hom}_R(A, \mathrm{Hom}_{\mathbb{Z}}(R, D)) \to \mathrm{Hom}_{\mathbb{Z}}(A, D)$$

given by $\Phi_A(\alpha)(a) = (\alpha(1))(a)$. Now given R-modules $A \subset B$ and a R-linear map $\alpha : A \to \text{Hom}_{\mathbb{Z}}(R, D)$, consider $\Phi(\alpha) : A \to D$ as a homomorphism of abelian groups. Since D is divisible and hence injective, we obtain an extension of this to a map $\tau : B \to D$. But then there is a unique $\beta : B \to \text{Hom}_{\mathbb{Z}}(R, D)$ such that $\Phi(\beta) = \tau$. It is easily checked that $\beta|_A = \alpha$ by the naturality of Φ.   ♠

Theorem 9.2.10 *Every R-module A is a submodule of an injective module.*

Proof: . We shall produce a monomorphism of R-modules $\beta : A \to \prod A_i$, where each A_i is a copy of $\text{Hom}_{\mathbb{Z}}(R, \mathbb{Q}/\mathbb{Z})$. Since $\mathbb{Q}/\mathbb{Z}$ is divisible, by the previous proposition, $\text{Hom}_{\mathbb{Z}}(R, \mathbb{Q}/\mathbb{Z})$ is injective. It is easily checked that the direct product of injective modules is injective. That will complete the proof.

To produce β, it is enough to produce a family $\beta_i : A \to \text{Hom}_{\mathbb{Z}}(R, \mathbb{Q}/\mathbb{Z})$ such that for each non zero element $a \in A$ there exists at least one i such that $\beta_i(a) \neq 0$. For then, we can take $\beta(x) = \prod_i \beta_i(x)$.

To this end, given $0 \neq a \in A$, if a is an element of finite order in the abelian group A say, $na = 0, n > 1$, then choose an element $y \in \mathbb{Q}/\mathbb{Z}$ of order n; otherwise, choose y to be any non zero element of $\mathbb{Q}/\mathbb{Z}$. Then we get a group homomorphism $\tau : (a) \to \mathbb{Q}/\mathbb{Z}$, where (a) denotes the subgroup generated by a in A. Since $\mathbb{Q}/\mathbb{Z}$ is injective, there is a group homomorphism $\tau' : A \to \mathbb{Q}/\mathbb{Z}$ which extends τ. As before, τ' corresponds to a R-linear map $\beta_a : A \to \text{Hom}_{\mathbb{Z}}(R, \mathbb{Q}/\mathbb{Z})$ i,e., $\Phi_A(\beta_a) = \tau'$. But then it also follows that $\beta_a(a) \neq 0$.   ♠

Before proceeding further with the sheaves, we need to strengthen this result with yet another basic result on injective modules. It is an immediate consequence of Theorem 9.2.10 that given any R-module M, there exists an exact sequence

$$0 \to M \to I^0 \to I^1 \to \cdots$$

wherein each I^i is injective. Such a sequence is called an *injective resolution* of M. We express this by writing $0 \to M \to I^*$. The homomorphisms $I^i \to I^{i+1}$ are all denoted by d, instead of decorating them such as d^i or d_i, etc. What we need is the following:

Proposition 9.2.11 *Given a short exact sequence of R-modules $0 \to M_1 \to M_2 \to M_3 \to 0$ there exist resolutions such that the following diagram is commutative with all the rows exact.*

$$
\begin{array}{ccccccccc}
0 & \longrightarrow & M_1 & \longrightarrow & M_2 & \longrightarrow & M_3 & \longrightarrow & 0 \\
  & & \downarrow{\scriptstyle \alpha_1} & & \downarrow{\scriptstyle \alpha_2} & & \downarrow{\scriptstyle \alpha_3} & & \\
0 & \longrightarrow & I_1^* & \longrightarrow & I_2^* & \longrightarrow & I_3^* & \longrightarrow & 0
\end{array}
$$

Proof: Note that if the sequences I_1^* and I_3^* are determined then, since I_1^* are all injective, it follows that I_2^* is a direct sum and so we are forced to take $I_2^j = I_1^j \oplus I_3^j$. So, we plan to construct these exact sequences of resolutions, by first fixing the injective resolutions I_1^* and I_3^*. Indeed, if we take $I_2^* = I_1^* \oplus I_3^*$ as a complex, it yields an exact complex. The problem is right at the topmost level, since M_2 need not be the direct sum of M_1 and M_3. So, we need first of all to select a monomorphism $\alpha_2 : M_2 \to I_1^0 \oplus I_3^0$ and then keep modifying the boundary homomorphisms (d_1, d_3) inductively so that
(i) the diagrams commute,
(ii) I_2^* is a complex, i.e, $d^2 = 0$ and
(iii) $0 \to M_2 \to I_2^*$ is a resolution. Luckily, the last part comes to us freely, merely by diagram chasing. (Note that you cannot prove (ii) by mere diagram chasing.) So, we shall achieve (i) and (ii) simultaneously.

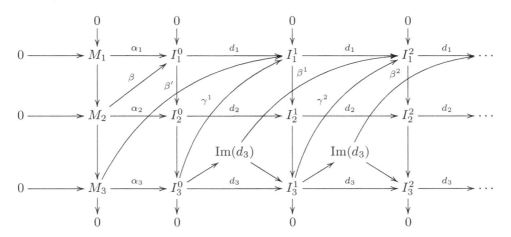

By injectivity of I_1^0, there exists $\beta : M_2 \to I_1^0$ which restricts to α_1 on M_1. We define $\alpha_2(x) = (\beta(m), \alpha_3(\bar{m}))$. Verify that α_2 is a monomorphism and it fits the diagram.

Let us now define $d_2 : I_2^0 \to I_2^1$. The obvious candidate satisfying (i) is (d_1, d_3). However,

$$(d_1, d_3)\alpha_2(m) = (d_1 \circ \beta(m), d_3 \circ \alpha_3(\bar{m})) = (d_1 \circ \beta(m), 0)$$

which may not be zero. So, we need to modify d_2 in the first slot. Note that, $d_1 \circ \beta$ vanishes on M_1 and hence factors down to define a homomorphism $\beta' : M_3 \to I_1^1$. By the injectivity of I_1^1 and α_3, there is a homomorphism $\gamma^1 : I_3^0 \to I_1^1$ such that $\gamma_1 \circ \alpha_3 = \beta'$. We can now define

$$d_2(x, y) := (d_1(x) - \gamma^1(y), d_3(y))$$

which clearly satisfies (i). It now satisfies (ii) as well:

$$d_2 \circ \alpha_2(m) = d_2(\beta(m), \alpha_3(\bar{m})) = (d_1 \circ \beta(m) - \gamma^1 \circ \alpha_3(\bar{m}), d_3\alpha_3(\bar{m})) = (0, 0).$$

The inductive step is similar except that, right in the beginning, we have to deal with the boundary map d_3 in place of α_3 and d_3 is not necessarily injective. We have a homomorphism $\gamma^i : I_3^{i-1} \to I_1^i$ such that $d_1 \circ \gamma^i \circ d_3 = d_1 \circ d_1 \circ \gamma^{i-1} = 0$ and hence defines a map $\beta^i : \mathrm{Im}(d_3) \to I_1^i$. Using the injectivity of I_1^{i+1}, since $\mathrm{Im}(d_3) \subset I_3^i$ we get an extension of this map to $\gamma^{i+1} : I_3^i \to I_1^{i+1}$. Clearly, $\gamma^{i+1} \circ d_3 = d_1 \circ \gamma^i$. Therefore, we define

$$d_2(x, y) = (d_1(x) + (-1)^i \gamma^i(y), d_3(y)).$$

Form now onward, the inductive step is similar. This completes the proof. ♠

We can now study injectivity of sheaves.

Theorem 9.2.12 *Every sheaf of R-modules is a subsheaf of an injective sheaf.*

Proof: Given a sheaf $\mathcal{F}$ of R-modules over a space X, for each $x \in X$, let I_x denote some R-module which is injective and contains $\mathcal{F}_x$ as a submodule. Consider the presheaf $\mathcal{I}$ which associates to an open subset U of X the module $\mathcal{I}(U) := \prod_{x \in U} I_x$, along with the obvious projection maps as restrictions. It is easily seen that this is actually a sheaf. The monomorphisms $\mathcal{F}_x \to I_x$ define a monomorphism $\prod_{x \in U} \mathcal{F}_x \to \mathcal{I}(U)$. Pre-composing this with the natural morphism $\mathcal{F}(U) \to \prod_{x \in X} \mathcal{F}_x$ which is a monomorphism, since $\mathcal{F}$ is sheaf, we get a monomorphism $\mathcal{F} \to \mathcal{I}$. We need to show that the sheaf $\mathcal{I}$ is injective. Consider the following diagram of sheaves and morphisms:

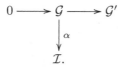

Passing onto stalk-level and composing with the projection $\mathcal{I}_x \to I_x$ and using the fact that I_x is injective, we obtain

$$
\begin{array}{ccc}
0 \longrightarrow \mathcal{G}_x \longrightarrow \mathcal{G}'_x \\
\quad\quad\Big\downarrow{\alpha'_x} \quad{}^{\beta_x} \\
I_x.
\end{array}
$$

This yields homomorphisms $\mathcal{G}'(U) \to \prod_{x \in U} \mathcal{G}'_x \to \mathcal{I}(U)$, compatible with the restrictions and hence a morphism $\beta : \mathcal{G}' \to \mathcal{I}$. Since for each x we have β_x restricts to α_x on $\mathcal{F}_x$, it follows that $\beta = \alpha$ on U just because $\mathcal{I}(U)$ is merely the product $\prod_{x \in U} I_x$. ♠

Definition 9.2.13 Let $\mathcal{J}^*$ be a complex of sheaves, i.e., for each $k \geq 0$ we have a morphism $d : \mathcal{J}^k \to \mathcal{J}^{k+1}$ such that $d \circ d = 0$. $\mathcal{J}^*$ is called a resolution of a sheaf $\mathcal{F}$ if we have a monomorphism $0 \to \mathcal{F} \to \mathcal{J}^0$ and the entire sequence

$$0 \to \mathcal{F} \to \mathcal{J}^0 \to \mathcal{J}^1 \to \cdots$$

is exact. We express this by writing $0 \to \mathcal{F} \to \mathcal{J}^*$. If each $\mathcal{J}^k$ is injective then this is called an injective resolution.

Theorem 9.2.14 *For every sheaf $\mathcal{F}$ there exists an injective resolution $0 \to \mathcal{F} \to \mathcal{I}^*$.*

Proof: Starting with $\mathcal{F}$ we put it inside an injective sheaf $\mathcal{I}^0$ as guaranteed by Theorem 9.2.12. We then take the quotient sheaf $\mathcal{I}^0/\mathcal{F}$ and put it in another injective sheaf $\mathcal{I}^1$. The composite morphism $\mathcal{I}^0 \to \mathcal{I}^0/\mathcal{F} \to \mathcal{I}^1$ is denoted by d^1 and fits into an exact sequence

$$0 \longrightarrow \mathcal{F} \longrightarrow \mathcal{I}^0 \overset{d^1}{\longrightarrow} \mathcal{I}^1.$$

This procedure can be continued indefinitely to produce the complex $\mathcal{I}^*$ as required. ♠

Indeed, if we start with an exact sequence of sheaves

$$0 \to \mathcal{F}_1 \to \mathcal{F}_2 \to \mathcal{F}_3 \to 0$$

by Proposition 9.2.11, for each x we get a commutative diagram of resolutions

$$
\begin{array}{ccccccccc}
0 & \longrightarrow & (\mathcal{F}_1)_x & \longrightarrow & (\mathcal{F}_2)_x & \longrightarrow & (\mathcal{F}_3)_x & \longrightarrow & 0 \\
& & \Big\downarrow{\alpha_1} & & \Big\downarrow{\alpha_2} & & \Big\downarrow{\alpha_3} & & \\
0 & \longrightarrow & (I_1)^*_x & \longrightarrow & (I_2)^*_x & \longrightarrow & (I_3)^*_x & \longrightarrow & 0.
\end{array}
$$

The sheaves $\mathcal{I}^i_1, \mathcal{I}^i_2, \mathcal{I}^i_3$ are defined by taking direct products and hence it follows that we have an exact sequence

$$0 \to \mathcal{I}^*_1 \to \mathcal{I}^*_2 \to \mathcal{I}^*_3 \to 0$$

For each open set U of X we obtain

$$0 \longrightarrow \prod_{x \in U}(\mathcal{F}_1)_x \longrightarrow \prod_{x \in U}(\mathcal{F}_2)_x \longrightarrow \prod_{x \in U}(\mathcal{F}_3)_x \longrightarrow 0 \qquad (9.3)$$

$$0 \longrightarrow \mathcal{I}_1^*(U) \longrightarrow \mathcal{I}_2^*(U) \longrightarrow \mathcal{I}_3^*(U) \longrightarrow 0.$$

We also have the canonical inclusions $\mathcal{F}_i(U) \to \prod_{x \in U}(\mathcal{F}_i)_x$ fitting the following diagram:

$$0 \longrightarrow \mathcal{F}_1^*(U) \longrightarrow \mathcal{F}_2^*(U) \longrightarrow \mathcal{F}_3^*(U) \longrightarrow 0 \qquad (9.4)$$

$$0 \longrightarrow \prod_{x \in U}(\mathcal{F}_1)_x \longrightarrow \prod_{x \in U}(\mathcal{F}_2)_x \longrightarrow \prod_{x \in U}(\mathcal{F}_3)_x \longrightarrow 0.$$

Combining (9.3) and (9.4) we obtain:

Theorem 9.2.15 *Given an exact sequence $0 \to \mathcal{F}_1 \to \mathcal{F}_2 \to \mathcal{F}_3 \to 0$ of sheaves there is an exact sequence of injective resolutions:*

$$\begin{array}{ccccccccc}
0 & \longrightarrow & \mathcal{F}_1 & \longrightarrow & \mathcal{F}_2 & \longrightarrow & \mathcal{F}_3 & \longrightarrow & 0 \\
& & \downarrow{\scriptstyle \alpha_1} & & \downarrow{\scriptstyle \alpha_2} & & \downarrow{\scriptstyle \alpha_3} & & \\
0 & \longrightarrow & \mathcal{I}_1^* & \longrightarrow & \mathcal{I}_2^* & \longrightarrow & \mathcal{I}_3^* & \longrightarrow & 0.
\end{array}$$

Definition 9.2.16 Two morphisms $\phi^*, \psi^* : \mathcal{I}^* \to \mathcal{J}^*$ are said to be *chain-homotopic* if there exists a sequence of homomorphisms $h^i : \mathcal{I}^i \to \mathcal{J}^{i+1}, i \geq 0$ such that

$$d \circ h^i + h^{i+1} \circ d = \phi^i - \psi^i. \qquad (9.5)$$

Two complexes $\mathcal{I}^*, \mathcal{J}^*$ over a given sheaf $\mathcal{F}$ are said to be chain (homotopy) equivalent if there are morphisms $\phi^* : \mathcal{I}^* \to \mathcal{J}^*, \psi^* : \mathcal{J}^* \to \mathcal{I}^*$ such that the composites $\phi^* \circ \psi^*, \psi^* \circ \phi^*$ are homotopic to the identity morphisms of the respective complexes.

Proposition 9.2.17 Given any resolution $0 \to \mathcal{F} \to \mathcal{J}^*$ of $\mathcal{F}$, an injective resolution $0 \to \mathcal{G} \to \mathcal{I}^*$, and any morphism $f : \mathcal{F} \to \mathcal{G}$ there is a morphism $\phi^* : \mathcal{J}^* \to \mathcal{I}^*$ which covers f, i.e., the following diagram is commutative:

$$\begin{array}{ccccc}
0 & \longrightarrow & \mathcal{F} & \longrightarrow & \mathcal{J}^* \\
& & \downarrow{\scriptstyle f} & & \downarrow{\scriptstyle \phi^*} \\
0 & \longrightarrow & \mathcal{G} & \longrightarrow & \mathcal{I}^*.
\end{array}$$

Moreover, if $\psi^* : \mathcal{J}^* \to \mathcal{I}^*$ is another morphism which covers f then ψ^* is chain homotopic to ϕ^*.

Proof: We shall leave the proof of the existence of ϕ^* as an exercise to the reader.

Let us construct the chain homotopy $h : \phi^* \sim \psi^*$. We start with $h^0 = 0$. Since on the image of $\mathcal{F}$, $\phi^0 = \psi^0$, we get a morphism $\alpha : \mathcal{J}^0/\mathcal{F} \to \mathcal{I}^0$ induced by $\phi^0 - \psi^0$. By the injectivity of $\mathcal{I}^1$ and the fact that $0 \to \mathcal{J}^0/\mathcal{F}^0 \to \mathcal{J}^1$, there is a homomorphism $h^1 : \mathcal{J}^1 \to \mathcal{I}^0$ such that restricted to $\mathcal{J}^0/\mathcal{F}^0$ it equals α. This is the same as saying $h^1 \circ d = \phi^0 - \psi^0$ which gives (9.5) for $i = 0$. Assuming that we have constructed h^j for $j \leq i$

as required, we check that $\phi^i - \psi^i - d \circ h^i$ vanishes on $d(\mathcal{J}^{i-1})$ and hence define a morphism $\alpha^{i+1} : \mathcal{J}^i/d(\mathcal{J}^{i-1}) \to \mathcal{I}^i$. We also have a monomorphism $\mathcal{J}^i/d(\mathcal{J}^{i-1}) \to \mathcal{J}^{i+1}$ induced by d.

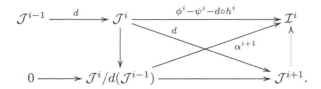

By injectivity of $\mathcal{I}^i$, there exists h^{i+1} (as indicated by the dotted arrow) which restricts to α^{i+1}. This is the same as saying

$$h^{i+1} \circ d = \phi^i - \psi^i - d \circ h^i$$

which is the same as (9.5). This completes the proof. ♠

The following corollary is immediate.

Corollary 9.2.18 Any two injective resolutions of a sheaf are chain equivalent.

Exercise 9.2.19 Given an exact sequence of sheaves $0 \to \mathcal{F} \to \mathcal{G} \to \mathcal{H} \to 0$, show that

(a) $\mathcal{F}$ and $\mathcal{G}$ are flabby $\implies \mathcal{H}$ is flabby;

(b) if X is paracompact, $\mathcal{F}$ and $\mathcal{G}$ are soft $\implies \mathcal{H}$ is soft.

9.3   Cohomology of Sheaves

We can now define the cohomology of a sheaf and establish some basic properties. We shall also relate it to the singular cohomology of the base space.

Note that a morphism $\phi^* : \mathcal{I}^* \to \mathcal{J}^*$ of complexes of sheaves defines a morphism of co-chain complexes $\phi_U^* : \mathcal{I}^*(U) \to \mathcal{J}^*(U)$. Even if $\mathcal{I}^*$ and $\mathcal{J}^*$ are exact, the complexes $\mathcal{I}^*(U), \mathcal{J}^*(U)$ need not be so. However, a chain homotopy of two morphisms ϕ^* and ψ^* yields a chain homotopy of ϕ_U^* and ψ_U^* and hence passing onto the homology, ϕ_U^*, ψ_U^* define the same sequence of homomorphisms. Thus in particular, chain homotopic resolutions define isomorphic homology modules. Thus the following definition makes sense.

Definition 9.3.1 Given a sheaf $\mathcal{F}$, taking an injective resolution $0 \to \mathcal{F} \to \mathcal{I}^*$ over a space X, we define
$$H^*(X; \mathcal{F}) := H^*(\mathcal{I}^*(X)) =: H^i(\mathcal{F}).$$

Remark 9.3.2

(a) Note that by definition, $H^0(X; \mathcal{F})$ is equal to the kernel of the morphism $\mathcal{I}^0(X) \to \mathcal{I}^1(X)$. By Proposition 9.1.32, the exactness of
$$0 \to \mathcal{F} \to \mathcal{I}^0 \to \mathcal{I}^1$$
implies the exactness of
$$0 \to \mathcal{F}(X) \to \mathcal{I}^0(X) \to \mathcal{I}^1(X)$$
and hence $H^0(X; \mathcal{F}) = \mathcal{F}(X)$.

(b) If $\mathcal{F}$ itself is injective, we can take the resolution to be $0 \to \mathcal{F} \to \mathcal{F} \to 0$ and then it follows that $H^i(X; \mathcal{F}) = 0$ for all $i > 0$.

(c) Given a morphism $f : \mathcal{F} \to \mathcal{G}$, Proposition 9.2.17 allows us to take any cover $\phi : \mathcal{I}^* \to \mathcal{J}^*$ of injective resolutions and then consider the induced homomorphisms $H^i(f) : H^i(X; \mathcal{F}) \to H^i(X; \mathcal{G})$ to be equal to the induced homomorphism $H^i(\phi(X)) : H^i(\mathcal{I}^*(X)) \to H^i(\mathcal{J}^*(X))$, independent of the resolutions as well as the covering map ϕ. It is routine to check that the assignment $\mathcal{F} \to H^*(\mathcal{F})$ defines a covariant functor.

(d) Given an exact sequence $0 \to \mathcal{F}_1 \to \mathcal{F}_2 \to \mathcal{F}_3 \to 0$ of sheaves, Theorem 9.2.15 yields an exact sequence of complexes

$$0 \to \mathcal{I}_1(X) \to \mathcal{I}_2(X) \to \mathcal{I}_3(X) \to 0,$$

which in turn gives an exact sequence of cohomology modules:

$$
\begin{array}{c}
0 \longrightarrow \mathcal{F}_1(X) \longrightarrow \mathcal{F}_2(X) \longrightarrow \mathcal{F}_3(X) \longrightarrow H^1(X; \mathcal{F}_1) \qquad (9.6) \\
H^1(X; \mathcal{F}_2) \longleftarrow H^1(X; \mathcal{F}_3) \longrightarrow H^2(X; \mathcal{F}_1) \longrightarrow \cdots
\end{array}
$$

One can easily check that this assignment defines a functor from the category of short exact sequences of sheaves to the category of long exact sequences of modules.

(e) Note that given an injective resolution $0 \to \mathcal{F} \to \mathcal{I}^*$ over X, for every open subset U of X, we get an injective resolution $0 \to \mathcal{F}_U \to \mathcal{I}^*_U$ of the restricted sheaf. This makes the assignment $U \mapsto H^*(U; \mathcal{F}(U))$ into a presheaf, which is, often, not a sheaf.

(f) Injective resolutions are often very unwieldy and so, it is desirable to be able to compute the cohomology modules using other resolutions. Thus, given any resolution $0 \to \mathcal{F} \to \mathcal{J}^*$, Proposition 9.1.32 yields an exact sequence

$$0 \to \mathcal{F}(X) \to \mathcal{J}^0(X) \to \mathcal{J}^1(X)$$

and hence $\mathcal{F}(X) = H^0(\mathcal{J}^*(X))$. Thus, to compute H^0, we need not bother about what kind of resolution we have. If we are interested in computing cohomology only up to a few dimensions, could there be resolutions which are more general than injective ones? The answer lies in property (b) above.

Proposition 9.3.3 Let $0 \to \mathcal{F} \to \mathcal{J}^*$ be a resolution of a sheaf $\mathcal{F}$ such that $H^i(X; \mathcal{J}^j) = (0), i + j \leq n, i > 0$. Then there are canonical isomorphisms

$$H^k(X; \mathcal{F}) \to H^k(\mathcal{J}^*(X))$$

for all $k \leq n$.

As an immediate corollary we get:

Corollary 9.3.4 Let $0 \to \mathcal{F} \to \mathcal{J}^*$ be a resolution of a sheaf $\mathcal{F}$ such that $H^i(X, \mathcal{J}^j) = (0)$ for all $i > 0$ and $j \geq 0$. Then there are canonical isomorphisms of $H^*(X; \mathcal{F})$ to the cohomology of the complex $\mathcal{J}^*(X)$.

Proof of Proposition 9.3.3: In Remark (f) above, we have already taken care of the case $n = 0$ (the hypothesis being vacuous). Inductively we shall assume that the proposition is true for all $n < m$ and prove it for m. Let us put $\mathcal{K} = \mathcal{J}^0/\mathcal{F}$, the cokernel of the inclusion $\mathcal{F} \to \mathcal{J}^0$. We then have an exact sequence

$$0 \to \mathcal{F} \to \mathcal{J}^0 \to \mathcal{J}^0/\mathcal{F} \to 0 \qquad (9.7)$$

and a resolution $0 \to \mathcal{K} \to \widehat{\mathcal{J}}^*$, where $\widehat{\mathcal{J}}^*$ is obtained merely by re-indexing of the terms in $\mathcal{J}^*$, viz., $\widehat{\mathcal{J}}^i = \mathcal{J}^{i+1}, i \geq 0$. Thus $H^i(\widehat{\mathcal{J}}^*(X)) = H^{i+1}(\mathcal{J}^*(X))$. Therefore, the hypothesis is true for the resolution $0 \to \mathcal{K} \to \widehat{\mathcal{J}}^*$ in the range $m - 1$. By induction hypothesis, we conclude that

$$H^i(\mathcal{K}) \approx H^i(\widehat{\mathcal{J}}^*(X)) = H^{i+1}(\mathcal{J}^*(X)). \qquad (9.8)$$

On the other hand, by Remark (d) above, the long exact sequence given by (9.7) yields exact sequences

$$H^{i-1}(\mathcal{J}^0) \to H^{i-1}(\mathcal{K}) \to H^i(\mathcal{F}) \to H^i(\mathcal{J}^0). \qquad (9.9)$$

Since $H^i(\mathcal{J}^0) = 0$ for $1 \leq i \leq m$, we have

$$H^i(\mathcal{F}) \approx H^{i-1}(\mathcal{K}), 2 \leq i \leq m$$

Combining this with (9.8) proves the proposition for $2 \leq i \leq m$.

The case $i = 0$, is taken care of by (a) of the above remark. For the case $i = 1$, we have to argue separately. We know that $H^0(\mathcal{K}) = \mathcal{K}(X)$ is equal to the kernel of $\widehat{\mathcal{J}}^0(X) \to \widehat{\mathcal{J}}^1(X)$ which is the same as the kernel of $\mathcal{J}^1(X) \to \mathcal{J}^2(X)$. In particular, $H^0(\mathcal{K})$ is a submodule of $\mathcal{J}^1(X)$. On the other hand, putting $i = 1$ in (9.9), since $H^1(\mathcal{J}^0) = 0$, we see that $H^1(\mathcal{F})$ is the quotient of $H^0(\mathcal{K})$ by the image of $H^0(\mathcal{J}^0) \to H^0(\mathcal{K})$ which is the same as the image of $\mathcal{J}^0(X) \to \mathcal{J}^0(\mathcal{K}) \subset \mathcal{J}^1(X)$. Therefore, $H^1(\mathcal{F})$ is isomorphic to the quotient of the kernel of $\mathcal{J}^1(X) \to \mathcal{J}^2(X)$ by the image of $\mathcal{J}^0(X) \to \mathcal{J}^1(X)$, which is nothing but $H^1(\mathcal{J}^*)$. ♠

An example where we can use the above proposition, we have:

Theorem 9.3.5 *Every flabby sheaf has vanishing cohomology. In particular, we can use a flabby resolution to compute the cohomology modules of a sheaf. Similarly, if X is paracompact, then we can use any soft resolution to compute the cohomology of a sheaf.*

Proof: Consider a flabby sheaf $\mathcal{F}$ and an exact sequence $0 \to \mathcal{F} \to \mathcal{I} \to \mathcal{G} \to 0$, where $\mathcal{I}$ is an injective sheaf. We have the long exact sequence of cohomology modules:

$$\to \mathcal{F}(X) \to \mathcal{I}(X) \to \mathcal{G}(X) \to H^1(\mathcal{F}) \to H^1(\mathcal{I}) \to \cdots$$

$$\cdots H^i(\mathcal{I}) \to H^i(\mathcal{G}) \to H^{i+1}(\mathcal{F}) \to H^{i+1}(\mathcal{I}) \to \cdots$$

In the first line, the last term $H^1(\mathcal{I}) = 0$. Since $\mathcal{F}$ is flabby, the morphism $\mathcal{I}(X) \to \mathcal{G}(X)$ is an epimorphism. Therefore we conclude that $H^1(\mathcal{F}) = 0$. Likewise, in the second line, $H^i(\mathcal{I}) = 0 = H^{i+1}(\mathcal{I})$ and hence $H^i(\mathcal{G}) \approx H^{i+1}(\mathcal{F}), i \geq 1$. We now appeal to the fact that an injective sheaf is flabby and the quotient of a flabby sheaf by a flabby subsheaf is flabby (see Exercise 9.2.19(a)). Therefore, $\mathcal{G}$ is flabby. Now, the theorem gets proved by an obvious induction.

The proof for soft sheaves on a paracompact space is similar. ♠

Example 9.3.6 de Rham complex of sheaves Let X be a smooth n-manifold. For $p \geq 0$, define the presheaf $\mathcal{D}^p$ on X by taking

$$\mathcal{D}^p(U) = \Omega^p(U),$$

the space of smooth p-forms on U, where $U \subset X$ is open. (For $p = 0$ this is nothing but $\mathcal{C}^\infty(U; \mathbb{R})$.) The exterior differentiation $d : \Omega^p(U) \to \Omega^{p+1}(U)$ then yields a chain complex of sheaves

$$\mathcal{D}^0 \xrightarrow{\ d\ } \mathcal{D}^1 \xrightarrow{\ d\ } \cdots \longrightarrow \mathcal{D}^n \longrightarrow 0.$$

This is called the *de Rham complex of sheaves*. By partition of unity, we know that every smooth p-form defined on a closed set can be extended to the whole of X. This means that the de Rham complex of sheaves is soft. Finally, by the Poincaré lemma, the above chain complex is exact except at the 0 level and the kernel of $d : \mathcal{D}^0 \to \mathcal{D}^1$ is the sheaf of constants $\mathbb{R}$. Therefore we have a resolution

$$0 \to \hat{\mathbb{R}} \to \mathcal{D}^0 \to \mathcal{D}^1 \to \cdots \to \mathcal{D}^n \to 0$$

of soft sheaves on the paracompact space X and hence the cohomology of the constant sheaf $\hat{\mathbb{R}}$ on X gives the de Rham cohomology. We could have as well continued to use the symbol Ω in place of $\mathcal{D}$; the purpose of the new symbol is to draw your attention to the fact that now we are thinking of Ω as a sheaf of chain complexes of differential forms and **not** just a chain complex of differential forms.

Let us now come to the study of relations between singular cohomology and sheaf cohomology.

Proposition 9.3.7 If $\mathcal{F}$ is a flabby presheaf satisfying F-II, then so is $\hat{\mathcal{F}}$.

Proof: Given $\sigma \in \hat{\mathcal{F}}(U)$, from Proposition 9.1.23, we obtain $s \in \mathcal{F}(U)$ such that $\Theta(s) = \sigma$. Since $\mathcal{F}$ is flabby, there is $\tau \in \mathcal{F}(X)$ such that $\tau|_U = s$. Now $\Theta(\tau)|_U = \Theta(\tau|_U) = \sigma$. ♠

Example 9.3.8 The presheaf $\mathcal{S}^*(-, A)$ of singular cochains satisfies F-II as seen in Example 9.1.24; it is also flabby for the same reason. Therefore, we conclude that the associated sheaf $\hat{\mathcal{S}}^*(-, A)$ is flabby.

Lemma 9.3.9 For any abelian group A, $\Theta : \mathcal{S}^*(-; A) \to \hat{\mathcal{S}}^*(-, A)$ of chain complexes induces an isomorphism of cohomology groups.

Proof: Let X be a topological space. For the sake of brevity, we shall not mention the coefficient group A.

Surjectivity: From Exercise 4.7.2 on small chains, it follows that for any open covering $\mathcal{U}$ of X, the natural restriction homomorphism $S^*(X) \to S^*_\mathcal{U}(X)$ induces an isomorphism in cohomology. Given an element $\mu \in H^k(\hat{\mathcal{S}}^*(X))$, represented by a cycle $c \in \hat{\mathcal{S}}^k(X)$, let $c' \in S^k(X)$ be such that $\Theta(c') = c$. (See Proposition 9.1.23.) Now $\Theta(dc') = d\Theta(c') = dc = 0$ implies that there is an open cover $\mathcal{U}$ of X such that $\Theta(dc')|_U = 0$, $U \in \mathcal{U}$. This means that in the complex $S^*_\mathcal{U}$ of small cochains, c' is a cocycle, which clearly maps onto c. Hence $H^k(\mathcal{S}_\mathcal{U}(X)) \to H^k(\hat{\mathcal{S}}(X))$ is surjective. By the above observation on small cochains, we conclude that $\Theta : H^k(S^*(X)) \to H^k(\hat{\mathcal{S}}(X))$ is surjective.

Injectivity Let $\Theta([c]) = 0$ for some cocycle $c \in S^k(X)$). This means that there is $c' \in \hat{\mathcal{S}}^{k-1}(X)$ such that $dc' = c$ in $\hat{\mathcal{S}}^k(X)$. Now there exists $c'' \in S^{k-1}(X)$ such that $\Theta(c'') = c'$. This means that $\Theta(c - dc'') = \Theta(c) - d\Theta(c'') = \Theta(c) - dc' = 0$. As before, this just means that there is an open covering $\mathcal{V}$ of X such that the class represented by $c - dc''$ is zero in $H^k(S^*_\mathcal{V}(X))$. Hence, $[c] = [c - dc'']$ must be zero in $H^k(S^*(X))$ itself. ♠

Remark 9.3.10 Though we have used the stronger hypothesis that every open set in X is paracompact, while applying Proposition 9.1.23, we only need to use the surjectivity of $S^*(X) \to \hat{\mathcal{S}}^*(X)$ in the proof of the above lemma. So, the hypothesis here can be weakened to just X being paracompact.

At last, we can now state and prove:

Theorem 9.3.11 *Let X be a locally contractible topological space in which every open subset is paracompact. Then there is a natural isomorphism of the singular cohomology groups of X with coefficients in an abelian group A with the cohomology of the constant sheaf $\hat{A}$.*

Proof: The complex $A \to \mathcal{S}^0 \to \mathcal{S}^1 \to \cdots$ of presheaves gives rise to a complex of sheaves

$$0 \to \hat{A} \to \hat{\mathcal{S}}^0 \to \hat{\mathcal{S}}^1 \to \cdots. \tag{9.10}$$

We shall claim that this is a resolution. By Example 9.3.8, each $\hat{\mathcal{S}}^j$ is flabby. Therefore, by Theorem 9.3.5, this resolution can be employed to compute the cohomology of $\hat{A}$. By the above lemma, this cohomology is isomorphic to the cohomology of the chain complex

$$0 \to A \to S^0(X; A) \to S^1(X; A) \to \cdots$$

which is nothing but $H^*(X; A)$.

So, we have to show that the complex of sheaves (9.10) is exact which is the same as saying that at the stalk level,

$$0 \to A \to \hat{\mathcal{S}}^0_x \to \hat{\mathcal{S}}^1_x \to \cdots \tag{9.11}$$

is exact for each $x \in X$. On the other hand, each $x \in X$ has a fundamental system of neighbourhoods $\{U_x\}$ which are contractible and hence we have

$$0 \to A \to S^0(U_x; A) \to S^1(U_x; A) \to \cdots$$

is exact. Upon passing to the direct limit, we obtain the exactness of (9.11). ♠

Combining with what we have seen in Example 9.3.6, we obtain another proof of de Rham's theorem:

Theorem 9.3.12 *For any smooth manifold, de Rham cohomology is canonically isomorphic to the singular cohomology with real coefficients.*

Exercise 9.3.13 Show that on a paracompact space, every soft sheaf has vanishing cohomology.

9.4 Čech Cohomology

We now come to the important goal of this chapter, viz., the Čech cohomology, which directly relates the topological behaviour of a space with various open coverings. As a simple consequence it is sensitive to the notion of covering dimension of a topological space. Čech homology is quite suitable when dealing with coefficients in a sheaf of modules. With additional assumptions such as existence of enough 'good' coverings, we establish that Čech cohomology of a space coincides with the singular cohomology.

Definition 9.4.1 Let $\mathcal{F}$ be a presheaf of modules over a space X. Let $\mathcal{U}$ be an open covering for X. We define a cochain complex $S^*(\mathcal{U}; \mathcal{F})$ as follows: For each $q \geq 0$, let $S^q(\mathcal{U}; \mathcal{F})$ be the module of functions ψ which assigns to each ordered $(q+1)$-tuple $(U_0, U_1, \ldots, U_q)$ of members of $\mathcal{U}$ an element in $\mathcal{F}(U_0 \cap \cdots \cap U_q)$. The coboundary operators $\delta : S^q(\mathcal{U}; \mathcal{F}) \to S^{q+1}(\mathcal{U}, \mathcal{F})$ are defined by the rule

$$\delta(\psi)(U_0, \ldots, U_{q+1}) = \sum_{i=0}^{q+1} (-1)^i \psi(U_0, \ldots, \hat{U}_i, \ldots, U_{q+1})|_{U_0 \cap \cdots \cap U_{q+1}}. \tag{9.12}$$

Verification that $\delta \circ \delta = 0$ is routine. The homology module of this cochain complex is denoted by $H^*(\mathcal{U}; \mathcal{F})$. If $\mathcal{V}$ is a refinement with a refinement function $\lambda : \mathcal{V} \to \mathcal{U}$ (i.e., $V \subset \lambda(V), V \in \mathcal{V}$) then we get a cochain map $\lambda^* : S^*(\mathcal{U}; \mathcal{F}) \to S^*(\mathcal{V}; \mathcal{F})$ defined by restriction. If $\mu : \mathcal{V} \to \mathcal{U}$ is another refinement map, then we verify that λ^* is chain homotopic to μ^* by the chain homotopy $D : S^q(\mathcal{U}; \mathcal{F}) \to S^{q-1}(\mathcal{V}, \mathcal{F})$ given by

$$D(\psi)(V_0, \ldots, V_{q-1}) = \sum_{j=0}^{q-1} (-1)^j \psi(\lambda(V_0), \ldots, \lambda(V_j), \mu(V_j), \ldots, \mu(V_{q-1}))|_{V_0 \cap \cdots \cap V_{q-1}}. \quad (9.13)$$

Thus there is a well defined homomorphism $H^*(\mathcal{U}; \mathcal{F}) \to H^*(\mathcal{V}; \mathcal{F})$ which is independent of the actual refinement map λ. Thus as $\mathcal{U}$ varies over all open covering of X, the family $\{H^*(\mathcal{U}; \mathcal{F})\}$ is a directed system. We define the Čech cohomology of X with coefficients in $\mathcal{F}$ as the direct limit of this directed system:

$$\check{H}^*(\mathcal{F}) = \varinjlim \{H^*(\mathcal{U}; \mathcal{F})\}. \quad (9.14)$$

Remark 9.4.2 If we take $\mathcal{F}$ to be the constant sheaf G, where G is an abelian group, we get the definition of $\check{H}(X; G) := \check{H}^*(G)$, the Čech cohomology of X with coefficients in G. Notice that, for any presheaf F, $H^0(\mathcal{U}; \mathcal{F}) = \mathcal{F}(\mathcal{U})$, the module of compatible $\mathcal{U}$-families of $\mathcal{F}$. Therefore, it follows that $\check{H}^0(\mathcal{F}) = \hat{\mathcal{F}}(X)$. In particular, when $\mathcal{F} = \Gamma$ is the sheaf of sections of a vector bundle, $\check{H}^0(\mathcal{F})$ is nothing but the module of global sections of the bundle. This example is perhaps enough to appreciate the importance of Čech cohomology.

It is easily checked that given a short exact sequence of presheaves:

$$0 \to \mathcal{F}' \to \mathcal{F} \to \mathcal{F}'' \to 0$$

for each open covering $\mathcal{U}$ of X, we get a short exact sequence of cochain complexes:

$$0 \to S^*(\mathcal{U}; \mathcal{F}') \to S^*(\mathcal{U}; \mathcal{F}) \to S^*(\mathcal{U}; \mathcal{F}'') \to 0. \quad (9.15)$$

This in turn yields a long exact sequence of cohomology modules:

$$\cdots \longrightarrow H^q(\mathcal{U}; \mathcal{F}') \longrightarrow H^q(\mathcal{U}; \mathcal{F}) \longrightarrow H^q(\mathcal{U}; \mathcal{F}'') \longrightarrow H^{q+1}(\mathcal{U}; \mathcal{F}') \longrightarrow \cdots \quad (9.16)$$

The collection of all these long exact sequences, as $\mathcal{U}$ varies over all open coverings of X is a directed system with respect to homomorphisms induced by refinement. Passing to the direct limit we obtain the following theorem.

Theorem 9.4.3 *Given a topological space X, there is a covariant functor from the category of short exact sequences of presheaves*

$$0 \to \mathcal{F}' \to \mathcal{F} \to \mathcal{F}'' \to 0$$

of modules over X to the category of long exact sequence of Čech cohomology modules of X:

$$\cdots \longrightarrow \check{H}^q(\mathcal{F}') \longrightarrow \check{H}^q(\mathcal{F}) \longrightarrow \check{H}^q(\mathcal{F}'') \longrightarrow \check{H}^{q+1}(\mathcal{F}') \longrightarrow \cdots \quad (9.17)$$

Definition 9.4.4 Let A be a subset of X and $\mathcal{F}$ be a presheaf on X. We define the cut-off presheaves $\mathcal{F}_A, \mathcal{F}^A$ as follows:

$$\mathcal{F}_A(U) = \begin{cases} \mathcal{F}(U), & U \cap A \neq \emptyset, \\ (0), & U \cap A = \emptyset. \end{cases}$$

$$\mathcal{F}^A(U) = \left\{ \begin{array}{ll} \mathcal{F}(U), & U \cap A = \emptyset, \\ (0), & U \cap A \neq \emptyset. \end{array} \right.$$

We define

$$\check{H}^*(A; \mathcal{F}) = \check{H}^*(\mathcal{F}_A); \quad \check{H}(X, A; \mathcal{F}) = \check{H}(\mathcal{F}^A).$$

Remark 9.4.5 It is clear that there is an exact sequence

$$0 \to \mathcal{F}_A \to \mathcal{F} \to \mathcal{F}^A \to 0.$$

From the above theorem we then have a long exact sequence

$$\cdots \longrightarrow \check{H}^q(A; \mathcal{F}) \longrightarrow \check{H}^q(X; \mathcal{F}) \longrightarrow \check{H}^q(X, A; \mathcal{F}) \longrightarrow \cdots$$

We shall now introduce a concept which helps us to conclude that the Čech cohomology is not 'disturbed' by the drawbacks of a presheaf as compared to a sheaf.

Definition 9.4.6 Let $\mathcal{F}$ be a presheaf of modules over X. We say it is locally zero, if for each $s \in \mathcal{F}(U)$, there is an over cover $\mathcal{V}$ of U such that $s|_V = 0$ for all $V \in \mathcal{V}$. This is the same as saying that $\mathcal{F}_x = (0)$; equivalently, the completion $\hat{\mathcal{F}}$ is the zero sheaf.

A morphism $\tau : \mathcal{F} \to \mathcal{F}'$ of presheaves is said to be a local isomorphism if $\operatorname{Ker} \alpha$ and $\operatorname{Coker} \alpha$ are both locally zero.

This is equivalent to saying that $\tau_x : \mathcal{F}_x \to \mathcal{F}'_x$ is an isomorphism for all x.

The following lemma is immediate since at the stalk level we have the identity map $\mathcal{F}_x \to \hat{\mathcal{F}}_x$.

Lemma 9.4.7 The canonical morphism $\Theta : \mathcal{F} \to \hat{\mathcal{F}}$ is a local isomorphism.

Theorem 9.4.8 *If X is paracompact and $\mathcal{F}$ is a locally zero presheaf, then $\check{H}^*(\mathcal{F}) = 0$.*

Proof: Let φ be a q-cochain on a locally finite open cover $\mathcal{U}$. For each $x \in X$, let W_x be a neighbourhood of x which intersects finitely many members of $\mathcal{U}$. Since $\mathcal{F}$ is locally trivial, we can choose a neighbourhood V_x of x such that $V_x \subset W_x$ and for all q-tuples of open sets $U_0, \ldots, U_q \in \mathcal{U}$, we have

$$\varphi(U_0, \ldots, U_q)|_{V_x} = 0.$$

Consider a common refinement $\mathcal{V}$ of both $\mathcal{U}$ and $\{V_x : x \in X\}$ and let $\alpha : \mathcal{V} \to \mathcal{U}$ be a refinement function. Then it follows that $\alpha^*(\varphi) = 0$ in $S^*(\mathcal{V}; \mathcal{F})$. Thus upon passing to the direct limit over all open covers, every cochain becomes zero. ♠

Corollary 9.4.9 *If $\tau : \mathcal{F} \to \mathcal{F}'$ is a local isomorphism of presheaves over a paracompact Hausdorff space, then $\tau_* : \check{H}^*(\mathcal{F}) \to \check{H}^*(\mathcal{F}')$ is an isomorphism.*

Proof: Observe that there are exact sequences

$$0 \longrightarrow \operatorname{Ker} \tau \longrightarrow \mathcal{F} \overset{\tau_1}{\longrightarrow} \operatorname{Im} \tau \longrightarrow 0$$

$$0 \longrightarrow \operatorname{Im} \tau \overset{\tau_2}{\longrightarrow} \mathcal{F}' \longrightarrow \operatorname{Coker} \tau \longrightarrow 0$$

so that $\tau = \tau_2 \circ \tau_1$. Now apply Theorems 9.4.3, 9.4.8 twice. ♠

Combining this with Lemma 9.4.7, we obtain:

Theorem 9.4.10 *If X is a paracompact Hausdorff space, the canonical morphism $\mathcal{F} \to \hat{\mathcal{F}}$ induces an isomorphism in cohomology: $\check{H}^*(\mathcal{F}) \approx \check{H}^*(\hat{\mathcal{F}})$.*

We shall now come to the special case when $\mathcal{F}$ is the constant sheaf $\hat{G}$, where G is an abelian group. The following statement is obvious.

Lemma 9.4.11 For any open covering $\mathcal{U}$ of X, let $N_{\mathcal{U}}$ denote the nerve of the covering. (See Example 2.6.2.(viii).) We have $S^*(\mathcal{U}; \hat{G}) = \operatorname{Hom}((C_*(N_{\mathcal{U}}); G)$ where C_* denotes the simplicial chain complex of the simplicial complex $N_{\mathcal{U}}$ with integral coefficients.

Remark 9.4.12 We say that a topological space is of covering dimension $\leq n$, and write $\dim X \leq n$ if for every open covering of X, there is a refinement whose nerve is a simplicial complex of dimension $\leq n$. If $\dim X \leq n$ but $\nleq n-1$ then we say $\dim X = n$. One can show that the covering dimension of $\mathbb{R}^n$ is n and every subspace of $\mathbb{R}^n$ is of dimension $\leq n$. (See [Hurewicz–Wallman, 1948] for more information.) It is an easy consequence of the above lemma that if $\dim X \leq n$, then for any presheaf $\mathcal{F}$ on X, we have $\check{H}^q(X; \mathcal{F}) = 0$ for $q > n$. As pointed out earlier in Remark 8.2.14, this is not the case with singular cohomology.

On the other hand, the above lemma points toward a certain relation between the Čech cohomology of an open covering of a space X with constant coefficients and the singular cohomology of X. There are several instances of this and several ways to carry out this plan. Below we shall present one way which seems to be the strongest of them all.

Definition 9.4.13 Given a topological space, an open covering $\mathcal{U}$ of X is said to be *good* if non empty finite intersections of members of $\mathcal{U}$ are all contractible.

Remark 9.4.14 Given a manifold M with a Riemannian metric, and an open covering $\mathcal{U}$ of X there is a good covering $\mathcal{V}$ which is a refinement of $\mathcal{U}$. In fact, members V of $\mathcal{V}$ have the property that between any two points of V there is a unique shortest path in M and this shortest path is contained in V. (Quite often, a student makes the mistake of thinking that any covering consisting of geodesically convex open sets of a manifold is good but that may not be so. See, for Example [Helgason, 1978].) Thus, for such spaces wherein the collection of good coverings forms a cofinal subfamily under refinement, good coverings can be used to compute the Čech cohomology, *without going to the direct limit*. The crux of the matter is that a good covering frees us from taking direct limits altogether. This is the content of the following discussion.

Theorem 9.4.15 *Let X be a paracompact Hausdorff space, $\mathcal{U} = \{U_j\}_{j \in J}$ be a good locally finite open covering, and $N_{\mathcal{U}}$ be the nerve of this covering. Then there is a homotopy equivalence $h_{\mathcal{U}} : X \to |N_{\mathcal{U}}|$, from X to the geometric realization of $N_{\mathcal{U}}$. Moreover, the homotopy equivalence is canonical with respect to taking refinements in the following sense: If $\mathcal{V}$ is another good covering and $\lambda : \mathcal{V} \to \mathcal{U}$ is a refinement function then there is a homotopy commutative diagram*

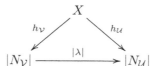

For any finite subset $A \subset J$, let us put $U_A = \cap_{j \in A} U_j$. Recall that the geometric realization $|K|$ of any simplicial complex K is a subspace of $\mathbb{I}^V$, (where V is the set of vertices of K,) consisting of $\theta : V \to \mathbb{I}$ with $\operatorname{supp} \theta \in K$ and such that $\sum_j \theta(j) = 1$. We put

$$\mathcal{D}(\mathcal{U}) = \{(\theta, x) \in |N_{\mathcal{U}}| \times X \ : \ x \in U_{\operatorname{supp} \theta}\}.$$

Let $p : \mathcal{D}(\mathcal{U}) \to N_{\mathcal{U}}$ and $q : \mathcal{D}(\mathcal{U}) \to X$ be the projection maps. The refinement function λ is a function on the indexing sets of the two coverings and defines a simplicial map $N_{\mathcal{V}} \to N_{\mathcal{U}}$

which we shall denote by λ itself. This in turn induces a map $\Lambda : \mathcal{D}(\mathcal{V}) \to \mathcal{D}(\mathcal{U})$ such that the following diagram is commutative:

$$
\begin{array}{ccc}
X \longleftarrow \mathcal{D}(\mathcal{V}) \longrightarrow |N_{\mathcal{V}}| \\
\Big\downarrow = \qquad \Big\downarrow \Lambda \qquad \Big\downarrow \lambda \\
X \longleftarrow \mathcal{D}(\mathcal{U}) \longrightarrow |N_{\mathcal{U}}|
\end{array}
$$

where all the horizontal arrows indicate appropriate projection maps. We claim that they are all homotopy equivalences from which the theorem follows. Indeed we shall prove the following two lemmas:

Lemma 9.4.16 If $\{\theta_j\}$ is a partition of unity subordinate to $\mathcal{U}$ then there is a section $\sigma : X \to \mathcal{D}(\mathcal{U})$ of the second projection q such that $\sigma(X)$ is a deformation retract of $\mathcal{D}(\mathcal{U})$.

Lemma 9.4.17 If $\mathcal{U}$ is a good cover then $\mathcal{D}(\mathcal{U})$ is a deformation retract of the mapping cylinder M_p of the projection $p : \mathcal{D}(\mathcal{U}) \to |N_{\mathcal{U}}|$.

The theorem follows immediately, since $|N_{\mathcal{U}}|$ is anyway a deformation retract of M_p.

In the proofs of the lemmas below, fixing the open cover $\mathcal{U}$ of X, we shall drop it from the notation $\mathcal{D}(\mathcal{U})$, $N_{\mathcal{U}}$, etc., and merely write $\mathcal{D}, N$, etc.

Proof of Lemma 9.4.16:

Define $\sigma(x) = (\Theta(x), x)$ where $\Theta(x)(j) = \theta_j(x)$. Clearly $\operatorname{supp} \Theta(x) \subset J$ is finite. Also, since $\{\theta_j\}$ is subordinate to $\mathcal{U}$, $\operatorname{supp} \theta_j \subset U_j$ and hence $x \in U_{\operatorname{supp} \Theta(x)}$. This implies $\operatorname{supp} \Theta(x) \in |N|$ and hence σ defines a function $X \to \mathcal{D}$. Clearly $q \circ \sigma(x) = x$. The continuity of σ follows from the fact that each θ_j is continuous and the family is locally finite. To show that $\sigma(X) \subset \mathcal{D}$ is a deformation retract, we note that given any $x \in X$, the set of all $j \in J$ such that $x \in U_j$ forms a simplex s in $N_{\mathcal{U}}$ and then we use the convexity of $|s|$. Thus, consider $H : \mathcal{D} \times \mathbb{I} \to \mathcal{D}$ defined by

$$H(\theta, x, t) = (t\Theta(x) + (1-t)\theta, x)$$

and verify that H is the homotopy as required.   ♠

Proof of Lemma 9.4.17:

Note that $\mathcal{U}$ is a good cover implies that the fibres of the first projection $p : \mathcal{D} \to |N|$ are all contractible. So, we may 'expect' p to be a homotopy equivalence. However, since there is no such general result, (see Exercise 1.9.18) we need to prove this carefully.

Let $N^{(k)}$ denote the k-skeleton of N; $\mathcal{D}^k = p^{-1}(N^{(k)})$ and $p^k = p|_{\mathcal{D}^k}$. We shall first show that $\mathcal{D}^k \cup M(p^{k-1})$ is a SDR of $M(p^k)$ for each $k \geq 0$. Clearly, this will imply that $\mathcal{D} \cup M(p^{k-1})$ is a SDR of $\mathcal{D} \cup M(p^k)$. An argument similar to the one employed in the proof of Theorem 2.5.3 can then be used to define a deformation retract of $M(p)$ to $\mathcal{D}$.

For $k = 0$, $\mathcal{D}^0$ is the disjoint union $\sqcup_{j \in J} \{j\} \times U_j$ and hence $M(p^0)$ is the disjoint union of the cones over these open sets. Since each U_j is contractible, there is a deformation retraction CU_j onto U_j (from Theorem 1.6.4). Putting them together gives a deformation retraction $\rho^0 : M(p^0) \to \mathcal{D}^0$. Figure 9.1 shows the part of $M(p^1)$ lying over a 1-simplex $\mathbb{I}$.

For $k \geq 1$, to construct a deformation retraction $M(p^k) \to \mathcal{D}^k \cup M(p^{k-1})$, it is enough to do this on each k-simplex of N separately and put them all together. Therefore, we may as well assume that N is a k-simplex σ, and $U_i = U_\sigma =: V$, say, for all $i \in \sigma$.

Now $\mathcal{D} = |\sigma| \times V$ and $p : |\sigma| \times V \to |\sigma|$ is the first projection and hence $M(p) = |\sigma| \times CV$, $M(p^{k-1}) = \partial|\sigma| \times CV$. As before, since V is contractible, V is a SDR of CV and hence

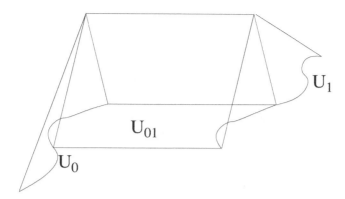

FIGURE 9.1. Nerve of a covering

$\partial|\sigma| \times V$ is a SDR of $\partial|\sigma| \times CV$ and hence $|\sigma| \times V \times 0$ is a SDR of $B = |\sigma| \times V \times 0 \cup \partial|\sigma| \times CV$. This shows that the latter space is contractible. Since $M(p) = |\sigma| \times CV$ is also contractible, we see that the inclusion map $B \hookrightarrow M(p)$ is a homotopy equivalence. Therefore, it is enough to show that the topological pair $(M(p), B)$ has HEP. By Theorem 1.6.9 again, we need to show that $M(p) \times \mathbb{I}$ retracts onto $M(p) \times 0 \cup B \times \mathbb{I}$. This follows from Exercise 1.6.15(ix) which indeed yields a strong deformation retraction of $M(p) \times 0 \cup B \times \mathbb{I}$, by taking $X = \partial|\sigma|$ and $Y = V$. ♠

Theorem 9.4.18 *Let X be a paracompact space which admits a cofinal family of good coverings. Then for any abelian group G, the Čech cohomology of X with constant sheaf $\hat{G}$ is isomorphic to the singular cohomology of X with coefficients in G, i.e., $\check{H}^*(X; \hat{G}) \approx H^*(X; G)$. Moreover, for any good covering $\mathcal{U}$ of X, we have $H^*(\mathcal{U}; \hat{G}) \approx H^*(X; G)$.*

Proof: By Lemma 9.4.9, for any open covering $\mathcal{U}$ of X, we have $H^*(\mathcal{U}; \hat{G})$ can be identified with $H^*(C_*(N_\mathcal{U}); G) = H^*(|N_\mathcal{U}|; G)$, the singular cohomology of the geometric realization of the simplicial complex $N_\mathcal{U}$. Since there is a cofinal family of good coverings, $\check{H}^*(X; \hat{G})$ can be computed by taking the direct limit over such a subfamily. Now from Theorem 9.4.15, we have a commutative diagram:

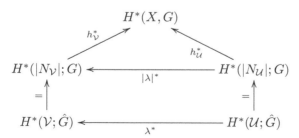

Therefore $\check{H}^*(X; \hat{G}) = \underset{\rightarrow}{\mathrm{dlim}}\, H^*(\mathcal{U}; \hat{G}) = \underset{\rightarrow}{\mathrm{dlim}}\, H^*(|N_\mathcal{U}|; G)$. Since the top triangle consists of all isomorphisms, it follows that the direct limit is canonically isomorphic to $H^*(X; G)$ and is isomorphic to any one of the members of this directed family. ♠

Corollary 9.4.19 The de Rham cohomology modules of a smooth manifold X are canonically isomorphic to the singular cohomology modules of X with constant coefficients $\mathbb{R}$.

Proof: As seen in the Example 9.3.6, the de Rham cohomology modules of X are the same as the sheaf cohomology of the constant sheaf $\hat{\mathbb{R}}$.

Exercise 9.4.20 Write down full details of continuity of σ in the proof of Lemma 9.4.16.

9.5 Miscellaneous Exercises to Chapter 9

1. Show that a presheaf $\mathcal{F}$ of modules satisfies (F-I) iff for every open set U, the morphism $\mathcal{F}(U) \to \hat{\mathcal{F}}(U)$ is injective, iff $\mathrm{Ker}\,[\mathcal{F} \to \hat{\mathcal{F}}] = (0)$. Make the corresponding statement for a presheaf of sets and prove it.

2. Let G be an abelian group, and $\mathcal{G}$ denote the corresponding constant presheaf on a topological space X. Show that $\mathrm{Sp\acute{e}}(\mathcal{G})$ is homeomorphic to $G \times X$, where G is given the discrete topology. Deduce directly from this the fact that the completion $\hat{\mathcal{G}}$ is precisely the sheaf $\hat{\mathcal{G}}$ as defined in Example 9.1.6(a).

3. Let X be a domain in $\mathbb{C}$ and $\mathcal{C}^\infty, \mathcal{H}$ denote the sheaf of $\mathcal{C}^\infty$ functions and the sheaf of holomorphic functions on X respectively.
   (a) Show that for each $x \in X$, the stalks $\mathcal{C}^\infty, \mathcal{H}_x$ are commutative rings.
   (b) Show that the canonical maps $s_x \mapsto s_x(x)$ define a surjective ring homomorphism $\mathcal{C}^\infty \to \mathbb{R}$ ($\mathcal{H}_x \to \mathbb{C}$, respectively) whose kernel is a maximal ideal.
   (c) Show that $\mathrm{Sp\acute{e}}(\mathcal{C}^\infty)$ is not Hausdorff whereas $\mathrm{Sp\acute{e}}(\mathcal{H})$ is Hausdorff.

4. Let A be a closed subspace of a paracompact Hausdorff space, $\mathcal{F}$ be a sheaf over X and s be a section of $\mathcal{F}$ over A. Show that there exists an open set U, such that $A \subset U \subset X$ and a section $s' \in \mathcal{F}(U)$ such that $s'|_A = s$. (Hint: Read the proof of Theorem 9.1.21 carefully.)

5. Given an open covering $\mathcal{U}$ of X and a family $\{\mathcal{F}_U\}$ where each $\mathcal{F}_U$ is a sheaf on $U \in \mathcal{U}$, write down a condition that ensures the existence of a sheaf $\mathcal{F}$ on X which when restricted to U is $\mathcal{F}_U$ for each U.

6. A presheaf $\mathcal{F}$ is said to be locally constant if there exists an open cover $\mathcal{U}$ of X such that for every $U \in \mathcal{U}$ and $x \in U$, the map $\mathcal{F}(U) \to \mathcal{F}_x$ is a bijection. Show that this condition is equivalent to saying that $\mathrm{Sp\acute{e}}(\mathcal{F}) \to X$ is a covering projection.

7. Every covering projection is a local homeomorphism and hence describes a sheaf. Describe the sheaf of sections of a universal covering space over a path connected space X.

8. Let $(\mathcal{C}^\infty)^\times$ denote the sheaf of continuous non vanishing complex valued functions on a manifold X. The exponential map $z \mapsto e^{2\pi i z}$, defines a morphism $\exp : \mathcal{C}^\infty \to (\mathcal{C}^\infty)^\times$. Show that there is an exact sequence of sheaves

$$0 \longrightarrow \mathcal{Z} \longrightarrow \mathcal{C}^\infty \xrightarrow{\ \exp\ } (\mathcal{C}^\infty)^\times \longrightarrow 0$$

where $\mathcal{Z}$ is the constant sheaf corresponding to the ring of integers.

9. Given a sheaf $\mathcal{F}'$ over Y and a map $f : X \to Y$ consider the sheaf of sections of the pullback of $\mathrm{Sp\acute{e}}(\mathcal{F}') \to Y$ under f. Show that this is the same as the sheaf $f^*(\mathcal{F}')$.

Chapter 10

Homotopy Theory

This chapter may be called the heart of the book. Throughout this chapter we shall be working with the category of pointed topological spaces. In Section 10.1, we begin with a general discussion on the homotopy sets $[X, Y]$ which leads us to the notion of H-spaces and their duals. In particular we get some canonical abelian group structures on $[\mathbb{S}^n, Y], n \geq 2$, which we call higher homotopy groups. In Section 10.2, we make an earnest beginning of the study of these groups. The central result here is the exact sequence of homotopy groups for pointed topological pairs and the same for fibrations. As a consequence we derive the non triviality of $\pi_3(\mathbb{S}^2)$ which is a milestone in homotopy theory and was a big surprise to the mathematical community when Hopf discovered it. Section 10.3 takes care of the study of the effect of change of base points on homotopy groups. In Section 10.4, we deal with a landmark result known as the Hurewicz isomorphism theorem, which relates homotopy groups with homology groups. As a consequence, we get a complete hold on the homotopy type of CW-complexes, a result due to J. H. C. Whitehead. It is high time now that we came back to one of the basic problems of extension and lifting. Sections 10.5—10.8 deal with these problems. This study goes under the generic name 'obstruction theory'. It has its own gems such as Eilenberg–Mac Lane spaces and representability of cohomology groups. We end this chapter with some computation of homotopy groups of classical groups, which in turn, have many diverse applications. In Section 10.9, we carry out some elementary computation of homotopy groups of classical groups. Section 10.10 contains a brief introduction to homology with local coefficients and relates it to the equivariant homology.

10.1 H-spaces and H'-spaces

Recall that by a pointed topological space, we mean a space X together with a specific point $x_0 \in X$ to be called the base-point of X and the whole thing being denoted by (X, x_0). A morphism $f : (X, x_0) \to (Y, y_0)$ is a map $f : X \to Y$ such that $f(x_0) = y_0$. One can easily check that this forms a category $\mathbf{Top_0}$ called the category of pointed spaces and base-point preserving maps. Corresponding to any full subcategory of $\mathbf{Top}$ such as the category of Hausdorff spaces, one can consider a full subcategory of pointed spaces as well, such as the category of pointed Hausdorff spaces, etc.

Likewise we also have the category $\mathbf{Top^2}$ of topological pairs (X, A) where $A \subset X$ with morphisms $f : (X, A) \to (Y, B)$ being maps $f : X \to Y$ such that $f(A) \subset B$. And then again, there is the category $\mathbf{Top_0^2}$ of pointed topological pairs and morphisms, with the additional condition that the base point of (X, A, x_0) is located inside A.

Given any of the above category of spaces, if we take the morphisms between any two objects B, C to be the set of all homotopy classes $[B, C]$ of maps from B to C we get yet another category. Algebraic topology is primarily interested in the study of such *homotopy categories*. Most often, we shall not mention the base point and merely say that X is a space when we actually mean (X, x_0), the base point x_0 being relegated to the background unless it is necessary to mention it. For instance $[X, Y]$ **will denote the set of all base-**

point preserving homotopy classes of base-point preserving maps from (X, x_0) **to** (Y, y_0).

The first problem that we consider is to put some binary operation on the set $[X, Y]$ of homotopy classes of maps. Our motivation comes from examples such as linear functions $V \to \mathbb{K}$ of vector spaces over a field $\mathbb{K}$, real (or complex) valued continuous (or $\mathcal{C}^\infty$) functions on manifolds, or simply the set $F(A, G)$ of all functions from a set A to a group G. All of them acquire a binary operation defined pointwise using the binary operation of the codomain.

More specifically, suppose G is a topological group. Then G^X acquires a group structure:

$$(f \cdot g)(x) = f(x)g(x).$$

This operation naturally goes down to define a group structure on $[X, G]$ with the constant function $x \mapsto 1 \in G$ playing the role of the identity for the operation. Now consider the converse problem:

Question A: For some space P, suppose the functor

$$\pi^P : X \rightsquigarrow [X, P]$$

takes values in the category of groups. Does this imply that P is a group?

Answer: Almost yes

This is what leads us to the notion of H-groups. Given a pointed space (P, p_0), let us denote by c the constant map $P \to P$ taking the value p_0. Let $T : P \times P \to P \times P$ denote the map $(x, y) \mapsto (y, x)$.

Definition 10.1.1 A pointed topological space (P, p_0) together with a binary operation $\mu : P \times P \to P$ is called an **H-space** if
(a) the two composite maps

$$P \xrightarrow{(c, Id)} P \times P \xrightarrow{\mu} P \qquad\qquad P \xrightarrow{(Id, c)} P \times P \xrightarrow{\mu} P$$

are homotopic to the $Id = Id_P$. In this case, the constant map c is referred to as the (unique) **homotopy identity** for μ.

The multiplication is respectively called **homotopy commutative** or **homotopy associative** according to whether the first or the second one of the following diagrams are homotopy commutative:

A map $\iota : P \to P$ is called a **homotopy inverse** if the two composites

$$P \xrightarrow{(Id, \iota)} P \times P \xrightarrow{\mu} P \qquad\qquad P \xrightarrow{(\iota, Id)} P \times P \xrightarrow{\mu} P$$

are both homotopic to c.

Theorem 10.1.2 *Given a pointed space P, the contravariant functor π^P on the category of pointed spaces takes values inside the category of semigroups (respectively, monoids, groups, abelian groups) iff P is an H-space (respectively, associative H-space, H-group, abelian H-group). Moreover, the algebraic structure on $[X, P]$ is the same as that induced by the corresponding homotopy binary operation on P.*

Theorem 10.1.3 *Let P, P' be any two H-groups. Then a map $f : P \to P'$ is a homomorphism iff f induces a natural transformation of π^P to $\pi^{P'}$ in the category of groups.*

We advise that the reader supply the proofs of these two theorems at her leisure. Let us discuss a few typical examples now.

Example 10.1.4 The Loop Spaces

For any pointed space (Y, y_0), let ΩY denote the space of all loops in Y based at y_0. The base point for ΩY is the constant loop at y_0. Define

$$\mu : \Omega Y \times \Omega Y \to \Omega Y$$

by the formula:

$$\mu(\omega, \omega')(t) = \begin{cases} \omega(2t), & 0 \leq t \leq 1/2, \\ \omega'(2t - 1), & 1/2 \leq t \leq 1. \end{cases}$$

We leave it to you to verify the continuity of μ. The verification that ΩY is an H-group is exactly the same as the verification that $\pi_1(Y, y_0)$ is a group and so we leave this also to you as a warming-up exercise. In summary, we have:

Theorem 10.1.5 *The loop functor is covariant from the category of pointed spaces to the category of H-groups and continuous homomorphisms.*

What makes the function space $\Omega(Y)$ acquire an H-group structure? What is so special about $\mathbb{S}^1$? This naturally leads us to a question 'dual' to Question A:

Question B For some pointed space Q, suppose the covariant functor $\pi_Q : X \rightsquigarrow [Q, X]$ on the category of pointed spaces takes values in the category of groups. Does this mean that Q is a co-group? Once again the answer is almost yes.

Definition 10.1.6 By an H'-space, (or a co-H-space) we mean a pointed topological space Q together with a co-multiplication $\nu : Q \to Q \vee Q$ such that
(a) the composites

$$Q \xrightarrow{\nu} Q \vee Q \xrightarrow{(c, Id)} Q \qquad\qquad Q \xrightarrow{\nu} Q \vee Q \xrightarrow{(Id, c)} Q$$

are homotopic to identity;
(b) the co-multiplication is said to be associative, if the square

$$
\begin{array}{ccc}
Q & \xrightarrow{\nu} & Q \vee Q \\
{\scriptstyle \nu}\downarrow & & \downarrow{\scriptstyle \nu \vee Id} \\
Q \vee Q & \xrightarrow{Id \vee \nu} & Q \vee Q \vee Q
\end{array}
$$

is homotopy commutative, i.e., ν is homotopy associative.

An associative H'-space Q satisfying the existence of homotopy inverse as formulated below will be called an H'-group (or an H-cogroup):
(c) Existence of homotopy inverse: there exists a map $\psi : Q \to Q$ (unique up to homotopy) such that the composites

$$Q \xrightarrow{\nu} Q \vee Q \xrightarrow{(Id, \psi)} Q \qquad\qquad Q \xrightarrow{\nu} Q \vee Q \xrightarrow{(\psi, Id)} Q$$

are homotopic to the constant map $c : Q \to Q$.

An H'-group Q as above is *abelian* if the following triangle is homotopy commutative:

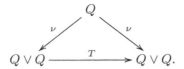

(Here T is the restriction of the permutation of coordinates in $Q \times Q$.

Given two H'-groups Q, Q', a map $\alpha : Q \to Q'$ is said to be a homomorphism if the following square is homotopy commutative:

$$
\begin{array}{ccc}
Q & \xrightarrow{\nu} & Q \vee Q \\
{\scriptstyle\alpha}\downarrow & & \downarrow{\scriptstyle\alpha\vee\alpha} \\
Q' & \xrightarrow{\nu'} & Q' \vee Q'
\end{array}
$$

Remark 10.1.7 If Q is an H'-group then for any pointed space X, the set $[Q, X]$ acquires a group structure given by $[f] * [g] = [(f, g) \circ \nu]$.

Theorem 10.1.8 π_Q *is a covariant functor taking values in the category of groups and homomorphisms iff Q is an H'-group. Moreover, in that case the group structure on $[Q, X]$ is the one induced by the H'-group structure on Q as in the above remark.*

The proof of this theorem is similar to that of Theorem 10.1.2 and likewise will be left to the reader as an exercise.

Example 10.1.9 Smash products and the reduced suspension
For any topological space X, we define $k(X)$ to be the space obtained by re-topologising X with the topology co-induced by the family of compact subspaces of X. (For more details see Exercise 1.9.45.) Given two pointed spaces X, Y their **smash product** $X \wedge Y$ is defined to be the quotient space of $k(X \times Y)$, wherein the subspace $X \times y_0 \cup x_0 \times Y$ is identified to a single point.

$$
X \wedge Y = \frac{X \times Y}{X \times y_0 \cup x_0 \times Y}
$$

with the class $[X \times y_0 \cup x_0 \times Y]$ taken as the base point. In particular, when $X = (\mathbb{S}^1, 1)$ then we call $X \wedge Y = \mathbb{S}^1 \wedge Y = SY$ the reduced suspension of Y. Let us denote the base point of SY by $[Y]$.

Inductively, we define $S^n Y = S(S^{n-1} Y)$, $S^n(Y, B) = (S^n Y, S^n B)$, $n \geq 2$.

We define $\nu : SX \to SX \vee SX$ as follows:

$$
\nu[x, t] = \begin{cases} ([x, 2t], [X]), & 0 \leq t \leq 1/2, \\ ([X], [x, 2t - 1]), & 1/2 \leq t \leq 1. \end{cases}
$$

The verification that ν makes SX into an H-cogroup is just 'dual' to the verification that ΩX is an H-group.

Indeed, this 'duality' is quite an important fact. Consider the two functors Ω and S. On the product category $\mathcal{T} \times \mathcal{T}$ consider the two associated functors:

$$
(X, Y) \rightsquigarrow \mathrm{Map}\,(SX, Y); \quad (X, Y) \rightsquigarrow \mathrm{Map}\,(X, \Omega Y).
$$

The exponential correspondence yields a natural bijection (exercise: verify this),

$$
\mathrm{Map}\,(SX, Y) \simeq \mathrm{Map}\,(X, \Omega Y)
$$

and so the two functors are equivalent. The natural equivalence passes down to homotopy category, both of free homotopies as well as pointed base-point homotopies. In particular, we have a natural equivalence

$$[SX; Y] \simeq [X, \Omega Y]. \tag{10.1}$$

(This is the same as saying that $S(-)$ is a left adjoint to $\Omega(-)$.)

We can use this duality to verify that SX is an H-cogroup. Thus for a fixed space X, the functor π_{SX} is naturally equivalent to the composite functor $\pi_X \circ \Omega$. Since Ω is a functor taking values in the category of H-groups, it follows that π_{SX} takes values in the category of groups. Therefore by Theorem 10.1.8, SX is an H-cogroup in such a way that the co-multiplication corresponds to the group structure on $[SX, X] \simeq [X, \Omega X]$ and hence up to homotopy, is the one that we have defined above. We summarise these observations in the following:

Theorem 10.1.10 *The reduced suspension is a covariant functor on the category of pointed spaces to the category of H-cogroups.*

Recall that the unreduced suspension $S'Y$ of a topological space Y is defined as the quotient $\mathbb{I} \times Y / 0 \times Y, 1 \times Y$. (See Exercises 1.9.27 and 1.9.28.) Thus you may think of the reduced suspension as the quotient of the unreduced suspension in which the image of $\mathbb{I} \times y_0$ is also identified to the single point, which is the base point.

Let $\kappa : S'Y \to SY$ denote this quotient map. Let $C'_{\pm}Y$ denote the subspaces of $S'Y$ which are the images of $Y \times [0, 1/2], Y \times [1/2, 1]$, respectively. Similarly let $C_{\pm}Y$ denote the image of $Y \times [0, 1/2]$ and $Y \times [1/2, 1]$ in SY. We shall identify Y with the image of the subspace $Y \times 1/2$ both in $S'Y$ as well as SY. We shall denote by $[y, t]'$ the image of (y, t) in $S'Y$.

Lemma 10.1.11 The quotient map κ induces a homotopy equivalence of pairs involving any two of the four spaces $S'Y, C'_+Y, C'_-Y, Y$ onto the corresponding pairs involving SY, C_+Y, C_-Y, Y, respectively.

Proof: This is a consequence of the non degeneracy of the point y_0: an immediate consequence being the fact that

$$Y \times \dot{\mathbb{I}} \cup y_0 \times \mathbb{I} \hookrightarrow Y \times \mathbb{I}$$

is a cofibration. We begin with the map

$$G : (Y \times \dot{\mathbb{I}} \cup y_0 \times \mathbb{I}) \times \mathbb{I} \cup Y \times \mathbb{I} \times 0 \to S'Y$$

given by:

$$\begin{aligned} G(y, 0, t) &= [y_0, t/2]'; & G(y, 1, t) &= [y_0, 1 - t/2]'; \\ G(y_0, s, t) &= [y_0, (1 - t)s + t/2]'; & G(y, s, 0) &= [y, s]'. \end{aligned}$$

Using the above cofibration, we can extend this to a homotopy $G : Y \times \mathbb{I} \times \mathbb{I} \to S'Y$. Clearly, this map factors down to define a homotopy $H : S'Y \times \mathbb{I} \to S'Y$ of the identity map with the map $\eta = H(-, 1)$ which collapses $y_0 \times \mathbb{I}$ to a single point. Therefore we obtain a map $g : SY \to S'Y$ such that $\eta = g \circ \kappa$. Also $H(SY_0 \times \mathbb{I}) \subset S'Y_0$. One can now easily verify that if A is any one of the subspaces $C'_{\pm}Y$ or Y, then $H(A \times \mathbb{I}) \subset A$ and g is a homotopy inverse of κ in all the cases. ♠

Corollary 10.1.12 If y_0 is a non degenerate base point of a path connected space, then the reduced suspension SY is simply connected.

Proof: Via the homotopy equivalence $\kappa : S'Y \to SY$ the problem is converted into proving that the unreduced suspension SY of Y is simply connected, which we have proved in Chapter 3.   ♠

Remark 10.1.13 These results will be put to use in a non trivial way in the next chapter while studying fibrations over suspensions.

Exercise 10.1.14

(i) Assume that all spaces are Hausdorff and compactly generated. Let $\mathrm{Map}\,(X, Y)$ denote the space Y^X re-topologised with the compactly generated topology $k(Y^X)$.
(a) The function spaces $\mathrm{Map}\,(X \wedge Y, Z)$ and $\mathrm{Map}\,(X, \mathrm{Map}\,(Y, Z))$ are naturally homeomorphic.
(b) The operation of composition induces a continuous map:

$$\mathrm{Map}\,(Y, X) \wedge \mathrm{Map}\,(X, Y) \to \mathrm{Map}\,(X, Z).$$

(c) The functors $\wedge Y$ and $\mathrm{Map}\,(Y, -)$ are adjoint to each other.
(d) The functors S and Ω are adjoint to each other, i.e., $\mathrm{Map}\,(SX, Y)$ and $\mathrm{Map}\,(X, \Omega Y)$ are naturally homeomorphic.

(ii) Show that X is an H-group then the set of path components of X forms a group.

(iii) Show that for any pointed space Y, $\pi_1(Y)$ is nothing but the set of path components of the H-group ΩY.

(iv) Show that for an H-group Y, $\pi_1(Y)$ is abelian.

10.2   Higher Homotopy Groups

In this section, we shall re-introduce the higher homotopy groups (which were introduced cursorily in Miscellaneous Exercise 1.9 earlier), and study some of their basic properties. We shall always use $0 \in \mathbb{I}$ as the base point for the interval $\mathbb{I} = [0, 1]$. Also $\dot{\mathbb{I}}$ will denote $\{0, 1\}$, the boundary of $\mathbb{I}$.

Definition 10.2.1 For $n \geq 1$ and a based space (X, x_0), we define

$$\pi_n(X, x_0) = [(S^n(\dot{\mathbb{I}}, 0); (X, x_0)] = [S^n(\dot{\mathbb{I}}); X] = [\dot{\mathbb{I}}; \Omega^n X].$$

For $n \geq 2$ and any based pair (X, A, x_0) we define

$$\pi_n(X, A) = [S^{n-1}(\mathbb{I}, \dot{\mathbb{I}}, 0); (X, A, x_0)] = [S^{n-1}(\mathbb{I}, \dot{\mathbb{I}}); (X, A)].$$

Remark 10.2.2

1. We shall denote by $\Omega^n(Y)$ the iterated loop space $\Omega(\Omega^{n-1}(Y))$, defined inductively. Also, given a topological pair (X, A), with a base point $x_0 \in A$, we define

$$\Omega(X, A) = \{\omega : \mathbb{I} \to X \;:\; \omega(0) = x_0, \omega(1) \in A\}$$

(We caution you not to confuse this with the topological pair $(\Omega(X), \Omega(A))$.) Also, for $n \geq 2$, we shall introduce the notation

$$\Omega^n(X, A) := \Omega^{n-1}(\Omega(X, A))$$

is the $(n-1)$-iterated loop space of the topological space $\Omega(X, A)$. From Exercise 10.1.14.(i)(d), we obtain

$$\begin{aligned} \text{Map}\,(S^{n-1}(\mathbb{I}, \dot{\mathbb{I}}, 0); (X, A, x_0)) &\approx \text{Map}\,(\mathbb{S}^{n-1}, \text{Map}\,((\mathbb{I}, \dot{\mathbb{I}}, 0), (X, A, x_0))) \\ &= \text{Map}\,(\mathbb{S}^{n-1}, \Omega(X, A)). \end{aligned}$$

It follows that $\pi_n(X, A, x_0) = [S^{n-1}; \Omega(X, A)] = \pi_{n-1}(\Omega(X, A)), n \geq 2$.

2. Thus, $\pi_n(X), n \geq 1$ (and $\pi_n(X, A), n \geq 2$) is the set of path components of the loop space $\Omega^n(X)$ (respectively, $\Omega^n(X, A)$) and hence is a group which is abelian if $n \geq 2$ (respectively, if $n \geq 3$). Similarly, one can of course define $\pi_1(X, A) = [(\mathbb{I}, \dot{\mathbb{I}}), (X, A)]$ but this is only a pointed set and does not have any natural group structure.

3. Also note that for $n \geq 1$, $S^n(\dot{\mathbb{I}})$ and $S^{n-1}(\mathbb{I}/\dot{\mathbb{I}})$ are homeomorphic. This induces a natural identification of the absolute group $\pi_n(X, x_0)$ with the relative group $\pi_n(X, \{x_0\}, x_0)$. Under this identification, the inclusion map $(X, \{x_0\}) \to (X, A)$ induces a homomorphism which we shall denote by $j_\# : \pi_n(X) \to \pi_n(X, A)$.

4. Since $S^n(\dot{\mathbb{I}})$ is path connected, if X_0 (respectively, if A_0) denotes the path component of X (resp. path component of A) containing x_0 then the inclusion maps $i : X_0 \to X$, (resp. $i : (X_0, A_0) \to (X, A)$ induce isomorphisms

$$i_\# : \pi_n(X_0) \to \pi_n(X, x_0); \quad i_\# : \pi_n(X_0, A_0) \to \pi_n(X, A).$$

For this reason, we can and shall assume, whenever it is convenient that X, A are path connected in the discussion of homotopy sets.

Example 10.2.3 In Section 1.2, we have seen that $\pi_1(\mathbb{S}^1) \approx \mathbb{Z}$, and $\pi_1(\mathbb{S}^n) = (1)$ for $n \geq 2$. In Chapter 2, as a simple application of cellular (or simplicial) approximation theorem, we have seen that $\pi_k(\mathbb{S}^{n+k}) = (0)$ for all $n, k \geq 1$. So far, we do not know the groups $\pi_n(\mathbb{S}^n), n \geq 2$, nor anything about $\pi_{n+k}(\mathbb{S}^n)$ for $n \geq 1$ and $k \geq 1$. It turns out that the computation of $\pi_n(\mathbb{S}^n) \approx \mathbb{Z}$ is not so difficult. We shall obtain this in Section 10.4 while proving a much stronger result. However, it turns out that computation of $\pi_{n+k}(\mathbb{S}^n)$ is a formidable task and only some partial results are known. These results will occupy a considerable amount of space in the rest of this book.

Lemma 10.2.4 Let (Y, B) be a topological pair and Y' be a closed subset of Y. Assume that there is a homotopy $H : (Y, B) \times \mathbb{I} \to (Y, B)$ such that
(a) $H(y, 0) = y, \; y \in Y$;
(b) $H(Y' \times \mathbb{I}) \subset Y'$;
(c) $H(Y' \times 1) = y_0$.
Then the collapsing map $k : (Y, B) \to (Y/Y', B/Y \cap B)$ is a homotopy equivalence. Furthermore,
(d) if $H(y_0 \times \mathbb{I}) = \{y_0\}$, then the inclusion map $(Y, B, y_0) \to (Y, B \cup Y', Y')$ is a homotopy equivalence.

Proof: Indeed, $f := H(-, 1) : Y \to Y$ factors down to define a map $\hat{f} : Y/Y' \to Y$ which happens to be a homotopy inverse to k. (Exercise: Check this.)

For the latter part, note that H itself is a homotopy of Id of $(Y, B \cup Y', Y')$ with $i \circ f$ as well as Id of (Y, B, y_0) with $f \circ i$. ♠

Remark 10.2.5 Consider the map $\Theta : S^{n-1}(\mathbb{I}) \to \mathbb{I} \times \mathbb{I}^{n-1}/(\mathbb{I} \times \partial(\mathbb{I}^{n-1}) \cup 0 \times \mathbb{I}^{n-1})$ given by

$$[\cdots[t, t_1]\cdots t_{n-1}] \mapsto [t, t_1, \ldots t_{n-1}].$$

Verify that Θ is a homeomorphism and $\Theta(S^{n-1}(\dot{\mathbb{I}})) \subset \partial(\mathbb{I}^n)/\mathbb{I} \times \partial(\mathbb{I}^{n-1}) \cup 0 \times \mathbb{I}^{n-1}$. Therefore,

$$\pi_n(X, A, x_0) = [(\mathbb{I}^n, \partial(\mathbb{I}^n), \mathbb{I} \times \partial(\mathbb{I}^{n-1}) \cup 0 \times \mathbb{I}^{n-1}); (X, A, x_0)].$$

Since $\mathbf{0} = (0, \ldots 0) \in \mathbb{I}^n$ is a SDR of $\mathbb{I} \times \partial(\mathbb{I}^{n-1}) \cup 0 \times \mathbb{I}^{n-1}$, it follows from the above lemma, that the inclusion map

$$(\mathbb{I}^n, \partial(\mathbb{I}^n), \mathbf{0}) \hookrightarrow (\mathbb{I}^n, \partial(\mathbb{I}^n), \mathbb{I} \times \partial(\mathbb{I}^{n-1}) \cup 0 \times \mathbb{I}^{n-1})$$

is a homotopy equivalence. Therefore

$$\pi_n(X, A, x_0) = [(\mathbb{I}^n, \partial(\mathbb{I}^n), \mathbf{0}); (X, A, x_0)].$$

Finally, given any point $p \in \mathbb{S}^{n-1}$, there is an obvious homeomorphism of the triples $(\mathbb{D}^n, \mathbb{S}^{n-1}, p_0) \to (\mathbb{I}^n, \partial(\mathbb{I}^n), \mathbf{0})$. Therefore, we have:

Theorem 10.2.6 $\pi_n(X, A, x_0) = [(\mathbb{D}^n, \mathbb{S}^{n-1}, p_0); (X, A, x_0)].$

The following theorem comes handy in a further interpretation of elements of $\pi_n(X, A, x_0)$.

Theorem 10.2.7 *A map* $\alpha : (\mathbb{D}^n, \mathbb{S}^{n-1}, p_0) \to (X, A, x_0)$ *represents the zero element of* $\pi_n(X, A, x_0)$ *iff* α *is homotopic relative to* $\mathbb{S}^{n-1}$ *to some map* $\beta : \mathbb{D}^n \to A$.

Proof: If $[\alpha] = 0$ then there is a homotopy

$$H : (\mathbb{D}^n, \mathbb{S}^{n-1}, p_0) \times \mathbb{I} \to (X, A, x_0)$$

such that $H(x, 0) = \alpha(x)$, and $H(x, 1) = x_0$ for all $x \in \mathbb{D}^n$. Define

$$H'(x, t) = \begin{cases} H\left(\frac{x}{1-t/2}, t\right), & 0 \le \|x\| \le 1 - t/2, \\ H\left(\frac{x}{\|x\|}, 2 - 2\|x\|\right), & 1 - t/2 \le \|x\| \le 1. \end{cases}$$

Check that H' is the required homotopy of α relative to $\mathbb{S}^{n-1}$. Conversely, if α is homotopic relative to $\mathbb{S}^{n-1}$ to a map $\beta : \mathbb{D}^n \to A$ then clearly $[\alpha] = [\beta]$. Therefore, it is enough to show that $[\beta] = 0$. The map $G(x, t) = \beta((1-t)x + tp_0)$ proves this. ♠

Definition 10.2.8 Let $n \ge 0$. We say (X, A) is n-connected, if for all $0 \le k \le n$, every map $\alpha : (\mathbb{D}^k, \mathbb{S}^{k-1}) \to (X, A)$ is homotopic relative to $\mathbb{S}^{k-1}$ to a map $\beta : \mathbb{D}^k \to A$.

Remark 10.2.9 In the above definition, $\mathbb{D}^0$ is a single point and $\mathbb{S}^{-1} = \emptyset$. Thus (X, A) is 0-connected iff every path component of X intersects A. Thus we obtain:

Corollary 10.2.10 A pair (X, A) is n-connected iff every path component of X intersects A and for every $1 \le k \le n$ and for every $x_0 \in A$, we have $\pi_k(X, A, x_0) = (0)$.

Remark 10.2.11 The sequence of groups $\{\pi_n(X)\}$ and $\{\pi_n(X, A)\}$ resemble the sequence of homology groups in many aspects. For example it is obvious that they satisfy the dimension axiom and the homotopy axiom. We shall see in the next section that they satisfy the exactness axiom. However, they do not satisfy the excision axiom to the full extent, the best known result going under the name Blakers–Massey theorem. On the other hand, under fibrations, homotopy groups are better behaved than the homology groups. In the rest of this section, we shall take up these properties one by one but not necessarily in this order.

Given a pointed pair of topological spaces, (X, A, x_0) we shall denote the inclusion maps by

$$i : (A, x_0) \to (X, x_0), \quad j : (X, x_0, x_0) \to (X, A, x_0).$$

To save some time and space we shall drop writing the base point unless it is very necessary to do otherwise. Restricting a map $(\mathbb{D}^n, \mathbb{S}^{n-1}) \to (X, A)$ to the boundary of $\mathbb{D}^n$, we also obtain the canonical homomorphism $\partial : \pi_n(X, A) \to \pi_{n-1}(A)$. These homomorphisms are neatly woven into a long exact sequence:

Theorem 10.2.12 (Homotopy exact sequence of a pair) *For any pointed pair (X, A), there is a canonical exact sequence*

$$\cdots \longrightarrow \pi_n(A) \xrightarrow{i_\#} \pi_n(X) \xrightarrow{j_\#} \pi_n(X, A) \xrightarrow{\partial} \pi_{n-1}(A) \longrightarrow \cdots$$

$$\cdots \longrightarrow \pi_1(X) \xrightarrow{j_\#} \pi_1(X, A) \xrightarrow{\partial} \pi_0(A) \xrightarrow{i_\#} \pi_0(X)$$

consisting of groups and homomorphisms except for the last three terms, which are based sets and base-point preserving functions.

Proof: The canonical property and the fact that the successive composites are trivial are easy to verify. The rest of the proof can be taken in three steps:
(1) $\operatorname{Ker} i_\# \subset \operatorname{Im} \partial$: Let $\alpha : \mathbb{S}^n \to A$ represent an element of $\pi_n(A)$ such that $i_\#[\alpha] = 0$. This means that there is a map $\alpha' : \mathbb{D}^{n+1} \to X$ such that $\alpha'|_{\mathbb{S}^n} = \alpha$. This means that $[\alpha'] \in \pi_{n+1}(X, A)$ and $[\alpha] = \partial[\alpha']$.
(2) $\operatorname{Ker} j_\# \subset \operatorname{Im} i_\#$: If $\alpha : (\mathbb{D}^n, \mathbb{S}^{n-1}) \to (X, x_0)$ is such that $j_\#([\alpha]) = 0$ then there is homotopy $H : \mathbb{D}^n \times \mathbb{I} \to X$ relative to $\mathbb{S}^{n-1}$ of α to a map $\beta : \mathbb{D}^n \to A$. It follows that $[\alpha] = i_\#[\beta]$.
(3) $\operatorname{Ker} \partial \subset \operatorname{Im} j_\#$: Suppose $\alpha : (\mathbb{D}^{n+1}, \mathbb{S}^n, 1) \to (X, A, x_0)$ is such that $\partial[\alpha] = 0$. This means that there is a null-homotopy $H : \mathbb{S}^n \times \mathbb{I} \to A$ of $\alpha|_{\mathbb{S}^n}$. By the homotopy extension property, there is an extension of H to a homotopy $G : \mathbb{D}^{n+1} \times \mathbb{I} \to X$ such that $G|_{\mathbb{D}^{n+1} \times 0} = \alpha$. Take $\beta = G|_{\mathbb{D}^{n+1} \times 1}$. Then $[\beta] \in \pi_{n+1}(X)$ and $j_\#[\beta] = [\alpha]$. ♠

Definition 10.2.13 A map $f : X \to Y$ is a called an *n-equivalence* $(n \geq 1)$ if
(a) f induces a bijection of path components of X with those of Y and
(b) for every $x_0 \in X$, the induced homomorphisms $f_\# : \pi_i(X, x_0) \to \pi_i(Y, f(x_0))$ is an isomorphism for $0 < q < n$ and an epimorphism for $q = n$.
    If f is an n-equivalence for all $n \geq 1$ then we say f is a *weak homotopy equivalence.*

Remark 10.2.14

1. Composite of n-equivalences is an n-equivalence.

2. Any map homotopic to an n-equivalence is an n-equivalence.

3. A homotopy equivalence is a weak homotopy equivalence. The converse is not true in general, but true for maps between CW-complexes, as we shall soon see.

4. Note that if $f : X \to Y$ is an inclusion map then by the homotopy exact sequence of the pair, f is an n-equivalence iff (Y, X) is n-connected. We would like to convert the above definition into something similar to Definition 10.2.8. As a first step, we see that $X \to Y$ is a n-equivalence iff (M_f, X) is n-connected, where M_f is the mapping cylinder of f (see Section 1.5).

The following lemma is an immediate consequence of the definition of n-connectivity.

Lemma 10.2.15 *Let B be obtained by attaching n-cells to A. Suppose (Y, X) is n-connected. Given maps $\alpha : A \to X$ and $\beta : B \to Y$ such that $\beta|_A = \alpha$, there is a map $\beta' : B \to X$ homotopic to β relative to A.*

Applying the lemma inductively, on the skeletons of (B, A) we obtain:

Theorem 10.2.16 *Let (B, A) be a relative CW-complex with $\dim (B, A) \leq n$. Suppose (Y, X) is n-connected. Given maps $\alpha : A \to X$ and $\beta : B \to Y$ such that $\beta|_A = \alpha$, there is a map $\beta' : B \to X$ homotopic to β relative to A.*

Theorem 10.2.17 *Let $f : X \to Y$ be an n-equivalence, $1 \leq n \leq \infty$. Let (B, A) be a relative CW-complex with $\dim (B, A) \leq n$. Then given $\alpha : A \to X$, $\beta : B \to Y$ such that $\beta|_A = f \circ \alpha$, there exists a map $\beta' : B \to X$ such that $\beta'|_A = \alpha$ and $f \circ \beta' \simeq \beta$ rel (A).*

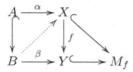

Proof: Replace Y by M_f and f by the inclusion map in the above diagram and then apply the previous theorem.

♠

Corollary 10.2.18 *Let $f : X \to Y$ be an n-equivalence, $1 \leq n \leq \infty$. Then for any CW-complex B, the function*
$$f_\# : [B; X] \to [B; Y]$$
is a surjection if $\dim B \leq n$ and a bijection if $\dim B \leq n - 1$.

Proof: Apply the previous theorem to the relative CW-complex $(B, \emptyset)$ to get the first part. For the second part, we have to work a little harder. This time we take the relative CW-complex $(B \times \mathbb{I}, B \times \partial\mathbb{I})$. Given two maps $\alpha_0, \alpha_1 : B \to X$ and a homotopy $\beta : B \times \mathbb{I} \to Y$ such that $\beta(x, 0) = \alpha_0(x), \beta(x, 1) = \alpha_1(x)$, since $\dim B \leq n$ there exists a map $\beta' : B \times \mathbb{I} \to X$ such that $f \circ \beta'$ is homotopic to β relative to $B \times \partial\mathbb{I}$. This means that β' is a homotopy of α_0 to α_1 in X. ♠

We can now prove one of the landmark results in homotopy theory.

Theorem 10.2.19 (Whitehead) *A map $: X \to Y$ where X and Y are CW-complexes is a homotopy equivalence iff f is a weak homotopy equivalence.*

Proof: We need to prove the 'if' part alone. So, let f be a weak homotopy equivalence. Then from the previous corollary (with $n = \infty$), we have bijections:
$$f_\#^Y : [Y; X] \to [Y; Y]; \quad f_\#^X : [X, X] \to [X; Y].$$
So, from the first bijection, we get a map $g : Y \to X$ such that $[f \circ g] = [Id_Y]$. This implies
$$f_\#^X[g \circ f] = [f \circ g \circ f] = [Id_Y \circ f] = [f] = f_\#^X[Id_X].$$
From the second bijection, this implies $[g \circ f] = [Id_X]$. This implies that f is a homotopy equivalence. ♠

Remark 10.2.20 In this context, one **cannot** loosely say that two complexes X, Y which have isomorphic homotopy groups are homotopy equivalent: it is important to have a map $f : X \to Y$ which induces all these isomorphisms. For instance, according to Theorem 3.5.11, it follows that $L(5,1)$ and $L(5,2)$ are not of same homotopy type. On the other hand, both have fundamental group isomorphic to $\mathbb{Z}_5$ and are covered by $\mathbb{S}^3$ and hence have all other homotopy groups isomorphic to that of $\mathbb{S}^3$. For an example of a pair of simply connected manifolds of this kind see the Misc. exercises at the end of the chapter.

Let us now take up the study of homotopy groups under fibrations.

Theorem 10.2.21 *Let* $p : E \to B$ *be a fibration,* $b_0 \in B' \subset B, E' := p^{-1}(B')$ *and* $e_0 \in E$ *be any point such that* $p(e_0) = b_0$. *Then* $p_\# : \pi_n(E, E', e_0) \to \pi_n(B, B', b_0)$ *is an isomorphism.*

Proof: To show $p_\#$ is surjective, consider a map $\alpha : (\mathbb{I}^n, \partial(\mathbb{I}^n), \mathbf{0}) \to (B, B', b_0)$ representing an element of $\pi_n(B, B', b_0)$. Take any contraction $H : \mathbb{I}^n \times \mathbb{I} \to \mathbb{I}^n$ to the single point $\mathbf{0}$. Then $\alpha \circ H : \mathbb{I}^n \times \mathbb{I} \to B$ defines a homotopy of α with the constant map $x \mapsto b_0$.

There is a homeomorphism of the pairs

$$\phi : (\mathbb{I}^n \times \mathbb{I}, \mathbb{I}^n \times 0 \cup \mathbb{I}^{n-1} \times 0 \times \mathbb{I}) \to (\mathbb{I}^n \times \mathbb{I}, \mathbb{I}^n \times 0).$$

By the HLP of p, there is a map $H' : \mathbb{I}^n \times \mathbb{I} \to E$ such that $p \circ H' = H \circ \phi^{-1}$ and $H'(\mathbb{I}^n \times 0) = e_0$. Take $\beta' = H' \circ \phi|_{\mathbb{I}^n \times 1}$. It follows that $\beta' : \mathbb{I}^n \to E$ is such that $p \circ \beta = \alpha$ and $\beta(\mathbf{0}) = e_0$. In particular, it follows that $\beta(\partial(\mathbb{I}^n)) \subset E' = p^{-1}(B')$. Clearly, then β represents an element of $\pi_n(E, E', e_0)$ and $p_\#([\beta]) = [\alpha]$.

To prove the injectivity, first consider the case $n > 1$. Suppose $\beta : (\mathbb{I}^n, \partial(\mathbb{I}^n), \mathbf{0}) \to (E, E', e_0)$ is such that $[p \circ \beta] = 0$ in $\pi_n(B, B', b_0)$. By Theorem 10.2.7, there is a homotopy $H : \mathbb{I}^n \times \mathbb{I}$ of $p \circ \beta$ relative to $\partial(\mathbb{I}^n)$ to a map $\alpha' : \mathbb{I}^n \to B'$. We want a homotopy $G : \mathbb{I}^n \times \mathbb{I} \to E$ of β relative to the boundary to a map $\beta' : \mathbb{I}^n \to E'$. So, take $F|_{\mathbb{I}^n \times 0} = \beta$ and $F(x, t) = \beta(x), x \in \partial \mathbb{I}^n, t \in \mathbb{I}$. Since the pair $(\mathbb{I}^n \times \mathbb{I}, \mathbb{I}^n \times 0 \cup \partial(\mathbb{I}^n) \times \mathbb{I})$ is homeomorphic to the pair $(\mathbb{I}^n \times \mathbb{I}, \mathbb{I}^n \times 0)$ by the homotopy lifting property there is a $G : \mathbb{I}^n \times \mathbb{I} \to E$ extending the above F and such that $p \circ G = H$. This H is the required homotopy.

For $n = 1$, note that there is no group structure. So, let $\beta, \beta' : (\mathbb{I}, \dot{\mathbb{I}}, 0) \to (E, E', e_0)$ be such that $[p \circ \beta] = [p \circ \beta']$. Let $H : \mathbb{I} \times \mathbb{I} \to B$ such that $H(x, 0) = p \circ \beta(x), H(x, 1) = p \circ \beta'(x),\ x \in \mathbb{I}$ and $H(0, t) = b_0, H(1, t) \in B'$ for all $t \in \mathbb{I}$. Take $F(x, 0) = \beta(x), F(x, 1) = \beta'(x)$ and $F(0, t) = e_0$. Now use the fact that the pair $(\mathbb{I} \times \mathbb{I}, \mathbb{I} \times 0 \cup \dot{\mathbb{I}} \times \mathbb{I})$ is homeomorphic to $(\mathbb{I} \times \mathbb{I}, \mathbb{I} \times 0)$ and the homotopy lifting property of p to obtain a homotopy $H : \mathbb{I} \times \mathbb{I} \to E$ as required. ♠

Given a fibration $p : E \to B,\ b_0 \in B, F = p^{-1}(b_0)$, etc., taking ∂' to be the composite of

$$\pi_n(B, b_0) \xrightarrow{p_\#^{-1}} \pi_n(E, F, e_0) \xrightarrow{\partial} \pi_{n-1}(F, e_0),$$

the homotopy exact sequence of the pair (E, F) immediately gives:

Theorem 10.2.22 (Homotopy exact sequence of a fibration) *Given a fibration* $p : E \to B$, *the sequence*

$$\cdots \longrightarrow \pi_n(F) \xrightarrow{i_\#} \pi_n(E) \xrightarrow{p_\#} \pi_n(B) \xrightarrow{\partial'} \pi_{n-1}(F) \longrightarrow \cdots$$

is exact.

Corollary 10.2.23 *Let* $p : \mathbb{S}^{2n+1} \to \mathbb{CP}^n$ *be the Hopf fibration. Then* $p_\#$ *is an isomorphism for* $n \geq 3$.

Proof: This follows from the exact sequence of p and the fact that $\pi_q(\mathbb{S}^1) = 0$ for $q > 1$.

Corollary 10.2.24 $\pi_3(\mathbb{S}^2) \approx \mathbb{Z}$.

Proof: Consider the Hopf fibration $p : \mathbb{S}^3 \to \mathbb{S}^2$ and the fact that $\pi_3(\mathbb{S}^3) \approx H_3(\mathbb{S}^3) \approx \mathbb{Z}$. ♠

Remark 10.2.25
(a) For a proof of the fact $\pi_3(\mathbb{S}^3) \approx H_3(\mathbb{S}^3)$, which we used in the above corollary, see (Hurewicz) theorem 10.4.3. More generally, one knows that $\pi_n(\mathbb{S}^n) \approx \mathbb{Z}$ which goes under the name Hopf's degree theorem. Hopf proved it using differential topological techniques especially the notion of degree (see [Shastri, 2011].) Since we know that the identity map $\xi_3 : \mathbb{S}^3 \to \mathbb{S}^3$ induces an isomorphism in homology, we may conclude that there is a surjection $\pi_3(\mathbb{S}^3) \to H_3(\mathbb{S}^3)$ and hence $\pi_3(\mathbb{S}^3)$ is nontrivial without appealing to the Hurewicz theorem or Hopf's degree theorem.
(b) To begin with, Hopf had only discovered that $\pi_3(\mathbb{S}^2)$ is non trivial, in quite a different way, though. This result itself was a surprise at that time. Hopf observed that the fibres of the so-called Hopf fibration $p : \mathbb{S}^3 \to \mathbb{S}^2$ are linked in a non trivial fashion and hence the fibration is non trivial. It then follows that the map p cannot be null homotopic and therefore $\pi_3(\mathbb{S}^2) \neq (0)$.
(c) Consider the case of the trivial fibration, viz., $\pi_1 : B \times F \to B$ is the projection to the first factor. It is fairly easy (see Exercise 1.9.38) to see that the homotopy groups of the product of two spaces are just the direct product of the corresponding homotopy groups of the two factor spaces. Thus it follows that for the trivial fibration, all the homomorphisms $\partial : \pi_n(B) \to \pi_{n-1}(F)$ are trivial. However, the converse is far from true. Nevertheless, one can say that the homomorphisms ∂ measure, to a large extent, how badly the fibres of a fibration are 'twisted'. These boundary homomorphisms are called *homotopy transgressions* of the fibration. Similarly there are homology and cohomology transgressions of a fibration which require the machinery of spectral sequence to understand (see Chapter 13).

Exercise 10.2.26

(i) Let $p = (0, 0, \ldots, 0) \in \mathbb{I}^n$. Given any map $f : (\mathbb{I}^n \times 0 \cup \partial(\mathbb{I}^n) \times \mathbb{I}, p) \to (Y, y_0)$ and an element $\alpha \in \pi_n(Y, y_0)$, there exists a map $g : (\partial(\mathbb{I}^{n+1}), p) \to (Y, y_0)$ which extends f and such that $[g] = \alpha \in \pi_n(Y, y_0)$.

(ii) Show that $(\mathbb{D}^{n+1}, \mathbb{S}^n)$ is n-connected.

(iii) Prove the converse of Lemma 10.2.15 and thereby the converse of Theorem 10.2.16 and that of Theorem 10.2.17.

(iv) Establish the homotopy exact sequence of a triple (X, A, B) similar to homology. (See Exercise 4.2.27.(vii).)

(v) **The double hexagon** of homotopy groups: Consider the diagram of topological spaces and inclusion maps:

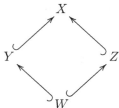

Assume that all the spaces are path connected and we have chosen the base point for all of them to be some point $x_0 \in W$. Verify that the five homotopy exact sequences (at any two consecutive dimensions $q, q+1$) of the five pairs

$$(X, Y), (X, Z), (X, W), (Y, W), (Z, W)$$

fit together into a double hexagon as in Figure 10.1, in which we also have portions of homotopy exact sequence of the triples (X, Y, W) and (X, Z, W) also. [Homomorphisms belonging to a particular exact sequence are indexed by the same suffix.]

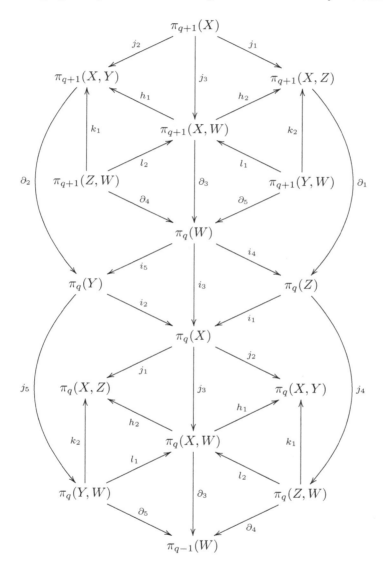

FIGURE 10.1. The double hexagon

Deduce that the inclusion induced homomorphisms k_1 and k_2 in the relative homotopy groups satisfy the following property: $k_1 : \pi_r(Z, W) \to \pi_r(X, Y)$ are isomorphisms for $r \leq q$ and an epimorphism for $r = q+1$ iff $k_2 : \pi_r(Y, W) \to \pi_r(X, Z)$ are isomorphisms and an epimorphism for $r = q+1$. [Hint: This is sheer diagram chasing! Enjoy it. We shall use this result in a later chapter.]

(vi) Given $A \subset X$, recall (see 1.7.10) that we have the mapping path fibration of the inclusion map $\iota : A \to X$. The total space of this fibration is

$$P(A, X) = \{\omega \in X^{\mathbb{I}} : \omega(0) \in A\}.$$

Then the fibration $p' : P(A, X) \to X$ is given by $p'(\omega) = \omega(1)$. Let $x_0 \in A$ be a base point. Let c denote the constant path at x_0. Let $F = p'^{-1}(x_0) = \{\omega \in P(A, X) : \omega(1) = x_0\}$ be the fibre of this fibration. Use the description of $\mathbb{D}^n$ as the reduced cone over $\mathbb{S}^{n-1}$ and the exponential correspondence to establish a canonical isomorphism $\pi_n(X, A, x_0) \approx \pi_{n-1}(F, c)$ for all n.

10.3 Change of Base Point

Often, we have dropped out writing down the base point in the notation of homotopy groups. However, we have been sufficiently apologetic about it. In this section, we shall bring out the necessity to keep track of the base point while studying homotopy groups.

We now appeal to you to reread Example 1.8.3.9 and work out the Exercise 3.9.15, if you have not done it already.

Definition 10.3.1 By a *bundle of groups* over a space X we mean a local system on X with values in the category of groups. Likewise we can speak about *a bundle of abelian groups* or a *bundle of R-modules* over X. We shall also use the terminology $\mathcal{P}_B$-group, $\mathcal{P}_B$-module, etc., to mean a bundle of groups, or a bundle of modules, etc.

Let us fix a topological pair (Y, B). Our first aim is to show that for $n \geq 2$,

$$b \rightsquigarrow \pi_n(Y, B, b)$$

defines a $\mathcal{P}_B$-group (abelian if $n \geq 3$) and

$$b \rightsquigarrow \pi_n(Y, b)$$

defines a $\mathcal{P}_Y$-group (abelian if $n \geq 2$). The main work is in defining the homomorphism $\pi_n(Y, B, b') \to \pi_n(Y, B, b)$ corresponding to a path from b to b' in B.

Let us treat this problem in a little more general context which requires no additional time or energy, which relates the set $[(X, x_0), (Y y_0)] =: [X; Y]$ (lazy notation!) of homotopy classes of base point-preserving maps with the set $[[X, Y]]$ of *free homotopy classes*.

Definition 10.3.2 Fix a topological pair (X, A) with a base point a and consider maps into other topological pairs (Y, B). Given a path ω from b_0 to b_1 in B, we say $\alpha : (X, A) \to (Y, B)$ *is ω-homotopic to* $\beta : (X, A) \to (Y, B)$ *if there is a homotopy* $H : (X, A) \times \mathbb{I} \to (Y, B)$ *from α to β such that $H(a, t) = \omega(t)$; and in that case we write* $\alpha \overset{\omega}{\sim} \beta$.

The following properties are verified in a straightforward way.

Lemma 10.3.3

(a) Suppose ω is the constant path at a point b. Then $\alpha \overset{\omega}{\sim} \beta$ iff $\alpha \sim \beta$ rel a.

(b) Suppose $\omega_1 * \omega_2$ is defined. Then $\alpha \overset{\omega_1}{\sim} \beta, \beta \overset{\omega_2}{\sim} \gamma$ implies $\alpha \overset{\omega_1 * \omega_2}{\sim} \gamma$.

(c) $\alpha \overset{\omega}{\sim} \beta$ implies $\beta \overset{\omega^{-1}}{\sim} \alpha$.

(d) For any $f : (Z, C) \to (X, A)$, $\alpha \overset{\omega}{\sim} \beta$ implies $\alpha \circ f \overset{\omega}{\sim} \beta \circ f$.

(e) For any $g : (Y, B) \to (Z, C)$, $\alpha \overset{\omega}{\sim} \beta$ implies $g \circ \alpha \overset{g \circ \omega}{\sim} g \circ \beta$.

(f) Suppose (X, A) possesses a co-multiplication ν with two-sided homotopy identity. If $H_i : \alpha_i \overset{\omega}{\sim} \beta_i$, then $(H_1 \vee H_2) \circ \nu$ defines an ω-homotopy of $(\alpha_1 \vee \alpha_2) \circ \nu$ with $(\beta_1 \vee \beta_2) \circ \nu$.

Definition 10.3.4 Given maps $\alpha, \beta : (X, A) \to (Y, B)$, we say α is *freely homotopic* to β if there is a map $H : (X, A) \times \mathbb{I} \to (Y, B)$ such that $H(x, 0) = \alpha(x)$ and $H(x, 1) = \beta(x)$. It follows easily that 'being freely homotopic' is an equivalence relation. Also α is freely homotopic to β iff there is a path ω in B such that α is ω homotopic to β. Thus 'being ω homotopic for some ω' is also an equivalence relation. However, if we fix a path ω then 'being ω-homotopic' is not an equivalence relation.

So far, the nature of the base point within the space was not of much concern. For going any further, we need to put some restriction on this.

Definition 10.3.5 Let (X, A) be a topological pair. Given a point $a \in A$ we say a is *non degenerate base point* for the pair (X, A) iff the inclusion map $(a, a) \subset (X, A)$ is a cofibration, i.e., given any map $f : (X, A) \to (Y, B)$ and a homotopy $\omega : a \times \mathbb{I} \to B$ such that $\omega(a, 0) = f(a)$, there is a homotopy $F : (X, A) \times \mathbb{I} \to (Y, B)$ such that $F(x, 0) = f(x), G(a, t) = \omega(t)$.

Recall that $A \subset X$ is a cofibration iff $X \times 0 \cup A \times \mathbb{I}$ is a retract of $X \times \mathbb{I}$. Similarly, it follows that $(a, a) \subset (X, A)$ is a cofibration iff $(X, A) \times 0 \cup a \times \mathbb{I}$ is a retract of $(X, A) \times \mathbb{I}$. (Exercise.) Also if $A \subset X$ and $\{a\} \subset A$ are cofibrations then it follows that $(a, a) \subset (X, A)$ is a cofibration. In particular, it follows that if (K, L) is a polyhedral pair then any point $p \in |L|$ is non degenerate for $(|K|, |L|)$. Thus further in particular, any point $p \in \mathbb{S}^{n-1}$ is non degenerate for the pair $(\mathbb{D}^n, \mathbb{S}^{n-1})$.

We shall from now on assume that a is a non degenerate base point of (X, A) throughout the rest of this section. Let us employ the notation $\overset{\cdot}{\sim}$ to indicate the homotopy relative to the base point a.

Lemma 10.3.6 Given any path ω in B from b_0 to b_1 and any map $\beta : (X, A, a) \to (Y, B, b_1)$ there is a map $\alpha : (X, A, a) \to (Y, B, b_0)$ which is ω homotopic to β. Moreover,
(a) the element $[\alpha]$ represented by α in $[(X, A, a), (Y, B, b_0)]$ depends only on the class $[\beta] \in [(X, A, a); (Y, B, b_1)]$ and the path-homotopy class $[\omega]$.
(b) If $\alpha' \overset{\omega}{\sim} \beta$, then $[\alpha'] = [\alpha]$.

Proof: Fix a retraction $r : (X, A) \times \mathbb{I} \to (X \times 1 \cup a \times \mathbb{I}, A \times 1 \cup a \times \mathbb{I})$.(See Figure 10.2.) Define $\lambda : X \times 1 \cup a \times \mathbb{I} \to Y$ by

$$\lambda(x, 1) = \beta(x), \quad \lambda(a, t) = \omega(t).$$

Take $\alpha(x) = \lambda \circ r(x, 0) = \lambda \circ r_0(x)$. Then $\lambda \circ r$ is a ω-homotopy of α to β.

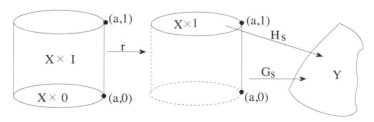

FIGURE 10.2. Homotopy invariance of ω-homotopy

(a) If $H : (X, A, a) \times \mathbb{I} \to (Y, B, b_1)$ is a base-point homotopy of β to β', and $G : \mathbb{I} \times \mathbb{I} \to B$ is a path-homotopy of ω to ω', consider the homotopy $\Lambda : (X \times 1 \cup a \times \mathbb{I}) \times \mathbb{I} \to Y$ defined by

$$\Lambda_s(x, 1) := \Lambda(x, 1, s) := H(x, s) =: H_s(x); \quad \Lambda_s(a, t) := \Lambda(a, t, s) := G(t, s) =: G(s(t).$$

Then $\Lambda_1 \circ r$ is a ω'-homotopy of α to β', whereas, $\Lambda_t \circ r_0$ is a base-point homotopy of α with α'.

(b) Let $F : (X \times 1 \cup a \times \mathbb{I}, A \times 1 \cup a \times \mathbb{I}) \times \mathbb{I} \to (X, A) \times \mathbb{I}$ be the homotopy defined by $F(x, t, s) = (x, ts)$. (Observe that this homotopy keeps $(a, 0)$ fixed.) If $F_s = F(-, -, s)$ and $r_t = r(-, t)$, then it follows that $F_0 \circ r_0 \overset{.}{\sim} F_1 \circ r_0$. Since $F_0(a \times \mathbb{I}) = (a, 0)$, it also follows that $F_0 \circ r_0 \overset{.}{\sim} F_0 \circ r_1$. Therefore $F_1 \circ r_0 \overset{.}{\sim} F_0 \circ r_1$.

Now suppose H is any ω homotopy of α' with β then $\alpha' = H \circ F_0 \circ r_1 \overset{.}{\sim} H \circ F_1 \circ r_0$. In particular, for the homotopy $H' = \lambda \circ r$, we get $\alpha \overset{.}{\sim} H' \circ F_1 \circ r_0$. Since

$$H|_{X \times 1 \cup a \times \mathbb{I}} = H'|_{X \times 1 \cup a \times \mathbb{I}},$$

it follows that $H \circ F_1 \circ r_0 = H' \circ F_1 \circ r_0$. This proves that $[\alpha'] = [\alpha]$.   ♠

Remark 10.3.7 It follows that there is a well-defined function

$$h_{[\omega]} : [(X, A, a); (Y, B, \omega(1))] \to [(X, A, a); (Y, B, \omega(0))]$$

characterised by the property that $h_{[\omega]}([\beta]) = [\alpha]$ iff α is ω-homotopic to β. Note that properties (d) and (e) in Lemma 10.3.3 imply that this construction is functorial both in (X, A) as well as in (Y, B); property (f) implies that if (X, A) is a suspension, then $h_{[\omega]}$ is a homomorphism and finally, property (a) and (b) yield the following:

Theorem 10.3.8 *Let a be a non degenerate base point for a pair (X, A). Then for any fixed topological pair (Y, B) the assignment*

$$b \rightsquigarrow [(X, A, a); (Y, B, b)]; \quad [\omega] \rightsquigarrow h_{[\omega]}$$

defines a covariant functor from the fundamental groupoid $\mathcal{P}_B$ to the category of pointed sets; if (X, A) is a suspension, this takes values in the category of groups and homomorphisms.

Corollary 10.3.9 *Let $a \in A$ be a non degenerate base point for (X, A). Then for all $b \in B$, $\pi_1(B, b)$ acts as a group of operators on the left on the pointed set $[(X, A, a), (Y, B, b)]$. If B is path connected then for any two points $b_0, b_1 \in B$, there is a 1-1 correspondence of the pointed sets $[(X, A, a), (Y, B, b_1)] \to [(X, A, a), (Y, B, b_0)]$ which is unique up to the action of $\pi_1(B, b_0)$.*

Proof: Clearly the map $([\omega], \alpha') \mapsto h_{[\omega]}([\alpha'])$ defines the action. If B is path connected, we can choose some path ω in B from b_0 to b_1 and get the 1-1 correspondence $[\alpha'] \mapsto h_{[\omega]}([\alpha'])$ the inverse of this being given by $h_{[\omega]^{-1}}$. Obviously, when we change the path, we may get different bijections. However, the rule $h_{[\omega]} = h_{[\omega * \omega'^{-1}]} \circ h_{[\omega']}$ proves the uniqueness part.   ♠

Remark 10.3.10

1. By taking $B = Y$ in the above discussions, we obtain the absolute versions of these results. Thus, $\pi_1(Y, b)$ acts as a group of operators on the left of the pointed set $[(X, a), (Y, b)]$ and every path from b to b' in Y defines a bijection of the pointed sets $[(X, a), (Y, b')]$ to $[(X, a), (Y, b)]$ and any two of these bijections are related by the action of an element in $\pi_1(Y, b)$.

2. Now suppose Y is an H-space. We know that $\pi_1(Y, b)$ is abelian. The same arguments now yield:

Theorem 10.3.11 *If (Y, B) is an H-space pair. Then the action of $\pi_1(B, b)$ is trivial.*

Proof: For any $[\omega] \in \pi_1(B, b)$ and any $\alpha : (X, A, a) \to (Y, B, b)$ we consider the homotopy

$$H(x, t) = \mu(\alpha(x), \omega(t))$$

which is a ω'-homotopy of α' with α'' where $\alpha \overset{.}{\sim} \alpha' \overset{.}{\sim} \alpha''$ and ω is path homotopic to ω'. This implies $h_\omega([\alpha]) = [\alpha]$. ♠

Remark 10.3.12 Relation with covering transformations Suppose B is simply connected. It follows that two maps $\alpha, \beta : (X, A, a) \to (Y, B, b)$ are base-point preserving homotopic iff they are freely homotopic. Then the natural map

$$\Theta := \Theta_b : [(X, A, a), (Y, B, b)] \to [[(X, A), (Y, B)]]$$

is a bijection for all b.

Now let us consider a more general situation, wherein Y, B are path connected and the inclusion map induces an isomorphism $\pi_1(B, b) \to \pi_1(Y, b)$. Suppose Y admits a simply connected covering space $p : \tilde{Y} \to Y$. Then $\tilde{B} = p^{-1}(B)$ is path connected and $p : \tilde{B} \to B$ is the simply connected covering of B. For each $\tilde{b} \in p^{-1}(b)$, we have an isomorphism

$$\Theta_{\tilde{b}} : [(X, A, a), (\tilde{Y}, \tilde{B}, \tilde{b})] \to [[(X, A), (\tilde{Y}, \tilde{B})]].$$

Let $g : \tilde{Y} \to \tilde{Y}$ be a covering transformation, $g(\tilde{b}') = g(\tilde{b})$ and ω be a path in $\tilde{B}$ from $\tilde{b}$ to $\tilde{b}'$. Then we have a commutative diagram:

$$
\begin{array}{ccccc}
[(X, A, a), (Y, B, b)] & \overset{p_\#}{\longleftarrow} & [(X, A, a), (\tilde{Y}, \tilde{B}, \tilde{b})] & \overset{\Theta}{\underset{\approx}{\longrightarrow}} & [[(X, A), (\tilde{Y}, \tilde{B})]] \\
\approx \downarrow h_{[p \circ \omega]} & & \approx \downarrow h_{[\omega]} \circ g_\# & & \approx \downarrow g_\# \\
[(X, A, a), (Y, B, b)] & \overset{p_\#}{\longleftarrow} & [(X, A, a), (\tilde{Y}, \tilde{B}, \tilde{b})] & \overset{\Theta}{\underset{\approx}{\longrightarrow}} & [[(X, A), (\tilde{Y}, \tilde{B})]]
\end{array}
$$

One can now specialize to the case when $B = Y$. Assume further that X and Y are connected locally path connected, and X is simply connected. It follows from the lifting theorem that $p_\#$ is also a bijection. Under the canonical isomorphism of the group of covering transformations of $p : \tilde{Y} \to Y$ with $\pi_1(Y, b)$, we can think of $\pi_1(Y)$ acting on the universal cover $\tilde{Y}$. Then the above commutative diagram just means that $p_\#$ is an equivariant function. We can summarize this as follows:

Theorem 10.3.13 *Let X be a simply connected locally path connected space with a non degenerate base point a. Let $\tilde{Y}$ be a simply connected covering space of a connected locally path connected space Y. Then there is a bijection from the free homotopy classes $[[X, \tilde{Y}]]$ to the pointed homotopy classes $[(X, a), (Y, b)]$ compatible with the action of the group $G(\tilde{Y}|Y)$ of covering transformations on the former and the group $\pi_1(Y, b)$ on the latter under the identification $\Psi : G(\tilde{Y}|Y) \to \pi_1(Y, b)$.*

Finally, we now specialize to the cases $(X, a) = S^n(\dot{\mathbb{I}}, 0)$.

Theorem 10.3.14 *For each $n \geq 1$ there is a covariant functor from $\mathcal{P}_Y$ to the category of groups (abelian if $n \geq 2$) and homomorphisms given by the assignments:*

$$y \rightsquigarrow \pi_n(Y, y); \quad [\omega] \rightsquigarrow h_{[\omega]} : \pi_n(Y, \omega(1)) \to \pi_n(Y, \omega(0)).$$

In particular, this defines an action of $\pi_1(Y, y)$ on $\pi_n(Y, y)$ on the left and each $[\omega]$ defines an isomorphism

$$h_{[\omega]} : \pi_n(Y, \omega(1)) \to \pi_n(Y, \omega(0));$$

two such isomorphisms being conjugate under the action of $\pi_1(Y, \omega(0))$. In case $n = 1$, this action is nothing but the conjugation in $\pi_1(Y, y)$.

Proof: Except for the last statement everything else is a special case of what we have seen already. The last statement, viz., $h_\omega[\alpha] = [\omega * \alpha * \omega^{-1}]$ for any two $[\alpha], \omega \in \pi_1(Y, b)$ follows from the ω-homotopy from $\omega * \alpha * \omega^{-1}$ to α depicted in Figure 10.3

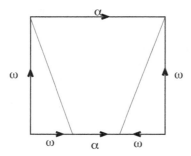

FIGURE 10.3. Conjugation action of π_1 on itself

Remark 10.3.15 This leads us to a simpler interpretation of $h_{[\omega]}(\alpha)$ for $[\alpha] \in \pi_n(Y, b)$. Let $\alpha : (\mathbb{D}^n, \mathbb{S}^{n-1}) \to (Y, b)$ any map. Consider $\alpha' : (\mathbb{D}^n, \mathbb{S}^{n-1}) \to (Y, b)$ defined by

$$\alpha'([x, t]) = \begin{cases} \alpha[x, 2t], & 0 \le t \le 1/2, \\ \omega(2t - 1), & 1/2 \le t \le 1. \end{cases}$$

Then $[\alpha'] = h_{[\omega]}([\alpha])$.

We leave it to the reader to draw a picture depicting an ω-homotopy of α' to α to justify this. This interpretation, in turn, leads us to a generalisation of the product

$$\pi_n(X, a) \times \pi_1(X, a) \to \pi_n(X, a)$$

into a product

$$\pi_n(X, a) \times \pi_m(X, a) \to \pi_{n+m-1}(X, a)$$

called the *Whitehead product* as introduced in Section 7.5.

The proof of the following theorem is totally obvious.

Theorem 10.3.16 *Let A be a path connected subspace of a space X, and $a \in A$. Then $\pi_1(A)$ acts on each term of the homotopy exact sequence of the pair (X, A) and all homomorphisms preserve this action. In other words these are $\pi_1(A)$-module homomorphisms.*

The non trivial action of the fundamental group causes many technical problems in homotopy theory. So, for future use, we shall introduce the notion of simple spaces.

Definition 10.3.17 A path connected topological space X (respectively, a topological pair (X, A) with A and X path connected) is said to be *n-simple* if for some $x_0 \in X$ (respectively, $x_0 \in A$) $\pi_1(X, x_0)$ (resp.$\pi_1(A, x_0)$) acts trivially on $\pi_q(X, x_0)$ (respectively, on $\pi_q(X, A, x_0)$) for all $q \le n$.

Remark 10.3.18 It then follows that $\pi_1(X, x)$ also acts trivially on $\pi_q(X, x), q \le n$ for all $x \in X$. Clearly any simply connected space is n-simple for all n. If X is n-simple for all n, we then say X is *simple*. For any simple space, $\pi_1(X, x_0)$ is abelian. The advantage of simple space is that we can really ignore the base point for all purposes— free homotopy classes and base point preserving homotopy classes become the same when a space is simple. We have seen that for an H-space, $\pi_1(X, x_0)$ is abelian. The same proof yields that every

H-space is simple. Likewise every H-space pair is also simple. Finally a map $f : X \to Y$ is called simple (n-simple) if $f_\# \pi_1(Y')$ is a normal subgroup of $\pi_1(Y)$ such that the quotient group is abelian and the relative mapping cylinder (M_f, Y) is simple (n-simple). We shall have some opportunity to use these concepts soon.

Exercise 10.3.19 Show that $\pi_1(\mathbb{S}^1)$ acts trivially on $\pi_2(\mathbb{D}^2, \mathbb{S}^1)$.

10.4 The Hurewicz Isomorphism

In Section 4.5 we have seen that the Hurewicz homomorphism $\varphi : \pi_1(X, x_0) \to H_1(X, \mathbb{Z})$ defines H_1 as the abelianization of π_1. The map itself easily generalizes to $\varphi : \pi_n(X, x_0) \to H_n(X, \mathbb{Z})$ though the result needs to be modified. First of all, since $\pi_i, i \geq 2$ are already abelian, there is no question of further abelianization. So, we ask the next best thing, viz., is φ then an isomorphism? And the answer is 'yes indeed' provided certain conditions are satisfied. In this section, we shall present this celebrated result of Hurewicz along with some closely related results.

We shall assume that members (X, A) of $\textbf{Top}_0^2$ occurring in this section are path connected.

Definition 10.4.1

1. Given $[\alpha] \in \pi_n(X, A, a)$, consider the homomorphism $\alpha_* : H_n(\mathbb{D}^n, \mathbb{S}^{n-1}) \to H_n(X, A)$. Put $\varphi[\alpha] = \alpha_*(\xi_n)$, where (ξ_n) is the homology element represented by the relative n-cycle given by the identity map. The function

$$\varphi : \pi_n(X, A, a) \to H_n(X, A)$$

is called the *Hurewicz map*. Note that if we take $A = \{a\}$, then we get the absolute version of this map which will also be denoted by φ.

2. The Hurewicz map is a functorial homomorphism which commutes with the restriction maps and hence with the boundary homomorphisms in the homotopy exact sequence of a pair. In particular, φ defines a homomorphism of the long exact homotopy sequence to the long exact homology sequence:

$$\begin{array}{ccccccccc}
\cdots \longrightarrow & \pi_n(A) & \longrightarrow & \pi_n(X) & \longrightarrow & \pi_n(X, A) & \longrightarrow & \pi_{n-1}(A) & \longrightarrow \cdots \\
& \downarrow{\scriptstyle\varphi} & & \downarrow{\scriptstyle\varphi} & & \downarrow{\scriptstyle\varphi} & & \downarrow{\scriptstyle\varphi} & \\
\cdots \longrightarrow & H_n(A) & \longrightarrow & H_n(X) & \longrightarrow & H_n(X, A) & \longrightarrow & H_{n-1}(A) & \longrightarrow \cdots
\end{array}$$

3. One knows that the commutator subgroup measures the amount of non abelianness of a group. Alternatively, you may interpret this as representing the effectiveness of conjugation action of a group on itself. In (the first Hurewicz) Theorem 4.5.2, it was obvious that φ completely ignores the conjugation action, merely being a homomorphism into an abelian group.

   This holds in higher dimensions as well. The Hurewicz map is insensitive to the action of π_1 on higher homotopy groups:

   $$\varphi(\alpha(g, \omega)) = \varphi(\omega).$$

   This is an easy consequence of the homotopy invariance of homology and the fact that ω-homotopy implies (free) homotopy.

For $n = 1$, we define $\pi_1'(X, x_0) = \pi_1^{ab}(X, x_0)$. For $n \geq 2$, we define the groups $\pi_n'(X, x_0)$ to be the quotient of $\pi_n(X, x_0)$ by the subgroup generated by

$$\{\alpha(g, \omega) - \omega \ : \ g \in \pi_1(X, x_0), \omega \in \pi_n(X, x_0)\}$$

We shall denote by $\varphi' : \pi_n'(X, x_0) \to H_n(X)$ the homomorphism induced by φ. Likewise, for $n \geq 2$, $\pi_n'(X, A, a)$ is defined to be the quotient of $\pi_n(X, A, a)$ by the normal subgroup generated by $\{\alpha(g, \omega)\omega^{-1} \ : \ g \in \pi_1(A, a), \omega \in \pi_n(X, A, a)\}$. Denote the induced homomorphism by $\varphi' : \pi_n'(X, A, a) \to H_n(X, A)$.

4. The main result that we are interested in is the following:

Theorem 10.4.2 (The relative Hurewicz isomorphism) *Let $a \in A \subset X$ be such that A, X are path connected.*
(a) *Suppose A, X are simply connected, and there is $n \geq 2$ such that $H_q(X, A) = 0$, for all $0 < q < n$. Then $\pi_q(X, A) = 0$ for $0 < q < n$ and*

$$\varphi : \pi_n(X, A) \to H_n(X, A)$$

is an isomorphism.
(b) *Conversely, suppose there is $n \geq 2$ such that $\pi_q(X, A, a) = 0$ for all $0 < q < n$. Then $H_q(X, A) = 0$ for all $0 < q < n$ and*

$$\varphi' : \pi_n'(X, A, a) \to H_n(X, A)$$

is an isomorphism.

The absolute version can be obtained by taking $A = \{a\}$ except that it does not include the case $n = 1$ which we have proved earlier. So, let us restate the absolute version to include the case $n = 1$ as well.

Theorem 10.4.3 (Absolute Hurewicz isomorphism)
(a) *Suppose X is simply connected and for some $n \geq 2$, $H_q(X) = 0$ for all $0 < q < n$. Then $\pi_q(X, x_0) = 0$ for all $q < n$ and $\varphi : \pi_n(X, x_0) \to H_n(X)$ is an isomorphism.*
(b) *Suppose there is some $n \geq 1$ such that $\pi_q(X, x_0) = (0)$ for all $0 < q < n$. Then $H_q(X) = 0$ for all $0 < q < n$ and $\varphi' : \pi_n'(X, x_0) \to H_n(X)$ is an isomorphism.*

Example 10.4.4 Consider the space $X = \mathbb{S}^1 \vee \mathbb{S}^2$. This is path connected, H_0, H_1 and H_2 are infinite cyclic and all other homologies vanish. Also the fundamental group is isomorphic to $\mathbb{Z}$ (van Kampen's theorem). What is $\pi_2(X)$?

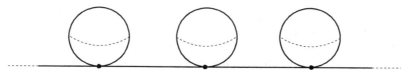

FIGURE 10.4. The universal cover of the one-point-union of a circle and a 2-sphere

The universal covering space $\tilde{X}$ of X can be visualized as the space obtained from the real line by attaching one copy of the 2-sphere at each of the integer points as in Figure 10.4. It is clear that $H_1(\tilde{X}) = 0$ and $H_2(\tilde{X})$ is the direct sum of countably infinite copies of $\mathbb{Z}$. Indeed, the group of covering transformation $T = \pi_1(X)$ acts on $H_2(\tilde{X})$ and one checks that $H_2(X) \approx \mathbb{Z}[T] = \mathbb{Z}[t, t^{-1}]$, as T-modules. The Hurewicz theorem for $\tilde{X}$ says that $\pi_2(\tilde{X}) \approx H_2(\tilde{X})$ only as abelian groups. On the other hand, notice that the absolute Hurewicz theorem is silent about any relation between $\pi_2(X)$ and $H_2(X)$.

On the other hand consider the relative version with $A = \mathbb{S}^1 \subset X$. Check that $\pi_q(X, A) = (0)$ for $q = 0, 1$. Therefore $\varphi' : \pi_2'(X, A) \to H_2(X, A) = \mathbb{Z}$ is an isomorphism. Once again, the relative version in the universal covering case gives us $\pi_2(X, A) = \pi_2'(X, A) \approx H_2(X, A) = \mathbb{Z}[T]$.

Sketch of the proof of the Hurewicz theorem:

There are quite a few technical hurdles one has to solve in the proof for $n \geq 2$ as compared to the case $n = 1$. The Eilenberg–Blakers homology groups were introduced to take care of the major hurdle. Nevertheless, the basic idea is quite similar to case $n = 1$, that we are familiar with.

Eilenberg–Blakers homology groups Fix an integer $r \geq 0$. Let $S(X, A, a)^{(r)}$ denote the subcomplex of the singular chain complex $S(X, A)$ generated by singular simplexes $\sigma : \Delta_n \to X$ such that $\sigma(\Delta_n^{(r)}) \subset A$ and $\sigma(\Delta_n^{(0)}) = \{a\}$. The Eilenberg–Blakers homology groups of degree r are defined to be

$$H_*^{(r)}(X, A, a) := H_*(S(X, A, a)^{(r)}, S(X, A, a)^{(r)} \cap S(A)).$$

Theorem 10.4.5 *Assume that A is path connected and (X, A) is r-connected, for some $r \geq 0$. Then the inclusion of the chain complexes $S(X, A, a)^{(r)} \subset S(X, A)$ is a chain-equivalence and hence induces an isomorphism in homology:*

$$H_*^{(r)}(X, A, a) \approx H_*(X, A).$$

Remark 10.4.6

(1) Notice that the Hurewicz map clearly factors as a composite

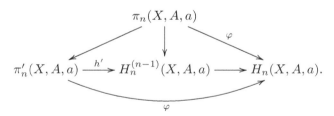

Thus the role of the Eilenberg–Blakers homology in the proof of the Hurewicz theorem is clear: *Under the hypothesis of the theorem, we can replace H_n by $H_n^{(n-1)}$ and prove the theorem for $H_n^{(n-1)}$.*

We now formulate three sequences of propositions A_n, B_n and C_n, which lead us to the proof of the Hurewicz theorems.

(2) **Homotopy addition theorem** Recall the face maps $F^r : \mathbb{R}^\infty \to \mathbb{R}^\infty$ are defined by

$$F^r(\mathbf{e}_i) = \begin{cases} \mathbf{e}_i, & i < r; \\ \mathbf{e}_{i+1}, & i \geq r. \end{cases}$$

and extended linearly. Clearly, $F^r(\Delta_n) \subset \dot{\Delta}_{n+1}$. In order to keep track of the dimension, we shall denote by $F_{n+1}^r := F^r|_{\Delta_n}$. It follows that we have the maps of the triples:

$$\left. \begin{array}{l} F_{n+1}^0 : (\Delta_n, \dot{\Delta}_n, \mathbf{e}_0) \to (\dot{\Delta}_{n+1}, \Delta_{n+1}^{(n-1)}, \mathbf{e}_1); \\ F_{n+1}^r : (\Delta_n, \dot{\Delta}_n, \mathbf{e}_0) \to (\dot{\Delta}_{n+1}, \Delta_{n+1}^{(n-1)}, \mathbf{e}_0), \end{array} \right\} \quad 0 < r \leq n + 1.$$

Recall also that for any two vertices $u, v \in \Delta_{n+1}$ we denote by $[uv]$ the linear path given by the edge u, v traced from u to v.

We define elements $b_n \in \pi_n(\dot{\Delta}_{n+1}, \Delta_{n+1}^{(n-1)}, \mathbf{e}_0), n \geq 2$ and $b_1 \in \pi_1(\dot{\Delta}_2, \mathbf{e}_0)$ by the formulae:

$$
\begin{cases}
b_1 & = [\mathbf{e}_0\mathbf{e}_1] * [\mathbf{e}_1\mathbf{e}_2] * [\mathbf{e}_2\mathbf{e}_0]; \\
b_2 & = \alpha([\mathbf{e}_0\mathbf{e}_1], [F_3^0])[F_3^2][F_3^1]^{-1}[F_3^3]^{-1}; \\
b_n & = \alpha([\mathbf{e}_0\mathbf{e}_1], [F_{n+1}^0]) + \left(\displaystyle\sum_{0 < r \leq n+1} (-1)^r [F_{n+1}^r] \right), \quad n \geq 3.
\end{cases}
$$

One of the forms of homotopy addition theorem is the following:

Proposition 10.4.7 $A_n(n \geq 1)$: The inclusion induced homomorphism

$$
j_\# : \pi_n(\dot{\Delta}_{n+1}, \Delta_{n+1}^{(n-1)}, \mathbf{e}_0) \to \pi_n(\Delta_{n+1}, \Delta_{n+1}^{(n-1)}, \mathbf{e}_0)
$$

maps b_n to zero.

Proposition 10.4.8 $B_n(n \geq 1)$: Let X be $(n-1)$-connected. Then

$$
h' : \pi_n'(X, x_0) \to H_n^{(n-1)}(X)
$$

is an isomorphism.

Proposition 10.4.9 $C_n, n \geq 2$: Let A be path connected and (X, A) be $(n-1)$-connected. Then

$$
h' : \pi_n'(X, A, x_0) \to H_n^{(n-1)}(X, A)
$$

is an isomorphism.

Remark 10.4.10 The proof of these three propositions will be done simultaneously in five steps as follows:

Step 1: proof of A_1 (already seen).

Step 2: $A_1 \Longrightarrow B_1$. (already done).

Step 3: For $n \geq 2$, $B_1, \ldots, B_{n-1} \Longrightarrow A_n$.

Step 4: For $n \geq 2$, $A_n \Longrightarrow C_n$.

Step 5: For $n \geq 2$, $C_n \Longrightarrow B_n$ (which is easy).

We shall postpone these proofs for the time being. The proof of the two Hurewicz theorems from Propositions B_n and C_n is direct. We shall now give an important consequence of Hurewicz theorems.

Theorem 10.4.11 (Whitehead) *Let X, Y be path connected spaces and let $f : (X, x_0) \to (Y, y_0)$ be a map.*

(a) If there is $n \geq 1$ such that $f_\# : \pi_q(X, x_0) \to \pi_q(Y, y_0)$ is an isomorphism for $q < n$ and an epimorphism for $q = n$ then $f_* : H_q(X) \to H_q(Y)$ is an isomorphism for $q < n$ and an epimorphism for $q = n$.

(b) Conversely, if X and Y are simply connected, and if there is $n \geq 2$ such that f_* is an isomorphism for $q < n$ and an epimorphism for $q = n$ then so is $f_\#$.

Proof: Consider the mapping cylinder M_f. Recall that there is a deformation retraction $r : M_f \to Y$ such that $f = r \circ i$, where $i : X \to M_f$ is the inclusion. It follows that we can replace Y in the statement of the theorem by M_f and the hypothesis and conclusion about $f_\#$ and f_* get converted into corresponding hypothesis and conclusion about $i_\#$ and

i_*. Now the homology and homotopy exact sequences complete the proof via the relative version of Hurewicz theorem.

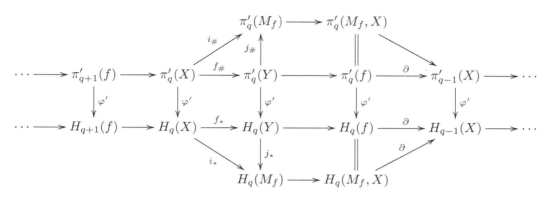

We can now give some applications of Whitehead's theorem in the computation of homotopy groups, etc.

Theorem 10.4.12 *Let $X = Y \cup_\alpha e^n, n \geq 2$, be obtained by attaching a n-cell to a connected, path connected, semi-locally simply connected space Y. Then the inclusion map $\iota : Y \to X$ induces an isomorphism in homotopy groups $\pi_k(Y) \to \pi_k(X)$ for $k < n - 1$ and a surjection $\pi_{n-1}(Y) \to \pi_{n-1}(X)$. Moreover,*
(a) for $n = 2$, $\pi_1(X)$ is isomorphic to the quotient of $\pi_1(Y)$ by the normal subgroup generated by $[\alpha] \in \pi_1(Y)$;
(b) for $n \geq 3$, $\pi_{n-1}(X)$ is the quotient of $\pi_{n-1}(Y)$ by the $\pi_1(Y)$-submodule generated by $[\alpha] \in \pi_{n-1}(Y)$.

Proof: The case $n = 2$ is a general consequence of the van Kampen theorem. (See Corollary 3.8.12.)

So we shall assume $n \geq 3$.

First consider the case when Y and hence X is simply connected. By Theorem 4.3.20, it follows that the inclusion map $\iota : Y \to X$ induces an isomorphism in $H_i(Y) \to H_i(X)$ for $i < n - 1$ and a surjection for $i = n - 1$. Hence the same is true for π_i. In particular, we also have that $\pi_{n-1}(X)$ is isomorphic to $\pi_{n-1}(Y)/\mathrm{Ker}\, \iota_\#$. The fact that $\mathrm{Ker}\, \iota_\# = \mathrm{Im}\, (\alpha_\#)$ follows from the commutative diagram below, wherein β, α, respectively, denote the characteristic map and the attaching map of the n-cell:

$$
\begin{array}{ccccc}
H_n(\mathbb{D}^n, \mathbb{S}^{n-1}) & \xleftarrow[\approx]{\varphi} & \pi_n(\mathbb{D}^n, \mathbb{S}^{n-1}) & \xrightarrow[\approx]{\partial} & \pi_{n-1}(\mathbb{S}^{n-1}) \\
{\scriptstyle \beta_*} \downarrow {\scriptstyle \approx} & & \downarrow {\scriptstyle \beta_\#} & & \downarrow {\scriptstyle \alpha_\#} \\
H_n(X, Y) & \xleftarrow[\varphi]{\approx} & \pi_n(X, Y) & \xrightarrow{\partial} & \pi_{n-1}(Y) \xrightarrow{i_\#} \pi_{n-1}(X).
\end{array}
$$

Finally in the general case, we first observe that $\iota_\# : \pi_1(Y) \to \pi_1(X)$ is an isomorphism. Let $p : \tilde{Y} \to Y$ be the universal covering space of Y. Let $\alpha_j : \mathbb{S}^{n-1} \to \tilde{Y}, j \in J$ be the collection of all possible distinct lifts of α. Put $\tilde{X} = \tilde{Y} \cup_{\alpha_j, j \in J} e_j^n$ be the space obtained by attaching n-cells to $\tilde{Y}$ along α_j. Clearly $p : \tilde{Y} \to Y$ extends to a covering projection $\tilde{p} : \tilde{X} \to X$ wherein for each j, interior of each of the n-cells attached to $\tilde{Y}$ is mapped homeomorphically onto the interior of the n-cell attached to Y. Generalizing the simply connected case from one cell to an arbitrary number of cells, we know that $\iota_\#$ is an isomorphism in dimensions $< n - 1$, and a surjection in dimension $n - 1$ with $\pi_{n-1}(\tilde{X})$ being the quotient of $\pi_{n-1}(\tilde{Y})$ by the submodule generated by the collection $\{[\alpha_j] : j \in J\}$. Under the covering projection, this submodule is precisely the $\pi_1(Y)$-module generated by the single element $[\alpha]$. This gives the desired conclusion. ♠

The arguments in the above theorem easily yield:

Corollary 10.4.13 For $n \geq 2$, let $X = Y \vee \{\mathbb{S}^n_j : J \in J\}$. Then $\pi_n(X, Y)$ is a free $[\pi_1(Y)]$-module over the base J, i.e., a direct sum of J-copies of $\mathbb{Z}[\pi_1(Y)]$.

Theorem 10.4.14 *Let π_1 be a group and $\{\pi_q : q \geq 2\}$ be a sequence of π_1-modules. Then there is a connected CW-complex X unique up to homotopy such that $\pi_1(X, x_0) = \pi_1$ and $\pi_q \approx \pi_q(X, x_0)$ as π_1-modules.*

Proof: The construction of X is as usual, by skeleton-by-skeleton. We take a single point x_0 and declare $X^{(0)} = x_0$. We choose a set A_1 of generators for π_1 and form the wedge sum $X^{(1)} = \vee_{\alpha \in A_1} \mathbb{S}^1_\alpha$, where each $\mathbb{S}^1_\alpha$ is a copy of $\mathbb{S}^1$, and the wedge sum is formed at the vertex x_0.

The construction of the 1-skeleton of X is over. Observe that there is a surjection $f_1 : \pi_1(X^{(1)}) \to \pi_1$ which sends the element represented by $\mathbb{S}^1_\alpha$ to $\alpha \in \pi_1$.

Let K_1 be the kernel of f_1. Now choose a set $B_1 \subset K_1$ which generates K_1 as a normal subgroup in $\pi_1(Y_1)$. Choose (base point preserving) maps $\phi_\beta : \mathbb{S}^1 \to Y_1$, to represent $\beta \in B_1$. Let $\phi_1 : \vee_\beta \mathbb{S}^1_\beta \to \vee_\alpha \mathbb{S}^1_\alpha$ be the map such that $\phi_1|_{\mathbb{S}^1_\beta} = \phi_\beta$. Take Y_2 to be the reduced mapping cone of ϕ_1. It follows that $\pi_1(Y_2)$ is isomorphic to π_1. Now take a set of generators A_2 for the π_1-module π_2 and put $X^{(2)} = Y_2 \vee (\vee_{\alpha \in A_2} \mathbb{S}^2_\alpha)$.

The construction of the 2-skeleton of X is over. We have $\pi_1(X^{(2)}) = \pi_1$ and there is a π_1-module surjection $f_2 : \pi_2(X^{(2)}) \to \pi_2$.

Inductively assume that we have constructed $X^{(n)}$:

$$X^{(0)} \subset \cdots \subset X^{(i)} \subset \cdots \subset X^{(n)},$$

a connected CW-complex of dimension n, such that $\pi_1(X^{(n)}) = \pi_1$, and for $1 < i < n$, $\pi_i(X^{(n)})$ is isomorphic to π_i as a π_1-module, and there is a surjection $f_n : \pi_n(X^{(n)}) \to \pi_n$ of π_1-modules. Then the kernel K_n of f_n is a π_1-module and we choose a set B_n of generators for it, represent them by maps $\phi_\beta : \mathbb{S}^n \to X^{(n)}$ for each $\beta \in B_n$, take their wedge sum

$$\phi_n : \vee_\beta \mathbb{S}^n_\beta \to X^{(n)}$$

and put Y_{n+1} equal to the mapping cone of ϕ_n. We then take a set of generators A_{n+1} for the π_1-module π_{n+1} and put

$$X^{(n+1)} = Y_{n+1} \vee \vee_{\alpha \in A_{n+1}} \mathbb{S}^{n+1}_\alpha.$$

This completes the inductive construction of $X^{(n+1)}$.

We now simply take $X = \cup_n X^{(n)}$. ♠

Closely related to this construction is the following result which can be viewed as a first step toward construction of maps between spaces.

Theorem 10.4.15 *Let $X = (\vee_i \mathbb{S}^n_i) \cup_j e^{n+1}_j$ be the space obtained by attaching $(n + 1)$-cells to a wedge of n-spheres, $n \geq 1$. Given any pointed space Y and a homomorphism $\phi : \pi_n(X, *) \to \pi_n(Y, *)$, there exists a map $f : (X, *) \to (Y, *)$ such that $\phi = f_\# : \pi_n(X, *) \to \pi_n(Y, *)$.*

Proof: To construct such a map f, first send the base point of X to the base of Y. Next for each index i, choose a map $f_i : \mathbb{S}^n \to Y$ representing the element $\phi[\eta_i] \in \pi_n(Y, *)$, where $\eta_i : \mathbb{S}^n \to X$ is the inclusion map into the ith sphere $\mathbb{S}_i^n$. Put $g : X^{(n)} \to Y$ equal to f_i on $\mathbb{S}_i^n$. We claim that g can be extended to a map $f : X \to Y$ as desired.

We know that $\pi_n(X^{(n)}) = \pi_n(\vee_i \mathbb{S}^n)$ is the free (abelian if $n \geq 2$) group generated by $\{[\eta_i]\}$ and if $\eta : X^{(n)} \to X$ is the inclusion map then $\eta_\# : \pi_n(X^{(n)}, *) \to \pi_n(X, *)$ is the quotient map. We thus have a commutative diagram

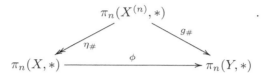

Therefore any extension $f : X \to Y$ of g will satisfy the condition $f_\# = \phi$.

Now let $\beta_j : \mathbb{S}^n \to X^{(n)}$ be the attaching map for the $(n+1)$-cell e_j^{n+1}. Then g extends over e_j^{n+1}, iff $f' \circ \beta_j$ is null homotopic in Y, iff $f'_\#[\beta_j] = 0$ in $\pi_n(Y)$ iff $\phi \eta_\#[\beta_j] = 0$. But we know that $\eta_\#[\beta_j] = 0$ is $\pi_n(X)$. Therefore, g can be extended over each e_j^{n+1}. Putting all these extensions together we get $f : X \to Y$. ♠

Modifications in the arguments of the above theorem can be used to prove various types of CW 'models' for a given topological space. (See Exercise 10.4.17.)

Theorem 10.4.16 *Given any topological pair (X, A) there exists a CW-pair (Y, B) and a weak homotopy equivalence $f : (Y, B) \to (X, A)$. If (X, A) is m-connected, then (Y, B) can be chosen to have cells of dimensions $n > m$ only.*

Proof: First obtain a CW-complex A and a weak homotopy equivalence $f_0 : B \to A$. Next, replace X by the mapping cylinder of $i \circ f_0 : B \to X$ and A by B. This amounts to assuming that A itself is a CW complex to begin with. We shall then prove a slightly stronger version of the above claim, viz., that there is a CW-complex Y containing A as a subcomplex and a homotopy equivalence $f : (Y, A) \to (X, A)$ such that $f|_A = Id_A$. The original claim is then recovered by composing with the deformation retraction $r : M_{i \circ f_0} \to X$.

The construction of Y and f, in this special case is done inductively, $Y_0 \subset Y_1 \cdots \subset Y_n \subset \cdots$ with $f_i : Y_i \to X$ be an i-equivalence such that $f_i|_{Y_{i-1}} = f_{i-1}$. We can then take $Y = \cup_i Y_i$ and $f|_{Y_i} = T_i$. We urge the reader to supply the details of inductive step. ♠

Completion of the proof of Hurewicz theorems

As observed in Remark 10.4.11, it remains to prove **Step 3**, **Step 4** and theorem 10.4.5. We shall take them up one by one.

Proof of Step 3: Consider the following commutative diagram

$$\pi_{n+1}(\Delta_{n+1}, \dot{\Delta}_{n+1}, v_0) \qquad (10.2)$$

$$\pi_n(\dot{\Delta}_{n+1}, v_0) \xrightarrow{i_\#} \pi_n(\dot{\Delta}_{n+1}, \Delta_{n+1}^{(n-1)}, v_0) \xrightarrow{\partial''} \pi_{n-1}(\Delta_{n+1}^{(n-1)}, v_0)$$

$$\pi_n(\Delta_{n+1}, \Delta_{n+1}^{(n-1)}, v_0)$$

in which the column is a part of the homotopy exact sequence of the triple

$(\Delta_{n+1}, \dot{\Delta}_{n+1}, \Delta_{n+1}^{(n-1)})$, and the row is a part of the homotopy exact sequence of the pair $(\dot{\Delta}_{n+1}, \Delta_{n+1}^{(n-1)})$. Clearly ∂ is an isomorphism. It follows that $\mathrm{Ker}\, j_{\#} = \mathrm{Ker}\, \partial''$.

Therefore, we shall prove that $\partial''(b_n) = 0$. The proofs for $n = 2$ and $n > 2$ have to be considered separately.

The case n=2. We have

$$\partial''(b_2) = \alpha([\mathbf{e}_0, \mathbf{e}_1], \partial''[F_3^0])\partial''[F_3^2]\partial''[F_3^1]^{-1}\partial''[F_3^3]^{-1}. \tag{10.3}$$

To compute $\partial''[F_3^i]$, consider the commutative diagram,

$$
\begin{array}{ccc}
\pi_2(\Delta_2, \dot{\Delta}_2, \mathbf{e}_0) & \xrightarrow{\ \partial\ } & \pi_1(\dot{\Delta}_2, \mathbf{e}_0) \\
\varphi \downarrow & & \downarrow \varphi \\
H_2(\Delta_2, \dot{\Delta}_2) & \xrightarrow{\ \partial\ } & H_1(\dot{\Delta}_2, \mathbf{e}_0).
\end{array}
$$

If $[\xi] \in \pi_2(\Delta_2, \dot{\Delta}_2, \mathbf{e}_0)$ denotes the element represented by the identity map, using the above commutative diagram, computing $\partial[\xi]$ in homotopy reduces to computing the same in homology (since B_1 is true). Therefore, we have

$$\partial[\xi] = [\mathbf{e}_0\mathbf{e}_1] \circ [\mathbf{e}_1\mathbf{e}_2] \circ [\mathbf{e}_2\mathbf{e}_0].$$

Temporarily reverting back to the notation F^i for F_3^i, by naturality, we have,

$$
\begin{aligned}
\partial''[F^i] &= \partial''(F^i)_{\#}[\xi] = (F^i|_{\dot{\Delta}_2})_{\#}\partial[\xi] \\
&= (F^i|_{\dot{\Delta}_2})_{\#}([\mathbf{e}_0\mathbf{e}_1] \circ [\mathbf{e}_1\mathbf{e}_2] \circ [\mathbf{e}_2\mathbf{e}_0]) \\
&= [F^i\mathbf{e}_0 F^i\mathbf{e}_1] \circ [F^i\mathbf{e}_1 F^i\mathbf{e}_2] \circ [F^i\mathbf{e}_2 F^i\mathbf{e}_0])
\end{aligned}
$$

Substituting this in (10.3), we get $\partial''(b_2) = 1$.

The case $n > 2$. Clearly $\pi_1(\Delta_{n+1}^{(n-1)}) = 0$ and for $q \le n - 2$, we have,

$$H_q(\Delta_{n+1}^{(n-1)}, \mathbf{v}_0) \approx H_q(\Delta_{n+1}) = 0.$$

Therefore, $B_1, \ldots, B_{n-2}$ imply that $\Delta_{n+1}^{(n-1)}$ is $(n-2)$-connected and by B_{n-1} we have

$$\varphi : \pi_{n-1}(\Delta_{n+1}^{(n-1)}) \approx H_{n-1}(\Delta_{n+1}^{(n-1)}).$$

Therefore it suffices to prove that $\varphi\partial''(b_n) = 0$. Since $\varphi\partial'' = \partial''\varphi$, we can pass onto the commutative diagram obtained by replacing homotopy groups by homology groups in the diagram (10.2). Since $\varphi(b_n) = \partial'\{\xi_{n+1}\} = i_*\partial\{\xi_{n+1}\}$, and since $\partial'' \circ i_* = 0$, we are through. This completes the proof of **Step 3**.

Proof of Step 4: This is similar to **Step 2**. The homomorphism φ' is indeed the composite of

$$\pi'_n(X, A, x_0) \xrightarrow{\ \varphi''\ } H_n^{(n-1)}(X, A, x_0) \xrightarrow{\ \approx\ } H_n(X, A).$$

The second map is an isomorphism by Theorem 10.4.5. Therefore it suffices to show that φ'' is an isomorphism. For this we directly construct its inverse.

Given a singular simplex $\sigma : (\Delta_n, \dot{\Delta}_n, (\Delta_n)^{(0)}) \to (X, A, x_0)$ we can consider the element $[\sigma]' \in \pi'_n(X, A, x_0)$ which is the image of $[\sigma] \in \pi_n(X, A, x_0)$ under the quotient homomorphism. If $\sigma(\Delta_n) \subset A$ then clearly, $[\sigma]' = 0$. Therefore there is a well-defined homomorphism

$$\psi : S_n^{(n-1)}(X, A, x_0)/S_n^{(n-1)}(X, A, x_0) \cap S_n(A) \to \pi'_n(X, A, x_0).$$

Let $\tilde{\sigma} : \dot{\Delta}_n, (\Delta_n)^{(0)} \to (A, x_0)$, denote the restriction of σ. The hypothesis A_n implies that

$$\psi \circ \partial(\sigma) = \sum_i (-1)^i [\sigma^i]' = \eta(\tilde{\sigma}_\#(b_n) = \eta \sigma_\# j_\#(b_n) = 0.$$

Therefore ψ factors down to define a homomorphism

$$\psi' : H_n^{(n-1)}(X, A, x_0) \to \pi'_n(X, A, x_0)$$

such that $\psi'\{\sigma\} = [\sigma]'$. Since $\varphi''[\sigma] = \{\sigma\}$, we are through with the proof of **Step 4**.

Proof of (Eilenberg–Blakers) Theorem 10.4.5 The idea is to use the prism construction Lemma 4.6.4. Toward this end, we need to construct the prisms $P\sigma$ as in the lemma with $C = S(X, A)^{(n-1)}$, the Eilenberg–Blakers complex. We carry it out by induction. For $q = 0$, since X and A are path connected, to each $x \in X$ we choose a path Px in X from x to x_0, being careful enough to choose this path inside A if $x \in A$ and of course, Px_0 is the constant path at x_0.

Assume now that $0 < q$ and we have defined $P\sigma$ in dimensions less than q, so as to satisfy the properties (a)-(e). Given a singular simplex $\sigma : \Delta_q \to X$, the maps $P(\sigma^i) : \mathbb{I} \times \Delta_{q-1} \to X$ fit together with σ to give a map $\lambda : 0 \times \Delta_q \cup \mathbb{I} \times \dot{\Delta}_q \to X$, viz.,

$$\lambda(0, x) = \sigma(x), \ x \in \Delta_q; \ \ \lambda(t, F^i x) = P(\sigma \circ F^i)(t, x), \ x \in \Delta_{q-1}.$$

Our $P\sigma$ is going to be an appropriate extension of λ so that (a) and (c) are automatically satisfied. If each of the map $P(\sigma^i)$ is stationary, we shall choose $P\sigma$ also stationary so that (b), (d) and (e) are also taken care of.

In the general case, notice that there is a (strong deformation) retraction map

$$r : \mathbb{I} \times \Delta_q \to 0 \times \Delta_q \cup \mathbb{I} \times \dot{\Delta}_q.$$

So, if we take $P\sigma = \lambda \circ r$ then it satisfies (e) also. Moreover, if $q > r$ then it satisfies (b) also.

In case $q \leq r$, then $\lambda(1 \times \dot{\Delta}_q) \subset A$. Since (X, A) is r-connected, we can choose an extension of λ to a map $P\sigma : \mathbb{I} \times \Delta_q \to X$ such that $P\sigma(1 \times \dot{\Delta}_q) \subset A$. For $q > r$, we can choose $P\sigma$ to be any extension of λ. This completes the proof of Theorem 10.4.5 and thereby also the proof of Hurewicz theorems. ♠

Exercise 10.4.17

(i) Given any topological space X and two subspaces X_1, X_2, show that there exist a CW-complex Y with two subcomplexes Y_1, Y_2 and a map $f : Y \to X$ which defines weak homotopy equivalences $Y \to X, (Y, Y_1) \to (X, X_1), (Y, Y_2) \to (X, X_2)$. [Hint: Start with a CW-complex Z and a weak homotopy equivalence $g : Z \to X$ and modify it using mapping cylinder technique.]

(ii) Show that the double combspace (see Exercise 1.9.17) has all its homotopy groups vanishing. (Hint: Prove π_1 is trivial by directly arguing as in van Kampen's theorem, with $U = Y \setminus [-1, 0] \times \{1\}$ and $V = Y \setminus \mathbb{I} \times \{-1\}$. Then appeal to Hurewicz Theorem and Exercise 5 in 4.7.) This provides an example of a space which is weak homotopy type of a singleton but not contractible.)

10.5 Obstruction Theory

So far, we have been dealing with 'invariants' such as homology and homotopy groups, which, in some sense, work only one-way, viz., two spaces X, Y are of the same homotopy type <u>only if</u> they have the same homology groups, homotopy groups, etc. Obstruction theory encompasses a large number of problems in which a topological construction can be carried out <u>if and only if</u> a certain algebraic quantity called 'obstruction' vanishes. At last, we can now consider one of the main problems in algebraic topology that we have proposed right in the first chapter. In this section, we shall make a beginning with obstruction theory for extending maps. Later, we shall use this to study obstruction theory for lifting maps. Obstruction theory takes many other forms which are spread out all over the literature: characteristic classes of a bundle, surgery obstructions, Whitehead torsion and so on, which we shall not be able to discuss.

Fix $n \geq 0$, and let X be obtained by attaching n-cells $\{e_j^n\}$ to A via the attaching maps $\phi_j : \mathbb{S}^{n-1} \to A$. Let $f : A \to Y$ be a given continuous map. Let us investigate the problem of extending the map f over X :

$$
\begin{array}{ccc}
X & & \\
\uparrow & \searrow & \\
\downarrow & & \searrow \\
A & \xrightarrow{\ f\ } & Y
\end{array}
$$

We begin with the following completely trivial observations.

Lemma 10.5.1 (a) Suppose $n = 0$. Then f can be extended continuously to a map $\hat{f} : X \to Y$.
(b) Suppose $n = 1$. Then f can be extended over X iff for each j, Im $f \circ \phi_j$ is contained in a single path component of Y. In particular, if Y is path connected then f can be extended.

Thus from now onward, we shall assume that $n \geq 2$ and Y is path connected. We shall also assume that A is path connected as the problem at hand can be studied by taking one path component at a time.

Lemma 10.5.2 Suppose $n \geq 2$. Then f can be extended over X iff each $f_{\#}([\phi_j]) \in \pi_{n-1}(Y)$ is trivial.

Notice that since we have not selected a base point for Y, the elements $[\phi_j]$ are defined only up to 'conjugacy', though there is no ambiguity about vanishing of these elements. To avoid such technicality, we shall from now on assume that Y *is a simple space, i.e., that the action of* $\pi_1(Y)$ *on all its homotopy groups is trivial. In particular, this assumption implies that* $\pi_1(Y)$ *is abelian.*

Definition 10.5.3 Let $n \geq 3$. Recall that $H_n(X, A)$ is a free abelian group with a generating set J which is in one-to-one correspondence with the set of n-cells attached. Consider the function $c(f) : H_n(X, A) \to \pi_{n-1}(Y)$ defined by $c(f)(j) = f_{\#}([\phi_j]), j \in J$, and extended linearly. This function $c(f)$ is called the *obstruction to extending f over X.*

The following observations are all easy.

Lemma 10.5.4

1. $c(f) = 0$ iff f can be extended over X.

2. If $f' : A \to Y$ is homotopic to f, then $c(f) = c(f')$.

3. If $g : Y \to Y'$ is any continuous map (and Y' is simple) then $c(g \circ f) = g_{\#}(c(f))$.

4. Given $h : A' \to A$, let X' be obtained by attaching n-cells to A' via $\phi_j \circ h$. Then $c(f \circ h) = h^*(c(f))$.

We shall now take up the general problem where (X, A) is a relative CW-complex and $f : X^{(n-1)} \to Y$ is a continuous function (Y is path connected and simple). We shall suppress mentioning A, since anyway $A \subset X^{(0)}$ and refer to X as a CW-complex.

Then the obstruction to extending f to $X^{(n)}$ will be denoted by $\tau^n(f)$ which is a homomorphism of abelian groups

$$\tau^n(f) : C_n^{CW}(X) := H_n(X^{(n)}, X^{(n-1)}) \to \pi_{n-1}(Y).$$

Thus we can think of $\tau^n(f)$ as a n-cochain on the CW-chain complex of X. The first non trivial observation is:

Lemma 10.5.5 $\tau^n(f)$ is a cocycle.

Proof: Consider the following commutative diagram:

$$
\begin{array}{ccc}
\pi_{n+1}(X^{(n+1)}, X^{(n)}) & \xrightarrow{\varphi_1} & H_{n+1}(X^{(n+1)}, X^{(n)}) \\
\downarrow{\scriptstyle \partial_1} & & \downarrow{\scriptstyle \partial_2} \\
\pi_n(X^{(n)}) & \xrightarrow{\varphi_2} & H_n(X^{(n)}) \\
\downarrow{\scriptstyle j_1} & & \downarrow{\scriptstyle j_2} \\
\pi_n((X^{(n)}, X^{(n-1)})) & \xrightarrow{\varphi_3} & H_n(X^{(n)}, X^{(n-1)}) \\
\downarrow{\scriptstyle \partial_3} & & \downarrow{\scriptstyle \tau^n(f)} \\
\pi_{n-1}(X^{(n-1)}) & \xrightarrow{f_\#} & \pi_{n-1}(Y)
\end{array}
$$

in which arrows labeled by φ_i indicate the Hurewicz homomorphisms, those with ∂_i indicate appropriate boundary homomorphisms, and those with j_i are induced by inclusions. Showing that $\tau^n(f)$ is a cocycle is the same as showing $\tau^n(f) \circ j_2 \circ \partial_2 = 0$. Since the Hurewicz homomorphism φ_1 is surjective, this amounts to showing that $\tau^n(f) \circ j_2 \circ \partial_2 \circ \varphi_1 = 0$. Now appeal to the above commutative diagram and the fact $\partial_3 \circ j_1 = 0$ from the homotopy exact sequence of the pair $(X^{(n)}, X^{(n-1)})$. ♠

The next step is to define obstructions for extending homotopies. Think of $\mathbb{I} = [0, 1]$ as a CW-complex with two 0-cells $\{0\}, \{1\}$ and one 1-cell denoted by $\mathbf{i}$. For any CW-complex (X, A) take the standard product CW-structure on $\hat{X} := \mathbb{I} \times (X, A)$. Note that $\hat{X}^{(n)} = \mathbb{I} \times X^{(n-1)} \cup \dot{\mathbb{I}} \times X^{(n)}$. Therefore, giving a map $F : \hat{X}^{(n)} \to Y$ is the same as giving two maps $f_0, f_1 : X^{(n)} \to Y$ and a homotopy $G : \mathbb{I} \times X^{(n-1)} \to Y$ from $f_0|_{X^{(n-1)}}$ and $f_1|_{X^{(n-1)}}$.

Definition 10.5.6 The *difference cochain of the data* (f_0, G, f_1) is defined to be an element $\Theta^n := \Theta^n(F) := \Theta^n(f_0, G, f_1) \in C_{CW}^n(X, A; \pi_n(Y))$ such that for each n-cell e_j^n we have

$$\Theta^n(F)(e_j^n) = (-1)^n \tau^{n+1}(F)(\mathbf{i} \times e_j^n)$$

An important special case of our interest will be when $f_0 = f_1$ on $X^{(n-1)}$ and the homotopy G is the constant, in which case, the notation need not involve G and so we shall use the notation $\Theta^n(f_0, f_1)$ for the difference cochain.

The naturality properties listed in Lemma 10.5.3 all hold for difference cochain as well.

Lemma 10.5.7

1. $\Theta^n = 0$ iff G can be extended to a homotopy of f_0 and f_1 on $X^{(n)}$.

2. If F, F' are homotopic then $\Theta^n(F) = \Theta^n(F')$.

3. If $g : Y \to Y'$ is any continuous map (and Y' is simple) then $\Theta^n(g \circ F) = g_\#(\Theta^n(F))$.

4. If (X', A') is a relative CW-complex and $h : (X', A') \to (X, A)$ is a cellular map, then $\Theta^n(F \circ h) = h^*(\Theta^n(F))$.

In general, the difference cochain need not be a cocycle. However, we have the following boundary formula:

Lemma 10.5.8 $\delta\Theta^n(f_0, G, f_1) = \tau^{n+1}(f_1) - \tau^{n+1}(f_0)$.

Proof: We have, $\delta\Theta^n = (-1)^n \Theta^n \circ \partial$. Therefore, for any $(n+1)$-chain c in (X, A) we have $\delta\Theta^n(c) = \tau^{n+1}(F)(\mathbf{i} \times \partial c)$. Since $\tau^{n+1}(F)$ is a cocycle, we have

$$
\begin{aligned}
0 &= \tau^{n+1}(F)(\partial(\mathbf{i} \times c)) \\
&= \tau^{n+1}(F)(1 \times c - 0 \times c - \mathbf{i} \times \partial c) \\
&= \tau^{n+1}(f_1)(c) - \tau^{n+1}(f_0)(c) - \tau^{n+1}(F)(\mathbf{i} \times \partial c)
\end{aligned}
$$

and hence the lemma follows. ♠

As an immediate corollary, we have:

Lemma 10.5.9 Let $f, f' : X^{(n)} \to Y$ be such that $f|_{X^{(n-1)}} = f'|_{X^{(n-1)}}$. Then the two cochains $\tau^{n+1}(f), \tau^{n+1}(f')$ are cohomologous.

Proof: Define $F : \hat{X}^{(n)} \to Y$ by $F(0, x) = f'(x)$; $F(1, x) = f(x), x \in X^{(n)}$; $F(t, x) = f(x) = f'(x), x \in X^{(n-1)}, t \in \mathbb{I}$. Then by the boundary formula in Lemma 10.5.8, we have

$$
\delta\Theta^n(F) = \tau^{n+1}(f) - \tau^{n+1}(f'|_{X^{(n)}}).
$$

The lemma follows. ♠

The next lemma tells you that every cochain is a difference cochain.

Lemma 10.5.10 Let $F_0 : 0 \times X^{(n)} \cup \mathbb{I} \times X^{(n-1)} \to Y$ be any map. Then any $d \in C^n_{CW}(X, A; \pi_n(Y))$ is the difference cochain of some extension $F : \hat{X}^{(n)} \to Y$ of F_0.

Proof: Since we can perform the extension cell-by-cell, the problem immediately reduces to the case when $X = X^{(n)}$ is obtained by attaching a single n-cell to $X^{(n-1)}$. This reduced problem is an immediate consequence of Exercise 10.2.26.(i). ♠

Lemma 10.5.11 A $(n+1)$-cocycle c which is cohomologous to an obstruction $\tau^{n+1}(f)$ is itself an obstruction $\tau^{n+1}(f')$ of $f|_{X^{(n-1)}}$.

Proof: Suppose λ is such that $\delta(\lambda) = \tau^{(n+1)}(f) - c$. By Lemma 10.5.10, there is a map $F : \hat{X}^{(n)} \to Y$ such that $\Theta^n(F) = \lambda$ and $F|_{0 \times X^{((n)}} = f$ and $F(t, x) = f(x), x \in X^{(n-1)}$. Now, taking $f' = F|_{1 \times X^{(n-1)}}$, we have

$$
\tau^{n+1}(f) - c = \delta(\lambda) = \delta(\Theta^n) = \tau^{n+1}(f) - \tau^{n+1}(f')
$$

and hence, $\tau^{n+1}(f') = c$. ♠

Lemma 10.5.12 (Addition formula for difference cochains) Suppose F', F'' : $\hat{X}^{(n)} \to Y$ are maps such that $F'(1, x) = F''(0, x), x \in X^{(n)}$. Define

$$F(t, x) = \begin{cases} F'(2t, x), & 0 \le t \le 1/2; \\ F''(2t - 1, x), & 1/2 \le t \le 1. \end{cases}$$

Then $\Theta^n(F) = \Theta^n(F') + \Theta^n(F'')$.

Proof: Here again, it is enough to prove this for the case when there is just one n-cell. This particular case follows from the addition law in homotopy. ♠

Theorem 10.5.13 *Let* $f : X^{(n)} \to Y$. *Then* $f|_{X^{(n-1)}}$ *can be extended over* $X^{(n+1)}$ *iff* $\tau^{n+1}(f)$ *is a coboundary.*

Proof: Suppose $f' : X^{(n+1)} \to Y$ is an extension of f. Define $F : \hat{X}^{(n)} \to Y$ as in Lemma 10.5.9. Then we have,

$$\delta\Theta^n(F) = \tau^{n+1}(f) - \tau^{n+1}(f'|_{X^{(n)}}).$$

Since, by Lemma 10.5.7.(1), $\tau^{n+1}(f'|_{X^{(n)}}) = 0$, $\tau^{n+1}(f)$ is a coboundary.

Conversely, suppose $\tau^{n+1}(f) = \delta\Theta$ for some $\Theta \in C^n_{CW}(X, A); \pi_n(Y))$. By Lemma 10.5.10, there is a map $F : \hat{X}^{(n)} \to Y$ such that

$$F(0, x) = f(x), x \in X^{(n)}; \quad F(t, x) = f(x), x \in X^{(n-1)}, t \in \mathbb{I}$$

and $\Theta^n(F) = -\Theta$. Define $f' : X^{(n)} \to Y$ by $f'(x) = F(1, x)$. Then

$$\tau^{n+1}(f) = \delta\Theta = -\delta\Theta^n(F) = \tau^{n+1}(f) - \tau^{n+1}(f').$$

Therefore, $\tau^{n+1}(f') = 0$ and by Lemma 10.5.7.(1) again, f' can be extended over $X^{(n+1)}$, which gives the required extension of $f|_{X^{(n-1)}}$. ♠

Definition 10.5.14 Given a map $f : A \to Y$ consider the set

$$\{\tau^{n+1}(f_n) \ : \ f_n : X^{(n)} \to Y \text{ is an extension of } f\}.$$

A priori, this set may be empty. However, we have seen that this set consists of $(n + 1)$-cocycles and is a (disjoint) union of cohomology classes (see Lemma 10.5.11). The image of this set in $H^{n+1}(X, A; \pi_n Y)$ is called the $(n + 1)$-dimensional obstruction set for f and is denoted by $\mathcal{O}^{n+1}(f)$.

Remark 10.5.15 Theorem 10.5.13 just says that f can be extended over $X^{(n+1)}$ iff $\mathcal{O}^{n+1}(f)$ contains the zero element. This is just the beginning of obstruction theory. There are finer versions of it for which you have to wait.

10.6 Homotopy Extension and Classification

Having made a beginning of obstruction theory in the previous section, let us quickly derive some very important results due to Eilenberg and Hopf–Whitney on homotopy classification of maps in certain special cases.

We assume $n \ge 1$ and Y is $(n - 1)$-connected; if $n = 1$, we continue to assume that Y is simple and hence in particular, $\pi_1(Y)$ is abelian. Temporarily, we use the notation Π for $\pi_n(Y)$. Let (X, A) be a CW-pair.

Lemma 10.6.1 Any map $f : A \to Y$ can be extended to a map $g : X^{(n)} \to Y$. Two such extensions g_0, g_1 restricted to $X^{(n-1)}$ are homotopic relative to A; moreover, $c^{n+1}(g_0) \sim c^{n+1}(g_1)$.

Proof: We have seen that since Y is path connected, f can be extended over $X^{(1)}$. Inductively assume that for $1 \le r \le n-1$ we have an extension $h : X^{(r)} \to Y$. Since $\pi_r(Y) = 0$ the obstruction to extending h to $X^{(r+1)}$ vanishes. Hence h can be extended over $X^{(r+1)}$.

Now suppose $g_0, g_1 : X^{(n)} \to Y$ are two extensions. Once again, in view of the fact that $\pi_i(Y) = 0$ for $i \le n-1$, inductive application of Lemma 10.5.6.(1) yields a homotopy relative to A of $g_0|_{X^{(n)}}$ with $g_1|_{X^{(n)}}$. Now Lemma 10.5.7 gives $c^{n+1}(g_0) \sim c^{n+1}(g_1)$.   ♠

Definition 10.6.2 The unique cohomology class $\gamma^{n+1}(f) = [c^{n+1}(g)] \in H^{n+1}(X, A; \Pi)$ given by the above lemma is called the *primary obstruction* to extending f.

Clearly this primary obstruction satisfies the naturality property: if $h : (X', A') \to (X, A)$ is a cellular map then $\gamma^{n+1}(f \circ h) = h^*\gamma^{n+1}(f)$. We now begin to realize that it may be possible to express all of these primary obstructions in a single way.

By Hurewicz isomorphism theorem, we have the isomorphism $\varphi : \Pi = \pi_n(Y) \to H_n(Y)$. By the universal coefficient theorem, we have an isomorphism

$$h : H^n(Y; \Pi) \to \mathrm{Hom}(H_n(Y), \Pi).$$

Let $\iota^n := \iota^n(Y) := h^{-1}(\varphi^{-1})$. The element ι^n is called the *n-characteristic* or simply the *characteristic element* for Y.

For instance, suppose Y is a CW-complex with $Y^{(n-1)} = *$, a singleton space. Then it is obvious that ι^n is represented by the n-cochain which sends each n-cell of Y to the element in $\Pi = \pi_n(Y)$, represented by the characteristic map of this n-cell.

In this background the following lemma looks completely natural:

Lemma 10.6.3 $\gamma^{n+1}(f) = (-1)^n \delta^* f^*(\iota^n)$.

Proof: Let $g : X^{(n)} \to Y$ be an extension of f. Consider the commutative diagram

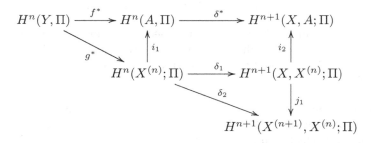

in which j_1 is a monomorphism and its image is the group $Z^{n+1}(X, A; \Pi)$ of CW-cocycles. Moreover, $i_2 \circ j_1^{-1} : Z^{n+1} \to H^{n+1}(X, A; \Pi)$ is nothing but the natural projection of the cocycles to cohomology classes. In particular, we have $\gamma^{n+1}(f) = i_2 j_1^{-1}(c^{n+1}(g))$. By the above commutative diagram the statement of the lemma is equivalent to showing that

$$\delta_2 g^*(\iota^n) = (-1)^n c^{n+1}(g). \tag{10.4}$$

Now consider another commutative diagram:

$$\begin{array}{ccccc}
H^n(Y;\Pi) & \xrightarrow{\;g^*\;} & H^n(X^{(n)};\Pi) & \xrightarrow{\;\delta_2\;} & H^{n+1}(X^{(n+1)},X^{(n)};\Pi) \\
\downarrow{\scriptstyle h_1} & & \downarrow{\scriptstyle h_2} & & \downarrow{\scriptstyle h_3} \\
\mathrm{Hom}(H_n(Y),\Pi) & \xrightarrow{\;g_1^*\;} & \mathrm{Hom}(H_n(X^{(n)}),\Pi) & \xrightarrow{\;\delta_3\;} & \mathrm{Hom}(H_{n+1}(X^{(n+1)},X^{(n)}),\Pi) \\
\downarrow{\scriptstyle \varphi_1^*} & & \downarrow{\scriptstyle \varphi_2^*} & & \downarrow{\scriptstyle \varphi_3^*} \\
\mathrm{Hom}(\pi_n(Y),\Pi) & \xrightarrow{\;g_2^*\;} & \mathrm{Hom}(\pi_n(X^{(n)}),\Pi) & \xrightarrow{\;\delta_4\;} & \mathrm{Hom}(\pi_{n+1}(X^{(n+1)},X^{(n)}),\Pi)
\end{array}$$

in which $\varphi_i's$ correspond to Hurewicz homomorphisms, $g_i's$ are induced by g, and δ_i^*'s are given by appropriate boundary homomorphisms. The homomorphisms h_i are given by the universal coefficient theorem. Since h_i is an isomorphism and φ_3^* is a monomorphism, (10.4) follows if we show that

$$(-1)^n \varphi_3^* h_3(c^{n+1}(g)) = \varphi_3^* h_3 \delta_2 g^*(\iota^n). \tag{10.5}$$

By definition of (ι^n) we have, $h_1(\iota^n) = \varphi_1^{-1}$. Therefore the right hand side of (10.5) is equal to

$$\delta_4 g_2^* \varphi_1^* h_1(\iota^n) = \delta_4 g_2^* = (-1)^n g_2 \circ \partial_4.$$

On the other hand, by the definition of $c^{n+1}(g)$ we have, $h_3 c^{n+1}(g) = g_2 \circ \delta_4 \circ \varphi_3^{-1}$. ♠

Here are a few immediate consequences:

Lemma 10.6.4 A map $f : A \to Y$ can be extended over $X^{(n+1)}$ iff $\delta^* f^*(\iota^n) = 0$.

Corollary 10.6.5 (Hopf–Whitney) If $\dim(X,A) \leq n+1$, and Y is $(n-1)$-connected and simple, a map $f : A \to Y$ can be extended over X iff $\delta^* f^*(\iota^n) = 0$.

Recall that a space Y is called q-simple if the action of $\pi_1(Y)$ on $\pi_j(Y)$ is trivial for $1 \leq j \leq q$.

The following generalization is immediate:

Theorem 10.6.6 (Eilenberg) *Let (X,A) be a CW-pair and Y be $(n-1)$-connected. Suppose further that for $n+1 \leq q < \dim(X,A)$, $H^{q+1}(X,A;\pi_q(Y)) = 0$ and Y is q-simple. Then a map $f : A \to Y$ can be extended over X iff $\gamma^{n+1}(f) = 0$.*

Proof: The 'only if' part is obvious. To prove the 'if' part, let $\gamma^{n+1}(f) = 0$. Then we know that there is an extension $f_{n+1} : X^{(n+1)} \to Y$. Inductively, if we have an extension $f_q : X^{(q)} \to Y$, for $n+1 \leq q$, then $\gamma^{q+1}(f_q) = 0$ because $\pi_q(Y) = 0$. Therefore, f_{q-1} can be extended to $f_{q+1} : X^{(q+1)} \to Y$. Finally $f : X \to Y$ is defined by $f|_{X^{(q)}} = f_q$, $q \geq n+1$. ♠

Corollary 10.6.7 If Y is n-simple and $\pi_i(Y) = 0$ for all $i \neq n$, then a map $f : A \to Y$ can be extended over X iff $\delta^* f^*(\iota^n) = 0$.

There are corresponding results for extending homotopies: given two maps $f_0, f_1 : X \to Y$ such that $f_0|_A = f_1|_A$, we shall merely state some results concerning extension of the stationary homotopy on A from f_0 to f_1 to a homotopy on X. Let $F : \mathbb{I} \times A \cup \dot{\mathbb{I}} \times X$ be defined as before. There is an obvious one-one correspondence

$$c_q \mapsto \mathbf{i} \times c_q$$

of q-cells in (X, A) with $(q+1)$-cells in $(\mathbb{I} \times X, \mathbb{I} \times A \cup \dot{\mathbb{I}} \times X)$ which induces an isomorphism of degree 1:

$$\zeta := \mathbf{i} \times \; : H^q(X, A) \to H^{q+1}(\mathbb{I} \times X, \mathbb{I} \times A \cup \dot{\mathbb{I}} \times X). \tag{10.6}$$

Consider the composite of

$$H^q(Y) \xrightarrow{F*} H^q(\mathbb{I} \times A \cup \dot{\mathbb{I}} \times X) \xrightarrow{\delta*} H^{q+1}(\mathbb{I} \times X, \mathbb{I} \times A \cup \dot{\mathbb{I}} \times X) \xrightarrow{\zeta^{-1}} H^q(X, A)$$

where $\delta*$ is the coboundary operator. Denote this composite by $(f_0, f_1)^*$.

Theorem 10.6.8 *The operator satisfies the following:*

1. $(f_0, f_0)^* = 0$.

2. $(f_0, f_1)^* + (f_1, f_2)^* = (f_0, f_2)^*$.

3. $(f_1, f_0)^* = -(f_0, f_1)^*$.

4. $j^*(f_0, f_1)^* = f_1^* - f_0^*$, *where* $j^* : H^q(X, A) \to H^q(X)$.

5. *For any* $g : Y \to Y'$, *we have,* $(f_0, f_1)^* \circ g^* = (g \circ f_0, g \circ f_1)^*$.

6. *If* $h : (X, A' \to (X, A)$ *is cellular, then* $(f_0 \circ h, f_1 \circ h)^* = h^* \circ (f_0, f_1)^*$.

7. $\delta^n(f_0, f_1) = (-1)^n (f_0, f_1)^*(\iota^n)$.

Theorem 10.6.9 (Eilenberg) *Suppose that* Y *is* q-*simple and that* $H^q(X, \pi_q(Y)) = 0$ *for all* q *such that* $n + 1 \le q < 1 + \dim(X, A)$. *Then* $f_0 \simeq f_1$ *(rel* A*) if* $(f_0, f_1)^*(\iota^n(Y)) = 0$.

We can now have the following classification theorems.

Theorem 10.6.10 (Eilenberg classification theorem) *Assume that*
(1) Y *is* q-*simple for* $n + 1 \le q < 1 + \dim(X, A)$;
(2) $H^q(X, A; \pi_q(Y)) = 0$, *for* $n + 1 \le q < 1 + \dim(X, A)$; *and*
(3) $H^{q+1}(X, A; \pi_q(Y)) = 0$, *for* $n + 1 \le q < \dim(X, A)$.
Then for any fixed map $f_0 : X \to Y$,

$$f \mapsto (f_0, f)^*(\iota^n)$$

defines a bijection of homotopy classes (rel A*) of extensions of* $f_0|_A$ *to* X *and the group* $H^n(X, A; \Pi)$.

Proof: Suppose $(f_0, f)^*(\iota^n) = (f_0, g)^*(\iota^n)$. Then by properties listed in Theorem 10.6.8, we have

$$(f, g)^*(\iota^n) = ((f, f_0)^* + (f_0, g)^*)(\iota^n) = 0.$$

By Theorem 10.6.9, $f \simeq g$ (rel A). This proves injectivity.

Given $z \in H^n(X, A; \Pi)$, let $\Delta \in Z^n(X, A; \Pi)$ be a representative cocycle. Let $F_0 : 0 \times X^{(n)} \cup \mathbb{I} \times X^{(n-1)} \to Y$ be the map

$$F_0(0, x) = f_0(x), x \in X^{(n)}; \quad F(t, x) = f_0(x), t \in \mathbb{I}, x \in X^{(n-1)}.$$

By Lemma 10.5.10, F_0 has an extension $F : X^{(n)} \to Y$ with $\Theta^n(F) = \Delta$. Let now $f(x) = F(1, x), x \in X^{(n)}$. By the coboundary formula in Lemma 10.5.8, we have,

$$0 = \delta\Delta = \delta\Theta^n(F) = c^{n+1}(f) - c^{n+1}(f_0|_{X^{(n)}}) = c^{n+1}(f).$$

Therefore, f can be extended to f_{n+1} over $X^{(n+1)}$. Now condition (3) implies that we can extend f_{n+1} to a map $f' : X \to Y$. But then $z = [\Delta] = \Theta^n(f_0, f') = (-1)^n (f_0, f')^*(\iota^n)$ which proves the surjectivity. ♠

By taking $A = \emptyset$ and $f_0 : X \to Y$ to be the constant map, we get

Corollary 10.6.11 Under the hypothesis of Theorem 10.6.10 on Y, $f \mapsto f^*(\iota^n)$ is a bijection $[X, Y] \to H^n(X, \Pi)$.

Theorem 10.6.12 (Hopf–Whitney) *Let Y be $(n-1)$-connected n-simple space. Then for any CW-complex X of dimension $\leq n$, we have*

$$[X, Y] \approx H^n(X; \pi_n(Y)).$$

Exercise 10.6.13 (See Exercise 4.7.12.)

(i) Show that all Moore spaces of type (G, n), where G is abelian and $n \geq 2$ belong to the same homotopy type. What can you say for $n = 1$?

(ii) Let $M(G, n)$ denote a Moore space of type (G, n), $n \geq 2$. Show that the inclusion map $M(\mathbb{Z}_p, n) \vee M(\mathbb{Z}_q, n) \to M(\mathbb{Z}_p, n) \times M(\mathbb{Z}_q, n)$ is a homotopy equivalence, where $n \geq 2$ and p, q are coprime.

10.7 Eilenberg–Mac Lane Spaces

Let us take a closer look at Corollary 10.6.7.

Having assigned a certain 'obstruction' element in the first non vanishing cohomology group of the codomain space Y, the vanishing of which guaranteed extension of certain maps over cells of that dimension, it is natural to look out for conditions on the space Y so that there is no more obstruction to extending maps over higher cells. Corollary 10.6.7 is an excellent instance of this. Perhaps, this is one of the motivations to come up with the notion of $K(\pi, n)$ spaces which have homotopy groups isolated in a single dimension. Later on we shall see how these 'atomic homotopy spaces' can be used to build all topological spaces and obtain a satisfactory obstruction theory for extensions as well as liftings.

Definition 10.7.1 Fix an integer $n \geq 1$ and a group G (abelian if $n \geq 2$). A topological space X is said to be *Eilenberg–Mac Lane space of type* (G, n) if it is path-connected and $\pi_q(X, x_0) = 0$ for all $q \neq n$ and $\pi_n(X, x_0) \approx G$. An *Eilenberg–Mac Lane space* is a space which is of type (G, n) for some G and n. Further if it has the homotopy type of a CW-complex also, then we say it is a $K(G, n)$-space.

Example 10.7.2

1. $\mathbb{S}^1$ is a $K(\mathbb{Z}, 1)$-space.

2. $\mathbb{P}^\infty$ is a $K(\mathbb{Z}_2, 1)$-space.

3. $\mathbb{CP}^\infty$ is a $K(\mathbb{Z}, 2)$-space.

4. $\mathbb{L}_n^\infty$, the infinite lens space is a $K(\mathbb{Z}_n; 1)$-space. (See Example 3.5.9.(iii).)

5. By taking finite products of the spaces above, we get some other Eilenberg–Mac Lane spaces.

6. $\Omega K(G, n)$ is an Eilenberg-Mac Lane space of type $(G, n-1)$. Though the loop space of a CW-complex fails to be a CW-complex, by a result of Milnor, it has the homotopy type of a CW-complex (see [Milnor, 1960]). In particular, $\Omega K(G, n)$ is a $K(G, n-1)$.

7. Two Eilenberg–Mac Lane spaces of the same type (G, n) are of the same *weak homotopy type*. The following two results remove whatever ambiguity there may be in the above definition.

Theorem 10.7.3 *Let n be a positive integer, Π be a group (abelian group if $n > 1$). Then there exists a $K(\Pi, n)$-space and any two of them have the same homotopy type.*

The existence is a consequence of a more general result about existence of a CW-complex with prescribed sequence of homotopy groups (see Theorem 10.4.14). The uniqueness is also a consequence of a more general result which we shall now describe.

Theorem 10.7.4 *Let G, G' be abelian groups, n a positive integer, and X, X' be some $K(G, n)$ and $K(G', n)$ spaces, respectively. Given any homomorphism $\eta : G \to G'$, there is a unique homotopy class of a map $f_\eta : X \to X'$ such that the induced homomorphism on the nth homotopy groups corresponds to η :*

$$\eta = \pi_n(f_\eta) : \pi_n(X) \to \pi_n(X').$$

Proof: By Corollary 10.6.11, the function $f \mapsto f^* \iota^n(X')$ is a bijection $[X, X'] \approx H^n(X; G')$. Consider $h^{-1}(\eta \circ \varphi^{-1}) \in H^n(X; G')$, where $\varphi : \pi_n(X) \to H_n(X)$ is the Hurewicz isomorphism and $h : H^n(X; G') \to \mathrm{Hom}(H_n(X); G')$ is the isomorphism given by the universal coefficient theorem. Let $f : X \to X'$ correspond to this element under the above correspondence. Then

$$\eta \circ \varphi^{-1} = h \circ f^* \iota^n(X') = h(\iota^n(X')) \circ H_n(f) = \varphi'^{-1} \circ H_n(f).$$

$$
\begin{array}{ccc}
\pi_n(X) & \xrightarrow{\ \eta\ } & \pi_n(X') \\
\varphi \downarrow & & \downarrow \varphi' \\
H_n(X) & \xrightarrow[H_n(f)]{} & H_n(X')
\end{array}
$$

Therefore, $\eta = \phi'^{-1} \circ H_n(f) \circ \phi = \pi_n(f)$.

If f, f' are maps such that $\pi_n(f) = \pi_n(f')$, then again by the above naturality of the Hurewicz homomorphisms, we have $H_n(f) = H_n(f')$ and $f^* = f'^* : H^n(X'; G') \to H^n(X; G')$. In particular, $f^* \iota^n(X') = f'^* \iota^n(X')$. By the uniqueness part in Corollary 10.6.11 $f \simeq f'$.   ♠

Theorem 10.7.5 *Let Y be a $K(G, n)$ space with a base point y_0 and n-characteristic $\iota_n \in H^n(Y, G)$ for some abelian group $G, n \geq 1$. Then for any relative CW-complex (X, A) the assignment $f \mapsto f^*(\iota_n)$ defines a functorial bijection*

$$\Psi : [(X, A, x_0); (Y, y_0)] \approx H^n(X; G). \tag{10.7}$$

Proof: When A is empty or a singleton this is immediate from Corollary 10.6.11. In the general case, use the collapsing map $k : (X, A) \to (X/A, x_0)$ and the following commutative diagram.

$$
\begin{array}{ccccc}
[(X, A); (Y, y_0)] & \xleftarrow[\ k_\#\ ]{\approx} & [(X/A, x_0), (Y, y_0)] & \xrightarrow{\ \approx\ } & [X/A; Y] \\
\Psi \downarrow & & \Psi \downarrow & & \approx \downarrow \Psi \\
H^n(X, A; G) & \xleftarrow[\ k^*\ ]{\approx} & H^n(X/A, x_0; G) & \xrightarrow{\ \approx\ } & H^n(X/A; G)
\end{array}
$$

There is a somewhat stronger result when $n = 1$.

Theorem 10.7.6 *Let Y be a $K(\pi, 1)$ for any group π. Then for any connected CW-complex X (with base point), the assignment $[f] \mapsto f_\#$ defines a functorial bijection*

$$\Psi : [(X, *); (Y, *)] \to \mathrm{Hom}(\pi_1(X, *), \pi).$$

Proof: The functoriality of Ψ is clear.

The homomorphism $f_\#$ on the fundamental group is completely determined by $f_\# : \pi_1(X^{(2)}) \to Y$. From Theorem 10.4.15 it follows that given a homomorphism $\phi : \pi_1(X) \to \pi_1(Y) = \pi$, there is a map $g : X^{(2)} \to Y$ such that $g_\# = \phi : \pi_1(X) \to \pi_1(Y)$. This map g can be extended to the whole of X, (Lemma 10.5.2), since $\pi_i(Y) = 0$ for $i \geq 2$. This proves surjectivity of Ψ.

Finally, suppose $f, g : X \to Y$ are such that $f_\# = g_\# : \pi_1(X) \to \pi_1(Y)$. We need to construct a homotopy $F : f \sim g$ which is constant on the base point x_0. Since $X \times \mathbb{I}$ has a cell structure consisting of cells from $X \times \{0, 1\}$ and cells of the type $e_i \times \mathbb{I}$ for each cell e_i of X, we need to define F on cells of the form $e_i \times \mathbb{I}$ since, any way, we must take $F|_{X \times 0} = f$ and $F|_{X \times 1} = g$.

For the simplicity of the exposition, let us first assume that the 0-skeleton of X consists of a single point. Put $F(x_0, t) = y_0$. This completes the construction of F on the 1-skeleton of $X \times \mathbb{I}$ which we shall denote by F_1.

Let us write $X^{(1)} = \vee_i \mathbb{S}_i^1$. Let f_1, g_1 denote the restriction of f to $X^{(1)}$. It then follows that $(f_1)_\# = (g_1)_\# : \pi_1(X^{(1)}) \to Y$. If β_i is the attaching map of the 2-cell $e_i \times \mathbb{I}$, it follows that $(F_1)_\#([\beta_i]) = (f_1)_\#([\beta_i])(h_{[\omega]}(f_2)_\#[\beta_i])^{-1}$, where $\omega(t) = F(x_0, t)$. Since this is a constant path, it follows that $(F_1)_\#([\beta_i]) = (1)$. Therefore we can extend F_1 to a map $F_2 : (X \times \mathbb{I})^{(2)} \to Y$.

Inductively having defined $F_n, n \geq 2$, we can extend it over cells of the form $e^n \times \mathbb{I}$ because $F_n \circ \gamma$ will be null homotopic since $\pi_n(Y) = 0$ for $n \geq 2$. This completes the definition of F, thereby establishing injectivity of Ψ.

We shall leave the details of the general case, viz., without the assumption that the 0-skeleton of X is a singleton space, to the reader as an exercise ♠

Remark 10.7.7 Now the uniqueness of a $K(G, n)$ up to homotopy follows. Because of the functoriality of $h : H^n(X; G) \to \text{Hom}(H_n(X); G)$, it follows that $h(f^*(\iota_n)) = f_* : H_n(X) \to H_n(K(G, n)) = G$, for any map $f : X \to K(G, n)$. This remark comes in handy later.

Theorem 10.7.5 goes under the name 'representability of the cohomology functor'. One can say this was sufficient motivation for inventing $K(G, n)$-spaces. There are quite a few by-products.

Combining these results with Theorem 10.1.2 we have:

Corollary 10.7.8 Every $K(G, n)$-space is an H-group, abelian if G is abelian. The group structure is unique up to homotopy and corresponds to the group structure of G in an obvious way.

Another consequence of Theorem 10.7.3 is:

Corollary 10.7.9 The homology and cohomology groups of a $K(G, n)$-space depend only on the group G and the integer n.

Because of the above result, it is natural to make the following definition.

Definition 10.7.10 Given a group G, we define $H_*(G, n), H^*(G, n)$ to be the integral homology and cohomology groups (respectively) of a $K(G, n)$ space.

Example 10.7.11 We know the homology groups of those spaces listed in Example 10.7.2. Using the Künneth formula, we can compute $H_*(G, 1)$ for any finitely generated abelian group. For instance, $H_2(\mathbb{Z}^2; 1) = \mathbb{Z}$. Also by Hurewicz theorem for any group G, $H_1(G, 1) = G/[G, G]$.

We end this section with yet another bonus of the notion of $K(G, n)$-spaces as a direct consequence of Theorem 10.7.5. Recall the definition of cohomology operations from Definition 7.5.1.

Theorem 10.7.12 (Serre) *There is a bijective correspondence* λ *between the set* $Op(p, p'; G, G')$ *of cohomology operations of type* $(p, p', G; G')$ *and the group* $H^{p'}(G, p; G')$ *given by* $\lambda : \Theta \mapsto \Theta(\iota_p)$ *where* $\iota_p \in H^p(K(G, p); G)$ *is the p-characteristic element.*

Proof: Since $H^{p'}(G, p; G') = H^{p'}(K(G, p); G')$, the assignment λ makes sense. Given an element u in $H^{p'}(G, p; G')$, we define $\Theta_u \in Op(p, p'; G, G')$ as follows. For any CW-complex X, and $x \in H^p(X, G)$, there is a unique homotopy class of a map $f : X \to K(G, p)$. We define $\Theta_u(x) = f^*(u)$. Verification that Θ_u is indeed an element of $Op(p, p'; G, G')$ is straight forward. We claim that $\Theta_u(\iota_p) = u$ and $\Theta_{\Theta(u)} = \Theta$. The first claim is obvious since the unique homotopy class of a map f corresponding to ι_p is nothing but the identity map. The second one follows because for any x and $[f]$ as above, we have

$$\Theta_{\Theta(u)}(x) = f^*(\Theta(u)) = \Theta(f^*(u)) = \Theta(x).$$

This shows that the assignment is a bijection.   ♠

Especially in light of the above theorem, it follows that any knowledge of the cohomology algebra of Eilenberg–Mac Lane spaces would be of some use. In the simplest cases, viz., $\mathbb{S}^1 = K(\mathbb{Z}, 1), \mathbb{P}^\infty = K(\mathbb{Z}_2, 1)$, and $\mathbb{CP}^\infty = K(\mathbb{Z}, 2)$, we have already witnessed this. Here is yet another instance — it suffices to prove that if R is an Adem's relation, then $R(\iota_n) = 0$ for the n-characteristic element ι_n of $K(\mathbb{Z}_2, n)$, for all n. For, given $y \in H^n(X; \mathbb{Z}_2)$, we have a map $f : X \to K(\mathbb{Z}_2, n)$ such that $f^*(\iota_n) = y$. Therefore by naturality of the Steenrod squares $R(y) = f^*(R(\iota_n)) = 0$. Indeed, we shall now present a theorem of Serre which gives a complete description of the cohomology algebra of $K(\mathbb{Z}_2, n)$ from which we shall be able to prove $R(\iota_n) = 0$.

We need a couple of definitions:

Definition 10.7.13 Let $I = (i_1, \ldots, i_r)$ denote a sequence of length $r \geq 1$ of positive integers. The *degree* $d(I)$ is the sum $\sum_k i_k$. The sequence I is said to be *admissible* if $i_k \geq 2i_{k+1}$ for all $0 \leq k < r$. The *excess* $e(I)$ of an admissible sequence is defined to be the number $e(I) = 2i_1 - d(I)$.

Remark 10.7.14 For any sequence I as above, we use the notation $Sq^I = Sq^{i_1} \cdots Sq^{i_r}$. Of course the empty sequence is allowed and we put $Sq^0 = Id$. Note that every sequence of length 1 is admissible.

Theorem 10.7.15 (Serre) *The cohomology algebra* $H^*(\mathbb{Z}_2, n; \mathbb{Z}_2)$ *is the polynomial ring over* $\mathbb{Z}_2$ *with generators* $\{Sq^I(\iota_n)\}$, *where* I *runs over all admissible sequences of excess less than* n.

The proof of this theorem is one of the few things that this exposition will miss. However, we shall now show how to prove Adem's relations using this theorem.

Proposition 10.7.16 For any 1-dimensional class u we have $Sq^i(u^j) = \binom{j}{i} u^{i+j}$.

Proof: Note that the total Steenrod operator $Sq : H^*(X; \mathbb{Z}_2) \to H^*(X, \mathbb{Z}_2)$ is a ring homomorphism. Since $Sq(u) = Sq^0(u) + Sq^1(u) = u + u^2$ it follows that $Sq(u^j) = u^j(1+u)^j$. Equating the terms of the corresponding degree on the either side, we get the result.   ♠

Let $Sym = \mathbb{Z}_2[\sigma_1, \ldots, \sigma_n]$ denote the subring of all symmetric polynomials in $\mathbb{Z}_2[x_1, \ldots, x_n] = H^*(P_n; \mathbb{Z}_2)$ where $P_n = (\mathbb{P}^\infty)^n$ and σ_i is the ith elementary symmetric function in $x_1, \ldots, x_n$. Let us consider an ordering on the set of monomials in σ_i's as follows: Given any monomial m, we write $m = \sigma_{j_1}^{k_1} \cdots \sigma_{j_r}^{k_r}$, where $j_1 > j_2 > \cdots > j_r$. Given two such monomials m, m' inductively we define $m < m'$ if $j_1 < j_1'$ or if $j_1 = j_1'$ then $m/\sigma_{j_1} < m'/\sigma_{j_1'}$.

Proposition 10.7.17 In $H^*(P_n; \mathbb{Z}_2)$, we have $Sq^i(\sigma_n) = \sigma_n \sigma_i, (1 \leq i \leq n)$. Moreover, $Sq^i(\prod_j \sigma_{i_j}) < \sigma_n$ if all $i_j < n$.

Proof: $Sq(\sigma_n) = Sq(x_1 \cdots x_n) = \prod_i Sq(x_i) = \prod(x_i + x_i^2) = \sigma_n \prod(1 + x_i)$ and again by comparing the degree terms we get the first part. The second part follows since the term is a polynomial in $\{\sigma_{i_k}\}_{i_k < n}$ and hence does not involve the 'variable' σ_n at all. ♠

Proposition 10.7.18 Let $I = (i_1, \ldots, i_r)$ such that $d(I) := \sum_j i_j \leq n$. Then $Sq^I(\sigma_n)$ can be written in the form $\sigma_n P_I$, where

$$P_I = \sigma_{i_1} \cdots \sigma_{i_r} + T$$

where T is a sum of monomials of order less than n.

Proof: We induct on the length of I. For $r = 1$ this follows from the previous proposition. Now write $I = i_1 J$ where J is a sequence of length $r - 1$. Assuming the result to be true for J we have

$$
\begin{aligned}
Sq^I(\sigma_n) &= Sq^{i_1} Sq^J(\sigma_n) \\
&= Sq^{i_1}(\sigma_n P_J) \\
&= \sum_0^{i_1} Sq^m(\sigma_n) Sq^{i_1 - m}(P_J) \\
&= \sigma_n \left(\sum_0^{i_1 - 1} \sigma_m Sq^{i_1 - m}(P_J) \right) + \sigma_n \sigma_{i_1} P_J
\end{aligned}
$$

where $P_J = \sigma_{i_2} \cdots \sigma_{i_r} +$ lower order terms. Since $Sq^j(\sigma_i) \leq \sigma_{i+j}$, and $Sq^i(\sigma_i) = \sigma_i^2$, the largest possible monomial from the bracketed sum is obtained by taking $m = i_1 - i_2 + 1$ and hence is equal to $y = \sigma_m Sq^{i_2 - 1}(\sigma_{i_2} \sigma_{i_3} \cdots \sigma_{i_r}) < \sigma_n$ as seen in Proposition 10.7.17. Of course the last term contributes a term of the form $\sigma_n(\sigma_{i_1} \sigma_{i_2} \cdots \sigma_{i_r}) +$ lower degree terms. The conclusion follows. ♠

Corollary 10.7.19 Let $f : P_n \to K(\mathbb{Z}_2, n)$ be a map such that $f^*(\iota_n) = \sigma_n$. Then $f^* : H^i(K(\mathbb{Z}_2, n); \mathbb{Z}_2) \to H^i(P_n; \mathbb{Z}_2)$ is injective for $i \leq 2n$.

Proof: As I runs through all admissible sequences of degree $\leq n$, the monomials $\sigma_I = \sigma_{i_1} \cdots \sigma_{i_r}$ are linearly independent elements of $Sym \subset \mathbb{Z}_2[\sigma_1, \ldots, \sigma_n] \subset H^*(P_n; \mathbb{Z}_2)$. If $\sum_I \alpha_I Sq^I(\iota_n) = 0$ for some coefficients α_I, then we get

$$0 = f^* \left(\sum_I \alpha_I Sq^I(\iota_n) \right) = \sum_I \alpha_I Sq^I(\sigma_n) = \sigma_n \left(\sum_I \alpha_I \sigma_I + \text{ lower order terms} \right).$$

This implies $\alpha_I = 0$ for I. It follows that $\{Sq^I(\iota_n)\}$ forms a linearly independent set in $H^*(K(\mathbb{Z}_2, n); \mathbb{Z}_2)$.

Now Serre's theorem says that $\{Sq^I(\iota_n)\}$ forms a generating set also. In particular, it follows that if $i \leq 2n$, then f^* is injective.

Proof of Adem's relations: Given $y \in H^n(X; \mathbb{Z}_2)$ and $i + j \leq n$, we want to show that $R(i, j)(y) = 0$. Choose a map $g : X \to K(\mathbb{Z}_2, n)$ such that $y = g^*(\iota_n)$. Then, by naturality, $R(y) = g^*(R(\iota_n))$. Therefore, it suffices to show that $R(\iota_n) = 0$. But we have shown that $R(\sigma_n) = 0$ in $H^*(P_n; \mathbb{Z}_2)$ (see Proposition 7.5.32). Now $R(\sigma_n) = R(f^*(\iota_n)) = f^*(R(\iota_n))$ and f^* is injective. The conclusion follows. ♠

Exercise 10.7.20

(i) We have given the proof of Theorem 10.7.6, only for the case when the 0-skeleton of X consists of a single vertex. How do you complete the proof of the general case?

(ii) Show that for $n \geq 2$, and for any abelian group G, $H_{n+1}(G, n) = (0)$.

(iii) Let X be a $(n-1)$-connected CW-complex such that $\pi_q(X) = 0$ for $n + 1 \leq q < m$. Prove that $H_q(X) \approx H_q(G; n)$, $q \leq m$ and $H_m(X)/\Sigma_m(X) \approx H_m(G, n)$, where $G = \pi_n(X)$, and $\Sigma_m(X)$ denotes the image of the Hurewicz map $\pi_m(X) \to H_m(X)$. Elements of $\Sigma_m(X)$ are called spherical homology classes.

(iv) Show that for any CW-complex X, $H_2(\pi_1(X), 1; \mathbb{Z}) = H_2(X)/\Sigma_2(X)$, where $\Sigma_2(X)$ is defined as in the above exercise.

10.8   Moore–Postnikov Decomposition

Eilenberg–Mac Lane spaces can be treated as 'building blocks' in homotopy theory somewhat like Euclidean cells were building blocks for homology theory. However, here the construction is not via gluing but via fibrations. The idea is to decompose, up to homotopy, any given map into a sequence of principal fibrations in a hope that the lifting problem through the given map gets decomposed into the lifting problems through these simpler fibrations, by converting a lifting problem into an extension problem. As a consequence, this gives an alternative approach to the extension problem itself. However, the real usefulness of Moore–Postnikov decomposition is that it naturally leads to a sweeping generalization of Serres' Theorem 10.7.12 to what are called *secondary cohomology operations, tertiary cohomology operations, etc.,* on the one hand and generalization of Eilenberg–Mac Lane spaces to spaces with nontrivial homotopy groups at two places, three places, etc., on the other. This in turn makes it possible to handle secondary obstructions, etc., in a similar (but obviously more complicated) way as we have handled primary obstruction. We shall not be presenting these aspects of Moore–Postnikov decomposition here. (See [Mosher–Tangora, 1968] for details.)

Definition 10.8.1 Given a sequence of maps $\{p_i : E_i \to E_{i-1}, i \geq 0\}$, we define their inverse limit

$$\lim_{\leftarrow}\{E_i, p_i\} = (E_\infty, \{a_i\})$$

where the space E_∞ and the maps $a_i : E_\infty \to E_i$ are defined by

$$E_\infty = \{(x_i) \in \times_i E_i \; : \; p_i(x_i) = x_{i-1}\}; \quad a_i((x_i)) = x_i,$$

i.e., a_i is the restriction of the projection to the ith factor.

Remark 10.8.2 Note that the inverse limit is characterised by the property that for any space X a map $f : X \to E_\infty$ is determined by a sequence of maps $f_i : X \to E_i$ such that $p_i \circ f_i = f_{i-1}$ under the correspondence $f_i = a_i \circ f$. Therefore, for the relative lifting problem,

$$
\begin{array}{ccc}
A & \xrightarrow{\;f''\;} & E_\infty \\
{\scriptstyle j}\downarrow & {\scriptstyle \tilde{f}}\nearrow & \downarrow {\scriptstyle a_0} \\
X & \xrightarrow{\;f'\;} & E_0,
\end{array}
$$

a lifting $\tilde{f} : X \to E_\infty$ is determined by a sequence of maps $\tilde{f}_i : X \to E_i$ such that $\tilde{f}_0 = f'$ and for each i we have a commutative square

$$
\begin{array}{ccc}
A & \xrightarrow{\;a_i \circ f''\;} & E_i \\
{\scriptstyle j}\downarrow & & \downarrow{\scriptstyle p_i} \\
X & \xrightarrow{\;\tilde{f}_i\;} & E_{i-1}.
\end{array}
$$

Thus a relative lifting problem for a pair $f : j \to a_0$ is 'factored' into a succession of relative lifting problems for maps $j \to p_i$. The idea is to obtain p_i so that the relative lifting problem for each $j \to p_i$ is hopefully simpler than the original lifting problem for $j \to a_0$. In short, the idea is to *replace* a given map $E \to B$ by a suitable inverse system of fibrations.

Definition 10.8.3 A sequence of fibrations

$$
E_0 \xleftarrow{\;p_1\;} E_1 \xleftarrow{\;p_2\;} E_2 \xleftarrow{\;p_3\;} E_3 \xleftarrow{\qquad} \cdots \tag{10.8}
$$

is said to be *convergent* if for any $n < \infty$, there is N_n such that p_i is an n-equivalence for $i > N_n$. By a *convergent factorization of a map* $f : E \to B$ we mean a sequence $\{p_i, E_i, f_i\}$ such that
(a) $E_0 = B$;
(b) for $i \geq 1$, $p_i : E_i \to E_{i-1}$ is a fibration;
(c) $f_i : E \to E_i$ are maps such that $f_0 = f$ and $f_i = p_{i+1} \circ f_{i+1}, i \geq 0$, and
(d) for each n, there exists N_n such that f_i is an n-equivalence for $i > N_n$.

Note that the first three conditions above imply that for each j, $f = p_1 \circ p_2 \circ \cdots \circ p_j \circ f_j$. Conditions (c) and (d) imply that the sequence (10.8) is convergent. That means that the infinite composition $p_1 \circ p_2 \circ \cdots$ makes sense. The following theorem, which is not very difficult to prove, makes this sense clearer.

Theorem 10.8.4 *Let* (10.8) *be convergent sequence of fibrations. Then*
(i) *the sequence* $\{p_i, E_i, a_i\}$ *is a convergent factorization of the map* $a_0 : E_\infty \to E_0$;
(ii) *if further,* $\{p_i, E_i, f_i\}$ *is a convergent factorization of a map* $f : E \to B = E_0$ *and* f' *is the inverse limit of* $f_i's$, *i.e.,* $f' : E \to E_\infty$ *be such that* $a_i \circ f' = f_i$, *then* f' *is a weak homotopy equivalence and* $f = a_0 \circ f'$.

Proof: (i) Conditions (a),(b) and (c) are checked easily. To prove (d), given n choose N so that p_i is an $(n+1)$-equivalence for $i \geq N$. We then claim that a_i is an n-equivalence for $i \geq N$. For this, it is enough to check that a_N is an n-equivalence, because $a_N = p_{N+1} \circ a_{N+1}$ and p_{i+1} is an $(n+1)$-equivalence and hence a_{N+1} is an n-equivalence and so on.

Using the converse of Theorem 10.2.17 (see Exercise 10.2.26.(iii)) it is enough to prove the following: Given a CW-pair (X, A) and maps $\alpha : A \to E_\infty, \beta : X \to E_N$ such that $\beta|_A = a_N \circ \alpha$, there exists $\beta' : X \to E_\infty$ such that $a_N \circ \beta' = \beta$ and $\beta'|_A = \alpha$. The map α corresponds to a sequence of maps $\alpha_i = a_i \circ \alpha : A \to E_i$ such that $\alpha_i = p_{i+1} \circ \alpha_{i+1}$. Likewise, getting a map $\beta' : X \to E_\infty$ with the desired property means getting a sequence of maps $\beta'_i : X \to E_i$ such that $\beta'_i|_A = \alpha_i, \beta'_i = p_{i+1} \circ \beta_{i+1}$ and $\beta'_N = \beta$.

For $i \leq N$, we take $\beta'_i = p_{i+1} \circ \cdots \circ p_N \circ \beta$. For $i > N$, we shall define it inductively. Assuming β'_i is defined, by applying Theorem 10.2.17 we first get $\beta'_{i+1} : X \to E_{i+1}$ such that $\beta'_{i+1}|_A = \alpha_{i+1}$ and $p_{i+1} \circ \beta'_{i+1} \simeq \beta_{i+1}$. Then using the fact that p_{i+1} is a fibration, we modify β'_{i+1} so that the homotopy becomes equality: $p_{i+1} \circ \beta'_{i+1} = \beta_{i+1}$. This proves (i).
(ii) By definition of $f = f_0 = a_0 \circ f'$. For each $1 \leq n < \infty$, there is $N > n$ such that both

a_N and f_N are n-equivalences. Since $a_n \circ f' = f_n$, it follows that f' is an n-equivalence for each $1 \leq n < \infty$.   ♠

Before proceeding further, the reader may recall some definition and notation introduced in Section 1.7 with respect to the notion of principal fibration.

Definition 10.8.5 By a *principal fibration of type* (G, n) *over* B we mean a principal fibration induced by a base point preserving map $f : B \to K(G, n) = K$, i.e., the pull-back $p_f : P_f \to B$ of the path fibration $PK \to K$ via f :

$$P_f = \{(b, \omega) \; : \; f(b) = \omega(1)\}; \quad p_f(b, \omega) = b.$$

Definition 10.8.6 A convergent sequence (10.8) is called a *Moore–Postnikov sequence of fibrations* if each p_i is a principal fibration of type (G_i, n_i) for some groups G_i and positive integers n_i. A Moore–Postnikov factorization of a map $f : Y' \to Y$ is a convergent factorization $\{p_i, E_i, f_i\}$ of f in which the sequence (10.8) is Moore–Postnikov. For a path connected space Y' by a Moore–Postnikov factorization of Y' we mean the same for the constant map $Y' \to y_0$.

The main result in this section is the following:

Theorem 10.8.7 *Let* $f : Y' \to Y$ *be a simple map between path-connected spaces. Then there is a Moore–Postnikov factorization* $\{p_i, E_i, f_i\}$ *of* f *such that*
(a) $E_0 = Y$, *each* E_i *is path-connected, and* $p_i : E_i \to E_{i-1}$ *is principal fibration of type* (π_i, i) *for some* π_i;
(b) *each* $(f_i)_\# : \pi_q(Y') \to \pi_q(E_i)$ *is an isomorphism for* $1 \leq q < i$ *and an epimorphism for* $q = i$; *and*
(c) *upon putting* $p'_i = p_1 \circ \cdots \circ p_i$, *we have,* $(p'_i)_\# : \pi_q(E_i) \to \pi_q(E_0) = \pi_q(Y)$ *is an isomorphism for* $q > i$ *and is a monomorphism for* $q = i$.

Two special cases of interest arise when Y is a simple, path connected space. The maps $Y \to \{y_0\}$ and the inclusion map $y_0 \subset Y$ are both simple and so we obtain two different types of Moore–Postnikov factorizations for Y.

Corollary 10.8.8 For any path connected simple space Y, there is a Moore–Postnikov factorization $\{p_i, E_i, f_i\}$, such that $\pi_q(E_i) = 0$, $q \geq i$ and $f_i : Y \to E_i$ is an i-equivalence.

Corollary 10.8.9 For any path connected simple space Y, there is a Moore–Postnikov sequence

$$Y \xleftarrow{\; p_1 \;} E_1 \xleftarrow{\; p_2 \;} E_2 \xleftarrow{\; p_3 \;} E_3 \xleftarrow{\quad\quad} \cdots$$

of fibrations such that E_i is i-connected and $p_1 \circ p_2 \circ \cdots \circ p_i : E_i \to Y$ induces an isomorphism $\pi_q(E_i) \to \pi_q(Y)$, for $q > i$.

Note that the last result is a kind of 'generalization' of the universal covering space. We shall postpone the proof of Theorem 10.8.7 to the end of this section. Right now let us see how this is going to be useful for us.

Let $f : B \to K = K(G, n + 1)$ be a map of pointed spaces. For any space W, a map $g : W \to P_f$ is nothing but a map $(g_1, g_2) : W \to B \times PK$ such that $f \circ g_1 = p \circ g_2$. Under the exponential correspondence g_2 corresponds to a homotopy $G : W \times \mathbb{I} \to K$ from the constant map at k_0 to $f \circ g_1$. Therefore, for any map $g_1 : W \to B$ we get a bijective correspondence between liftings $g : W \to E_f$ of g_1 and homotopies $G : W \times \mathbb{I} \to K$ of the constant map to $f \circ g_1$.

Let us now consider the situation where we are given a map of pairs $g : j \to p_f$ represented by the diagram

$$
\begin{array}{ccc}
A & \xrightarrow{g_2} & E_f \\
{\scriptstyle j}\downarrow & & \downarrow{\scriptstyle p_f} \\
X & \xrightarrow{g_1} & B,
\end{array}
$$

in which (X, A) is a relative CW-complex with a base point. In particular, there is a homotopy $h : A \times \mathbb{I} \to K$ from the constant map at k_0 to $f \circ g \circ j$. Let $f(g) : (A \times \mathbb{I} \cup X \times \dot{\mathbb{I}}, X \times 0) \to (K, k_0)$ be defined by

$$
f(g)(x, 0) = k_0; \quad f(g)(x, 1) = f \circ g_1(x), \quad x \in X; \quad f(g)(a, t) = h(a, t), \ a \in A, \ t \in \mathbb{I}.
$$

From Corollary 10.6.5, it follows that $f(g)$ can be extended over $X \times \mathbb{I}$ iff $\delta(f(g))^*(\iota^{n+1}) = 0 \in H^{n+2}((X, A) \times (\mathbb{I}, \dot{\mathbb{I}}); G)$, where $\iota^{n+1} \in H^{n+1}(K, k_0; G)$ is the $(n+1)$-characteristic for K. Note that we treat $\mathbb{I}$ as a CW-complex with $\{0, 1\}$ as the vertex set and a 1-cell $\mathbf{i}$ such that $\partial(\mathbf{i}) = \{1\} - \{0\}$.

Now we use the isomorphism $\zeta : H^{n+1}(X, A; G) \to H^{n+2}((X, A) \times (\mathbb{I}, \dot{\mathbb{I}}); G)$ as given in (10.6). Define $c(g) \in H^{n+1}(X, A; G)$ by the formula:

$$
\delta(f(g))^*(\iota^{n+1}) = (-1)^{n+1}\zeta(c(g)) \tag{10.9}
$$

and call it the *obstruction to lifting g through* p_f.

Remark 10.8.10
(a) In case $A = \emptyset$, i.e., when we have an absolute lifting problem of a map $g = g_1 : X \to B$, we have $f(g) : X \times \dot{\mathbb{I}} \to K$ is nothing but $f(g)(x, 0) = k_0$ and $f(g)(x, 1) = f \circ g_1(x)$. It follows that

$$
\delta(f(g))^*(\iota^{n+1}) = (-1)^{n+1}g_1^* f^*(\iota^{n+1}) = (-1)^{n+1}\zeta g_1^* f^*(\iota^{n+1})
$$

Therefore, $c(g) = c(g_1) = g_1^* f^*(\iota^{n+1})$.
(b) We have the functoriality of the obstruction with respect to mapping of the pairs: given $h : (X', A') \to (X, A)$, we have $h^*(c(g)) = c(g \circ h)$.
(c) Let us temporarily write $c(g)$ as $c(f, g)$ to indicate the role of f. Suppose $f = f_1 \circ f_2$ where $f_1 : B' \to K(G, n+1)$ and $f_2 : B \to B'$. Then $c(f_1, f_2 \circ g) = c(f, g)$. These properties are verified directly by going through the definition.

Suppose we are given two liftings $\bar{g}_i, i = 0, 1$ of g. Consider the map-pair $G = (G_1, G_2)$:

$$
G_1(x, t) = g_1(x); \ G_2(x, 0) = \bar{g}_0(x); \ G_2(x, 1) = \bar{g}_1(x); \ G_2(a, t) = g_2(a), \ a \in A, x \in X, t \in \mathbb{I}.
$$

The map-pair G is represented in the following commutative diagram:

$$
\begin{array}{ccc}
A \times \mathbb{I} \cup X \times \dot{\mathbb{I}} & \xrightarrow{G_2} & E_f \\
{\scriptstyle j'}\downarrow & {\scriptstyle \hat{G}}\nearrow & \downarrow{\scriptstyle p_f} \\
X \times \mathbb{I} & \xrightarrow{G_1} & B
\end{array}
$$

Now, $\bar{g}_0$ is homotopic to $\bar{g}_1$ relative to g iff G can be lifted, i.e., there is a map $\hat{G}$ represented by the dotted line and fitting the above commutative diagram. The obstruction for this is an element $c(G) \in H^n((X,A) \times (\mathbb{I}, \dot{\mathbb{I}}); G)$. We define the *difference element* $d(\bar{g}_0, \bar{g}_1) \in H^n(X, A; G)$ by the formula

$$c(G) = (-1)^{n+1} \zeta(d(\bar{g}_0, \bar{g}_1)).$$

Thus, for instance, $\delta f(G)^*(\iota^{n+1}) = -\zeta \circ \zeta(d(\bar{g}_0, \bar{g}_1))$.

It also follows that $\bar{g}_0$ is homotopic to $\bar{g}_1$ relative to g iff $d(\bar{g}_0, \bar{g}_1) = 0$. It is routine to verify that the difference is functorial and satisfies:

$$d(\bar{g}_0, \bar{g}_1) + d(\bar{g}_1, \bar{g}_2) = d(\bar{g}_0, \bar{g}_2)$$

for any three lifts $\bar{g}_i, i = 0, 1, 2$ of g. The proofs of the following facts are also similar to that of 10.5.10 and will be left as an exercise to the reader.

Theorem 10.8.11 *Given a map pair $g : j \to p_f$ and a lift $\bar{g}_0$ of g and an element $u \in H^{n-1}(X, A; G)$, there is a lift $\bar{g}_1$ of g such that $d(\bar{g}_0, \bar{g}_1) = u$.*

Remark 10.8.12 Now consider the constant map $c : E \to E_0 = \{*\}$. Then the lifting problem $f : j \to c$ is the same as the extension problem for maps $g : A \to E$ over X. Therefore the Moore–Postnikov decomposition of $E \to \{*\}$ decomposes this problem into a sequence of obstructions in $H^{n+1}(X, A; \pi_n(Y))$. Assuming Y is $(n-1)$-connected, the primary obstruction is a single element $c(g) \in H^{n+1}(X, A; \pi_n(Y))$. It can be easily seen that this obstruction coincides with $\gamma^{n+1}(g)$ that we have defined earlier. This is precisely what we meant by saying that Moore–Postnikov decomposition gives an alternative approach to obstructions for extension problems. Various results that we have proved in Section 10.6 can be reproved using this technique.

The rest of this section will be devoted to the proof of Theorem 10.8.7, which the reader may choose to study at her leisure.

Definition 10.8.13 An *n-factorization* of a map $f : E \to B$, where E, B are path connected spaces, we mean a factorization $f = p' \circ g$ where
(a) $p' : E' \to B$ is a fibration with E' path connected, $g : E \to E'$ is such that $f = p' \circ g$;
(b) $g_\# : \pi_q(E) \to \pi_q(E')$ is an isomorphism for $q < n$ and an epimorphism for $q = n$; and
(c) $p'_\# : \pi_q(E') \to \pi_q(B)$ is an isomorphism for $q > n$ and a monomorphism for $q = n$.

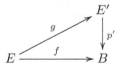

Example 10.8.14 For $n = 1$, this would mean that $p'_\# : \pi_1(E') \approx f_\# \pi_1(E)$. We can achieve this by taking the covering $p' : E' \to B$ of B corresponding to the subgroup $f_\# \pi_1(E) \subset \pi_1(B)$. Then from the theorem on lifting criterion 3.3.12, there exists $g : E \to E'$ such that $p \circ g = f$. Conditions (b) and (c) are automatically satisfied. For $n > 1$, we would like to imitate this construction. However, there seems to be some need to forego certain 'over-simplification' that we could afford in the case of $n = 1$. Recall that a covering space over a space was constructed as a space of homotopy classes of paths starting at a base point. For $n > 1$, we need to work directly in the space of paths and not in the homotopy classes.

We shall now need the generalization of the notion of an n-characteristic element for pairs of topological spaces (X, A).

Definition 10.8.15 An element $u \in H^n(X, A; G)$ is called an *n-characteristic* if the following holds:

(a) If $n = 1$ then $j_\# \pi_1(A)$ is a normal subgroup of $\pi_1(X)$ such that the composite is an isomorphism:

$$\pi_1(X)/j_\# \pi_1(A) \xrightarrow{\varphi} H_1(X)/j_*(H_1(A)) \longrightarrow H_1(X, A) \xrightarrow{h(u)} G.$$

(b) If $n > 1$, then the following composite is an isomorphism:

$$\pi_n(X, A) \xrightarrow{\varphi} H_n(X, A) \xrightarrow{h(u)} G.$$

In the rest of this section, G will denote $\pi_1(X)/j_\# \pi_1(A)$ for $n = 1$ and $\pi_n(X, A)$ for $n > 1$.

Lemma 10.8.16 Let $j : A \subset X$ denote the inclusion of path connected spaces where (X, A) is $(n-1)$-connected. Assume that j is simple.
(a) There exists $u \in H^n(X, A; G)$ which is an *n*-characteristic for (X, A).
(b) There is an *n*-factorization of $j : A \hookrightarrow X$, i.e., $j = p^X \circ g$, where $p^X : E^X \to X$ is a principal fibration of type (G, n).

Proof: (a) For $n = 1$, the kernel of the composite

$$\pi_1(X) \xrightarrow{\varphi} H_1(X) \longrightarrow H_1(X)/j_*H_1(A) \longrightarrow 0$$

is the normal subgroup generated by $j_\# \pi_1(A)$. Since we are assuming that $j_\# \pi_1(A)$ is normal, we have an isomorphism

$$\pi_1(X)/j_\# \pi_1(A) \to H_1(X)/j_*H_1(A).$$

By the universal coefficient theorem for cohomology, we also have an isomorphism $h : H^1(X, A; G) \approx \operatorname{Hom}(H_1(X, A), G)$. Therefore, by taking $G = \pi_1(X)/j_\# \pi_1(A)$, we get the element u as desired. For $n > 1$, the same argument works except that we need to appeal to the relative Hurewicz isomorphism Theorem.

(b) By Theorem 10.7.5, there is a unique homotopy class of a map $\phi : (X, A) \to (Y, y_0)$ where Y is a $K(G, n)$ space with a base point y_0 such that $\phi^*(\iota_n) = u$ is a characteristic element. Let $p : PY \to Y$ be the path fibration and $p^X : E^X \to X$ be the pull-back fibration $\phi^*(p)$.

Recall that $PY = \{\omega : \mathbb{I} \to Y : \omega(0) = y_0\}, p(\omega) = \omega(1)$. We also know that

$$E^X = \{(x, \omega) \in X \times PY : \omega(1) = \phi(x)\}, \quad p^X(x, \omega) = x.$$

Also each fibre is homeomorphic to ΩY. We shall identify the fibre $F = x_0 \times \Omega Y$ over x_0 with ΩY itself. Since $\phi(A) = \{y_0\}$, it follows that $E^A := (p^X)^{-1}(A) = A \times \Omega Y$. Let c be the constant path at y_0, and take $s(a) = (a, c)$, and $g = j' \circ s$, where $j' : E^A \to E^X$ is the inclusion. Then we have, $p' \circ g = j$. Let $\psi : E^A \to \Omega Y$ be defined by $\psi(a, \omega) = \omega$. Observe that ψ restricts to the identity map $F = \Omega Y \to \Omega Y = F$.

We claim that this gives an *n*-factorization of j.

Let us first see why E^X is path connected. For $n > 1$ this is an easy consequence of the fact that ΩY is path connected and X is path connected. For $n = 1$, again because X is path connected, it is enough to show that any member (x_0, ω) of the fibre F over x_0 can be joined to (x_0, c) by a path in E^X. This is where you have to use the fact that $\phi_\# : \pi_1(X) \to \pi_1(Y)$ is surjective. (See Remark 10.7.7.) We leave the details to you.

By the homotopy exact sequence of the fibration, since ΩY is of type $(G, n-1)$, p^X satisfies condition (c). We need to prove condition (b), viz., $g_\#$ is an isomorphism for $i < n$ and an epimorphism for $i = n$.

If $\eta^A : \Omega Y \to E^A$ is the inclusion map $\omega \mapsto (x_0, \omega)$, then we have a direct sum decomposition

$$\pi_i(E^A) \approx \eta_\#^A \pi_i(\Omega Y) \oplus g_\# \pi_i(A). \tag{10.10}$$

(Strictly speaking, for $i = 1$ this is a direct product. However, we can use the additive notation in moderation.)

The following commutative diagrams of maps sum up the entire situation.

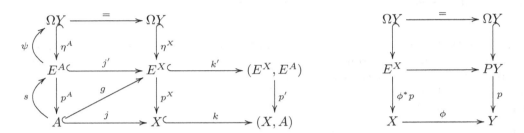

Consider the composite of homomorphisms

$$\pi_i(X, A) \xrightarrow[\approx]{p_\#'^{-1}} \pi_i(E^X, E^A) \xrightarrow{\partial} \pi_{i-1}(E^A) \xrightarrow{\psi_\#} \pi_{i-1}(\Omega Y)$$

and call it λ. We then have a diagram

$$\cdots \longrightarrow \pi_i(A) \xrightarrow{j_\#} \pi_i(X) \xrightarrow{k_\#} \pi_i(X, A) \xrightarrow{\partial} \pi_{i-1}(A) \longrightarrow \cdots \tag{10.11}$$
$$\qquad\quad\; g_\# \downarrow \qquad\quad = \downarrow \qquad\quad \lambda \downarrow \qquad\quad g_\# \downarrow$$
$$\cdots \longrightarrow \pi_i(E^X) \xrightarrow{p_\#^X} \pi_i(X) \xrightarrow{\bar{\partial}} \pi_{i-1}(\Omega Y) \xrightarrow{\eta_\#^X} \pi_{i-1}(E^X) \longrightarrow \cdots$$

in which, it is easily checked that the first two squares are commutative.

In the third square, we claim that $g_\# \circ \partial = -j_\# \circ \lambda$. For $i = 1$, note that both the sides are trivial maps. Consider the case $i > 1$. First of all, we have the following commutative diagram

$$\pi_i(E^X, E^A) \xrightarrow{\partial} \pi_{i-1}(E^A) \xrightarrow{\psi_\#} \pi_{i-1}(\Omega Y) \xrightarrow{\eta_\#^X} \pi_{i-1}(E^X)$$
$$\approx \downarrow p_\#' \qquad\qquad s_\# \left(\begin{array}{c} \\ \end{array}\right) p_\#^A \qquad g_\#$$
$$\pi_i(X, A) \xrightarrow{\partial} \pi_{i-1}(A)$$

from which it follows that $g_\# \circ \partial = \eta_\#^X \circ \lambda$. From (10.10), we have for $z \in \pi_{i-1}(E^A)$,

$$z = \eta_\#^A \psi_\# z + s_\# p_\#^A z.$$

Therefore for $w \in \pi_i(E^X, E^A)$, taking $z = \partial w$, we get

$$0 = j_\#' \partial w = j_\#'(\eta_\#^A \psi_\# \partial w + s_\# p_\#^A \partial w) = \eta_\#^X \psi_\# \partial w + g_\# \partial \circ p_\#^X w.$$

By the definition of λ, we have $\lambda \circ p'_{\#} = \psi_{\#} \circ \partial$. Therefore,

$$0 = (\eta^X_{\#} \circ \lambda + g_{\#} \circ \partial) \circ p'_{\#}(w).$$

Since $p'_{\#}$ is an isomorphism, the claim is established.

Finally, we claim that λ is an isomorphism in dimensions $\leq n$. Since Y is a $K(G, n)$ and $(X < A)$ is $(n - 1)$-connected, we need to consider the case $i = n$ alone. Note that there is a commutative diagram:

$$
\begin{array}{ccccc}
\pi_i(X, A) & \xrightarrow{\phi_{\#}} & \pi_i(Y) & \xrightarrow[\approx]{\bar{\partial}} & \pi_{i-1}(\Omega Y) \\
\approx \downarrow \varphi & & \approx \downarrow \varphi & & \\
H_i(X, A) & \xrightarrow{\phi_*} & H_i(Y) & & \\
h(u) \downarrow \approx & & \approx \downarrow h(\iota_i) & & \\
G & =\!\!=\!\!= & G & &
\end{array}
$$

in which the various φ denote the corresponding Hurewicz map, and h denotes the homomorphism occurring in the universal coefficient theorem. (Recall that $u = \phi^*(\iota_n)$.) The composite of all maps in the first row is easily seen to be equal to λ and therefore λ is an isomorphism.

Going back to diagram (10.11) and using Five lemma, it follows that $g_{\#}$ is an isomorphism for $i < n$ and an epimorphism for $i = n$ as desired. ♠

Exercise 10.8.17 In the proof of part (b) of Lemma 10.8.16, complete the proof that E^X is path connected.

10.9 Computation with Lie Groups and Their Quotients

We shall use the notation $\mathbb{K}$ to denote the fields $\mathbb{R}, \mathbb{C}$ or the skew field of the quaternions $\mathbb{H}$. Accordingly $c = \dim_{\mathbb{R}} \mathbb{K}$ will denote the number $1, 2$ or 4, the vector space dimension of $\mathbb{K}$ over $\mathbb{R}$. We shall denote by $\mathbb{K}^{\infty}$ the countable infinite direct sum of copies of $\mathbb{K}$. The elements

$$\mathbf{e}_i = (0, \ldots, 0, 1, 0, \ldots), \quad i = 0, 1, 2, \ldots$$

with all entries zero except in the ith place, wherein it is equal to 1, form a basis for $\mathbb{K}^{\infty}$. Notice that for each n, $\mathbb{K}^n$ is identified with the subspace of $\mathbb{K}^{\infty}$ spanned by $\{\mathbf{e}_0, \ldots, \mathbf{e}_{n-1}\}$. With each $\mathbb{K}^n$ carrying the Cartesian product topology, we topologised $\mathbb{K}^{\infty}$ with the weak topology with respect to the family $\{\mathbb{K}^n, n \geq 1\}$, viz., a subset $F \subset \mathbb{K}^{\infty}$ is closed iff each $F \cap \mathbb{K}^n$ is closed in $\mathbb{K}^n$. Note that this topology (which is called weak topology) is finer than the product topology. Indeed, it is easily seen that $\mathbb{K}^{\infty}$ can be triangulated. In particular, the topology is compactly generated also. As a consequence, it follows that every compact subset of $\mathbb{K}^{\infty}$ is contained in some $\mathbb{K}^n$. Moreover, the standard inner product on each of $\mathbb{K}^n$ extends to an inner product on $\mathbb{K}^{\infty}$ as well. However the metric topology does not cocoincide with the weak topology and indeed is weaker than the weak topology (see Exercise 2.2.23). The unit sphere in $\mathbb{K}^{\infty}$ is denoted by $\mathbb{S}^{\infty}_{\mathbb{K}}$. Under the identification of $\mathbb{K}$ with $\mathbb{R}^c$, all these three infinite spheres are topologically the same (and so we can drop the lower suffix $\mathbb{K}$). The CW-topology that we have given earlier to $\mathbb{S}^{\infty}$ coincides with the subspace topology on $\mathbb{S}^{\infty} \subset \mathbb{R}^n$. As an immediate consequence, we obtain the following result.

Theorem 10.9.1 *For $n \geq 0$, $\pi_n(\mathbb{S}^{\infty}) = (0)$. In particular, $\mathbb{S}^{\infty}$ is contractible.*

Exercise 10.9.2 Indeed, show that $\mathbb{S}^\infty$ is contractible by writing down a specific contraction.

Let $\mathcal{O}_n := \mathcal{O}_n(\mathbb{K})$ denote the orthogonal group with respect to the standard inner product on $\mathbb{K}^n$. We then have

$$\cdots \mathcal{O}_n \subset \mathcal{O}_{n+1} \subset \cdots$$

and we define

$$\mathcal{O} := \cup_n \mathcal{O}_n$$

to be the infinite orthogonal group over $\mathbb{K}$. This is topologised again with the weak topology. With the usual multiplication, it becomes a topological group. It is customary to denote

$$\mathcal{O}_n(\mathbb{R}) := O(n); \mathcal{O}_n(\mathbb{C}) = U_n; \mathcal{O}_n(\mathbb{H}) = Sp(n).$$

Similarly, we shall denote by O, U, Sp, the three infinite orthogonal groups. Also, for the case $\mathbb{K} = \mathbb{R}, \mathbb{C}$ we have the special orthogonal groups

$$SO(n) = \{A \in O(n) \ : \ \det A = 1\}; \quad SU(n) = \{A \in U(n) \ : \ \det A = 1\}.$$

For $n = 1$, we have $SO(1) = SU(1) = (1); O(1) = \mathbb{Z}_2 = \{-1, 1\}$, and $U(1) = SO(2) = \mathbb{S}^1$ as a group of unit complex numbers and $Sp(1) = SU(2) = \mathbb{S}^3$, the group of unit quaternions.

We also have two exact sequences:

$$SO(n) \xrightarrow{\hspace{3cm}} O(n) \xrightarrow{\quad \det \quad} \mathbb{Z}_2;$$

$$SU(n) \xrightarrow{\hspace{3cm}} U(n) \xrightarrow{\quad \det \quad} \mathbb{S}^1.$$

We shall use matrix representations for elements of these groups with respect to the standard basis. For instance, I_k will denote the identity matrix of size $k \times k$.

Steifel Varieties Fix $1 \le k \le n$. Consider the smooth action of $\mathcal{O}_n$ on $\mathbb{K}^{n \times k}$ (the space of $n \times k$ matrices over $\mathbb{K}$), given by the left multiplication:

$$(g, A) \mapsto gA.$$

The orbit of the element $p = (\mathbf{e}_1, \ldots, \mathbf{e}_k)$ is called the Steifel variety $V_{n,k}(\mathbb{K})$. This can be described as the set of all orthonormal k-frames in $\mathbb{K}^n$, i.e.,

$$\{(\mathbf{v}_1, \ldots, \mathbf{v}_k) \text{ such that } \langle \mathbf{v}_i, \mathbf{v}_j \rangle = \delta_{ij}\}.$$

The quotient map $\alpha_n : \mathcal{O}(n) \to V_{n,k}$ given by $\alpha_n(g) = (g(\mathbf{e}_0), \ldots, g(\mathbf{e}_{k-1}))$ is then a locally trivial fibre bundle and hence a fibration. The isotropy group of the element p is nothing but the subgroup of $\mathcal{O}_n$ consisting of elements of the form

$$\begin{bmatrix} I_k & 0 \\ 0 & A \end{bmatrix},$$

where $A \in \mathcal{O}_{n-k}$. We shall denote this subgroup by $\mathcal{O}(, n-k)$. The inclusions

$$\cdots \subset \mathbb{K}^n \subset \mathbb{K}^{n+1} \subset \cdots$$

induce compatible inclusions of Steifel varieties

$$\cdots \subset V_{n,k}(\mathbb{K}) \subset V_{n+1,k}(\mathbb{K}) \subset \cdots$$

and of the corresponding fibrations

$$\mathcal{O}(k) \subset \mathcal{O}(n) \xrightarrow{\alpha} V_{n,k}(\mathbb{K})$$

as well and thereby yields a fibration

$$\mathcal{O}_k(\mathbb{K}) \hookrightarrow \mathcal{O}(\mathbb{K}) \xrightarrow{\alpha} V_k(\mathbb{K})$$

over infinite Steifel variety of k-frames in $\mathbb{K}^\infty$.

Example 10.9.3
(a) For $k = n$, we have $\mathcal{O}(n) = V_{n,n}(\mathbb{K})$.
(b) For $k = 1$, we have $V_{n,1}(\mathbb{K}) = \mathbb{S}^{cn-1}, V_{n,1}(\mathbb{R}) = \frac{O(n)}{O(n-1)} = \frac{SO(n)}{Id_1 \times SO(n-1)}$, and $V_{n,1}(\mathbb{C}) = \frac{U(n)}{Id_1 \times U(n-1)} = \frac{SU(n)}{Id_1 \times SU(n-1)}$.
(c) For $k = n - 1$, $SO(n) = V_{n,n-1}(\mathbb{R}^n)$; $SU(n) = V_{n,n-1}(\mathbb{C}^n)$.

Grassmann Varieties Fix integers $1 \leq k \leq n$. Consider the set $G_{n,k}(\mathbb{K})$ of all k-dimensional, $\mathbb{K}$-linear subspaces of $\mathbb{K}^n$. The group $\mathcal{O}(n)$ acts on this set on the left as follows:

$$(g, V) \mapsto g(V) \subset \mathbb{K}^n.$$

Clearly, this action is transitive and hence $G_{n,k}$ can be identified with the homogeneous space $O(n)/H$ of left-cosets of some closed subgroup H. The subgroup H is the isotropy group of the subspace, say, $\mathbb{K}^k \times 0 \subset \mathbb{K}^n$ and is easily seen to be equal to $\mathcal{O}(k) \times \mathcal{O}(n-k)$ which we shall denote by $\mathcal{O}(k, n-k)$. Thus we have the following commutative diagram:

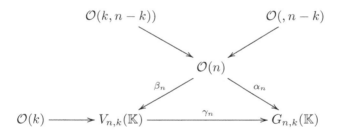

where $\alpha_n, \beta_n, \gamma_n$ are all locally trivial fibre bundle maps. As seen before, these are compatible with the inclusions

$$\cdots \subset \mathbb{K}^n \subset \mathbb{K}^{n+1} \subset \cdots$$

and hence upon taking the union, give the fibre bundle

$$\mathcal{O}(k) \hookrightarrow V_k(\mathbb{K}) \to G_k(\mathbb{K})$$

over the infinite Grassmann.

Remark 10.9.4

1. Put $\tau = \begin{bmatrix} 0 & I_k \\ I_{n-k} & 0 \end{bmatrix}$. The conjugation $A \mapsto \tau A \tau^t$ on $\mathcal{O}(n)$ takes the subgroup $\mathcal{O}(k, n-k)$ to $\mathcal{O}(n-k, k)$ and hence induces a homeomorphism of $G_{n,k}(\mathbb{K})$ with $G_{n,n-k}(\mathbb{K})$. (This is the reason why some authors use the 'symmetric' notation $G_{n-k,k}$ in place of our notation $G_{n,k}$.) Therefore, while computing homotopy groups, etc., we can always assume that $n \geq 2k$.

2. Assume $n \geq 2k$. The homotopy

$$(A, t) \mapsto \begin{bmatrix} (\cos \pi t/2)A \\ (\sin \pi t/2)I_k \\ 0 \end{bmatrix}, \quad 0 \leq t \leq 1,$$

shows that the inclusion map $\mathcal{O}(k) \to V_{n,k}(\mathbb{K})$ is null-homotopic. From the homotopy exact sequence of the fibration

$$\mathcal{O}(k) \to V_{n,k}(\mathbb{K}) \to G_{n,k}(\mathbb{K})$$

we obtain:

Theorem 10.9.5 *If $n \geq 2k$, we have*

$$\pi_q(G_{n,k}(\mathbb{K})) \approx \pi_q(V_{n,k}(\mathbb{K})) \oplus \pi_{q-1}(\mathcal{O}(k)), \quad q \geq 1.$$

It follows that the homotopy groups of the Grassmann varieties are completely determined by those of Steifel varieties and orthogonal groups. We shall now extract information on the homotopy groups of these spaces.

Let us consider the action of $\mathcal{O}(n+1)$ on the unit sphere $\mathbb{S}^{c(n+1)-1}$ given by: $(A, \mathbf{v}) \mapsto A\mathbf{v}$. Given any $v \in \mathbb{S}^{c(n+1)-1}$, there exists $A \in \mathcal{O}(n+1)$ whose last column is equal to v, i.e., $A\mathbf{e}_{n+1} = v$. This just means that the action is transitive. Clearly the isotropy group of $\mathbf{e}_{n+1}$ is the subgroup $\mathcal{O}(n)$. Thus we have the locally trivial fibre bundle map

$$\mathcal{O}(n) \subset \mathcal{O}(n+1) \to \mathbb{S}^{c(n+1)-1}.$$

The long homotopy exact sequence of this fibration therefore yields:

Theorem 10.9.6 *The inclusion induced homomorphism $\pi_q(\mathcal{O}(n)) \to \pi_q(\mathcal{O}(n+1))$ is an isomorphism for $q < (c(n+1) - 2$ and an epimorphism for $q = c(n+1) - 2$ (equivalently, $(\mathcal{O}(n+1), \mathcal{O}(n))$ is $(c(n+1) - 2)$-connected.*

Corollary 10.9.7 *The pairs $(SO(n+1), SO(n))$ are $(n-1)$-connected.*

Remark 10.9.8 Thus fixing an integer $q > 0$, we choose $n \geq \lceil (q+2)/c \rceil$. It follows that the inclusion induced homomorphism $\pi_q(\mathcal{O}(n)) \to \pi_q(\mathcal{O})$ is an isomorphism for all such n. In other words, the homotopy group π_q has stabilized at the stage $n = \lceil q+2/c \rceil$. This is called the stable range for $\pi_q(\mathcal{O}(n))$. For instance, in the case of real orthogonal groups, the stable range for π_q is $[q+2, \infty)$. We state this as follows:

Corollary 10.9.9 $(\mathcal{O}, \mathcal{O}(n))$ is $c(n+1) - 2$ connected.

Now, using the fibration

$$\mathcal{O}(n) \to \mathcal{O}(n+r) \to V_{n+r,r}(\mathbb{K})$$

we get

Theorem 10.9.10 $\pi_q(V_{n+r,r}(\mathbb{K})) = (0)$ *for $q \leq c(n+1) - 2$.*

This of course implies immediately that $\pi_i(V_k(\mathbb{K})) = 0$ for all i, upon taking the limit as $n \to \infty$.

Remark 10.9.11 Consider the inclusion map $j : V_{n,k}(\mathbb{K}) \to V_{n+1,k+1}(\mathbb{K})$ given by

$$(\mathbf{v}_1, \ldots, \mathbf{v}_k) \mapsto (\mathbf{v}_1, \ldots, \mathbf{v}_k, \mathbf{e}_{n+1}).$$

This is actually the inclusion map of the fibration

$$q : V_{n+1,k+1}(\mathbb{K}) \to \mathbb{S}^{c(n+1)-1}$$

given by the projection to the last coordinate vector. The homotopy exact sequence of this yields:

Theorem 10.9.12 *The inclusion induced homomorphism*

$$j_\# : \pi_i(V_{n,k}(\mathbb{K})) \to \pi_i(V_{n+1,k+1}(\mathbb{K}))$$

is an isomorphism for $i \leq c(n+1) - 2$ and an epimorphism for $i = c(n+1) - 1$.

So far, we have not used anything more than the general homotopy properties of the fibrations over a sphere. Let us take a closer look at some of the special cases. In the fibration

$$SO(n) \hookrightarrow SO(n+1) \xrightarrow{p_n} \mathbb{S}^n,$$

the first non trivial and interesting thing is happening at the nth level. The kernel of the inclusion induced epimorphism $\pi_{n-1}(SO(n)) \to \pi_{n-1}(SO(n+1))$ is a cyclic group generated by the element $\omega_{n-1} := \partial[\xi_n]$, where $[\xi_n] \in \pi_n(\mathbb{S}^n)$ is the homotopy class of the identity map and $\partial : \pi_n(\mathbb{S}^n) \to \pi_{n-1}(SO(n))$ is the 'boundary' homomorphism of the long exact sequence. Let us get an explicit description of ω_{n-1}.

Consider the function $f : \mathbb{S}^n \to O(n+1)$ given by

$$f(x)(y) = y - 2\langle x, y \rangle x, \tag{10.12}$$

i.e., $f(x)$ is the reflection in the hyperplane perpendicular to x. The 'adjoint' of this map f is the map $\hat{f} : \mathbb{S}^n \times \mathbb{S}^n \to \mathbb{S}^n$ given by

$$\hat{f}(x, y) = f(x)y.$$

Fixing x or y we get two maps

$$_x\hat{f} := \hat{f}(x, -), \hat{f}_y := \hat{f}(-, y) : \mathbb{S}^n \to \mathbb{S}^n.$$

Since $\mathbb{S}^n$ is connected, it follows that the degree of $_x\hat{f}$ is independent of x. Similarly for $\hat{f}_y$. The pair $(deg\, \hat{f}_y, deg\, _x\hat{f})$ is called the *bi-degree* of $\hat{f}$.

Lemma 10.9.13 The bi-degree of $\hat{f}$ is equal to $(1 + (-1)^{n+1}, -1)$.

Proof: Since $_x\hat{f}$ is the reflection in a hyperplane, clearly its degree is -1. So, we need to compute the degree of $\hat{f}_y$. We are free to choose $y \in \mathbb{S}^n$ and let us choose $y = -\mathbf{e}_n$. Write

$$g(x) = f(x)(-\mathbf{e}_n) = 2\langle x, \mathbf{e}_n \rangle x - \mathbf{e}_n.$$

We have to show that $\deg g = 1 + (-1)^{n+1}$. Clearly $g(x) = g(-x)$. (This immediately says that g factors through the double covering

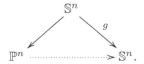

Hence its degree must be 0 if n is even and some even number if n is odd. So, we may concentrate on the case when n is odd. Moreover, $g(\mathbb{S}^{n-1}) = \{-\mathbf{e}_n\}$. Therefore, the degree of g is equal to the sum of the degrees of g restricted to the upper and lower hemispheres $\mathbb{D}^n_\pm$. Put $g_\pm = g|_{\mathbb{D}^n_\pm}$. Then $g_- = g \circ \alpha$, where $\alpha : \mathbb{S}^n \to \mathbb{S}^n$ is the antipodal map. We conclude that

$$\deg g = (1 + (-1)^{n+1})\deg g_+ = 2\deg g_+.$$

It is easily checked that $g_+ : \operatorname{int}\mathbb{D}^n_+ \to \mathbb{S}^n \setminus \{-\mathbf{e}_n\}$ is a diffeomorphism. Therefore $\deg g_+ = \pm 1$.

Now note that $g(x) \neq -x$ for any $x \in \mathbb{D}^n_+$. Therefore, there is a unique shortest path from $g(x)$ to x which gives a homotopy of g_+ with the inclusion map. Therefore $\deg g_+ = 1$. Therefore, $\deg g = 2$. This completes the proof of the theorem. ♠

We define $f_n : \mathbb{S}^n \to SO(n+1)$ by:

$$f_n(x) = f(x)f(\mathbf{e}_0). \tag{10.13}$$

Observe that $f_n|_{\mathbb{S}^{n-1}} = f_{n-1}. : \mathbb{S}^{n-1} \to SO(n)$. Moreover, for all $x \in \mathbb{S}^n$, we have

$$p_n \circ f_n(x) = f_n(x)(\mathbf{e}_n) = f(x)f(\mathbf{e}_0)(\mathbf{e}_n) = f(x)(\mathbf{e}_n) = -g(x).$$

Therefore $p_n \circ f_n : (\mathbb{D}^n_+, \mathbb{S}^{n-1}) \to (\mathbb{S}^n, \mathbf{e}_n)$ which equals $\alpha \circ g_+$ has degree $(-1)^{n+1}$. This just means that $(p_n)_\#([f_n]) = (-1)^{n+1}[\iota_n]$. Note that the boundary homomorphism in the homotopy exact sequence of the fibration p is given by

$$\partial' = \partial \circ (p_n)_\#^{-1} : \pi_n(\mathbb{S}^n, \mathbf{e}_n) \to \pi_{n-1}(SO(n)),$$

where, $\partial([f_n] - [f_n|_{\mathbb{S}^{n-1}}]) = [f_{n-1}]$. We put $\omega_{n-1} = \partial'[\iota_n]$. Then

$$(-1)^{n+1}\omega_{n-1} = (-1)^{n+1}\partial'[\iota_n] = \partial[f_n] = [f_{n-1}]. \tag{10.14}$$

We conclude:

Theorem 10.9.14 *The map $f_{n-1} : \mathbb{S}^{n-1} \to SO(n)$ represents the element $(-1)^{n+1}\omega_{n-1} \in \pi_{n-1}(SO(n))$.*

We have seen that $p_n \circ f_n = -g$. Therefore, from Lemma 10.9.13, it follows that $p_n \circ f_n : \mathbb{S}^n \to \mathbb{S}^n$ has degree $(-1)^{n+1}(1 + (-1)^{n+1}) = 1 + (-1)^{n+1}$. Therefore, for n even the degree of $p_{n-1} \circ f_{n-1} : \mathbb{S}^{n-1} \to \mathbb{S}^{n-1}$ is 2.

As an immediate corollary we have:

Theorem 10.9.15 *For n even, $\omega_{n-1} \in \pi_{n-1}(SO(n))$ generates an infinite cyclic sub-group.*

Let us put this information to some immediate use. Since $SO(2) = \mathbb{S}^1$, the exact sequence of the fibration $p_3 : SO(3) \to \mathbb{S}^2$ gives:

$$0 \longrightarrow \pi_2(SO(3)) \xrightarrow{(p_3)_\#} \pi_2(\mathbb{S}^2) \xrightarrow{\partial'} \pi_1(SO(2)) \xrightarrow{\iota_\#} \pi_1(SO(3)) \longrightarrow 0.$$

It follows that ∂' is injective and hence $(p_3)_\# = 0$. Hence $\pi_2(SO(3)) = (0)$. Also since the image of ∂' is equal to $2\mathbb{Z} \subset \mathbb{Z} = \pi_1(SO(2))$, it follows that $\pi_1(SO(3)) = \mathbb{Z}_2$. Since $(SO(n), SO(2))$ is 1-connected and $(SO(n), SO(3))$ is 2-connected, it follows that:

Corollary 10.9.16 *For all $n \geq 3$, $\pi_1(SO(n)) = \mathbb{Z}_2$, and $\pi_2(SO(n)) = (0)$.*[1]

_____

[1]There is a general result due to E. Cartan that says $\pi_2(G) = 0$ for any Lie group G. Cartan proved this using deep results in Lie theory. This was generalized by Browder [Browder, 1965] for any H-space G using only homological arguments.

To extract information on $\pi_3(SO(n))$ is a little more difficult. Recall that representing $\mathbb{S}^3$ as the space of unit quaternions, and taking the conjugation action on the space of purely imaginary quaternions, we get a homeomorphism of the 3-dimensional real projective space with $SO(3)$. Indeed, if $\tau(x)$ denotes the conjugation

$$y \mapsto x^{-1}yx,$$

then $\tau(x) \in SO(4)$, and $\tau(x)(\mathbf{e}_0) = \mathbf{e}_0$. Therefore, $\tau(x) \in 1 \times SO(3) \subset SO(4)$. Moreover, it is easily seen that $\tau : \mathbb{S}^3 \to SO(3)$ is a homomorphism with its kernel equal to $\{-1, 1\}$. Therefore it induces an injective homomorphism $\mathbb{P}^3 \to SO(3)$, which is a submersion and hence an open mapping. Since $\mathbb{P}^3$ is compact and $SO(3)$ is connected, it follows that this is surjective as well. In particular, we know that $\tau_\# : \pi_3(\mathbb{S}^3) \to \pi_3(SO(3))$ is an isomorphism. On the other hand, the fibration

$$SO(3) \hookrightarrow SO(4) \xrightarrow{p_3} \mathbb{S}^3$$

is actually a 'trivial' fibration, i.e., there is a homeomorphism

$$h : SO(4) \to SO(3) \times \mathbb{S}^3$$

such that $\pi \circ h = p_3$. (This follows from general considerations about principal G-bundles, once we have a cross section for p_3, viz., a continuous map $s : \mathbb{S}^3 \to SO(4)$ such that $p_3 \circ s = Id_{\mathbb{S}^3}$. However, here we can directly establish this by elementary considerations, as follows.

Recall that we denote the conjugation of an element $x \in \mathbb{H}$ by $\bar{x}$. Let $\mathbf{k}$ denote the element $(0, 0, 0, 1)$. Define

$$s(x) = L_{x\bar{\mathbf{k}}}$$

Then $s(x)$ a norm preserving linear map $\mathbb{R}^4 \to \mathbb{R}^4$. Check that $p_3(s(x)) = x\bar{\mathbf{k}}\mathbf{k} = x$. Therefore s is a section of p_3.

Now every element $A \in SO(4)$ can be expressed as

$$A = L_{(A\mathbf{k})\bar{\mathbf{k}}} \circ (L_{\mathbf{k}\overline{A\mathbf{k}}} \circ A),$$

in a unique way, so that $L_{\mathbf{k}\overline{A\mathbf{k}}} \circ A \in SO(3)$ and $L_{A(\mathbf{k})\bar{\mathbf{k}}} = s(A\mathbf{k}) \in s(\mathbb{S}^3)$. Now check that the function

$$A \mapsto (A\mathbf{k}, L_{\mathbf{k}\overline{A\mathbf{k}}} \circ A)$$

defines the required homeomorphism $h : SO(4) \to \mathbb{S}^3 \times SO(3)$.

It follows that

$$\pi_3(SO(4)) \approx \pi_3(SO(3) \times \mathbb{S}^3) = \pi_3(SO(3)) \times \pi_3(\mathbb{S}^3) = \mathbb{Z} \times \mathbb{Z}.$$

Indeed, if $\xi = \tau_\#[\iota_3]$ and $\eta = s_\#[\iota_3]$, then $\{i_\#(\xi), \eta\}$ is a basis for the free abelian group $\pi_3(SO(4))$. Let us write

$$h_\#(-\omega_3) = [h \circ f_3] = ai_\#(\xi) + b\eta.$$

We want to determine the integers a, b. For this we notice that

$$f_3(x) = L_x \circ R_x$$

where L_x and R_x denote the left and right multiplication by the quaternion x. (Verify this directly by writing down the 4×4 matrices of all the transformations involved.) In particular, we have $f_3(x)(y) = xyx = x^2(x^{-1}yx) = (L_{x^2} \circ \tau(x))(y)$, i.e., $f_3(x) = L_{x^2} \circ \tau(x)$. Therefore $h \circ f_3(x) = (x^2, \tau(x))$. It immediately follows that $a = 1$. Since, $[p_3 \circ f_3] = 2[\iota_3]$, it follows that $b = 2$. Thus we have proved:

Theorem 10.9.17 $h_{\#}(\omega_3) = -(i_{\#}(\xi) + 2\eta)$.

It follows that $\{\omega_3, \eta\}$ is also a basis for the free abelian group $\pi_3(SO(4))$.

This allows us to conclude that $\pi_3(SO(5)) \approx \mathbb{Z}$, generated by the appropriate inclusion of the section $s : \mathbb{S}^3 \to SO(4) \subset SO(5)$. Therefore, $\pi_3(SO(n)) \approx \mathbb{Z}$ for $n \geq 5$.

To summarize the results so far, we have the following table:

	$SO(2)$	$SO(3)$	$SO(4)$	$SO(n)$, $n \geq 5$
π_1	$\mathbb{Z}$	$\mathbb{Z}_2$	$\mathbb{Z}_2$	$\mathbb{Z}_2$
π_2	0	0	0	0
π_3	0	$\mathbb{Z}$	$\mathbb{Z} \oplus \mathbb{Z}$	$\mathbb{Z}$

We now consider the case n odd. We have seen that $2\omega_{n-1} = \partial'[2\iota_n] = \partial \circ (p_n)_{\#}(\omega_n) = 0$. Therefore ω_{n-1} is either the zero element or an element of order 2. Now $\omega_{n-1} = 0$ iff $(p_n)_{\#} : \pi_n(SO(n+1)) \to \pi_n(\mathbb{S}^n)$ is surjective. This means that $Id : \mathbb{S}^n \to \mathbb{S}^n$ has a homotopy lift across p_n and since p_n is a fibration this is equivalent to saying that Id has an actual lift, i.e., p_n has a cross section. Thus we have:

Theorem 10.9.18 $\omega_{n-1} = 0$ *iff the fibration* $p_n : SO(n+1) \to \mathbb{S}^n$ *is trivial.*

We have seen that this is the case $n = 1, 3$. Indeed, this is also the case for $n = 7$, using the Cayley numbers. The converse part is a deep result which we shall not be able to prove here (see [Adams, 1960]):

Theorem 10.9.19 $\omega_{n-1} = 0$ *iff* $n = 1, 3, 7$.

Exercise 10.9.20

1. Consider the maps f_n as defined in (10.13).
   (i) $f_n(-x) = f_n(x)$ and hence f_n defines a map $g_n : \mathbb{P}^n \to SO(n+1)$.
   (ii) g_n is an embedding.
   (iii) $g_{n+1}|_{\mathbb{P}^n} = g_n$ and hence there is an embedding $g : \mathbb{P}^\infty \hookrightarrow SO$.
   (iv) $p_n \circ g_n(\mathbb{P}^{n-1}) = \{e_n\}$.
   (v) The maps $p_n \circ g_n : (\mathbb{P}^n, \mathbb{P}^{n-1}) \to (\mathbb{S}^n, e_n)$ are relative homeomorphisms.

2. Show that $V_{r+2,2}(\mathbb{R})$ is diffeomorphic to the total space of the unit tangent bundle to $\mathbb{S}^{r+1}$. Show that this space is $(r-1)$-connected.

3. Compute the image of $\partial'(\xi_{r+1})$ in $\pi_r(V_{r+1,1}) = \pi_r(\mathbb{S}^r)$, where ∂' is the connecting homomorphism of the homotopy exact sequence of the fibration $V_{r+2,2}(\mathbb{R}) \to \mathbb{S}^{r+1}$ using the following commutative diagram:

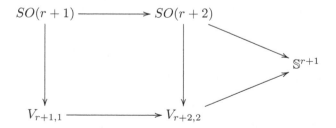

Conclude the following theorem:

Theorem 10.9.21 *If* r *is even then* $\pi_r(V_{r+k,k}(\mathbb{R}))$ *is infinite cyclic. If* r *is odd then* $\pi_r(V_{r+1,1}(\mathbb{R}))$ *is infinite cyclic and* $\pi_r(V_{r+k,k}(\mathbb{R})) = \mathbb{Z}_2, k \geq 2$.

10.10 Homology with Local Coefficients

This section is just like an appendix, in which we introduce the concept of homology and cohomology with local coefficients and relate it to a more general concept of equivariant homology and cohomology. The treatment is just skeletal and the reader is expected to supply most of the proofs by herself.

Definition 10.10.1 Let X be a topological space and $\mathcal{G}$ be a bundle of abelian groups on X which is the same as a local system of abelian groups. (See Definition 10.3.1). Let for each $n \neq 0$, $S_n(X; \mathcal{G})$ denote the abelian group of finite formal sums

$$c = \sum_{i=1}^{k} g_i \sigma_i$$

where each $\sigma_i : \Delta_n \to X$ is a singular n-simplex and $g_i \in \mathcal{G}(\sigma_i(\mathbf{e}_0))$, under the usual law of addition. Members of $S_n(X; \mathcal{G})$ are called singular n-chains in X with coefficients in $\mathcal{G}$. A chain of the form $g\sigma$ where σ is a singular n-simplex and $g \in G(\sigma(\mathbf{e}_0))$ is called an elementary n-chain.

Note that if $\mathcal{G}$ were the constant bundle G, then $S_n(X; \mathcal{G}) = S_n(X; G)$ the usual free module of singular n-chains with coefficients in G. We wish to define the boundary operators $\partial : S_n(X; \mathcal{G}) \to S_{n-1}(X; \mathcal{G})$ so as to make $S_*(X, \mathcal{G})$ into a chain complex in such a way that if $\mathcal{G}$ is the constant bundle, then ∂ coincides with the usual boundary operator of the singular chain complex. It is enough to define ∂ on elementary n-chains and then extend the definition by linearity over all n-chains. We are tempted to take $\partial(g\sigma) = g(\partial(\sigma))$ where $\partial(\sigma) = \sum_{r=0}^{n}(-1)^r \sigma \circ F^r$. But we note that $g \in G(\sigma(\mathbf{e}_0))$ and hence $g\sigma \circ F^0$ does not make sense, because we can multiply $\sigma \circ F^0$ with elements of $G(\sigma \circ F^0(\mathbf{e}_0)) = G(\sigma(\mathbf{e}_1))$. Of course, all the remaining terms make sense. So, we remedy this situation by using the isomorphism $\mathcal{G}(\sigma(\mathbf{e}_0)) \to \mathcal{G}(\sigma(\mathbf{e}_1))$ of the local coefficient system which corresponds to the path $\sigma([\mathbf{e}_0, \mathbf{e}_1])$ from $\sigma(\mathbf{e}_0)$ to $\sigma(\mathbf{e}_1)$. Let us denote this isomorphism σ_{01}. Thus we define:

$$\partial(g\sigma) = \sigma_{01}(g)\sigma \circ F^0 + g\sum_{r=1}^{n}(-1)^r \sigma \circ F^r. \tag{10.15}$$

The next task is to show that $\partial \circ \partial = 0$. We shall merely write down the proof for the case $n = 2$, the proof for the general case being similar. We shall also use the notation $[u_i, u_j]$ to denote the singular simplex $\sigma[\mathbf{e}_i, \mathbf{e}_j]$, etc. We have

$$
\begin{aligned}
\partial^2(g\sigma) &= \partial(\sigma_{01}(g)[u_1, u_2] - g[u_0, u_2] + g[u_0, u_1]) \\
&= \sigma_{12} \circ \sigma_{01}(g)[u_2] - \sigma_{01}(g)[u_1] - (\sigma_{02}(g)[u_2] - g[u_0]) + (\sigma_{01}(g)[u_1] - g[u_0])
\end{aligned}
$$

Since the paths $[u_0, u_1] * [u_1, u_2]$ and $[u_0, u_2]$ are homotopic relative to the end-points, it follows that the isomorphism $\sigma_{12} \circ \sigma_{01} : \mathcal{G}(\sigma(\mathbf{e}_0)) \to \mathcal{G}(\sigma(\mathbf{e}_2))$ is equal to the isomorphism σ_{02}. Therefore the above expression vanishes.

The homology groups $H_*(X; \mathcal{G})$ of the chain complex $S_*(X; \mathcal{G})$ are called the homology of X with local coefficients in $\mathcal{G}$.

It is clear that if $\mathcal{G} = G$ is the constant coefficient system, then this homology coincides with the usual homology $H_*(X; G)$. For a subspace $A \subset X$ we shall denote the restriction of the coefficient system $\mathcal{G}$ on X to A also by $\mathcal{G}$. It then follows easily that the chain complex $S_*(A; \mathcal{G})$ is a subchain complex of $S_*(X; \mathcal{G})$. The relative chain complex $S_*(X, A; \mathcal{G})$ is then defined as the quotient and we have an exact sequence of chain complexes:

$$0 \longrightarrow S_*(A; G) \longrightarrow S_*(X; \mathcal{G}) \longrightarrow S_*(X, A; \mathcal{G}) \longrightarrow 0.$$

This in turn induces a long exact sequence of homology groups:

$$\cdots \longrightarrow H_{q+1}(X,A;\mathcal{G}) \xrightarrow{\partial} H_q(A;\mathcal{G}) \xrightarrow{i_*} H_q(X;\mathcal{G}) \xrightarrow{j_*} H_q(X,A;\mathcal{G}) \xrightarrow{\partial} \cdots$$

Given a morphism $\Phi : \mathcal{G} \to \mathcal{G}'$ of bundles of abelian groups, there is an obvious homomorphism $\Phi_* : S_*(X,A;\mathcal{G}) \to S_*(X,A;\mathcal{G}')$ which respects the boundary homomorphism and hence is a chain map. We shall denote the induced homomorphism on homology also by

$$\Phi_* : H_*(X,A;\mathcal{G}) \to H_*(X,A;\mathcal{G}').$$

These homomorphisms are themselves categorical, viz., if we have another morphism $\Phi' : \mathcal{G}' \to \mathcal{G}''$ of local systems, then $(\Phi' \circ \Phi)_* = \Phi'_* \circ \Phi_*$ etc. Similar statements hold in the space slots as well. That is, if $f : (X,A) \to (Y,B)$ is a map and $\mathcal{G}$ is a bundle of abelian groups on Y, we then have the pull-back bundle $f^*\mathcal{G}$ on X and we have a homomorphism $f_* : H_*(X,A;f^*(\mathcal{G})) \to H_*(Y,B;\mathcal{G})$ with functorial properties.

Thus, the correct settings for discussion on homology (and similarly for cohomology) with local coefficient systems is the category $\mathcal{L}$ whose objects are pairs $((X,A),\mathcal{G})$ of topological pairs (X,A) and local systems $\mathcal{G}$ on them. We shall use the simplified notation $(X,A;\mathcal{G})$ for $((X,A),\mathcal{G})$. (Of course, we may restrict these topological pairs within those which are compactly generated or within CW-pairs, etc.) A morphism $\alpha : (X,A;\mathcal{G}) \to (Y,B;\mathcal{G}')$ in this category consists of a map $\alpha_1 : (X,A) \to (Y,B)$ and a functor $\alpha_2 : \mathcal{G} \to f^*\mathcal{G}'$. We define $\mathbb{I} \times (X,A;\mathcal{G})$ to be the object $(\mathbb{I} \times (X,A);p^*\mathcal{G})$ where $p : \mathbb{I} \times (X,A) \to (X,A)$ is the projection map. Given two morphisms $\alpha,\beta : (X,A;\mathcal{G}) \to (Y,B;\mathcal{G}')$, a homotopy

$$F : \mathbb{I} \times (X,A;\mathcal{G}) \to (Y,B;\mathcal{G}')$$

from α to β is a pair $F = (F_1,F_2)$, consisting of a homotopy $F_1 : \mathbb{I} \times (X,A) \to (Y,B)$ from $\alpha_1 \to \beta_1$ and a homomorphism $F_2 : p^*\mathcal{G} \to F_1^*\mathcal{G}'$ such that when restricted to $0 \times (X,A)$ and $1 \times (X,A)$ it is equal to α_2 and β_2, respectively. If such a homotopy exists, we write $\alpha \simeq \beta$. It is a matter of routine verification that $\simeq$ is an equivalence relation and that composites of homotopic morphisms are homotopic. Given two objects $(X_i,A_i;\mathcal{G}_i) \in \mathcal{L}, i = 1,2$, such that $X_1 \subset X_2, A_1 \subset A_2$ and $\mathcal{G}_2|_{X_1} = \mathcal{G}_1$, we shall denote by $k : (X_1,A_1;\mathcal{G}_1) \hookrightarrow (X_2,A_2;\mathcal{G}_2)$ the morphism which is the inclusion of $(X_1,A_1) \hookrightarrow (X_2,A_2)$ and the identity map on $\mathcal{G}_1 \to \mathcal{G}_2|_{X_1}$. There is the obvious restriction functor $R : \mathcal{L} \to \mathcal{L}$ given by $R(X,A;\mathcal{G}) := (A,\emptyset;i^*\mathcal{G}) = (A;\mathcal{G})$ with morphisms, etc., also being restrictions. We can now formulate a list of properties of homology with local coefficients, similar to the Eilenberg–Steenrod axioms for ordinary homology.

Theorem 10.10.2 *There are given a sequence of functors $H_q : \mathcal{L} \rightsquigarrow \mathbf{Ab}$ and a natural transformation $\partial : H_q \to H_{q-1} \circ R$ satisfying the following properties:*
(a) **Exactness:** *For every object $(X,A);\mathcal{G}$ in $\mathcal{L}$ there is an exact sequence:*

$$\cdots \longrightarrow H_{q+1}(X,A;\mathcal{G}) \xrightarrow{\partial_{q+1}(X,A;\mathcal{G})} H_q(A;\mathcal{G}) \xrightarrow{H_q(i)} H_q(X;\mathcal{G}) \xrightarrow{H_q(j)} H_q(X,A;\mathcal{G}) \longrightarrow \cdots$$

(b) **Homotopy:** *If $\alpha,\beta : (X,A;\mathcal{G}) \to (Y,B;\mathcal{G}')$ is a homotopy in $\mathcal{L}$ then $H_q(\alpha) = H_q(\beta)$ for all q.*
(c) **Excision:** *If $X = \mathrm{int}\,X_1 \cup \mathrm{int}\,X_2$ then for any local coefficient system $\mathcal{G}$ of abelian groups $\mathcal{G}$ on X, we have the inclusion induced homomorphism $H_q(X_1,X_1 \cap X_2;\mathcal{G}) \to H_q(X,X_2;\mathcal{G})$ are isomorphisms for all q.*
(d) **Dimension:** *$H_q(*;\mathcal{G}) = 0, q > 0$ and $H_0(*;\mathcal{G}) = \mathcal{G}(*)$ where $*$ denotes any singleton space.*

The proof of the above theorem is just analogous to those for the case of ordinary homology and so will be left to the reader. We can also derive the following additive property and direct limit properties in the usual way.

Theorem 10.10.3 *If (X, A) is a disjoint union of a family of $\{(X_j, A_j)\}$ of topological spaces then the inclusion induced homomorphisms $H_*(X_j, A_j; \mathcal{G}) \to H_*(X, A; \mathcal{G})$ define $H_*(X, A; \mathcal{G})$ as a direct sum. More generally, if (X, A) is a union of subspaces (X_j, A_j) with the direct limit topology, then the inclusion induced homomorphisms above define $H_*(X, A; \mathcal{G})$ as a direct limit.*

Theorem 10.10.4 *If $\{X_\alpha\}$ are path components of a space X, then $H_*(X, \mathcal{G}) = \oplus_\alpha H_*(X_\alpha; \mathcal{G})$.*

More or less similarly we can consider the cohomology with local coefficients. The difference is in setting up the correct category $\mathcal{L}^*$. As before the objects are triples $(X, A; \mathcal{G})$ of topological pairs (X, A) and a bundle of abelian groups $\mathcal{G}$ over X. However, a morphism $\alpha : ((X, A); \mathcal{G}) \to ((Y, B); \mathcal{G}')$ consists of a map $f : (X, A) \to (Y, B)$ and a morphism $\alpha_2 : f^*\mathcal{G}' \to \mathcal{G}$. Verification that this actually forms a category is straightforward. The singular cochain complex $S^n(X; \mathcal{G})$ is defined to be the abelian group of all functions c which assign, to each singular n-simplex σ in X an element $c(\sigma) \in \mathcal{G}(\sigma(\mathbf{e}_0))$ under the usual point-wise addition and the coboundary operator $\delta : S^n(X; \mathcal{G}) \to S^{n+1}(X; \mathcal{G})$ given by

$$(-1)^n (\delta c)(\sigma) = \mathcal{G}(\sigma_{0,1})^{-1}(c(\sigma \circ F^0)) + \sum_{r=1}^{n+1} c(\sigma \circ F^r)$$

for each singular $(n + 1)$-simplex σ. This time we shall leave the verification that $\delta^2 = 0$ completely to the reader. If A is a subspace of X then the inclusion map $i : A \hookrightarrow X$ induces a chain homomorphism $i^* : S^*(X; \mathcal{G}) \to S^*(A; \mathcal{G})$ by restriction and the kernel is therefore a subchain complex which we define to be $S^*(X, A; \mathcal{G})$. The homology of groups of these chain complexes are defined to be the singular cohomology groups with local coefficient system $\mathcal{G}$ and are respectively denoted by $H^*(X; \mathcal{G})$ and $H^*(X, A; \mathcal{G})$. We then have results parallel to those in Theorem 10.10.2 for these cohomology groups. Also, a result similar to that of Theorem 10.10.4 holds for cohomology. Of course, cohomology does **not** commute with direct limit as usual.

There is a more general theory than the theory of homology with local coefficients. We shall briefly introduce this theory of equivariant (co)homology and relate it to homology with local coefficients. These have far better applicability. The only reason then for introducing homology with local coefficients is that it gives a slightly better geometric picture of equivariant theory, albeit in a special case.

Let X be a space on which a group Π acts on the left. This in turn induces an action of Π on the singular chain groups $S_*(X; \mathbb{Z})$ of X : given a singular simplex $\sigma : \Delta_n \to X$, and an element $\xi \in \Pi$, take $(\xi\sigma)(t) = \xi\sigma(t)$. This action is clearly seen to respect the boundary homomorphism and hence $S_*(X; \mathbb{Z})$ becomes a left Π-module.

Now suppose we have a right Π-module G. We can then form the tensor product $G \otimes_\Pi S_*(X)$ which is nothing but the quotient of the usual tensor product $G \otimes_\mathbb{Z} S_*(X)$ of abelian groups modulo the subgroup generated by elements of the form $g\xi \otimes \sigma - g \otimes \xi\sigma$. (If you have not seen tensor product of modules over non commutative rings, and you wish to go ahead with the study of equivariant homology, it is time you learnt it.) Once again, $Id \otimes \partial$ makes sense and is a boundary operator on $G \otimes_\Pi S_*(X)$. The homology of this chain complex is denoted by $E_*(X; G)$ and is called the equivariant homology of X with coefficients in the Π-module G.

While considering cohomology, we have to be a little more careful since the group ring $\mathbb{Z}\Pi$ may be non commutative. So now, let G be a left Π-module. We can then take the group of Π-linear homomorphisms $\operatorname{Hom}^{\Pi}(S_*(X); G)$ which is clearly a subcomplex of the cochain complex $\operatorname{Hom}(S_*(X); G)$ of all group homomorphisms.

The homology of modules of the cochain complex $\operatorname{Hom}^{\Pi}(S_*(X); G)$ are denoted by $E^*(X; G)$ and are called the equivariant cohomology of X with coefficients in the Π-module G.

A special case that we are interested in is when E is the universal covering space of a space B, $\Pi = \pi_1(B, b_0)$ and the action of Π on X is the usual action via covering transformation. We already know that giving a bundle $\mathcal{G}$ of abelian groups over B is equivalent to giving a Π-module $G = \mathcal{G}(x_0)$. The proof of the following result is completely canonical, though not all too trivial, and is left to the reader as an exercise.

Theorem 10.10.5 (Eilenberg): *The homology (cohomology) groups $H_*(B; \mathcal{G})$, (respectively, $H^*(B; \mathcal{G})$) with respect to the local system $\mathcal{G}$ are canonically isomorphic with the equivariant homology $E_*(X; G)$, (respectively, cohomology $E^*(X; G))$), where $G = \mathcal{G}(x_0)$.*

10.11   Miscellaneous Exercises to Chapter 10

1. We have seen, using the Hurewicz theorem or otherwise that $\pi_n(\mathbb{S}^n) \approx \mathbb{Z}$ and the Hurewicz homomorphism $\varphi : \pi_n(\mathbb{S}^n) \to H_n(\mathbb{S}^n)$ is an isomorphism. For $n \geq 2$, we also know that, if $p : \mathbb{S}^n \to \mathbb{P}^n$ is the double cover, then $p_\# : \pi_n(\mathbb{S}^n) \to \pi_n(\mathbb{P}^n)$ is an isomorphism. We also have the antipodal map $\alpha : \mathbb{S}^n \to \mathbb{S}^n$ which has the property $p \circ \alpha = p$. Does this imply that $\alpha_\# = Id$? Does this imply $\alpha_* = Id : H_n(\mathbb{S}^n) \to H_n(\mathbb{S}^n)$?

2. Show that a connected closed triangulated n-manifold has a CW-structure with one n-cell.

3. Choosing $\{1, \mathbf{j}\}$ as a $\mathbb{C}$-basis for $\mathbb{H}$, the right multiplication by an element $z + w\mathbf{j}$ on $\mathbb{H}$ is represented by the matrix

$$\begin{bmatrix} z & w \\ -\bar{w} & z \end{bmatrix} \in M(2, \mathbb{C}).$$

   More generally, the $\mathbb{C}$-isomorphism $\mathcal{Q}_n : \mathbb{H}^n \to \mathbb{C}^{2n}$ given by

$$(z_1 + w_1\mathbf{j}, \ldots, z_n + w_n\mathbf{j}) \mapsto (z_1, w_1, \ldots, z_n, w_n)$$

   gives an identification of $M(n, \mathbb{H})$ with the subspace

$$\{A \in M(2n, \mathbb{C}) \ : \ A J_{2n} = J_{2n} \bar{A}\}$$

   where $J_{2n} = diag(J_2, \ldots J_2)$ and

$$J_2 = \begin{bmatrix} 0 & 1 \\ -1 & 0 \end{bmatrix}.$$

   Prove that there is an isomorphism $Sp(1) \to SU(2)$. Also show that $Sp(1) = \mathbb{S}^3$.

4. Show that $\pi_i(U(2)) \approx \pi_i(\mathbb{S}^3), i \geq 2$.

5. Using the fact that $\pi_{k+1}(\mathbb{S}^k) \approx \mathbb{Z}_2, k \geq 3$, compute the groups $\pi_4(U(2)), \pi_4(SU(2))$, and $\pi_4(SO(k)), \pi_4(Sp(k))$, for $1 \leq k \leq 3$.

Chapter 11

Homology of Fibre Spaces

As indicated in the introduction to Chapter 1, the concept of fibration plays a central role in algebraic topology. However, apart from studying some special cases, viz., the covering projection in Chapter 2, a little bit about vector bundles in Chapter 5 and establishing the homotopy exact sequence of a fibration in Chapter 7, we have not studied much about fibrations so far. In this chapter, let us take up the ever-so-important study of fibrations again and go one step further. In Section 11.1, we shall deal with some generalities of fibrations. In Section 11.2, we shall discuss fibrations with fibres homotopy type of a sphere and establish Thom isomorphism theorem and, as an immediate consequence, the Gysin sequences. In Section 11.3, we begin the study of fibrations over suspensions, establish Wang homology and cohomology exact sequences by specialising to the case of fibrations over spheres. As a bonus, we derive the well-known suspension theorem due to Freudenthal. We shall then apply this study to compute the cohomology groups of some classical groups in Section 11.4. We also include some basic facts about Hopf algebras.

The reader is urged to revisit Section 1.7, before proceeding with this chapter.

11.1   Generalities about Fibrations

In this section, we shall begin with some generalities about fibrations. An important technical result here combines homotopy extension and homotopy lifting properties together. At the end of the section, through step-by-step exercises, we have indicated how one can prove Serre's theorem on homology sequence of a fibration, the celebrated Blakers–Massey theorem and a version of Freudenthal's suspension theorem.

Definition 11.1.1 Given two fibrations $p_i : E_i \to B$ over the same base space, by a *fibre map* $f : E_1 \to E_2$ we mean a map f such that $p_2 \circ f = p_1$. Two fibre maps $f_i : E_1 \to E_2, i = 0, 1$ are said to be fibre homotopic if there exists a homotopy $F : E_1 \times \mathbb{I} \to E_2$ such that $p_2 \circ F(x, t) = p_1(x)$, for all t and x and $F_0 = f_0$ and $F_1 = f_1$. In case $p_1 = p_2 = p$, we express this by writing

$$f_1 \underset{\tilde{p}}{\simeq} f_2.$$

We say p_1 and p_2 have the *same fibre homotopy type* or p_1 *is fibre homotopy equivalent to* p_2 if there exist fibre maps $f : E_1 \to E_2$ and $g : E_2 \to E_1$ such that

$$g \circ f \underset{\tilde{p_1}}{\simeq} Id_{E_1}; \quad f \circ g \underset{\tilde{p_2}}{\simeq} Id_{E_2}.$$

There is a category $\mathbf{Fib}_B$, whose objects are fibrations over a given space B and morphisms are fibre maps. One can change the morphisms to fibre homotopy classes to get another category and that is the one which we are presently interested in.

We are interested in the study of lifting problems. Recall that covering projections are a very special class of fibrations which satisfy the unique path lifting property. In general, we do not have such uniqueness. However, the following result gives us uniqueness up to 'fibre homotopy'. Indeed, this result combines both homotopy extension and lifting properties, in some sense. You can simply ignore the parameter X, and appeal to the fact that there is a homeomorphism of the topological pairs:

$$(\mathbb{I} \times \mathbb{I}, 0 \times \mathbb{I} \cup \mathbb{I} \times \dot{\mathbb{I}}) \to (\mathbb{I} \times \mathbb{I}, \mathbb{I} \times 0).$$

(See Exercise 1.5.19.(iii).) Then the theorem below is an immediate consequence of the homotopy lifting property.

Theorem 11.1.2 Homotopy Extension-Lifting Property *Let $p : E \to B$ be a fibration. Given any homotopy $H : X \times \mathbb{I} \times \mathbb{I} \to B$ and a map $G : X \times \mathbb{I} \times \dot{\mathbb{I}} \cup X \times 0 \times \mathbb{I} \to E$ such that $p \circ G = H \circ \iota$ where $\iota : X \times \mathbb{I} \times \dot{\mathbb{I}} \cup X \times 0 \times \mathbb{I} \to X \times \mathbb{I} \times \mathbb{I}$ is the inclusion map, there is a homotopy $H' : X \times \mathbb{I} \times \mathbb{I} \to E$ such that $p \circ H' = H$, and $H' \circ \iota = G$.*

$$
\begin{array}{ccc}
X \times 0 \times \mathbb{I} \cup X \times \mathbb{I} \times \dot{\mathbb{I}} & \xrightarrow{\quad G \quad} & E \\
{\scriptstyle \iota} \downarrow & {\scriptstyle H'} \nearrow & \downarrow {\scriptstyle p} \\
X \times \mathbb{I} \times \mathbb{I} & \xrightarrow{\quad H \quad} & B
\end{array}
$$

The data can be interpreted as a partial lift of a homotopy of homotopies and the conclusion as an extension of this lift. Or, you may think of the data as an 'initial point' preserving homotopy of two lifts and the conclusion as an extension of this homotopy to a full homotopy of the two lifts.

Following this we make another definition:

Definition 11.1.3 Let $p : E \to B$ be a fibration. Two maps $f, g : X \to E$ such that $p \circ f = p \circ g$ are said to be *fibre homotopic* to each other if there exists a homotopy $H : f \simeq g$ which respects the fibration p, i.e., $p \circ H(x, t) = pf(x), x \in X$.

The corollary below is obtained by taking H, G to be constant homotopies in the above theorem.

Corollary 11.1.4 *Given a fibration $p : E \to B$, any two liftings $F_0, F_1 : X \times \mathbb{I} \to E$ of the same map $p \circ F = p \circ G$ such that $F_0|_{X \times 0} = F_1|_{X \times 0}$ have the same fibre homotopy type.*

Given a fibration $p : E \to B$ and a path $\mathbb{I} : \omega \to B$ from $b_0 = \omega(0)$ to $b_1 = \omega(1)$, consider the following diagram:

$$
\begin{array}{ccc}
F_{b_0} \times 0 & \hookrightarrow & E \\
\downarrow & {\scriptstyle \Omega} \nearrow & \downarrow {\scriptstyle p} \\
F_{b_0} \times \mathbb{I} & \xrightarrow{\;\widehat{\omega}\;} & B
\end{array}
$$

where $\widehat{\omega}(e, t) = \omega(t)$, the HLP gives the homotopy $\Omega : F_{b_0} \times \mathbb{I} \to E$ such that $p \circ \Omega(e, t) = \omega(t)$ and $\Omega(e, 0) = e$. Let us define

$$f_\omega : F_{b_0} \to F_{b_1}$$

by $f_\omega(e) = \Omega(e, 1)$. If $\omega \simeq \omega'$ in B, it follows from the previous theorem, by taking $X = F_{b_0}$, that $f_\omega \simeq f_{\omega'}$. We define $h[\omega] = [f_\omega] \in [p^{-1}(\omega(0)); p^{-1}(\omega(1))]$.

We summarise this in the following:

Theorem 11.1.5 *Given a fibration $p : E \to B$, the assignment*

$$b \rightsquigarrow p^{-1}(b) =: F_b; \qquad [\omega] \rightsquigarrow h[\omega],$$

defines a contravariant functor from the fundamental groupoid $\mathcal{P}(B)$ of B to the homotopy category whose objects are fibres of E. In particular, if B is path connected, all fibres of p have the same homotopy type.

Remark 11.1.6 It follows that there is a well-defined homomorphism $\pi_1(B, b_0) \to \mathcal{H}(F)$, the group of homotopy equivalences of the fibre $F = F_{b_0}$. We can interpret this as a 'homotopy-action' of $\pi_1(B)$ on F which passes down to define an action on the homology groups of F. Note that in general, it does not define an action on the homotopy groups of F since the base-point of F may not be preserved by the homotopy equivalences $f_{[\omega]} : F \to F$. However, it gives an action on the groups $\pi'_q(F)$ (see Definition 10.4.1).

The action of $\pi_1(B)$ on the homology of the fibre is an important ingredient whenever we need to study a particular fibration in detail. So, we shall study this a little more here in the general set-up itself.

Consider the mapping cylinder M_p of p and let $\hat{p} : M_p \to B$ be the associated strong deformation retraction (see 1.5.17). Put $\hat{F} = \hat{p}^{-1}(*)$. Clearly,

$$\hat{F} = \frac{F \times \mathbb{I}}{F \times 1}$$

can be thought of as the cone over F. Since p is a fibration, it follows that the homomorphism $p_\# : \pi_q(E, F) \to \pi_q(B, b_0)$ is an isomorphism for all q, whereas $\hat{p}_\# : \pi_q(M_p, \hat{F}) \to \pi_q(B, b_0)$ are isomorphisms since $\hat{p}$ is a SDR. Therefore, from the commutativity of the diagram

$$\pi_q(E, F) \xrightarrow{\quad k_1 \quad} \pi_q(M_p, \hat{F})$$
$$p_\# \searrow \qquad \swarrow \hat{p}_\#$$
$$\pi_q(B, b_0)$$

where k_1 is the inclusion induced homomorphism, it follows that k_1 is an isomorphism. We now appeal to the double-hexagon Exercise 10.2.26(v) with $X = M_p, Y = E, Z = \hat{F}$ and $W = F$. We conclude that

Lemma 11.1.7 The inclusion map induces an isomorphism $k_2 : \pi_q(\hat{F}, F) \to \pi_q(M_p, E)$ for all q.

Let Θ be the composite isomorphism

$$\pi_{q+1}(M_p, E) \xrightarrow{k_2^{-1}} \pi_{q+1}(\hat{F}, F) \xrightarrow{\partial} \pi_q(F).$$

(Note that the boundary ∂ is an isomorphism here.)

Now, there is the standard action of $\pi_1(E)$ on the relative homotopy groups $\pi_{q+1}(M_p, E)$. Also, via the homomorphism $p_\# : \pi_1(E) \to \pi_1(B)$, we obtain an action of $\pi_1(E)$ on $\pi'_q(F)$. These two actions turn out to be compatible under Θ.

Theorem 11.1.8 *Let $p : E \to B$ be a fibration with a path connected fibre and non degenerate base point $e \in F = p^{-1}(b_0)$. Then the composite Θ' of the homomorphisms*

$$\pi_{q+1}(M_p, E) \xrightarrow{\Theta} \pi_q(F) \to \pi'_q(F)$$

is a $\pi_1(E)$-module homomorphism.

Proof: Given $\alpha \in \pi_{q+1}(M_p, E)$ and $\xi \in \pi_1(E)$, we have to show that

$$\Theta'(\phi_\xi(\alpha)) = (h_{p_\#(\xi)})_\#(\Theta'(\alpha)).$$

We need to make good use of the 'isomorphism' Θ. So, we begin with an element $k_2[\lambda] = \alpha$, for some $\lambda : (\mathbb{D}^{q+1}, \mathbb{S}^q) \to (\hat{F}, F)$. Let $\omega : (\mathbb{I}, \dot{\mathbb{I}}) \to (F, e)$ represent ξ. We then obtain

$\beta : (\mathbb{D}^{q+1}, \mathbb{S}^q) \to (\hat{F}, F)$ such that $\beta \overset{\omega}{\sim} \lambda$ in (M_p, E). This will imply that $\phi_\xi(\alpha) = k_2([\beta])$. We then plan to show that $\partial[\beta] = h_{[p \circ \omega]}(\partial[\lambda])$. This will complete the proof.

By HLP, there is a map $H : F \times \mathbb{I} \to E$ such that

$$H(x, 1) = x; \ H(e, t) = \omega(t); \ p \circ H(x, t) = p \circ \omega(t), \ x \in F, t \in \mathbb{I}.$$

We extend this map to $\hat{H} : \hat{F} \times \mathbb{I} :\to M_p$ by the formula $\hat{H}([x, s], t) = [H(x, t), s]$. Since for a fixed t, $p \circ H(x, t) \subset p^{-1}(p(\omega(t)))$, it follows that $\hat{H}$ is well defined and continuous.

Observe $\hat{H}([x, s], 0) \in \hat{F}$. Put $\beta := \hat{H}_0 \circ \lambda : (\mathbb{D}^{q+1}, \mathbb{S}^q) \to (\hat{F}, F)$. Then $\hat{H} \circ \lambda$ defines a ω-homotopy from β to λ in M_p.

Finally, $\partial[\beta]$ is represented by $\beta|_{\mathbb{S}^q} = H_0 \circ \lambda|_{\mathbb{S}^q}$. Therefore $\partial[\beta] = h_{[p \circ \omega]}[\partial\lambda]$. ♠

Corollary 11.1.9 The composite $\varphi \circ \Theta : \pi_{q+1}(M_p, E) \to H_q(F)$ is a $\pi_1(E)$-module homomorphism, where $\varphi : \pi_q(F) \to H_q(F)$ denotes the Hurewicz map.

Example 11.1.10 Let F be a connected orientable closed n-manifold and $h : F \to F$ be a homeomorphism. Consider the space $M = F_f = F \times \mathbb{I}/(x, 0) \sim (f(x), 1)$. There is the obvious map $p : M \to \mathbb{S}^1$ given by $p([x, t]) = e^{2\pi i t}$. It is easy to see that p is a fibre bundle and hence a fibration. The action of the generator $[\xi] \in \pi_1(\mathbb{S}^1)$ on the fibre F is easily seen to be via the homeomorphism f. Thus the action of $\pi_1(\mathbb{S}^1)$ on $H_n(F; \mathbb{Z})$ is via the isomorphism $f_* : H_n(F) \to H_n(F)$ which is trivial iff f is orientation preserving.

Returning to Definition 11.1.3, fixing a base and the fibre up to homotopy, what are all possible fibrations $p : E \to B$ up to fibre homotopy equivalence? We shall give a partial answer to this here.

Proposition 11.1.11 Let $p : E \to B \times \mathbb{I}$ be a fibration. For each $t \in \mathbb{I}$ consider the fibration $p_t : E_t = p^{-1}(B \times t) \to B$ given by $p_t(e) = \pi(p(e))$ where $\pi : B \times \mathbb{I} \to B$ is the projection. All the fibrations p_t are fibre homotopy equivalent to each other.

Proof: It is enough to prove that p_0, p_1 are equivalent. The HLP applied to the following diagram for $t = 0, 1$,

$$\begin{array}{ccc} E_t \times t & \hookrightarrow & E \\ \downarrow & \overset{H_t}{\nearrow} & \downarrow p \\ E_t \times \mathbb{I} & \overset{p_t \times Id}{\longrightarrow} & B \times \mathbb{I} \end{array}$$

yields homotopies $H_t : E_t \times \mathbb{I} \to E$. Put $f_0(e) = H_0(e, 1); f_1(e) = H_1(e, 0)$. Then $f_0 : E_0 \to E_1$ and $f_1 : E_1 \to E_0$ are such that $p_1 \circ f_0 = p_0$ and $p_0 \circ f_1 = p_1$. We claim f_0 and f_1 are fibre homotopy inverses of each other.

For this we use Theorem 11.1.2 as follows: To get a fibre homotopy from Id_{E_0} to $f_1 \circ f_0$, we consider the diagram

$$\begin{array}{ccc} E_0 \times 1 \times \mathbb{I} \cup E_0 \times \mathbb{I} \times \dot{\mathbb{I}} & \overset{G}{\longrightarrow} & E \\ \downarrow & \overset{H'}{\nearrow} & \downarrow p \\ E_0 \times \mathbb{I} \times \mathbb{I} & \overset{H}{\longrightarrow} & B \times \mathbb{I} \end{array}$$

where $H(e_0, s, t) = (p_0(e_0), s)$ and

$$G(e_0, s, t) = \begin{cases} f_0(e_0), & s = 1; \\ H_0(e_0, s), & t = 0; \\ H_1(f_0(e_0), s), & t = 1. \end{cases}$$

Check that $H'(e_0, 0, t)$ is the required homotopy. By symmetry, we can also get a homotopy from Id_{E_1} to $f_0 \circ f_1$. ♠

Recall that given functions $f : B' \to B$ and $p : E \to B$ their fibred product is defined to be

$$E' = \{(b', e) \in B' \times E \ : \ f(b') = p(e)\}$$

together with the projection map $p' : E' \to B'$ given by $p'(b', e) = b'$. The map p' is also called the *pullback* of p via f and is denoted by $f^*(p)$. There is a canonical map $\hat{f} : E' \to E$ given by $\hat{f}(b', e) = e$. (See Section 1.7.)

Theorem 11.1.12 *Let $p : E \to B$ be a fibration. Then for any map $f : B' \to B$ the pullback $f^*(p) : E' \to B'$ is a fibration. Moreover, if $g : B' \to B$ is homotopic to f then $g^*(p)$ is fibre homotopic to $f^*(p)$.*

Proof: The fibration property is easily verified. If $H : B' \times \mathbb{I} \to B$ is the homotopy between f and g, then the fibration $H^*(p) \to B' \times \mathbb{I}$ restricts to $f^*(p)$ and $g^*(p)$ over $B \times 0$ and $B \times 1$, respectively, and hence Proposition 11.1.11 is applicable. ♠

Corollary 11.1.13 *Any fibration over a contractible space B is fibre homotopic to the trivial fibration $p_1 : B \times F \to B$.*

Corollary 11.1.14 *Let $p : E \to B$ be a fibration. If $f : B' \to B$ is a homotopy equivalence then the canonical map $\hat{f} : E(f^*(p)) \to E$ is a fibre homotopy equivalence.*

Remark 11.1.15

(i) Corollary 11.1.13 can be viewed as a partial converse to the celebrated theorem of Hurewicz which says that every (locally trivial) fibre bundle map over a paracompact space is a fibration. We can at least say that if the base space B is locally contractible, (e.g., if B is a manifold) then every fibration over B is locally fibre homotopy trivial.

(ii) Also, a parallel result to Proposition 11.1.11 for locally trivial fibrations holds provided we assume B is paracompact, viz., a locally trivial bundle over $B \times \mathbb{I}$ where B is paracompact has the property that restrictions to each $B \times \{t\}$ are all isomorphic to each other (see [Husemoller, 1994]).

(iii) In particular, any fibration over the interval $\mathbb{I}$ is fibre-homotopy equivalent to a trivial fibration. Suppose now that the fibration $p : E \to B \times \mathbb{I}$ is such that

$$p : p^{-1}(b_0 \times \mathbb{I}) \to b_0 \times \mathbb{I}$$

is fibre homotopic to the trivial fibration $F \times \mathbb{I} \to b_0 \times \mathbb{I}$. Then in the proof of Proposition 11.1.11, it follows that $f_0(F \times 0) \subset F \times 1$ and similarly, $f_1(F \times 1) \subset F \times 0$. Thus the fibre homotopies of $f_1 \circ f_0$, etc., give us the additional information that restricted to F they are homotopic to the identity map of F. Consequently, if B is deformable to a point $b_0 \in B$, then for any fibration $p : E \to B$, there is a fibre homotopy equivalence $H : F \times B \to E$, where $F = p^{-1}(b_0)$ such that $H : F \times 0 \to F$ is homotopic to the identity map.

(iv) The above results describe when two pullback fibrations are equivalent. There is a result that says that given a space F there is a 'universal fibration' $\gamma_F : E_F \to B_F$ such that all fibrations over a paracompact space B and fibre homotopy type of F are obtained as pullback of γ_F via a map $f : B \to B_F$. Such a space B_F is unique up to homotopy type and is called the classifying space for F. Thus the set of homotopy classes $[B, B_F]$ classifies all possible fibrations up to fibre homotopy equivalence. We shall not go into much detail here and refer the reader to [Husemoller, 1994] for further reading.

Exercise 11.1.16

(i) Let $p : E \to B$ be a fibration. Let A be a strong deformation retract of a space X. Given $f : A \to E, g : X \to B$ such that $p \circ f = g|_A$, show that there is a map $\hat{g} : X \to E$ such that $p \circ \hat{g} = g$ and $\hat{g}|_A$ is fibre homotopic to f.

(ii) Let $p : E \to B \times \mathbb{I}$ be a fibration. Prove that p is fibre homotopy equivalent to the fibration $p_0 \times Id : E_0 \times \mathbb{I} \to B \times \mathbb{I}$, where $p_0 = p|_{p^{-1}(B \times 0)}$.

(iii) Prove the following theorem due to Serre by completing the details in the following five steps:

Serre's theorem: Let $E \to B$ be a fibration with fibre type F such that $\tilde{H}_q(F) = 0$ for all $q < n, (n \geq 1)$. Let $B_0 \neq \emptyset$ be a subspace of B such that (B, B_0) is $(m-1)$-connected. Put $E_0 = p^{-1}(B_0)$. Then $p_* : H_q(E, E_0) \to H_q(B, B_0)$ is an isomorphism $q < m + n$ and an epimorphism for $q = m + n$.

Step 1 May assume that (B, B_0) is a CW-pair having cells of dimension $\geq m$.

Step 2 The theorem is true if p is homotopically a trivial fibration.

Step 3 The theorem is true if B is obtained by attaching cells of dimension k for some fixed $k \geq m$, to B_0.

Step 4 By induction, the theorem is true for $(B, B_0)^{(k)}$.

Step 5 By taking direct limit, the proof is completed.

(iv) Give examples to show that the above result is sharp.

(v) Can we take $B_0 = \emptyset$ in the above theorem?

(vi) Obtain a proof of the following theorem by supplying details in the steps that follow:

Blakers–Massey homotopy excision theorem: Let $X = A \cup B$ satisfy the following conditions:

(a) X, A, B, and $C = A \cap B$ are all simply connected.

(b) (X, A) is $(m-1)$-connected, (X, B) is $(n-1)$-connected, $m, n \geq 2$.

(c) The inclusion induced homomorphism $H_q(B, C) \to H_q(X, A)$ is an isomorphism for $q < m + n - 2$ and an epimorphism for $q = m + n - 2$. Then the inclusion induced homomorphism $\pi_q(B, C) \to \pi_q(X, A)$ is an isomorphism for $q < m + n - 2$ and an epimorphism $q = m + n - 2$.

Step 1 Consider mapping path fibration $p : P = P(B, X) \to X$ of the inclusion map $\iota : B \to X$ (see Exercise 10.2.26.(vi)). Let F be its fibre over the chosen base point $x_0 \in C$. Then $\pi_q(F) \approx \pi_{q+1}(X, B)$ and hence F is $(n-2)$-connected.

Step 2 Let $P_0 = p^{-1}(A)$. Use Exercise (iii) above to conclude that

$$p_* : H_q(P, P_0) \to H_q(X, A)$$

is an isomorphism for $q \leq m + n - 2$ and an epimorphism for $q = m + n - 1$.

Step 3 Recall that there is a homotopy equivalence $s : B \to P$ such that $p \circ s = j : B \to X$ is the inclusion. Conclude that

$$s_* : H_q(B, C) \to H_q(P, P_0)$$

is an isomorphism for $q < m + n - 2$ and an epimorphism for $q = m + n - 2$ AND $s_* : H_*(B) \to H_*(P)$ is an isomorphism for all q.

Step 4 Appeal to Five lemma to conclude

$$(s|_C)_* : H_q(C) \to H_q(P_0)$$

is an isomorphism for $q < m + n - 3$ and an epimorphism for $q = m + n - 3$.

Step 5 Use the following diagram to conclude that $\pi_1(P_0)$ is abelian.

$$
\begin{array}{ccccc}
\pi_2(P, P_0) & \xrightarrow{\partial} & \pi_1(P_0) & \longrightarrow & \pi_1(P) = (0) \\
\approx \downarrow p_\# & & \downarrow p_\# & & \\
\pi_2(X) \longrightarrow & \pi_2(X, A) & \longrightarrow & \pi_1(A) = (0).
\end{array}
$$

Step 6 Combine Steps 4 and 5 to conclude P_0 is simply connected.

Step 7 Appeal to Whitehead's theorem to conclude that

$$(s|_C)_\# : \pi_q(C) \to \pi_q(P_0)$$

is an isomorphism for $q < m + n - 2$ and an epimorphism for $q = m + n - 2$.

Step 8 Combine this with the fact that $s : B \to P$ is a homotopy equivalence using Five lemma to arrive at the result.

(By interchanging A and B, we can also conclude that the inclusion induced homomorphism $\pi_q(A, C) \to \pi_q(X, B)$ is an isomorphism for $q < m+n-2$ and an epimorphism for $= m+n-2$. (Compare Exercise 10.2.26(v).)

(vii) By filling in the details in the following steps, obtain a proof of the following weaker version of

Freudenthal's suspension theorem Suppose that X is n-connected for some $n \geq 1$. Then $S : \pi_q(X) \to \pi_{q+1}(SX)$ is an isomorphism for $q \leq 2n$ and an epimorphism for $q = 2n + 1$.

[For the stronger version and an alternative proof see Theorem 11.3.7.]

Step 1: As in Theorem 4.2.21, show that the suspension homomorphism

$$S : [\mathbb{S}^q; X] \to [\mathbb{S}^{q+1}, SX]$$

is given by the composite

$$\pi_q(X) \xrightarrow[\approx]{\partial^{-1}} \pi_q(CX, X) \xrightarrow{\eta_\#} \pi_{q+1}(SX, X)$$

where $\partial : \pi_{q+1}(CX, X)$ is the boundary homomorphism of the exact homotopy sequence of the pair (CX, X) and η is the composite of the quotient map and the obvious homeomorphism $CX \to CX/X \to SX$.

Step 2 The quotient map $\eta : (CX, X) to (SX, \star)$ factors as the composite of an inclusion, a quotient and a homeomorphism:

$$(CX, X) \to (SX, C_-(X) \to (SX/C_-X, \{C_-X\}) \to (SX, \star).$$

(See Exercise 1.9.28.)

Step 3 Use the previous exercise to conclude that $j_\# : \pi_i(CX, X) \to \pi_i(SX, C_-X)$ is an isomorphism for $i \leq 2n + 1$ and an epimorphism for $q = 2n + 1$.

(viii) Let $n \geq 2, m \geq 1$. Let $A \subset X$ be a cofibration, A is m-connected and X is n-connected. Then the collapsing map $k : X \to X/A$ induces an isomorphism $k_\# : \pi_i(X, A) \to \pi_i(X/A, A/A)$ for $i \leq m+n$ and an epimorphism for $i = m+n+1$. [Hint: Use the cone CA and apply Exercise (vi) above.]

11.2   Thom Isomorphism Theorem

In this section, you will learn the proofs of two celebrated results: Thom isomorphism theorem and the Leray–Hirsch theorem. You shall also see the derivation of Gysin homology exact sequence for sphere bundles.

Fix a commutative ring R. We shall be considering homology and cohomology groups with coefficients in modules over R. For simplicity, we shall assume that the base space B is path connected. We are going to study fibrations over B with fibres having homotopy type of a sphere $\mathbb{S}^n$. The motivation for this comes from geometry of vector bundles E over a manifold M. When E is equipped with a Riemannian metric, the space of unit vectors in E becomes a sphere bundle over M, the study of which obviously has bearing on the study of the vector bundle. The fundamental results needed can be derived in the wider class of fibrations with fibres as homotopy spheres without any extra effort and are useful within algebraic topology.

Since the case $n = 0$ is somewhat too special (the statements need to be modified slightly and the conclusions trivial), we shall assume that $n \geq 1$.

Definition 11.2.1 Let $p : E \to B$ be a fibration with the fibre $F = p^{-1}(*)$ having homotopy type of $\mathbb{S}^n$, $n \geq 1$. Consider the mapping cylinder M_p of p and let $\hat{p} : M_p \to B$ be the associated strong deformation retraction (see 1.5.17). Put $\hat{F} = \hat{p}^{-1}(*)$, the cone over F. An element $u \in H^{n+1}(M_p, E; R)$ is called a *Thom class* or *an orientation class* for p if for every $b \in B$, $(\iota_b)^*(u) \in H^{n+1}(\hat{F}_b, F_b)$ is a generator. The fibration is called *orientable* if there exists an orientation class as above.

Remark 11.2.2 The above definition leads us to some natural questions: Why not consider similar definitions in dimensions lower than $n + 1$? How to ensure the existence of an orientation class? Is it unique if it exists? All these can be answered very satisfactorily in a fashion similar to the answers for similar questions about orientability of manifolds.

Clearly, $(\hat{F}, F)$ has homotopy type of $(\mathbb{D}^{n+1}, \mathbb{S}^n)$. Now, Lemma 11.3.2 implies that (M_p, E) is n-connected and $\pi_{n+1}(M_p, E)$ is an infinite cyclic group. Appealing to Hurewicz isomorphism theorem we easily deduce that $H_i(M_p, E) = (0)$ for $i \leq n$ and $H_{n+1}(\hat{F}, F; \mathbb{Z}) \to H_{n+1}(M_p, E; \mathbb{Z})$ is surjective. Moreover, this is an isomorphism iff $\pi_1(B) = \pi_1(M_p)$ operates trivially on $H_n(F; \mathbb{Z})$ via the isomorphisms

$$\pi_{n+1}(M_p, E) \approx \pi_{n+1}(\hat{F}, F) \approx H_{n+1}(\hat{F}, F; \mathbb{Z}) \approx H_n(F; \mathbb{Z}).$$

Since the cohomology modules of the pair $(\hat{F}, F)$ vanish below the dimension $n + 1$, we do not have to bother about lower dimensions. Since $H^{n+1}(\hat{F}, F)$ is infinite cyclic, existence of u implies that the inclusion induced map $H^{n+1}(M_p, E) \to H^{n+1}(\hat{F}_b, F_b)$ is surjective. Since $H^{n+1}(M_p, E)$ itself is infinite cyclic, surjectivity of the inclusion induced map gives us that it is actually an isomorphism and hence we can always take one of the generators of $H^{n+1}(M_p, E)$ to be a Thom class. This answers the uniqueness part: Thom class if it exists is unique up to sign. (Path connectivity of B has played its role here.) Finally the existence of Thom class is ensured if the inclusion induced homomorphism $H_{n+1}(\hat{F}, F) \to H_{n+1}(M_p, E)$ is an isomorphism which in turn holds iff the action of $\pi_1(B)$ on $H_n(F)$ is trivial. Do you notice any analogy between this and orientability of a manifold?

The Thom class satisfies certain naturality properties: Let $f : B' \to B$ be a map and $p' : E' \to B'$ be the pullback fibration:

$$
\begin{array}{ccc}
E' & \longrightarrow & E \\
{\scriptstyle p'}\downarrow & & \downarrow{\scriptstyle p} \\
B' & \xrightarrow{\ f\ } & B
\end{array}
$$

It follows that we have a commutative diagram

$$
\begin{array}{ccc}
(M_{p'}, E') & \xrightarrow{\ \tilde{f}'\ } & (M_p, E) \\
{\scriptstyle \hat{p}'}\downarrow & & \downarrow{\scriptstyle \hat{p}} \\
(B', B') & \xrightarrow{\ f\ } & (B, B).
\end{array}
\tag{11.1}
$$

Lemma 11.2.3 *If $u \in H^{n+1}(M_p, E)$ is a Thom class so is $v = (\tilde{f}')^*(u) \in H^{n+1}(M_{p'}, E')$ and we have the following commutative diagram*

$$
\begin{array}{ccc}
H_q(M_{p'}, E'; R) & \xrightarrow{\ (\tilde{f}')_*\ } & H_q(M_p, E; R) \\
{\scriptstyle \frown v}\downarrow & & \downarrow{\scriptstyle \frown u} \\
H_{q-n-1}(M_{p'}; R) & \xrightarrow{\ (\tilde{f}')_*\ } & H_{q-n-1}(M_p; R).
\end{array}
\tag{11.2}
$$

In particular, if $B_0 \subset B$ then every Thom class for $p : E \to B$ restricts to a Thom class of $p_0 : p^{-1}(B_0) \to B_0$.

Proof: The commutativity of the above diagram is just the naturality of the cap product. Note that the fibres of p' can be identified with the fibres of p, we have commutative diagrams:

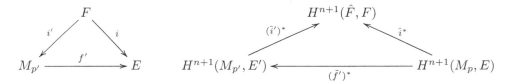

The surjectivity of $(\hat{i}')^*$ follows from that of $\hat{i}^*$. ♠

For any subspace $B_0 \subset B$, let us denote by p_0 the restricted fibration $p : p^{-1}(B_0) \to B_0$. Then the mapping cylinder M_{p_0} can be identified with the subspace $\hat{p}^{-1}(B_0)$ and the map $\hat{p}_0$ with the restriction of $\hat{p}$ to M_{p_0}. The main result in this section is the following theorem:

Theorem 11.2.4 (Thom) *Let $u \in H^{n+1}(M_p, E; R)$ be a Thom class where $p : E \to B$ is a n-spherical fibration. Then for all closed subspaces B_0 of B and all R-modules G we have isomorphisms:*

$$
\begin{aligned}
&\smile u : H^q(M_p, M_{p_0}; G) \to H^{q+n+1}(M_p, M_{p_0} \cup E; G); \\
&\frown u : H_q(M_p, M_{p_0} \cup E; G) \to H_{q-n-1}(M_p, M_{p_0}; G).
\end{aligned}
\tag{11.3}
$$

Proof: By taking $B_0 = \emptyset$, and pre-composing with the isomorphisms $p^* : H^*(B; G) \to H^*(M_p; G)$ or post-composing with the isomorphisms $p_* : H_*(M_p; G) \to H_*(B, G)$ we obtain the more popular version of the Thom isomorphisms:

$$
\begin{aligned}
&\Phi^* : H^q(B; G) \overset{\approx}{\to} H^{q+n+1}(M_p, E; G); \\
&\Phi_* : H_q(M_p, E; G) \overset{\approx}{\to} H_{q-n-1}(B; G).
\end{aligned}
\tag{11.4}
$$

We shall make a sequence of observations, each of which reduces the theorem to simpler situations until the final situation wherein the proof becomes completely easy.

First of all we observe that:

Lemma 11.2.5 If $\frown u : H_q(M_p, E; R) \to H_{q-n-1}(M_p; R)$ is an isomorphism for all q then the absolute version of the above theorem holds, viz., for all R-modules G and for all q,

$$\smile u : H^q(M_p; G) \to H^{q+n+1}(M_p, E; G);$$
$$\frown u : H_q(M_p, E; G) \to H_{q-n-1}(M_p, ; G).$$

are isomorphisms.

Proof: We fix a representative, $U \in Z^{n+1}(M_p, E; R)$ for the Thom class. Then $c \mapsto c \frown U$ defines a chain map

$$\alpha : A_* := C_*(M_p, E; R) \to C_*(M_p, ; R) =: B_*$$

of degree $-(n+1)$. By naturality of the exact sequences in the universal coefficient theorem we have a commutative diagram of exact sequences:

$$
\begin{array}{ccccccccc}
0 & \longrightarrow & H_q(A_*) \otimes G & \longrightarrow & H_q(A_*; G) & \longrightarrow & H_{q-1}(A_*) \star G & \longrightarrow & 0 \\
 & & \downarrow{\scriptstyle \alpha_* \otimes 1} & & \downarrow{\scriptstyle (\alpha \otimes 1)_*} & & \downarrow{\scriptstyle \alpha_* \star 1} & & \\
0 & \longrightarrow & H_{q-n-1}(B_*) \otimes G & \longrightarrow & H_{q-n-1}(B_*; G) & \longrightarrow & H_{q-n-2}(B_*) \star G & \longrightarrow & 0
\end{array}
$$

The hypothesis that $\frown u : H_q(M_p, E; R) \to H_{q-n-1}(M_p; R)$ is an isomorphism for all q is the same as saying that the 1st and the 3rd vertical arrows are isomorphisms which yields that the middle arrows are also isomorphisms for all q. This gives the conclusion in the homology.

Similarly, the cochain map $\mathrm{Hom}(\alpha, G) : C^*(M_p; G) \to C^*(M_p, E; G)$ induces isomorphisms in homology. But these are nothing but the cohomology homomorphisms given by $\smile u$. (See (h) of Theorem 7.3.5.)   ♠

So, in the rest of the proof, we are working with $G = R$ and hence suppress mentioning it for the sake of brevity. Also, we need only to prove the homology isomorphisms in (11.3).

Put $E_0 = p^{-1}(B_0)$ and $p_0 = p|_{E_0}$. We shall use the notation u itself to denote the image of u under the inclusion induced map $(M_{p_0}, E_0) \subset (M_p, E)$.

Lemma 11.2.6 If any two of the homomorphisms

$$\frown u : H_q(M_{p_0}, E_0) \to H_{q-n-1}(M_{p_0})$$
$$\frown u : H_q(M_p, E) \to H_{q-n-1}(M_p)$$
$$\frown u : H_q(M_p; M_{p_0} \cup E) \to H_{q-n-1}(M_p, M_{p_0})$$

are isomorphisms, (for every p), so is the third. In particular, it is enough to prove the theorem in the absolute case, i.e., with $B_0 = \emptyset$.

Proof: For the homology statement, consider the following commutative diagram shown in Figure 11.1 in which the two columns are part of the homology exact sequences of the triple $(M_p, M_{p_0} \cup E, E)$ and the pair (M_p, M_{p_0}), arrows labelled by η are excision isomorphisms, φ is defined so that $\varphi \circ \eta = \frown u$ and all arrows without any marking represent maps induced by appropriate inclusions. The commutativity of the entire diagram is obvious except perhaps the third rectangle involving the boundary homomorphisms. To see that this rectangle is also commutative, we use Theorem 7.3.5(e). Since the homomorphisms η are isomorphisms, saying that $\frown u : H_q(M_{p_0}, E_0) \to H_{q-n-1}(M_{p_0})$ are isomorphisms for all p is equivalent to saying that φ are isomorphisms.

We now appeal to Five lemma to conclude the result.   ♠

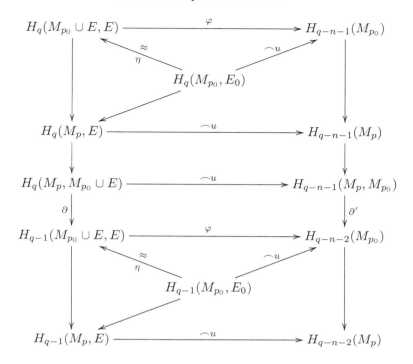

FIGURE 11.1. Reduction step in Thom isomorphism

Lemma 11.2.7 Given a weak homotopy equivalence $f : B' \to B$, let $v = (\tilde{f}')^*(u)$. Then the theorem is true if it is true for the induced fibration p' with v in place of u.

Proof: Let $p' : E' \to B'$ be the induced fibrations, etc. From the ladder of long homotopy exact sequences, and the Five lemma, it easily follows that $f' : E' \to E$ is a weak homotopy equivalence. In the commutative diagram (11.1), since the vertical arrows are homotopy equivalences, it follows that $f' : E' \to E$, and $\tilde{f}' : M_{p'} \to M_p$ are weak homotopy equivalences. Therefore by Whitehead's Theorem, we have isomorphisms:

$$f'_* : H_*(E';\mathbb{Z}) \to H_*(E;\mathbb{Z}); \quad \tilde{f}'_* : H_*(M_{p'};\mathbb{Z}) \to H_*(M_p;\mathbb{Z}).$$

Since $\tilde{f}'|_{E'} = f'$, by Five lemma, it follows that

$$\tilde{f}'_* : H_*(M_{p'}, E';\mathbb{Z}) \to H_*(M_p, E;\mathbb{Z})$$

are isomorphisms. By universal coefficient theorem, this is true with $\mathbb{Z}$ replaced by R. Now the commutative diagram (11.2) completes the proof of the lemma. ♠

Lemma 11.2.8 It is enough to prove the theorem when B is finite CW-complex.

Proof: Since every space is the weak homotopy type of a simplicial complex (see Theorem 10.4.16 and Exercise 2.11.11), the previous lemma reduces the proof of the theorem to the case when B is a simplicial complex. So, we assume B is a simplicial complex, and write

$$B = \varinjlim B_\alpha,$$

where B_α is the family of finite subcomplexes of B directed by the inclusions. Let $u_\alpha \in$

$H^{n+1}(M_{p_\alpha}, E_\alpha)$ be the restriction of the Thom class u. By naturality of the cap product it follows that

$$\frown u = \varinjlim \frown u_\alpha.$$

Therefore, in order to prove that $\frown u$ is an isomorphism, it suffices to prove this for the case when B is a finite simplicial complex. ♠

Lemma 11.2.9 Suppose Theorem 11.2.4 is true when B is a simplex, then it is true when B is any finite simplicial complex.

Proof: We shall prove this by induction on the number of simplexes in B using Lemma 11.2.6. The induction starts with the hypothesis here. Suppose now that the theorem is true for simplicial complexes with total number of simplexes $< N$. Let B be a simplicial complex which has N simplexes in it and let F be a maximal simplex in it. Take A to be the subcomplex of B in which only F is absent, put $B_0 = |A|, B_1 = |F|, B_2 = B_0 \cap B_1$ and $p_i = p|_{B_i}$. Then by induction hypothesis, the theorem is true for the fibration $p_i, i = 0, 1, 2$. By Lemma 11.2.6, we need to prove that

$$\frown u : H_q(M_p, M_{p_0} \cup E) \to H_{q-n-1}(M_p, M_{p_0})$$

are isomorphisms, which follows from the commutative diagram

$$
\begin{array}{ccc}
H_q(M_{p_1}, M_{p_2} \cup E_1) & \xrightarrow{\ \frown u_1\ } & H_{q-n-1}(M_{p_1}, M_{p_2}) \\
\downarrow & & \downarrow \\
H_q(M_p, M_{p_0} \cup E) & \xrightarrow{\quad \frown u \quad} & H_{q-n-1}(M_p, M_{p_0})
\end{array}
$$

in which the vertical arrows are excision isomorphisms. The top horizontal arrow is an isomorphism by hypothesis, because, B_1 is a single simplex. ♠

Since the underlying space of a single simplex is contractible, and since any fibration over a contractible space is fibre homotopy equivalent to a trivial fibration, the following lemma will complete the proof of Theorem 11.2.4.

Lemma 11.2.10 Let $p : \mathbb{S}^n \times B \to B$ be the projection to the second factor. Then for any Thom class $u \in H^{n+1}(M_p, E)$, $\frown u$ defines isomorphisms.

Proof: Note that $E = \mathbb{S}^n \times B, M_p = \mathbb{D}^{n+1} \times B$. Therefore by the Künneth formula, we have

$$
\begin{array}{llll}
H^{n+1}(M_p, E) & = & H^{n+1}((\mathbb{D}^{n+1}, \mathbb{S}^n) \times B) & \approx & H^{n+1}(\mathbb{D}^{n+1}, \mathbb{S}^n) \otimes H^0(B); \\
H_q(M_p, E) & = & H_q((\mathbb{D}^{n+1}, \mathbb{S}^n) \times B) & \approx & H_{n+1}(\mathbb{D}^{n+1}, \mathbb{S}^n) \otimes H_{q-n-1}(B); \\
H_{q-n-1}(M_p) & = & H_{q-n-1}(\mathbb{D}^{n+1} \times B) & \approx & H_0(\mathbb{D}^{n+1}) \otimes H_{q-n-1}(B).
\end{array}
$$

Under these isomorphisms u corresponds to $w^* \times 1$, where w^* is a generator of the free module $H^{n+1}(\mathbb{D}^{n+1}, \mathbb{S}^n)$. Let $w \in H_{n+1}(\mathbb{D}^{n+1}, \mathbb{S}^n)$ be the element dual to w^*, i.e., $\langle w^*, w \rangle = w \frown w^* = 1$. Then w generates the free module $H_{n+1}(\mathbb{D}^{n+1}, \mathbb{S}^n)$ and every element of $H_q(M_p, E)$ can be written as $w \times c$ for a unique $c \in H_{n-p-1}(B)$. Now

$$(w \times c) \frown u = (w \times c) \frown (w^* \times 1) = \pm (w \frown w^*) \times (c \frown 1) = \pm c.$$

Therefore $\frown u$ is an isomorphism. ♠

This completes the proof of Thom isomorphism Theorem 11.2.4. ♠

A ready to use consequence of Thom isomorphism is the following long homology sequence called the Gysin sequence of the sphere-bundle.

Theorem 11.2.11 (Gysin sequence) *Let R be a principal ideal domain and $\dot{E} \to B$ be a R-oriented n-spherical fibration with the Thom class $w \in H^{n+1}(M_p, E; R)$. Then for any closed subspace B_0 of B, and any coefficient module G there are long exact sequences of homology (and cohomology) groups with coefficients in G (which is suppressed):*

$$\cdots \longrightarrow H_{q-n}(B, B_0) \xrightarrow{\beta} H_q(E, E_0) \xrightarrow{p_*} H_q(B, B_0) \xrightarrow{\gamma} H_{q-n-1}(B, B_0) \longrightarrow \cdots$$

$$\cdots \longrightarrow H^{q-n-1}(B, B_0) \xrightarrow{\gamma^*} H^q(B, B_0) \xrightarrow{p^*} H^q(E, E_0) \xrightarrow{\beta^*} H^{q-n}(B, B_0) \longrightarrow \cdots$$

where the homomorphisms γ, γ^* are given by $\gamma(x) = x \frown w$; $\gamma^*(x) = w \smile x$ with $w = (\hat{p}^*)^{-1}j^*(u) \in H^{n+1}(B; R)$. (Here $j^* : H^{n+1}(M_p, E; R) \to H^{n+1}(M_p; R)$ is the inclusion induced homomorphism.)

Proof: Let us establish the exact homology sequence, the proof of the cohomology sequence being identical. We consider the long homology exact sequence of the triple $(M_p, M_{p_0} \cup E, M_{p_0})$. First thing to do is to replace the groups $H_*(M_{p_0} \cup E, M_{p_0})$ by groups $H_*(E, E_0)$ under excision isomorphisms. Next we replace $H_*(M_p, M_{p_0})$ by $H_*(B, B_0)$ under $\hat{p}_*$ since $\hat{p} : (M_p, M_{p_0}) \to (B, B_0)$ is a homotopy equivalence. Finally, we use Thom isomorphism followed by the isomorphism $\hat{p}_*$ to replace the groups $H_q(M_p, M_{p_0} \cup E)$ by the groups $H_{q-n-1}(B, B_0)$. This gives the homology exact sequence, as above. To prove the statement that $\gamma(x) = x \frown w$, consider the diagram

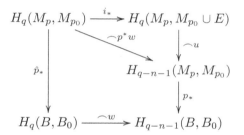

the commutativity of which follows from the naturality of the cap product. Now γ is, by definition, the homomorphism obtained by going up in the reverse of the first vertical arrow, going to the right and then coming all the way down along the two vertical arrows. ♠

The element w has a nice obstruction theoretic interpretation. We shall merely state this and refer the reader to Chapter VII Section 5 of [Whitehead, 1978].

Theorem 11.2.12 *Let $u \in H^{n+1}(M_p, E; R)$ be a Thom class. Then $w = (\hat{p}^*)^{-1}j^*(u)$ is equal to the image of the primary obstruction $\gamma^{n+1} \in H^{n+1}(B; R)$ under the canonical coefficient extension $\mathbb{Z} \to R$ of the primary obstruction to finding a cross section to the fibration $p : E \to B$.*

We are tempted to give a sort of a generalization of Thom's theorem, when the fibre is allowed to have a little more complicated homology than that of a sphere. Instead, we shall now revert to the more geometric situation and restrict ourselves to fibre bundles rather than fibrations. *The Leray–Hirsch Theorem* is a popular result here, which can be thought of as a generalization of Thom isomorphism.

Definition 11.2.13 A fibre bundle pair consists of a base B, a total pair $(\hat{E}, E)$, a fibre pair $(\hat{F}, F)$ and a projection map $p : \hat{E} \to B$ such that the following local triviality condition holds: There is an open covering $\{U_j\}$ of B and homeomorphisms $\phi_j : U_j \times (\hat{F}, F) \to$

$(p^{-1}(U_j), p^{-1}(U_j) \cap E)$, such that $p \circ \phi_j = p_1$,the projection to the first factor. For each $b \in B$, we put $(\hat{F}_b, F_b) = (p^{-1}(b), p^{-1}(b) \cap E)$ and call it the fibre pair over b. Fixing the base space, one can define the notion of a morphism from one bundle pair to another in an obvious way and this will give us the category of bundle pairs over the base space B.

Example 11.2.14

(i) Any product pair $B \times (\hat{F}, F)$ with the projection to the second factor is clearly an example of a fibre bundle pair. This is called the trivial fibre bundle pair. Indeed, any bundle pair which is equivalent to a product bundle pair will be called a trivial bundle pair.

(ii) Given a smooth manifold M with a Riemannian metric, we can consider the space DM of all tangent vectors of length less than or equal to 1 and the space of UM of unit tangent vectors. Then along with the restriction of the projection $p : TM \to M$ we obtain a fibre bundle pair $p : (DM, UM) \to M$ with fibre pair $(\mathbb{D}^n, \mathbb{S}^{n-1})$.

(iii) Indeed, given any fibre bundle $p : E \to B$ with fibre F, the mapping cylinder $\hat{p} : (M_p, E) \to B$ is a fibre bundle pair with fibre pair $(\hat{F}, F)$. Note that in the previous example, for the unit tangent bundle $p : UM \to M$, the map $\hat{p}$ is nothing but the corresponding disc-bundle $DM \to M$.

Definition 11.2.15 By a cohomology extension of the fibre of a bundle pair $p : (\hat{E}, E) \to B$ with fibre pair $(\hat{F}, F)$, we mean a degree 0 graded homomorphism $\theta : H^*(\hat{F}, F; R) \to H^*(\hat{E}, E; R)$ such that for every $b \in B$, the composite

$$H^*(\hat{F}, F; R) \xrightarrow{\theta} H^*(\hat{E}, E; R) \to H^*(\hat{F}_b, F_b)$$

is an isomorphism.

Remark 11.2.16

(i) It is easily seen that for the product bundle pair $p_1 : B \times (\hat{F}, F) \to B$, we have $\theta = p_1^*$ is a cohomology extension of the fibre.

(ii) In case of the mapping cylinder of a spherical bundle pair, one readily checks that having a Thom class is equivalent to having a cohomology extension of the fibre.

(iii) For any map $f : B' \to B$ and a fibre bundle pair $p : (\hat{E}, E) \to B$, consider the pullback fibre bundle pair $p' : (\hat{E}', E') \to B'$. Let $\hat{f} : (\hat{E}', E') \to (\hat{E}, E)$ be the map that covers f, i.e., $p \circ \hat{f} = f \circ f$. If θ is a cohomology extension of the fibre for p, then $\hat{f}^* \circ \theta$ is a cohomology extension of the fibre for p'.

(iv) Write $B = \sqcup B_j$ where B_j are path components of B. Then giving a cohomology extension θ for p over B is the same as giving a cohomology extension θ_j for the restricted bundles over B_j for each j.

Let $M = \oplus_r M_r$ be a finitely generated free graded module over R with a basis $\{m_i\}$ consisting of homogeneous elements and $M^* = \mathrm{Hom}(M; R)$. Given a map $f : X \to Y$ of topological spaces a subspace $A \subset X$, and a degree 0 homomorphism $\theta : M^* \to H^*(X, A; R)$, let us define

$$\Theta_f : H_*(X, A; G) \to H_*(Y; G) \otimes M; \quad \Theta_f^* : H^*(Y; G) \otimes M^* \to H^*(X, A; G)$$

by the formula

$$\Theta_f(z) = \sum_i f_*(\theta(m_i^*) \frown z) \otimes m_i; \quad \Theta_f^*(u \otimes m_i^*) = f^* u \frown \theta(m_i^*) \tag{11.5}$$

where $\{m_i^*\}$ are the dual basis elements.

We can now state:

Theorem 11.2.17 (Leray–Hirsch) *Let $p : (\hat{E}, E) \to B$ be a fibre bundle pair with a cohomology extension of the fibre $\theta : H^*(\hat{F}, F; R) \to H^*(\hat{E}, E; R)$ where $H_*(\hat{F}, F)$ is a finitely generated and free graded R-module. Then for any coefficient R-module G, we have the isomorphisms*

$$\Theta = \Theta_p : H_*(\hat{E}, E; G) \to H_*(B, G) \otimes H_*(\hat{F}, F; G); \tag{11.6}$$

$$\Theta^* = \Theta_p^* : H^*(B; G) \otimes H^*(\hat{F}, F; G) \to H^*(\hat{E}, E; G) \tag{11.7}$$

given by

$$\Theta(z) = \sum_i p_*(\theta(m_i^*) \frown z) \otimes m_i; \quad \Theta^*(u \times v) = p^*(u) \smile \theta(v). \tag{11.8}$$

The proof of this theorem is exactly parallel to the proof of Thom's theorem, with a few technical modifications. The first step is the analogue of Lemma 11.2.5 with a similar proof, which is purely algebraic.

Lemma 11.2.18 Suppose that Θ_f is an isomorphism for $G = R$. Then for R-modules G Θ_f, Θ_f^* are isomorphisms.

Proof: Let the degree of m_i be q_i. For each i, let U_i^* be the cocycles representing $\theta(m_i^*)$. Consider the chain map of degree 0

$$\tau : S_.(X)/S_.(A) \to S_.(Y) \otimes M$$

given by

$$\tau(c) = \sum_i f_*(U_i^* \frown c) \otimes m_i.$$

One readily checks that in homology and cohomology, we have $\tau_* = \Phi_f$ and $\tau^* = \Phi_f^*$. Now by the universal coefficient theorem for homology and the same for cohomology, the result follows. ♠

Thus the theorem is reduced to proving the homology part of the theorem and for $G = R$. Next, the reduction to the case when the fibre bundle is a product, is exactly the same as that for Theorem 11.2.4. However, for the trivial case, we must include the relative version of the theorem as well. So, we shall now prove the theorem for the trivial fibre bundle pair $B \times (\hat{F}, F)$.

Lemma 11.2.19 Theorem (11.2.17) holds for the product bundle pair $B \times (\hat{F}, F)$.

Proof: Remember that we need to prove only the homology isomorphisms and only for coefficients $G = R$, which we suppress. We can also assume that B is path connected. We then have the Künneth formula

$$H_*(B \times (\hat{F}, F)) \approx H_*(B) \otimes H_*(\hat{F}, F).$$

Therefore we can view Θ as an endomorphism of the graded module $H_*(B) \otimes H_*(\hat{F}, F)$.

Consider the graded submodules $N_s = \oplus_q (N_s)_q$ of $H_*(B) \otimes H_*(\hat{F}, F)$ given by

$$(N_s)_q = \oplus_{i+j=q, j \geq s} H_i(B) \otimes H_j(\hat{F}, F).$$

Clearly

$$H_*(B) \otimes H_*(\hat{F}, F) = N_0 \supset N_1 \supset \cdots \supset N_s \cdots$$

and $N_s = 0$ for all large s because $H_*(\hat{F}, F)$ is finitely generated. We claim that
(i) $\Theta(N_s) \subset N_s$ and hence induces endomorphisms of the quotients

$$\Theta_s : N_s/N_{s+1} \to N_s/N_{s+1}.$$

(ii) For each s, Θ_s is an isomorphism.
Once these claims have been established, because of the short exact sequences

$$0 \to N_{s+1} \to N_s \to N_s/N_{s+1} \to 0$$

by reverse induction, it would follow that Θ is an isomorphism.

Proof of Claim (i): First we observe that $p_*(N_1) = 0$. Now, consider an element of the form $z \times z' \in N_s$ where $z' \in H_q(\hat{F}, F), q \geq s$. Then using the distributive property of cap product over the cross product (see Theorem 7.3.5 (f)), it follows that for any $\lambda \in H_*(B) \times H_*(\hat{F}, F)$ of degree $< s$, $\lambda \cap (z \times z') \in N_1$. Therefore in the summation

$$\Theta(z \times z') = \sum_i p_*(\theta(m_i^*) \frown (z \times z')) \otimes m_i$$

only those terms with $\deg m_i \geq s$ survive, which means that $\Theta(z \times z') \in N_s$. Therefore $\Theta(N_s) \subset N_s$.

Proof of Claim (ii): Given $u \in H^s(\hat{F}, F)$, we can write $\theta(u) = 1 \times \gamma(u) + u'$ where $u' \in \oplus_{i+j=s, j<s} H^i(B) \otimes H^j(\hat{F}, F)$. Clearly for any $b \in B$, $u'|_{b \times (\hat{F}, F)} = 0$. Therefore, $\theta(u)|_{b \times (\hat{F}, F)} = 1 \times \gamma(u)$. Thus, by hypothesis it follows that the assignment $u \mapsto \gamma(u)$ defines an automorphism of $H^*(\hat{F}, F)$. Let γ^* be the dual automorphism of $H_*(\hat{F}, F)$. We claim that for $z' \in H_s(\hat{F}, F)$,

$$\Theta(z \times z') = z \otimes \gamma^*(z')$$

modulo N_{s+1}. Since elements of N_s/N_{s+1} can be represented by sums of the form $\sum_j z_j \times z_j'$ with $z_j' \in H_s(\hat{F}, F)$, this will imply that Θ_s is an isomorphism. So, modulo N_{s+1}, we have

$$
\begin{aligned}
\Theta(z \times z') &= \textstyle\sum_{\deg m_i = s} p_*[(1 \times \gamma(m_i^*) + m_i^{*\prime}) \frown z \times z'] \otimes m_i \\
&= p_*(1 \times \gamma(m_i^*) \frown z \times z') \otimes m_i; \quad (\text{because } m_i^{*\prime} \frown (z \times z') \in N_1) \\
&= z \otimes \langle \gamma(m_i^*), z' \rangle m_i \quad (\text{use (f) of Theorem 7.3.5}) \\
&= z \otimes \gamma_*(z')
\end{aligned}
$$

Along with the proof of lemma, this completes the proof of Theorem 11.2.17.   ♠

11.3   Fibrations over Suspensions

This section will contain another milestone result in homology of fibre spaces, viz., the Wang sequence for homology as well as cohomology of a fibre space over a suspension and in particular over a sphere. As an easy consequence, we obtain a proof of the stronger form of Freudenthal's homotopy suspension theorem.

Let us begin with a space Y with a non degenerate base point y_0. Recall the definition of unreduced and reduced suspensions from Example 10.1.9 and denote them, respectively by $S'Y, SY$. Let $C'_{\pm} Y$ denote the subspaces of $S'Y$ which are the images of $Y \times [0, 1/2], Y \times [1/2, 1]$, respectively. Similarly let $C_{\pm} Y$ denote the image of $Y \times [0, 1/2]$ and $Y \times [1/2, 1]$ in SY. Let $\kappa : S'Y \to SY$ denote the quotient map which collapses $y_0 \times \mathbb{I}$ to a single point. We shall identify Y with the image of the subspace $Y \times 1/2$ both in $S'Y$ as well as SY.

We consider the suspension map

$$S : [X; Y] \to [SX; SY]$$

given by $S([f]) = [Sf]$. Composing this with the exponential correspondence

$$\eta : [SX; SY] \approx [X; \Omega SY],$$

we obtain a map

$$\hat{S} : [X; Y] \to [X; \Omega SY].$$

Let us define $\rho : Y \to \Omega SY$ by

$$\rho(y)(t) = [y, t]. \tag{11.9}$$

It follows easily that $\hat{S} = \eta \circ S = \rho_\#$. This just means that $\hat{S}$ is actually induced by the map $\rho : Y \to \Omega SY$ and hence the study of $\hat{S}$ is converted into the study of $\rho_\#$. But then we notice that the space ΩSY is the fibre of the path fibration $PSY \to SY$. This leads us to the study of the homology properties of fibrations over the suspensions. We begin with another corollary to Lemma 10.1.11.

Corollary 11.3.1 Let $y_0 \in Y$ be a non degenerate base point. For any fibration $p : E \to SY$, put $E_\pm = p^{-1}(C_\pm Y)$. Then $\{E_+, E_-\}$ is an excisive couple in E for the singular homology.

Proof: Let $p' : E' \to SY$ be the pullback fibration $\kappa^*(p)$. From Lemma 10.1.11, there are homotopies of pairs involving E, E'_+, E'_- with the corresponding pairs involving E, E_+, E_-, etc., induced by the homotopy equivalence $\bar{\kappa} : E' \to E$. Therefore we need to prove the same statement over the unreduced suspension, viz., $\{E'_+, E'_-\}$ is an excisive couple in E'. If $U_+Y = CY \setminus [Y \times 1], U_-Y = CY \setminus [Y \times 0]$, then we have $U_+Y \cap C_-Y = Y \times [1/2, 1)$ and $U_-Y \cap C_+Y = Y \times (0, 1/2]$. Therefore p restricted over these sets are fibre homotopy equivalent to products $p' \times Id$ where $p' : p^{-1}(Y) \to Y$ is the restriction of p. It follows that $E'_\pm$ are strong deformation retracts of $p^{-1}(U_\pm)$, respectively. Since $\{E_\pm\}$ forms an open cover for E, the conclusion follows. ♠

Definition 11.3.2 Let $y_0 \in Y$ be a non degenerate base point, $p : E \to SY$ be a fibration. Put $F = p^{-1}([y_0, 1/2])$. Let $f_- : C_-Y \times F \to E_-$, $g_+ =: E_+ \to C_+Y \times F$ be fibre homotopy equivalences such that $z \mapsto f_-([y_0, 1/2], z)$ and $z \mapsto \pi_2 \circ g(z)$ are homotopic to the identity maps (see Remark 11.1.15.(iii). Let $\varphi : Y \times F \to F$ be defined by the equation

$$g_+ \circ f_-(y, z) = (y, \varphi(y, z)).$$

We call φ a *clutching function* for the fibration p.

Remark 11.3.3

1. It follows that we can assume and do assume that $\varphi|_{y_0 \times F}$ is homotopic to the map $(y_0, z) \mapsto z$. Indeed, if and when necessary, we can assume $\varphi_{y_0 \times F}$ is actually equal to the map $(y_0, z) \mapsto z$.

2. Conversely, given a function $\phi : Y \times F \to F$ such that $\phi|_{y_0 \times F}$ is homotopic to the map $(y_0, z) \mapsto z$, we can construct a fibration $q : X \to SY$ where X is the quotient space

$$\frac{C_+Y \times F \coprod C_-Y \times F}{(y, z) \sim (y, \phi(y, z))}$$

and q is induced by the projection maps on each of the two parts. It follows that for a path connected space Y, fibre homotopy classes of fibrations over SY with fibre F are in one-one correspondence with the homotopy classes of maps $Y \to \mathcal{H}_e(F)$, where $\mathcal{H}_e(F)$ denotes the space of homotopy equivalences of F which are homotopic to Id_F.

Consider the following commutative diagram

$$\cdots \longrightarrow H_q(E) \xrightarrow{\ j_* \ } H_q(E,F) \xrightarrow{\ \partial \ } H_{q-1}(F) \xrightarrow{\ i_* \ } \cdots$$

$$\approx \Big\downarrow j_1 \qquad\qquad j_2 \Big\downarrow \approx$$

$$H_q(E,E_+) \xrightarrow{\ \partial \ } H_{q-1}(E_+)$$

$$\approx \Big\uparrow \eta_1 \qquad\qquad \eta_2 \Big\uparrow \approx$$

$$H_q(E_-, E_+ \cap E_-) \xrightarrow{\ \partial \ } H_{q-1}(E_+ \cap E_-)$$

$$\approx \Big\uparrow f_{-*} \qquad (f_-|_{Y\times F})_* \Big\uparrow \approx$$

$$H_q(C_-Y, Y) \times F \xrightarrow{\ \partial \ } H_{q-1}(Y \times F)$$

where the top row is the homology exact sequence of the pair (E, F) and j_1, j_2, η_1, η_2 are induced by appropriate inclusion maps. Since j_1, j_2 and f_{-*} are induced by homotopy equivalences, and η_1 is an excision map, we have the corresponding arrows marked as isomorphisms. We wish to replace $H_q(E, F)$ by $H_q((C_-Y, Y) \times F)$ in the long homology exact sequence. So, we take

$$\alpha_* = f_{-*}^{-1} \circ \eta_1^{-1} \circ j_1 \circ j_*; \quad \beta_* = j_2^{-1} \circ \eta_2 \circ (f_-|_{Y\times F})_* \circ \partial$$

to get the exact sequence

$$\cdots \longrightarrow H_q(E) \xrightarrow{\ \alpha_* \ } H_q(C_-Y, Y) \times F \xrightarrow{\ \beta_* \ } H_{q-1}(F) \xrightarrow{\ i_* \ } \cdots$$

The homomorphism α_* is fairly easy to understand, being the composite of various inclusion induced maps. We now wish to give a precise description of β_*. First we observe that there is a homotopy equivalence $g_+ : E_+ \to C_+Y \times F$ with $g_+|_F$ being homotopic to the projection $z \mapsto (y_0, z)$. Therefore the composite isomorphism

$$H_{q-1}(E_+) \xrightarrow{\ (g_+)_* \ } H_{q-1}(C_+Y \times F) \to H_{q-1}(F)$$

is equal to j_*^{-1}. By definition φ is equal to the composite of

$$Y \times F \xrightarrow{\ f_-|_{Y\times F} \ } E_+ \cap E_- \lhook\joinrel\longrightarrow E_+ \xrightarrow{\ g_+ \ } C_+Y \times F \longrightarrow F.$$

Therefore $\beta_* = \varphi_* \circ \partial$.

Similar discussion holds with cohomology modules as well.

Thus we have proved:

Theorem 11.3.4 (Wang) *Given a fibration $p : E \to SY$, $F = p^{-1}(y_0)$, where y_0 is a non degenerate base point of Y, there are exact sequences*

$$\cdots \longrightarrow H_q(E) \xrightarrow{\ \alpha_* \ } H_q((C_-Y, Y) \times F) \xrightarrow{\ \varphi_* \circ \partial \ } H_{q-1}(F) \xrightarrow{\ i_* \ } H_{q-1}(E) \longrightarrow \cdots$$

$$\cdots \longrightarrow H^q(E) \xrightarrow{\ i^* \ } H^q(F) \xrightarrow{\ \delta \circ \varphi^* \ } H^{q+1}((C_-Y, Y) \times F) \xrightarrow{\ \alpha^* \ } H^{q+1}(E) \longrightarrow \cdots$$

where $\varphi : Y \times F \to F$ is a clutching function for p and homology and cohomology modules are taken with any coefficient module.

Let us specialize to the case when $Y = \mathbb{S}^{n-1}$, i.e., when we have a fibration $p : E \to \mathbb{S}^n$. Clearly, $(C_-Y, Y) = (\mathbb{D}^n, \mathbb{S}^{n-1})$. Let ξ_n denote a generator of $H_n(\mathbb{D}^n, \mathbb{S}^{n-1})$. Then the map $u \mapsto \xi_n \times u$ defines an isomorphism

$$\tau : H_{q-n}(F) \to H_q((C_-Y, Y) \times F).$$

Put $\gamma_* = \tau^{-1} \circ \alpha_*$ and $\theta_* = \varphi_* \circ \partial \circ \tau$, and replace the term $H_q(C_-Y, Y \times F)$ by $H_{q-n}(F)$ in the Wang sequence. By similar consideration with cohomology as well, we obtain:

Theorem 11.3.5 (Wang) *For any fibration $p : E \to \mathbb{S}^n$ with fibre F we have exact sequences:*

$$\cdots \xrightarrow{i_*} H_q(E) \xrightarrow{\gamma_*} H_{q-n}(F) \xrightarrow{\theta_*} H_{q-1}(F) \xrightarrow{i_*} H_{q-1}(E) \longrightarrow \cdots$$

$$\cdots \xrightarrow{\gamma^*} H^q(E) \xrightarrow{i^*} H^q(F) \xrightarrow{\theta^*} H^{q+1-n}(F) \xrightarrow{\gamma^*} H^{q+1}(E) \longrightarrow \cdots$$

where γ^* has the following linearity property,

$$\gamma^*(x \smile i^*(y)) = \gamma^*(x) \cup y, \quad x \in H^{q-n}(F), y \in H^q(E). \tag{11.10}$$

Moreover, when the coefficient module is the commutative ring R with a unit, θ^* is a derivation, i.e.,

$$\theta^*(u \smile v) = \theta^*(u) \smile v + (-1)^{(n-1)\deg u} u \smile \theta^*(v); \quad \text{and } \gamma^*(1) = p^*(\xi_n^*). \tag{11.11}$$

Proof: We need to prove only (11.10) and (11.11). We have the commutative diagram

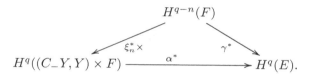

Therefore,

$$\gamma^*(x \smile i^*(y)) = \alpha^*(\xi_n^* \times (x \smile i^*(y))) = \alpha^*((\xi_n^* \times x) \smile i^*(y)) = \alpha^*(\xi_n^* \times x) \smile \alpha^* i^*(y) = \gamma^*(x) \smile y,$$

the last equality holds because the fibre homotopy equivalence $f : (C_-Y, Y) \times F \to E_-$ is such that on the fibre over the base point it is homotopic to the identity map. This proves (11.10).

Next, the homomorphism θ is the composite of

$$H^q(F) \xrightarrow{\delta \circ \varphi^*} H^{q+1}((\mathbb{D}^n, \mathbb{S}^{n-1}) \times F) \longrightarrow H^{q+1-n}(F).$$

The homomorphism φ^* has the property that

$$\varphi^*(u) = 1 \times u + s^* \times \theta(u)$$

where $s^* \in H^{n-1}(\mathbb{S}^{n-1})$ is a suitable generator. Therefore,

$$\begin{aligned}
&1 \times (u \smile v) + s^* \times \theta^*(u \smile v) \\
=\ &\varphi^*(u \smile v) = \varphi^*(u) \smile \varphi^*(v) \\
=\ &[1 \times u + s^* \times \theta^*(u)] \smile [1 \times v + s^* \times \theta^*(v)] \\
=\ &1 \times (u \smile v) + s^* \times [\theta^*(u) \smile v + (-1)^{(n-1)\deg u} u \smile \theta^*(v)].
\end{aligned}$$

From this the first part of (11.11) follows.

From (11.10), it follows that the image of γ^* is an ideal in $H^*(E)$, such that the product of any two elements is zero, i.e., $\gamma^*(a) \frown \gamma^*(b) = \gamma^*(a \frown i^* \gamma^*(b)) = 0$, since $i^* \circ \gamma^* = 0$. Again, because of (11.10), it becomes important to know what is $\gamma^*(1)$, where $1 \in H^0(F) \approx R$ is the unit. For this, consider the diagram

$$((\mathbb{D}^n, \mathbb{S}^{n-1}) \times F) \xrightarrow{f_-} (E_-, E_+ \cap E_-) \xrightarrow{\eta_1} (E, E_+) \xleftarrow{j_1} (E, F) \longleftarrow E$$

where p_1 is the projection to the first factor and we have denoted various restrictions of $p : E \to \mathbb{S}^n$ by p itself. The triangle is homotopy commutative and the rest of the diagram is commutative. Denoting the generator of $H^n(\mathbb{D}^n, \mathbb{S}^{n-1})$ by ξ_n^* and the corresponding generator of $H^n(\mathbb{S}^n)$ by the same symbol $\bar{\xi}_n^*$, by the homotopy commutativity of the triangle, it follows that $(f_-)^* p^*(\bar{\xi}_n^*) = p_1^*(\bar{\xi}_n^*)$. Therefore

$$
\begin{aligned}
\gamma^*(1) &= \alpha^* \tau^*(1) \\
&= \alpha^*(\bar{\xi}_n^* \times 1) \\
&= \alpha^* p_1^*(\bar{\xi}_n^*) \\
&= j^* \circ j_1^* \circ (\eta_1^*)^{-1} \circ ((f_-)^*)^{-1}(p_1^* \bar{\xi}_n^*) \\
&= j^* \circ j_1^* \circ (\eta_1^*)^{-1} p^*(\bar{\xi}_n^*) \\
&= p^*(\xi_n).
\end{aligned}
$$

This completes the proof of the theorem.   ♠

Next, we specialize to the path fibration $p : PSY \to SY$ with fibre ΩSY. The characteristic function $\rho : Y \to \Omega SY$ now enters the description of a clutching function for the fibration p.

Lemma 11.3.6 Let $\rho : Y \to \Omega SY$ be defined as in (11.9). Then the function

$$\varphi(y, \omega) = \omega * \rho(y)$$

is a clutching function for the path fibration $p : PSY \to SY$.

Proof: Define

$$s_+(y)(t) = [y, t/2], \ \ s_-(y)(t) = [y, (t-1)/2], \ \ 0 \le t \le 1, \ \ y \in Y.$$

Then $s_\pm : Y \to PSY$ are such that $p \circ s_\pm(y) = [y, t/2] = y$. Indeed both $s_\pm$ can be extended to sections of p over $C_\pm Y$, respectively, viz.,

$$s_+([y, s])(t) = [y, st], \ \ 0 \le s \le 1/2; \ \ s_-([y, s])(t) = [y, s(t-1)/2], 1/2 \le s \le 1, \ \ 0 \le t \le 1.$$

We can use $s_\pm$ to define fibre homotopy equivalences:

$$f_- : C_- Y \times \Omega SY \to p^{-1}(C_- Y); \ \ g_+ : p^{-1}(C_+ Y) \to C_+ Y \times \Omega SY$$

given by

$$f_-([y, s], \omega) = \omega * s_-([y, s]); \ \ g_+(\omega) = (p(\omega), \omega * s_+(p(\omega))^{-1}).$$

The maps f_+ and g_- are similarly defined and one can easily verify that $g_- \circ f_-, f_- \circ g_-$ are fibre homotopic to the respective identity maps and hence f_- is a fibre homotopy equivalence. Likewise, g_+ is also a fibre homotopy equivalence. Finally, $s_-(y_0)(t) = s_-([y_0, 0])(t) =$

$[y_0, 0] = y_0$ and hence $f_-(y_0, \omega) = \omega * s_-(y_0)$ is homotopic to the map $(y_0, \omega) \mapsto \omega$. Similar property holds for g_+ also. Therefore, the composite

$$Y \times \Omega SY \xrightarrow{f_-} p^{-1}(Y) \xrightarrow{g_+} Y \times \Omega SY \longrightarrow \Omega SY$$

is a clutching function which is nothing but φ defined above. ♠

Theorem 11.3.7 *Let Y be n-connected $(n \geq 0)$ and having a non degenerate base point. Then the characteristic function $\rho : Y \to \Omega SY$ of the path fibration $p : PSY \to SY$ induces isomorphisms*

$$\rho_* : H_q(Y) \to H_q(\Omega SY), \quad 0 \leq q \leq 2n + 1.$$

Proof: Since $E = PSY$ is contractible, the homology exact sequence of Theorem 11.3.4 yields

$$\varphi_* \partial : H_q((C_-Y, Y) \times \Omega SY) \approx \tilde{H}_{q-1}(\Omega SY).$$

By homology suspension theorem $H_q(SY) = H_{q-1}(Y) = 0, 1 \leq q \leq n + 1$. Since SY is simply connected as well, by the Hurewicz theorem, SY is $(n + 1)$-connected. Therefore ΩSY is n-connected. Moreover, (C_-Y, Y) is $(n + 1)$-connected. Let ω_0 denote the constant loop at y_0. By the Künneth formula, the inclusion map $(C_-Y, Y) \times \omega_0 \subset (C_-Y, Y) \times \Omega SY$ induces isomorphisms

$$H_q(C_-Y, Y) \times \omega_0) \approx H_q((C_-Y, Y) \times \Omega SY), \quad q \leq 2n + 2.$$

By Lemma 11.3.6, we can choose the clutching function $\varphi : Y \times \Omega SY \to \Omega SY$ to be such that $\varphi(y, \omega) = \omega * \rho(y)$ and hence $\varphi(y, \omega_0) = \omega_0 * \rho(y)$ is homotopic to ρ. It follows that we have a commutative diagram:

$$
\begin{array}{ccccc}
H_q(C_-Y, Y) & \xrightarrow{\approx} & H_q((C_-Y, Y) \times \omega_0) & \xrightarrow{\approx} & H_q((C_-Y, Y) \times \Omega SY) \\
\approx \downarrow \partial & & \approx \downarrow \partial & & \approx \downarrow \partial \\
\tilde{H}_{q-1}(Y) & \xrightarrow{\approx} & \tilde{H}_{q-1}(Y \times \omega_0) & \xrightarrow{\approx} & H_{q-1}(Y \times \Omega SY) \\
& \rho_* \searrow & \downarrow & \swarrow \varphi_* \\
& & \tilde{H}_{q-1}(\Omega SY).
\end{array}
$$

The result follows. ♠

As an immediate corollary we obtain:

Theorem 11.3.8 (Freudenthal) *Let Y be an n-connected space $(n \geq 1)$ with a non degenerate base point. Then for any CW-complex X the suspension map*

$$S : [X, Y] \to [SX; SY]$$

is surjective if $\dim X \leq 2n + 1$ and bijective for $\dim X \leq 2n$.

Proof: As seen just after (11.9), the problem gets converted into proving similar assertions about $\tilde{S} = \rho_\#$. Since Y and ΩSY are simply connected, it follows from the above theorem and Whitehead's theorem 10.4.11 that ρ is a $(2n + 1)$-equivalence. Now appeal to Theorem 10.2.16. ♠

Often the following corollary goes under the name Freudenthal suspension theorem, which we have proved in Exercise 11.1.16.(vii) by a different method.

Corollary 11.3.9 For any space X which is n-connected ($n \geq 1$) the suspension homomorphism $S : \pi_q(X) \to \pi_{q+1}(SX)$ is an isomorphism for $q \leq 2n$ and surjective for $q = 2n + 1$.

We shall end this section with an example of a useful computation in which we make full use of cohomology Wang sequence:

Example 11.3.10 Let X be a $K(\mathbb{Z}, 3)$-space and $\eta : \mathbb{S}^3 \to X$ be a map representing a generator of $\pi_3(X)$. Let $p : E = M^\eta \to \mathbb{S}^3$ be the principal fibration induced by η. Let us compute $H_*(E)$ by first computing $H^*(E)$ for which we shall use the Wang sequence of the fibration p. The fibre F of p is of type $K(\mathbb{Z}, 2) \approx \mathbb{CP}^\infty$. Therefore, $H^*(F) \approx \mathbb{Z}[x]$, the polynomial algebra generated by a single element of degree 2.

Since η is the generator, it follows that $\partial : \pi_3(\mathbb{S}^3) \to \pi_2(F)$ is an isomorphism and hence E is 3-connected. Therefore $H^1(E) = H^2(E) = H^3(E) = (0)$. By the cohomology Wang exact sequence of the fibration,

$$0 = H^2(E) \longrightarrow H^2(F) \xrightarrow{\ \theta^*\ } H^0(F) \longrightarrow H^3(E) = 0$$

it follows that $\theta^* : H^2(F) \to H^0(F)$ is an isomorphism.

We pick up a generator u for $H^2(F) \approx \mathbb{Z}$ such that $\theta^*(u) = 1$. By the derivation property of θ^*, it follows that

$$\theta^*(u^n) = n u^{n-1}.$$

Now, again by the Wang exact sequence, it follows that $H^q(E) = 0$ for q odd and for $q = 2n, n \geq 2$, $H^q(E) = \mathbb{Z}_n$. By appealing to the cohomology universal coefficient theorem, we conclude that

$$H_q(E) = \begin{cases} \mathbb{Z}, & q = 0; \\ 0, & q \text{ odd}; \\ \mathbb{Z}_n, & q = 2n. \end{cases}$$

These computations will become handy in establishing Serre's theorem on homotopy groups of spheres.

11.4   Cohomology of Classical Groups

Throughout this section X will denote a connected space with a non degenerate base point. Homology and cohomologies will be taken with coefficients over R which is a PID. As usual, we shall suppress R from the notation. We shall first give some generalities about Hopf algebras and state the structure theorem for Hopf algebras due to Borel. We shall then compute the cohomology algebras of some classical groups.

Recall that $H_*(X)$ is said to be of *finite type* (over R) if all $H_i(X)$ are finitely generated, and *free of finite type* if each $H_i(X)$ is finitely generated and free (over R). Assume for a while this is the case. Then by the Künneth formula, we have the isomorphism given by the cross product

$$H_*(X) \otimes H_*(X) \to H_*(X \times X)$$

with which we identify these two modules. Then the diagonal map $\Delta : X \to X \times X$ can be thought of as inducing a homomorphism

$$\Delta_* : H_*(X) \to H_*(X) \otimes H_*(X)$$

which has properties akin to the H'-spaces that we have discussed. So, we can call such a structure a co-algebra.

Further, assume for a while that there is an H-space structure $\mu : X \times X \to X$. Then we get a homomorphism

$$\mu_* : H_*(X \times X) \to H_*(X).$$

Combining this with the homology cross product we get a pairing

$$H_*(X) \otimes H_*(X) \to H_*(X)$$

which makes $H_*(X)$ into a graded R-algebra. This is called the *Pontrjagin Algebra* of X. This algebra structure and the co-algebra structure have a certain inter-relation as well which we shall soon see. This motivates the following definitions. We follow the book [Whitehead, 1978] closely, but not exactly.

Definition 11.4.1 A *Hopf algebra* over R is a graded R-module $H = \oplus_{n \geq 0} H_n$, together with R-module homomorphisms

$$\varphi : H \otimes H \to H; \qquad \eta : R \to H$$
$$\psi : H \to H \otimes H; \qquad \epsilon : H \to R$$

fitting the following two commutative diagrams:

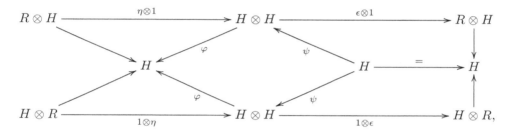

wherein all the unlabelled arrows indicate the canonical isomorphism $r \otimes x \mapsto rx$, and τ is the factor-interchanging map (graded version)

$$\tau(x \otimes y) = (-1)^{(\deg x)(\deg y)}(y \otimes x)$$

Because of the top two commutative triangles in the above diagram, H becomes an algebra and because of the two commutative quadrilaterals there, it is a co-algebra. The maps η and ϵ are called unit of the product and the co-unit of the co-product, respectively. The bottom rectangle tells you that ψ is an endomorphism of the algebra or equivalently, φ is an endomorphism of the co-algebra. (The twisting factor along with the sign is something which may take some time to digest.) We shall follow the normal practice of writing xy for $\varphi(x \otimes y)$.

Just as we have defined associativity, commutativity, etc., of H-spaces and H'-spaces (see Section 10.1), we define associativity and commutativity of φ and ψ. We leave this to you as an exercise. (Do not forget to include the twisting τ in commutativity and co-commutativity.)

Definition 11.4.2 We say H is connected if $\eta : R \to H_0$ is an isomorphism (equivalently, $\epsilon : H_0 \to R$ is an isomorphism).

Remark 11.4.3 If X is a connected space and $H^*(X)$ is free of finite type, then it follows that $H^*(X)$ is a connected associative and commutative algebra (with the cup product). Likewise it also follows that H_* is a connected co-associative and co-commutative co-algebra. Moreover, if X is an H space, then both $H_*(X)$ and $H^*(X)$ become connected associative, commutative, co-associative and co-commutative Hopf algebras. More generally, if H_* is a free and finite type Hopf algebra, then $H^* = \text{Hom}(H_*, R)$ is again a Hopf algebra called to dual of H_*. No prize for guessing that the product of the one induces the coproduct of the other, etc. We shall assume from now onward, that all Hopf algebras are connected. We can then treat the submodule H^+ of all elements of positive degree as a quotient of H as well, with $q : H \to H^+$ being the projection.

Definition 11.4.4 Let $x \in H$ be an element of positive degree. We say x is a *primitive* if $(q \otimes q) \circ \psi(x) = 0$; x is said to be *decomposable* if it is of the form $x = \sum_i a_i b_i$, with $a_i \in H_{p_i}, b_i \in H_{n-p_i}$ for $0 < p_i < n$.

Remark 11.4.5 Clearly, the set of primitive elements $P(H)$ of H forms a graded submodule of H. Check that $x \in P(H)$ iff $\psi(x) = x \otimes 1 + 1 \otimes x$. (Notice the similarity in the definition of primitive elements in the homology that we have considered in 6.4.5.) Similarly, the decomposable elements also form a graded submodule denoted by $D(H)$. Elements of the quotient module $Q(H) = H^+/D(H)$ are called *indecomposables*.

Before going further with the algebraic aspects of Hopf algebras, let us consider a few results to see why it is important from the topological point of view.

Theorem 11.4.6 *The n-characteristic element $\iota_n \in H^n(K(G;n);G)$ is a primitive. Moreover, for any connected CW-complex X, the bijection*

$$\Psi : [X, K(G;n)] \to H^n(X;G)$$

of Theorem 10.7.5 is an isomorphism.

Proof: The first statement is an easy consequence of the Künneth theorem and the remark above that ι_n is a primitive iff $\mu^*(\iota_n) = \iota_n \otimes 1 + 1 \otimes \iota_n$. Note that the H-space structure $\mu : K \times K \to K$ where $K = K(G;n)$ induces an addition on $[X;K]$. If $f, g : X \to K$ are any two maps then we must show that $\Psi([f] + [g]) = \Psi[f] + \Psi([g])$ which is the same as showing that

$$[\Delta \circ (f \times g) \circ \mu]^*(\iota_n) = f^*(\iota_n) + g^*(\iota_n).$$

This is easily verified.   ♠

Theorem 11.4.7 *Under Serre's correspondence $\lambda : Op(G, p; G', p') \to H^{p'}(G, p; G')$ (see Theorem 10.7.12) given by $\Theta \mapsto \Theta(\iota_p)$, Θ is additive iff $\Theta(\iota_p)$ is a primitive.*

Proof: We shall leave this as Exercise 11.5.3 for the reader.   ♠

Definition 11.4.8 Let H be an associative Hopf algebra. The height $h = \text{ht}\, x$ of an element $x \in H^+$ is defined to be one plus the supremum of all k such that $x^k \neq 0$. Thus $h = \text{ht}\, x$, iff $x^h = 0$ and $x^k \neq 0$ for $k < h$. Note that the height of an element can be infinite.

Lemma 11.4.9 Let H be an associative and commutative Hopf algebra over a field R of characteristic $p \geq 0$. Let $x \in H_n$ be a non zero primitive element of height h. Then the following hold:
(i) If n is odd and $p \neq 2$ then $h = 2$.
(ii) If n is even and $p = 0$ then $h = \infty$.
(iii) If n is even and p is odd then $h = \infty$ or $h = p^r$, for some r.
(iv) If $p = 2$ then $h = \infty$ or $h = 2^r$ for some r.

Proof: Since H is commutative, $x^2 = (-1)^{n^2} x^2$. Therefore if n is odd, $2x^2 = 0$ which implies $x^2 = 0$ unless $p = 2$. This proves (i).

Now assume that n is even or $p = 2$. Since x is a primitive, $\psi(x) = x \otimes 1 + 1 \otimes x$. Then

$$\psi(x^h) = (\psi(x))^h = (x \otimes 1 + 1 \otimes x)^h$$

and by binomial theorem, we get

$$\psi(x^h) = \sum_{i=0}^{h} \binom{h}{i} x^i \otimes x^{h-i} = x^h \otimes 1 + 1 \otimes x^h + \sum_{i=1}^{h-1} \binom{h}{i} x^i \otimes x^{h-i}. \qquad (11.12)$$

Now if h is the height of x and $h < \infty$, then LHS above is zero and the first two terms on the RHS are zero. Since the rest of the terms are all in different graded components, it follows that each one must be zero. Since $x^i \otimes x^{h-i} \neq 0, 0 < i < h$, it follows that each of the binomial coefficient must be zero in the field R. This implies that $p < \infty$ and h is a power of p. This proves all three statements (ii), (iii) and (iv). ♠
As an immediate consequence, we have the following two theorems.

Theorem 11.4.10 *Let H be an associative Hopf algebra of finite dimension over a field of characteristic $p = 0$. Then there are no non zero primitive elements of even degree in H.*

Theorem 11.4.11 *Let H be an associative, commutative, connected Hopf algebra generated by a single element $x \in H_n$ (i.e., H is monogenic) over a field R of characteristic p. Then x is a primitive and the following statements hold:*
(i) If n is odd and $p \neq 2$ then H is the exterior algebra $\Lambda(x)$.
(ii) If n is even and $p = 0$ then H is the polynomial algebra $R[x]$.
(iii) If n even and p is an odd prime then either $H \approx R[x]$ or the truncated polynomial algebra: $H \approx R[x]/(x^{p^r})$.
(iv) If $p = 2$, then $H \approx R[x]$ or $R[x]/(x^{2^r})$.
In all these cases the coproduct is given by (11.12).

Note that given a (countable) family of Hopf algebras over the same field, we can form their tensor product, which is again a Hopf algebra in an obvious way. The following theorem of Borel completes the picture. Recall that a field R is *perfect* if the characteristic $p = 0$ or $R = R^p$, i.e., $y = x^p$ is solvable in R for all $x \in R$. Also note that if R is finite then since the Frobenius map $x \mapsto x^p$ is injective, it follows that R is perfect. We can now state Borel's theorem.

Theorem 11.4.12 (Borel) *Let H be a connected, associative, commutative Hopf algebra of finite type over a perfect field R. Then H is isomorphic as a Hopf algebra, to the tensor product of monogenic Hopf algebras.*

Our main interest is in the following corollaries:

Corollary 11.4.13 Let X be a connected H-space with its homology having finite type over a field R of characteristic p. Then the cohomology ring $H^*(X, R)$ is a tensor product $\otimes_{i=1}^{\infty} B_i$, where each B_i is generated by one element x_i of degree n_i. Moreover, we have:
(i) if n_i is odd and $p \neq 2$ then $B_i = \Lambda(x_i)$;
(ii) if n_i is even and $p = 0$, then $B_i = R[x_i]$;
(iii) if n_i is even and p is an odd prime then $B_i = R[x_i]$ or $R[x_i]/(x_i^{p^{r_i}})$;
(iv) if $p = 2$ then $B_i = R[x_i]$ or $R[x_i]/(x_i^{2^{r_i}})$.

Proof: We apply the above theorem first for a perfect field. Now if R is an arbitrary field of characteristic $p > 0$, we have the result for the perfect field $\mathbb{Z}_p$ and then $H^*(X, R) = H^*(X, \mathbb{Z}_p) \otimes R$ and so the result follows for R as well. ♠

Corollary 11.4.14 Let X be a connected H-space such that $H^*(X, R)$ is finite dimensional, where characteristic of R is zero. Then $H^*(X; R) \approx \Lambda(x_1, \ldots, x_k)$ generated by elements x_i of odd degree.

Corollary 11.4.15 If X is a connected compact Lie group, and R is a field of characteristic 0, then $H^*(X, R)$ is an exterior algebra generated by primitive elements of odd degree.

We are not going to prove Borel's theorem here, which is quite algebraic in nature. Interested readers may look into [Milnor–Moore, 1965] for the proof.

Instead, let us workout some specific examples such as $H^*(SO(N))$, $H^*(U(n))$, etc. We would like to do this over a PID R (not necessarily a field). But then the existence of torsion in the homology becomes a big hurdle.

We follow the notation as in Section 10.9. In particular, recall that there are fibrations

$$SO(n) \hookrightarrow SO(n+1) \to \mathbb{S}^{n+1}$$
$$U(n) \hookrightarrow U(n+1) \to \mathbb{S}^{2n+1}$$
$$Sp(n) \hookrightarrow Sp(n+1) \to \mathbb{S}^{4n+1}$$

So, we want to appeal to the Wang sequence to give more information. This is definitely the case. Just like the case of cellular homology (cohomology) computation of complex projective space was easier than the same for real projective space, it turns out that the cases $U(n)$ and $Sp(n)$ are easier than that of $SO(n)$.

Theorem 11.4.16 *The Hopf algebra $H^*(U(n), R)$ is the exterior algebra $\Lambda(x_1, \ldots, x_n)$ with primitive elements $x_i \in H^{2n-1}(U(n); R)$.*

Theorem 11.4.17 *The Hopf algebra $H^*(Sp(n), R)$ is the exterior algebra $\Lambda(x_1, \ldots, x_n)$ with primitive elements $x_i \in H^{4n-1}(Sp(n); R)$.*

Proof: We use induction on n. For $n = 1$, U_1 is the group of unit complex numbers $\mathbb{S}^1$ and the statement is obviously true. Assume the result for $n \geq 1$ and let us prove it for $n + 1$. The Wang sequence (see Theorem 11.3.5) in this case is

$$\cdots \longrightarrow H^q(U(n+1)) \xrightarrow{i^*} H^q(U(n)) \xrightarrow{\theta^*} H^{q-2n}(U(n)) \xrightarrow{\gamma^*} H^{q+1}(U(n+1)) \longrightarrow \cdots$$

wherein θ^* is a derivation. Since it has degree $-2n$, it vanishes on the primitive elements x_i of $H^*(U(n))$. Therefore $\theta^* \equiv 0$. This means that the above long exact sequence breaks-up into a sequence of short exact sequences:

$$0 \longrightarrow H^{q-2n}(U(n)) \xrightarrow{\alpha^*} H^{q+1}(U(n+1)) \xrightarrow{i^*} H^{q+1}(U(n)) \longrightarrow 0.$$

Since $H^{q+1}(U(n))$ is a free module the above sequence is split exact. Moreover, for $q < 2n$, since $H^{q-2n}(U(n)) = 0$, we have, $i^* : H^{q+1}(U(n+1)) \approx H^{q+1}(U(n))$. So, let us denote $(i^*)^{-1}(x_i)$ also by $x_i, i = 1, 2, \ldots, n$. Let us denote the image of $1 \in H^0(U(n))$ under γ^* by x_{n+1}. The monomials $X_J = x_{i_1} x_{i_2} \cdots x_{i_q}, (J = (n \geq i_1 > i_2 > \cdots > i_q \geq 1)$ form a module basis for $H^q(U_n)$. And we have

$$\gamma^*(x_j) = \gamma^* i^*(x_j) = \gamma^*(1 \smile i^*(x_j)) = \gamma^*(1) \smile x_j = x_n x_j.$$

This proves that $H^*(U(n+1))$ is $\Lambda(x_1, \ldots, x_{n+1})$. The elements $x_i, i < n+1$ are primitive since i^* is an isomorphism in those dimensions and by induction hypothesis they are primitives in $H^*(U(n))$. The primitivity of x_{n+1} follows from a slightly more general result which we state and prove below as a lemma. This completes the proof of the theorem. ♠

Lemma 11.4.18 Let $p : E \to \mathbb{S}^n$ be a fibration with fibre F. Suppose that (E, F) is an H-pair under the homotopy multiplication $\mu : E \times E \to E$, i.e., in addition to being an H-space structure, the restriction maps μ_i of μ fit into the following homotopy commutative diagrams:

$$
\begin{array}{ccc}
E \times F & \xrightarrow{\mu_1} & E \\
{\scriptstyle p_1}\downarrow & & \downarrow{\scriptstyle p} \\
E & \xrightarrow{p} & \mathbb{S}^n
\end{array}
\qquad
\begin{array}{ccc}
F \times E & \xrightarrow{\mu_2} & E \\
{\scriptstyle p_2}\downarrow & & \downarrow{\scriptstyle p} \\
E & \xrightarrow{p} & \mathbb{S}^n.
\end{array}
$$

Then the homomorphism $\mu_1^* : H^*(E) \to H^*(E \times F)$ sends $u := p^*(\xi_n^*)$ to the element $u \times 1$. In particular, u is a primitive.

Proof: Clearly, $\mu_1^*(u) = \mu_1^* p^*(\xi_n^*) = (p \circ p_1)^*(\xi_n^*) = p_1^* u$. Likewise the homomorphism $\mu_2^* : H^*(E) \to H^*(F \times E)$ will send u to $p_2^*(u)$. It follows that under $\mu^* : H^*(E) \to H^*(E \times E)$ we have $\mu^*(u) = p_1^*(u) + p_2^*(u)$ which means, by definition, that u is a primitive. ♠

This completes the proof of Theorem 11.4.16. The proof of Theorem 11.4.17 is identical.

By duality, there are similar results about the Pontrjagin algebra. One can appeal to certain general results about dual algebras to conclude the following. Since we have not treated any such results, we shall give some details of the proofs of the following theorem, the details in the next one being similar.

Theorem 11.4.19 *The Pontrjagin algebra $H_*(U(n); R)$ is isomorphic to the exterior algebra $\Lambda(x_1', \ldots, x_n')$ with primitive elements $x_i' \in H_{2i-1}(U(n); R)$. Hence, in particular, it is commutative.*

Theorem 11.4.20 *The Pontrjagin algebra $H_*(Sp(n); R)$ is isomorphic to the exterior algebra $\Lambda(x_1', \ldots, x_n')$ with primitive elements $x_i' \in H_{4i-1}(Sp(n); R)$. Hence, in particular, it is commutative.*

Proof of 11.4.19: Let $\{x_i'\}$ form the basis of $H_n = H_*(U(n))$ which is dual to the basis $\{x_I\}$ of $H^n = H^*(U(n))$. Let I, J be two sequences such that $I \cap J = \emptyset$. Let us denote by I+J, the sequence obtained by arranging $I \cup J$ in the increasing order. Then we have

$$\mu^*(x_K) = \sum_{I+J=K} \eta(I, J) x_I \times x_J,$$

where $\eta(I, J)$ is the signature of the permutation

$$
\begin{pmatrix}
k_1 & \cdots & k_r & k_{r+1} & \cdots & k_n \\
i_1 & \cdots & i_r & j_1 & \cdots & j_s
\end{pmatrix}
$$

Fix two sequences P, Q. Let us compute $x_P' \cdot x_Q'$ by computing the effect of x_K on it. We have

$$\begin{aligned}
\langle x_K, x_P' \cdot x_Q' \rangle &= \langle x_K, \mu_*(x_P' \times x_Q') \rangle \\
&= \langle \mu^*(x_K), x_P' \times x_Q' \rangle \\
&= \sum_{I+J=K} \eta(I, J) \langle x_I \times x_J, x_P' \times x_Q' \rangle.
\end{aligned}$$

Since $\langle x_I \times x_J, x_P' \times x_Q' \rangle = (-1)^{pq} \langle x_I, x_P' \rangle \langle x_J, x_Q' \rangle \neq 0$ iff $(I, J) = (P, Q)$ and since $I \cap J = \emptyset$, we obtain

$$\langle x_K, x_P' \cdot x_Q' \rangle = (-1)^{pq} \eta(P, Q).$$

Therefore

$$x_P' \cdot x_Q' = \begin{cases} (-1)^{pq} \eta(P, Q) x_{P+Q}, & \text{if } P \cap Q = \emptyset, \\ 0, & \text{otherwise.} \end{cases}$$

In particular, the Pontrjagin algebra H_n is commutative and $x_i^2 = 0$ for all i. Next we claim that the monomials $x_{i_1}' \cdots x_{i_k}'$ where $i_1 < i_2 < \cdots < i_k$ form a basis. In fact, we shall prove that if $I = (i_1, \ldots, i_k)$ then $x_{i_k}' \cdots x_{i_1}' = x_I'$. (Pay attention to the reversed order!) For $k = 1$, this is clearly true. Put $J = (i_2, \ldots, i_k)$ and by induction suppose $x_J' = x_{i_k} \cdots x_{i_2}$. Then

$$x_J' \cdot x_{i_1}' = (-1)^{k-1} \eta(J, i_1) x_I' = (-1)^{k-1} (-1)^{k-1} x_I' = x_I'.$$

It remains to prove that x_i' are primitive, i.e., $\Delta_*(x_i') = x_i' \times 1 + 1 \times x_i'$. Once again we compute the Kronecker indices:

$$\langle x_I \times x_J, \Delta_*(x_i') \rangle = \langle \Delta^*(x_I \times x_J), x_i' \rangle = \langle x_I \cdot x_J, x_i' \rangle.$$

Now if $I \cap J \neq \emptyset$ then $x_I \cdot x_J = 0$. Otherwise, we have $x_I \cdot x_J = x_{I+J}$. In that case, the Kronecker index is non zero iff $I + J = \{i_1\}$. Therefore the only non zero indices are:

$$\langle x_i \times 1, \Delta_*(x_i') \rangle = \langle x_i, x_i' \rangle = 1; \quad \langle 1 \times x_i, \Delta_*(x_i') \rangle = \langle x_i, x_i' \rangle = 1.$$

This proves that $\Delta_*(x_i') = x_i' \times 1 + 1 \times x_i'$, as required. ♠

Remark 11.4.21 Notice that the commutativity of the above Pontrjagin algebras is non trivial, since the multiplication in the corresponding Lie groups are highly non commutative; however, it is analogous in spirit to the commutativity of their fundamental group.

Almost similarly, we have the results for the groups $SU(n)$ as well. However, the induction here starts at $SU(2) = \mathbb{S}^3$ (because $SU(1) = (1)$.) Thus we have:

Theorem 11.4.22 *The Hopf algebra $H^*(SU(n); R)$ is isomorphic to $\Lambda(x_2, \ldots, x_n)$, the exterior algebra generated by primitives $x_i \in H^{2n-1}(SU(n); R)$.*

Theorem 11.4.23 *The Pontrjagin algebra $H_*(SU(n); R)$ is isomorphic to $\Lambda(x_2', \ldots, x_n')$, the exterior algebra generated by primitives $x_i' \in H_{2n-1}(SU(n); R)$.*

The case of $SO(n)$ is more complicated due to the fact that we have to deal with spheres of all dimensions unlike the cases $U(n)$ and $Sp(n)$ wherein the base spaces of successive fibrations were spheres of odd dimension. This had indirectly helped us in two different ways. First of all, primitive elements of odd dimensions have the property that the squares are zero. Secondly, there was 'enough room' between successive primitive elements, which implies that the Wang sequence broke-up into a sequence of short exact sequences. Here, we need the embeddings $g_n : \mathbb{P}^n \to SO(n+1)$ given in Exercise 10.9.20.(i) to bail us out.

We propose to handle this situation (at least) in the case $R = \mathbb{Z}_2$, which will bring some simplifications. Even here, we cannot immediately say that an odd dimensional primitive has vanishing square—for, all that we get is $2x^2 = 0$, which is useless, because we are working mod 2. So, we have to take our steps carefully. We begin with a technical definition.

Definition 11.4.24 Let A be a commutative graded algebra over $\mathbb{Z}_2$. We say a set (finite or infinite) $\{x_1, x_2, \ldots\}$ of homogeneous elements of A forms a simple system of generators for A iff the monomials $\{x_I\}$ form an additive basis for A.

Remark 11.4.25 The point is that now it follows that each x_i^2 is a linear combination of monomials $\{x_I\}$ and once these linear combinations are known, we know the entire algebra structure of A.

Theorem 11.4.26 *The Hopf algebra $H^*(SO(n+1); \mathbb{Z}_2)$ has a simple system of primitive generators $\{x_1, \ldots, x_n\}$ such that*

$$x_i^2 = \begin{cases} x_{2i}, & 2i \leq n; \\ 0, & 2i > n. \end{cases} \tag{11.13}$$

The embedding $g_n : \mathbb{P}^n \hookrightarrow SO(n+1)$ induces a homomorphism g_n^* which maps the module M^n generated by $\{x_1, \ldots, x_n\}$ isomorphically onto $H^*(\mathbb{P}^n; \mathbb{Z}_2)$.

Theorem 11.4.27 *The Pontrjagin algebra $H_*(SO(n+1); \mathbb{Z}_2)$ is an exterior algebra $\Lambda(x_1', \ldots, x_n')$ with x_i' homogeneous of degree i. The primitive submodule M_n is generated by $\{x_{2i-1} : 1 \leq i \leq n\}$ as a basis.*

We prove these two theorems simultaneously by induction. Temporarily, let us have the notation $H^n := H^*(SO(n+1))$, $H_n = H_*(SO(n+1))$. For $n = 1$ there is nothing much to prove. Now assume both the theorems to be true for $n-1$ and let us proceed toward the proof for the case n.

Just as in the case of $U(n)$, since θ^* is of degree $-n+1$, it follows that $\theta^*(x_i) = 0$ in dimension $< n-1$. We claim $\theta^*(x_{n-1}) = 0$. The only other possibility is that $\theta^*(x_{n-1}) = 1$. But then $0 = \gamma^*(1) = p_n^*(\xi^*)$. On the other hand, since $p_n \circ g_n : (\mathbb{P}^n, \mathbb{P}^{n-1}) \to (\mathbb{S}^n, *)$ is a relative homeomorphism, it follows that $0 \neq g_n^* p_n^*(\xi_n^*) = 0$ which is absurd. Therefore, $\theta^*(x_{n-1}) = 0$. Since θ is a derivation and since H^{n-1} is generated by $x_1, \ldots, x_{n-1}$, we get $\theta^* \equiv 0$ on the whole of H^{n-1}. Therefore we have short exact sequences

$$0 \longrightarrow H^{q-n}(SO(n)) \overset{\gamma^*}{\longrightarrow} H^q(SO(n+1)) \overset{i^*}{\longrightarrow} H^q(SO(n)) \longrightarrow 0.$$

Let us denote the inverse-images of x_i under the isomorphism i^*, again by x_i for $q < n$. Put $x_n = \gamma^*(1)$. It follows that $\{x_1, \ldots, x_n\}$ forms a simple system of generators of $H^*(SO(n+1))$. Since $g_n^*(\xi_n)$ is the non zero element of $H^n(\mathbb{P}^n)$ as seen above, and since $g_{n-1} = g_n|_{\mathbb{P}^{n-1}}$, it follows inductively that $g_n^*(x_q) = u^q, 1 \leq q \leq n$. The primitivity of $x_i, i < n$ follows by induction hypothesis, whereas the primitivity of x_n follows from Lemma 11.4.17.

It remains to verify (11.4.27). In $SO(3)$, it is not hard to see $x_1^2 = x_2$ for the simple reason that $g_2^*(x_2) = u^2 = (g_2^*(x_1))^2$, since x_2 is the lone generator of H^2. However, for $SO(n), n > 3$, it becomes more complicated and we need to appeal to the duality and the induction hypothesis on the theorem on homology. Let $\{x_I'\}$ be the dual basis for H_n. It follows, as in the unitary case, that the Pontrjagin algebra H_n is the exterior algebra on $\{x_1', \ldots, x_n'\}$ with $x_I' = x_{i_1}' \cdots x_{i_k}'$. (We still do not know about the coproduct here.) Since the module of primitive elements M^n in H^n is dual to $Q_n = H_n/D_n$ and M_n is dual to $Q^n = H^n/D^n$, where D_n, D^n are spaces of decomposable elements, respectively, in H_n, H^n. Since H_n is an exterior algebra, D_n is spanned by the unit 1 and $\{x_I' : \#(I) \geq 2\}$. This means that Q_n has one basis element, viz., x_q' in each dimension q. By duality, M^n has one basis element in each dimension q which is clearly x_q. Since $g_n^*(x_q) = u^q$, g_n^* is an isomorphism of M^n onto $H^*(\mathbb{P}^n)$.

From (11.12), it follows that $x_q^2 \in M^n$ and is homogeneous of degree $2q$. If $2q \leq n$, since $g_n^*(x_{2q}) = u^{2q} = g_n^*(x_q^2)$, it follows that $x_q^2 = x_{2q}$. On the other hand if $2q > n$, since M^n has no components of degree bigger than n, $x_q^2 = 0$. This completes the proof of Theorem 11.4.26.

In particular, Q^n is spanned by $\{x_{2i-1} : 1 \leq i \leq n\}$. By duality, it follows that M_n is spanned by $\{x'_{2i-1}\}$. This completes the proof of Theorem 11.4.27. ♠

11.5   Miscellaneous Exercises to Chapter 11

1. For $n = 1, 3, 7$, consider the map $\phi_k : \mathbb{S}^n \to \mathbb{S}^n$ given by $\phi_k(x) = x^k$, where we are using the multiplication of complex numbers, quaternions or Cayley numbers, respectively. Compute the degree of ϕ_k.

2. Let p, q be any integers. Consider $f_{p,q} : \mathbb{S}^3 \times \mathbb{S}^3 \to \mathbb{S}^3 \times \mathbb{S}^3$ given by $f_{p,q}(a, b) = (a, a^p b a^q)$. Let $M_{p,q}$ be the quotient space of the disjoint union two copies of $\mathbb{D}^4 \times \mathbb{S}^3$ by the identification $(a, b) \sim f_{p,q}(a, b)$ for $(a, b) \in \mathbb{S}^3 \times \mathbb{S}^3$.
   (a) Compute the homomorphism $(f_{i,j})_* : H_3(\mathbb{S}^3 \times \mathbb{S}^3; \mathbb{Z}) \to H_3(\mathbb{S}^3 \times \mathbb{S}^3; \mathbb{Z})$.
   (b) Show that $M_{p,q}$ is a connected, simply connected, closed topological manifold. (It can be shown that this manifold has a unique smooth structure compatible with the smooth structures on the two copies of $\mathbb{D}^4 \times \mathbb{S}^3$.)
   (c) Compute the homology modules of $M_{p,q}$ using (a).
   (d) Conclude that if $p + q = \pm 1$, then $M_{p,q}$ is a homotopy sphere (and hence by the solution of Poincaré conjecture, $M_{p,q}$ is homeomorphic to $\mathbb{S}^7$.)
   (e) Show that there is a fibre bundle $\mathbb{S}^3 \to M_{p,q} \to \mathbb{S}^4$ in general, and compute the homology and cohomology modules using this fact.

   (These manifolds were discovered by Milnor, some of them as (first) examples of **Exotic Spheres**, i.e., topological spheres with smooth structures not diffeomorphic to the standard $\mathbb{S}^n$. (See [Milnor, 1956] for more details.)

3. Show that a cohomology operation $\Theta \in Op(G, p; G', p')$ is additive iff $\Theta(\iota_p) \in H^{p'}(K, G')$ is a primitive.

Chapter 12

Characteristic Classes

One of the central topics of interest in topology which is common to algebraic topology, differential geometry and algebraic geometry, etc., is the characteristic classes. In this introductory chapter, we make an attempt to present some rudimentary facts about characteristic classes. For further study, the reader may consult books such as [Milnor–Stasheff, 1974], [Husemoller, 1994], etc.

In Section 12.1, we discuss the orientation of vector bundles and the Euler class and establish the relation between the Euler class and the Euler characteristic, from which the name 'characteristic class' is derived. In Section 12.2, we begin the study of Steifel–Whitney classes and Chern classes, simultaneously, by actually constructing them. In Section 12.3, we study some basic properties of these classes and give some important applications. In Section 12.4, we prove the uniqueness of Steifel–Whitney classes and Chern classes. In the last section, we introduce Pontrjagin classes and indicate indicate some applications, especially in the study of oriented cobordism.

The reader is advised to revisit Section 5.4, before going further with this chapter.

12.1   Orientation and Euler Class

We begin with a stronger version of Lemma 8.1.3.

Theorem 12.1.1 *A vector bundle ξ is orientable iff there is an element $\mu \in H^n(E, E_0; \mathbb{Z})$ whose restriction to any fibre is a generator of $H^n(\xi_b, (\xi_b)_0; \mathbb{Z})$.*

Proof: The 'if' part is obvious. We need to prove the 'only if' part here.

Reduction to the Finite Type Case Suppose we have proved the statement when the bundle is of finite type, viz., when there is a finite cover of B which trivializes p. In particular, this would prove the statement when the base space is compact. We can then appeal to Thom isomorphism Theorem 11.2.4 to conclude that $H_{n-1}(E, E_0;) \approx H_{-1}(E) = 0$, whenever the base space is compact. By taking the direct limit, it then follows that $H_{n-1}(E, E_0) = 0$ in the general case also. But then by the universal coefficient theorem, it follows that $H^n(E, E_0) \approx \operatorname{Hom}(H_n(E, E_0); \mathbb{Z})$. This means that $H^n(E, E_0)$ is isomorphic to the inverse limit of the groups

$$\{H^n(p^{-1}(K), p^{-1}(K)_0), \ K \subset B \text{ compact}\}.$$

The statement of the theorem then follows from the corresponding statement when the base is compact.

Proof when p Is of Finite Type The statement is trivially verified in the case when the entire bundle is trivial. Now it suffices to prove the theorem for the case when B is covered by two open sets V_1, V_2, over each of which the bundle is trivial. Put

$$E_i = p^{-1}(V_i), i = 1, 2, E' = p^{-1}(V_1 \cap V_2), (E_i)_0 = E_i \cap E_0, i = 1, 2, E'_0 = E' \cap E_0.$$

Let $\mu_i \in H^k(E_i; (E_i)_0; \mathbb{Z}), i = 1, 2$ be such that restricted to each fibre they give the pre-orientation class. Now in the exact Mayer–Vietoris sequence

$$H^{k-1}(E', E_0') \to H^k(E, E_0) \to H^k(E_1, (E_1)_0) \oplus H^k(E_2, (E_2)_0) \to H^k(E', E_0') \quad (12.1)$$

since the bundle over $V_1 \cap V_2$ is trivial, it is easily verified that the element $\mu_1 - \mu_2$ is mapped to zero. Therefore there exists $\mu \in H^k(E, E_0)$ which maps onto (μ_1, μ_2). By the canonical property of cohomology elements, it follows that μ restricts to the generator of $H^k(\xi_b, (\xi_b)_0)$ for each $b \in B$. This completes the proof of the theorem. ♠

Definition 12.1.2 Let $\xi = (E, p, B)$ be an oriented real k-plane bundle with an orientation class $\mu \in H^n(E, E_0; \mathbb{Z})$. Then

$$e(\xi) := (p^*)^{-1} i^*(\mu) \in H^n(B; \mathbb{Z})$$

is called the Euler class of ξ. If ξ is a complex k-plane bundle, then $\xi_{\mathbb{R}}$ has a canonical orientation with respect to which we take the Euler class, i.e., $e(\xi) = e(\xi_{\mathbb{R}})$.

Remark 12.1.3
(i) Let ξ and ξ' be oriented k-plane bundles and $(f, \bar{f}) : \xi \to \xi'$ is a map which is an isomorphism on each fibre and preserves orientations. Then by naturality property of cohomology classes, it follows that $e(\xi) = f^*(e(\xi'))$.
(ii) If we change the orientation on ξ, then it follows that $e(\xi)$ also changes its sign. Thus if we agree to denote the bundle with opposite orientation by $-\xi$ then we have $e(-\xi) = -e(\xi)$.
(iii) Combining (i) and (ii), it follows that if k is odd, then $2e(\xi) = 0$. [To see this, consider the automorphism which sends each vector to its negative. This is orientation reversing if the dimension k is odd. Therefore we have $(Id, \eta) : \xi \to -\xi$ given by $v \mapsto -v$. From (i), it follows that $e(\xi) = Id^*(e(-\xi)) = e(-\xi) = -e(\xi)$.]

Example 12.1.4 If ξ is a trivial bundle then $e(\xi) = 0$. To see this first of all note that $B \times \mathbb{R}^k$ is orientable and an orientation class looks like $\mu = \pi_2^*(\nu)$ where $\nu \in H^k(\mathbb{R}^k; \mathbb{R}_0^k; \mathbb{Z})$ is a generator, where π_2 is the second projection. Since $\pi_2 \circ i$ is a constant map, it follows that $i^*(\mu) = 0$ in $H^k(B \times \mathbb{R}^k; \mathbb{Z})$.

Example 12.1.5 Consider the canonical complex line bundles $\gamma_{\mathbb{C}}^1$ over the complex projective space $\mathbb{P}_{\mathbb{C}}^1 = \mathbb{S}^2$. From the definition of the projective space, it is clear that the associated sphere-bundle, viz., the bundle restricted to the subspace of norm one vectors is the Hopf-bundle $p : \mathbb{S}^3 \to \mathbb{S}^2$ with fibres $\mathbb{S}^1$. Clearly the bundle restricted to $V_i = \mathbb{S}^2 \setminus \{P_i\}$ is trivial where P_i denote the north and south pole and we can orient each piece with the orientation coming from the complex structure on the fibre $\mathbb{R}^2 = \mathbb{C}$. The compatibility of this choice over $V_1 \cap V_2$ follows since the transition function $([z_2, z_2], z) \mapsto ([z_1, z_2], z \cdot z_1/z_2)$ is
and hence orientation preserving. Therefore, the bundle is orientable. Indeed, since $\mathbb{S}^3 \subset E_0$ is a deformation retract, we have $H^1(E_0) = 0 = H^2(E_0)$ and hence

$$H^2(E, E_0) \approx H^2(E) \approx H^2(\mathbb{S}^2) \approx \mathbb{Z}.$$

Therefore, the orientation class μ has to be a generator of $H^2(E, E_0)$ which is mapped onto a generator of $H^2(\mathbb{S}^2)$. This just means $e(\gamma_{\mathbb{C}}^1) = (p^*)^{-1}(i^*(\mu))$ is a generator of $H^2(\mathbb{S}^2)$.
Since $\mathbb{S}^2 = \mathbb{P}_{\mathbb{C}}^1$ is a complex manifold, it has a canonical orientation which fixes a generator $z \in H^2(\mathbb{P}_{\mathbb{C}}^1; \mathbb{Z})$ and hence $e(\gamma_{\mathbb{C}}^1) = \pm z$. We need to determine the sign.
This is best understood if we work inside $\mathbb{P}_{\mathbb{C}}^2$ in which there is a natural embedding of the total space of γ^1 as an open subspace:

$$E(\gamma^1) = \{[z_0, z_1, z_2] : z_0 \neq 0\} \subset \mathbb{P}_{\mathbb{C}}^2.$$

Note that the zero section of the bundle is identified with the line at infinity in $\mathbb{P}^2_{\mathbb{C}}$.

There is an orientation on the total space $E(\gamma^1)$ which is obtained by taking the orientation on the base followed by the orientation on the fibre. With respect to this orientation, it is clear that the intersection number of each fibre with the zero section is equal to 1. This implies that $p^*(z) \smile \mu$ is the generator of $H^4(\mathbb{P}^2_{\mathbb{C}}; \mathbb{Z})$. On the other hand, we also have $p^*(z) \smile p^*(z)$ is equal to this generator. Therefore, it follows that $p^*(z) = \mu$. Inside E this just implies that

$$e(\gamma^1_{\mathbb{C}}) = z. \tag{12.2}$$

Theorem 12.1.6 *The Euler class of the Cartesian product is the cross product of the Euler classes; also the Euler class of a Whitney sum is the cup product of the Euler classes.*

$$e(\xi_1 \times \xi_2) = e(\xi_1) \times e(\xi_2); \quad e(\xi_1 \oplus \xi_2) = e(\xi_1) \smile e(\xi_2).$$

Proof: Note that for any two oriented vector spaces V, W, $V \times W$ is oriented by first taking the basis of V followed by the basis of W. It follows easily from this that $\mu(\xi \times \xi) = \mu(\xi) \times \mu(\xi')$ from which the first property follows. But then $\xi_1 \oplus \xi_2 = \Delta^*(\xi_1 \times \xi_2)$ and the second property follows from the above Remark (i). ♠

Corollary 12.1.7 *Let M be a smooth oriented manifold such that $e(M) \neq 0$. Then the tangent bundle $\tau(M)$ does not admit any subbundle of odd rank.*

Proof: Suppose ξ is an oriented subbundle of odd rank. Fixing a Riemannian metric on τ, we can take the orthogonal complement $\xi^\perp$ and orient it in such a way that the direct sum orientation coincides with that of τ. Therefore,

$$2e(M) = 2e(\tau(M)) = 2e(\xi) \smile e(\xi^\perp) = 0 \in H^n(M, \mathbb{Z}) \approx \mathbb{Z}$$

which contradicts the hypothesis $e(M) \neq 0$. ♠

Now, in the situation of the above corollary, it may happen that ξ is not orientable. From the earlier example, there is a double cover $\phi : \tilde{M} \to M$ such that $\phi^*(\xi)$ is orientable. On the other hand, it is a subbundle of $\phi^*(\tau(M)) = \tau(\tilde{M})$. Now we are in the orientable case. ♠

Exercise 12.1.8 Extend the notion of orientability to the sphere bundles $\dot{\eta} = (\dot{E}, p, B)$. Show that a sphere bundle is orientable iff there is $\mu \in H^{k-1}(\dot{E}; \mathbb{Z})$ whose restriction to each fiber is a generator of $H^{k-1}(\dot{\eta}_b; \mathbb{Z})$.

Exercise 12.1.9 Fix a Riemannian metric on a k-plane bundle ξ. Let $\dot{E}$ denote the subspace of unit vectors in E. Then $p : \dot{E} \to B$ defines a (locally trivial) $\mathbb{S}^{k-1}$-bundle over B.

Theorem 12.1.10 *Let ξ be a k-plane bundle with a metric and $\dot{\xi}$ be the sphere bundle. Let η be the tautological line bundle over the total space of $\dot{\xi}$. Of the three bundles $\xi, \dot{\xi}, \eta$, if two of them are orientable then the third is also orientable.*

Proof: Easy.

Theorem 12.1.11 *For any oriented vector bundle ξ, $e(\xi) = 0$ if ξ admits a nowhere vanishing section.*

Proof: Suppose ξ admits a nowhere vanishing section. Then we can write $\xi \cong \xi' \oplus \Theta^1$. It follows that $e(\xi) = e(\xi') \smile e(\Theta^1) = 0$. ♠

Lemma 12.1.12 Let M^m be closed submanifold of a manifold N^{m+k}. Then for any coefficient ring R, there is a canonical isomorphism

$$H^*(E, E_0; R) \to H^*(N, N \setminus M; R)$$

where E is the total space of the normal bundle of M in N and E_0 is the subspace of E consisting of non zero normal vectors.

Proof: Fixing a Riemannian metric on N recall that for some suitable $\epsilon > 0$ the exponential map $\exp : E(\tau(N)) \to N$ restricts to a diffeomorphism

$$E(\epsilon) \to N(\epsilon)$$

of the space of all vectors in $\nu(M)$ of norm less than ϵ to a tubular neighbourhood $E(\epsilon)$ of M in N. This diffeomorphism is identity on M. On the other hand we have the excision map

$$(N(\epsilon), N \setminus M) \hookrightarrow (N, N \setminus M)$$

inducing an isomorphism in cohomology. Combining this with $(\mathrm{Exp})^*$ gives the required isomorphism. ♠

Remark 12.1.13 The isomorphism does not depend upon the choice of $\epsilon > 0$, nor on the choice of the Riemannian metric. This is so because the homotopy type of the tubular neighbourhood is independent of such choices. Now suppose that the normal bundle of M is oriented. Then the image of the fundamental cohomology class μ in $H^k(N, N \setminus M)$ under the above isomorphism will be denoted by μ'. Since the diffeomorphism Exp is Identity on M it follows that under the inclusion induced maps $M \to N$ followed by $N \to (N, N \setminus M)$ the element μ' is mapped onto $e(\nu(M))$. As an immediate consequence we have:

Theorem 12.1.14 *Let M^m be a closed manifold embedded in $\mathbb{R}^{m+k}$ so that the normal bundle $\nu(M)$ is orientable. Then $e(\nu(M)) = 0$.*

Proof: You have to only notice that $H^k(\mathbb{R}^{m+k}) = (0)$. ♠

Finally, we come to the result that justifies the name of this particular characteristic class:

Theorem 12.1.15 *Let M be a closed oriented m-dimensional manifold. Then*

$$e(M) = \chi(M)\overline{[M]}, \tag{12.3}$$

where $\overline{[M]} \in H^m(M; \mathbb{Z})$ is the dual to the fundamental class $[M] \in H_m(M, \mathbb{Z})$.

In Remark 12.1.3, we have seen that $2e(\xi) = 0$ for any odd rank bundle ξ. In particular, $e(M) = 0$ for odd dimensional manifolds. We have also seen that $\chi(M) = 0$ in this case, as an easy consequence of Poincaré duality. Therefore, in the proof of the above theorem we need to consider the case when m is even.

We shall present two proofs of this wonderful theorem, both the proofs have their own educative value. The first proof is more algebraic topological and is presented with slight variations in several expositions. The second one is more differential topological and brings out Euler class as an obstruction for existence of non vanishing sections.

Toward the first proof, we shall begin with an alternative description of the tangent bundle and thereby obtain a useful alternative description of $e(M)$.

Let M^m be a closed, connected, smooth manifold and $\Delta : M \to M \times M$ be the diagonal embedding. We shall identify M with $\Delta(M) =: \Delta_M$. First note that the tangent bundle $\tau(\Delta_M)$ is canonically isomorphic to the normal bundle $\nu(\Delta_M)$ in $M \times M$ via

$$(\mathbf{v}, \mathbf{v}) \mapsto (\mathbf{v}, -\mathbf{v}).$$

Since both Δ_M and $M \times M$ are oriented, the normal bundle ν_Δ is also oriented. Note that $\Delta_*[M]$ is not zero since under any one of the coordinate projection, it is mapped on the fundamental class $[M]$. By Poincaré duality, there exists a unique element $\mathcal{T} \in H^m(M \times M)$ such that

$$\mathcal{T} \frown [M \times M] = \Delta_*[M].$$

Lemma 12.1.16 There is a unique $\mu' \in H^m(M \times M, M \times M \setminus \Delta_M)$ which is mapped onto $\mathcal{T}$ under the inclusion induced homomorphism $j^* : H^m(M \times M, M \times M \setminus \Delta_M) \to H^m(M \times M)$. Moreover, this unique element is the fundamental class of ν_Δ and we have

$$e(M) := e(\tau(M)) := e(\nu(M)) = \Delta^*(\mathcal{T}). \tag{12.4}$$

Proof: If U is a tubular neighbourhood of Δ_M in $M \times M$ and $M \times M = \bar{U} \cup \bar{V}$, with $\bar{U} \cap \bar{V} = \partial \bar{U} = \partial \bar{V}$, using the property $\mathcal{T} \frown [M \times M] = \Delta_*[M]$ it follows that $i^*(\mathcal{T}) \frown [\bar{V}] = 0$. By Poincaré duality, this implies $i^* \mathcal{T} = 0$. By the cohomology exact sequence it follows that there is $\mu' \in H^m(M \times M, M \times M \setminus \Delta_M)$ such that $j^*(\mu') = \mathcal{T}$.

We shall show that any μ' with the property that $j^*(\mu') = \mathcal{T}$ is the fundamental class of ν from which uniqueness part will also follow. That means we have to prove that for each $x \in M$ if $j_x : (M, M \setminus x) \to (M \times M, M \times M \setminus \Delta_M)$ denotes the inclusion then $j_x^*(\mu')$ is a generator of $H^m(M, M \setminus x)$.

First suppose for some x, $j_x^*(\mu') = 0$. By local triviality of the bundle and the Künneth formula, it follows that $j_U^*(\mu') = 0$ where U is a coordinate neighbourhood of x and $j_U : \pi^{-1}(U) \to M \times M$ is the inclusion. The same is true if $j_x^*(\mu) \neq 0$ also. Therefore, by connectivity of M, we conclude that $j_x^*(\mu') = 0$ for some $x \in M$ iff it is true for one x. Moreover, since M is compact, by Mayer–Vietoris argument, it follows that $j_x^*(\mu') = 0$ iff $\mu' = 0$.

The above argument is available for any coefficient module $\mathbb{Z}_p$ as well. Therefore, we can also conclude that for any prime p, $j_x^*(\mu')$ is divisible p for some x, iff μ' is divisible by p. But that would imply that $\Delta_*[M]$ is divisible p which is not true. We conclude that $j_x^*(\mu')$ is a generator. Equation (12.4) follows. ♠

Our next task is to get a description of the Thom class $\mathcal{T}$.

Lemma 12.1.17 Let M be a closed, connected oriented manifold of even dimension m. Let $\{b_{ir}\}$ be a basis for $H^r(M)$. Let $\{b_{i,m-r}^\#\}$ be the dual basis for $H^{m-r}(M)$ under Poincaré duality, viz.,

$$\langle b_{i,r}^\# \smile b_{j,m-r}, [M] \rangle = \delta_{i,j}, \text{ for all } i, j, \text{ and for each fixed } r. \tag{12.5}$$

Then in $H^m(M)$, we have,

$$\mathcal{T} = \sum_r \sum_i (-1)^i b_{i,r}^\# \times b_{i,m-r}. \tag{12.6}$$

In particular, (12.3) holds.

Proof: By the Künneth theorem, it follows that

$$\mathcal{T} = \sum_{i,j,r} \alpha_{i,j,r} b_{i,r}^\# \times b_{j,m-r}.$$

We shall compute the coefficients $\alpha_{i,j,r}$ by evaluating the cup product of $\mathcal{T}$ with the generators $b_{i,r} \times b_{j,m-r}^\#$ in two different ways. For a fixed degree r and any two fixed indices p, q, put $b = b_{p,r}, b^{\#} = b^{\#}_{q,m-r}$. Then

$$
\begin{aligned}
\langle (b \times b^{\#}) \smile \mathcal{T}, [M \times M] \rangle &= (-1)^{r(m-r)} \langle (b^{\#} \times b), \mathcal{T} \frown [M \times M] \rangle \\
&= (-1)^{r(m-r)} \langle (b^{\#} \times b), \Delta_{*}[M] \rangle \\
&= (-1)^{r(m-r)} \langle \Delta^{*}(b^{\#} \times b), [M] \rangle \\
&= (-1)^{r(m-r)} \langle b^{\#} \smile b, [M] \rangle \\
&= (-1)^{r(m-r)} \delta_{p,q}.
\end{aligned}
$$

On the other hand,

$$
\begin{aligned}
(b \times b^{\#}) \smile \mathcal{T} &= (b \times b^{\#}) \smile (\sum_{i,j,s} \alpha_{i,j,s}(b^{\#}_{i,s} \times b_{j,m-s})) \\
&= \sum_{i,j} (-1)^{r(m-r)} \alpha_{i,j,r}(b \smile b^{\#}_{i,r}) \times (b^{\#} \smile b_{j,m-r})
\end{aligned}
$$

Notice that the terms for which the degree s is not equal to r have dropped out in the last sum above because one of the terms $b \smile b^{\#}_{i,s}$ or $b^{\#} \smile b_{j,m-s}$ is of degree bigger than m and hence equals zero. Therefore,

$$
\begin{aligned}
&\langle (b \times b^{\#}) \smile \mathcal{T}, [M \times M] \rangle \\
&= (-1)^{r(m-r)}(-1)^{m^2} \sum_{i,j} \alpha_{i,j,r} \langle b \smile b^{\#}_{i,r}, [M] \rangle \langle b^{\#} \smile b_{j,m-r}, [M] \rangle \\
&= (-1)^{r(m-r)}(-1)^{m}(-1)^{r(m-r)} \sum_{i,j} \alpha_{i,j,r} \delta_{p,i} \delta_{q,j} \\
&= (-1)^{m} \alpha_{p,q,r}.
\end{aligned}
$$

Now using the fact that m is even, it follows that $\alpha_{p,q,r} = (-1)^{r} \delta_{p,q}$.

This establishes formula 12.6. ♠

Proof of theorem 12.1.15 Now using (12.5) we have

$$
e(M) \frown [M] = \Delta^{*}(\mathcal{T}) \frown [M] = \sum_{r} \sum_{i} (-1)^{r}(b_{i,r} \smile b^{\#}_{i,m-r}) \frown [M] = \sum_{r} (-1)^{r} \mathrm{rk}\, H^{r}(M) = \chi(M).
$$

This proves that $e(M) = \chi(M)\overline{[M]}$. ♠

Alternative proof of theorem 12.1.15

We shall prove two lemmas, from which the theorem would follow immediately.

Lemma 12.1.18 Let M be a smooth oriented closed n-manifold. If $s : M \to \tau(M)$ is a (smooth) section of the tangent bundle of a manifold with finitely many zeros, then the index $\iota(s)$ of s is equal to $e(M) \frown [M]$ where $[M] \in H_n(M; \mathbb{Z})$ denotes the fundamental class of M.

Lemma 12.1.19 Let K be a triangulation of a closed oriented manifold M. Then there exists a (smooth) vector field s on M, with finitely many zeros such that $\iota(s) = \chi(K)$.

For other avatars of Euler characteristic of a smooth closed manifold see [Shastri, 2011].

Proof of Lemma 12.1.18 Let $z_1, \ldots, z_k \in M$ be the zeros of s. Choose disjoint disc neighbourhoods $B_1, \ldots, B_k$ around the zeros $z_1, \ldots, z_k$, respectively. Put $P = \cup_i B_i$ and $Q = M \setminus \mathrm{int}\, P$, $E' = p^{-1}(Q)$, where $p : E \to M$ is the projection of the tangent bundle. Put $E_0 = E \setminus s_0(M)$ where $s_0 : M \to E$ is the zero section, and $E'_0 = E' \frown E_0$. Let $\mu \in H^n(E, E_0)$ denote the orientation class of the tangent bundle. Then it follows that $s : Q \to E'$ factors through $s' : Q \to E_0 \frown E' \to E'$ and therefore, $s^{*}(\mu)|_Q = 0$. Therefore the computation of $e(M) \frown [M] = \langle s^{*}(\mu), [M] \rangle$ can be effectively carried out by restricting attention on P. In other words, $s^{*}(\mu)$ can be thought of as a relative n-cochain on $(P, \partial P)$ and

$$
e(M) \frown [M] = \sum_{i} s^{*}(\mu) \frown [B_i, \partial B_i]. \tag{12.7}
$$

Now on each $B_i = B$, the bundle is trivial and fixing some trivialization on each of them we can write, $s(x) = (x, \sigma(x))$, where $\sigma : B \to \mathbb{R}^n$ is a (smooth) map such that $\sigma^{-1}(0) = \{z_i\}$. Also over B we can represent μ as $1 \times v$, where $v \in H^n(\mathbb{R}^n, \mathbb{R}^n \setminus 0)$ and $1 \in H^0(B)$. Therefore $s^*[M] \frown [B, \partial B] = \sigma^*[v] \frown [B, \partial B]$. This is nothing but the winding number of $\sigma|_{\partial B}$ around the point 0 which is also equal to the degree d_i of the map $\sigma/\|\sigma\| : \partial B \to \mathbb{S}^{n-1}$. The index of s at the point $z = z_i$ is equal to d_i. Now (12.7) implies $\iota(s) = \sum_i d_i = e(M) \frown [M]$. ♠

Proof of Lemma 12.1.19 Let K', K'' denote respectively the first and second barycentric subdivision of K. Then note that the vertex set of K' is the set of all barycentres $\beta(\tau)$ where τ is a simplex of K. Also, each vertex v of K'', as an element of $|K|$ belongs to the interior of a unique simplex $s \in K$. Treating s as a vertex of K', this assignment gives a vertex map $\varphi : V(K'') \to V(K')$. Check that φ is actually a simplicial map $\varphi : K'' \to K'$. Also, check that on the vertices of K', φ is identity. In fact, we claim that the fixed points of $|\varphi|$ are precisely vertices of K'. To prove this, let $x \in \langle \tau \rangle$ for $\tau = \{v_0, \ldots, v_k\} \in K''$. Then $x = \sum_i t_i v_i$, with $t_i \neq 0$. Therefore, $|\varphi|(x) = x$ implies $\sum_i t_i v_i = \sum_i t_i \varphi(v_i)$. The two convex combinations on either side are taken inside a single simplex of K and hence it follows that $\{v_0, \ldots, v_p\} = \{\varphi(v_0), \ldots, \varphi(v_k)\}$. This is possible iff $k = 0$ and v_0 is a vertex of K'.

[With respect to the CW-structure of K', $|\varphi|$ is a cellular map such that under the canonical orientation of the simplices τ of K', $|\varphi| : |\tau| \to |\tau|$ defines a degree 1 map. Therefore it follows that $|\varphi|$ induces the identity map of the CW chain complex of K'. Hence the Lefschetz number of $|\varphi|$ is equal to $\chi(K') = \chi(M)$. This fact also follows from the general theory since φ is a simplicial approximation to Id_M.]

We shall now construct a vector field s on M such that the zero set of the section s is precisely $V(K')$, the vertex set of K', so that its index at $\beta(\tau) \in K'$ is equal to $(-1)^{\dim \tau}$. Taking the sum of all these indices it would follow that $\iota(s) = \chi(K) = \chi(M)$. Figure 12.1 depicts such a vector field.

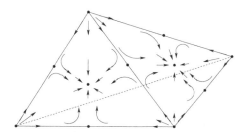

FIGURE 12.1. Vector field on a triangulated manifold

For points $x \in U = |K| \setminus V(K')$ the line segment $[x, \varphi(x)]$ in $|K|$ defines a smooth parameterised curve. The tangent vector $\alpha(x)$ to this curve at x, defines a non vanishing continuous vector field on $|K| \setminus V(K')$. In fact, it is smooth wherever $|\varphi|$ is smooth. We choose disjoint disc neighbourhoods B_s around each $\beta(s)$ and take a smooth approximation α_1 to the vector field α on the compact space, $|K| \setminus \cup_s \text{int } B_s$ so that α_1 is nowhere vanishing. Choosing a trivialization of the tangent bundle of M on each B_s, the section $\alpha_1|_{\partial B_s}$ corresponds to a smooth map $f_s : \partial B_s \to \mathbb{R}^n \setminus \{0\}$. This map can be extended to a smooth map $g_s : B_s \to \mathbb{R}^n$ so that $g_s^{-1}(0) = \beta(s)$. This in turn gives a smooth extension of $\alpha_1|_{B_s}$ to a smooth vector field α_2 of M with its zeros precisely at $\beta(s), s \in K$. The index of α_2 is equal to the sum of the indices of α_2 at $\beta(s), s \in K$.

On the other hand, the index at a given $\beta(s)$ is determined by $\alpha_2|_{\partial B_s} = \alpha_1|_{\partial B_s}$. Since α_1 is homotopic to α, we need to determine the index of α around the point $\beta(s)$. (It may be noted here that the smoothness of the vector field is not essential in determining the index.)

Suppose $v = \beta(\tau)$ for some simplex $\tau \in K$ of dimension d. The vector field α is pointing towards v at each point of $|\tau| \setminus \{v\}$. On the other hand, if τ' is the dual $(n-d)$-cell in K'', then α is pointing away from v at all the points $|\tau'| \setminus \{v\}$. Therefore the same property holds of α_2 also. This means the number of eigen values of dg_v which are less than 1 is equal to d. Hence, the index of α_2 at $v = \beta(s)$ is equal to $(-1)^d$. (See Section 7.7 of [Shastri, 2011].) ♠

Corollary 12.1.20 Let M be a smooth closed manifold. Then M has a smooth nowhere vanishing section iff $e(M) = 0$.

Proof: We have already seen the 'only if' part. Suppose now that $e(M) = 0$. We may assume that M is connected. We can take any vector field s with finitely many zeros and then $\iota(s) = 0$. We can assume that all the zeros of s are contained in the interior of a single disc D in M (by an ambient isotopy). Via a trivialization of the tangent bundle of M restricted to the disc D, the vector field $s|_{\partial D}$ now corresponds to a map $\sigma : \partial D \to \mathbb{R}^n \setminus \{0\}$ whose degree (equivalently, winding number around 0) is zero. Therefore σ can be extended to a map $D \to \mathbb{R}^N \setminus \{0\}$. This means that the the vector field $s|_{M \setminus \text{int } D}$ can now be extended to a non vanishing vector field all over M.   ♠

Remark 12.1.21 In subsequent sections, we shall introduce three more types of characteristic classes of vector bundles. Following Theorem 12.1.15, it is a standard practice to call the characteristic class of the tangent bundle of a smooth manifold M as the corresponding characteristic class of the manifold itself.

12.2   Construction of Steifel–Whitney Classes and Chern Classes

In this section, we shall introduce the so-called Steifel–Whitney classes and Chern classes.
Convention:

c	$F = F_c$	$K = K_c$
1	$\mathbb{R}$	$\mathbb{Z}_2$
2	$\mathbb{C}$	$\mathbb{Z}$

The table above represents a convention that we are going to follow in the rest of this chapter, viz., whenever we are dealing with real vector bundles we shall use homology and cohomology modules with $\mathbb{Z}_2$ coefficients and similarly, when we are dealing with complex bundles, we shall use integer coefficients. The number c then denotes the $\mathbb{R}$-dimension of K.

We shall consider an F-vector bundle $\xi = (E, p, B)$ of rank n. One easy fall-out of this convention is that our bundles are always K-orientable.

Let E_0 denote the open space consisting of non zero vectors in E and let $p_0 = p|_{E_0}$. Let $P(\xi) = (P(E), q, B)$ denote the projectivized bundle. We have the natural projection $E_0 \to P(E)$ and q is the factorization of p_0 through this map. Note that elements of $P(E)$ are lines in $p^{-1}(b), b \in B$.

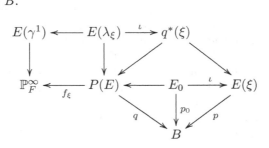

Consider the pullback bundle $q^*(\xi)$ over $P(E)$:

$$q^*(\xi) = \{(L, v) \in E' \times E \ : \ q(L) = p(v)\}$$

This bundle has a natural line subbundle given by

$$\lambda_\xi = \{(L, v) \ : \ v \in L\}.$$

Also observe that, for each $b \in B$, the fiber $q^{-1}(b)$ can be identified with the projective space $\mathbb{P}_F^{n-1}$ and if $j_b : q^{-1}(b) \to P(E)$ is the inclusion map, then $j_b^*(\lambda_\xi)$ is the canonical line bundle on the projective space $\mathbb{P}_F^{n-1}$. From the classification theory, it follows that there is a unique homotopy class $f_\xi : E(P(\xi)) = P(E) \to \mathbb{P}_F^\infty = G_1(F^\infty)$ such that $\lambda_\xi = f_\xi^*(\gamma_F^1)$ where γ_F^1 is the universal line bundle over $\mathbb{P}_F^\infty$. Also recall:

Theorem 12.2.1 *The cohomology ring $H^*(\mathbb{P}_{F_c}^\infty, K_c)$ is the polynomial algebra $K_c[z]$, where* $\deg z = c, c = 1, 2$. *The rings $H^*(\mathbb{P}_{F_c}^n; K_c)$ are got by putting one relation $z^{n+1} = 0$ under the map induced by the inclusion $\mathbb{P}_{F_c}^n \subset \mathbb{P}_{F_c}^\infty$.*

Remark 12.2.2 In case of real projective spaces, $(c = 1)$, there is no ambiguity in the choice of a generator $z \in H^2(\mathbb{P}_\mathbb{R}^\infty; \mathbb{Z}_2) \approx \mathbb{Z}_2$. In case of complex projective spaces, first of all note that there is a canonical choice for the orientation class of $\mathbb{P}_\mathbb{C}^1$ being a 1-dimensional compact complex manifold, which is taken as the generator z for $H^2(\mathbb{P}_\mathbb{C}^1; \mathbb{Z})$. Under the inclusion induced map this defines the choice of the generator for all $H^2(\mathbb{P}_\mathbb{C}^n; \mathbb{Z})$ $(c = 2)$.

Let $f_\xi : P(E(\xi)) \to \mathbb{P}_F^\infty$ be the classifying map for the line bundle λ_ξ as constructed in the above diagram. We denote

$$a_\xi = -f_\xi^*(z) \in H^c(P(\xi)) \tag{12.8}$$

which depends on the homotopy class of f_ξ and not on the specific choice of f_ξ. The following theorem is an easy consequence of Theorem 11.2.17, since our bundles are orientable.

Theorem 12.2.3 *The mapping $q^* : H^*(B) \to H^*(P(\xi))$ is a monomorphism and the elements $\{1, a_\xi, \ldots, a_\xi^{n-1}\}$ form a $H^*(B)$ base for $H^*(P(\xi))$.*

In particular consider the universal n-plane bundle $\xi = \gamma^n$ over $G_n(F^\infty)$. It follows that the element $a_\xi^n \in H^{nc}(P(\xi))$ can be expressed in a unique way

$$a_\xi^n = \sum_{i=1}^n (-1)^{i-1} x_i(\gamma^n) a_\xi^{n-i}, \tag{12.9}$$

for some $x_i(\gamma^n) \in H^{ci}(G_n; K_c)$, which are called the *universal characteristic classes for n-plane bundles*. Now, for an arbitrary n-plane bundle $\xi = g^*(\gamma^n)$ over B, where $g : B \to G_k(F^\infty)$, we define the characteristic classes

$$x_i(\xi) = g^*(x_i(\gamma^n)), \quad i \le n, \text{ and } x_i(\xi) = 0, \quad i > n. \tag{12.10}$$

We put

$$x(\xi) = 1 + x_1(\xi) + \cdots + x_n(\xi) + 0 + \cdots \tag{12.11}$$

For $F = \mathbb{R}$ (respectively, $F = \mathbb{C}$), the element $x_i(\xi) \in H^i(B, \mathbb{Z}_2)$ respectively, $\in H^{2i}(B; \mathbb{Z})$ is called the ith *Steifel–Whitney class (Chern class)* of ξ denoted by $w_i(\xi)$ (respectively, $c_i(\xi)$. Also, $x(\xi)$ is called the *total Steifel–Whitney class* denoted by $w(\xi)$ (*total Chern class* denoted by $c(\xi)$.)

12.3 Fundamental Properties

The following four properties of characteristic classes are so fundamental that they have been upgraded to the status of being called axioms.

(A1) If ξ and η are isomorphic bundles over B then $x(\xi) = x(\eta)$.
(A2) If $g : B' \to B$ is a continuous map, then $x(g^*(\xi)) = g^*(x(\xi))$.
(A3) For any two vector bundles ξ and η over B, we have,

$$x(\xi \oplus \eta) = x(\xi) \smile x(\eta).$$

(A4) $x(\gamma^1) = 1 + z$, where $\gamma^1 := \gamma_F^1$ is the universal line bundle over $\mathbb{P}_F^\infty$.

The properties (A1), (A2) are verified easily. To verify property (A4), we note that when $\xi = \gamma^1$ then $P(\xi) = G_1(F^\infty) = \mathbb{P}_F^\infty$ and q is the identity map. Therefore $\lambda_\xi = \gamma^1$ and $f = Id$. Therefore $a_\xi = z$ is the generator of $H^c(\mathbb{P}_F^\infty)$. On the other hand the identity (12.9) reduces to the identity $a_\xi = x_1(\gamma^1)$. This proves (A4). Property (A3) is the one which will take some effort to verify. We postpone the proof this for a while. First, let us derive some easy and beautiful consequences of these axioms.

Corollary 12.3.1
(1) $x(\Theta^k) = 1$.
(2) If η and ξ are stably equivalent, then $x(\eta) = x(\xi)$.
(3) If B is stably parallelizable manifold, then $w(B) := w(\tau(B)) = 1$.
In particular $w(\tau(\mathbb{S}^n)) = 1$.
(4) $x(\mathbb{P}_F^n) := x(\tau(\mathbb{P}_F^n)) = (1 + z)^{n+1}$.

Proof: (1) The trivial bundle Θ^k is induced by the constant map $f : B \to G_k(F^\infty)$.
(2) We have, by property (A3), $x(\eta) = x(\eta \oplus \Theta^k) = x(\xi \oplus \Theta^k) = x(\xi)$.
(3) follows from (2).
(4) If γ^1 denotes the canonical line bundle over $\mathbb{P}_F^n$ then we have seen that the tangent bundle $\tau(\mathbb{P}_F^n)$ is stably equivalent to $(n + 1)\gamma^1$. (See Theorem 5.4.25.) Now use (A3) and (A4). ♠

Corollary 12.3.2 (Steifel) The class $w(\mathbb{P}^n) := w(\tau(\mathbb{P}^n))$ is equal to 1 iff $n + 1$ is a power of 2. Thus the only projective spaces which can be parallelizable are $\mathbb{P}^{2^k - 1}, k \geq 1$.

It is known that $\mathbb{P}^n$ is parallelizable iff $n = 1, 3, 7$. This requires digging deeper into the properties of characteristic classes which we shall not deal with here. (See [Milnor–Stasheff, 1974] or [Husemoller, 1994] for further references.)

Some Applications

(a) Division algebras

Theorem 12.3.3 *Suppose there is a bilinear map* $\beta : \mathbb{R}^n \times \mathbb{R}^n \to \mathbb{R}^n$ *which has no zero divisors. Then* $\mathbb{P}^{n-1}$ *is parallelizable. In particular,* n *is a power of* 2.

Proof: Let $\{e_j\}_{1 \leq j \leq}$ denote the standard basis for $\mathbb{R}^n$. That there are no divisors for β means that

$$\alpha_j : x \mapsto \beta(x, e_j)$$

define isomorphisms for each j. Also note that for any fixed $x \neq 0$, the set $\{\alpha_j(x)\}$ is linearly independent. To see this, note that

$$0 = \sum_j r_j \alpha_j(x) = \beta(x, \sum_j r_j \mathbf{e}_j)$$

implies that $\sum_j r_j \mathbf{e}_j = 0$ which means each $r_j = 0, j = 1, 2, \ldots, n$.

Putting $v_j = \alpha_1^{-1} \circ \alpha_j$, we get, for each non zero $x \in \mathbb{R}^n$, a linearly independent set $\{v_1(x) = x, v_2(x), \ldots, v_n(x)\}$. If p_x denotes the projection on the plane $x^\perp$, orthogonal to x, then it follows that $\{p_x v_2(x), \ldots, p_x v_n(x)\}$ forms a basis of $x^\perp$. This then defines a trivialization of the bundle $Hom(\gamma_{n-1}^1, \gamma_{n-1}^{1}{}^\perp) \cong \tau(\mathbb{P}^{n-1})$.

Remark 12.3.4 Note that by applying the Gram–Schmidt process to $\{v_1(x) = x, v_2(x), \ldots, v_n(x)\}$ we obtain trivialization of $\tau(\mathbb{S}^{n-1})$ also.

Remark 12.3.5 Of course, it is known that only $\mathbb{R}, \mathbb{R}^2, \mathbb{R}^4, \mathbb{R}^8$ admit bilinear forms without any zero divisors but we cannot prove this with the techniques developed so far.

(b) **Steifel–Whitney numbers and un-oriented cobordism**

Given any closed (i.e., compact and without boundary) n-dimensional manifold M, one knows that $H_n(M; \mathbb{Z}_2) = \mathbb{Z}_2$. A generator of this group is called a fundamental class μ_M (or orientation class) of M. We shall write $w_i(M)$ for the ith Steifel–Whitney class of the tangent bundle of M.

Now consider a sequence of variables $T_1, \ldots, T_k$ where we give (weighted) degree $\deg(T_i) = i$. Then each monomial $m(T) = T_1^{r_1} \cdots T_k^{r_k}$ is of total degree

$$\deg(m(T)) = \sum_i r_i i.$$

For each $m(T)$ of total degree n we get a number

$$SW(T) = [w_1(M)^{r_1} \cdots w_n(M)^{r_n}] \frown \mu_M \in \mathbb{Z}_2.$$

The collection $\{SW(T)\}$ where T varies over all the monomials of total degree n is referred to as the collection of Steifel–Whitney(SW) numbers of M.

Example 12.3.6 Let us prove that all SW numbers of $\mathbb{P}^{2n-1}$ vanish. We know $w(\mathbb{P}^m) = (1 + a)^{m+1}$. Putting $m = 2n - 1$, we see that $w(\mathbb{P}^{2n-1}) = (1 + a^2)^n$. In particular, $w_{2i}(\mathbb{P}^{2n-1}) = 0$ for all i. Now any monomial of total degree odd will have at least one of the variables of odd degree, the conclusion follows.

On the other hand, for $m = 2n$, $w_{2n}(\mathbb{P}^{2n}) = (2n+1)a^{2n} = a^{2n} \neq 0$. Similarly, $w_1(\mathbb{P}^{2n}) = (2n+1)a = a$ and hence $w_1^{2n} = a^{2n} \neq 0$. So, there are at least two of them which are non zero.

In the special case when $m = 2^n$, we have $w(\mathbb{P}^m) = 1 + a + a^m$ and so there are no other non zero S-W numbers.

This computation may not be so impressive. However, the following two theorems due to two great topologists take the cake.

Theorem 12.3.7 (Pontrjagin) *If M is the total boundary of a compact manifold W, then the Steifel–Whitney numbers of M are all zero.*

Proof: Let $\partial : H_{i+1}(W, M) \to H_i(M)$ and $\delta : H^i(M) \to H^{i+1}(W, M)$ denote the canonical connecting homomorphism in the respective long homology (cohomology) exact sequence of the pair (W, M). The relative fundamental class $\mu_W \in H_{n+1}(W, M; \mathbb{Z}_2)$ has the property that $\partial(\mu_W) = \mu_M$. By the property of cap product under boundary homomorphism (see Theorem 7.3.5(e)), we have, for any $v \in H^n(M)$

$$v \frown \mu_M = v \frown (\partial \mu_W) = (\delta v) \frown \mu_M. \tag{12.12}$$

We know that the tangent bundle $\tau(W)$ restricted to $\partial W = M$ has the tangent bundle $\tau(M)$ of M as a subbundle. Moreover, the normal bundle of M in W is a trivial 1-dimensional bundle with, for example, a strictly outward normal drawn at each point of M. Thus we have

$$\tau(W)|_M \cong \tau(M) \oplus \Theta^1.$$

This then means that $\iota^* w(W) = w(M)$. Therefore, each class $w_1(M)^{r_1} \cdots w_n(M)^{r_n} \in H^n(M), (\sum r_i = n)$, is in the image of $\iota^* : H^n(W) \to H^n(M)$. By the long exact sequence

$$H^n(W) \xrightarrow{\iota^*} H^n(M) \xrightarrow{\delta} H^{n+1}(W, M)$$

it follows that $\delta(w_1(M)^{r_1} \cdots w_n(M)^{r_n}) = 0$. The result sow follows from (12.12). ♠

The converse of this theorem is a very deep result due to Thom (see [Thom, 1954]):

Theorem 12.3.8 (Thom) *If all the Steifel–Whitney numbers of a closed manifold vanish then M is the total boundary of some manifold.*

Definition 12.3.9 Let $M_i, i = 1, 2$ be any two closed manifolds. We say M_1 is cobordant to M_2 if there exists a compact manifold W with $\partial W = M_1 \sqcup M_2$.

Remark 12.3.10 It can be shown that being cobordant is an equivalence relation on the diffeomorphism class of all closed n-dimensional manifolds. Under disjoint union, these classes form an abelian group. Under the Cartesian product, this abelian group becomes a graded commutative ring. It is clear that the SW numbers of a disjoint union $M_1 \sqcup M_2$ are the sum of the SW numbers of M_i. Thus the theorem 12.3.7 means that the SW-numbers are invariant under cobordism. And theorem 12.3.8 means that if all the SW numbers of equal for M_1, M_2 then M_1 is cobordant to M_2.

For strengthening these results, another type of characteristic classes called *Pontrjagin Classes* were invented. We shall briefly study them in the next section.

(c) **Immersions and embeddings** For any immersed manifold M in $\mathbb{R}^N$ we can talk about the normal bundle $\nu(M)$ and then we have

$$\tau(M) \oplus \nu(M) \cong \Theta^N.$$

It follows that

$$w(M)w(\nu(M)) = 1.$$

That is $w(\nu(M))$ is the multiplicative inverse of $w(M)$ in the cohomology algebra $H^*(M; \mathbb{Z}_2)$. In particular, the total Steifel–Whitney class of the normal bundle $\nu(M)$ is independent of the dimension of the immersion. On the other hand, since $\nu(M)$ is a $(N-n)$-plane bundle, we know that $w_i(\nu(M)) = 0$ for $i > N - m$. This then puts an obvious lower bound for the immersion dimension, provided we can compute the inverse of $w(M)$.

This is where we use the formal graded-algebra approach. Consider, for any connected space M,

$$H^{\Pi}(M) = K + H^1(M; K) + \cdots + H^n(M, K) + \cdots$$

the direct product of $H^i(M), i \geq 0$. (We are using the additive notation so that we can think of them as power series as well.) An element of this direct product is a finite or infinite sum

$$a_0 + a_1 + \cdots$$

where $a_i \in H^i(M, K)$. One defines component-wise addition and Cauchy product as multiplication to make it into a K-algebra. It follows easily that $H^{\Pi}(M)$ is a graded-commutative algebra in which an element of the above form is invertible iff $a_0 \neq 0$. In particular, let us compute the inverse of $w(\mathbb{P}^9)$. Indeed

$$w(\mathbb{P}^9) = (1+a)^{10} = 1 + a^2 + a^8$$

and hence

$$w(\mathbb{P}^9)^{-1} = 1 + a^2 + a^4 + a^6.$$

Therefore we cannot immerse $\mathbb{P}^9$ in $\mathbb{R}^N$ for $N < 9 + 6 = 15$. For $n = 2^r$, we actually get a sharp result. Here

$$w(\mathbb{P}^n) = (1+a)^{n+1} = 1 + a + a^n$$

and therefore

$$w(\nu) = w(\mathbb{P}^n)^{-1} = 1 + a + \cdots + a^{n-1}.$$

This then implies that we cannot immerse $\mathbb{P}^n$ in $\mathbb{R}^N$ for $N < 2n - 1$. On the other hand, by a celebrated theorem of Whitney, any manifold can be immersed in $\mathbb{R}^{2n-1}$.

12.4 Splitting Principle and Uniqueness

Let $\xi = (E, p, B)$ be any vector bundle. Call a map $f : B_1 \to B$ a splitting map for ξ if $f^* : H^*(B, K) \to H^*(B_1, K)$ is injective and $f^*(\xi)$ is a direct sum of line bundles. Clearly
(1) If $f : B_1 \to B$ is a splitting map for ξ and $g : B_2 \to B_1$ is such that $g^* : H^*(B_1, K) \to H^*(B_2, K)$ is injective then $f \circ g$ is also a splitting map for ξ.
(2) Any map $f : B_1 \to B$ such that $f^* : H^*(B; K) \to H^*(B_1; K)$ is injective is a splitting map for all line bundles over B.
(3) Given any ξ over B let $q : P(\xi) \to B$ be the associated projective bundle. Then $H^*(B, K) \to H^*(P(\xi); K)$ is injective. Moreover, $q^*(\xi) = \lambda_\xi \oplus \sigma_\xi$, where λ_ξ is the canonical line subbundle and σ is some complementary subbundle.
(4) Thus by induction on the rank of ξ, it follows that there is a splitting map for each ξ.
(5) Indeed, by repeated application of (1) it also follows that for any finitely many bundles ξ_i over B, there is a common splitting map.

Theorem 12.4.1 Uniqueness of Characteristic Classes *If $x(\xi), y(\xi)$ satisfy the properties (A1)–(A4), then $x = y$.*

Proof: To begin with, for the universal line bundle γ^1, from (A4), we have $x(\gamma^1) = 1 + z = y(\gamma^1)$. Therefore for any line bundle λ, by (A2) it follows that $x(\lambda) = y(\lambda)$.

Now, given any n-plane bundle ξ, let $f : B_1 \to B$ be a splitting map. To show that $x(\xi) = y(\xi)$, it is enough to show that $f^*(x(\xi)) = f^*(y(\xi))$. But we have

$$
\begin{aligned}
f^*(x(\xi)) = x(f^*(\xi)) &= x(\lambda_1 \oplus \cdots \oplus \lambda_n) = x(\lambda_1)x(\lambda_2)\cdots x(\lambda_n) \\
&= y(\lambda_1)y(\lambda_2)\cdots y(\lambda_n) \\
&= y(\lambda_1 \oplus \cdots \oplus \lambda_n) \\
&= y(f^*(\xi) = f^*(y(\xi)).
\end{aligned}
$$

This completes the proof of the theorem.   ♠

Verification of Property (A3)

We shall now complete these results by proving axiom (A3).

Lemma 12.4.2 Let $\xi = \lambda_1 \oplus \cdots \oplus \lambda_k$, a direct sum of line bundles. Then

$$x(\xi) = x(\lambda_1) \cdots x(\xi_k) = (1 + x_1(\lambda_1)) \cdots (1 + x_1(\lambda_k)). \qquad (12.13)$$

Proof: Let $q : P(\xi) \to B$ be the projective bundle. Consider the line subbundle λ_ξ of $q^*(\xi) = \oplus_i q^*(\lambda_i)$. Proving (12.13) is the same as proving that the product

$$(a_\xi + x_1(\lambda_1)) \cdots (a_\xi + x_1(\lambda_k)) = 0. \qquad (12.14)$$

Let λ_ξ^* be the multiplicative inverse of the line subbundle of λ_ξ. Tensoring $q^*(\xi)$ with this yields a trivial subbundle of $\oplus_{i=1}^n [\lambda_\xi^* \otimes q^*(\lambda_i)]$. This is the same as having a nowhere vanishing section s of the direct sum. When projected to any of the summands, this yields a cross section s_i of the line bundle $\lambda_\xi^* \otimes q^*(\lambda_i)$. Let $V_i \subset P(\xi)$ be the open set on which s_i does not vanish. This means that restricted to V_i, $\lambda_\xi^* \otimes q^*(\lambda_i)$ is trivial. Therefore $a_\xi + q^* x_1(\lambda_i) = x_1(\lambda_\xi) + x_1(q^*(\lambda_i)) = 0$ on V_i. Hence the LHS of (12.14) vanishes on V_i. Since $\cup_i V_i = P(\xi)$, (12.14) follows.   ♠

We can now prove (A3): Given ξ, η on B, let $f : B_1 \to B$ be a common splitting. Let

$$\oplus_{i=1}^k \lambda_i = f^*(\xi); \qquad \oplus_{j=1}^l \lambda_{n+j} = f^*(\eta).$$

Then

$$
\begin{aligned}
f^* x(\xi \oplus \eta) &= x(f^*(\xi \oplus \eta)) = x(\lambda_1 \oplus \cdots \oplus \lambda_{m+n}) \\
&= [x(\lambda_1) \cdots x(\lambda_n)][x(\lambda_{n+1}) \cdots x(\lambda_{n+m}) \\
&= x(f^*(\xi)) x(f^*(\eta)) = f^*(x(\xi) x(\eta)).
\end{aligned}
$$

Since f^* is injective, we are through.   ♠

We end this section with a result that relates the Euler class and Chern class.

Theorem 12.4.3 *For any complex k-plane bundle η over a paracompact space, we have $e(\eta) = c_k(\eta)$.*

Proof: By the splitting principle, and the product property of Euler class and Chern class for Whitney sums, it is enough to prove this for line bundles. By the classification of line bundles, it is enough to prove this for the universal line bundle γ^1 over $\mathbb{P}_{\mathbb{C}}^\infty$. Since the inclusion $\mathbb{P}_{\mathbb{C}}^1 \hookrightarrow \mathbb{P}_{\mathbb{C}}^\infty$ induces an isomorphism in second cohomology, it is enough to prove this for $\gamma_{\mathbb{P}^1}^1$. For this case, in Example 12.1.5, we have verified that $e(\gamma^1) = z$, the canonical generator and by definition (A4) $c_1(\gamma^1) = z$.   ♠

Exercise 12.4.4 Show that for any complex line bundle η over a paracompact space, $w_2(\eta)$ is equal to the mod 2 reduction of $c_1(\eta)$. [Hint: First show that the restriction of canonical line bundle over $\mathbb{P}_{\mathbb{C}}^{n-1}$ to $\mathbb{P}_{\mathbb{R}}^{2n-1}$ is the complexification of the canonical line bundle over $\mathbb{P}_{\mathbb{R}}^{2n-1}$.]

12.5   Complex Bundles and Pontrjagin Classes

Definition 12.5.1 Let V be a $\mathbb{R}$-vector space of even dimension. By a complex structure on V we mean an $\mathbb{R}$-linear isomorphism $J : V \to V$ such that $J \circ J = -Id$.

Given a complex vector space F, we shall denote by $F_\mathbb{R}$ the underlying real vector space of dimension $2(\dim_\mathbb{C} F)$.

Remark 12.5.2 This is an example of a *forgetful functor*. Here it forgets the complex structure retaining only the real vector space structure and the orientation. Clearly the map $J(\mathbf{v}) = \imath \mathbf{v}$ gives a complex structure on $F_{\mathbb{R}}$ which is $\mathbb{C}$-isomorphic to F.

Definition 12.5.3 By the complexification of a real vector space V we mean taking $V \otimes_{\mathbb{R}} \mathbb{C}$. The complex structure is defined by $J(\mathbf{v} \otimes 1) = \mathbf{v} \otimes \imath$. Under the identification $V \otimes \mathbb{C} \to V \oplus V$ given by

$$(\mathbf{u} \otimes 1 \mapsto (\mathbf{u}, 0); \quad \mathbf{u} \otimes \imath \mapsto (0, \mathbf{u})$$

the complex structure takes the form:

$$J(\mathbf{u}, \mathbf{v}) = (-\mathbf{v}, \mathbf{u}).$$

Definition 12.5.4 Given a complex vector space F by the conjugate complex vector space $\bar{F}$ we mean the underlying real vector space together with the complex structure $J(\mathbf{v}) = -\imath \mathbf{v}$.

Lemma 12.5.5 (i) We have for any real vector space V, $V \otimes \mathbb{C}$ is canonically isomorphic to the conjugate $\mathbb{C}$-vector space $\overline{V \otimes \mathbb{C}}$.
(ii) Given a complex vector space F we have a canonical isomorphism $F_{\mathbb{R}} \otimes \mathbb{C} \to F \oplus \bar{F}$.
(iii) Both (i) and (ii) hold for vector bundles as well.

Proof: (i) Define $\Theta(\mathbf{u} + \imath \mathbf{v}) = \mathbf{u} - \imath \mathbf{v}$ and verify that Θ is as required.
(ii) Consider the following two maps $f, g : F \to F_{\mathbb{R}} \otimes \mathbb{C}$ given by

$$g(\mathbf{u}) = (\mathbf{u}, -\imath \mathbf{u}); \quad h(\mathbf{u}) = (\mathbf{u}, \imath \mathbf{u}).$$

Verify that g is complex linear and h is conjugate linear and both are injective. Moreover, images of the two maps span the entire $F_{\mathbb{R}} \otimes \mathbb{C}$. Therefore, we can identify $F_{\mathbb{R}} \otimes \mathbb{C}$ with $F \oplus \bar{F}$.
(iii) Since this isomorphism is canonical, we get the same statement for vector bundles as well. ♠

Lemma 12.5.6 $c_1(\overline{\gamma^1}) = -c_1(\gamma_1)$ where γ^1 is the canonical line bundle over $\mathbb{CP}^{\infty}$.

Proof: Note that it is enough to prove the statement for the canonical line bundle over $\mathbb{CP}^1$. Consider the map $j : \mathbb{CP}^1 \to \mathbb{CP}^1$ defined by $z \mapsto \bar{z}$. Then $j * (\gamma_1) \cong \overline{\gamma^1}$. Therefore $c(\overline{\gamma_1}) = j^*(1 + z) = 1 - z$. ♠

Theorem 12.5.7 $c_i(\bar{\xi}) = (-1)^i c_i(\xi)$.

Proof: Use splitting principle. ♠

Thus for a real bundle ξ, if the total Chern class is

$$c(\xi \otimes \mathbb{C}) = 1 + c_1 + \cdots + c_k$$

then

$$c(\overline{\xi \otimes \mathbb{C}}) = 1 - c_1 + c_2 - + \cdots + (-1)^k c_k.$$

From Lemma 12.5.5 (iii), it follows that $2c_{2i-1}(\xi \otimes \mathbb{C}) = 0$. So, we concentrate our attention on the even degree terms.

Definition 12.5.8 The ith Pontrjagin class of a real vector bundle ξ is defined as

$$p_i(\xi) := (-1)^i c_{2i}(\xi \otimes \mathbb{C}) \in H^{4i}(B; \mathbb{Z}) \qquad (12.15)$$

and the total Pontrjagin class

$$p(\xi) := 1 + p_1(\xi) \cdots + p_{[n/2]}(\xi). \qquad (12.16)$$

Remark 12.5.9

(a) The sign on the RHS of (12.15) is introduced so that, elsewhere, some formula becomes nicer. (See Corollary 12.5.13.) You may ignore the sign (like some authors), which is OK, provided you are consistent.

(b) All the four fundamental properties of the Chern classes hold for Pontrjagin classes as well except that the product formula is valid only up to order 2 terms, i.e.,

$$2[p(\xi \oplus \eta) - p(\xi)p(\eta)] = 0.$$

However, in a special case, we have the stronger result

$$p(\xi \oplus \Theta^1) = p(\xi)$$

which follows directly from the corresponding result for Chern classes. In particular, $p(\tau(\mathbb{S}^n)) = 1$.

Theorem 12.5.10 *For any complex k-plane bundle ω we have*

$$1 - p_1 + p_2 - + \cdots + (-1)^k p_k = (1 - c_1 + c_2 - + \cdots + (-1)^k c_k)(1 + c_1 + \cdots + c_k).$$

In particular,

$$p_j(\omega) = c_j^2 - 2c_{j-1}c_{j+1} + - \cdots + (-1)^k 2c_{2k}.$$

Proof: Exercise.

Corollary 12.5.11 $p(\mathbb{CP}^k) = (1 + a^2)^{k+1}$.

Proof: Exercise.

Lemma 12.5.12 Given any $\mathbb{R}$-bundle of rank k, there is a $\mathbb{R}$-isomorphism $\xi \oplus \xi \to (\xi \otimes \mathbb{C})_{\mathbb{R}}$ which is orientation preserving iff $k(k-1)/2$ is even.

Proof: Exercise.

Corollary 12.5.13 If ξ is an oriented real bundle of rank $2k$, then $p_k(\xi)$ is equal to the square of the Euler class $e(\xi)$.

Proof: We have $p_k(\xi) = (-1)^k c_{2k}(\xi \otimes \mathbb{C}) = (-1)^k e(\xi \otimes \mathbb{C})$. On the other hand, $e(\xi \otimes \mathbb{C}) = (-1)^{2k(2k-1)/2} e(\xi \oplus \xi) = (-1)^{k(2k-1)} e(\xi)^2 = (-1)^k e(\xi)^2$. ♠

Theorem 12.5.14 *The cohomology ring of the oriented infinite real Grassmann manifold $\tilde{G}_{2n+1}$ (respectively, $\tilde{G}_{2n}$) is, up to 2-torsion, isomorphic to the polynomial ring generated by the Pontrjagin classes $p_1, \ldots, p_n$ (respectively, $p_1, \ldots, p_n$, and $e(\gamma^{2n})$) of the canonical oriented $(2n + 1)$-plane bundle (respectively $(2n)$-plane bundle) over $\tilde{G}_{2n+1}$ (resp. $\tilde{G}_{2n}$).*

We end this chapter with a very brief introduction to oriented cobordism theory.

Definition 12.5.15 Let M_1, M_2 be two oriented, m-dimensional closed manifolds. We say M_1 is oriented cobordant to M_2 if there exists an oriented compact manifold W of dimension $m + 1$ such that $\partial W = M_1 \sqcup M_2$ with the induced orientation.

Remark 12.5.16 The oriented cobordism defines an equivalence relation among the class of all closed manifolds of a fixed dimension m. The set of equivalence classes is denoted by Ω^m. The disjoint union of manifolds defines an abelian binary operation with the empty manifold as the additive identity. Recall that given an oriented manifold M there is an obvious orientation on $M \times \mathbb{I}$ so that $\partial(M \times \mathbb{I}) = M \times 0 \sqcup M \times 1$ wherein the induced orientation on $M \times 1$ coincides with that of M and the on $M \times 0$ is the opposite orientation. This allows us to write $\partial M \times \mathbb{I} = (-M \sqcup M)$. Thus the manifold with the opposite orientation is its negative in the group structure of cobordism classes. Clearly $\Omega^0 = \mathbb{Z}$. Let $\Omega = \oplus_{m \geq 0} \Omega^m$. Then the Cartesian product defines a graded ring structure on Ω. This is called the oriented cobordism ring. We state without proof:

Theorem 12.5.17 Thom *The oriented cobordism group Ω_n is finite for $n \not\equiv 0 \mod 4$ and is a finitely generated group of rank equal to the number of partitions of r for $n = 4r$.*

Corollary 12.5.18 Let M be a smooth closed oriented manifold. Then some positive multiple of M is oriented null cobordant iff all the Pontrjagin numbers of M vanish.

12.6   Miscellaneous Exercises to Chapter 12

1. Prove Theorem 12.5.10, Corollary 12.5.11 and Lemma 12.5.12.

2. Show that two complex line bundles over a given Reimann surface are isomorphic iff their Chern classes are the same.

3. Consider the quadratic surface S over $\mathbb{C}$ given by

$$S = \{([z_1, z_2, z_3], [w_1, w_2]) \in \mathbb{CP}^2 \times \mathbb{CP}^1 \; : \; z_1 w_2 = z_2 w_1\}$$

The surface S is called the blow-up of $\mathbb{CP}^2$ at the point $[0, 0, 1]$. Let $\varphi_j : S \to \mathbb{CP}^2$ denote the restriction of the two projection maps. Show that
(1) $\varphi^{-1}([0, 0, 1]) = \mathbb{CP}^1$, and $\varphi : S \setminus \varphi^{-1}([0, 0, 1]) \to \mathbb{CP}^2 \setminus \{[0, 0, 1]\}$ is a diffeomorphism.
(2) S is diffeomorphic to $\mathbb{CP}^2 \# (-\mathbb{CP}^2)$.
(3) $\varphi_2 : S \to \mathbb{CP}^1$ is a $\mathbb{CP}^1$-fibre bundle.
(4) The pull-back of the above bundle via the Hopf fibration $h : \mathbb{S}^3 \to \mathbb{CP}^1$ is a product bundle.
(5) $\mathbb{S}^2 \times \mathbb{S}^2$ and S have isomorphic homotopy groups, but nonisomorphic cohomology algebras, in particular S, is not homotopy equivalent to $\mathbb{S}^2 \times \mathbb{S}^2$.

Chapter 13

Spectral Sequences

Spectral sequences were invented by Leray in the context of study of cohomology of sheaves. Soon it was taken up by other authors who realized its potential in application to a wide class of situations. At present, there are more than three dozen various spectral sequences. In this chapter, we introduce the basics of this wonderful tool with a single example, viz., the Leray–Serre spectral sequence of a fibration. We give both homology and cohomology versions, and include a variety of applications such as the generalised Wang sequence, Gysin sequence, generalised Hurewicz theorem and Whitehead theorem. As a culmination of all this and several other results developed in this book, we present Serre's result on homotopy groups of spheres, in the last section.

Within mathematics, perhaps the earliest use of the word 'spectrum' is in analysis wherein we study 'spectrum of an operator' and 'spectral theorem' for eigenvalues of operators. Very closely related is the usage 'spectrum of a ring' in algebra and algebraic topology. For a lucid account of this you may read [Taylor, 1985]. One may wonder why Leray chose the name 'spectral sequences'. In advanced homotopy theory, you will come across the terminology such as 'Ω-spectrum', etc. The simple reason for this usage is in the counter-questions: *What do we see in a rainbow? What does a prism do to a ray of light?* At the end of the article [Taylor, 1985], the author expresses his view about the use of the word spectrum and its derivatives elsewhere as follows: *These uses* (of the word 'spectrum' and its derivatives) *tend to conform to the common meaning of 'a range of inter-related values or objects, etc.' and are not connected with eigen-values of operators.*

13.1 Warm-up

The Mayer–Vietoris principle has been of great use in the study of homological properties of topological spaces. It had some partial success in the homotopical study as well (e.g., Blakers–Massey excision theorem). Let us take a closer look at this principle again.

Recall how we have constructed the CW-homology of a CW-complex X (see Definition 4.3.24). We consider the graded module

$$C_n^{CW}(X) := H_n(X^{(n)}, X^{(n-1)}; \mathbb{Z})$$

(we shall suppress the coefficients) and make it into a chain complex by taking ∂ to be the composite of

$$H_n(X^{(n)}, X^{(n-1)}) \xrightarrow{d_n} H_{n-1}(X^{(n-1)}) \xrightarrow{j_n} H_{n-1}(X^{(n-1)}, X^{(n-2)})$$

which is nothing but the connecting homomorphism of the long homology exact sequence of the triple $(X^{(n)}, X^{(n-1)}, X^{(n-2)})$. Homology of this chain complex was then shown to be isomorphic to the singular homology of X. In order to understand this a little better, we shall reconstruct the '*cellular singular homology*' (which we have seen in Chapter 4), in a more general context.

Definition 13.1.1 Let $F = \{X_s\}$ be an increasing sequence of subspaces of a topological space X and $S_.(X)$ denote the singular chain complex of X. Define $S^F(X)_n$ to be the submodule of $S_n(X)$ generated by singular simplexes $\sigma : \Delta_n \to X$ such that $\sigma(\Delta_n^{(k)}) \subset X_k$ for each $0 \leq k \leq n$. Then it is easily checked that $S^F(X) = \oplus_{n \geq 0} S_n^F(X)$ is indeed a chain subcomplex of $S_.(X)$. In case $X_s = X^{(s)}, s \geq 0$, is the sth-skeleton of a relative CW-complex X, we call this the cellular singular chain complex and denote it by $C_.^{cell}(X, A)$. (See Definition 4.3.30.)

Theorem 13.1.2 *Suppose $\{X_s\}$ is an increasing sequence of subspaces of X such that (X, X_s) is $(s - 1)$-connected for each $s \geq 0$. Then the inclusion map $S^F(X) \subset S_.(X)$ is a chain equivalence.*

Proof: We appeal to retraction-operator Lemma 4.6.4. For this, we need to construct, for each singular simplex $\sigma : \Delta_q \to X$, the prisms $P\sigma : \mathbb{I} \times \Delta_q \to X$ satisfying the hypothesis of this lemma, with $S^F(X)$ in place of C. As usual, the construction of P is done by induction. For $q = 0$, if $\sigma(\Delta_0) \in X_0$, define $P\sigma(t, \mathbf{e}_0) = \sigma(\mathbf{e}_0), t \in \mathbb{I}$. If $\sigma(\Delta_0) \notin X_0$, choose a path $\omega : \mathbb{I} \to X$ from $\sigma(\Delta_0)$ to some point in X_0. (This is possible because (X, X_0) is 0-connected.) Then take $P\sigma(t, \mathbf{e}_0) = \omega(t)$.

Inductively, assume that $q > 0$ and we have defined $P\tau$ for all singular simplexes τ of dimension $< q$. Let now σ be a singular simplex of dimension q. If $\sigma(\Delta_q^{(k)}) \subset X_k$, for all k, define $P\sigma(v, t) = \sigma(v), t \in \mathbb{I}$. In the general case, we can define $P\sigma$ on $0 \times \Delta_q \cup \mathbb{I} \times \dot{\Delta}_q$ by using (a) and (c) of Lemma 4.6.4, and using (b), we ensure that $P\sigma(1 \times (\dot{\Delta}_q)^{(k)}) \subset X_k$ for all k. It remains to extend this $P\sigma$ all over $\mathbb{I} \times \Delta_q$ appropriately. This is where we have to use the $(q - 1)$-connectivity of (X, X_q). Since there is a homeomorphism of the topological triples

$$(\mathbb{I} \times \Delta_q, 1 \times \Delta_q, 0 \times \Delta_q \cup \mathbb{I} \times \dot{\Delta}_q) \to (\mathbb{I} \times \mathbb{D}^q, \mathbb{I} \times \mathbb{S}^{q-1} \cup 1 \times \mathbb{D}^q, 0 \times \mathbb{D}^q),$$

Theorem 10.2.7 yields an extension of $P\sigma$ as desired.   ♠

As an immediate corollary we have:

Corollary 13.1.3 For any relative CW-complex X, the inclusion map $C_.^{cell}(X, A) \to S_.(X, A)$ is a chain equivalence.

Indeed, if (X', A) is a subcomplex of (X, A), then $C_.^{cell}(X', A) = C_.^{cell}(X, A) \cap S_.(X', A)$. Then by Five lemma, it follows that the inclusion map $C_.^{cell}(X, X') \to S_.(X, X')$ induces an isomorphism of the homology. From Theorem 6.1.6, we conclude that it is a chain equivalence. Thus in particular, the inclusion map

$$C_.^{cell}(X^{(s)}, X^{(s-1)}) \subset S_.(X^{(s)}, X^{(s-1)})$$

is a chain equivalence. Hence Lemma 4.3.22 is valid if we replace the singular homology everywhere by cellular singular homology.

Consider the filtration $F_s(C_.^{cell}(X, A)) := C_.^{cell}(X^{(s)}, A) \subset C_.^{cell}(X, A)$. Corresponding to this filtration, define the chain complex $(Q_., d)$ as follows:

$$Q_s = H_*(C_.^{cell}(X^{(s)}, X^{(s-1)})) = H_*(X^{(s)}, X^{(s-1)})$$

and $d : Q_s \to Q_{s-1}$ is the boundary operator of the corresponding triples. We then have Theorem 4.3.26 valid for cellular singular homology:

Theorem 13.1.4 $H_*(Q_., d) \approx H_*(C^{CW}(X))$.

Thus, the homology of the homology of a filtration gives the homology of the original chain complex.

Spectral sequence is a sophisticated machinery which generalizes this concept of iterating homology. It turns out that filtrations on chain complexes are not the only sources for spectral sequences. There is a concept which is somewhat more general and primitive and which is the parent of all spectral sequences. This is the content of the next section, which a beginner may afford to skip and go directly to Section 13.3.

13.2   Exact Couples

In this section, we shall study the algebraic notion of an exact couple which is more fundamental than spectral sequences and which produces 'all' spectral sequences. For the sake of completeness, we begin with some relevant category theory.

Definition 13.2.1 Let $\mathcal{C}$ be any category. A morphism $f \in M(A, B)$ is said to be a *monomorphism* (respectively an *epimorphism*) if for every $g_1, g_2 : M \to A$ in $\mathcal{C}$ (respectively, for every $h_1, h_2 : B \to N$ in $\mathcal{C}$)

$$f \circ g_1 = f \circ g_2 \Longrightarrow g_1 = g_2; \quad (\text{respectively } h_1 \circ f = h_2 \circ f \Longrightarrow h_1 = h_2).$$

Remark 13.2.2 In many of the familiar small categories, 'monomorphism' is the same as 'injective' and 'epimorphism' is the same as 'surjective', but this is not always true. For example, neither of the above is true in the category $\mathcal{T}$ of topological spaces and homotopy class of maps.

Definition 13.2.3 Let $\mathcal{C}$ be a category which has zero objects (see Definition 1.8.15). Then for any morphism $f : A \to B$, the *kernel* of f is a morphism $k : K \to A$ such that
(i) $f \circ k = 0_{K,B} = 0$ and
(ii) if $l : L \to A$ is such that $f \circ l = 0_{L,B}$, then there is a unique $l' : L \to K$ such that $l = k \circ l'$.

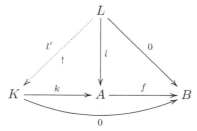

Remark 13.2.4
(i) (In other words, (K, k) is a terminal object in an appropriate category.) It follows that up to equivalences in $\mathcal{C}$, the kernel is unique, if it exists. Often we call the domain K of the kernel as the kernel of f, the morphism k being understood. It also follows that k is a monomorphism due to the uniqueness of l'.
(ii) Exactly similarly, in a dual fashion, we define cokernel as well. It can be shown that if a morphism has a cokernel, then it is unique up to equivalence.

Definition 13.2.5 Sums and Products Let A, B be any two objects in a category $\mathcal{C}$. By a product of A and B in $\mathcal{C}$ we mean an object C in $\mathcal{C}$ together with morphisms $p_A : C \to A$

and $p_B : C \to B$ satisfying the following universal property: given any two morphisms $q_A : D \to A$ and $q_B : D \to B$, there is a unique morphism $q : D \to C$ such that $q_A = p_A \circ q$ and $q_B = p_B \circ q$. Once again, due to the uniqueness of q, it can be shown that (C, p_A, p_B) is unique up to equivalence. So, if it exists, we write $A \times B$ for C. The morphisms p_A, p_B are called the projections.

Remark 13.2.6 Similarly, there is also a notion of the 'coproduct' or the 'direct sum' which is dual to the notion of product; we denote it by $A \oplus B$. It is easily shown that product exists iff direct sum exists. The notion gets easily generalised to product (or sum) of finitely many objects. Indeed, they can be generalised to arbitrary families of objects, but then products and sums will no longer be the 'same'.

Definition 13.2.7 By an additive category, we mean a category $\mathcal{C}$ with zero objects such that
(i) For any two objects A, B in $\mathcal{C}$, the product $A \times B$ exists.
(ii) For any two objects A, B in $\mathcal{C}$, the morphism set $M(A, B)$ is an abelian group.
(iii) For any three objects A, B, C in $\mathcal{C}$, the composition $M(A, B) \times M(B, C) \to M(A, C)$ is bilinear.

Example 13.2.8 The foremost example of an additive category is **Ab** of abelian groups. Indeed, the definition of additive category is modelled on these properties of **Ab**. Likewise the foremost example of a category which is not additive (but has zero objects) is **Gr**. The categories **Vect$_k$**, **FVect$_k$**, **R-mod** (see Section 1.8) and the category **Ch$_R$** (see Section 4.1) are all additive categories.

Definition 13.2.9 An additive category $\mathcal{C}$ is called an abelian category if
(i) every morphism in $\mathcal{C}$ has kernel and cokernel;
(ii) every monomorphism is the kernel of its cokernel, and every epimorphism is the cokernel of its kernel; and
(iii) every morphism f is the composite, $f = \alpha \circ \beta$, where β is an epimorphism and α is a monomorphism.

Example 13.2.10 All the additive categories mentioned in Example 13.2.8 above are abelian categories. The category of free abelian groups is additive but not abelian. Presently, we are interested in a few more examples. Given a ring R, consider modules such as $A = \oplus_{n \in \mathbb{Z}} A_n$, where each A_i is an R-module. These are called $\mathbb{Z}$-graded modules. A morphism $f : A \to B$ of $\mathbb{Z}$-graded modules is said to be homogeneous if there exists k such that $f(A_n) \subset B_{n+k}$ for all n. The integer k is called the degree of the morphism f. Graded modules together with homogeneous morphisms form an additive category. Moreover, if we restrict ourselves to degree 0 morphisms then we get an abelian category.

Example 13.2.11 Likewise we can take modules which are graded over $\mathbb{Z}^r$, $r \geq 2$ with homogeneous morphisms. They form an additive category and if we take morphisms of homogeneous multi-degree $(k_1, \ldots, k_r) = (0, \ldots, 0)$ only, then we get an abelian category. Of course, like the category of chain complexes **Ch$_R$** which is a subcategory of the category of $\mathbb{Z}$-graded modules, there are subcategories of these categories which are additive or abelian as the case may be.

Definition 13.2.12 A differential object (C, d) in an additive category $\mathcal{C}$ is an object C in $\mathcal{C}$ together with a morphism $d : C \to C$ such that $d \circ d = 0$. Often, we may drop mentioning the differential d and just say C is a differential object.

Remark 13.2.13 The foremost example of this is a chain complex. However, what we are aiming at are $\mathbb{Z}^2$-graded modules with a differential of degree $(-r, r-1)$. In any case, we can borrow standard terminology from homology theory such as cycles, boundaries, etc. To each differential object (C, d) we may assign the graded group $H(C, d) = \frac{\operatorname{Ker} d}{\operatorname{Im} d}$ and abbreviate it as $H(C)$.

Definition 13.2.14 An exact couple $C = (D, E, i, j, k)$ in an abelian category $\mathcal{C}$, consists of two objects D, E and three morphisms,

$$i : D \to D; \quad j : D \to E, \quad k : E \to D,$$

in $\mathcal{C}$ such that

$$\operatorname{Ker} i = \operatorname{Im} k; \quad \operatorname{Ker} j = \operatorname{Im} i; \quad \operatorname{Ker} k = \operatorname{Im} j.$$

We represent this by the following triangle

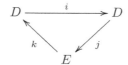

which is not to be confused with a commutative triangle.

If $C_1 = (D_1, E_1, i_1, j_1, k_1)$ is another exact couple, a morphism $\phi : C \to C_1$ consists of a pair $\phi = (\alpha, \beta)$ of morphisms $\alpha : D \to D_1$ and $\beta : E \to E_1$ such that in the following diagram all three parallelograms are commutative:

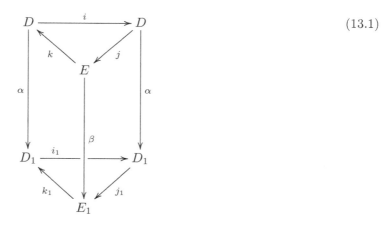

(13.1)

It follows easily, that there is a category of exact couples in any given abelian category.

Put $d = j \circ k : E \to E$, then $d \circ d = 0$. The homology module $H(E, d) = \operatorname{Ker} d / \operatorname{Im} d$ fits into another exact couple

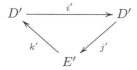

where, $D' = i(D), E' = H(E, d)$, i' is the restriction of i to $i(D)$ and j', k' are given by

$$j'(ix) = jx + jkE, \quad k'(y + jkE) = ky, y \in E \text{ such that } jky = 0.$$

Check that j', k' are well-defined morphisms and the above triangle is exact. This is called the derived couple of C and is denoted by C'. Given a morphism $\phi : C_1 \to C_2$ of two exact couples it is fairly obvious that there is an induced morphism $f' : C_1' \to C_2'$ of derived couples. Indeed, check that $C \rightsquigarrow C'$ defines a covariant functor on the category of exact couples.

Iteratively, we define $C^0 = C, C^1 = C', \ C^r = (C^{r-1})', r \geq 2$. The sequence of modules $\{C^r\}$ is called the spectral sequence associated to the exact couple C. Let us denote $C^r = (D^r, E^r, i_r, j_r, k_r)$. We then have a category of spectral sequences in which a morphism τ is a collection of morphisms $\tau^r : C_1^r \to C_2^r$ of exact couples such that $\tau_{r+1} = \tau_r'$. It follows that the association $C \rightsquigarrow \{C^r\}_{r \geq 1}$ itself is a covariant functor.

Clearly, the modules D^r and E^r, etc., are subquotients of subquotients $\cdots$ of the modules D, E with which we begin. The following lemma tells us that instead of subquotients of subquotients and so on, we can work with submodules of D and quotients of E alone.

Lemma 13.2.15 Put $i^r = i \circ i \circ \cdots \circ i, (r$ times$)$. Then

$$D^r = i^r(D); \quad E^r = \frac{k^{-1}(i^r(D))}{j(\operatorname{Ker} i^r)}.$$

Moreover, the morphism i_r is the restriction of i to $\operatorname{Im} i^r$ and j_r and k_r are given by

$$j_r(i^r x) = j(x) + j(\operatorname{Ker} i^r); \quad k_r(y + j\operatorname{Ker} i^r) = k(y), \ y \in k^{-1}(\operatorname{Im} i^r).$$

Proof: The proof is by a very simple induction on r. ♠

Putting $Z^r = k^{-1}(i^r(D))$ and $B^r = j(\operatorname{Ker} i^r)$ we have,

$$0 = B^0 \subset B^1 \subset B^2 \subset \cdots \subset Z^2 \subset Z^1 \subset Z_0 = E.$$

We put $E^\infty := \cap_r Z^r / \cup_r B^r$. Clearly,

Lemma 13.2.16 $\lim_r \lim_s Z^r / B^s = \lim_s \lim_r Z^r / B^s = E^\infty$.

Remark 13.2.17 We could go on with this kind of generality for some time but by the very nature of being general, the results that we may get will be weak. So, it is time to look out for specific situations. Let us now take a look at some situations which give rise to exact couples.

Example 13.2.18 Bockstein spectral sequence

For any prime p, consider the exact sequence

$$0 \longrightarrow \mathbb{Z} \xrightarrow{\ \cdot p\ } \mathbb{Z} \longrightarrow \mathbb{Z}_p \longrightarrow 0.$$

Tensoring with the chain complex C of abelian groups, we get an exact sequence

$$0 \longrightarrow C \xrightarrow{\ \times p\ } C \longrightarrow C \otimes \mathbb{Z}_p \longrightarrow 0.$$

This in turn yields a long exact homology sequence which can be thought of as an exact couple

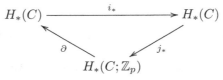

Here i_* is the multiplication by p, j_* is obtained by going modulo (p) and ∂ is the connecting homomorphism of the long homology exact sequence. The spectral sequence associated to this exact couple is called the *Bockstein spectral sequence*. (Unfortunately, we are not going to discuss much about this very interesting spectral sequence.)

The next example is of our principal source of spectral sequences, which was also the motivation behind the invention of spectral sequences.

Definition 13.2.19 Let (C, d) be a differential graded object in an abelian category $\mathcal{C}$. By an increasing filtration on (C, d) we mean a sequence

$$\mathcal{F} : \cdots \subset C^{(p-1)} \subset C^{(p)} \subset \cdots \subset C$$

of subobjects of (C, d), i.e., $d(C^{(p)}) \subset C^{(p)}$ for each p.

There is an obvious category of filtered differential objects; a morphism $\phi : (C, d, \mathcal{F}) \to (C', d', \mathcal{F}')$ is a morphism ϕ of the differential graded objects which preserves the filtrations; $\phi(C^{(p)}) \subset C'^{(p)}$ for each p.

We shall now construct a functor from this category to the category of exact couples in $\mathcal{C}$. Given an object C in $(\mathcal{C}, d, \mathcal{F})$, for each p we have the short exact sequence

$$0 \longrightarrow C^{(p-1)} \longrightarrow C^{(p)} \longrightarrow C^{(p)}/C^{(p-1)} \longrightarrow 0$$

which, in turn, yields a long exact sequence of homology groups. All of these sequences fit into the following ladder in which the vertical arrows indicate inclusion induced homomorphisms.

We can split up this ladder into a sequence of staircases as indicated by the thick arrows. Thus we define $D = \oplus_{p,q} D_{p,q}, E = \oplus_{p,q} E_{p,q}$ where,

$$D_{p,q} = H_{p+q}(C^{(p)}), \quad E_{p,q} = H_{p+q}(C^{(p)}/C^{(p-1)}). \tag{13.2}$$

Then $(D, E, i_*, j_*, \partial)$ is an exact couple associated with $(C, d, \mathcal{F})$. Given a morphism $\phi : (C, d, \mathcal{F}) \to (C', d', \mathcal{F}')$ there are induced morphisms of bigraded modules $\alpha(\phi) : D \to D'$ and $\beta(\phi) : E \to E'$ such that the diagram (13.1) commutes. This completes the construction of the functor from the category of filtered differential objects to the category of exact couples.

As seen before, the exact couple so constructed in turn gives rise to a spectral sequence, which we call the spectral sequence associated to the filtered differential object. It turns out that we get to know these spectral sequences better in terms of the filtered differential object (the grandparents) rather than those of the associated exact couples (the parents).

Pay attention to the way we choose the indexing in (13.2). Each of i_*, j_*, ∂ is homogeneous of bidegrees $(1, -1), (0, 0), (-1, 0)$ and hence $d = j_* \circ \partial$ is homogeneous of degree $(-1, 0)$. It follows that in Lemma 13.2.15, $\deg i_r = (1, -1), \deg j_r = (1 - r, r - 1)$ and $\deg k_r = (-1, 0)$ and hence d^r is homogeneous of bidegree $(-r, r - 1)$. Putting

$$Z_{s,t}^r := \partial^{-1}(i_*^{r-1}(H_{s+t-1}(C^{(s-r)}))); \quad B_{s,t}^r := j_*(\mathrm{Ker}\,[i_*^{r-1} : H_{s+t}(C^{(s)}) \to H_{s+t}(C^{(s+r-1)})]),$$

it also follows that $E_{s,t}^r = Z_{s,t}^r / B_{s,t}^r$.

Indeed, to a large extent, we can completely bypass all mention of the exact couples and just work with the filtered differential object as done classically. Since most of our examples arise with a filtration on a chain complex, it is worth getting more familiar with this special case, even at the cost of some repetition. That is what we are going to do in the next section.

13.3   Algebra of Spectral Sequences

In this section, we shall develop the basic algebra of spectral sequences independent of the exact couples and what you may have read in the previous section. We shall fix R to denote a principal ideal domain. All modules will be taken over R.

Definition 13.3.1 By a bigraded module E we mean a direct sum

$$E = \oplus_{(s,t) \in \mathbb{Z} \times \mathbb{Z}} E_{s,t}$$

of modules $E_{s,t}$. A homomorphism $f : E \to E'$ of two bigraded modules is called *homogeneous*, if there are integers (p, q) such that $f(E_{s,t}) \subset E'_{s+p,t+q}$. Then f is called a morphism of *bidegree* (p, q) and *total degree* $p + q$.

We shall consider only homogeneous homomorphisms, without specifically mentioning it. There is a category of bigraded modules and homogeneous morphisms.

Definition 13.3.2 Let E be a bigraded module. By a *differential* on E we mean an endomorphism $d : E \to E$ of total degree -1 such that $d \circ d = 0$. It is a common practice to denote the degree of d by $(-r, r - 1)$ (note that there is a unique such r). Thus d can be thought of as a direct sum of homomorphisms

$$d : E_{s,t} \to E_{s-r,t+r-1}.$$

The pair (E, d) is then called a differential bigraded module. Just like in the case of chain complexes, a morphism between two differential graded objects is supposed to commute with the differentials. We then have a category of differential bigraded modules.

We may think of a differential bigraded module as a sheet of graph paper on which you have placed modules at all the lattice points. You put arrows from one module to another, all arrows having same slope $\frac{r}{1-r}$ and are of the same length. (See Figure 13.1.)

Definition 13.3.3 Given a differential bigraded module (E, d), we define $H(E)$, the homology of (E, d), to be the bigraded module

$$H_{s,t}(E, d) := \mathrm{Ker}\,[d : E_{s,t} \to E_{s-r,t+r-1}]/d(E_{s+r,t-r+1}).$$

E^1 E^2

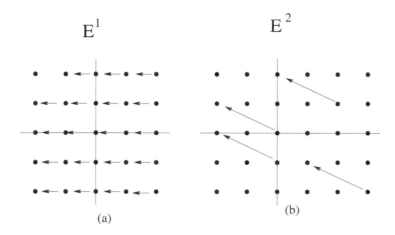

(a)   (b)

FIGURE 13.1. Bigraded modules

Definition 13.3.4 Associated to (E, d), we can define a chain complex $\{E_q, \partial\}$ by taking $E_q = \oplus_{s+t=q} E_{s,t}$ and ∂ to be the sum of the corresponding components of d. Clearly, since homology commutes with the direct sum we have

$$H_q(E_*, \partial) = \oplus_{s+t=q} H_{s,t}(E, d).$$

Definition 13.3.5 Let $k \geq 1$ be an integer. A sequence $\{(E^r, d^r)\}_{r \geq k}$ of differential bigraded modules is called an E^k-spectral sequence if
(i) The bidegree of d^r is $(-r, r-1)$ for each $r \geq k$; and
(ii) $E^{r+1} = H(E^r), r \geq k$.

The role of k is just to indicate where we begin. Strangely, here is a situation wherein we are dealing with sequences and even the very first term and how you index it makes a difference. Unfortunately, there are slightly different conventions (see [Hilton–Stommbach, 1970]) which can be fatal if you do not pay proper attention to the indexing. Luckily, in application, we encounter only two cases, viz., $k = 1, 2$.

Remark 13.3.6 Note that the modules at the $(r+1)^{\text{th}}$ stage are completely determined by the terms and the differential at the rth stage but the differentials at the $(r+1)^{\text{th}}$ stage are completely arbitrary except for its bidegree. For instance, suppose the differentials at the rth stage are all zero. This means $E^{r+1} = E^r$ but the differentials d^r and d^{r+1} are definitely different being of different bidegree. In particular, we cannot interpolate a given spectral sequence with some additional terms.

Definition 13.3.7 Given an E^k-spectral sequence, for each $r \geq k$, let $\mathcal{Z}(E^r), \mathcal{B}(E^r)$ be the bigraded modules respectively given by:

$$\mathcal{Z}_{s,t}(E^r) = \text{Ker}\,[d^r : E^r_{s,t} \to E^r_{s-r,t+r-1}]; \quad \mathcal{B}_{s,t}(E^r) = d^r(E^r_{s+r,t-r+1})$$

Clearly $\mathcal{B}(E^r) \subset \mathcal{Z}(E^r)$ and $\mathcal{Z}(E^r)/\mathcal{B}(E^r) = E^{r+1}$. We put $Z^k = \mathcal{Z}(E^k)$ and $B^k = \mathcal{B}(E^k)$. For $r > k$, let $q^r : E^k \to E^r$ denote the composite of the quotient maps $E^k \to E^{k+1} \to \cdots \to E^r$. Put $Z^{r+1} = (q^r)^{-1}(\mathcal{Z}(E^r)), B^{r+1} = (q^r)^{-1}(\mathcal{B}(E^r))$. It follows that we have a sequence of bigraded modules

$$B^k \subset B^{k+1} \subset \cdots \subset B^r \subset \cdots \subset Z^r \subset \cdots Z^{k+1} \subset Z^k$$

such that for each $r \geq k$, we have $E^{r+1} = Z^r/B^r$. We define

$$Z^\infty = \cap_r Z^r; \quad B^\infty = \cup_r B^r; \quad E^\infty = Z^\infty/B^\infty.$$

Definition 13.3.8 A morphism $\phi : \{E^r\}_{r \geq k} \to \{E'^r\}_{r \geq k}$ from one spectral sequence to another is a collection of morphisms $\phi^r : E^r \to E'^r$ (i.e., which commutes with the differentials, and) such that for all $r \geq k$, $\phi^{r+1} = H(\phi^r)$. Clearly, we have a category of E^k-spectral sequences.

Remark 13.3.9 It follows easily that if a morphism ϕ is such that ϕ^r is an isomorphism for some r then ϕ^s is an isomorphism for all $s \geq r$. It then follows that ϕ induces an isomorphism of the E^∞ terms as well. Take care to note that if we have a morphism $\phi : E^k \to E'^k$ of just the E^k-terms, it may not induce a morphism of E^{k+1} terms though we do get a morphism $H(\phi)$ of the underlying modules of the E^{k+1}-terms. There is no guarantee that this $H(\phi)$ will commute with the differentials of the E^{k+1}-term.

We would like to call the bigraded module E^∞ the 'limit' of the E^k spectral sequence $\{E^r\}$. But most often we will not be able to justify this term. So, we shall merely call it the infinity term of the spectral sequence. As you may expect, analogous to pointwise convergence and uniform convergence, etc., of sequences of functions, there are notions of convergence of spectral sequences to suit one's purpose. We consider three of them here.

Definition 13.3.10

(i) We say the spectral sequence is *convergent* if for every (s,t), there exists an integer $r(s,t)$ such that for $r \geq r(s,t)$, the homomorphism $d^r : E^r_{s,t} \to E^r_{s-r,t+r-1}$ is zero. In this case, for each $r \geq r(s,t)$, $E^{r+1}_{s,t}$ is a quotient of $E^r_{s,t}$ and $E^\infty_{s,t}$ is the direct limit of these quotients:

$$E^{r(s,t)}_{s,t} \to E^{r(s,t)+1}_{s,t} \to \cdots$$

(ii) We say $\{E^r\}$ is *strongly convergent* if for some r there exist integers N, N' such that $E^r_{s,t} = 0$ for $s < N$ or $t < N'$. Then for all $r' \geq r$ the same will be true. Under this situation, if we choose $r' > \sup\{s - N, t - N' + 1, r\}$ then in the sequence

$$E^{r'}_{s+r',t-r'+1} \xrightarrow{d^{r'}} E^{r'}_{s,t} \xrightarrow{d^{r'}} E^{r'}_{s-r',t+r'-1}$$

the two end modules are zero because $t - r' + 1 < N'$ and $s - r' < N$. Therefore, for each $E_{s,t}$ for large r we have

$$E^r_{s,t} = E^{r+1}_{s,t} = \cdots = E^\infty_{s,t}.$$

This notion is quite useful, because there are a lot of spectral sequences which are *first quadrant*, i.e., for some r, $E^r_{s,t} = 0$ for $s < 0$ or $t < 0$. Such spectral sequences are strongly convergent and have the additional property that for any fixed q, there are only finitely many non zero terms $E^\infty_{s,t}$ such that $s + t = q$.

(iii) We say a spectral sequence $\{E^r\}_{r \geq k}$ *collapses* at the nth term if the differentials $d^r = 0$ for $r \geq n$. It then follows that $E^r = E^{r+1} = \cdots = E^\infty$. For example, it may happen that there are numbers p, q such that $E^r_{s,t} = 0$ for $s > p$ or $t > q$.

We now return to an elaborate description of the case of the bigraded differential (E, d) with a filtration, which is after all, the primary source of spectral sequences.

Definition 13.3.11 An increasing filtration F on a module A is a sequence of submodules $F_s := F_s A$ such that $F_s \subset F_{s+1}$. If A is $\mathbb{Z}$-graded then this filtration is supposed to be compatible with the gradation, viz., if $A = \oplus_{t \in \mathbb{Z}} A_t$, then $F_s = \oplus F_{s,t} = \oplus_t F_s \cap A_t$, for all s. The associated graded module $G = G_F(A)$ is defined as a bigraded module $G_{s,t} = F_{s,t}/F_{s-1,t}$. Pay attention to the indexing pattern. Here s is the *filtration degree*, $s+t$ is the *total degree* and t is called the *complementary degree*. The filtration F is said to be *convergent* if $\cap_s F_s = (0)$ and $\cup_s F_s = A$.

Remark 13.3.12 When we have a convergent filtration F on a bigraded module A, the associated graded module $G_F(A)$ is expected to contain 'all' the information on A and yet more accessible. This turns out to be more or less the case when we are working over a field and certainly not the case over general rings, due to the module extension problem. Thus, the study of spectral sequence, in general, may give all the information on the graded module $G_F(A)$ rather than on A itself. This being the case, it makes sense to redefine our target of study to be $G_F(A)$ rather than A itself.

Definition 13.3.13 By a filtration F on a chain complex $C.$, we mean a filtration on the graded module $C.$ which is compatible with the differential. (Thus, each filtration submodule of C is a subchain complex as well. With such a filtration of $C.$, there is an induced filtration on the homology $H_*(C.)$ denoted by

$$F_s(H_*(C)) = \mathrm{Im}\,[H_*(F_s(C)) \to H_*(C)]. \tag{13.3}$$

Here our initial target of study is $H_*(C)$ and the modestly modified new target is the bigraded module associated with the filtration (13.3). Various definitions of convergence considered above should be understood from this point of view.

Since the homology commutes with the direct limit, if F is a convergent filtration, then it follows that $\cup_s F_s(H_*(C)) = H_*(C)$. Once again, caution is necessary: It is not true that if $\cap_s F_s = (0)$ then $\cap_s F_s(H_*(C)) = (0)$. Thus, we need to have a stronger hypothesis on the original filtration:

Definition 13.3.14 A filtration on a graded module A is bounded below if for each t there is $s(t)$ such that $F_{s(t),t} = 0$.

It follows easily that if we have a convergent filtration which is bounded below on a chain complex then the induced filtration on the homology is convergent.

Theorem 13.3.15 *Let F be a filtration which is convergent and bounded below on a chain complex C. Then there is a convergent E^1-spectral sequence with*

$$E^1_{s,t} = H_{s+t}(F_s C/F_{s-1} C); \quad E^\infty \approx G_F H_*(C)$$

where d^1 is the boundary operator of the homology exact sequence of the triple $(F_s C, F_{s-1} C, F_{s-2} C)$.

Proof: For any $r \geq 1$, put

$$Z^r_s = \partial^{-1} F_{s-r} C, \quad Z^\infty_s = \mathrm{Ker}\, \partial \cap F_s C.$$

Note that the gradation on C induces a gradation on each of Z^r_s and Z^∞_s. Indeed,

$$Z^r_{s,t} = \{c \in F_s C_{s+t} \; : \; \partial c \in F_{s-r}\}; \quad Z^\infty_{s,t} = \{c \in F_s C_{s+t} \; : \; \partial c = 0\}.$$

Therefore, we have:

$$\cdots \subset \partial Z^{-1}_{s-1} \subset \partial Z^0_s \subset \partial Z^1_{s+1} \subset \cdots \subset \partial C \cap F_s C \subset Z^\infty_s \subset \cdots \subset Z^1_s \subset Z^0_s = F_s C.$$

Put $B^r_{s,t} := Z^{r-1}_{s-1} + \partial Z^{r-1}_{s+r-1}$ and define

$$E^r{}_{s,t} := Z^r_{s,t}/B^r_{s,t}; \quad E^\infty_{s,t} = Z^\infty_{s,t}/(Z^\infty_{s-1,t+1} + \partial C \cap F_s C).$$

Since $\partial(Z_s^r) \subset Z_{s-1}^r, \partial(B_s^r) = \partial(Z_{s-1}^{r-1} + \partial Z_{s+r-1}^{r-1}) \subset \partial Z_{s-1}^{r-1} \subset B_{s-r}^r$, it induces a homomorphism $d^r : E_s^r \to E_{s-r}^r$, viz.,

$$d^r(x + B_{s,t}^r) = \partial(x) + B_{s-r,t+r-1}^r. \tag{13.4}$$

Therefore (E^r, d^r) is a differential graded module, where the bidegree of d^r is $(-r, r-1)$. Note that for $r \le 0$, $Z_s^r = F_s(C)$, and for $r < 0$, $d^r = 0$. Hence $E_s^r = F_s(C)/F_{s-1}C$. In particular, we have,

$$E_{s,t}^0 = F_s C_{s+t}/F_{s-1}C_{s+t} = G_F(C)_{s,t}.$$

Also, d^0 is the boundary operator of the quotient chain complex $F_s C/F_{s-1}C$. Now, by Noether's isomorphism theorem,

$$E_{s,t}^1 = Z_{s,t}^1/(Z_{s-1,t+1}^0 + \partial Z_{s-1,t+1}^0) = \frac{Z_{s,t}^1/Z_{s-1,t+1}^0}{(Z_{s-1,t+1}^0 + \partial Z_{s-1,t+1}^0)/Z_{s-1,t+1}^0}.$$

Clearly $Z_{s,t}^1/Z_{s-1,t+1}^0$ is the module $Z_{s+t}(F_s C/F_{s-1}C)$ of $(s+t)$-cycles of the chain complex $F_s C/F_{s-1}C$ whereas $[Z_{s-1,t+1}^0 + \partial Z_{s-1,t+1}^0]/Z_{s-1,t+1}^0$ is the module $B_{s+t}(F_s C/F_{s-1}C)$ of $(s+t)$-boundaries of $F_s C/F_{s-1}C$. Therefore,

$$E_{s,t}^1 = H_{s+t}(F_s C/F_{s-1}C).$$

Since the boundary operator of the triple $(F_c C, F_{s-1}C, F_{s-2}C)$ is nothing but the composite of the connecting homomorphism and the inclusion induced map

$$H_{s+t}(F_s C/F_{s-1}C) \xrightarrow{\quad d \quad} H_{s+t-1}(F_{s-1}C) \xrightarrow{\quad j_* \quad} H_{s+t-1}(F_{s-1}C/F_{s-2}C) \,,$$

it follows that it is equal to the homomorphism d^1 induced by ∂.

The proof that $H_*(E^r) = E^{r+1}$ is similar for $r \ge 1$. Thus $\{E^r\}$ is a spectral sequence.

We shall now prove that this spectral sequence 'converges'. Since the filtration is bounded below, given s, t, there exists $\sigma(s+t)$ such that $F_{s'}C_{s+t} = 0$ for all $s' \le \sigma(s+t)$. Therefore for every fixed s, t, if $r > s - \sigma(s+t)$, then $Z_{s,t}^r = Z(C_{s+t})$. Hence we have surjective mappings

$$E_{s,t}^r \longrightarrow E_{s,t}^{r+1} \longrightarrow \cdots \tag{13.5}$$

That is what is meant by convergence of the spectral sequence.

Moreover, we have

$$E_s^\infty = Z_s^\infty/(Z_{s-1}^\infty + \partial C \cap F_s C) = (Z^\infty + F_{s-1}C)/(F_{s-1}C + \partial C \cap F_s C). \tag{13.6}$$

Since $\cup_s F_s C = C$, it follows that $\partial C \cap F_s C = \cup_r Z_{s+r-1}^{r-1}$. Since for each fixed t, $F_s C_t = 0$ for small t, it follows that $\cap_r Z_{s,t}^r = Z_{s,t}^\infty$. Therefore, carrying further with (13.6), we get

$$E^\infty = (\cap_r Z_s^r + F_{s-1}C)/(F_{s-1}C + \cup_r Z_{s+r-1}^{r-1}) = \cap_r(Z_s^r + F_{s-1}C)/ \cup_r (F_{s-1}C + \partial Z_{s+r-1}^{r-1}).$$

Since the numerator is the intersection of a decreasing sequence of modules, whereas the denominator is the union of an increasing sequence of modules, it follows that E^∞ is the direct limit of the sequence (13.5).

Finally, it remains to describe the E^∞ term. We have seen that for each fixed $s+t$, $Z_{s,t}^r = Z(C_{s+t})$ for large r. Therefore, every $(s+t)$-cycle in C is represented by an element in Z_s^∞. This just means that

$$F_s H_*(C) := \mathrm{Im}\,[H_{s+t}(F_s C) \to H_{s+t}(C)] = Z_s^\infty/\partial C \cap F_s C$$

and therefore

$$G_F H_*(C) = F_s H_*(C)/F_{s-1}H_*(C) = \frac{Z_s^\infty/\partial C \cap F_s C}{Z_{s-1}^\infty/\partial C \cap F_{s-1}C} = \frac{Z_s^\infty}{Z_{s-1}^\infty + \partial C \cap F_s C} = E_s^\infty.$$

This completes the proof of the theorem.   ♠

13.4 Leray–Serre Spectral Sequence

Let us now consider some examples of topological situations in which we get filtered differential graded objects. Naturally, we must look at various chain complexes and look out for situations wherein they get a filtration structure. Once again, the most obvious one is when the topological space itself is filtered.

Example 13.4.1 An increasing filtration on a topological pair (X, A) is a sequence of subsets X_s of X such that

$$A = X_{-1} \subset X_0 \subset X_1 \subset \cdots \subset X_s \subset X_{s+1} \subset \cdots \qquad (13.7)$$

for each s and $\cup_s X_s = X$. Depending on situations, we may impose further topological conditions on X_s such as X_s is closed, or open, etc. For instance, in many situations in analysis, you may desire that each X_s is (relatively) compact. Suppose we want to study the singular homology of (X, A). Then it is natural to assume that the topology on X is the weak topology with respect to the family $\{X_s\}$. It follows that, corresponding to the filtration (13.7), we immediately get a filtration

$$\cdots \subset S(X_s, A) \subset S(X_{s+1}, A) \subset \cdots \subset S(X, A)$$

which is convergent and bounded below, so that we can apply the above theorem.

As a specific case, let (B, A) be a connected relative CW-complex and $p : E \to B$ be a fibration. Put $E_A = p^{-1}(A), E_s = p^{-1}((B, A)^{(s)}), s \geq 0$ and $E_s = E_A$ for $s < 0$ so that $\{E_s\}$ is an increasing filtration on E with $E_1 = E_A$ and $\cup_s E_s = E$. Observe that every compact subset of E is contained in E_s for some s. Therefore we can conclude:

Theorem 13.4.2 *Let (B, A) be a relative CW-complex and $p : E \to B$ be a fibration. There is a convergent first quadrant E^1-spectral sequence with $E^1_{s,t}$ isomorphic to the singular homology $H_{s+t}(E_s, E_{s-1}; G)$, d^1 the boundary operator of the homology exact sequence of the triple (E_s, E_{s-1}, E_{s-2}) and E^∞ the bigraded module associated to the filtration $H_*(E, E_A; G)$, viz.,*

$$F_s H_*(E, E_A; G) = \mathrm{Im}\left[H_*(E_s, E_A; G) \to H_*(E, E_A; G)\right].$$

We need to take Theorem 13.4.2 a step further in order to make some use of it by studying the E^1-terms further and get hold of a description of the E^2 terms. Just like in the study of sphere bundles, we shall restrict ourselves to fibrations, which satisfy a certain orientability condition. (It is possible to carry out the study in more general situations, using 'homology with local coefficients' which we have discussed in Section 10.10. We shall leave this to the reader who is familiar with the contents of Section 10.10.) So, we need to extend the concept of orientability of sphere bundles to arbitrary fibrations before we go further with the description of the E^2 term of the spectral sequence.

Throughout the rest of this section, $p : E \to B$ denotes a fibration over path connected space, $A \subset B$, $b_0 \in A$ and $F = p^{-1}(b_0)$ is path connected. Recall from Theorem 11.1.5 and the subsequent remark, that there is an action of $\pi_1(B)$ on the homology of the fibre F.

Definition 13.4.3 Let $p : E \to B$ be a fibration over a path connected base space. We say it is orientable if the action of $\pi_1(B)$ on the homology of the fibre is trivial.

Clearly any fibration over a simply connected base is orientable. Also any pullback of an orientable fibration is orientable. It follows that if $[\omega]$ is any path class in B then the homomorphism $h[\omega] : H_*(F_{\omega(0)}) \to H_*(F_{\omega(1)})$ is independent of the path class. Such homomorphisms play an important role in converting the orientability of the fibration into a stronger result on homology. So, we need to give them some attention.

Definition 13.4.4 Let $p : E \to B$ be a fibration. A map $f : p^{-1}(b_0) \to p^{-1}(b_1)$ is called admissible if there exists a path class $[\omega]$ from b_0 to b_1 such that $[f] = h[\omega]$.

Remark 13.4.5 It is easy to verify each of the following properties.
(a) Every admissible map is a homotopy equivalence.
(b) Composite of admissible maps is admissible.
(c) A homotopy inverse of an admissible map is admissible.
(d) Over a path connected space, there is always an admissible map between any two fibres.
(e) If $p : E \to B$ is orientable then any two admissible maps from one fibre F_0 to another fibre F_1 induce the same homomorphism $H_*(F_0) \to H_*(F_1)$.

Lemma 13.4.6 Let $p : E \to B$ be a fibration, $b_0 \in B$, $F = p^{-1}(b_0)$. Let X be a path connected space, $\alpha : X \to B$ and $\hat{\alpha} : X \times F \to E$ be maps such that $p \circ \hat{\alpha}(x, e) = \alpha(x), x \in X, e \in F$. Then the following are equivalent:
(a) For each $x \in X$, the map $f_x : e \mapsto \hat{\alpha}(x, e)$ from $F \to p^{-1}(\alpha(x))$ is admissible.
(b) There is some $x_1 \in X$ such that $f_{x_1} : e \mapsto \hat{\alpha}(x_1, e)$ is admissible.

Proof: The implication (a) $\implies$ (b) is obvious, we need to prove (b) *Lrw* (a) only. Let $\omega : \mathbb{I} \to B$ be a path from b_0 to $\alpha(x_1)$ such that $[f_{x_1}] = h[\omega]$. Choose a path $\lambda : \mathbb{I} \to X$ from x_1 to some $x \in X$. The proof is completed if we show that

$$[f_{x_1}] = h[\omega * (\alpha \circ \lambda)].$$

Let $\hat{\omega} : \mathbb{I} \times F \to E$ be a map such that $p \circ \hat{\omega}(t, z) = \omega(t)$ and $\hat{\omega}(0, z) = z, z \in F$. Put $g(z) = \hat{\omega}(1, z)$. Then by definition, we have $h[\omega] = [g]$. Now define $\sigma : \mathbb{I} \times F \to E$ by the formula:

$$\sigma(t, z) = \begin{cases} \hat{\omega}(2t, z), & 0 \le t \le 1/2; \\ \hat{\alpha}(\lambda(2t - 1), z), & 1/2 \le t \le 1. \end{cases}$$

Then $\sigma(0, z) = \hat{\omega}(0, z) = z$ and $p \circ \sigma = \omega * (\alpha \circ \lambda)$. Moreover, $\sigma(1, z) = \hat{\alpha}(x, z) = f_x$ and hence by definition $h[\omega * (\alpha \circ \lambda)] = [f_{x_1}]$. ♠

Definition 13.4.7 Given $\alpha : X \to B$, a map $\hat{\alpha}$ satisfying the condition in the above lemma is called an admissible lift of α.

Now let us specialize to the case where $p : E \to B$ is an orientable fibration over a path connected relative CW-complex (B, A), $F = p^{-1}(b_0)$, for some $b_0 \in A$.

Fix an integer $s \ge 0$. Let $\{e_\alpha\}$ denote the collection of s-cells of $(B, A)^{(s)}$.

Lemma 13.4.8 The inclusion induced homomorphisms

$$(i_\alpha)_* : H_*((p^{-1}(e_\alpha), p^{-1}(\dot{e}_\alpha)) \to H_*(E_s, E_{s-1})$$

define a direct sum representation of $H_*(E_s, E_{s-1})$:

$$H_n(E_s, E_{s-1}) \approx \oplus_\alpha H_n(p^{-1}(e_\alpha), p^{-1}(\dot{e}_\alpha)).$$

Proof: This is an easy consequence of the fact that $(B, A)^{(s-1)}$ is a neighbourhood deformation retract in $(B, A)^{(s)}$ and excision. ♠

Lemma 13.4.9 Let $p : E \to B$ be an orientable fibration. For each singular simplex $\sigma : (\Delta_s, \dot{\Delta}_s) \to ((B, A)^{(s)}, (B, A)^{(s-1)})$, there is $\tilde{\sigma} : (\Delta_s, \dot{\Delta}_s) \times F \to (E_s, E_{s-1})$, which is an admissible lift of σ and the induced homomorphism

$$\tilde{\sigma}_* : H_*((\Delta_s, \dot{\Delta}_s) \times F) \to H_*(E_s, E_{s-1})$$

is independent of the lift.

Proof: Using Exercise 11.1.16(i), since $\mathbf{e}_0$ is a strong deformation retract of Δ_s, there exists $\tilde{\sigma} : (\Delta_s, \dot{\Delta}_s) \times F \to (E_s, E_{s-1})$ such that $p \circ \tilde{\sigma}(t, z) = \sigma(t), t \in \Delta_s, z \in F$ and $\tilde{\sigma}(\mathbf{e}_0, z) : F \to p^{-1}(\sigma(\mathbf{e}_0))$ is an admissible map. From Lemma 13.4.6, it follows that $\tilde{\sigma}$ is an admissible lift.

Given two admissible lifts $\tilde{\sigma}_i, i = 0, 1$ of σ, there is an admissible map $f : F \to F$ such that $\tilde{\sigma}_0|_{\mathbf{e}_0 \times F}$ is homotopic to $(\tilde{\sigma}_1|_{\mathbf{e}_0 \times F}) \circ f$. Put

$$Y = (\Delta^s \times F \times \dot{\mathbb{I}}) \cup (\mathbf{e}_0 \times F \times \mathbb{I})$$

and consider $g : Y \to E_s$ defined by

$$g(x, z, 0) = \tilde{\sigma}_0(x, z), \quad g(x, z, 1) = \tilde{\sigma}_1(x, f(z)).$$

Take $G : \Delta^s \times F \times \mathbb{I} \to B$ to be $G(x, z, t) = \sigma(x)$. Then G is the extension of $p \circ g$. Since Y is a strong deformation retract of $\Delta_s \times F \times \mathbb{I}$, again it follows from Exercise 11.1.16(i) that $g|_{\Delta_s \times F \times 0} \simeq_p g|_{\Delta_s \times F \times 1}$. Therefore $\tilde{\sigma}_0$ is homotopic to the composite

$$(\Delta_s, \dot{\Delta}_s) \times F \xrightarrow{\ 1 \times f\ } (\Delta_s, \dot{\Delta}_s) \times F \xrightarrow{\ \tilde{\sigma}_1\ } (E_s, E_{s-1}).$$

Since p is orientable, f_* is identity on homology. Therefore it follows that $(\tilde{\sigma}_0)_* = (\tilde{\sigma}_1)_*$. ♠

Remark 13.4.10 What can we say, if p were not orientable? Well, in obtaining the homomorphism (13.8) as in the next lemma, orientability is crucial. On the other hand, by replacing the first term in (13.8) by what are known as 'homology with local coefficients', the same statement would hold, without any further effort. Moreover, the rest of the theory would also hold, if we make the corresponding replacements for the homology of the base with ordinary coefficients with the local coefficients. As suggested earlier, the reader may figure out the details on her own or consult the book [Whitehead, 1978].

Lemma 13.4.11 Let $p : E \to B$ be an orientable fibration. There is a homomorphism

$$\psi : H_s((B, A)^{(s)}, (B, A)^{(s-1)}; H_n(F; G)) \to H_{n+s}(E_s, E_{s-1}; G) \tag{13.8}$$

with the property that for any cellular singular simplex $\sigma : \Delta_s \to (B, A)^{(s)}$ and for any $w \in H_n(F; G)$, we have

$$\psi(\sigma \otimes w) = \tilde{\sigma}_*(\xi_s) \times w, \tag{13.9}$$

where $\tilde{\sigma}$ is an admissible lift of σ.

Proof: Note that the identity map ξ_s of Δ_s is a cycle modulo $\dot{\Delta}_s$ and the homology class $\{\xi_s\}$ is a generator of $H_s(\Delta_s, \dot{\Delta}_s; R)$. Clearly, for each fixed singular s-simplex σ, the assignment

$$w \mapsto \tilde{\sigma}_*(\{\xi_s\}) \times w$$

defines a homomorphism $H_n(F; G) \to H_{n+s}(E_s, E_{s-1}; G)$. Since cellular singular s-simplexes σ's form a basis for the free module $C_s^{cell}((B, A)^{(s)})$, we get a homomorphism

$$\psi : C_{\cdot}^{cell}((B, A)^{(s)}) \otimes H_n(F; G) \to H_{n+s}(E_s, E_{s-1}; G)$$

satisfying (13.9). We claim that this homomorphism itself induces the required homomorphism. For this, it suffices to prove that ψ vanishes on the boundary elements.

Let $\tau : \Delta_{s+1} \to (B, A)^{(s)}$ be a cellular singular $(s+1)$-simplex and let $\tilde{\tau}$ be an admissible lift of τ. Put $\tau^i = \tau \circ F^i$ where $F^i : \Delta_s \to \Delta_{s+1}$ are the face maps $0 \le i \le s+1$. It follows that the composites

$$\tilde{\tau}^i : \ \Delta_s \times F \xrightarrow{F^i \times Id} \Delta_{s+1} \times F \xrightarrow{\tilde{\tau}} E_s$$

are admissible lifts of τ^i. Therefore

$$\psi(\tau^i \times w) = \tilde{\tau}^i_*(\{\xi_s\} \times w) = \tilde{\tau}_*(F^i \times Id)_*(\{\xi_s\} \times w) = \tilde{\tau}'_*(\{F^i\} \times w),$$

where $\{F^i\} \in H_s(\dot{\Delta}_{s+1}, \Delta_{s+1}^{(s-1)})$ and $\tilde{\tau}'$ is the restriction of $\tilde{\tau}$ to $\dot{\Delta}_{s+1} \times F$. Therefore

$$\psi(\partial\tau \times w) = \tilde{\tau}_* \left(\left\{ \sum_i (-1)^i F^i \right\} \times w \right).$$

However, in the singular chain complex $S(\Delta_{s+1})$, we have $\partial \xi_{s+1} = \sum_i (-1)^i F^i$. If

$$j : (\dot{\Delta}_{s+1}, \Delta_{s+1}^{(s-1)}) \to (\Delta_{s+1}, \Delta_{s+1}^{(s-1)})$$

is the inclusion map, then $j_*(\{\sum_i (-1)^i F^i\}) = 0$. Since $\tilde{\tau}' = \tilde{\tau} \circ (j \times Id)$, we have

$$\tilde{\tau}'_*(\{F^i\} \times w) = \tilde{\tau}_* j_* \left(\left(\sum_i (-1)^i F^i \right) \times w \right) = 0.$$

This completes the proof of the lemma.   ♠

We shall now be able to compute the E^1-terms of the spectral sequence in Theorem 13.4.2.

Lemma 13.4.12 (a) For each $s \ge 0$, there is an isomorphism

$$\psi_* : H_s((B, A)^{(s)}, (B, A)^{(s-1)}; H_n(F; G)) \to H_{n+s}(E_s, E_{s-1}; G)$$

(b) For each $s \ge 0$, there is a commutative diagram:

$$
\begin{array}{ccc}
H_s((B, A)^{(s)}, (B, A)^{(s-1)}; H_n(F; G)) & \xrightarrow{\ \psi_*\ } & H_{n+s}(E_s, E_{s-1}; G) \\
\Big\downarrow{\partial} & & \Big\downarrow{\partial} \\
H_{s-1}((B, A)^{(s-1)}, (B, A)^{(s-2)}; H_n(F; G)) & \xrightarrow{\ \psi_*\ } & H_{n+s-1}(E_{s-1}, E_{s-2}; G)
\end{array}
$$

Proof: (a) Using the cellular chain complex for the term on the left hand side and using Lemma 13.4.8, we have direct sum decompositions on either side, and ψ_* respects this decomposition. Therefore, it is enough to prove that for any s-cell e of $B \setminus A$,

$$\psi_* : H_s(e, \dot{e}; H_n(F; G)) \to H_{n+s}(p^{-1}(e), p^{-1}(\dot{e}); G)$$

is an isomorphism. Let $f : (\mathbb{D}^s, \mathbb{S}^{s-1}) \to (e, \dot{e})$ be the characteristic map of the cell. Let $q : (X, Y) \to (\mathbb{D}^s, \mathbb{S}^{s-1})$ be the pullback of p via f and let $f' : (X, Y) \to (p^{-1}(e), p^{-1}(\dot{e}))$ be the map that covers f.

$$
\begin{array}{ccc}
(X, Y) & \xrightarrow{\ f'\ } & (p^{-1}(e), p^{-1}(\dot{e})) \\
\Big\downarrow{q} & & \Big\downarrow{p} \\
(\mathbb{D}^s, \mathbb{S}^{s-1}) & \xrightarrow{\ f\ } & (e, \dot{e})
\end{array}
$$

There is a commutative diagram

$$
\begin{array}{ccc}
H_{n+s}(X, Y; G) & \xrightarrow{\ q^*\ } & H_{n+s}(p^{-1}(e), p^{-1}(\dot e); G) \\[4pt]
\Big\uparrow{\psi_*} & & \Big\uparrow{\psi_*} \\[4pt]
H_s(\mathbb{D}^s, \mathbb{S}^{s-1}; H_n(F; G)) & \xrightarrow{\ f_*\ } & H_s(e, \dot e; H_n(F; G))
\end{array}
$$

in which the horizontal arrows are isomorphisms induced by relative homeomorphisms. Therefore, proving that the right-side vertical arrow is an isomorphism is the same as proving that the left-side vertical arrow is an isomorphism.

Thus the problem is reduced to the case when the fibration is trivial and then this is just the Künneth theorem.

(b) This is similar to the proof that ψ_* vanishes on the boundary, the main idea being that the restrictions of admissible lifts to faces are admissible lifts of the restrictions to faces. ♠

Theorem 13.4.13 *Let $p : E \to B$ be an orientable fibration with path connected base B and path connected fibre $F = p^{-1}(b_0)$, $b_0 \in B$. Let $b_0 \in A \subset B$, $E_A = p^{-1}(A)$. For the singular homology, and for any coefficient module G, there is a convergent first quadrant E^2-spectral sequence such that:*

(a) $E^2_{s,t} \approx H_s(B, A; H_t(F; G))$.

(b) E^∞ is the bigraded module associated to some filtration on $H_*(E, E_A; G)$.

The spectral sequence is functorial in G as well as on the category of orientable fibrations and fibre-preserving maps.

Proof: We shall first prove the result for the case when (B, A) is a relative CW-complex. Of course in this special case, the functoriality should be taken on the category of orientable fibrations (over a fixed pair (B, A) or over CW-pairs and cellular maps) and fibre-preserving maps. On the other hand we can make conclusion (b) more specific:

(b') E^∞ is the bigraded module associated to the filtration

$$
F_s H_*(E, E_A; G) = \mathrm{Im}\,[H_*(E_s, E; G) \to H_*(E, E_A; G)]
$$

where $E_s = p^{-1}((B, A)^{(s)})$.

Because $H_s((B, A)^{(s)}, (B, A)^{(s-1)})$ is a free module, from the universal coefficient theorem, we conclude that

$$
\begin{aligned}
H_s((B, A)^{(s)}, (B, A)^{(s-1)}; H_n(F; G)) &\approx H_s((B, A)^{(s)}, (B, A)^{(s-1)}) \otimes H_n(F; G) \\
&= C_s^{CW}(B, A) \otimes H_n(F; G).
\end{aligned}
$$

Moreover, it is easy to see that under this isomorphism the boundary operator of the triple corresponds to

$$
\partial \otimes 1 : C_s^{CW}(B, A) \otimes H_n(F; G) \to C_{s-1}^{CW}(B, A) \otimes H_n(F; G).
$$

Thus, Lemma 13.4.12 implies that ψ_* induces an isomorphism of bigraded chain complex $C_*(B, A) \otimes H_*(F; G)$ with the E^1-term of the spectral sequence of Theorem 13.4.2. Since the E^2-term is the homology of the E^1-terms, the conclusion follows in the case of relative CW-complex (B, A).

In the general case, we take a relative CW-complex (B', A') with a weak homotopy equivalence $f : (B', A') \to (B, A)$. We apply the result to the induced fibration $p' : (E', E'_{A'}) \to (B', A')$. The covering map $f' : (E', E'_{A'}) \to (E, A)$ is easily seen to be a weak homotopy equivalence. Then f' induces isomorphism of homology sequence $(E', E'_{A'})$ with that of (E, E_A). In particular, it defines a filtration on the $H_*(E, E_A; G)$.

If $g : (B'', A'') \to (B, A)$ is another CW-approximation, then there is a cellular homotopy equivalence $h : (B'', A'') \to (B', A')$ such that $g \simeq f \circ h$. The map h induces isomorphisms of the two spectral sequences, and hence takes one filtration on the homology $H_*(E, E_A; G)$ onto the other. Thus there is a well-defined filtration on $H_*(E, E_A)$ independent of the choice of the approximation.

The functoriality also follows from similar consideration, because maps between spaces can be realized by maps at the approximation level also.   ♠

13.5   Some Immediate Applications

Theorem 13.5.1 *Let R denote a field. Let $p : E \to B$ be an R-orientable fibration with base B and fibre F path connected. If two of the three Euler characteristics $\chi(E), \chi(F), \chi(B)$ are defined then the third is defined and we have*

$$\chi(E) = \chi(B)\chi(F).$$

Proof: For any finitely generated bigraded vector space G, (over the field R), we define

$$\chi(G) = \sum_{s,t} (-1)^{s+t} \dim G_{s,t}.$$

Assume that $\chi(B)$ and $\chi(F)$ are defined. This implies by the Künneth formula that $E^2 = \{E_{s,t}^2\} = \{H_s(B) \otimes H_t(F)\}$ is finitely generated and we have $\chi(E^2) = \chi(B)\chi(F)$. Just as in Theorem 4.1.19, it follows that

$$\chi(E^2) = \chi(E^3) = \cdots = \chi(E^r).$$

Again since E^2 is finitely generated, $E_r \approx E^\infty$ for some large r. Therefore $\chi(E^2) = \chi(E^\infty)$. But then E^∞ is the graded module associated to the filtration of $H_*(E)$ and hence

$$\dim H_n(E) = \sum_{s+t=n} \dim E_{s,t}^\infty.$$

Therefore $\chi(E^\infty) = \chi(E)$. The other two cases now follow from Corollary 13.8.12 applied to the Serre class of finite dimensional vector spaces over R. (See Section 13.8.)   ♠

Example 13.5.2 Let $p : E \to B$ be an orientable fibration over a path connected space with a path connected fibre.

(a) **Fibrations of $\mathbb{R}^n$** Let us take homology with a field coefficient k. Assume that E is acyclic. Further assume that there is some r such that $H_i(B) = 0 = H_i(F)$ for $i > r$. If p is the least with the property that $H_i(B) = 0$, for all $i > p$ and q is the least such that $H_i(F) = 0$ for all $i > q$, it follows that $E_{r,s}^2 = H_r(B) \otimes H_s(F) = 0$ for $r > p$ or $s > q$. Thus the non zero terms of E^2 are contained in the shaded rectangle, in Figure 13.2.

It follows that the same is true for E^r for all $r \geq 3$ as well. Therefore, $E_{p,q}^r$ term is never disturbed by any non zero differential, which means $E_{p,q}^2 = E_{p,q}^3 = \cdots = E_{p,q}^\infty$. (In such a situation, we say $E_{p,q}^2$ persists.) But then for $s + t = p + q$, the only non zero term $E_{s,t}^\infty$ is $E_{p,q}^\infty$. Since we are working with field coefficients, $H_{p+q}(E) = \oplus_{s+t=n} E_{s,t}^\infty = E_{p,q}^\infty = E_{p,q}^2 = H_p(B) \otimes H_q(F) \neq (0)$. Since we have assumed that E is acyclic, this implies $p + q = 0$. Therefore both B and F are acyclic. This is not a fictitious situation. Here is an example how this can be put to use.

Theorem 13.5.3 (Borel) *Let $F \to \mathbb{R}^n \to B$ is a fibre bundle with connected fibres. Then both F, B are acyclic.*

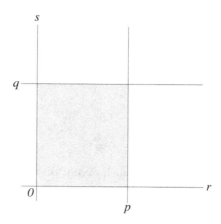

FIGURE 13.2. E_2-terms vanish outside the shaded region

Proof: Note that from the homotopy exact sequence, it follows that $\pi_1(B) = (1)$ and hence the fibration is orientable. Therefore, in order to apply the above arguments, all that we need to prove is that the homology of F and B vanish beyond some dimension. For F, this follows because p is a locally trivial fibration implying that F is a retract of an open subset of $\mathbb{R}^n$, and hence has homology vanishing beyond dimension n.

For B, the argument is subtler. B has an open cover $\{U_\alpha\}$ such that $p^{-1}(U_\alpha)$ is open in $\mathbb{R}^n$ and is homeomorphic to $U_\alpha \times F$. So, to begin with each U_α has homology vanishing in dimension beyond n. So, if B' denotes the union of some finitely many U_α's, then by Mayer–Vietoris argument, we can say that there is some m beyond which the homology of B' vanishes. We can now use the arguments above with the spectral sequence for the fibration $p^{-1}(B') \to B'$ to conclude first that the homology of B' vanishes beyond the dimension n, because that is what happens to $p^{-1}(B')$ which is an open subset of $\mathbb{R}^n$. Now since homology commutes with direct limit, we can conclude the same for B.

Thus, the argument above is applicable to the fibration $p : \mathbb{R}^n \to B$ and this time we conclude that both F and B are acyclic. ♠

Example 13.5.4 Fibred spheres over spheres

Let us now assume that E is a homology sphere and B is also a homology sphere. What can we say about the homology of the fibre? Say E has homology of $\mathbb{S}^n$ and B has homology of $\mathbb{S}^p$. It follows that the E^2-terms are non zero only on two vertical lines $r = 0$, and $r = p$ in Figure 13.3.

As in the previous case, it follows that $p + q = n$ where q is the maximum for which $H_i(F) \neq 0$ and $H_q(F) = H_p(B) = H_n(E) = k$. It follows that

$$E^2 = E^3 = \cdots = E^p; \quad E^{p+1} = \cdots = E^\infty.$$

Only at the E^p level, there is just one possibility for a non zero differential, viz., $d^p : E^p_{p,0} \to E^p_{0,p-1}$. If this differential is also zero, then $H_p(B) = E^2_{p,0} = E^p_{p,0} = E^{p+1}_{p,0} = E^\infty_{p,o}$ will be a non trivial direct summand of $H_p(E)$ and hence $p = n$ and F is acyclic. If this differential is non zero, then of course, $E^p_{0,p-1} \neq (0)$. But $H_{p-1}(F) = E^2_{0,p-1} = E^p_{0,p-1} \neq (0)$. This implies $p - 1 \leq q < n$. It follows that for all $i \neq 0, p - 1$, $H_i(F) = 0$, and hence $q = p - 1$. Therefore F is a homology $(p - 1)$-sphere and $n = 2p - 1$. Of course, we know examples of this type,

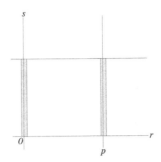

FIGURE 13.3. E_2-terms vanish except on two vertical lines

viz., Hopf fibrations

$$\mathbb{S}^1 \subset \mathbb{S}^3 \to \mathbb{S}^2; \quad \mathbb{S}^3 \subset \mathbb{S}^7 \to \mathbb{S}^4.$$

Example 13.5.5 Generalized homology Wang sequence

Assuming that B is a homology p-sphere (and without the assumption that E is a homology sphere), we can still say something. The only possible non zero differentials are $d^p : E_{p,t}^p \to E_{0,t+p-1}^p$ and hence we have $E^2 = E^3 = \cdots = E^p$ and $E^{p+1} = H(E^p, d^p) = E^{p+1} = \cdots = E^\infty$. Moreover, $d^p = 0$ on the s-axis and hence $E_{0,s}^{p+1} = E_{0,s}^p/\mathrm{Im}\, d^p$. Similarly, $E_{p,s}^{p+1} = \mathrm{Ker}\, d^p$. Therefore we have exact sequences

$$0 \longrightarrow E_{p,s}^\infty \longrightarrow E_{p,s}^2 \overset{d^p}{\longrightarrow} E_{0,s-p+1}^2 \longrightarrow E_{0,s-p+1}^\infty \longrightarrow 0 \tag{13.10}$$

Since $E_{r,s}^\infty$ is non trivial only for $r = 0, p$, writing H_k for $H_k(E)$, we also have,

$$H_{p+s} = F_0 H_{p+s} = \cdots = F_p H_{p+s} \supset F_{p+1} H_{p+s} = \cdots = F_{p+s} H_{p+s} \supset (0)$$

and $E_{p,s}^\infty = H_{p+s}/F_{p+1}H_{p+s} = H_{p+s}/E_{0,p+s}^\infty$. Therefore, we have exact sequences

$$0 \to E_{0,p+s}^\infty \to H_{p+s}(E) \to E_{p,s}^\infty \to 0. \tag{13.11}$$

When we splice these with the sequences (13.10)

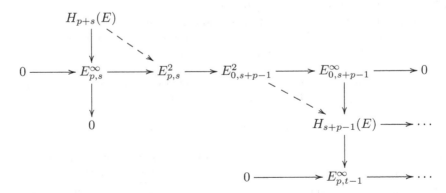

we get a long exact sequence

$$\cdots \longrightarrow H_{p+s}(E) \longrightarrow E_{p,s}^2 \overset{d^p}{\longrightarrow} E_{0,s+p-1}^2 \longrightarrow H_{s+p-1}(E) \longrightarrow \cdots$$

We can now use isomorphisms

$$E^2_{p,s} \approx H_p(B) \otimes H_s(F) \approx H_s(F); \quad E^2_{0,t+p-1} \approx H_0(B) \otimes H_{t+p-1}(F) \approx H_{t+p-1}(F),$$

in the above sequence to obtain a long exact sequence

$$\cdots \longrightarrow H_{p+t}(E) \longrightarrow H_t(F) \longrightarrow H_{t+p-1}(F) \longrightarrow H_{t+p-1}(E) \longrightarrow \cdots \quad (13.12)$$

It remains to identify the homomorphisms in this sequence. We leave it to the reader to figure it out that (13.12) is indeed the Wang sequence (11.3.4).

Similarly, we can get a **generalized Gysin homology sequence** from the spectral sequence associated to an orientable fibration with fibre, a homology sphere.

Example 13.5.6 Serre exact sequence of a fibration

Now suppose, complementary to the case (a) that we have $H_i(B) = 0$, $0 < i < p$ and $H_i(F) = 0$, $0 < i < q$. (See Figure 13.4.)

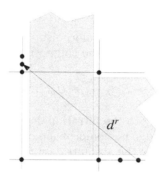

FIGURE 13.4. E^2-terms vanish inside the shaded rectangle

It follows that in the range $s + t < p + q$, the only possible non zero differentials are of the form

$$d^r : E^r_{r,0} \to E^r_{0,r-1}, \quad r \geq \max\{p, q+1\}.$$

Proceeding exactly as in (c), for $r \leq p + q$, we get exact sequences

$$0 \to E^\infty_{s,0} \to E^2_{s,0} \xrightarrow{d^s} E^2_{0,s-1} \to E^\infty_{0,s-1} \to 0$$

and

$$0 \to E^\infty_{r,0} \to H_r(E) \to E^\infty_{0,r} \to 0.$$

When we splice them up, and replace $E^2_{s,0} \approx H_s(B)$, $E^2_{0,s-1}$ by $H_{s-1}(F)$, etc., we get an exact sequence called the Serre homology sequence.

$$H_{p+q-1}(F) \to \cdots \to H_s(F) \xrightarrow{i_*} H_s(E) \xrightarrow{p_*} H_s(B) \xrightarrow{d^s} H_{s-1}(F) \to \cdots \to H_0(B). \quad (13.13)$$

13.6 Transgression

In this section, we shall derive some easy consequences of the naturality property of the Serre spectral sequence, coupled with the fact that it is a first quadrant spectral sequence. As a consequence, we shall show that the homology transgression corresponds to the differentials.

Let $p : E \to B$ be an orientable fibration with base B, where the fibre $F = p^{-1}(b_0)$ is path connected, $b_0 \in B' \subset B$, and $E' = p^{-1}(B')$. We shall denote the corresponding maps of pairs by $p_0 : (E, E') \to (B, B')$. Homology groups are taken with the coefficients in a ring R which we shall suppress from the notation.

To begin with, we have $E_{p,0}^2 = H_p(B, B'; H_0(F)) \approx H_p(B, B')$. Since $E_{p,q}^r = 0$ for $q < 0$, we also have:

$$E_{p,0}^3 = \operatorname{Ker} d^2 : E_{p,0}^2 \to E_{p-2,1}^2, \ E_{p,0}^4 = \operatorname{Ker} d^3 : E_{p,0}^3 \to E_{p-3,2}^3, \cdots$$

Moreover, since $E_{t,s}^2 = 0$ for $t \le 0$, $d^r(E_{p,q}^r) = 0$ if $r \ge p$, we have

$$H_p(B, B') = E_{p,0}^2 \supset \cdots \supset E_{p,0}^p = \cdots = E_{p,0}^\infty.$$

On the other hand, putting $F_{p,q}$ equal to the image of $i_* : H_{p+q}(E_p, E') \to H_{p+q}(E, E')$, we have,

$$H_p(E, E') = F_{p,0} \supset F_{p-1,1} \supset \cdots$$

and $E_{p,0}^\infty = F_{p,0}/F_{p-1,1}$. Thus we have an epimorphism $\alpha_p : H_p(E, E') \to E_{p,0}^\infty$ and an inclusion map $\kappa_p : E_{p,0}^\infty \to H_p(B, B')$. We call the composite

$$H_p(E, E') \xrightarrow{\alpha_p} E_{p,0}^\infty \xrightarrow{\kappa_p} H_p(B, B') \tag{13.14}$$

the *edge homomorphism* corresponding to the filtration degree, p.

Exactly like this, we get another edge homomorphism corresponding to the complementary degree, except that we need to take the absolute case $B' = \emptyset$ only, so that to begin with we have $E_{0,q}^2 = H_0(B; H_q(F)) = H_q(F)$. The rest of the arguments are identical to the above and we get an epimorphism $\lambda_q : H_q(F) \to E_{0,q}^\infty$ and a monomorphism $\tau_q : E_{0,q}^\infty \to H_q(E)$. The composite of these two is called the edge homomorphism corresponding to the complementary degree, q.

Theorem 13.6.1 *The filtration degree edge homomorphism is nothing but $(p_0)_*$ induced by the fibre map $p_0 : (E, E') \to (B, B')$; the complementary degree edge homomorphism is nothing but i_*, induced by the inclusion $i : F \to E$. In particular we have:*
(a) the kernel of $(p_0)_*$ is equal to the image of $i_* : H_p(E_{p-1}, E') \to H_p(E, E')$; the image of $(p_0)_*$ is equal to the image of $E_{p,0}^\infty$.
(b) the image of $i_* : H_q(F) \to H_q(E)$ (which is equal to the kernel of $j_* : H_q(E) \to H_q(E; F)$ is equal to the image of $E_{0,q}^\infty$ in $H_q(E)$.

Proof: First consider the case of filtration degree. Consider the identity map $Id : B \to B$ as a fibration with fibre a single point. We then have a filtration preserving map of the two fibrations:

$$\begin{array}{ccc}
E & \xrightarrow{\ p\ } & B \\
\downarrow{\scriptstyle p} & {\scriptstyle Id} & \downarrow{\scriptstyle Id} \\
B & \xrightarrow{\ Id\ } & B
\end{array}$$

which induces a map of the corresponding (exact couples and thereby) spectral sequences. This gives rise to the following commutative diagram:

$$
\begin{array}{ccccccc}
H_p(B, B') & = & E^2_{p,0}(p) & \xleftarrow{\;\kappa_p\;} & E^\infty_{p,0}(p) & \xleftarrow{\;\alpha_p\;} & H_p(E, E') \\
\downarrow{\scriptstyle Id} & & \downarrow{\scriptstyle Id} & & \downarrow & & \downarrow{\scriptstyle (p_0)_*} \\
H_p(B, B') & = & E^2_{p,0}(Id) & \longleftarrow & E^\infty_{p,0}(Id) & \longleftarrow & H_p(B, B')
\end{array}
$$

where the composite in each row is the corresponding edge homomorphism. In the bottom row, the edge homomorphism is the identity map. Therefore, $\kappa_p \circ \alpha_p = (p_0)_*$. The conclusion (a) is obvious from this.

For the case of filtration degree, we consider the constant map $F \to b_0$ as a fibration with the inclusion map

$$
\begin{array}{ccc}
F & \hookrightarrow & E \\
\downarrow & & \downarrow{\scriptstyle p} \\
\{b_0\} & \hookrightarrow & B
\end{array}
$$

The rest of the argument is similar to the above case. ♠

Lemma 13.6.2 The following diagram is commutative:

$$
\begin{array}{ccc}
H_n(E; F) & \xrightarrow{\;\partial_*\;} & H_{n-1}(F) \\
\downarrow{\scriptstyle \alpha_n} & & \downarrow{\scriptstyle \lambda_{n-1}} \\
E^n_{n,0} & \xrightarrow{\;d^n\;} & E^n_{0,n-1}
\end{array}
\tag{13.15}
$$

Proof: Since B is connected, we may assume that B has only one vertex, $B_0 = \{b_0\}$. Then $E_0 = F$ and $H_{n-1}(F) = H_{n-1}(E_0) = H_{n-1}(E_0, E_{-1}) = E^1_{0,n-1}$. Now in the commutative diagram

$$
\begin{array}{ccc}
H_n(E, F) & \xrightarrow{\hspace{5.5cm}\partial_*\hspace{5.5cm}} & H_{n-1}(F) \\
\uparrow{\scriptstyle i_2} & & \downarrow{\scriptstyle i_1} \\
H_n(E_n, F) \xrightarrow{\;j_1\;} H_n(E_n, E_{n-1}) & \xrightarrow{\;\partial_1\;} H_{n-1}(E_{n-1}) & \xrightarrow{\;j_2\;} H_{n-1}(E_{n-1}, F)
\end{array}
$$

i_2 is an epimorphism. Therefore, it is enough to prove that

$$
d^n \circ \alpha_n \circ i_2 = \lambda_{n-1} \circ \partial_* \circ i_2.
$$

To see this we merely need to recall how the domains and codomains of these homomorphisms are defined and how d^n itself is defined as in (13.4). We have $E^n_{n,0} = Z^n_{n,0}/B^n_{n,0}$, $E^n_{0,n-1} = Z^n_{0,n-1}/B^n_{0,n-1}$, and $d^n(x + B^n_{n,0}) = \partial(x) + B^n_{0,n-1}$. On the other hand, for any n-cycle c representing an element $[c] \in H_n(E_n, F)$ we have,

$$
\alpha_n \circ i_2([c]) = c + F_{n-1,1} = c + B^n_{n,0} \in F_{n,0}/F_{n-1,1} = E^\infty_{n,0} = E^n_{n,0}.
$$

Therefore

$$
d^n \circ \alpha_n \circ i_2([c]) = \partial(c) + B^n_{0,n-1}.
$$

On the other hand, $\lambda_{n-1}([y]) = y + B^n_{0,n-1}$ for any $(n-1)$-cycle in F and $\partial_* \circ i_2([c]) = [\partial(c)]$. Therefore,

$$
\lambda_{n-1} \circ \alpha_n \circ i_2([c]) = \partial(c) + B^n_{0,n-1}.
$$

This completes the proof of the lemma. ♠

Definition 13.6.3 Given any fibration $p : E \to B$, with the base B and the fibre $F = p^{-1}(b_0)$, path connected. Let $j_* : H_*(B) \to H_*(B, b_0)$ be inclusion induced homomorphism and $(p_0)_* : H_*(E, F) \to H_*(B, b_0)$ be the homomorphism induced by the projection p, and let $\partial : H_q(E, F) \to H_{q-1}(F)$ be the boundary homomorphism.

$$
\begin{array}{ccc}
H_q(E, F) & \xrightarrow{\ \partial\ } & H_{q-1}(F) \\
\downarrow{\scriptstyle (p_0)_*} & & \\
H_q(B) & \xrightarrow{\ j_*\ } & H_q(B, b_0)
\end{array}
$$

We define the *homology transgressions* to be the homomorphisms from a subgroup of $H_q(B)$ to a quotient group of $H_{q-1}(F)$, viz., $\tau_q : j_*^{-1}(\operatorname{Im} p_*) \to H_{q-1}(F)/\partial(\operatorname{Ker}(p_0)_*)$ as follows: $z \in H_q(B)$ be such that $j_*(z) = (p_0)_*(u)$ for some $u \in H_q(E, F)$. Put $\tau_q(z) = \partial(u) + \partial(\operatorname{Ker}(p_0)_*)$. That this gives a well-defined homomorphism is obvious.

(Such homomorphisms are called additive relations (see [Whitehead, 1978] appendix B). The importance of this stems from the following.

Theorem 13.6.4 *For a fibration $p : E \to B$ with base and fibre connected, and the associated Leray–Serre spectral sequence, we have:*
(a) $E_{n,0}^n \approx j_*^{-1}(\operatorname{Im}(p_0)_*) \subset H_n(B)$.
(b) $E_{0,n-1}^n \approx H_{n-1}(F)/\partial(\operatorname{Ker}(p_0)_*)$.
(c) *Under the above isomorphisms, $d^n : E_{n,0}^n \to E_{0,n-1}^n$ corresponds to the transgression τ_n.*

Proof: All that we need is the following commutative diagram in which the rows are exact. Using earlier results of this section, a reader may like to supply the rest of the arguments.

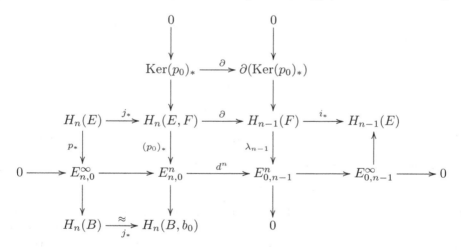

Exercise 13.6.5 Let τ denote the transgression homomorphisms of an oriented fibration $p : E \to B$. If $z \in H_q(B)$ is spherical then show that $\tau(z)$ is spherical.

13.7   Cohomology Spectral Sequences

In Sections 13.2 and 13.3, we had restricted ourselves to discussing spectral sequences of 'homological type'. By merely changing the indexing (p, q) to $(-p, -q)$ everywhere, we would obtain the theory of spectral sequences of cohomological type. Alternatively, we can retain the indexing but reverse all the arrows. In practice, however, we also interchange lower and upper indices (which is not a logical necessity but conforms to existing practice). Thus spectral sequence of cohomological type is a bigraded sequence of modules $\{E_r^{p,q}\}$ together with differentials d_r of bidegree $(r, 1-r)$. Everything that we have done in Sections 13.2 and 13.3 applies verbatim if we only take care to reverse the arrows. However, when we come to topological applications as in Section 13.4, we need to be extra careful. For the singular cochain complex is defined as the dual to the singular chain complex and if we have convergent filtration on the singular chain complex then the induced filtration on the cochain complex may not converge. Having said that, we add that all results in the previous section about the Leray–Serre homology spectral sequence of a fibration have analogous results for the cohomology spectral sequence also.

The main idea of going to cohomology is that we can exploit the rich algebraic structure in it. In what follows, we shall consider all modules over a commutative ring R.

Definition 13.7.1 An algebra A over R is an R module together with a product

$$\eta : A \otimes A \to A.$$

The product is said to be associative if the following diagram is commutative:

$$
\begin{array}{ccc}
A \otimes A \otimes A & \xrightarrow{\psi \otimes Id} & A \otimes A \\
{\scriptstyle Id \otimes \psi}\downarrow & & \downarrow{\scriptstyle \psi} \\
A \otimes A & \xrightarrow{\psi} & A.
\end{array}
$$

The product is said to have a two-sided unit if there is a mapping $\eta : R \to A$ such that the following diagram commutes:

$$
\begin{array}{ccccc}
R \otimes A & \xrightarrow{\eta \otimes Id} & A \otimes A & \xleftarrow{Id \otimes \eta} & A \otimes R \\
{\scriptstyle \approx}\downarrow & & \downarrow{\scriptstyle \psi} & & \downarrow{\scriptstyle \approx} \\
A & = \!\!=\!\!= & A & = \!\!=\!\!= & A
\end{array}
$$

Definition 13.7.2 The tensor product of two differential graded modules $(A, d_A), (B, d_B)$ is defined to be the graded differential module $(A \otimes B, d_\otimes)$ as follows:

$$(A \otimes B)^n = \oplus_{p+q=n} A^p \otimes B^q; \quad d_\otimes(a \otimes b) = d_A(a) \otimes b + (-1)^{\deg a} a \otimes d_B(b).$$

Definition 13.7.3 By a graded algebra we mean an algebra A together with a $\mathbb{Z}$-gradation on $A = \oplus A^n$, such that $A^n \cdot A^m \subset A^{m+n}$. By a differential graded algebra we mean a graded algebra in which the differential d satisfies the Leibnitz' rule

$$d(xy) = d(x)y + (-1)^{\deg x} x d(y).$$

We shall refer to this property by saying d is a *derivation*. The example of a differential graded algebra that we should keep in mind is the singular co-cochain complex of a topological space with the usual cup product. Because of the Leibnitz' rule, it follows that the cohomology $H^*(A, d)$ acquires a graded algebra structure. Associativity, existence of unit, etc., for the product operation are defined in an obvious way with appropriate commutative diagrams.

Likewise, given two differential bigraded modules $(E^{*,*}, d_E), (\bar{E}^{*,*}, d_{\bar{E}})$, their tensor product $(E \otimes \bar{E}, d_\otimes)$ is defined by

$$(E \otimes \bar{E})^{p,q} = \bigoplus_{\substack{r+t=p \\ s+u=q}} E^{r,s} \otimes \bar{E}^{t,u};$$

and $d_\otimes(e \otimes \bar{e}) = d_E(e) \otimes \bar{e} + (-1)^{r+s} e \otimes d_{\bar{E}} \bar{e}$, for $e \in E^{r,s}$ and $\bar{e} \in \bar{E}^{t,u}$.

A *spectral sequence of algebras* over R is a spectral sequence $\{(E_r^{*,*}, d_r)\}$ of cohomological type, together with algebra structures $\psi_r : E_r \otimes E_r \to E_r$ such that ψ_{r+1} is equal to the composite of the following:

$$E_{r+1} \otimes E_{r+1} \xrightarrow{\approx} H(E_r) \otimes H(E_r) \xrightarrow{p} H(E_r \otimes E_r) \xrightarrow{H(\psi_r)} H(E_r) \xrightarrow{\approx} E_{r+1}$$

where $p([x] \otimes [y]) = [x \otimes y]$.

Theorem 13.7.4 *Let $p : E \to B$ be an orientable fibration with the base space B path connected and fibre $F = p^{-1}(b_0), b_0 \in B$. Then there is a first quadrant spectral sequence of algebras $\{(E_r^{*,*}, d_r)\}$ converging to $H^*(E; R)$ as an algebra with the following properties:*
(i) $E_2^{p,q} \approx H^p(B; H^q(F; R))$;
(ii) *If $\smile_2$ denotes product in E_2 then under the above isomorphism, we have*

$$u \smile_2 v = (-1)^{p'q} u \smile v, \quad u \in E_2^{p,q}, v \in E_2^{p',q'}.$$

(iii) *The spectral sequence is natural with respect to fibre preserving maps between fibrations.*

As remarked earlier, except when the algebra structure is involved, the rest of the proof is similar to the case of homology spectral sequence. The proof of the algebra structure is complicated. We skip it. Interested readers may refer to the seminal paper of Serre [Serre, 1951] or [Spanier, 1966].

Example 13.7.5 Let us work out the cohomology algebra of $\mathbb{CP}^\infty$ using the spectral sequence so as to get familiar with the spectral sequence technique itself. For this we can use the fibration

$$\mathbb{S}^1 \to \mathbb{S}^\infty \to \mathbb{CP}^\infty.$$

Since $\mathbb{CP}^\infty$ is simply connected, this fibration is orientable and hence we can apply Theorem 13.7.4 with the coefficient ring $\mathbb{Z}$. We have,

$$E_2^{p,q} = H^p(\mathbb{CP}^\infty; H^q(\mathbb{S}^1)) = H^p(\mathbb{CP}^\infty) \otimes H^q(\mathbb{S}^1),$$

(by UCT), since $H^*(\mathbb{S}^1)$ is torsion free. It follows that the terms of the spectral sequence vanish except perhaps on the two horizontal lines $q = 0, 1$, as depicted in Figure 13.5.

It also follows that all differentials $d_j = 0$ for $j \geq 3$ and hence $E_\infty^{p,q} = E_3^{p,q}$. Since $\tilde{H}^*(\mathbb{S}^\infty)$ vanishes, it follows that the graded module associated to the filtration on the cohomology is trivial, which means $E_\infty^{p,q} = 0$ for all p, q.

Now consider the differentials $d_2 : E_2^{p,1} \to E_2^{p+2,0}$. It follows that $E_3^{p,0} = \text{Coker } d_2$ and $E_3^{p,1} = \text{Ker } d_2 \subset E_2^{p,1}$. Since both these groups vanish, we conclude that $d_2 : E_2^{p,1} \to E_2^{p+2,0}$ are isomorphisms. Now

$$E_2^{0,1} = H^0(\mathbb{CP}^\infty) \otimes H^1(\mathbb{S}^1) = \mathbb{Z} \otimes H^1(\mathbb{S}^1) \approx H^1(\mathbb{S}^1)$$

whereas $E_2^{1,1} = 0$ since $\mathbb{CP}^\infty$ is simply connected. We choose a generator $v \in H^1(\mathbb{S}^1)$. It

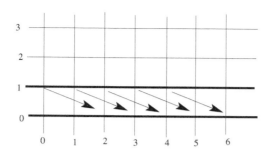

FIGURE 13.5. Computing the cohomology of $\mathbb{CP}^\infty$

follows that $H^{2n+1}(\mathbb{CP}^\infty) = 0$ and $H^{2n}(\mathbb{CP}^\infty) \approx \mathbb{Z}$ and is generated by u_{2n}, where we define u_{2n} inductively, by $u_2 = d_2(v)$ and $u_{2n} = d_2(u_{2n-2} \otimes v)$. Since d_2 vanishes on $E_2^{0,q}$, by the derivation property of d_2, it follows that

$$u_{2n} = d_2(u_{2n-2}) \smile v + u_{2n-2} \smile d_2(v) = 0 + u_{2n-2} \smile u_2 = \cdots = u_2^n.$$

Therefore, we conclude that the cohomology algebra of $\mathbb{CP}^\infty$ is isomorphic to the polynomial algebra $\mathbb{Z}[u_2]$ in one variable of degree 2.

We can now discuss a few applications promised in the beginning of this section.

(A) Cohomology Wang sequence We have discussed how to get the generalized Wang homology sequence. The discussion for cohomology is identical except that we would like to bring in the multiplicative property as well.

Theorem 13.7.6 *Let $p : E \to B$ be a fibration with fibre F. Suppose B is simply connected and $H^*(B; R) \approx H^*(\mathbb{S}^n; R)$ for some $n \geq 2$. Then there is an exact sequence*

$$\cdots \longrightarrow H^q(E; G) \xrightarrow{i^*} H^q(F; G) \xrightarrow{\Theta} H^{q-n+1}(F; G) \longrightarrow H^{q+1}(E; G) \longrightarrow \cdots \quad (13.16)$$

where Θ satisfies the following derivation property:

$$\Theta(u \smile v) = \Theta(u) \smile v + (-1)^{(n+1)\deg u} u \smile \Theta(v). \quad (13.17)$$

Proof: Since B is a cohomology sphere, it follows that $E_2^{s,t} = H^s(B) \otimes H^t(F; G) = 0$ unless $s = 0$ or n. From this, as seen in Example 13.5.2 (c), we obtain an exact sequence

$$\cdots \longrightarrow H^t(E; G) \longrightarrow E_2^{0,t} \xrightarrow{d_n} E_2^{n,t-n+1} \longrightarrow H^{t+1}(E; G) \longrightarrow \cdots$$

If $1 \in H^0(B), w \in H^n(B)$ denote the generators, the map $u \mapsto 1 \otimes u$ is an isomorphism of $H^t(F; G)$ with $E_2^{0,t}$ and the map $v \mapsto v \otimes w$ is an isomorphism of $H^{t-n+1}(F; G)$ with $E_2^{n,t-n+1}$. We take $\Theta : H^t(F; G) \to H^{t-n+1}(F; G)$ by the formula

$$d_n(1 \otimes u) = w \otimes \Theta(u).$$

The exact sequence (13.16) is obtained by replacing the E_2 terms via the above isomorphisms. To see that Θ satisfies (13.17), we use the property of mutual distribution of cup product and cross product (see (f) of Theorem 7.3.5) and the fact that d_n is a derivation:

$$
\begin{aligned}
w \otimes \Theta(u \smile v) &= d_n(1 \otimes (u \smile v)) = d_n(1 \otimes u \smile 1 \otimes v) \\
&= d_n(1 \otimes u) \smile 1 \otimes v + (-1)^{\deg u} 1 \otimes u \smile d_n(1 \otimes v) \\
&= w \otimes \Theta(u) \smile 1 \otimes v + (-1)^{\deg u} 1 \otimes u \smile w \otimes \Theta(v) \\
&= (w \smile 1) \otimes (\theta(u) \smile v) + (-1)^{(n+1)\deg u}(1 \smile w) \otimes (u \smile \theta(v)) \\
&= w \otimes [\Theta(u) \smile v + (-1)^{(n+1)\deg u} u \smile \Theta(v)].
\end{aligned}
$$

This completes the proof of the theorem. ♠

(B) **Cohomology of loop spaces:** We begin with the following application of the generalized Wang cohomology sequence.

Theorem 13.7.7 *Let R be a PID. Let X be a simply connected space with $H^*(X; R) \approx H^*(\mathbb{S}^n; R)$ where $n > 1$ is odd. Then $H^*(\Omega X; R)$ has a basis consisting of $\{1, u_1, u_2, \dots\}$ with $\deg u_k = k(n-1)$ and such that*

$$
p! \, q! \, u_p \smile u_q = (p+q)! \, u_{p+q}. \tag{13.18}
$$

In particular, we have $u_1^p = p! \, u_p$.

Proof: Since the total space of $p : PX \to X$ is contractible, in the Wang cohomology exact sequence (13.16), we have $\Theta : H^q(\Omega X) \to H^{q-n+1}(\Omega X)$ is an isomorphism, $q > 0$. We choose $u_0 = 1$. Inductively, for $k \geq 1$, we take $u_k \in H^{k(n-1)}(\Omega X)$ such that $\Theta(u_k) = u_{k-1}$. It follows that $\{1, u_1, u_2, \cdots\}$ is a basis for $H^*(\Omega X)$. Since n is odd (13.17) yields

$$
\Theta(u_p \smile u_q) = \Theta(u_p) \smile u_q + u_p \smile \Theta(u_q).
$$

Formula (13.18) is easily derived by a double induction on p, q. ♠

Remark 13.7.8 For $r = \mathbb{Q}$, it follows that $H^*(\Omega X; \mathbb{Q}) \approx \mathbb{Q}[u_1]$, the polynomial ring with a generator of degree $n-1$. However, for $R = \mathbb{Z}$, we do not exactly get a polynomial ring; what we get is called a divided factorial polynomial ring because of the property $u_1^p = p! \, u_p$.

Toward the converse of the above result, we need a technical lemma:

Lemma 13.7.9 Let X be a simply connected space. Suppose for some $n \geq 2$, there is $u \in H^n(X; R)$ such that for some $m \geq 2$, $u^{m-1} \neq 0$ and $\{1, u, u^2, \dots, u^{m-1}\}$ forms a R-module basis for $H^*(X; R)$ in degrees $< mn$. Then there is an element $v \in H^{n-1}(\Omega X; R)$ such that $\{1, v\}$ forms a R-module basis for $H^*(\Omega; R)$ in degree $< mn - 2$.

Proof: We appeal to the path fibration $p : PX \to X$ with fibre ΩX. Since PX is contractible, we have $E_\infty^{p,q} = 0$ unless $p = 0 = q$. Since X has no R-torsion in degree $< mn$ we have $E_2^{p,q} = H^p(X) \otimes H^q(\Omega X), p < mn$. In particular, for $p < mn$, $E_2^{p,q} = 0$ unless $p = kn, k = 0, 1, \dots, m-1$. Therefore the same holds for E_r terms also for all $r \geq 2$. Since the bidegree of d_r is $(r, 1-r)$, it follows that for $p < mn$, $d_r : E_r^{p,q} \to E_r^{p+r,q-r+1}$ is zero unless $r = kn, k = 1, \dots m-1$. Therefore $E_n^{p,q} \approx E_2^{p,q}$ for $p < mn$.

Now if $0 < q < n-1$, we have $H^q(\Omega X) \approx E_2^{0,q} \approx E_\infty^{0,q} = 0$. We also have the exact sequence

$$
0 \longrightarrow E_\infty^{0,n-1} \longrightarrow H^0(X) \otimes H^{n-1}(\Omega X) \xrightarrow{d_n} H^n(X) \otimes H^0(\Omega X) \longrightarrow E_\infty^{n,0} \longrightarrow 0.
$$

Since both the E_∞ terms in the sequence are zero, it follows that there is $v \in H^{n-1}(\Omega X)$ such that $d_n(1 \otimes v) = u \otimes 1$. Using the derivation property of d_n we conclude that

$$
d_n(u^k \otimes v) = (-1)^{k(n-1)} u^{k+1} \otimes 1.
$$

Combining this with the hypothesis on $H^*(X)$, we conclude that $d_n : E_n^{p-n,n-1} \to E_n^{p,0}$ are isomorphisms for $p < mn$. Since $d_n \circ d_n = 0$, it follows that $d_n : E_n^{p-2n,2n-2} \to E_n^{p-n,n-1}$ are zero for $p < mn$. Also it follows that $E_{n+1}^{p-n,n-1} = 0 = E_{n+1}^{p,0}, p < mn$. Thus we have

$$\left. \begin{array}{rcll} E_r^{p,q} & = & 0, & p < mn, \ q < n-1, \ r \geq n+1, \\ E_{n+1}^{0,2n-2} & = & E_2^{0,2n-2}. & \end{array} \right\} \tag{13.19}$$

We need to show that $H^t(\Omega X) = 0, \quad n-1 < t < mn-1$. If this is not true then there is some $n-1 < t < mn-2$ for which $H^t(\Omega X) \neq 0$ and $H^s(\Omega X) = 0, \ n-1 < s < t$. Then $E_n^{0,t} \approx E_2^{0,t} \approx H^t(\Omega X)$. We claim that $E_n^{0,t} = E_\infty^{0,t}$ which will be a contradiction.

It suffices to show that $E_n^{0,t} \approx E_{n+1}^{0,t}$ for all the other differentials beyond this stage are zero in the range $p < mn, q < t$. Now, if $t = 2n-2$, this follows from the fact $E_{n+1}^{0,2n-2} \approx E_2^{0,2n-2}$, as observed above. If $t \neq 2n-2$, then $H^{t-n+1}(\Omega X) = 0$ and hence $E_r^{p,t-n+1} = 0$ for p, r. In particular, the differential $d_n = 0 : E_n^{0,t} \to E_n^{n,t-n+1}$. Thus, $E_n^{0,t} \approx E_{n+1}^{0,t}$ and the conclusion follows. ♠

As an immediate corollary, letting $m \to \infty$, we have:

Theorem 13.7.10 *Let X be a simply connected space such that $H^*(X; R)$ is a polynomial algebra over a generator of degree n. Then $H^*(\Omega X; R) \approx H^*(\mathbb{S}^{n-1}; R)$.*

Note that in the context of the above theorem, the number n is necessarily even.

13.8 Serre Classes

We are familiar with the notion of 'getting rid' of troublesome factors in the study of abelian groups, e.g., by tensoring over $\mathbb{Q}$ or $\mathbb{R}$ to get rid of torsion or by tensoring with $\mathbb{Z}_p$ to get rid of torsions coprime to p and so on. Serre has systematized this with the machinery which, nowadays goes under the name 'Serre class of abelian groups'. In this section, after introducing the basics, we shall give an application of this to generalize Hurewicz theorem and Whitehead's theorem. In the next section, we shall use it again to arrive at Serre's results on homotopy groups of spheres.

Definition 13.8.1 A non empty class $\mathcal{C}$ of abelian groups is called a *Serre class* if it is closed under three term short exact sequences: i.e., whenever $A \to B \to C$ is an exact sequence of abelian groups with $A, C \in \mathcal{C}$ then $B \in \mathcal{C}$.

The following lemma is immediate and gives some ready-to-use properties of Serre classes.

Lemma 13.8.2 A class of abelian groups $\mathcal{C}$ is a Serre class iff the following properties are all true:
(a) $\mathcal{C}$ contains a trivial group.
(b) $A \in \mathcal{C}, A \approx B \Longrightarrow B \in \mathcal{C}$.
(c) $A \subset B, B \in \mathcal{C} \Longrightarrow A \in \mathcal{C}$ and $B/A \in \mathcal{C}$.
(d) If $0 \to A \to B \to C \to 0$ is a short exact sequence with $A, C \in \mathcal{C}$, then $B \in \mathcal{C}$.

Example 13.8.3 The following are some commonly used examples of Serre classes. Verification that these are actually Serre classes is left to the reader as an exercise.
(i) The class of all abelian groups.
(ii) The class of all finitely generated abelian groups.
(iii) The class of all finite abelian groups.

(iv) The class of torsion abelian groups.

(v) The class of p-groups where p is a prime number.

(vi) The class of all torsion abelian groups containing no element of order equal to a power of p for a given prime p.

(vii) The class of trivial groups.

Remark 13.8.4 Note that because of (a) and (b), it follows that a Serre class can never be a set. Also a trivial example of a class which is **not** a Serre class is the class of all free abelian groups.

The idea of introducing Serre class $\mathcal{C}$ is to study properties of homomorphisms $f : A \to B$ of abelian groups modulo the class $\mathcal{C}$. This leads us to make the following definitions:

Definition 13.8.5 A homomorphism $f : A \to B$ is a $\mathcal{C}$-monomorphism (respectively, $\mathcal{C}$-epimorphism) if Ker $f \in \mathcal{C}$ (respectively, Coker $f \in \mathcal{C}$.) We say f is a $\mathcal{C}$-isomorphism if it is both $\mathcal{C}$-monomorphism and a $\mathcal{C}$-epimorphism. If such f exists then we write $A \approx_{\mathcal{C}} B$. It is not hard (exercise) to prove that $\approx_{\mathcal{C}}$ is an equivalence relation on $\mathcal{C}$.

Remark 13.8.6 Tensor product and torsion product play a central role in controlling the torsions. Thus we need to put some additional conditions on $\mathcal{C}$ if we want to make any meaningful statement about the behaviour of $\approx_{\mathcal{C}}$ under these products. This is the motivation for the following definitions:

Definition 13.8.7 We say a Serre class $\mathcal{C}$ is a *ring of abelian groups*, if $A, B \in \mathcal{C}$ implies $A \otimes B, A \star B \in \mathcal{C}$. It is called an *ideal of abelian groups*, if $A \in \mathcal{C}$ implies $A \otimes B, A \star B \in \mathcal{C}$ for all abelian groups B.

Remark 13.8.8 Obviously an ideal of abelian groups is also a ring of abelian groups. Classes occurring in Example 13.8.3 are all rings of abelian groups and except (ii) and (iii), all are ideals of abelian groups. In what follows we shall state, simultaneously, results which are valid for rings of abelian groups and somewhat stronger results valid for ideals of abelian groups. We shall prove the former ones and leave the proof of the latter, which need only minor modifications, to the reader. We begin with the following generalization of Serre's result given in Exercise 11.1.16(ii).

Theorem 13.8.9 *Let* $p : E \to B$ *be an orientable fibration with fibre* F *and base* B, *both path connected. Let* $\emptyset \neq B' \subset B, E' = p^{-1}(B')$. *Let* $\mathcal{C}$ *be a Serre class of abelian groups such that* $H_i(B, B'; R) \in \mathcal{C}, 0 \leq i < n$ *and* $H_j(F; G) \in \mathcal{C}, 0 < j < m$.
(a) *If* $\mathcal{C}$ *is a ring of abelian groups, assume further that* $H_1(B, B'; R) = 0$ *and take* $r = \inf\{n, m+1\}$.
(b) *If* $\mathcal{C}$ *is an ideal of abelian groups, then take* $r = m + n - 1$.
Then the homomorphism $p_* : H_q(E, E'; G) \to H_q(B, B'; G)$ *is a* $\mathcal{C}$-*monomorphism for* $q \leq r$ *and a* $\mathcal{C}$-*epimorphism for* $q \leq r + 1$.

Proof: Recall the spectral sequence of Theorem (13.4.13) wherein we have:

$$E_{s,t}^2 = H_s(B, B'; H_t(F; G)) = H_s(B, B'; R) \otimes H_t(F; G) \oplus H_{s-1}(B, B'; R) \star H_t(F; G).$$

Suppose $s + t \leq r$, and $t \geq 1$.

Consider the situation (a). Since $H_0(B, B'; R) = 0 = H_1(B, B'; R)$, it follows that $E_{s,t}^2 = 0$ for $s = 0, 1$. If $s > 1$, then $t < m, s < n$ and hence $H_t(F; G), H_s(B, B'; R), H_{s-1}(B, B'; R)$ are all in $\mathcal{C}$. Therefore, $E_{s,t}^2 \in \mathcal{C}$.

Since we have a first quadrant sequence, from (b) of Theorem 13.4.13, for each $q \geq 0$, we get a finite filtration

$$0 \subset D_0 \subset D_1 \subset \cdots \subset D_q = H_q(E, E'; G)$$

with $D_i/D_{i-1} \approx E^\infty_{i,q-i}$. In particular, we have

$$H_q(E, E'; G)/D_{q-1} \approx E^\infty_{q,0} \subset E^2_{q,0} = H_q(B, B'; G).$$

This implies that the kernel of $p_* : H_q(E, E'; G) \to H_q(B, B'; G)$ is isomorphic to the kernel of the quotient map $D_q \to D_q/D_{q-1}$ which in turn is equal to D_{q-1}. Thus, if we show that $D_{q-1} \in \mathcal{C}$, then it would follow that p_* is a $\mathcal{C}$-monomorphism. Now this can be achieved by a simple induction, provided we show that $E^\infty_{s,t} \in \mathcal{C}$ for $s + t \leq r$ and $t \geq 1$. We have already proved this for the E^2-term. Since the spectral sequence is a first quadrant one, it follows that E^∞-terms are subquotients of E^2 terms and so they are all in $\mathcal{C}$ for the range $s + t \leq r, t > 1$.

To show that p_* is a $\mathcal{C}$-epimorphism for $q \leq r + 1$, we must show that $E^2_{q,0}/E^\infty_{q,0} \in \mathcal{C}$ for $q \leq r + 1$. Here we have

$$E^2_{q,0} \supset E^3_{q,0} \supset \cdots \supset E^{q+1}_{q,0} = E^\infty_{q,0}.$$

Therefore it suffices to show that the successive quotients $E^k_{q,0}/E^{k+1}_{q,0} \in \mathcal{C}, q \leq r + 1, k > 1$. Since $E^{k+1}_{q,0} = \mathrm{Ker}\,[d^k : E^k_{q,0} \to E^k_{q-k,k+1}]$, the required quotients are isomorphic to some submodules of $E^k_{q-k,k+1}$, and hence it suffices to show that $E^k_{q-k,k+1} \in \mathcal{C}$ in the said range. Once again this follows from the same fact proved for the E^2-term. ♠

Definition 13.8.10 A topological space X is said to be $\mathcal{C}$-acyclic if its reduced integral homology groups $\tilde{H}_q(X) \in \mathcal{C}$. Similarly a topological pair (X, A) with $A \neq \emptyset$ is said to be $\mathcal{C}$ acyclic if $H_q(X, A) \in \mathcal{C}$ for all q.

The following corollary is immediate, by taking $B' = \{*\}$ in the above theorem.

Corollary 13.8.11 Let $p : E \to B$ be a fibration with a path connected total space and a simply connected base B. Assume that E is $\mathcal{C}$-acyclic and $H_i(B) \in \mathcal{C}$ for $0 < i < n$, where $\mathcal{C}$ is a ring of abelian groups. Then F is path connected and $H_i(F) \in \mathcal{C}$ for $0 < i < n - 1$, and $H_{n-1}(F) \approx_\mathcal{C} H_n(B)$. Further, if $\mathcal{C}$ is an ideal of abelian groups, then $H_i(F) \approx_\mathcal{C} H_{i+1}(B)$ for $i < 2n - 2$.

Proof: That F is path connected is a standard exercise (see 1.7.14(ii)). We shall now induct on n. Assume that we have proved $H_i(F) \in \mathcal{C}$ for $0 < i < m - 1$ for $m < n - 1$. We have to show that $H_m(F) \in \mathcal{C}$. By part (a) of the above theorem, we have $p_* : H_i(E, F; G) \to H_i(B, *)$ is a $\mathcal{C}$-isomorphism for $i \leq m + 1$. Since E is $\mathcal{C}$-acyclic, we have $H_i(E, F) \approx_\mathcal{C} H_{i-1}(F)$ for all i and hence in particular, this implies $H_i(F) \in \mathcal{C}, i \leq n - 1$ and $H_{n-1}(F) \approx_\mathcal{C} H_n(B)$. We leave the proof of part (b) to the reader. ♠

Corollary 13.8.12 Let $p : E \to B$ be a fibration with path connected base B and fibre F. Let p be orientable over R. Then for any ring $\mathcal{C}$ of R-modules, if two of the three spaces E, B, F are $\mathcal{C}$-acyclic then the third one is also $\mathcal{C}$-acyclic.

Proof: Assume that B, F are $\mathcal{C}$-acyclic. Put $B' = \{b_0\}$, so that $E' = F$. By Theorem 13.8.9 with $m = n = \infty$, we get (E, F) is $\mathcal{C}$-acyclic. Since F is $\mathcal{C}$-acyclic, it follows that E is also so. Now consider the case when E, B are $\mathcal{C}$-acyclic. Now apply Corollary 13.8.11 for each n to conclude that F is $\mathcal{C}$-acyclic. Finally, suppose E and F are $\mathcal{C}$-acyclic. If B is not $\mathcal{C}$-acyclic there is a smallest integer $n > 0$ such that $H_n(B) \notin \mathcal{C}$. Again by Corollary 13.8.11, $H_{n-1}(F) \notin \mathcal{C}$ which is a contradiction. ♠

The following corollary is easy to prove:

Corollary 13.8.13 Let X be a simply connected space and $\mathcal{C}$ be a ring of abelian groups. Then X is $\mathcal{C}$-acyclic iff its loop space ΩX is $\mathcal{C}$-acyclic.

Definition 13.8.14 We say a Serre class $\mathcal{C}$ is an *acyclic class* if whenever $\pi \in \mathcal{C}$ then any space of type $(\pi, 1)$ is also $\mathcal{C}$-acyclic, i.e., $H_*(\pi) \in \mathcal{C}$.

Clearly the class of all abelian groups is acyclic. We have seen that the class of finite abelian groups is acyclic. Indeed, it is not difficult to show that all the Serre classes mentioned in Example 13.8.3 are acyclic. The following is an immediate corollary to the above corollary once we note that the loop space of a space of type (π, n) is a space of type $(\pi, n-1)$.

Corollary 13.8.15 Let $\mathcal{C}$ be an acyclic ring of abelian groups. Then $\pi \in \mathcal{C}$ implies that any space of type (π, n) for any $n \geq 1$ is $\mathcal{C}$-acyclic.

We can now build on this by using a Postnikov tower and prove that any simply connected space with all its homotopy groups belonging to an acyclic ring $\mathcal{C}$ of abelian groups is itself $\mathcal{C}$-acyclic. However, we can also deduce this from the following generalization of the Hurewicz theorem, which also uses a Postnikov tower.

Theorem 13.8.16 *Let $\mathcal{C}$ be an acyclic ring of abelian groups and X be any simply connected space. Then for each $q \geq 2$, the following statements are equivalent:*
(A_q) $\pi_i(X) \in \mathcal{C}$, *for* $2 \leq i < q$.
(B_q) $H_i(X) \in \mathcal{C}$, *for* $2 \leq i < q$.
Moreover, either of them implies that the Hurewicz map $\varphi : \pi_i(X) \to H_i(X)$ is $\mathcal{C}$-isomorphism for $i \leq q$.

Proof: As in the proof of the Hurewicz theorem, it is enough to prove that (B_q) implies that $\varphi : \pi_q(X) \approx_{\mathcal{C}} H_q(X)$. For $q = 2$, this follows from the absolute Hurewicz theorem. Let $q \geq 3$. Assume that we have proved the equivalence of (A_{q-1}) and (B_{q-1}), etc. Consider a sequence of fibrations:

$$E_q \xrightarrow{p_q} E_{q-1} \xrightarrow{p_{q-1}} \cdots \xrightarrow{p_2} E_1 = X$$

where each p_j is a j-connected fibration with fibre F_j being a space of type $(\pi_j(X), j-1)$ (as in Theorem 10.8.7). Induction hypothesis combined with Corollary 13.8.15, implies that each F_j is $\mathcal{C}$-acyclic, $2 \leq i < q-1$.

Let us prove that $(p_j)_* : H_i(E_j) \to H_i(E_{j-1})$ are $\mathcal{C}$-isomorphisms for $j \leq q$, and for all $i \leq q$. Since F_j is $\mathcal{C}$-acyclic, this would follow if we prove $(p_j)_* : H_i(E_j, F_j) \to H_i(E_{j-1}, *)$ are $\mathcal{C}$-isomorphisms. But this is a consequence of Theorem 13.8.9 (a).

Putting $f = p_q \circ \cdots \circ p_2$, it follows that $f_* : H_i(E_q) \approx_{\mathcal{C}} H_i(X)$. Consider the commutative diagram

$$
\begin{array}{ccc}
\pi_q(E_q) & \xrightarrow{\varphi} & H_q(E_q) \\
{\scriptstyle f_\#} \downarrow & & \downarrow {\scriptstyle f_*} \\
\pi_q(X) & \xrightarrow{\varphi} & H_q(X)
\end{array}
$$

wherein the top horizontal arrow is an isomorphism. The vertical arrows are $\mathcal{C}$-isomorphisms and hence the bottom horizontal one is also a $\mathcal{C}$-isomorphism.   ♠

Corollary 13.8.17 Let X be a simply connected space. Then $\pi_i(X)$ are finitely generated for all i iff $H_i(X)$ are finitely generated.

Corollary 13.8.18 All the homotopy groups of $\mathbb{S}^n$ are finitely generated.

Remark 13.8.19 Easy examples such as $\mathbb{S}^1 \vee \mathbb{S}^2$ show that without the simple connectivity condition Corollary 13.8.17 is false. Nevertheless, there are far-reaching generalisations for spaces such as 'nilpotent spaces' which we shall not be able to touch. (See [Hilton–Mislin–Roitberg, 1975].) However, we shall need one more step of generalization of this result to what are called strongly simple spaces.

Definition 13.8.20 Let X be a path connected space. Consider an embedding of X in a space B of type $(\pi_1(X), 1)$ such that the inclusion $j : X \to B$ induces an isomorphism $j_\# : \pi_1(X) \to \pi_1(B)$. (Such a space can be constructed by simply attaching cells of dimension ≥ 2 successively to kill all the higher homotopy groups.) Let $p = p^j : M^j \to B$ be the mapping path fibration (see Definition 1.7.8). We say X is *strongly simple* if $\pi_1(X)$ is abelian and the fibration p^j is orientable over $\mathbb{Z}$.

Lemma 13.8.21 A pointed space (X, x_0) is strongly simple if for each element $[\omega] \in \pi_1(X, x_0)$, there is a map $\hat{\omega} : \mathbb{S}^1 \times X \to X$ such that $x \mapsto \hat{\omega}(1, x) = x$ is homotopic to Id_X, and $\theta \mapsto \hat{\omega}(\theta, x_0)$ represents $[\omega]$.

Proof: Let us call this condition on (X, x_0) (SC). First of all notice that in this condition, we can as well say $x \mapsto \hat{\omega}(1, x)$ is actually the identity map. Next, observe that if (Y, y_0) is homotopy equivalent to (X, x_0), then the (Y, y_0) also satisfies (SC).

Now let us prove that (SC) implies (X, x_0) is strongly simple. Given $[\omega] \in \pi_1(X)$ and $[\lambda] \in \pi_n(X)$ consider the map $H : \mathbb{S}^1 \times \mathbb{S}^n \to X$ given by $H(\theta, x) = \hat{\omega}(\theta, \lambda(x))$, where $\hat{\omega} : \mathbb{S}^1 \times X \to X$ satisfies the condition above. From this it follows that $h[\omega]([\lambda]) = 0$. Therefore, in particular, $\pi_1(X)$ is abelian (and its action on $\pi_n(X), n \geq 2$ is trivial).

To see that the action of $\pi_1(B)$ on the homology of the fibre F of $p^j : M^j \to B$ is trivial, we use the fact that $j_\# : \pi_1(X) \to \pi_1(B)$ is an isomorphism and X is a deformation retract of M^j. It follows that M^j satisfies (SC) and $(p^j)_\# : \pi_1(M^j) \to \pi_1(B)$ is an isomorphism.

Given $[\omega] \in \pi_1(M^j)$ consider a map $\hat{\omega} : \mathbb{S}^1 \times M^j \to M^j$ such that $\hat{\omega}(1, \lambda) = \lambda$ and $\hat{\omega}(\theta, \lambda_0) = \omega(\theta)$. Putting $p^j \circ \omega = \omega'$, this implies that the map $h_{\omega'} : F \to F$ is actually the identity map. Since every element of $\pi_1(B)$ is of the form $[p^j \circ \omega]$ for some $\omega : \mathbb{S}^1 \to M^j$, it follows that the action of $\pi_1(B)$ on the fibre of p^j is homotopically trivial. In particular, p^j is orientable. ♠

Corollary 13.8.22 Every H-space is strongly simple. In particular, every loop space is strongly simple.

Exercise 13.8.23 Show that if $\mathbb{S}^1 \vee X$ is a retract of $\mathbb{S}^1 \times X$, then X is strongly simple.

Lemma 13.8.24 Let X be a strongly simple space and $\mathcal{C}$ be an acyclic ring of abelian groups. Suppose $\pi_1(X) \in \mathcal{C}$ and $H_i(X) \in \mathcal{C}$, $0 < i < q$ for some $q \geq 2$. If F is the fibre of the mapping path fibration $M^j \to B$, then $H_i(F) \to H_i(M^j)$ are $\mathcal{C}$-isomorphisms for $0 \leq i \leq q$.

Proof: Proof is by induction. Assume that the result is true for $1 \leq m < q$. Since X and M^j have the same homotopy type, $H_i(M^j) \in \mathcal{C}$, $0 < i < q$. By induction hypothesis, $H_i(F) \in \mathcal{C}$, for $0 < i \leq m$. From this we shall conclude that $H_{m+1}(F) \approx_\mathcal{C} H_{m+1}(M^j)$, which will complete the induction step.

We can apply Theorem 13.4.13. There is a filtration

$$0 \subset D_0 \subset D_1 \subset \cdots \subset D_{m+1} = H_{m+1}(M^j)$$

where $D_s/D_{s-1} \approx E^\infty_{s,m+1-s}$ and $D_0 = \text{Im}\,[H_{m+1}(F) \to H_{m+1}(M^j)]$. Therefore, in order to show that $H_{m+1}(F) \to H_{m+1}(M^j)$ is a $\mathcal{C}$-epimorphism, it suffices to show that $E^\infty_{s,m+1-s} \in \mathcal{C}$ for each $1 \leq s \leq m$. For this, it suffices to show that $E^2_{s,m+1-s} \in \mathcal{C}$. But we have

$$E^2_{s,m+1-s} \approx H_s(B) \otimes H_{m+1-s}(F) \oplus H_{s-1}(B) * H_{m+1-s}(F)$$

wherein $H_s(B) \in \mathcal{C}, s > 0$ because $\mathcal{C}$ is acyclic. By induction hypothesis $H_{m+1-s}(F) \in \mathcal{C}$ for $s > 0$ (and since $H_0(B) = \mathbb{Z}$), we conclude that $E^2_{s,m+1-s} \in \mathcal{C}$ since $\mathcal{C}$ is a ring of abelian groups.

Next we want to show that $H_{m+1}(F) \to H_{m+1}(M^j)$ is a $\mathcal{C}$-monomorphism. This homomorphism is actually the composite of

$$H_{m+1}(F) \approx E^2_{0,m+1} \to E^3_{0,m+1} \to \cdots \to E^{m+2}_{0,m+1} = E^\infty_{0,m+1} \approx D_0 \subset D_{m+1} = H_{m+1}(M^j).$$

Therefore, it suffices to prove that $\text{Ker}\,[E^r_{0,m+1} \to E^{r+1}_{0,m+1}] \in \mathcal{C}$ for $r \geq 2$. This is the same as showing that $d^r(E^r_{r,m+2-r}) \subset E^r_{0,m+1}$ is in $\mathcal{C}$ for all $r \geq 2$. This will be the case if $E^2_{r,m+2-r} \in \mathcal{C}$ for $r \geq 2$. But then

$$E^2_{r,m+2-r} \approx H_r(B) \otimes H_{m+2-r}(F) \oplus H_{r-1}(B) \star H_{m+2-r}(F)$$

wherein $H_{m+2-r}(F) \in \mathcal{C}$ by induction hypothesis. Since $\mathcal{C}$ is a ring of abelian groups, we are done. ♠

Now the proof of the following strengthened version of Theorem 13.8.16 is easy.

Theorem 13.8.25 *Let X be a strongly simple space and $\mathcal{C}$ be an acyclic ring of abelian groups. If $\pi_1(X) \in \mathcal{C}$, the following are equivalent:*
(a) $\pi_i(X) \in \mathcal{C}, 2 \leq i < q$.
(b) $H_i(X) \in \mathcal{C}, 2 \leq i < q$.
Moreover, either of them implies that $\varphi : \pi_i(X) \to H_i(X)$ is a $\mathcal{C}$-isomorphism for $i \leq q$.

Proof: As seen earlier, it suffices to prove that (b) implies that $\varphi : \pi_q(X) \approx_\mathcal{C} H_q(X)$. Following the notation in Definition 13.8.20, let F be the fibre of the fibration $M^j \to B$. Since X and M^j have the same homotopy type, there is a map $f : F \to X$ which is equivalent to the inclusion $F \hookrightarrow M^j$. It follows that $f_\# : \pi_i(F) \to \pi_i(X)$ is an isomorphism $i \geq 2$. Also, by the above lemma, $f_* : H_i(F) \approx_\mathcal{C} H_i(X)$. Therefore, F satisfies the same hypothesis as X does, as in the theorem. In addition, it is simply connected and hence from Theorem 13.8.16, it follows that $\varphi : \pi_q(F) \approx_\mathcal{C} H_q(F)$. Since $\varphi \circ f_\# = f_* \circ \varphi$, the same conclusion follows from X also. ♠

We shall now embark upon proving the relative version of the above theorem.

Lemma 13.8.26 *Let (X, A) be a topological pair such that both X and A are simply connected. Let $x_0 \in A$ be the base point. Then the relative loop space*

$$\Omega(X, A) = \{\omega : \mathbb{I} \to X \text{ such that } \omega(0) = x_0, \omega(1) \in A\}$$

is strongly simple.

Proof: The map $\psi : \Omega X \times \Omega(X, A) \to \Omega(X, A)$ given by composition

$$(\omega, \lambda) \mapsto \omega * \lambda$$

is continuous. Also, we know that the inclusion induces an isomorphism $\pi_1(\Omega X) \to \pi_1(\Omega(X, A))$. Therefore, the condition of Lemma 13.8.21 is satisfied by $\Omega(X, A)$. ♠

We are now in a position to state and prove the following relative version of Theorem 13.8.16.

Theorem 13.8.27 *Let (X, A) be a topological pair such that A, X are simply connected. Let C be an acyclic ideal of abelian groups. Then the following two statements are equivalent:*
(a) $\pi_i(X, A) \in C$, $2 \le i < q$.
(b) $H_i(X, A) \in C$, $2 \le i < q$.
Moreover, either of them implies that the Hurewicz map
(c) $\varphi : \pi_q(X, A) \to H_q(X, A)$ *is a C-isomorphism.*

Proof: As usual, it suffices to prove that (b) implies (c) which we shall do by induction on q. For $q = 2$ this is just the relative version of the Hurewicz isomorphism theorem. So assume that $q \ge 3$, (b) holds and (c) holds for $i < q - 1$. This implies that $\pi_i(X, A) \in C$ for $i < q$.

Let $x_0 \in A$ be the base point, PX be the space of paths in X starting at x_0 and $p : PX \to X$ be the fibration given by $p(\omega) = \omega(1)$. The fibre $p^{-1}(x_0)$ is the loop space ΩX. We know that $p_\# : \pi_i(PX, p^{-1}(A)) \to \pi_i(X, A), i \ge 1$, is an isomorphism. Since PX is contractible, we also know that for $i \ge 2$, $\pi_i(PX, p^{-1}(A)) \approx \pi_{i-1}(p^{-1}(A))$ and $H_i(PX, p^{-1}(A)) \approx H_{i-1}(p^{-1}(A))$. Consider the following commutative diagram

$$
\begin{array}{ccccc}
\pi_{i-1}(p^{-1}(A)) & \xleftarrow[\partial]{\approx} & \pi_i(PX, p^{-1}(A)) & \xrightarrow[\approx]{p_\#} & \pi_i(X, A) \\
\downarrow{\scriptstyle\varphi} & & & & \downarrow{\scriptstyle\varphi} \\
H_{i-1}(p^{-1}(A)) & \xleftarrow[\partial]{\approx} & H_i(PX, p^{-1}(A)) & \xrightarrow{p_*} & H_i(X, A).
\end{array}
$$

Since X is simply connected, we have that ΩX is path connected and the fibration is orientable. Therefore Theorem 13.8.9(b) is valid, with $n = q$ and $m = 1$. Therefore, we conclude that $p_* : H_i(PX, p^{-1}(A)) \approx_C H_i(X, A), i \le q$. Therefore, all the horizontal arrows in the above diagram are C-isomorphisms. Thus it suffices to show that the first vertical arrow is a C-isomorphism for $i = q - 1$. Now, combining Lemma 13.8.26 and Theorem 13.8.25, we are through. ♠

Now by taking the mapping cylinder, as usual, we directly obtain the following generalization of Whitehead's theorem.

Theorem 13.8.28 *Let C be an acyclic ideal of abelian groups and let $f : X \to Y$ be a mapping of simply connected spaces. Then for any $n \ge 1$ the following are equivalent:*
(a) $f_\# : \pi_i(X) \to \pi_i(Y)$ *is a C-isomorphism for $i < n$ and a C-epimorphism for $i = n$.*
(b) $f_* : H_i(X) \to H_i(Y)$ *is a C-isomorphism for $i < n$ and a C-epimorphism for $i = n$.*

Exercise 13.8.29 We have computed the homology modules of $K(\mathbb{Z}, 3)$ in Example 11.3.10 using the Wang sequences. Compute the cohomology algebra of $K(\mathbb{Z}, 3)$ directly from the cohomology spectral sequence of the fibration

$$\mathbb{CP}^\infty \subset \mathbb{S}^3 \to K(\mathbb{Z}, 3)$$

and thereby compute the homology modules of $K(\mathbb{Z}, 3)$.

13.9   Homotopy Groups of Spheres

In this last section, we use many deep results that we have seen so far and present the classical result due to Serre on the homotopy groups of spheres.

Odd dimensional case

As an immediate corollary to Theorem 13.8.16 and our computation in Example 11.3.10 of the homology of the total space of the principal fibration of $\psi : \mathbb{S}^3 \to K(\mathbb{Z}, 3)$ we have:

Corollary 13.9.1 The p-primary component of $\pi_i(\mathbb{S}^3)$ vanishes if $3 < i < 2p$ and is equal to $\mathbb{Z}_p$ for $i = 2p$.

Proof: Let $\mathbb{CP}^\infty \subset E \to \mathbb{S}^3$ be the principal fibration induced by a map $\eta : \mathbb{S}^3 \to K(\mathbb{Z}, 3)$ which represents a generator of π_3. In the Example 11.3.10, we have seen that

$$H_q(E) = \begin{cases} \mathbb{Z}, & q = 0; \\ 0, & q \text{ odd}; \\ \mathbb{Z}_n, & q = 2n. \end{cases}$$

Given a prime p, let $\mathcal{C}_p$ denote the Serre class of abelian groups having no element of order a power of p, then $H_i(E) \in \mathcal{C}$ for $0 < i < 2p$. This implies $\pi_i(E) \in \mathcal{C}$ for the same range by Theorem 13.8.16. Since $p_\# : \pi_i(E) \to \pi_i(\mathbb{S}^3)$ is an isomorphism for $i > 3$, we are home. ♠

Theorem 13.9.2 *If n is odd, $\pi_m(\mathbb{S}^n)$ is finite for all $m \neq n$.*

We induct on n, the truth of the statement being familiar for $n = 1$. For $n = 3$, if E is as in Example 11.3.10, taking $\mathcal{C}$ to be the class of finite abelian groups, we know that $H_i(E) \in \mathcal{C}$ for all $i > 0$. Since E is simply connected also, by Theorem 13.8.16, it follows that $\pi_i(E) \in \mathcal{C}$. Since $\pi_i(E) \approx \pi_i(\mathbb{S}^3), i > 3$, $\pi_i(\mathbb{S}^3)$ are finite for $i \neq 3$. The induction starts at this stage.

Assume now $n > 3$ and is odd and $\pi_m(\mathbb{S}^{n-2})$ are finite for $m \neq n - 2$. The rational cohomology algebra of $\Omega\mathbb{S}^n$ is the polynomial algebra $\mathbb{Q}[x]$ with degree of x equal to $n - 1$, (see Theorem 13.7.7), whereas that of $\Omega^2(\mathbb{S}^n)$ is the same as $H^*(\mathbb{S}^{n-2}; \mathbb{Q})$ (see Theorem 13.7.10). By the universal coefficient theorem, the integral homology group $H_i(\Omega^2\mathbb{S}^n)$ is a finite group for $i \neq n - 2$ and is of rank 1 for $i = n - 2$. Since $\Omega^2\mathbb{S}^n$ is $(n-3)$-connected, $\varphi : \pi_{n-2}(\Omega^2\mathbb{S}^n) \approx H_{n-2}(\Omega^2\mathbb{S}^n)$. Let $\alpha : \mathbb{S}^{n-2} \to \Omega^2\mathbb{S}^n$ be a generator of the infinite cyclic component. Then it follows that $\alpha_* : H_*(\mathbb{S}^{n-2}) \approx_\mathcal{C} H_*(\Omega^2\mathbb{S}^n)$. By the generalized Whitehead Theorem 13.8.28, this implies that $\alpha_\# : \pi_i(\mathbb{S}^{n-2}) \approx_\mathcal{C} \pi_i(\Omega^2\mathbb{S}^n) \approx \pi_{i+2}(\mathbb{S}^n)$. This completes the inductive step. ♠

The Even dimensional case

Let us denote by V, the real Steifel manifold $V_{2,2n+1}$ of orthonormal pairs $(\mathbf{v}_1, \mathbf{v}_2)$ in $\mathbb{R}^{2n+1}$.

Lemma 13.9.3 The integral homology groups $H_*(V; \mathbb{Z})$ are all finite, except that $H_0(V; \mathbb{Z})$ and $H_{4n-1}(V, \mathbb{Z})$ which are infinite cyclic.

Proof: Note that $V := V_{2,2n+1}$ is a closed orientable manifold of dimension $4n-1$. Consider the map $p : V \to \mathbb{S}^{2n}$ given by $p(\mathbf{v}_1, \mathbf{v}_2) = \mathbf{v}_1$. This is indeed a fibre bundle map (the projection of the unit tangent bundle of $\mathbb{S}^{2n}$) with fibre $\mathbb{S}^{2n-1}$. From the homotopy exact sequence of the fibration, for $i \leq 2n - 2$, we obtain exact sequences,

$$\begin{array}{ccccc} \pi_i(\mathbb{S}^{2n-1}) & \longrightarrow & \pi_i(V) & \longrightarrow & \pi_i(\mathbb{S}^{2n}) \\ \| & & & & \| \\ 0 & & & & 0 \end{array}$$

and

$$\pi_{2n}(V) \xrightarrow{p_\#} \pi_{2n}(\mathbb{S}^{2n}) \xrightarrow{\partial} \pi_{2n-1}(\mathbb{S}^{2n-1}) \longrightarrow \pi_{2n-1}(V) \longrightarrow 0.$$

In particular, we conclude that $\pi_i(V) = 0$, $i \leq 2n - 2$.

We know by the hairy ball Theorem 4.4.7 that there is no map $f : \mathbb{S}^{2n} \to \mathbb{S}^{2n}$ such that $f(x)$ is orthogonal to x everywhere. This is the same as saying that the fibration $p : V \to \mathbb{S}^{2n}$ has no section. This then implies that the generator of $\pi_{2n}(\mathbb{S}^{2n})$ is not in the image of $p_{\#} : \pi_{2n}(V) \to \pi_{2n}(\mathbb{S}^{2n})$. This in turn implies that the boundary homomorphism $\partial : \mathbb{Z} = \pi_{2n}(\mathbb{S}^{2n}) \to \pi_{2n-1}(\mathbb{S}^{2n-1}) = \mathbb{Z}$ has non trivial image and hence is a monomorphism. This also implies that $\pi_{2n-1}(V)$ is a finite group. By the Hurewicz theorem, $\tilde{H}_i(V; \mathbb{Z}) = 0, i \leq 2n - 2$ and $H_{2n-1}(V; \mathbb{Z})$ is a finite group. By the universal coefficient theorem, we see that $H^i(V, \mathbb{Z}) = 0$ for $0 < i \leq 2n - 1$. Therefore by Poincaré duality, $H_{4n-1-i}(V; \mathbb{Z}) = 0$ for $0 < i \leq 2n - 1$. Thus $H_i(V)$ is finite for $0 < i < 4n - 1$, and $H_0(V), H_{4n-1}(V)$ are infinite cyclic. ♠

Theorem 13.9.4 $\pi_m(\mathbb{S}^{2n})$ *is finite except for* $m = 2n$ *and* $m = 4n - 1$ *and* $\pi_{2n}(\mathbb{S}^{2n})$ *is infinite cyclic whereas* $\pi_{4n-1}(\mathbb{S}^{2n})$ *is a direct sum of an infinite cyclic group with a finite group.*

Proof: Let $\mathcal{C}$ be the acyclic ideal of abelian groups consisting of torsion groups. By Lemma 13.9.3, for $V = V_{2,2n+1}$, $H_i(V; \mathbb{Z}) \in \mathcal{C}$ for $0 < i < 4n - 1$. By generalized Hurewicz Theorem 13.8.16, $\pi_{4n-1}(V) \approx_{\mathcal{C}} H_{4n-1}(V; \mathbb{Z})$. In any case, $\pi_{4n-1}(V)$ is finitely generated (by Corollary 13.8.17) and $H_{4n-1}(V) \approx \mathbb{Z}$. Therefore $\pi_{4n-1}(V)$ is a direct sum of an infinite cyclic group with a finite group. As before, if we take $\alpha : \mathbb{S}^{4n-1} \to V$ to represent a generator of the infinite cyclic summand, and appeal to generalized Whitehead Theorem 13.8.28, we get $\alpha_{\#} : \pi_i(\mathbb{S}^{4n-1}) \approx_{\mathcal{C}} \pi_i(V)$. By Theorem 13.9.2, it follows that $\pi_i(V)$ is finite for $i \neq 4n - 1$. By the exact homotopy sequence of the fibration $V \to \mathbb{S}^{2n}$, the conclusion follows. ♠

Definition 13.9.5 Given an integer $n \geq 0$, we define the nth *stable homotopy group*

$$\Pi_n := \lim_{k \to \infty} \pi_{n+k}(\mathbb{S}^k)$$

the direct limit of the directed system $S : \pi_{n+k}(\mathbb{S}^k) \to \pi_{n+k+1}(\mathbb{S}^{k+1}), k \geq 0$, where S is the suspension homomorphism.

From Freudenthal's suspension Theorem 11.3.7, it follows that $\Pi_n = \pi_{2n+2}(\mathbb{S}^{n+2})$. Clearly $\Pi_0 = \mathbb{Z}$. As an immediate consequence of the above two Theorems 13.9.2 and 13.9.4 we have:

Corollary 13.9.6 Π_n *is finite for all* $n \geq 1$.

Hints and Solutions

§1.1.19 (page no. 11)

(i). If H is the homotopy from the identity map to a constant map at x_0, then $t \mapsto H(x,t)$ gives a path from x to x_0.

(ii). There are many such topological properties though giving examples may not be easy. Here are six of them for which the examples are easy.
(a) compactness; (b) being locally Euclidean of the same dimension; (c) locally Euclidean; (d) locally path connected; (e) locally connectedness; (f) Hausdorffness. For (a) and (b) you can use the fact that $\mathbb{R}^n$ is homotopy equivalent to a single point for all n. For (c), you can use the union of two axes which is contractible. For (d) and (e) you can use the comb space which is not locally connected but contractible. For (f) take the example given in remark 1.3.5 (b) and check that it is contractible.

§1.2.33 (page no. 21)

(i). $\omega \simeq \tau \implies \omega * \tau^{-1} \simeq \tau * \tau^{-1} \simeq c$. Conversely, $\omega * \tau^{-1} \simeq c \implies \omega * \tau^{-1} * \tau \simeq c * \tau \implies \omega \simeq \tau$.

(ii). Any homomorphism $\alpha : \mathbb{Z} \to \mathbb{Z}$ is given by multiplication by the integer $\alpha(1)$. We treat $\mathbb{Z}$ as the fundamental group $\pi_1(\mathbb{S}^1, 1)$ with the generator $[\xi]$, where $\xi : \mathbb{S}^1 \to \mathbb{S}^1$ is the identity map. Now for any map $f : (\mathbb{S}^1, 1) \to (\mathbb{S}^1, 1)$, the induced homomorphism $f_\#$ is completely determined by the element $f_\#([\xi])$. Taking $f(z) = z^n$, we see that $f_\#[\xi] = [f \circ \xi]$. But the loop $f \circ \xi$ is nothing but $z \mapsto z^n$ and hence represents the element $n[\xi]$. If two functions $f, g : (\mathbb{S}^1, 1) \to (\mathbb{S}^1, 1)$ are such that $f_\# = g_\#$, then as loops, $[f] = [f \circ \xi] = [g \circ \xi] = [g]$. Hence f and g are homotopic.

(iii). The first part is straightforward. For the second part, use Exercise (i).

(iv). If p, q are the projection maps from $X \times Y$ to X, Y, respectively, then a map $f : Z \to X \times Y$ is completely determined by the two maps $p \circ f, q \circ f$. Keep using this repeatedly. It follows that the induced homomorphisms $p_\#, q_\#$ on the fundamental groups, have the same property for homomorphism of groups $G \to \pi_1(X \times Y)$. This means that $\pi_1(X \times Y)$ is a direct product of $\pi_1(X)$ and $\pi_1(Y)$.

(v). By the previous exercise, and since $\pi_1(\mathbb{S}^1)$ is abelian, we may regard $\pi_1(\mathbb{S}^1 \times \mathbb{S}^1)$ as a direct sum $\pi_1(\mathbb{S}^1 \times \mathbb{S}^1) = \mathbb{Z} \times \mathbb{Z}$ generated by $(1,0)$ and $(0,1)$ which are the elements represented by the loops $\xi_1 = \mathbb{S}^1 \times 1$ and $\xi_2 = 1 \times \mathbb{S}^1$. Since $f(z_1, 1) = (z_1, 1)$ and $f(1, z_2) = (z_2, z_2)$, it follows that $f_\#(1,0) = (1,0)$ and $f_\#(0,1) = (1,1)$.

(vi). Consider the map $\rho : \mathbb{I} \times \mathbb{I} \times \mathbb{I} \to \mathbb{I} \times \mathbb{I}$ defined by $(x, y, t) \mapsto (x, t/2 + (1-t)y)$.

(vii). Note that the boundary circle of $\ddot{\mathrm{M}}$ is the image of $I \times \{0, 1\}$. Under the deformation retract both these arcs traverse the entire central circle and hence the boundary circle wraps onto the central circle twice. Therefore on π_1, the induced homomorphism is given by multiplication by 2.

(viii). Let $\omega_i, i = 1, 2$ be two loops based at a point x_0 in a space X such that in $\pi_1(X, x_0)$, $[\omega_1]$ does not commute with $[\omega_2]$. Consider the map $\Theta : \mathbb{S}^1 \vee \mathbb{S}^1 \to X$ given by $\Theta(x, 1) = \omega_1(x)$ and $\Theta(1, x) = \omega_2(x)$. If $\xi_j : \mathbb{S}^1 \to \mathbb{S}^1 \vee \mathbb{S}^1$ denotes the two inclusion maps, then $\Theta_\#[\xi_j] = [\omega_j], j = 1, 2$. Therefore $[\xi_1], [\xi_2]$ do not commute.

(ix). If p is not in the image of f, then f can be thought of as a map $f : X \to \mathbb{S}^n \setminus \{p\}$, and $\mathbb{S}^n \setminus \{p\}$ is contractible.

(x). For any two points $x, y \in \mathbb{S}^n$ such that $y \neq -x$, there is a unique shortest geodesic path (great circle) joining them. Use them to define the homotopy.

(xi). Let U be a disc-like neighbourhood of q disjoint from p. Then $f^{-1}(q)$ is a compact subset of $\mathbb{S}^1$ covered by the connected components of the open set $f^{-1}(U)$ and hence is covered by finitely many open arcs. Under f the boundary points of these arcs are mapped onto the boundary circle of U. Join them by arcs inside the boundary of U. This way you get the loop f modified to avoid the point q. Since U is contractible, there is a path homotopy of each modification. Put them together to get the required homotopy.

(xiii). If not, this will imply $\mathbb{R}^n \setminus \{0\}$ is homeomorphic to $\mathbb{R}^2 \setminus \{0\}$. The first one has the homotopy type of $\mathbb{S}^{n-1}$ and the latter that of $\mathbb{S}^1$. They have non isomorphic fundamental groups, which is a contradiction.

(xiv). Let $H : X \times I \to Y$ be a homotopy from f to g. Put $\lambda(t) = H(x_0, t)$. Take $\phi = h_{[\lambda]}$.

(xv). (Caution: You have to keep track of the base points while writing down a solution to this problem.) If $f : X \to Y$, $g : Y \to X$ are homotopy inverses of each other, from the previous exercise, it follows that $(g \circ f)_\# : \pi_1(X, x_0) \to \pi_1(X, gf(x_0))$ is an isomorphism for every $x_0 \in X$. In particular, this implies that $f_\# : \pi_1(X, x_0) \to \pi_1(Y, f(x_0))$ is injective and $g_\# : \pi_1(Y, f(x_0)) \to \pi_1(X, gf(x_0))$ is surjective for all $x_0 \in X$. Reversing the role of f, g, we conclude that $g_\# : \pi_1(Y, y_0) \to \pi_1(X, g(y_0))$ is injective for every $y_0 \in Y$. Therefore $g_\# : \pi_1(Y, f(x_0)) \to \pi_1(X, gf(x_0))$ is an isomorphism. Since $g_\# \circ f_\#$ is an isomorphism, this implies $f_\# : \pi_1(X, x_0) \to \pi_1(Y, f(x_0))$ is an isomorphism.

§1.3.11 (page no. 26)

(i). Consider $h(x, y) = \frac{|x^2 - y^2|}{x^2 + y^2}$ for $(x, y) \neq (0, 0)$ and $h(0, 0) = 1$. Taking the limits along the lines $y = mx, x \to 0$ we get different values for different m.

(ii). The map $z \mapsto z^2$ from $G \to \mathbb{D}^2$ is surjective and factors down to define a homeomorphism $Y \to \mathbb{D}^2$.

(iii). (a) Writing $(x_1, x_2, x_3) = (r \cos \theta, x_2, r \sin \theta)$, where $r = \sqrt{1 - x_2^2}$, consider the map $(r \cos \theta, x_2, r \sin \theta) \mapsto (r \cos 2\theta, x_2, r \sin 2\theta), \ -\pi/2 \leq \theta \leq \pi/2$.

(b) This quotient map is the restriction of the standard quotient map $\mathbb{S}^2 \to \mathbb{P}^2$. We can now appeal to Theorem 1.3.6.

(iv). Think of $\mathbb{S}^1 \times \mathbb{S}^1$ as the quotient space of $\mathbb{I} \times \mathbb{I}$ in which opposite sides are identified appropriately, viz., $(0, y) \sim (1, y); (x, 0) \sim (x, 1)$, for all $x, y \in \mathbb{I}$. Now the additional identification by the group action renders $[1/2, 1] \times \mathbb{I}$ redundant. So, the resulting space is obtained by identifying the edges of the rectangle $[0, 1/2] \times \mathbb{I}$ as in Figure 13.6 Now use Exercises (ii) and (iii) above. (Look into Exercise 1.5.19.(ii) also.)

FIGURE 13.6. Torus modulo the diagonal $\mathbb{Z}_2$-action

(v). Let $q : X \to Z$ and Y be as in Munkres example. Let us denote by $Z \times_q Y$, the space $Z \times Y$ with the quotient topology whereas $Z \times Y$ will have the product topology. If the evaluation map $E : (Z \times_q Y)^Y \times Y \to Z \times_q Y$ is continuous then using Remark 1.3.2, it would follow that the identity map $Z \times Y \to Z \times_q Y$ is continuous which will contradict the conclusion in Example 1.3.10.

(vi). First prove the theorem for the case $A = q^{-1}(B)$ and when B is open or closed in Y. The rest of the argument is easy.

§1.5.19 (page no. 35)

(i). There is a homeomorphism of the cone $C\mathbb{S}^{n-1}$ which is identity on the base sphere and maps the apex to any given interior point of the disc $\mathbb{D}^n$.

(ii). (e) By (d) we may assume that A_1, A_2, A_3 are $(0,1), (0,0), (0,-1)$, respectively, on the boundary of G. Now use (ii).

(iii). Use Exercise 1.5.19.(ii) (b).

(iv). (a) $C_f \setminus Y$ is contractible.

(b) Since $X \times Y$ is locally contractible, the only points that need to be taken care of are the points of Y. For any open subset U of Y there are neighbourhoods in C_f which contain U as a deformation retract.

(v). Use Theorem 1.5.5.

(vi). Consider $\exp(t) = e^{2\pi\imath t} : [0,1] \to \mathbb{S}^1$. Then every loop $\omega : [0,1] \to X$ based at a corresponds to a map $f : \mathbb{S}^1 \to X$ under $f \mapsto f \circ \exp = \omega$. Moreover, a homotopy $H : \mathbb{S}^1 \times \mathbb{I} \to X$ relative to 1 corresponds to a path homotopy $G : \mathbb{I} \times \mathbb{I} \to X$ given by $G(t,s) = H(\exp(t), s)$

(vii). Consider the following chain of implications:

$f : \mathbb{S}^1 \to X$ is null homotopic for every f

$\iff g : (\mathbb{S}^1, 1) \to (X, x_0)$ is null homotopic relative 1 for every g and every $x_0 \in X$

$\iff$ every loop $\omega : [0,1] \to X$ at x_0 is null homotopic (as a path) for every $x_0 \in X$

$\iff \pi_1(X, x_0) = (1)$ for every $x_0 \in X$.

The only thing that you need to prove yet is the first implication, which is a consequence of Theorem 1.5.5.

(viii). Define $\Theta : \pi_1(X, x_0) \to [[\mathbb{S}^1, X]]$ as follows. Represent an element of $\pi_1(X, x_0)$ by a map $\mathbb{S}^1 \to X$ such that $f(1) = x_0$. Take $\Theta[f] = [[f]]$. Clearly, Θ is well defined. Given any $g : \mathbb{S}^1 \to X$ join x_0 to $g(1)$ by a path λ and then $[\lambda * g * \lambda^{-1}] \in \pi_1(X, x_0)$ and $\Theta[\lambda * g * \lambda^{-1}] = [[g]]$. This shows surjectivity. If $[f_1], [f_2] \in \pi_1(X, x_0)$ are freely homotopic in X, say $H : \mathbb{S}^1 \times I \to X$ is such a homotopy, then put $\lambda(t) = H(1, t)$. Then λ is a loop at x_0 and we have $\lambda * f_2 * \lambda^{-1} \simeq f_1$ which means $[f_1], [f_2]$ are conjugates in $\pi_1(X, x_0)$. Now for the last part, $\pi_1(X, x_0)$ is trivial iff the set of conjugacy classes in $\pi_1(X, x_0)$ is a singleton iff $[[\mathbb{S}^1, X]]$ is a singleton iff every map $f : \mathbb{S}^1 \to X$ is null homotopic.

(ix). $h_{[\tau]} = h_{[\lambda]} \iff h_{[\tau]} \circ h_{[\lambda]}^{-1} = Id \iff h_{[\tau * \lambda^{-1}]} = Id \iff [\tau * \lambda^{-1}]$ is in the centre of $\pi_1(X, a)$. Taking $[\lambda] = [c_a]$, this implies that every element of $\pi_1(X, a)$ is in the centre. Converse is easy.

§**1.6.15** (page no. 40)

(i). A retract of a Hausdorff space is closed.

(iii). If $r : X \times \mathbb{I} \to X \times 0 \cup A \times \mathbb{I}$ is a retraction, consider $Id_Z \times r$.

(v). Given the HED for $B \hookrightarrow C$, by restriction, you get an HED for $A \cap B \hookrightarrow A$. This yields the extension on $A \times \mathbb{I}$ which can be patched up with the map on $B \times \mathbb{I}$ to get the map on $C \times \mathbb{I}$.

(vi). This is a special case of the previous exercise.

(vii). If M_f is Hausdorff, the subspaces X, Y are also Hausdorff. For the converse part, note that since $X \times \mathbb{I}$ is Hausdorff, there is no problem if the two points are of the form $[x_1, t_1], [x_2, t_2]$ for $t_1, t_2 < 1$. Now suppose the second point is $[y]$ for some $y \in Y$. Then, under the quotient map, the image of $X \times [0, r)$ and $X \times (r, 1] \cup Y$ will give required disjoint neighbourhoods. Similarly, for points of the form $[y_1], [y_2]$ where $y_1 \neq y_2 \in Y$, first choose disjoint neighbourhoods U_i containing y_i and take the images of $f^{-1}(U_1) \times [0, 1] \cup U_1$ and $f^{-1}(U_2) \times [0, 1] \cup U_2$.

(viii). Hint: Use the following fact: Given any subset A of X and point $z \in Z \setminus X$, there exist disjoint open sets U, V of Z such that $A \subset U$, and $z \in V$.

§**1.7.14** (page no. 43)

(i). Use exponential correspondence (see Theorem 1.3.1).

(ii). Given any two points $e_1, e_2 \in p^{-1}(b)$, take a path $\omega : \mathbb{I} \to E$ joining them, take a null homotopy $H : \mathbb{I} \times \mathbb{I} \to B$ of $p \circ \omega$ and lift it.

(iii). Take $H(\omega, t)(s) = \omega(ts)$.

§1.8.20 (page no. 53)

(iv). In **Ens**, the pushout is just the quotient set of the disjoint union $X \sqcup Y$ by the relation $f(a) \sim g(a)$, $a \in A$. In **Ab**, it is the quotient of $X \oplus Y$ by the subgroup generated by $\{f(a) - g(a), \quad a \in A\}$. In **Gr**, it is the quotient of the free product $X * Y$ by the normal subgroup generated by $\{f(a)g(a)^{-1}, \quad a \in A\}$.

(v). If $C_n \approx \mathbb{Z}$ denotes the nth group in the system, we define the homomorphisms $f_n : C_n \to \mathbb{Z}[1/2]$ by the formula $f_n(1) = \frac{1}{2^n}$. Let $\alpha_n : C_n \to C_{n+1}$ be the multiplication by 2. Check that $f_{n+1} \circ \alpha_n = f_n$. Now given any abelian group G and a sequence of homomorphisms $g_n : C_n \to G$ such that $g_{n+1} \circ \alpha_n = g_n$, the only homomorphism $g : \mathbb{Z}[1/2] \to G$ such that $g \circ f_n = g_n$ is given by

$$g\left(\frac{m}{2^n}\right) = g_n(m).$$

§1.9 (page no. 54)

1. (a) $\mathbf{x} \mapsto \frac{\mathbf{x}}{\|\mathbf{x}\|}$.

(b) Every element of $\mathbb{I}^n \setminus \{p\}$ is uniquely expressed as $t(\mathbf{x} - p) + p$ where, $\mathbf{x} \in \partial \mathbb{I}^n$ and $0 < t \leq 1$. The map $t(\mathbf{x} - x) + p \mapsto \mathbf{x}$ gives a strong deformation retract.

(c) Use the above exercise for $n = 2, p = (1/2, 1/2)$ and the quotient map $(t, s) \mapsto (e^{2\pi \imath t}, e^{2\pi \imath s})$.

(d), (e) Just as in (c), in each case, consider the appropriate quotient map.

5. Polar Decomposition Theorem (See Theorem 9.1.10) in [Shastri, 2011])

Every element of $GL(n, \mathbb{C})$ can be written in a unique way as a product $A = UH$, where U is unitary and H is positive definite Hermitian. The decomposition defines a diffeomorphism $\varphi : GL(n, \mathbb{C}) \to U(n) \times \mathbb{R}^{n^2}$. Furthermore, if A is real then U is real orthogonal and H is real symmetric and hence φ restricts to a diffeomorphism $\varphi : GL(, \mathbb{R}) \to O(n) \times \mathbb{R}^{n(n-1)/2}$.

It follows that $p \circ \varphi : GL(n, \mathbb{R}) \to O(n)$ is a deformation retraction.

6. This exercise can be solved in several ways: Using the above exercise, it is enough to prove that $SO(n)$ is path connected. Induct on n. Given $A \in SO(n)$ suppose first that $A(e_n) = e_n$. Then it follows that A can be considered as an element in $SO(n-1)$ and by induction, there is a path in $SO(n-1)$ from A to Id. In the general case, put $Ae_n = v$ which is a unit vector. Let $R_{t\theta}$ be the rotation in the plane spanned by $\{e_n, v\}$ (or if $v = -e_n$, in any plane containing e_n) through an angle $t\theta$ from v to e_n. Then $R_{t\theta} \circ A$ is a path in $SO(n)$ joining A to $B = R_\theta A$ where $Be_n = e_n$. Clearly $B \in SO(n-1)$. Fill in the details yourself.

Alternatively, use induction and the fact that if G is a topological group and H is a connected subgroup such that the quotient space is connected then G is connected.

7. Use the above exercise. A path in $O(n)$ defines a homotopy $\mathbb{S}^{n-1} \times \mathbb{I} \to \mathbb{S}^{n-1}$.

8. There is a unique great circular arc joining two points on $\mathbb{S}^{n-1}$ which are not anti-podal.

12. Given $a \in A$ and an open subset U of X such that $x \in U$, we can find a neighbourhood V of x in X such that $V \subset U$ and V is (path) connected. This implies $V \cap A$ is a neighbourhood of x in A such that $V \cap A \subset U \cap A$. Now if $r : X \to A$ is a retraction, we can choose V such that $r(V) \subset A \cap U$. But $r(V)$ is (path) connected and $V \cap A \subset r(V)$ and hence $r(V)$ is a neighbourhood of x in A.

14. The combspace is not locally (path) connected. Now use Corollary 1.6.3.

15. Let $H : X \times \mathbb{I} \to X$ be a strong deformation retraction onto p. Since $\{p\} \times I$ is contained in the open set $H^{-1}(U)$, it follows that there is an open neighbourhood V of p such that $V \times \mathbb{I} \subset H^{-1}(U)$ (Wallman's theorem). This means in particular, $V \subset U$ and $H|_V : V \times \mathbb{I} \to U$ is a homotopy of the inclusion with the constant map.

17. Suppose $H : D \times \mathbb{I} \to D$ is a homotopy of the Id_D with some constant map. WLOG, we may assume that $H(y, 1) = p$ for all $y \in D$. Consider the points $q = (0, 1), r = (0, -1)$

and put

$$h_p(t) = H(p, t); \quad A = h_p^{-1}\{q\}; \quad B = h_p^{-1}\{r\}.$$

Claim I: $A \neq \emptyset \neq B$.

Claim II: Inf A = inf B.

But then $(0, 1) = h_p(\text{Inf } A) = h_p(\text{Inf } B) = (0, -1)$ which is absurd.

Try to prove these two claims yourself. Only after you have given enough time read further.

Proof of Claim I: Put $h_n(t) = H((1/n, 0), t)$. Then $h_n(0) = (1/n, 0)$ and $h_n(1) = p$. For $n > 0$ these two points are in different path components of $D \setminus \{r\}$. Therefore, there exist $t_n \in I$ such that $h_n(t_n) = r$. Passing to a subsequence, if necessary, we may assume that $t_n \to t_0$. Then $((1/n, 0), t_n) \to (p, t_0)$ and hence $H(p, t_0) = \lim H(1/n, 0, t_n) = r$. This proves that $B \neq \emptyset$. The argument is similar for $n < 0$ and gives $A \neq \emptyset$.

This should give you some idea how to prove Claim II. Once again, try to prove Claim II by yourself and read further only after enough trials.

Proof of Claim II: It is enough to prove that given $t_0 \in A$ there exists $t_1 \in B$ such that $t_1 < t_0$. It follows that Inf $B \leq$ Inf A. By symmetry, this will prove II.

So, let $t_0 \in A$. By continuity, for large n, $h_n(t_0) \in B_{1/2}(q)$. For similar reasons as in Claim I, there exist $s_n < t_0$ such that $h_n(s_n) = r$. Passing to a subsequence, etc., as before, this gives $t_1 \leq t_0$ such that $h_p(t_1) = r$.

19. Put $A(1/n) = \{(x, \sin \frac{\pi}{x}) \in A_1 : x \leq 1/n.\}$. First claim that given any path $\omega : I \to A_1$ there exists n such that $\omega(\mathbb{I}) \subset C \setminus A(n)$. This will first of all prove that L is not path connected. The same thing will also imply that given any path $\omega : \mathbb{I} \to C$ there is n such that $\omega(\mathbb{I}) \subset C \setminus A(n)$. Since $C \setminus A(n)$ is easily seen to be contractible, it follows that $\pi_1(C) = (1)$. Finally, the above claim is the same as saying that there is n such that $(1/m, \sin m\pi) \notin \omega(\mathbb{I})$ for $m \geq n$. If this is not the case, we can find a sequence of points $t_n \in \mathbb{I}$ monotonically converging to say $t_0 \in$ such that $\omega(t_n) = (1/m_n, \sin m_n \pi)$ with $m_n \to \infty$. Using the intermediate value theorem you can now show that ω cannot be continuous at t_0.

20. In fact the map $z \mapsto z/|z|$ is a strong deformation retraction.

21. The map $\bar{q}_1 : \mathbb{R} \to \mathbb{R}$ given by $t \mapsto nt$ is a lift of q_1. Therefore $\deg q_1 = \bar{q}_1(1) = n$. Likewise $\deg q_2 = -1$.

22. Every polynomial can have at most as many roots as its degree. Therefore, we can choose $r >> 0$ such that all the possible roots of the polynomial $p(z) - z^n$ are inside the disc $\{z : |z| < r/2\}$. This means that on the circle S_r the map $(z, t) \mapsto (1 - t)p(z) + tq(z)$ defines the required homotopy.

23. From the previous exercise, it follows that $q|_{S_r} \sim p|_{S_r}$ is null homotopic. This implies $q|_{\mathbb{S}^1}$ is also null homotopic. This contradicts the fact that the degree of q is $n > 0$.

24. (a) A lift $g : \mathbb{R} \to \mathbb{R}$ of f will have the property that $g(t + 1/2) - g(t) = (2n + 1)/2$ for all t. Therefore

$$g(t + 1) - g(t) = g(t + 1/2 + 1/2) - g(t + 1/2) + g(t + 1/2) - g(t) = 2n + 1.$$

25. Put $f_i(x) = d(x, F_i)$. Apply Borsuk–Ulam to conclude that there exist $x \in \mathbb{S}^2$ with

$$(d(x, F_1), d(x, F_2)) = (d(-x, F_1), d(-x, F_2)).$$

If $\pm x \in F_1$ or $\pm x \in F_2$ then we are done. If not, this implies both $x, -x \in F_3$.

26. Notice that the size, shape or taste of the pieces of ham and bread does not matter to us. Even the way these three objects are placed in the 3-dimensional Euclidean space is immaterial provided the knife is large enough. Thus, the mathematical model of the problem is the following: Given three bounded domains (connected) in $\mathbb{R}^3$ does there exist a plane which cuts each of them into equal halves?

Given any solid (bounded measurable set) in $\mathbb{R}^3$ and a unit vector v, there is a unique $r \in \mathbb{R}$ such that the plane perpendicular to v and passing through rv cuts the solid into two equal halves. To prove this claim use intermediate value theorem. Next using elementary calculus, prove that the assignment $v \mapsto r$ is continuous.

Now, let B_1, B_2, B_3 denote (the two pieces of bread and the ham) the three solids and let f_1, f_2, f_3 be the three maps $\mathbb{S}^2 \to \mathbb{R}$ so that the plane perpendicular to v and passing through $f_i(v)v$ cuts B_i into two equal halves. Observe that $f_i(-v) = -f_i(v)$ for $i = 1, 2, 3$. Apply Borsuk–Ulam to the map $v \mapsto (f_1(v) - f_2(v), f_1(v) - f_3(v))$. Fill in the rest of the details.

29. Use (d) of Exercise 28 repeatedly along with the associativity of the join operation.

30. (a) First of all we can show that the Id_X is homotopic to the constant map $x \mapsto p$. Let $H : X \times \mathbb{I} \to X$ be such a homotopy. Consider the loop $t \mapsto H(p, t)$ in X. Since X is contractible, we get a homotopy $G : \mathbb{I} \times \mathbb{I} \to X$ such that $G(t, 0) = h(p, t), G(t, 1) = G(0, s) = G(1, s) = p$ for all $t, s \in \mathbb{I}$. Define $f : p \times \mathbb{I} \times \mathbb{I} \cup X \times \mathbb{I} \times 0 \to X$ to be such that $f(p, t, s) = G(t, s)$ and $f(x, t, 0) = H(x, t)$. Then f is continuous. Now $\{p\} \hookrightarrow X$ is a cofibration implies so is $\{p\} \times \mathbb{I} \hookrightarrow X \times \mathbb{I}$. Therefore there is a retraction $r : X \times \mathbb{I} \times \mathbb{I} \to p \times \mathbb{I} \times \mathbb{I} \cup X \times \mathbb{I} \times 0$. Define $F : X \times \mathbb{I} \to X$ as follows:

$$G(x, t) = \begin{cases} f \circ r(x, 0, 3t), & 0 \le t \le 1/3; \\ f \circ r(x, 3t - 1, 1), & 1/3 \le t \le 2/3; \\ f \circ r(x, 1, 3 - 3t), & 2/3 \le t \le 1. \end{cases}$$

Verify that G is a homotopy of Id_x with the constant map $x \mapsto p$ relative to $\{p\}$.

(b) We have seen in Example 1.4.6 that $(0, 1)$ is not a SDR of the comb space and the combspace is contractible. Now appeal to the above exercise.

(c) The double combspace (Exercise 17 has one degenerate point $(0, 0)$. In the zig-zag zebra (Exercise 16), every point is degenerate. On the topologists' sign loop (Exercise19), every point on the line segment $0 \times [-1, 1]$ is degenerate. There are other examples such as Hawaiian ring, etc., which you will see later. The proofs of all these use just one argument.: A non degenerate point is a DR of a neighbourhood (see Remark 1.6.8) and from Exercise 1.9.30.(a), it is a SDR of its neighbourhood. Combine this with Exercise 1.9.15.

31. ((i) The contraction $([x, t], s) \mapsto [x, ts]$ of the unreduced cone factors down to define a contraction of the reduced cone.

(ii) Take $(\mathbf{v}, t) \mapsto t\mathbf{v} + (1 - t)p$ to get a homeomorphism of the reduced cone $(C(\mathbb{S}^{n-1}, p)$ with $(\mathbb{D}^n, p)$. Working with upper and lower hemispheres separately gives the corresponding statement for reduced suspensions.

(iii) Just as in Theorem 1.5.5, use (ii).

41. Given any two points $x, y \in \underset{\to}{\mathrm{dlim}} X_j$, there exist p, q and $x_p \in X_p, y_q \in X_q$ such that $\alpha_p(x_p) = x, \alpha_q(y_q) = y$. Choose $r > p, q$ and a path ω in X_r joining $f_{p,r}(x_p)$ and $f_{q,r}(y_q)$. Then $\alpha_r(\omega)$ is a path in $\underset{\to}{\mathrm{dlim}} X_j$ from x to y.

42. ((b) Any sequence $\{a_n\}$ of positive integers in which, given any prime p and a $m \ge 1$, there exist $n = n(p, m)$ such that p^m divides a_n. For example, if q_j denotes the jth prime, then $a_n = (q_1 q_2 \cdots q_n)^n$ will do.

§2.1.37 (page no. 74)

Put $M = \{x_1, \ldots, x_n\}$. There exist real numbers α_i not all zero such that $\sum_i \alpha_i = 0$ and $\sum_i \alpha_i x_i = 0$. Put $M_1 = \{x_i \in M : \alpha_i > 0\}$ and $M_2 = M \setminus M_1$. If t is the sum of all those α_i which are positive, put $x = (\sum_{x_i \in M_1} \alpha_i x_i)/t$. Check that x belongs to conv $M_1 \cap$ conv M_2.

§2.1.44 (page no. 82)

We have the regular n-simplex Δ_n inscribed inside the sphere in the plane $L := \sum_i x_i = 1$ with centre $\frac{1}{n+1}(1, 1, \ldots, 1)$ and radius $\sqrt{\frac{n}{n+1}}$. Translate this centre to the origin, rotate the plane L onto the plane $x_n = 0$. This amounts to writing down the inverse of an orthogonal matrix with the last row $(\frac{1}{\sqrt{n+1}}, \ldots, \frac{1}{\sqrt{n+1}})$. Finally, scale-up by a factor of $\sqrt{\frac{n+1}{n}}$.

§2.2.23 (page no. 89)

(i). For any closed set F in the metric topology of $\mathbb{R}^\infty$, $\mathbb{R}^n \cap F$ is closed in the metric topology of $\mathbb{R}^n$ for all n. Hence F is closed in the weak topology in $\mathbb{R}^\infty$. Take $\{\mathbf{e}_1, \mathbf{e}_2/2, \ldots, \mathbf{e}_n/n, \ldots, \}$ to see the converse does not hold.

(ii). Choose a locally finite refinement $\mathcal{V}$ of $\mathcal{U}$ and a partition of unity $\{\eta_j : j \in J\}$ subordinate to it. Choose a well order on the indexing set J. Define $\theta_j(x) = \sum_{i \leq j} \eta_i(x)$.

§2.4.7 (page no. 95)

(i). Given any point $x \in X$ and an open subset U of X containing x, take the open cell E such that $x \in E$. Since E is locally compact, you can take a neighbourhood A of x in $E \cap U$ such that $\bar{A}$ is compact. Now in extending this neighbourhood A to a neighbourhood W of x in X itself, in the Proposition 2.2.5, you have to deal with only finitely many cells that meet E. In each of these steps, you get an extended neighbourhood which has compact closure and hence the final extended neighbourhood W also has compact closure.

Converse is a bit subtle. Assume X is locally compact. Let E be a closed cell. Cover E with finitely many open sets K_i whose closures are compact. Now for every closed cell $\bar{e}$ which meets E, the open cell e should meet $\overline{\cup_i K_i}$ which is compact. Therefore there can be only finitely many such cells e.

(i). Use cellular approximation.

(ii). Fix a vertex v_0. Let $A_{0,n}$ be the set of all vertices of X which can be joined to v_0 by an arc in $X^{(1)}$ which intersects at most n, 1-cells. Local finiteness implies each $A_{0,n}$ is finite. By the previous exercise, $X^{(0)} = \cup_n A_{0,n}$. Hence $X^{(0)}$ is countable. Inductively, this implies $X^{(1)}, X^{(2)}, \cdots$ are countable and hence X is countable.

§2.8.11 (page no. 112)

(i). Take the standard simplex $\Delta_n \subset \mathbb{R}^{n+1}$ and project radially from the barycentre c onto the sphere

$$\{x \in \mathbb{R}^n : \sum_i x_i = 1, \ \|x - c\| = 1\}.$$

(ii). In this case, the nerve is isomorphic to K itself.

(iii)b. $f_i(K \star v) = f_i(K) + f_{i-1}(K), i \geq 1$. Now take the alternating sum.

(iii)c. $St_K(F) = Lk_K(F) \star F$ which is a cone.

(iii)d. In view of Exercise (iii)a. above, it is enough to prove this for $K = \Delta_n$. Now $\mathrm{sd}\,(\Delta_n) = (\mathrm{sd}\,\dot{\Delta}_n) \star \{\beta\}$ and hence from Exercise (iii)b above, $\chi(\mathrm{sd}\,(\Delta_n)) = 1 = \chi(\Delta_n)$.

§2.9.21 (page no. 119)

(i). Note that both $f(\alpha)$ and $|\phi|(\alpha)$ belong to $|F_2|$ which is convex. If F_2' is another such face of K_2 then both $f(\alpha)$ and $|\phi|(\alpha)$ would belong to $|F_2 \cap F_2'|$, $F_2 \cap F_2'$ is a simplex of K_2 and the convex combinations have the same meaning in all three simplexes. By the property of the weak topology, any function $h : |K| \times \mathbb{I} \to |K_2|$ is continuous iff it is so restricted to $|F| \times \mathbb{I}$ for each simplex F in K.

(iii). Let K be a simplicial complex such that $|K| = \mathbb{S}^1$ and $\varphi : K \to K$ be a simplicial approximation to $f : \mathbb{S}^1 \to \mathbb{S}^1$. Let v be a vertex of K. Now there are two points u_1, u_2 such that $f(u_1) = v = f(u_2)$. If φ is a simplicial approximation to f then $|\varphi|(u_i) = v$. If u_i is a vertex of K, it is fine. Otherwise, it belongs to the open simplex $\langle s \rangle$ for some 1-simplex $s = \{v_1, v_2\}$ of K. But then it follows that $\phi(v_1) = \phi(v_2) = v$. Thus, in any case, we conclude that there are at least two distinct vertices mapped onto v by φ. Therefore φ is mapping of the vertex set of K onto itself which is not one-one. Therefore, the vertex set of K is infinite, which is absurd.

§2.10.8 (page no. 122)

Of course $St_{St(F)}(G) \subset St(F) \cap St(G)$. The equality does not hold in general: take $K = \dot{\Delta}_2$ and $F = \{0\}, G = \{1\}$.

§2.11 (page no. 122)

1. (a) Use induction on the number of vertices. Given a convex polytope P with vertex set V, choose any vertex $v_0 \in V$, put $V' = V \setminus \{v_0\}$, and consider the convex polytope $P' = \operatorname{conv} V'$. By induction P' has a triangulation with the vertex set V'. (The visible complex $\mathcal{V}(P', v_0)$ gets a triangulation as a subcomplex. (It is indeed a triangulated facet of P', though this fact is not needed.) Check that P is the union of P' and the cone over $\mathcal{V}(P', v_0)$ with apex v_0 to get a triangulation of P with the vertex set V.

(b) In order to distinguish the two factors, let us denote the vertices of the first Δ_m by $U = \{u_0, \ldots, u_m\}$ and those of the second Δ_n by $V = \{v_0, \ldots, v_n\}$. Choose the strict order on $U \times V$ viz., $(u, v) < (u', v')$ iff $u < u'$ and $v < v'$. Consider the set of all maximal chains in $U \times V$ as the maximal faces of a simplicial complex K. We claim that the identity map $U \times V \to U \times V$, extends linearly on each simplex of $|K|$ (with respect to the standard linear structure on $|\Delta_m| \times |\Delta_m|$) and defines a homeomorphism onto $|\Delta_m| \times |\Delta_n|$. Try to prove this yourself and only after enough trials, read further.

The first thing you have to verify is that every maximal chain in $U \times V$ is affinely independent in $|\Delta_m| \times |\Delta_n|$. This is easy. Now given an element $x = \sum_{ij} s_{ij}(u_i, v_j) \in |F|$ where F is a maximal simplex in K, consider the matrix (s_{ij}). Let $C_1, \ldots, C_n$ be the column sums and $R_1, \ldots, R_m$ be the row-sums. Consider the point $h(x) = (\sum_i R_i u_i, \sum_j C_j v_j)$. Check that this is the linear extension of the inclusion map of the vertices of F. It remains to see why h is one-one and onto. Instead one can describe the inverse map directly. So, let $y = (\sum_i \alpha_i u_i, \sum_j \beta_j v_j)$ be an arbitrary point of $|\Delta_m| \times |\Delta_n|$. We shall first construct the maximal chain in $U \times V$ to which this point has to be sent: All maximal chains start at (u_0, v_0) and end at (u_m, v_n). (They are in one-one correspondence with the lattice paths from (u_0, v_0) to (u_m, v_n).) To begin with check whether $\alpha_0 \leq \beta_0$. If so, take the next vertex in the chain to be (u_1, v_0), otherwise take (u_0, v_1). Having arrived at some vertex (u_k, v_l) check whether $\sum_{i \leq k} \alpha_i \leq \sum_{j \leq l} \beta_j$. If so, take the next vertex to be (u_{k+1}, v_l); otherwise take the vertex (u_k, v_{l+1}). Of course, when you have arrived at (u_m, v_n) you stop. The affine coordinates of the point $h^{-1}(y) = \sum_{ij} s_{ij}(u_i, v_j)$ are obtained by solving the system of linear equations given by comparing the coefficients of u_i and v_j. (Carry out the case when $m = 2$ and $n = 1$.)

(c) Order the vertex sets U, V of K, L. For $|K| \times |L|$ choose the vertex set to be $U \times V$ with the strict order: $(u, v) \leq (u', v')$ iff $u \leq u'$ and $v \leq v'$. Now declare a finite subset $F \subset U \times V$ as a simplex in $K \times L$ iff its projections on U and V are simplexes in K and L, respectively, and F is a chain. The rest of the proof follows from the previous exercise.

4. (a) The simplicial structure of Δ_2 thought of as a CW-structure quotients down to a CW-structure on D.

(b) Take the second barycentric subdivision of Δ_2 and the induced structure on the quotient space.

(c) The attaching map of the 2-cell for the CW-structure in (a) is homotopic to the identity map of $\mathbb{S}^1$. So, we you can appeal to Theorem 1.6.10.

(d) This is very difficult for me.

6. The answer is 'not necessarily'. In Example 2.8.4 take $f = |\varphi|$ and see that

$$f\left(\frac{2}{7}\mathbf{e}_1 + \frac{2}{7}\mathbf{e}_2 + \frac{3}{7}\mathbf{e}_3\right) = \frac{4}{7}\mathbf{e}_1 + \frac{3}{7}\mathbf{e}_2 = \frac{1}{7}\mathbf{e}_1 + \frac{6}{7}\left(\frac{\mathbf{e}_1 + \mathbf{e}_2}{2}\right),$$

which is in the closed simplex $|\{\mathbf{e}_1, \widetilde{\{\mathbf{e}_1, \mathbf{e}_2\}}\}|$ whereas,

$$|\operatorname{sd} \varphi| \left(\frac{2}{7}\mathbf{e}_1 + \frac{2}{7}\mathbf{e}_2 + \frac{3}{7}\mathbf{e}_3 \right) = |\operatorname{sd} \varphi| \left(\frac{6}{7}\frac{\mathbf{e}_1 + \mathbf{e}_2 + \mathbf{e}_3}{3} + \frac{1}{7}\mathbf{e}_3 \right) = \frac{3}{7}\mathbf{e}_1 + \frac{4}{7}\mathbf{e}_2,$$

is not in $|\{\mathbf{e}_1, \widetilde{\{\mathbf{e}_1, \mathbf{e}_2\}}\}|$.

7. First, observe that under the given hypothesis, we have $f|_{\partial \mathbb{D}^n} : \partial \mathbb{D}^n \to \mathbb{D}^n \setminus \{0\}$ is homotopic to the inclusion map. Now if the conclusion does not hold, then it means that some point $p \in B_{1-\epsilon}(0)$ is not in the image. This means $f|_{\partial \mathbb{D}}$ is null homotopic in $\mathbb{D}^n \setminus \{p\}$. This in turn is equivalent to saying that $Id_{\mathbb{S}^{n-1}}$ is null homotopic, which is absurd.

11. Let us construct the simplicial complex K inductively. Clearly X_0 is a discrete space and so can be thought of as a 0-dimensional simplicial complex K_0. Now all the attaching maps are simplicial and hence by the mapping cone construction, $X^{(1)}$ is triangulable so that $X^{(0)}$ is a subcomplex but not necessarily the whole of 0-dimensional skeleton. Now replace all the attaching maps by simplicial ones homotopic to the original ones and let X_2' be the space obtained by attaching 2-cells via these simplicial maps. Then $(X_2', X^{(1)})$ is homotopy equivalent to $(X^{(2)}, X^{(1)})$ and by mapping cone argument X_2' can be triangulated extending the triangulation on $X^{(1)}$. Iteration of this step yields a sequence of simplicial complexes,

$$X^{(0)} = K_0 \subset K_1 \subset \cdots \subset K_r \subset$$

and maps $(f_i, f_{i-1}) : (|K_i|, |K_{i-1}|) \to (X^{(i)}, X^{(i-1)})$ which are homotopy equivalences. Take $K = \cup_i K_i$.

13a. Consider the right-shift map $\rho : \mathbb{R}^\infty \to \mathbb{R}^\infty$ given by $\rho(\mathbf{e}_i) = \mathbf{e}_{i+1}$. Consider the map

$$(x, t) \mapsto (\cos \pi t/2)x + (\sin \pi t/2)\rho(x).$$

This defines a homotopy (isotopy) of Id and ρ on $\mathbb{S}^\infty$.

13b. Take $T = (\rho^k, \ldots, \rho^k) : (\mathbb{S}^\infty)^k \to (\mathbb{S}^\infty)^k$, where ρ is as in the previous exercise and ρ^k is the composite of ρ with itself taken k-times. This T is homotopic to $Id_{\mathbb{S}^\infty}$ by a homotopy defined as in the previous exercise. Indeed, the same homotopy restricts to a homotopy of $T : V_k \to V_k$ with the identity. Finally, let $E = (\mathbf{e}_1, \ldots, \mathbf{e}_k) \in V_k$ and consider

$$X \mapsto (\cos \theta)T(X) + (\sin \theta)E$$

to get a homotopy of T with the constant map.

14i. We have already seen that each element X of G_k belongs to a unique Schubert cell $E(X(\sigma))$. Since $\bar{E}'(\sigma)$ is compact, it follows that $\psi(\bar{E}'(\sigma)) = \bar{E}(\sigma)$. We need to show that $\psi\partial(\bar{E}'(\sigma))$ is contained in the union of Schubert cells of lower dimensions. Let $L \in \bar{E}(\sigma) \setminus E(\sigma)$. Then there exists a basis $X = (x_1, \ldots, x_k) \in \bar{E}'(\sigma)$ for L. This means $x_i \in \mathbb{R}^{\sigma_i}$ and $\dim L \cap \mathbb{R}^{\sigma_i} \geq i$. Therefore, the Schubert symbol $\sigma' := \sigma(L)$ associated with L satisfies $\sigma_i' \leq \sigma_i$. Since X is not in $E(\sigma)$ at least one of the inequality is strict and hence $d(\sigma') < d(\sigma)$. Finally, since we know that the topology on G_k is compactly generated, it coincides with the weak topology with respect to the closed cover $\{\bar{E}(\sigma)\}$.

15. Inductively it suffices to prove that $|\dot{F}| \setminus |G|$ is a SDR of $|F|$. Use construction as in Example 1.5.11 to 'push' through the face G.

16. Of course the dunce hat is triangulable but there will not be any free face so that we can begin collapsing.

17. Starting with a cube (preferably made up of play-dough) keep pushing one finger each on two opposite faces and carve out the duplex igloo. Reversing this argument, we have just proved that the duplex-igloo $X \subset \mathbb{I}^3$ is a deformation retract which means X is contractible.

§**3.1.7** (page no. 129)

(i). If $a, b \in \overline{X}$ are such that $p(a) \neq p(b)$, then we can take disjoint open sets U, V in X such that $a \in p^{-1}(U), b \in p^{-1}(V)$. If $p(a) = p(b)$ then we can take an evenly covered open set around $p(a) = p(b)$ and get disjoint open sets V_1, V_2 such that $a \in V_1, b \in V_2$. For the converse part, assuming that p is a finite cover, given $x_1 \neq x_2 \in X$, let U_i be an evenly covered neighbourhood of x_i. Using Hausdorffness of $\overline{X}$, we can get open sets U_{ij} around each point of $p^{-1}(x_i)$ such that $p(U_{ij}) \subset U_i$, $p|_{U_{ij}}$ is one-one and $U_{1,j} \cap U_{2,k} = \emptyset$ for every j, k. Take $V_i = \cap_j p(U_{ij})$, $i = 1, 2$.

(ii). See Definition 3.5.6 and Theorem 3.5.8. Verify that we have an action of $\mathbb{Z}$ which is even. (For a point (x, y) with $x \neq 0$, choose $0 < \epsilon < |x|/3$, and the neighbourhood $W = (x - \epsilon, x + \epsilon) \times V$, where V is any open interval around y. Then verify the $gW \cap W = \emptyset$ for $g \neq 1$. Similarly for $y \neq 0$. This proves that the quotient map is a covering projection.) For any neighbourhood U of $[(1, 0)]$, there is n_0 such that $q^{-1}(U)$ will contain all points of the form $(1, 1/2^n)$ for $n > n_0$. Similarly, for any neighbourhood V of $[(0, 1)]$, there is some m_0 such that $q^{-1}(V)$ contains all points of the form $(1/2^n, 1)$ for $n > m_0$. So, for $n > \max\{m_0, n_0\}$ we see that $(1, 1/2^n) \sim (1/2^n, 1) \in q^{-1}(U) \cap q^{-1}(V)$ which implies $U \cap V \neq \emptyset$.

(iv). Given $\bar{x} = s(x)$, first choose a neighbourhood U of $\bar{x} \in \overline{X}$ such that $p : U \to p(U)$ is a homeomorphism. By continuity of s, it follows that there is a neighbourhood V of x such that $s(V) \subset U$. But then $s(V) = p^{-1}(V) \cap U$ and hence is open in $\overline{X}$. This proves that $s(X)$ is open in $\overline{X}$ in both the cases (a), (b). To show that $s(X)$ is closed, start with $z \in \overline{X} \setminus s(X)$ put $x = p(z), \bar{x} = s(x)$. Find disjoint neighbourhoods U, U' of $\bar{x}, z$ respectively, so that $p : U \to W$ and $p : U' \to W$ are homeomorphisms (possible under either of the hypotheses (a), (b)). As before, by continuity of s, there is an open set $V \subset W$ such that $s(V) \subset U$. Check that $z \in p^{-1}(V) \cap U'$, and $p^{-1}(V) \cap U' \cap s(X) = \emptyset$.

§**3.2.4** (page no. 132)

Indeed, using lifting criterion, first prove that there is a map $\Phi : p^{-1}(X \times 0) \times \mathbb{I} \to Y$ such that $p \circ \Phi(z, t) = (p(z), t)$. Verify that Φ is a homeomorphism.

§**3.3.15** (page no. 135)

Keep using homotopy lifting property.

§**3.4.25** (page no. 142)

(i). By lifting criterion, the identity map $X \to X$ can be lifted, i.e., there is a continuous section $s : X \to \overline{X}$ for p. Now use Exercise 3.1.7.(iv).

ii.(c). Recall that $\mathbb{S}^n$, $n \geq 2$ is simply connected (see Corollary 1.2.32). Now use Corollary 3.4.8.

ii.(e). For any map $f : \mathbb{P}^n \to \mathbb{S}^1$, $n \geq 2$, $f_{\#} : \mathbb{Z}_2 = \pi_1(\mathbb{P}^n) \to \pi_1(\mathbb{S}^1) = \mathbb{Z}$ is trivial and hence by lifting criterion, there is a lift $\hat{f} : \mathbb{P}^n \to \mathbb{R}$ such that $\exp \circ \hat{f} = f$. Since $\mathbb{R}$ is contractible, $\hat{f} \simeq 0$ which implies $f \simeq 0$.

(iv). Infinite product of covering projections is a covering projection iff all but finitely many of them are trivial. This is because the basic open subsets of the product $\prod_j X_j$ are of the form $\prod_j U_j$ wherein U_j is open in X_j for all j and all but finitely many U_j are equal to the whole space X_j.

(v). At the two apex points local contractibility is obvious. At others it follows from the same properties in $X \times (0, 1)$. From this it follows that there is a simply connected covering $p : Y \to SX$. If $C_{\pm} X \subset SX$ are the two cones, the inclusion maps can be lifted and the lifts patch-up together to give a section $s : SX \to Y$ of p. By Exercise 3.1.7.(iv), p is a homeomorphism.

(vii). More generally, if $q : Y \to Z$ is an open quotient map of Hausdorff spaces, then it is easily seen that Y is CG implies that so is Z. Therefore X' is CG implies so is X. To see the converse, use the fact that if $\mathcal{U}$ is an open cover of X by evenly covered open sets, then $\{p^{-1}(U) \ : \ U \in \mathcal{U}\}$ is an open cover of X' such that each member of this cover is CG being a disjoint union of spaces each of which is homeomorphic to some open subset of X.

(viii). Use lifting property to establish a r to 1 relation between n-cells of $\bar{X}$ and n cells of X for each n.

(ix). Use lifting criterion and the fact that $\mathbb{S}^n, n \geq 2$ are simply connected.

§**3.5.23** (page no. 152)

(i). The order of $G(p)$ has to be either 1 or 3 and order 3 implies that the covering has to be normal, which is not true in this case.

(ii). Each orbit is dense.

(iii). We have seen (see 3.1.7.(iv)) that existence of a section implies that the covering is trivial, i.e., there is a homeomorphism $h : E \to B \times G$ such that $p_2 \circ h = p$. Also there is an automorphism $\alpha : G \to G$ such that $h(ge) = \alpha(g)(p(e), g)$. Take $h_1 : B \times G \to B \times G$ to be $h_1(b, g) = \alpha(g^{-1})(b, g)$. The composite $h_1 \circ h$ is the required G-isomorphism.

(iv) Clearly, multiplication by ω is a covering transformation which generates a group of order 3. Since the order of $\mathbf{G}(p)$ cannot exceed 3 here, we conclude $\mathbf{G}(p) \approx \mathbb{Z}_3$.

§**3.9** (page no. 165)

2. Go through the proof of Lemma 3.4.6 carefully.

5. Let $X = \times_{\alpha \in A} \mathbb{S}^1_\alpha$, where each $\mathbb{S}^1_\alpha$ denotes a copy of $\mathbb{S}^1$ and A is an infinite set. Suppose we have a covering $p : X' \to X$. Let U be an evenly covered open subset of X. Then there exists a finite subset $B \subset A$ and a path connected open subset V of $\times_{\alpha \in B} \mathbb{S}^1_\alpha$ such that $W = V \times (\times_{\alpha \in A \setminus B} \mathbb{S}^1_\alpha) \subset U$. Fix $\beta \in A \setminus B$ and let $p_\beta : X \to \mathbb{S}^1$ be the projection map onto the β-factor. If $\eta : W \to X$ is the inclusion map then $(p_\beta \circ \eta)_\#$ is a surjection on the fundamental groups. Let $W' \subset \tilde{X}$ be such that $p : W' \to W$ is a homeomorphism. If $\eta' : W' \to X'$ is the inclusion, then $(p_\beta \circ p \circ \eta')_\#$ is a surjection on the fundamental groups. So is $(p_\beta \circ p)_\# \pi_1(X') \to \pi_1(\mathbb{S}^1)$. Let now $q : X'' \to X$ be the covering corresponding to the kernel of $p_{\beta\#}$. Then it follows that there is no map $g : X' \to X$ such that $q \circ g = p$.

6. Figure 3.7 represents the space $\hat{W}$. Notice that in each cluster of circles there is one extra circle at the bottom than at the top, which is of decreasing radius as we proceed away from the y-axis. To get the 2-fold covering $q : \hat{W} \to W$, fold along the line $y = 1$ except for the outermost circles $C((4m, 1), 1)$ which should be wound twice along the extra circles at the bottom $C((4m, 0), 1/|m| + 2)$.

§**4.1.24** (page no. 176)

(i). (a) Tensoring with $\mathbb{Q}$ we may assume that A, B, C, etc., are finite dimensional vector spaces over $\mathbb{Q}$. We can then assume that $B = A \oplus C$ and write matrices for f, g, h after fixing bases for A and C. It follows that the matrices M_f, M_g, M_h of the homomorphisms f, g, h are related by

$$M_g = \begin{bmatrix} M_f & \star \\ 0 & M_h \end{bmatrix}$$

The result follows.

(b) Let $z(f_n) : Z_n(C) \to Z_n(C), b(f_n) : B_n(C) \to B_n(C)$ denote the restrictions of $f_n : C_n \to C_n$ to Z_n, B_n, respectively, and $h(f_n) : H_n(C) \to H_n(C)$ denote the homomorphism induced by f. Then, from the previous exercise, we have

$$tr(f_n) = tr(z(f_n)) + tr(b(f_{n-1})); \quad tr(z(f_n)) = tr(h(f_n)) + tr(b(f_n)).$$

Taking the alternative sum, the result follows.

§**4.2.27** (page no. 186)

(ii). We shall discuss the case $X = \mathbb{S}^n_1 \vee \mathbb{S}^n_2$ leaving the general case to the reader. Let $a_i \in \mathbb{S}^n_i$

be any point other than the base point. Then X is the union of two open sets $X_i = X \setminus \{a_i\}$ and hence you can use Mayer–Vietoris. But both $\mathbb{S}_i^n \hookrightarrow X_j, i \neq j$ are deformation retracts and hence induce isomorphisms in the homology. Moreover, $X_1 \cap X_2$ is contractible. Putting this information together, we obtain $H_n X \approx H_n(X_1) \oplus H_n(X_2), n > 0$ and $H_0(X) = \mathbb{Z}$.

(iii). Clearly, $(\alpha_0)_0(g_0) = -g_0$. Inductively, we have

$$(\alpha_n)(g_n) = (\alpha_n)_*(Sg_{n-1}) = -S((\alpha_{n-1})_*(g_{n-1})) = (-1)^{n+1}S(g_{n-1}) = (-1)^{n+1}g_n.$$

Therefore $\deg \alpha_n = (-1)^{n+1}$.

(iv). Under the conjugation, g_1 is mapped onto $-g_1$ and so the degree of the conjugation is -1. To compute the degree of $z \mapsto z^n$, we need to subdivide the S-triangulation properly. For instance taking the barycentric subdivision of the S triangulation we can check that the degree of the map $z \mapsto z^2$ is equal to 2. In general, we need to consider the triangulation of the domain $\mathbb{S}^1$ with vertex set equal to the $2n^{\text{th}}$-roots of unity.

(v). These are similar to exercise (i) above with the help of S-triangulation, the degree of the reflection being -1.

(vi). Take a regular value x_0 for f. Then there is a neighbourhood A of x_0 diffeomorphic to $\mathbb{D}^n$ such that $f^{-1}(A_0)$ is the disjoint union of a finite number of subspaces A_i such that $f_{A_i} : A_i \to A_0, \ i > 0$ are diffeomorphisms. Under excision and with the induced orientations, the inclusion induced homomorphisms $H_n(A_,, \partial A_i; \mathbb{Z}) \to H_n(\mathbb{S}^n, \star; \mathbb{Z})$ map the generator to the generator. Therefore, it follows that the algebraic degree of f is equal to the sum of the algebraic degrees of f_{A_i}. On the other hand, the geometric degree of f is defined to be the number of times $f : A_i \to A_0$ is orientation preserving minus the number of times $f : A_i \to A_0$ is orientation reversing. Therefore the two degrees coincide.

(vii). Use the functorial exact sequence

$$0 \to S_.(A)/S_.(B) \to S_.(X)/S_.(B) \to S_.(X)/S_.(A) \to 0.$$

(viii). Let U be a neighbourhood of A in X such that A is DR of U. (Remember that A is a closed subset of X by definition.) Then $\{X \setminus A, U\}$ and $\{Y \setminus q(A), q(U)\}$ are excisive couples in X, Y, respectively.

Use the commutative diagram

$$
\begin{array}{ccccccccc}
0 & \longrightarrow & S_.(A) & \longrightarrow & S_.(X) & \longrightarrow & S_.(X, A) & \longrightarrow & 0 \\
 & & \downarrow & & \downarrow & & \downarrow & & \\
0 & \longrightarrow & S_.(U) & \longrightarrow & S_.(X) & \longrightarrow & S_.(X, U) & \longrightarrow & 0
\end{array}
$$

pass on to the long homology exact sequences and appeal to Five lemma to conclude that the inclusion $i_1 : (X, A) \to (X, U)$ induces isomorphism in homology. Similarly for the inclusion $j_1 : (Y, q(A)) \to (Y, q(U))$. Consider the following diagram

$$
\begin{array}{ccccc}
(X, A) & \xrightarrow{\ i_1\ } & (X, U) & \xleftarrow{\qquad i_2 \qquad} & (X \setminus a, U \setminus A) \\
\downarrow{\scriptstyle q} & & \downarrow{\scriptstyle q} & & \downarrow{\scriptstyle q} \\
(Y, q(A)) & \xrightarrow{\ j_1\ } & (Y, q(U)) & \xleftarrow{\qquad j_2 \qquad} & (Y \setminus q(A), q(U) \setminus q(A))
\end{array}
$$

in which all the horizontal arrows are inclusion maps. The inclusions i_2, j_2 are excisions and therefore all horizontal arrows induce isomorphism in homology. The third vertical arrow q is a homeomorphism of the pairs. Therefore, the first vertical arrow also induces isomorphism in homology.

§**4.3.19** (page no. 193)

(i). Let v denote the apex of cK. Then the oriented k-simplices of cK consists of those in K and those of the form $v * \sigma$, where σ is an oriented $(k-1)$ simplex of K. Therefore

$$C_k(cK) = C_k(K) \oplus C_{k-1}(K),$$

wherein for the sake of clarity we shall write elements of the second summand in the form $v * \sigma$. Let $\hat{\partial}$ denote the boundary operator of $C.(cK)$. Then we have

$$\hat{\partial}(c) = \partial(c), \ c \in C_k(K), \quad \hat{\partial}(v * c') = c' - v * \partial(c') \ c' \in C_{k-1}(K).$$

Now suppose $\hat{\partial}(c + v * c') = 0$. This implies that $c' = -\partial c$. Therefore $\hat{\partial}(v * c) = c - v * \partial(c) = c + v * c'$, proving the exactness.

(ii). The simplicial chain complex of $\dot{\Delta}_n$ coincides with $C.(\Delta_n)$ except in the n-term which is zero. Since $C_n(\Delta_n) \approx \mathbb{Z}$, $\partial_n : C_n(\Delta_n) \to C_{n-1}(\Delta_n)$ is injective, and since $C.(\Delta_n)$ is exact we get $\mathrm{Ker}\,(\partial_{n-1}) = \mathbb{Z}$. Therefore, $H_{n-1}(\dot{\Delta}_n) = \mathrm{Ker}\,\partial_{n-1} \approx \mathbb{Z}$.

§**4.3.32** (page no. 198)

(i). We use the cell structure of $\mathbb{S}^n$ with one 0-cell and one n-cell and then give the product cell structure on $\mathbb{S}^p \times \mathbb{S}^q$. The case when $p, q \geq 2$ poses no problem at all since all the boundary maps in the CW-chain complex vanish. Therefore if $p, \neq q$ then

$$H_i(\mathbb{S}^p \times \mathbb{S}^q) = \begin{cases} \mathbb{Z}, & i = 0, p, q, p+q \\ 0, & \text{otherwise,} \end{cases}$$

and if $p = q$ then

$$H_i(\mathbb{S}^p \times \mathbb{S}^q) = \begin{cases} \mathbb{Z}, & i = 0, p+q \\ \mathbb{Z} \oplus \mathbb{Z}, & i = p; \\ 0, & \text{otherwise.} \end{cases}$$

Now consider the case when we have p or q is equal to 1. Say, for the sake of definiteness that $q = 1$. This time, there is a p-cell and a $(p+1)$-cell and so, we need to show that $\partial_{p+1} : C_{p+1} \to C_p$ is zero. So, we need to closely examine the attaching map $f : \mathbb{S}^p \to \mathbb{S}^p \vee \mathbb{S}^1$ of the $(p+1)$-cell. Let $\psi_1 : \mathbb{D}^1 \to \mathbb{S}^p \vee \mathbb{S}^1$, and $\psi_2 : \mathbb{D}^p \to \mathbb{S}^p \vee \mathbb{S}^1$ be the characteristic maps of the 1-cell and the p-cell, respectively. Then $f : \partial(\mathbb{D}^1 \times \mathbb{D}^p) \to \mathbb{S}^p \vee \mathbb{S}^1$ is given by

$$f(x, y) = \begin{cases} \psi_1(x), & y \in \partial\mathbb{D}^p; \\ \psi_2(y), & x \in \partial\mathbb{D}^1. \end{cases}$$

We now consider the case $p = 1$ and $p > 1$ separately. For $p = 1$, we note that the element represented by f in $\pi_1(\mathbb{S}^1 \vee \mathbb{S}^1)$ is precisely equal to the commutator $[\psi_1][\psi_2][\psi_1]^{-1}[\psi_2]^{-1}$. Passing onto the homology, this becomes trivial. Therefore $f_* = 0$. For $p > 1$, we know that the projection map $\eta : \mathbb{S}^p \vee \mathbb{S}^1 \to \mathbb{S}^p$ induces an isomorphism $\eta_* : H_p(\mathbb{S}^p \vee \mathbb{S}^1) \to H_p(\mathbb{S}^p)$. Therefore, it is enough to compute the map $(\eta \circ f)_*$. But $\eta \circ f$ maps $\mathbb{S}^1 \times \partial\mathbb{D}^p$ to the basepoint and hence defines a map $g : \partial(\mathbb{D}^1 \times \mathbb{D}^p)/\mathbb{D}^1 \times \partial\mathbb{D}^p \to \mathbb{S}^p$. The domain of g is nothing but the wedge of two copies of $\mathbb{S}^p$ and g restricted to one of these spheres is equal to the identity map whereas, on the other sphere, it is the negative of the identity map. (This is because the orientations induced on $1 \times \mathbb{D}^p$ and $-1 \times \mathbb{D}^p$ from that $\mathbb{D}^1 \times \mathbb{D}^p$ are opposite of each other.) Therefore $(\eta \circ f)_* = 0$. The rest of the argument is the same as before.

[In Section 10.4, while discussing the Hurewicz homomorphism, we shall have an elegant proof of the triviality of the homomorphism induced by the attaching map of product cells.]

(ii). This follows if we use CW-homology instead of singular homology.

(iii). Use the previous exercise and the product CW-structure.

(iv). Use Mayer–Vietoris. In case $n = 1$, the situation is somewhat different. If $\varphi : \mathbb{S}^0 \to$

Y takes values in the same path component of Y then $H_0(X) \approx H_0(Y)$ and $H_1(X) \approx H_1(Y) \oplus \mathbb{Z}$. On the other hand, if φ takes values in different path components of Y, then $H_0(Y) \approx H_0(X) \oplus \mathbb{Z}$, and $H_1(X) \approx H_1(Y)$. In either case $H_i(X) \approx H_i(Y)$ for $i > 1$.

§4.4.17 (page no. 204)

(i). A null homotopic map has Lefschetz number equal to 1.

(ii). Choose any path ω joining e with $g \neq e$. Then the right multiplication map $R_g : X \to X$ has no fixed point. On the other hand it is homotopic to $Id = R_e$ via $H(x,t) = x\omega(t) = R_{\omega(t)}(x)$. Therefore, $\chi(X) = L(Id) = Lf(R_g) = 0$.

(iii). $\chi(\mathbb{S}^{2n}) = 2$ and so previous exercise is applicable.

(iv). Same argument as in (ii) above.

§4.7 (page no. 212)

1.

(i)a. Given a directed system $\{G_\alpha, \{\phi_{\alpha\beta}\}\}$ of abelian groups, the following construction of the direct limit comes in handy in dealing with many problems with direct limits, viz., the quotient group of the direct sum $\oplus G_\alpha$ by all the relations of the form $g_\alpha \sim \phi_{\alpha\beta}(g_\alpha)$, for all $g_\alpha \in G_\alpha$ and for all $\beta \geq \alpha$. Now given $a \in A$ represented by $a_\alpha \in A_\alpha$, $f(a) = 0$ iff there is some $\beta \geq \alpha$ such that $\psi_{\alpha\beta}(f_\alpha(a_\alpha)) = 0$ iff $f_\beta\psi_{\alpha\beta}(a_\alpha) = 0$ iff $a \in \text{dlim}_{\to\alpha}\text{Ker}f_\alpha$. Similarly you can argue for images also.

(i)b. It is enough to show the following: Suppose we have a directed system of abelian groups and homomorphism

$$A_\alpha \xrightarrow{f_\alpha} B_\alpha \xrightarrow{g_\alpha} C_\alpha$$

such that for each α, $\text{Ker}\,g_\alpha = \text{Im}\,f_\alpha$. If

$$A \xrightarrow{f} B \xrightarrow{g} C$$

is the direct limit, then $\text{Ker}\,g = \text{Im}\,f$. This follows easily from the above exercise.

(i)c. The image of every singular simplex is compact. So, a singular chain, being a finite linear combination, has its support inside a compact subset.

(i)d. Using (i)a, this is similar to (i)b.

3. Under the given hypothesis, show that $X = \text{int}\,X_1 \cup \text{int}\,X_2$.

4. Use simplicial approximation.

5. Put $D_1 = D \cap \{(x,y) : y < 1/3\}, D_2 = D \cap \{(x,y) : y > -1/3\}$. Then we can apply the Mayer–Vietoris sequence. Note that $D_1 \cap D_2$ contains a totally disconnected subspace as a SDR and hence we can use the previous exercise to compute its homology. It follows immediately that $H_i(D) = (0), i > 1$. Also, since D is connected $H_0(D) \approx \mathbb{Z}$. So, we have an exact sequence

$$0 \longrightarrow H_1(D) \xrightarrow{\partial} H_0(D_1 \cap D_2) \xrightarrow{(j_1,j_2)_*} H_0(D_1) \oplus H_0(D_2) \longrightarrow H_0(D) \to 0.$$

We claim that $(j_1, j_2)_*$ is injective and hence $H_1(D) = 0$. For this, we can replace D_1, D_2 by $Z = D_1 \cap D_2 \cap \mathbb{R}$ which is a DR of $D_1 \cap D_2$. Let $\sigma = \sum_i n_i v_i$ be a finite where $v_i \in Z$, be such that $(j_1, j_2)_*(\sigma) = 0$. This implies that $(j_1)_*(\sigma) = 0 = (j_2)_*(\sigma)$. The first one implies that all the v_i are $\mathbb{R}^+$ and the second one implies that they are in $\mathbb{R}^-$. Therefore $\sigma = n_0 v_0$, where $v_0 = 0$. But then $(j_1)_*(\sigma) = 0$ implies $n_0 = 0$.

7. Consider the two 3-subsets $\{1, \omega, \omega^2\} \times \{0\}, \{0\} \times \{1, \omega, \omega^2\}$ of $\mathbb{C} \times \mathbb{C}$. There is an embedding of $K_{3,3}$ in $Y \times Y \subset \mathbb{C} \times \mathbb{C}$ with these six points as vertices and the image of this embedding is a deformation retraction of $Y \times Y \setminus \{(0,0)\}$. If we had an embedding of $Y \times Y$ in $\mathbb{R}^3$ we can use this to construct an embedding of $K_{3,3}$ inside $\mathbb{S}^2$, which will be absurd.

§**5.1.12** (page no. 218)

(**i**) A manifold is locally path connected.

(**ii**). These spaces are certainly non Euclidean, nor they are locally Euclidean : No non empty open set U of the Zariski topology is homeomorphic to an open subset of a Euclidean space even though each of them is an open subset of the Euclidean space. There are many more open subsets in the Euclidean topology than in the Zariski topology. On the other hand any topological space X with cofinite topology and with cardinality of $\mathbb{R}$ is contractible. (The usual maps such as $(x, t) = xt$ will not do because no polynomial maps $X \times \mathbb{I} \to X$ other than constants are **not** continuous.) On the other hand, you can take any finite-to-one map $H : X \times (0, 1) \to X$ (possible because of the cardinality assumption on X) and extend to $H : X \times \mathbb{I} \to X$ by defining $H(x, 0) = x_0, H(x, 1) = x$ for all $x \in X$ to get a homotopy of Id_X with a constant function. Note that for any space Z, any function $Z \to X$ is continuous iff its fibres are closed.

§**5.1.20** (page no. 220)

(**i**). There are many subspaces of $\mathbb{R}^n$ which do not satisfy the local euclidean condition: $\mathbb{Q} \subset \mathbb{R}, \mathbb{R} \setminus \mathbb{Q}$, union of any two coordinate axes in $\mathbb{R}^n$, etc.

(**ii**). The Grassmann space of k-dimensional subspaces of $\mathbb{R}^n$. (See Section 5.4.)

(**iii**). Choose a path from p to q, and cover it with finitely many inter-locking balls. In each of these balls use Exercise 1.5.19.(i) and then take the composite.

(**iv**). Use Exercise (iii) above and induction. Here you need to use the fact that $M \setminus A$ is connected for any finite subset A. If dim $X = 1$, then you need to consider the two cases $X = \mathbb{R}, \mathbb{S}^1$ separately. For $\mathbb{R}$ the result is true iff $k \leq 2$, viz., there is actually an (unique) affine linear transformation $f(t) = \frac{(b_2 - b_1)t + a_2 b_1 - a_1 b_2}{a_2 - a_1}$; and if we choose $a_1 < a_2 < a_3$ and $b_1 < b_3 < b_2$ then there is no homeomorphism which $g : \mathbb{R} \to \mathbb{R}$ such that $f(a_i) = b_i$. In case $X = \mathbb{S}^1$, we can do this iff $k \leq 3$. This statement can be deduced from the case $X = \mathbb{R}$.

(**v**). If dim $X \geq 2$ we can use the previous exercise to find a homeomorphism which maps F inside a n-cell C and then take $E = f^{-1}(C)$. If dim $X = 1$ then it follows from the classification of 1-dimensional manifolds.

(**vii**). Under Hausdorffness, paracompactness is equivalent to the existence of partition of unity. Now use theorem 5.1.7.

§**5.3.19** (page no. 240)

(**i**). **aabbcc**, **aa**, **aabb**.

(**ii**). Take $\left[(x^2 + (y - g - 1)^2 - (g + 1)^2 \prod_{k=1}^{g} (x^2 + (y - k)^2 - 1/9) \right] + z^2 = 0$. (For more details, see [Shastri, 2011] page 235.)

§**5.4.18** (page no. 246)

Note that $V^* = \wedge^1 V$. Let $A(V, W)$ denote the image of the exterior product map

$$\wedge : \wedge^1(V) \times \wedge^1(W) \to \wedge^2(V \oplus W),$$

then we see that $(v, w) \mapsto v \wedge w$ defines a canonical isomorphism of $V^* \otimes W^* \to A(V, W)$. Since

$$\wedge^2(V \oplus W) = \wedge^2(V) \oplus \wedge^2(W) \oplus A(V, W)$$

the conclusion follows.

§**5.4.23** (page no. 247)

There are projection maps $p_i : \xi_2 \oplus \xi_3 \to \xi_i, i = 2, 3$. The required isomorphism is given by $f \mapsto (p_2 \circ f, p_3 \circ f)$.

§**5.4.24** (page no. 247)

For each $b \in B$, a homomorphism $f : \eta_b \to \eta_b$ is determined by a unique element $\lambda \in \mathbb{R}$

by the rule $f(v) = \lambda v$ (does not depend on the choice of the basis element for η_b. Thus the function $f \mapsto (x, \lambda)$ is well defined on $E(\mathrm{Hom}\,(\eta, \eta)) \to B \times \mathbb{R}$. Continuity of this function, etc., is easily checked using local triviality condition. Note that the same holds for line bundles over $\mathbb{C}$ also.

§**5.4.31** (page no. 249)
Consider the map $A \mapsto$ the linear span of first k columns of A. By Gram–Schmidt, this is surjective and is smooth. If $H \in O(k, -) \times O(-, n - k)$, then for any $A \in O(n)$, we have $f(HA) = f(H)$ and conversely.

§**5.5** (page no. 251)
1. These are easy consequences of Brouwer's invariance of domain.
2. We appeal to Theorem 5.4.37. If $f_\xi : X \to G_{n+1}$ is the classifying map, by simplicial approximation theorem, we may assume that f_ξ takes values inside the nth-skeleton $G_{n+1}^{(n)}$ of G_{n+1}. Now $\xi \oplus \theta^r \approx \theta^{n+r+1}$ implies that $\eta \circ f_\xi : X \to G_{n+r+1}$ is null homotopic, where $\eta : G_{n+1} \to G_{k+r+1}$ is the standard inclusion. If $H : X \times I \to G_{n+k+1}$ is such a homotopy, then again by simplicial approximation theorem, we get a null homotopy of f_ξ inside the $(n+1)^{th}$-skeleton of G_{n+k+1} which is nothing but $G_{n+1}^{(n+1)}$.
3. (a) To begin with we have the standard embedding of $\mathbb{S}^{k_1} \subset \mathbb{R}^{1+k_1}$. Its normal bundle is trivial. Take the composite embedding

$$\mathbb{S}^{k_1} \subset \mathbb{R}^{1+k_1} = \mathbb{R}^{1+k_1} \times 0 \subset \mathbb{R}^{k_1+1} \times \mathbb{R}^{k_2}$$

and take the unit normal bundle which is a trivial bundle. That gives an embedding of $\mathbb{S}^{k_1} \times \mathbb{S}^{k_2} \subset \mathbb{R}^{1+k_1+k_2}$ with its normal bundle trivial. Proceed iteratively.
(b) follows from (a).
(c) Here we use the fact that every odd dimensional sphere admits a nowhere vanishing vector field, e.g., take

$$(x_0, \ldots x_{2n+1}) \mapsto (x_0, -x_1, x_2, -x_3, \ldots, x_{2n}, -x_{2n-1}).$$

We shall prove that if M is s-parallelizable manifold of dimension $m \geq 1$, then $M \times \mathbb{S}^k$ is parallelizable where k is odd. Here we appeal to Exercise 2 above to see that $\tau_M \oplus \Theta^1$ itself is trivial. Write $\tau_{\mathbb{S}^k} = \xi \oplus \Theta^1$. Let $q_1 : M \times \mathbb{S}^k \to M, q_2 : M \times \mathbb{S}^k \to \mathbb{S}^k$ be the projections. It follows that

$$
\begin{aligned}
\tau_{M \times \mathbb{S}^k} &= q_1^*(\tau_M) \oplus q_2^*(\tau_{\mathbb{S}^k}) &&= q_1^*(\tau_M) \oplus q_2^*(\Theta^1 \oplus \xi) \\
&= q_1^*(\tau_M) \oplus \Theta^1 \oplus q_2^*(\xi) &&= q_1^*(\tau_M \oplus \Theta^1) \oplus q_2^*(\xi) &&= q_1^*(\Theta^{m+1}) \oplus q_2^*(\xi) \\
&= \Theta^{m+1} \oplus q_2^*(\xi) &&= \Theta^{m-1} \oplus q_2^*(\Theta^2 \oplus \xi) &&= \Theta^{m-1} \oplus q_2^*(\Theta^1 \oplus \tau_{\mathbb{S}^k}) \\
&= \Theta^{m-1} \oplus q_2^*(\Theta^{k+1}) &&= \Theta^{m-1} \oplus \Theta^{k+1} &&= \Theta^{m+k}.
\end{aligned}
$$

4. See [Shastri, 2011] Section 7.5.
5. Since f is surjective, M_f is the quotient of $\mathbb{S}^1 \times \mathbb{I}$ under the relation $(z_1, 1) \sim (z_2, 1)$ iff $z_1^n = z_2^n$. So, consider the map $g : \mathbb{S}^1 \times \mathbb{I} \to \mathbb{D}^2 \times \mathbb{S}^1$ given by $g(z, t) = ((1 - t)z, z^n))$. This map then factors through the quotient map and defines an embedding $\bar{g} : M_f \to \mathbb{D}^2 \times \mathbb{S}^1$. Now extend this embedding on the mapping cone via $t[z, 0] \mapsto t\bar{g}([z, 0])$ inside $\mathbb{D}^2 \times \mathbb{D}^2$.
9. Apply Van Kampen's theorem.
10. Apply Whitney embedding theorem.
11. First observe that $\pi_1(\mathbb{S}^1 \times \mathbb{D}^4) \approx \mathbb{Z}$. By taking the boundary connected sum of k copies of this, we obtain a 5-manifold X with $\pi_1(X) \approx F_k$, a free group of rank k. If $N = \partial X$, then it follows that the inclusion map induces an isomorphism: $\pi_1(N) \approx \pi_1(X)$. Note that N is orientable. In fact N is parallelizable also.

Now let G be a finitely presented group with k generators and l relations. This means we have a normal subgroup H in F_k generated by l elements and $G = F_k/H$. Represent a generator of H by an embedded loop ω in N. Since N is orientable, it follows that the tubular neighbourhood U of ω in N is a product, i.e., U diffeomorphic to $\mathbb{S}^1 \times \mathbb{D}^3$. Attach $\mathbb{D}^2 \times \mathbb{D}^3$ to X along U to obtain a manifold X_1. The boundary of X_1 will have the fundamental group isomorphic to the quotient K_1 of F_k by the normal subgroup generated by $[\omega]$. We have $G = K_1/H_1$ where H_1 is now generated by $l - 1$ elements in K_1. Inductively, in l steps by attaching 2-handles, the fundamental group of the resulting manifold will become isomorphic to G.

12. Moore space $M := M(G, n)$: Every abelian group G is the quotient of a free abelian group F by a free abelian subgroup H. Let $\{\alpha\} \subset F$ and $\{\beta\} \subset H$ be their bases. Put $Y = \vee_\alpha \mathbb{S}^n_\alpha$ and let $f_\beta : \mathbb{S}^n \to Y$ be maps which represent the elements $\beta \in H_n(Y; \mathbb{Z})$. Let M be the space obtained by attaching $(n+1)$-cells $\{e^{n+1}_\beta\}$ to Y via f_β. Use CW-homology to check that M is as required.

13. Since every map $\sigma : \Delta^n \to X$ is a constant, it follows that $S_.(X)$ is a direct sum of chain complexes of each singleton subspace of X. Therefore $H_*(X) = \oplus_{x \in X} H_*(\{x\})$.

§6.2.20 (page no. 262)
(i). Using the resolution

$$0 \longrightarrow \mathbb{Z} \xrightarrow{\cdot m} \mathbb{Z} \longrightarrow \mathbb{Z}_m \to 0$$

and using the isomorphism $\mathbb{Z} \otimes \mathbb{Z}_n \approx \mathbb{Z}_n$ this amounts to determining the kernel of the homomorphism $\mathbb{Z}_n \xrightarrow{\cdot m} \mathbb{Z}_n$, which is nothing but the subgroup $\mathbb{Z}_d \subset \mathbb{Z}_n$, given by $\bar{1} \mapsto \overline{n/d}$, where $d = gcd(m, n)$.

(ii). Consider the case when A'' is torsion free. If $A' \otimes B \to A \otimes B$ is not injective, there is a finitely generated submodule of M of A such that $(M \cap A') \otimes B \to M \otimes B$ is not injective. By considering the exact sequence $0 \to M \cap A' \to M \to M/M \cap A' \to$ not that $M \cap A' \subset of A''$ and hence we may as well assume A'' itself is finitely generated. But then A'' being torsion free, is a free module and hence the sequence is split-exact. Upon tensoring with any B, it remains exact. Now consider the case when B is torsion free. As before we can reduce the case when B is finitely generated and then being a torsion free module it is a free module. Tensoring with B is then the same as taking direct sum of so many copies and hence the sequence remains exact.

(iii). Use the previous exercise.

(iv). Let $0 \to C_1 \to C_0 \to M \to 0$ be a free presentation of M. This gives us a free chain complex which has vanishing homology modules in dimension ≥ 2. The long homology exact sequence given by the previous exercise is the required six-term exact sequence.

(v). Let $0 \to C_1 \to C_0 \to B \to 0$ be a free presentation of B. Apply the six-term exact sequence to this short exact sequence and use the canonical isomorphism $M \otimes N \to N \otimes N$ to conclude the result.

(vi). Use the exact sequences $0 \to \operatorname{Tor} A \to A \to A/\operatorname{Tor} A \to 0$ and $0 \to \operatorname{Tor} B \to B \to B/\operatorname{Tor} B \to 0$ and use the corresponding six-term exact sequences twice.

(vii). This is an easy consequence of Theorem 6.2.4.

(viii). This is also an easy consequence of Theorem 6.2.4, since for any field $\mathbb{K}$, we have $C \star \mathbb{K} = 0$.

§6.4 (page no. 271)
1. Write B and C as direct sum of unital graded modules, prove the statement for unital graded modules and use the fact that tensor product commutes with direct sum.

2. (4.7) yields the following exact sequence

$$0 \longrightarrow \frac{S_.(X_1 \cap X_2)}{S_.(A)} \longrightarrow \frac{S_.(X_1)}{S_.(A)} \oplus \frac{S_.(X_2)}{S_.(A)} \longrightarrow \frac{S_.(X_1) + S_.(X_2)}{S_.(A)} \longrightarrow 0.$$

Similarly, we have an exact sequence

$$0 \longrightarrow \frac{S.(Y)}{S.(X_1 \cap X_2)} \longrightarrow \frac{S.(Y)}{S.(X_1)} \oplus \frac{S.(Y)}{S.(X_2)} \longrightarrow \frac{S.(Y)}{S.(X_1)+S.(X_2)} \longrightarrow 0.$$

They respectively give the implications (a)$\Longrightarrow$ (b) and (a)$\Longrightarrow$(c). To show (b)$\Longrightarrow$(a), put $A = X_1 \cap X_2$ and to show (c)$\Longrightarrow$(a), put $Y = X_1 \cup X_2$.

3. The homology of the product is the tensor product of the homology of those many spheres. In the second statement, the 'if' part is obvious. For the converse, we first observe that a stronger statement, viz., if $H_*(X \times \mathbb{S}^r) \approx H_*(Y \times \mathbb{S}^r)$, then $H_*(X) \approx H_*(Y)$. This is just an application of the previous result. Now suppose $H_*(\times_k \mathbb{S}^{n_k}) \approx H_*(\times_l \mathbb{S}^{m_l})$. If $n = \min, \{n_k\}$ and $m = \min \{m_l\}$, then by the Künneth formula it follows that the smallest non trivial module in the positive dimensions on either side are $H_m m, H_n = n$, respectively, and hence $m = n$. Inductively, it would follow that the set of exponents $\{n_k\}$ is equal to $\{m_k\}$.

4. Recall that (see Exercise 1.9.30) saying x_0 is a non degenerate base point is the same as saying $\{x_0\} \subset X$ is a cofibration. Hence by Exercise 1.6.15.(v), it follows that the two inclusion maps $j_t : X \hookrightarrow X \vee X, t = 1, 2$ are cofibrations. From Theorem 4.2.19, it follows that we can use Mayer–Vietoris sequences. The rest of the exercise is routine.

§7.4.13 (page no. 291)

(i). We shall prove that any two coordinate inclusions, $\eta_i : \mathbb{S}^m \to \mathbb{S}^n$ are isotopic through orthogonal transformations: First take a permutation P on $n+1$ letters such that $P\eta_1 = \eta_2$, where $\eta_i : \mathbb{R}^{m+1} \to \mathbb{R}^{n+1}$ are some coordinate inclusions. By changing the sign at one of the coordinates $x_j, j > m$, if necessary, we can modify P so that it is of determinant 1, i.e., $P \in SO(n + 1)$ and $P\eta_1 = \eta_2$. Take a path $\omega : \mathbb{I} \to SO(n + 1)$ such that $\omega(0) = Id$ and $\omega(1) = P$. Put $H(x,t) = \omega(t)(x)$. Then $H \circ (\eta_1 \times Id)$ is the required isotopy from η_1 to η_2. This isotopy factors down to an isotopy on the projective spaces as well, because $H(-x,t) = -H(x,t)$.

(iii). The claim for CW-structure on X is obvious. Note that the quotient map $q : \mathbb{S}^n \times \mathbb{S}^n \to X$ is a relative homeomorphism on the $2n$-cells and send the two n-spheres homeomorphically onto the single n-sphere. Let u, v denote some 'standard' generators of $H^n(X), H^{2n}(X)$, and x, y, and z those of $H^n(\mathbb{S}^n \times \mathbb{S}^n)$, and $H^{2n}(\mathbb{S}^n \times \mathbb{S}^n)$, we have $q^*(u) = x + y$ and $q^*(v) = z$. Clearly $u \smile u = 0$ if n is odd. However, for n even we have $q^*(u \smile u) = (x + y) \smile (x + y) = 2(x \smile y) = 2z = q^*(2v)$. Therefore, $u \smile u = 2v$.

§8.1.21 (page no. 311)

(i). Let X be a connected manifold $x_0 \in X$ be any point. First of all one proves that if any two loops at x_0 are homotopic then X is orientable along one loop iff it is so along the other. Now assign $+1$ to a loop along which the manifold is orientable and assign -1 otherwise, to obtain a homomorphism $\Theta : \pi_X, x_0) \to \mathbb{Z}_2$. This homomorphism is onto iff X is not orientable. The conclusion follows. Of course, manifolds such as $\mathbb{S}^1$ are easy examples of orientable ones, wherein the fundamental group may have subgroups of index 2.

(ii). We use Theorem 8.1.13. One can prove a statement analogous to it by replacing $\mathbb{Z}$ by any commutative ring. In particular, X is orientable over $\mathbb{Q}$ iff there is a continuous section for the bundle $X_{\mathbb{Q}} \to X$. which is nowhere vanishing. By continuity it follows that such a section is either positive everywhere or negative everywhere on X and hence can be normalized to obtain a nowhere vanishing section of $X_{\mathbb{Z}}$. Note that $q : X_{\mathbb{Z}} \to X$ is just the restriction of $q : X_{\mathbb{Q}} \to X$.

(iii). For any number $r > 0$ consider the family of open balls of radius r with centres inside K and let $\mathcal{U}_r$ denote a chosen finite subcover. Let $l(r)$ denote the Lebesgue number of this cover. Inductively, we define a sequence s_n as follows: $s_1 = 1, s_n = \min \{l(s_{n-1}), 1/n\}$. Put

K_n =union of closures of members of $\mathcal{U}_{s_n}$. Then K_n is a finite union of closed (convex) balls and $K = \cap_n K_n$.

§8.2.19 (page no. 319)

(i). Any open subset of an orientable manifold is again an orientable manifold. So, after removing finitely many points, an orientable manifold remains orientable. To see the converse, we may assume that the dimension is > 1. Now an orientation on $\mathbb{D}^n \setminus \{0\}$ extends uniquely to an orientation on the whole of $\mathbb{D}^n$. Given a finite subset $A \subset M$, we can choose disjoint neighbourhoods U_i of $x_i \in A$, and homeomorphism to $f_i : \mathbb{D}^n \to U_i$ such that $f(0) = x_i$. Given an orientation on $M \setminus A$, we can then extend the induced orientation on $U_i \setminus \{x_i\}$ to an orientation on U_i and then put all of this together to get an orientation on M.

(ii). M is orientable by Exercise 8.1.21.(i). Then $H_{n-1}(M) \approx H^1(M) \approx \text{Hom}\,(H_1(X); \mathbb{Z})$ which is clearly a torsion free abelian group.

(iii). Use (8.7) to see that $\bar{H}^{n-1}(A) \approx H_1(X, X \setminus A)$ and then use the exact sequence of the pair $(X, X \setminus A)$:

$$H_1(X) \longrightarrow H_1(X, X \setminus A) \longrightarrow H_0(X \setminus A) \longrightarrow H_0(X). \tag{13.20}$$

(iv). We shall prove this only in the smooth case. We first show that $X \setminus A$ is disconnected. If not, we can find a smooth loop ω in X which intersects A transversely in a single point. Working with $\mathbb{Z}_2$ coefficients, this just means that the mod 2 intersection number of ω and A is non zero and hence ω represents a non trivial element in $H_1(X)$ which is a contradiction. Now we use the above exact sequence (13.20) from which we can conclude that $H_1(X, X \setminus A) \approx \mathbb{Z}$. Therefore, $\bar{H}^{n-1}(A) \approx \mathbb{Z}$. Since A is a smooth manifold, it has a fundamental system of tubular neighbourhoods and hence $\bar{H}^*(A) = H^*(A)$. Therefore $H^{n-1}(A) \approx \mathbb{Z}$ which means A is orientable.

(v). Appeal to the previous exercise.

(vi). Since $K = f(\mathbb{S}^{n-2})$ is a smooth embedding, by using tubular neighbourhoods as before, it follows that $\bar{H}^*(K)$ and $H^*(K)$ coincide. By duality Theorem (8.7), we have

$$H_i(\mathbb{S}^n, \mathbb{S}^n \setminus K) = \begin{cases} 0, & i \neq 2, n; \\ \mathbb{Z} & i = 2, n. \end{cases}$$

From this it follows that $H_i(\mathbb{S}^n \setminus K) = 0$, for $i \neq 0, 1$ and $\mathbb{Z}$ if $i = 0, 1$.

§8.3.17 (page no. 324)

All oriented surfaces occur as the boundary of a handle-body in $\mathbb{R}^3$. As seen in Example 8.3.15, for any manifold $M \# M$ is always the boundary of a manifold. An unoriented surface of genus $2g$ is homeomorphic to $P_g \# P_g$ and hence is a boundary.

§8.5 (page no. 327)

1. Assume that such a trivializing cover exists, we can orient each fibre in an unambiguous way by the rule that the homeomorphisms $h_i : p^{-1}(U_i) \to U_i \times \mathbb{R}^k$ are all orientation preserving. Conversely, if we have an orientation of ξ, then it follows that for each i, j, $p_i \circ p_j^{-1} : (U_i \cap U_j) \times \mathbb{R}^n \to (U_i \cap U_j) \times \mathbb{R}^n$ is orientation preserving on each fibre, which is the same as saying that the transition functions take values inside matrices of positive determinant.

2. The transition functions associated with the exterior powers are all obtained via those of the original bundle and hence will be of positive determinant.

3. If a line bundle is orientable, then it has a nowhere vanishing section and hence is trivial. Of course a trivial line bundle is orientable.

4. Recall that $\Lambda^k V = \mathbb{R}$ for a vector space of dimension k. Therefore $\Lambda^k \xi$ of a k-plane bundle is a line bundle. ξ is orientable iff the transition functions take positive values. The transition functions for the determinant bundle are nothing but the determinant of the transition functions of the given bundle.

5. Any complex linear isomorphism is automatically orientation preserving. Alternatively, the transition functions take values inside $GL(n, \mathbb{C}) \subset SL_{2n}(\mathbb{R})$.

§9.5 (page no. 356)

1. For each open set U, given $t, s \in \mathcal{F}(U)$, $\Theta(t) = \Theta(s)$ iff there is an open cover $\{U_i\}$ of U such that $s|_{U_i} = t|_{U_i}$ for all U_i. By F-I, this is the same as saying $s = t$. This is, in turn, the same as Ker $\Theta(U) = 0$. This takes care of the first part. For the second part, note that the Ker $\Theta(U)$ still makes sense, for each open set U, $\mathcal{F}(U)$ contains the zero element $\mathcal{F}(\emptyset)$ and hence Ker $\Theta(U)$ consists of those elements $s \in \mathcal{F}(U)$ such that there exists an open cover $\{U_i\}$ of U with the property that $s|_{U_i} = \mathcal{F}(\emptyset)$. Then the first part of the statement holds. However, in the latter part, we get only one implication, viz., the condition implies that Ker $\Theta = 0$ but we cannot go in the reverse direction.

2. For any open set U in X, $\mathcal{G}(U) = U \times G$. Therefore, Spé($\mathcal{G}$) is the quotient of disjoint union of all (U, g) where U is open in X and $g \in G$. The claims follow.

3. (a) and (b) are trivial. To prove (c), note that the collection of open sets $\{q(U_s)\}$ form a base for the topology of Spé in each case. Now suppose $\sigma \neq \tau \in$ Spé($\mathcal{H}$). If $z = \pi(\sigma) \neq \pi(\tau) = w \in \mathbb{C}$ then we can choose disjoint neighbourhoods U, V of z, w, respectively, in $\mathbb{C}$. Let (U, s) and (V, t) represent σ and τ. Then clearly, $q(U_s), q(V_t)$ form disjoint neighbourhoods of σ, τ, respectively. Next consider the case when $\pi(\sigma) = \pi(\tau) = z$. Let D be an open disc centred at z and contained in $U \cap V$. Then $q(D_s)$ and $q(D_t)$ are neighbourhoods of σ, τ, respectively. We claim that $q(D_s) \cap q(D_t) = \emptyset$. If this is not true, there exists $z' \in D$ and a neighbourhood W such that $z' \in W \subset D$ and $s|_W = t_W$. By the principle of analytic continuation, it follows that $s|_D = t_D$. This implies $\sigma = q(s) = q(t) = \tau$ a contradiction.

The fact that Spé($\mathcal{C}^\infty$) is not Hausdorff must be clear now, because of the lack of analytic continuation in this situation. Indeed, it is not hard to see that for any open set U and any two $\mathcal{C}^\infty$ functions $s, t : U \to \mathbb{R}$ we have $q(U_s) \cap q(U_t) \neq \emptyset$, so that no points in the same stalk can be separated by disjoint open sets.

5. For any two members $U_1, U_2 \in \mathcal{U}$, we must have $\mathcal{F}_{U_1}|_{U_1 \cap U_2} = \mathcal{F}_{U_2}|_{U_1 \cap U_2}$.

6. Claim: $\mathcal{U}$ is evenly covered by $\pi : \text{Spé}(\mathcal{F}) \to X$. For, $\pi^{-1}(U) = q(\sqcup\{U_s \ : \ s \in \mathcal{F}(U)\}) = q(U \times \mathcal{F}_x) = \sqcup\{q(U \times \{s\}) \ : \ s \in \mathcal{F}_x\}$ and $\pi : q(U \times \{s\}) \to U$ is clearly a homeomorphism.

7. By Exercise 5 it suffices to describe this sheaf on an even covering $\mathcal{U}$ for the covering projection. On members of $\mathcal{U}$ the sheaf of sections clearly satisfies the property in Exercise 6. Notice that $\mathcal{F}_{x_0} = \pi^{-1}(x_0) = \pi_1(X, x_0)$. So, we conclude that the sheaf of sections of a covering projection is nothing but a sheaf of locally constant functions into $\pi_1(X, x_0)$.

8. At the stalk level, we clearly have such an exact sequence.

9. The pullback of $\pi' : \text{Spé}(\mathcal{F}') \to Y$ is clearly a local homeomorphism $\pi : Z \to X$ with its sheaf of sections isomorphic to the sheaf of sections of π under the inverse-correspondence $s \mapsto f^*(s)$ where $f^*(s)(x) = (x, s(f(x)))$. On the other hand, for any open set $U \in Y$ by definition $f^*(\mathcal{F}')(f^{-1}(U)) = \mathcal{F}'(U)$. So, the claim follows.

§10.1.14 (page no. 362)

(i). The crux of the matter is that Theorem 1.3.1 is valid if we replace the Cartesian product $X \times Y$ with $X \times_w Y = k(X \times Y)$ without the additional assumption of certain spaces being locally compact. For instance, in the proof of part (a) of the Theorem 1.3.1, we used the fact that X is locally compact. In this modified situation, the corresponding step easily reduces to the case when the whole of X is compact.

(ii). If C_x denotes the path component of X containing the point x, then

$$C_x \cdot C_y = C_{x \cdot y}$$

makes sense. The rest of the details are easy.

(**iv**). Use Exercise 1.9.36.

§10.2.26.(i). (page no. 368)

Find a retraction $r : \partial(\mathbb{I}^{n+1}) \times \mathbb{I} \to \mathbb{I}^n \times 0 \cup \partial(\mathbb{I}^n) \times \mathbb{I} \cup p \times \mathbb{I} \cup \mathbb{I}^{n+1} \times 1$.

§10.3.19 (page no. 375)

Consider $\varphi : (\mathbb{D}^2, \mathbb{S}^1) \times \mathbb{I} \to (\mathbb{D}^2, \mathbb{S}^1)$ given by $(r, \theta_1, \theta_2) \mapsto (r, \theta_1 + \theta_2)$, where we are using polar coordinate representation of elements of $\mathbb{D}^2$. Check that ϕ is a ω-homotopy of $Id : (\mathbb{D}^2, \mathbb{S}^1) \to (\mathbb{D}^2, \mathbb{S}^1)$ to itself, where $\omega : \mathbb{S}^1 \to \mathbb{S}^1$ is the identity loop. The claim follows.

§10.6.13 (page no. 391)

(**i**). We have constructed a Moore space Y of type (G, n) which has cells just in dimension $0, n, n+1$ alone, in Exercise 4.7.12. The idea is to show that any other Moore space X of type (G, n) is homotopy equivalent to this one. By definition, $\pi_i(X) = (0)$ for $i < n$ and $\pi_n(X) = G$. Therefore, first of all we can assume that X is a CW-complex with $X^{(n-1)}$ equal to a singleton and $X^{(n)}$ a wedge of spheres. Therefore, the inclusion map $X^{(n)} \to X$ induces a surjection $\pi_n(X^{(n)}) \to G = \pi_n(Y)$. This surjection can be realized by a map $f_n : X^{(n)} \to Y$. The obstruction to extending f_n to a map $f_{n+1} : X^{(n+1)} \to Y$ vanishes because $\pi_n(X^{(n+1)}) = G$. Moreover f_{n+1} now induces an isomorphism on π_n and hence on H_n. Since $H^{n+k}(X, M) = (0)$ for all $k \geq 2$ and all modules M, the obstructions to extending f_{n+k} to f_{n+k+1} vanish for all $k \geq 2$ and we get a map $f : X \to Y$ which induces isomorphisms of homology groups and hence on homotopy groups. By Whitehead's theorem f is a homotopy equivalence.

(**ii**). By the Van Kampen theorem and the Mayer–Vietoris, it follows that $M(\mathbb{Z}_p, n) \vee M(\mathbb{Z}_q, n)$ is a Moore space of $(\mathbb{Z}_p \oplus \mathbb{Z}_q; n)$. By universal coefficient theorem, the same is true for $M(\mathbb{Z}_p, n) \times M(\mathbb{Z}_q, n)$, since the torsions vanish.

§10.7.20 (page no. 396)

(**i**). There is a tree T contained in the 1-skeleton of X. It follows from Theorem 1.6.4 that the quotient map $q : X \to X/T$ is a homotopy equivalence and X/T has a CW structure with a single vertex.

(**ii**),(**iii**), (**iv**). All these three exercises follow from the construction of an appropriate CW-complex representing the corresponding $K(G, n)$. For instance the $(n + 1)$ skeleton of any $K(G, n)$ can be chosen to be the mapping cone of a map $f : A \to B$ where both A, B are wedge of copies of n-spheres and f induces a monomorphism in H_n. Therefore, it follows that $H_{n+1}(X^{(n+1)}) = 0$ and hence $H_{n+1}(X) = 0$.

§10.8.17 (page no. 403) We need to produce a path in E^X from (x_0, ω) to (x_0, c) where ω is a loop in Y at y_0. Since $\phi_\#$ is surjective, there is a homotopy $H : I \times I \to Y$ such that $H(t, 0) = \omega(t), H(t, 1) = \phi \circ \tau(t)$ and $H(0, s) = H(1, s) = y_0$, where τ is a loop in X at x_0. This gives a path in E^X from (x_0, ω) to $(x_0, \phi \circ \tau)$. Now, for each t, put $\tau_t(s) = \tau(st)$. Then $t \mapsto (\tau(t), \phi \circ \tau_t)$ is a path in E^X which joins (x_0, c) with $(x_0, \phi \circ \tau)$.

§10.11 (page no. 414)

1. Recall that the antipodal map is of degree $(-1)^{n+1}$. Therefore at the homology level we have $\alpha_* = (-1)^{n+1} Id$. From this we may try to conclude that the same is true at π_n level also via the Hurewicz map. But then $p \circ \alpha = p$ implies that $p_\# \circ \alpha_\# = p_\#$ which in turn implies that $\alpha_\# = Id$, which is absurd in case n is even.

This is all because we ignore the base points. Since $\alpha(x_0) = -x_0$ the two isomorphisms $p_\# : \pi_n(\mathbb{S}^n, \pm x_0) \to \pi_n(\mathbb{P}^n, z_0)$ are not the same.

2. Use Theorem 5.2.6.

3. See [Shastri, 2011] Chapter 9.

4. Use the fibration $SU(2) \to U(2) \xrightarrow{\det} \mathbb{S}^1$.

5. From the fibration $\mathbb{S}^1 \to U(2) \to SU(2) = \mathbb{S}^3$ we obtain $\pi_4(U(2)) = \pi_4(SU(2) = \pi_4(Sp(1)) = \pi_4(\mathbb{S}^3) \approx \mathbb{Z}_2$.

$SO(1) = (1), SO(2) = \mathbb{S}^1$ and so there is nothing to compute. There is a double cover $\mathbb{S}^3 \to SO(3)$ and hence $\pi_i(SO(3)) \approx \pi_i(\mathbb{S}^3), i \geq 2$.

Since there is a homeomorphism $SO(4) \to \mathbb{S}^3 \times SO(3)$, it follows that $\pi_4(SO(4)) = \mathbb{Z}_2 \oplus \mathbb{Z}_2$. Finally from 10.9.6, it follows that $(Sp(k), Sp(1))$ is at least 6-connected and hence $\pi_i(Sp(1)) \approx \pi_i(Sp(k)), k \geq 2, i \leq 5$.

§11.5 (page no. 444)

1. More generally, if X is any Lie group then the degree of the map $f_k(x) = x^k$ is equal to k. To see this use the following facts: If $\mu : X \times X \to X$ is the multiplication, and $f, g : X \to X$ are any two maps $h = \mu \circ (f \times g) \circ \Delta$ then $h_* = f_* + g_*$ and $\deg f_* + \deg g_* = \deg h_*$.

2. (a) This amounts to computing the degree of the composites $p_i \circ f_{p,q} \circ \eta_j$ for $i, j = 1, 2$ where η_j denote the coordinate inclusions $\mathbb{S}^3 \to \mathbb{S}^3 \times \mathbb{S}^3$ and $p_i : \mathbb{S}^3 \times \mathbb{S}^3 \to \mathbb{S}^3$ are projections. The answer is $(f_{p,q})_*(m, n) = (m, m(p+q) + n)$.

(c) We use the Mayer–Vietoris sequence. Need to concentrate on dimensions 3 and 4:

$$\cdots 0 \to H_4(M) \xrightarrow{\partial} H_3(\mathbb{S}^3 \times \mathbb{S}^3) \xrightarrow{\alpha} H_3(\mathbb{D}^4 \times \mathbb{S}^3) \oplus H_3(\mathbb{D}^4 \times \mathbb{S}^3) \to H_3(M) \to 0 \cdots$$

The homomorphism α is $((i_1)_*, (i_2)_*)$. Using the fact that $(i_2)_* = (f_{p,q})_*$ check that α is injective if $p + q \neq 0$ and the image of α is equal to $(p+q)\mathbb{Z} \oplus \mathbb{Z}$. It follows that $H_4(M) = (0)$ and $H_3(M) \approx \operatorname{Coker} \alpha \approx \mathbb{Z}_{a+b}$.

(d) (You have only to observe that by Van Kampen theorem, $M_{p,q}$ is simply connected also.

(e) One can use Wang sequence 11.3.5 to compute the homology of the fibre space $M \to \mathbb{S}^4$ with fibre $\mathbb{S}^3$. This amounts to computing $\theta_* = \varphi_* \circ \partial \circ \tau$ as given just before Theorem 11.3.5. One checks that θ_* is the multiplication by $p + q$.

3. Let $K = K(G, p), \mu : K \times K \to K$ be the homotopy multiplication and $p_i : K \times K \to K$ be the two projections. Then we know that $p_1^*(x) = x \otimes 1, p_2^*(x) = 1 \otimes x$. Let Θ and u be such that $u = \Theta(\iota_p)$. Suppose Θ is additive. Then

$$
\begin{aligned}
\mu^*(u) &= \mu^* \Theta(\iota_p) = \Theta(\mu^*(\iota_p)) \\
&= \Theta(p_1^*(\iota_p) + p_2^*(\iota_p)) = \Theta p_1^*(\iota_p) + \Theta p_2^*(\iota_p) \\
&= p_1^* \Theta(\iota_p) + p_2^* \Theta(\iota_p) = u \otimes 1 + 1 \otimes u.
\end{aligned}
$$

This proves u is a primitive.

Conversely, suppose that u is a primitive. Given $x, y \in H^p(X, Y)$, let $f, g : X \to K$ be such that $f^*(\iota_p) = x, g^*(\iota_p) = y$. Put $h = \mu \circ (f \times g) \circ \Delta$. Then $h^*(\iota_n) = f^*(\iota_n) + g^*(\iota_n) = x + y$. Now

$$
\begin{aligned}
\Theta(x + y) &= \Theta h^*(\iota_n) = h^* u \\
&= \Delta^* \circ (f \times g)^* \circ \mu^*(u) = \Delta^*(f \times g)^*(u \otimes 1 + 1 \otimes u)) \\
&= \Delta^*(f^*(u) \otimes 1 + 1 \otimes g^*(u)) = f^*(u) + g^*(u) = \Theta(x) + \Theta(y).
\end{aligned}
$$

Therefore Θ is additive.

§12.6 (page no. 461)

3.

(4) The pullback bundle has the total space given by

$$E = \{(x, [z], [w]) \in \mathbb{S}^3 \times S : h(x) = [w]\}$$

$$= \{(x, [z], [x]) \in \mathbb{S}^3 \times \mathbb{CP}^2 \times \mathbb{CP}^1 : x_1 z_2 = x_2 z_1\}.$$

and the bundle map is the projection to the first factor. Consider the map $f : \mathbb{S}^3 \times \mathbb{CP}^1 \to E$ given by

$$((x_1, x_2), [y_1, y_2]) \mapsto ((x_1, x_2), [y_1 x_1, y_1 x_2, y_1 + y_2], [x])$$

which is clearly an injective immersion and because the dimensions are the same is a diffeomorphism. Since this respects the first projection, this proves (4).

(5) Both spaces serve as the base for a $\mathbb{S}^1$-fibration on the same total space $E = \mathbb{S}^3 \times \mathbb{CP}^1$. The first fibration is got by the symmetric nature of the pullback and since h is a $\mathbb{S}^1$-fibration. The second fibration is just the product $h \times Id_{\mathbb{CP}^1}$. Therefore, from the homotopy exact sequences of the respective fibrations, it follows that both the base spaces have the same homotopy groups. The cup product pairing on $\mathbb{S}^2 \times \mathbb{S}^2$ is positive definite whereas on S, it is indefinite because of (2).

Bibliography

[Adams, 1960] J. F. Adams, On the non-existence of Hopf invariant one, *Ann. Math.* (2) 72, pp. 20–104.

[Adams, 1974] J. F. Adams, *Stable Homotopy and Generalized Homology,* Chicago Lecture Series, University of Chicago Press.

[Ahlfors–Sario, 1960] L. V. Ahlfors and L. Sario, *Riemann Surfaces,* Princeton Univesity Press.

[Alexander, 1924] J.W. Alexander, An example of a simply connected surface bounding a region which is not simply connected, *Proc. Nat. Acad. Sc. of USA,* 10 (1), pp. 8–10.

[Armstrong, 2004] M. A. Armstrong, *Basic Topology,* Springer International Edition.

[Barratt–Milnor, 1962] M. G. Barratt, and J. Milnor, An example of anamalous singular homology, *Proc. Amer. Math. Soc.* 13, pp. 293–297.

[Bredon, 1977] G. E. Bredon, *Topology and Geometry,* Springer-Verlag.

[Brondsted, 1982] A. Brondsted, *An Introduction to Convex Polytopes*, Springer-Verlag.

[Browder, 1965] W. Browder, *Surgery on Simply Connected Manifolds,* Springer Verlag.

[Brown, 1968] R. Brown, *Elements of Modern Topology,* Mc–Graw-Hill, London.

[Chinn–Steenrod, 1966] W. G. Chinn and N. Steenrod, *First Concepts of Topology,* Random House.

[Cohen, 1973] M. Cohen, *A Course in Simple-Homotopy Theory,* Springer-Verlag, GTM 10.

[Coxeter, 1973] H. S. M. Coxeter, *Regular Polytopes,* Dover Publications, Inc., New York.

[Croom, 1978] F. H. Croom, *Basic Concepts of Algebraic Topology,* Springer-Verlag.

[Datta, 2007] Basudeb Datta, Minimal triangulation of manifolds, *Journal of Indian Inst. Sc.* 87, pp. 429–449.

[Deo, 2003] Satya Deo, *Algebraic Topology–A Primer*, TRIM- 27, Hindustan Book Agency.

[Diedonne, 1989] J. Dieudonne, *A History of Algebraic and Differential Topology,* Birkhauser.

[Dold, 1972] A. Dold, *Lectures on Algebraic Topology,* Springer-Verlag.

[Dowker, 1951] C. H. Dowker, Topology of metric complexes, *Amer. J. Math.* 4, pp. 555–577.

[Dugundji, 1990] J. Dugundji, *Topology,* Universal Book Stall, New Delhi.

[Eilenberg–Steenrod, 1952] S. Eilenberg and N. Steenrod, *Foundations of Algebraic Topology*, Princeton University Press.

[Engleking, 1968] R. Engleking, *Outline of General Topology*, North-Holland Pub. Co., Amsterdam.

[Farrell, et al., 2002] F.T. Farrell, L. Göttsche and W. Lück, *Topology of High-Dimensional Manifolds,* ICTP Lecture Notes Series 9, Part 1 and 2.

[Froster, 1981] O. Froster, *Riemann Surfaces,* Springer-Verlag.

[Fulton, 1995] W. Fulton, *Algebraic Topology, A First Course,* Springer International Edition, GTM 163.

[Gamelin–Greene, 1997] T. W. Gamelin and R. E. Greene, *Introduction to Topology,* II edition, Dover, Mineola, New York.

[Ghorpade et al., 2013] S. R. Ghorpade, A. R. Shastri, M. K. Srinivasan, and J. K. Verma, *Combinatorial Topology and Algebra*, Proceedings of the Instructional Conference on Combinatorics, Topology, and Algebra, held in Dec. 1993 at I. I. T. Bombay, Ramanujan Mathematical Society Lecture Notes, Vol. 18.

[Godement, 1958] R. Godement, *Topologie Algébrique et Théory des Faisceaux*, Hermann, Paris.

[Greenberg–Harper, 1981] M. Greenberg and J. Harper, *Algebraic Topology, A First Course*, Benjamin/Cummings.

[Grunbaum, 1967] B. Grünbaum, *Convex Polytopes*, Interscience.

[Hardt, 1977] R. Hardt, Triangulation of subanalytic sets and proper light subanalytic maps, *Invent. Math.* 38, pp. 207–217. MR 56:12302.

[Hartshorne, 1977] R. Hartshorne, *Algebraic Geometry*, GTM 52, Springer.

[Hatcher, 2002] A. Hatcher, *Algebraic Topology*, Cambridge University Press.

[Helgason, 1978] S. Helgason, *Differential Geometry, Lie Groups and Symmetric Spaces,* Academic Press.

[Higgin–Neumann–Neumann, 1949] G. Higgin, B. H. Neumann, Hanna Neumann, Embedding theorems for groups, *J. London. Math. Soc.* s1-24(4) pp. 247–254.

[Hilton–Mislin–Roitberg, 1975] P. J. Hilton, G. Mislin, and J. Roitberg, *Localization of nilpotent groups and spaces*, North Holland Pub. Co.

[Hilton–Stommbach, 1970] P. J. Hilton and U. Stommbach, *A Course in Homological Algebra*, GTM 4, Springer-Verlag.

[Hilton–Wylie, 1960] P. J. Hilton and S. Wylie, *Homology Theory*, Cambridge Univerity Press.

[Hudson, 1969] J. F. P. Hudson, *Piecewise Linear Topology,* Benjamin Inc.

[Hurewicz–Wallman, 1948] W. Hurewicz and H. Wallman, *Dimension Theory*, Princeton University Press.

[Husemoller, 1994] D. Husemoller, *Fibre Bundles,* III edition, GTM 20, Springer-Verlag.

[Joshi, 1983] K. D. Joshi, *Introduction to General Topology*, Wiley Eastern Ltd., New Delhi.

[Jungerman–Ringel, 1940] M. Jungerman and G. Ringel, Minimal triangulations on orientable surfaces, *Acta Mat.* 145, pp. 121–154.

[Kühnel, 1990] W. Kühnel, Triangulations of manifolds with few vertices, *Advances in Diff. Geom. Topology,* World Scientific, pp. 59–119.

[Lefschetz, 1930] S. Lefschetz, *Topology,* AMS Colloq. Publications No. 12, New York.

[Lefschetz, 1942] S. Lefschetz, *Algebraic Topology,* AMS Colloq. Publications No. 27, New York.

[Lovasz, 1978] L. Lovasz, Kneser's conjecture, chromatic number and homotopy, *J. Comb. Theory,* Ser. A 25, pp. 319–324.

[Lundell–Weingram, 1969] A. T. Lundell and S. Weingram, *The Topology of CW Complexes*, Van Nostrand Reinhold Company, New York.

[Massey, 1977] W. Massey, *Algebraic Topology: An Introduction*, Springer-Verlag.

[Mac Lane, 1961] S. Mac Lane, *Homology*, Addison Wesley.

[Matousek, 2003] J. Matousek, *Using Borsuk-Ulam theorem: Lectures on Topological Methods in Combinatorics and Geometry*, Berlin: Springer.

[Maunder, 1970] C. R. F. Maunder, *Algebraic Topology*, Van Nostrand.

[McCleary, 2001] John McCleary, *A User's Guide to Spectral Sequences*, II edition, Cambridge University Press.

[Milnor, 1956] J. W. Milnor, On manifolds homeomorphic to the 7-sphere, *Ann. Math.* 64, pp. 399–405.

[Milnor, 1959] J. W. Milnor, Differentiable Structures on Spheres, *Amer. Journ. Math.* 81, pp. 962–972.

[Milnor, 1960] J. W. Minor, On spaces having the homotopy type of a CW-complex, *Trans. Amer. Math. Soc.* 90, pp. 272–280.

[Milnor, 1961] J.W. Milnor, Two complexes which are homeomorphic but combinatorially distinct, *Ann. of Math.* 74, No. 3. pp. 575–590.

[Milnor–Moore, 1965] John W. Milnor and John C. Moore, On the structure of Hopf Algebras, *Ann. Math.* (2) 81, pp. 211–264.

[Milnor–Stasheff, 1974] John W. Milnor and James D. Stasheff, *Characteristic Classes*, Annals of Math. Studies, 76, Princeton University Press.

[Miyazaki, 1952] H. Miyazaki, Paracompactness of CW complexes, *Tohoku Math. J.* 4, pp. 309–313.

[Moise, 1977] E.E. Moise, *Geometric Topology in Dimension 2 and 3.* Springer -Verlag.

[Mosher–Tangora, 1968] R. E. Mosher and M. C. Tangora, *Cohomology Operations and Applications in Homotopy Theory*, Dover Books in Mathematics.

[Munkres, 2008] J. R. Munkres, *Topology*, II edition, Fourth impression. Pearson Education, Prentice Hall.

[Munkres, 1984(1)] J. R. Munkres, *Elements of Algebraic Topology*, Addison-Wesley, 1984.

[Munkres, 1984(2)] J. R. Munkres, Topological results in combinatorics, *Michigan Math. J.*, 31, pp. 113–128.

[Nevanlina, 1973] R. Nevanlinna, *Uniformiseirung*, Springer-Verlag.

[Novikov, 1955] P.S. Novikov, On the algorithmic unsolvability of the word problem in group theory, *Proceedings of the Steklov Institute of Mathematics*, 44: 1V143 (Russian).

[Ramanan, 2004] S. Ramanan, *Global Calculus*, Graduate Studies in Mathematics, 65, American Mathematical Society, Providence, Rhode Island.

[de Rham, 1969] G. de Rham, Lecture Notes on *Introduction to Algebraic Topology*, notes by V. J. Lal, www.math.tifr.res.in/ publ/In/tifr44.pdf.

[Rudin, 1976] W. Rudin, *Principles of Mathematical Analysis*, III edition, McGraw-Hill International Editions Mathematics Series.

[Seifert–Threlfall, 1990] H. Seifert and W. Threlfall, *A Textbook of Topology*, Academic Press Inc.

[Serre, 1951] J.-P. Serre, Homologie singulière des espaces fibrés, *Ann. of Math.* 54, pp. 425–505.

[Serre, 1980] J.-P. Serre, *Trees*, (English translation by J. Stillwell) Springer-Verlag, Berlin-New York.

[Shastri, 2009] A. R. Shastri, *Basic Complex Analysis of One Variable*, Macmillan India Ltd, Delhi.

[Shastri, 2011] A. R. Shastri, *Elements of Differential Topology*, CRC Press, Boca Raton.

[Spanier, 1966] E. H. Spanier, *Algebraic Topology*, McGraw-Hill. Delhi.

[Stanely, 1975] R. Stanley, The Upperbound Conjecture and Cohen-Macaulay Rings, *Studies in Applied Math.*, 54, pp. 135-142.

[Steenrod, 1951] N. E. Steenrod, *Topology of Fibre Bundles* Princeton University Press, Princeton, NJ.

[Steenrod, 1967] N. E. Steenrod, A convenient category of topological spaces, *Michigan Math. J.* 14, pp. 133–152.

[Taylor, 1985] D.E. Taylor, *Fourier Techniques and Applications*, (Kensington 1983) pp. 221–224. Plenum Press, New York.

[Thom, 1954] René Thom, Quelques propriértés globales des variétés différentiables, *Comment. Math. Helv.*, 28, pp. 17–86.

[Thomassen, 1992] C. Thomassen, The Jordan-Schönflies Theorem and the Classification of Surfaces, *Amer. Math. Monthly*, 99, pp. 116–130.

[Varadarajan, 1966] K. Varadarajan, Groups for which Moore spaces $M(\pi, 1)$ exist, *Ann. Math. II series*, 84, pp. 368–371.

[Whitehead, 1940] J. H. C. Whitehead, On C^1-complexes, *Ann. Math.* 41, pp. 809–824.

[Whitehead, 1939] J. H. C. Whitehead, Simplicial Spaces, Nuclei and m-groups, *Proc. Lond. Math. Soc.* (2) 45, pp. 243–327. [Also in *Collected Works, Vol.* II]

[Whitehead, 1939] J. H. C. Whitehead, On incidence matrices, nuclei and homotopy types, *Ann. of Math.*(2) 42, pp. 1197–1239. [Also in *Collected Works* Vol. II]

[Whitehead, 1950] G. W. Whitehead, A generalization of Hopf invariant, *Ann. of Math.* 51, pp. 192–237.

[Whitehead, 1953] G. W. Whitehead, On Freudenthal theorems, *Ann. Math.* 57, pp. 209–228.

[Whitehead, 1978] G. W. Whitehead, *Elements of Homotopy Theory*, Springer-Verlag.

[Whitney, 1938] H. Whitney, On products in a complex, *Ann. Math.* 39, pp. 397–432.

Index